A Balanced Approach

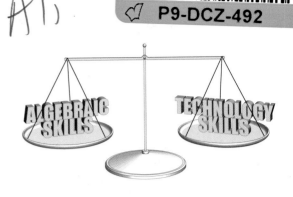

Solve: $x^2 - 3x = 4$.

Algebraic Solution

We have

$$x^2 - 3x = 4$$
$$x^2 - 3x - 4 = 0$$
$$(x + 1)(x - 4) = 0$$
$$x + 1 = 0 \quad or \quad x - 4 = 0$$
$$x = -1 \quad or \quad x = 4.$$

A table set in ASK mode can be used to check the solutions.

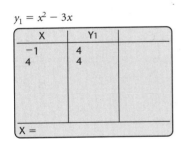

$y_1 = x^2 - 3x$

X	Y1
−1	4
4	4

X =

The solutions are −1 and 4.

Graphical Solution

INTERSECT METHOD

We graph $y_1 = x^2 - 3x$ and $y_2 = 4$ and find the first coordinates of the points of intersection.

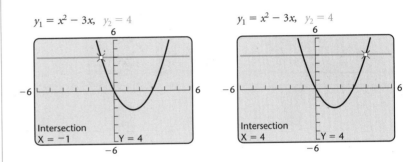

The solutions are −1 and 4. We could also use the Zero method.

ZERO METHOD

We graph $y = x^2 - 3x - 4$ and find the zeros of $f(x) = x^2 - 3x - 4$.

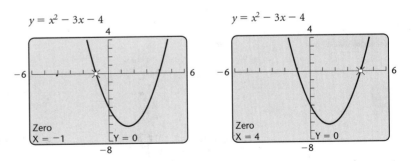

The zeros of the function $f(x) = x^2 - 3x - 4$ are the solutions of the equation $x^2 - 3x = 4$. They are −1 and 4.

College Algebra

GRAPHS & MODELS

SECOND EDITION

College Algebra

GRAPHS & MODELS

SECOND EDITION

Marvin L. Bittinger
Indiana University–Purdue University at Indianapolis

Judith A. Beecher
Indiana University–Purdue University at Indianapolis

David Ellenbogen
Community College of Vermont

Judith A. Penna
Indiana University–Purdue University at Indianapolis

Addison
Wesley

Boston   San Francisco   New York
London   Toronto   Sydney   Tokyo   Singapore   Madrid
Mexico City   Munich   Paris   Cape Town   Hong Kong   Montreal

Publisher	Jason A. Jordan
Project Manager	Kari Heen
Assistant Editor	Suzanne Alley
Managing Editor	Ron Hampton
Production Supervisor	Kathleen A. Manley
Editorial and Production Services	Martha K. Morong/Quadrata, Inc.
Art Editor	Geri Davis/The Davis Group, Inc.
Marketing Managers	Brenda Bravener and Laura Potter
Illustrators	Scientific Illustrators and Barbara Harmon
Technical Art Consultants	Scott Silva and Joe Vetere
PrePress Buyer	Caroline Fell
Compositor	The Beacon Group
Cover Designer	Abby Stoltz
Design Direction	Susan Carsten Raymond
Original watercolor on cover	*Nauset Light* © 2001 Gina Hagen
Print Buyer	Evelyn Beaton

Photo Credits

134, PF Sports Images **140 (left),** SuperStock **165,** Tom and DeeAnn McCarthy, The Stock Market **169,** DigitalSTOCK **170, 186, 203,** PhotoDisc **206,** Ryan McVay, PhotoDisc **321,** UPI Newsphotos **347,** SuperStock **349,** The Image Bank **352,** © 1985 Topps Chewing Gum, Inc. **375,** PhotoDisc **377,** PhotoLink, PhotoDisc **378,** Buccina Studios, PhotoDisc **387,** The Stock Market **394,** SW Productions, PhotoDisc **410,** PhotoLink, PhotoDisc **414,** Robert Shaw, The Stock Market **536,** Mary Dyer

Reprinted with corrections.

Library of Congress Cataloging-in-Publication Data

College algebra: graphs & models/Marvin L. Bittinger . . . [et al.].—2nd ed.
     p. cm.
   Includes index.
   ISBN 0-201-61672-6
   1. Algebra. I. Bittinger, Marvin L.
QA154.2.C643 2000
512.9—dc21   00-026613

4 5 6 7 8 9 10—QWT—040302

Contents

Preface

Note to Students

Our challenge and our goal when writing this edition of *College Algebra: Graphs and Models* was to do everything possible to help you learn the concepts and skills contained between its covers. Every improvement we made and every feature we added was done with this objective in mind. We realize that your time is both valuable and limited, so we communicate in a highly visual way that allows you to focus easily and learn quickly and efficiently. Take advantage of the side-by-side algebraic and graphical solutions, the numerous graphing calculator windows, the Connecting the Concepts features, and the Study Tips. We included them to enable you to make the most of your study time and to be successful in this course.

Best wishes for a positive learning experience,

Marv Bittinger
Judy Beecher
Dave Ellenbogen
Judy Penna

Content Features

College Algebra: Graphs and Models, Second Edition, covers college-level algebra and is appropriate for a one-term course in precalculus mathematics. Our approach is more interactive and more visual than that of most college algebra texts and our goal is to enhance the learning process through the use of technology. Although a course in intermediate algebra is a prerequisite for using this text, Chapter R provides sufficient review to unify the diverse mathematical backgrounds of most students.

New!

• **New Topics** In response to reviewers' requests, we have added the following topics to the Second Edition: variation in Section 1.6, zeros of linear functions in Section 2.1, the leading-term test for polynomials in Section 3.1, and Descartes' Rule of Signs in Appendix A.

• **Integrated Technology** In order to increase students' understanding of the course content through a visual means, we completely integrate graphing calculator technology throughout. In this text, the term "grapher" refers to all graphing calculator technology. The use of the grapher is woven throughout the text's exposition, exercise sets, and testing program without sacrificing algebraic skills. Grapher technology is included in order to enhance, not replace, students' mathematical skills and to alleviate the tedium associated with certain procedures. We assume that each student is required to have a grapher (or at least access to one) while enrolled in this course. (See pp. 183, 266, and 408.)

• **Optional Review Chapter** Chapter R provides an optional review of intermediate algebra. We purposely placed the *Introduction to Graphs and the Graphing Calculator* before this chapter to allow the grapher to be used in the review. Incorporating technology gives these topics a fresh perspective and sets the tone for the rest of the course. (See pp. 38, 46, and 55.) Chapter R can also be used as a convenient source of information for students who need a quick review of a particular topic.

New!

• **Increased Visualization** We use technology to make this edition much more visual than the first edition. Believing the adage "a picture is worth a thousand words," we include many new graphing calculator windows that illustrate algebraic concepts and skills. By displaying graphs, tables, or computation, we make it possible for students, particularly visual learners, to gain quick understanding of these concepts and skills. The additional windows also clearly illustrate how the graphing calculator can be used in combination with algebra to solve problems. (See pp. 74 and 266.)

• **Function Emphasis** Since graphing calculator technology enables students to visualize a concept immediately, we believe it supports the early presentation of functions. We introduce graphing and functions in the first section of Chapter 1 and enhance and streamline the study of the family of functions (linear, quadratic, higher-degree polynomial, rational, exponential, and logarithmic) with the inclusion of the grapher. In addition, in order to amplify and add relevance to the study of functions, we incorporate applications with graphs throughout. (See pp. 64, 229, 282, and 297.)

New!

• **Zeros, Solutions, and *x*-Intercepts Theme** We find that when students understand the connections among *the real zeros of a function, the solutions of its associated equation,* and *the x-coordinates of the x-intercepts of its graph,* a door opens to a new level of mathematical comprehension that increases the probability of success in this course. We emphasize zeros, solutions, and *x*-intercepts throughout the text by using consistent, precise terminology and including exceptional graphics. (See pp. 189, 210, and 344.)

New!

• **Connecting the Concepts** Combining design and art, this feature highlights the importance of connecting concepts. When students are presented with concepts in a visual form (as in an outline or a chart) rather than merely in paragraphs of text, their *comprehension* is streamlined and their *retention* maximized. The visual aspect of this feature invites students to stop and check their understanding of how concepts work together in one section or in several sections. This concept check in turn enhances student performance on homework assignments. (See pp. 164, 238, and 312.)

- **Interactive Discoveries** The grapher provides an exciting teaching opportunity and encourages students to discover and further investigate mathematical concepts. Our unique Interactive Discovery feature introduces new topics and provides a vehicle for students to see a concept in its visual form quickly. This feature reinforces the idea that grapher technology is an integral part of the course as well as an important learning tool. It invites students to develop analytic and reasoning skills while taking an active role in the learning process. (See pp. 89, 228, and 360.)

- **Side-by-Side Algebraic and Graphical Solutions** When students are exposed to a variety of approaches to finding a solution, their skill in solving mathematical problems grows. To highlight both the algebraic- and graphical-solution approaches to solving equations, we use a two-column solution format in numerous examples (see pp. 172, 272, and 339) and we have increased the number of these two-column solutions in the Second Edition. In the algebraic/graphical side-by-side features, each method provides a complete solution. Side-by-side solutions emphasize that there is more than one way to obtain a result and illustrate the comparative efficiency and accuracy of the two methods.

- **Real-Data Applications** We encourage students to see and interpret the mathematics that appears every day in the world around them. Throughout the writing process, we conducted an energetic search for real-data applications and the result is a variety of examples and exercises that connect the mathematical content with the real world. Most of these applications feature source lines and frequently include charts and graphs. Many are drawn from the fields of health, business and economics, life and physical sciences, social science, and areas of general interest such as sports and daily life. (See pp. 101, 169, 206, and 321.)

- **Regression** We introduce the use of regression or curve fitting to model data in Chapter 1 with linear functions and continue this visual theme with quadratic, cubic, quartic, higher-degree polynomial, exponential, logarithmic, and logistic functions. (See pp. 203–204 and 232–233.) Although the theoretical aspects of curve fitting cannot be developed in this course, the power of the grapher is very apparent in this area as the technique is applied to real data. Students can quickly make the "Aha!" connection between real data and the interpolated and extrapolated results of the curve fitting, giving them a better conceptual understanding of the material.

New!

- **Study Tips** Appearing in the text margins, Study Tips provide helpful study hints and briefly remind students to use the electronic and print supplements that accompany the text. (See pp. 108, 233, and 311.)

New!

- **Review Icons** Placed next to the concept that a student is currently studying, a review icon references another text section or sections in which the student can find and review more content related to the concept at hand. (See pp. 163, 227, and 301.)

• **Technology Learning Aids**   In order to minimize the amount of valuable class time needed to teach students how to use a grapher, we provide several aids that shorten the learning curve while increasing students' knowledge of graphing calculator fundamentals. The first learning aid is the "Introduction to Graphs and the Graphing Calculator" found at the beginning of the text. This section introduces students to the basic functions of the grapher. The other aids are the *Graphing Calculator Manual* packaged with each copy of the text (see p. xiv) and the series of Graphing Calculator Instructional Videos (see p. xvi). All of these technology learning aids have been written and produced specifically for this text.

Pedagogical Features

• **Chapter Openers**   Each chapter opens with an application relevant to the chapter content and is illustrated with both a technology window and situational art. The chapter openers also include a table of contents listing section titles. (See pp. 224 and 359.)

• **Section Objectives**   Content objectives are listed at the beginning of each section. Together with subheadings throughout the section, these objectives provide a useful outline for both instructors and students. (See pp. 160 and 310.)

• **Use of Color**   This text uses full color in an extremely functional way, as seen in the design elements and artwork on nearly every page. The choice of color has been carried out in a precise manner so that its use carries a consistent meaning, which enhances the readability of the text for both student and instructor. (See pp. 73, 133, and 289.)

• **Art Package**   The text contains over 1000 art pieces including a form of art called photorealism. Photorealism superimposes mathematics on a photograph and encourages students to see mathematics in familiar settings (see pp. 83 and 197). The exceptional situational art and statistical graphs throughout the text highlight the abundance of real-world applications while helping students visualize the mathematics (see pp. 84 and 167). The design and use of color with the grapher windows exemplifies the impact that technology has in today's mathematical curriculum (see pp. 129, 292, and 363).

• **Annotated Examples**   Over 560 examples fully prepare the student for the exercise sets. Learning is carefully guided with numerous color-coded art pieces and step-by-step annotations, with substitutions and annotations highlighted in red. (See pp. 194 and 382.)

• **Five-Step Problem-Solving Process**   The basis for problem solving is a distinctive five-step process established early in the text (Section 2.1) to help students learn strategic ways to approach and solve applied problems. This process is then used consistently throughout the text to give students a consistent framework for problem solving. (See pp. 164–165 and 364–367.)

- Variety of Exercises   There are over 4100 exercises in this text. The exercise sets are enhanced with real-data applications and source lines, detailed art pieces, and technology windows that include both tables and graphs, and feature the following elements as well.

  Technology Exercises reflect the integration of the grapher throughout this text, and exercise sets include both grapher and nongrapher exercises. In some cases, detailed instruction lines indicate the approach that the student is expected to use. In others, we ask the student to choose the approach that seems best, thereby encouraging critical thinking. (See pp. 190 and 340.)

  Discussion and Writing Exercises encourage students to both consider and write about key mathematical ideas in each chapter. Many of these exercises are open-ended, making them particularly suitable for use in class discussions or as collaborative activities. (See pp. 137 and 237.)

  Skill Maintenance Exercises review concepts previously taught in the text and provide excellent review for a final examination. Answers to all Skill Maintenance exercises appear along with a section reference in the answer section at the back of the book. (See pp. 152 and 387.)

  Synthesis Exercises appear at the end of each exercise set and encourage critical thinking by requiring students to synthesize concepts from several sections or to take a concept a step further than in the general exercises. (See pp. 121 and 370.)

- Highlighted Information   Certain material is presented and organized for efficient learning and review. Important definitions, properties, and rules are displayed in screened boxes, and summaries and procedures are listed in boxes outlined in color. (See pp. 178 and 181.)

- Summary and Review   The Summary and Review at the end of each chapter provides an extensive set of review exercises along with a list of important properties and formulas covered in that chapter. This feature provides excellent preparation for chapter tests and the final examination. Answers to all review exercises appear in the text along with section references that direct students to the appropriate material to reexamine if they have difficulty with a particular exercise. (See pp. 153 and 356.)

Supplements for the Instructor

For more information on these and other helpful instructor supplements, please contact your Addison Wesley Longman sales representative.

Annotated Instructor's Edition *(ISBN 0-201-70990-2, bundled with the Graphing Calculator Manual)*

New!

This specially bound version of the student edition contains answers to both even- and odd-numbered exercises at the back of the text.

Instructor's Solutions Manual *(ISBN 0-201-70879-5)*

The *Instructor's Solutions Manual* by Judith A. Penna contains worked-out solutions to all exercises in the exercise sets, including the Discussion and Writing exercises. It also includes a sample test with answers for each chapter and answers to the exercises in the appendixes. These sample tests are also included in the *Student's Solutions Manual.*

Printed Test Bank/Instructor's Resource Guide *(ISBN 0-201-70880-9)*
Prepared by Laurie Hurley, the *Printed Test Bank/Instructor's Resource Guide* contains the following:

- 4 free-response test forms for each chapter, following the format and with the same level of difficulty as the tests in the *Student's Solutions Manual.*
- 2 multiple-choice test forms for each chapter.
- 6 forms of a final examination, 4 with free-response questions and 2 with multiple-choice questions.
- Index to the Content Videos.
- Index to the Graphing Calculator Instructional Videos.
- Conversion Guide from the First Edition to the Second Edition.

TestGen-EQ with QuizMaster-EQ *(ISBN 0-201-70882-5)*
Available on a dual-platform Windows/Macintosh CD-ROM, this fully networkable software enables instructors to build, edit, print, and administer tests using a computerized test bank of questions organized according to the chapter content of the text. Tests can be printed or saved for on-line testing via a network or the Web, and the software can generate a variety of grading reports for tests and quizzes.

InterAct Math Plus *(Windows ISBN 0-201-63555-0;*
Macintosh ISBN 0-201-64805-9)
This networkable software provides course management and on-line administration for Addison Wesley Longman's InterAct Math Tutorial Software (see "Supplements for the Student"). InterAct Math Plus enables instructors to create and administer on-line tests, summarize students' results, and monitor students' progress in the tutorial software, providing an invaluable teaching and tracking resource.

Supplements for the Student

For more information on these and other helpful student supplements, please contact your bookstore.

Graphing Calculator Manual *(ISBN 0-201-70878-7)*
The *Graphing Calculator Manual* by Judith A. Penna, with the assistance of Daphne A. Bell, contains keystroke level instruction for the Texas Instruments TI-82®, TI-83®, TI-83+®, TI-85®, TI-86®, and TI-89®. Modules for other models are available on request. Contact your local Addison Wesley Longman sales consultant for details. Bundled free with every copy of the text, the *Graphing Calculator Manual* uses actual examples and exercises from *College Algebra: Graphs and Models,* Second Edition, to help teach students to use the graphing calculator. The order of topics in the *Graphing Calculator Manual* mirrors that of the text, providing a just-in-time mode of instruction.

Student's Solutions Manual *(ISBN 0-201-66237-X)*

The *Student's Solutions Manual* by Judith A. Penna contains completely worked-out solutions with step-by-step annotations for all the odd-numbered exercises in the text, with the exception of the Discussion and Writing exercises. It also includes a self-test with answers for each chapter and a final examination.

InterAct Math Tutorial CD-ROM *(ISBN 0-201-70883-3, stand-alone)*

This interactive tutorial software provides algorithmically generated practice exercises that correlate at the objective level to the odd-numbered exercises in the text. Each practice exercise is accompanied by an example and guided solution designed to involve students in the solution process. The software recognizes common student errors and provides appropriate feedback. Instructors can use InterAct Math Plus course management software to create, administer, and track on-line tests and monitor student performance during practice sessions.

New!

World Wide Web Supplement   www.awl.com/BBEP

This free Web site provides additional practice and learning resources such as chapter quizzes, collaborative activities, and InterAct Math tutorial exercises.

New!

AWL Math Tutor Center *(ISBN 0-201-44461-5, stand-alone)*

The AWL Math Tutor Center is staffed by qualified mathematics instructors who provide students with tutoring on examples and exercises from the textbook. Tutoring is available via toll-free telephone, fax, or e-mail five days a week, seven hours a day.

Videotapes *(ISBN 0-201-70839-6)*

Developed and produced especially for this text, these videotapes feature an engaging team of instructors presenting material and concepts from every section of the text in a format that stresses student interaction. The lecturers' presentations include examples and problems from the text and support an approach that emphasizes the use of technology, visualization, and problem solving.

New!

Digital Video Tutor *(ISBN 0-201-70950-3, stand-alone)*

The videos for this text are now available on CD-ROM, making it easy and convenient for students to watch video segments from a computer at home or on campus. The complete digitized video set, now affordable and portable for students, is ideal for distance learning or supplemental instruction.

New!

InterAct MathXL *(12-month registration ISBN 0-201-71630-5, stand-alone)*
www.mathxl.com

This Web-based diagnostic testing and tutorial system allows students to take practice tests correlated to the textbook and receive customized study plans based on their results. Each time a student takes a practice test, the resulting study plan identifies areas for improvement and links to appropriate practice exercises and tutorials generated by InterAct Math. A course-management feature allows instructors to view students' test results, study plans, and practice work.

Graphing Calculator Instructional Videos *(ISBN 0-201-87352-4)*
Designed and produced specifically for *College Algebra: Graphs and Models*, the Graphing Calculator Instructional Videos take students through procedures on the graphing calculator using content from the text. These videos include most topics covered in the *Graphing Calculator Manual*, as well as several additional topics. Every video section uses actual text examples or odd-numbered exercises from the text to help motivate students while they learn to use the graphing calculator. The videos are correlated to the sections of the text. A complete set of Graphing Calculator Instructional Videos is free to qualifying adopters.

Acknowledgments

We wish to express our genuine appreciation to a number of people who have contributed in special ways to the development of this textbook. Jason Jordan, Publisher, and Kari Heen, Executive Project Manager, at Addison Wesley Longman, shared our vision and provided encouragement and motivation. In addition, the production and marketing departments of Addison Wesley Longman brought to the project their unsurpassed commitment to excellence. The unwavering support from the Higher Education Group has been a continuing source of strength for this author team. For this we are most grateful. We are also indebted to Daphne Bell, Joan Bookbinder, Shane Goodwin, Laurie Hurley, and Barbara Johnson for their accuracy checking of the manuscript.

We would also like to thank the following reviewers for their invaluable contribution to the development of this text:

Jamie Ashby, *Texarkana College*

Brian Balman, *Johnson County Community College*

Ray Battee, *Cochise College*

Daphne Bell, *Motlow State Community College*

Susan Bradley, *Angelina College*

Beverly Broomell, *Suffolk County Community College*

Terry Cheng, *Irvine Valley College*

Al Coons, *Pima Community College*

Cynthia Coulter, *Catawba Valley Community College*

Paul Cox, *Ricks College*

Lisa DeLong Cuneo, *Pennsylvania State University— DuBois Campus*

Phil DeMarois, *William Rainey Harper College*

Jo Dobbin, *Fayetteville Technical Community College*

Rob Farinelli, *Community College of Allegheny County— Boyce Campus*

Joy Fett, *Moraine Valley Community College*

James Galloway, *Collin County Community College*

William Grimes, *Central Missouri State University*

Deborah Hanus, *Brookhaven College*

Daniel Harned, *Michigan State University*

Brian Hickey, *East Central College*

Robert Houston, *Rose State College*

Heidi Howard, *Florida Community College—Jacksonville*

Miles Hubbard, *St. Cloud State University*

Helen Kolman, *Central Piedmont Community College*

LuAnn Linton, *Arapahoe Community College*

Linda Long, *Ricks College*

Judy Marwick, *Prairie State College*

Marsha May, *Midwestern State University*

Jean McArthur, *Joliet Junior College*

Timothy McLendon, *East Central College*

Sue Neal, *Wichita State University*

Nancy Olson, *Johnson County Community College*

Kathy Rodgers, *University of Southern Indiana*

Mike Rosenborg, *Canyonville Christian Academy*

Randy Ross, *Morehead State University*

Bob Sweetland, *Redlands Community College*

Peter Uluave, *Utah Valley State College*

Lucio Della Vecchia, *Daytona Beach Community College—North*

Bette Warren, *Eastern Michigan University*

Jon Weerts, *Triton College*

Brenda Wood, *Florida Community College—Jacksonville*

M.L.B.
J.A.B.
D.J.E.
J.A.P.

Feature Walkthrough

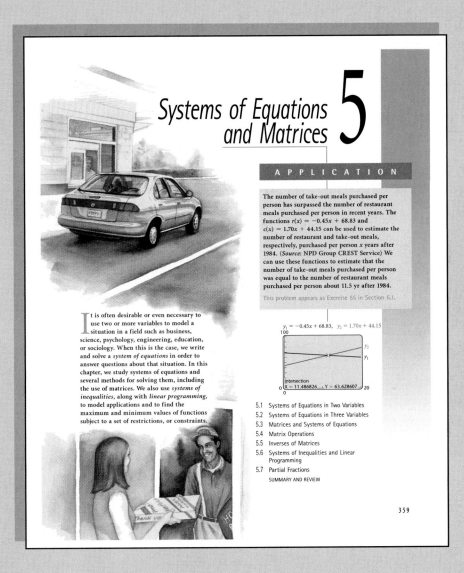

Systems of Equations and Matrices 5

A P P L I C A T I O N

The number of take-out meals purchased per person has surpassed the number of restaurant meals purchased per person in recent years. The functions $r(x) = -0.45x + 68.83$ and $c(x) = 1.70x + 44.15$ can be used to estimate the number of restaurant and take-out meals, respectively, purchased per person x years after 1984. (*Source*: NPD Group CREST Service) We can use these functions to estimate that the number of take-out meals purchased per person was equal to the number of restaurant meals purchased per person about 11.5 yr after 1984.

This problem appears as Exercise 55 in Section 5.1.

$y_1 = -0.45x + 68.83, \quad y_2 = 1.70x + 44.15$

Intersection
X = 11.486826 Y = 63.628607

It is often desirable or even necessary to use two or more variables to model a situation in a field such as business, science, psychology, engineering, education, or sociology. When this is the case, we write and solve a *system of equations* in order to answer questions about that situation. In this chapter, we study systems of equations and several methods for solving them, including the use of matrices. We also use *systems of inequalities*, along with *linear programming*, to model applications and to find the maximum and minimum values of functions subject to a set of restrictions, or constraints.

359

Applied Chapter Openers

Each chapter begins with an application showing students how chapter concepts are used in the real world. This application is modeled with a function and illustrated with a graph. The chapter openers are not only motivating to students but also help students develop their critical-thinking skills.

Zeros, Solutions, and x-Intercepts Theme

There is an increased emphasis on visualizing and connecting the following concepts:

- the real zeros of a function;
- the solutions of the associated equation;
- the x-coordinates of the x-intercepts of the graph of the function.

Seeing the connection between these concepts increases student understanding. This theme is reinforced visually throughout the text.

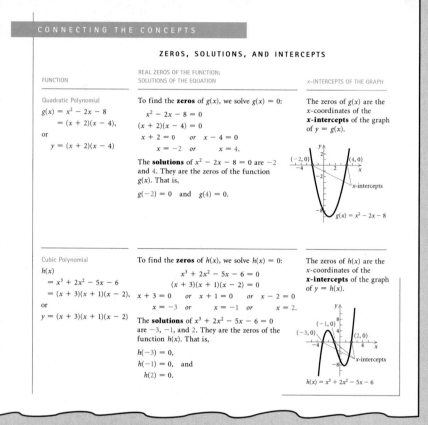

CONNECTING THE CONCEPTS

ZEROS, SOLUTIONS, AND INTERCEPTS

FUNCTION	REAL ZEROS OF THE FUNCTION; SOLUTIONS OF THE EQUATION	x-INTERCEPTS OF THE GRAPH
Quadratic Polynomial $g(x) = x^2 - 2x - 8$ $= (x + 2)(x - 4),$ or $y = (x + 2)(x - 4)$	To find the **zeros** of $g(x)$, we solve $g(x) = 0$: $x^2 - 2x - 8 = 0$ $(x + 2)(x - 4) = 0$ $x + 2 = 0$ or $x - 4 = 0$ $x = -2$ or $x = 4.$ The **solutions** of $x^2 - 2x - 8 = 0$ are -2 and 4. They are the zeros of the function $g(x)$. That is, $g(-2) = 0$ and $g(4) = 0.$	The zeros of $g(x)$ are the x-coordinates of the **x-intercepts** of the graph of $y = g(x)$.
Cubic Polynomial $h(x)$ $= x^3 + 2x^2 - 5x - 6$ $= (x + 3)(x + 1)(x - 2),$ or $y = (x + 3)(x + 1)(x - 2)$	To find the **zeros** of $h(x)$, we solve $h(x) = 0$: $x^3 + 2x^2 - 5x - 6 = 0$ $(x + 3)(x + 1)(x - 2) = 0$ $x + 3 = 0$ or $x + 1 = 0$ or $x - 2 = 0$ $x = -3$ or $x = -1$ or $x = 2.$ The **solutions** of $x^3 + 2x^2 - 5x - 6 = 0$ are -3, -1, and 2. They are the zeros of the function $h(x)$. That is, $h(-3) = 0,$ $h(-1) = 0,$ and $h(2) = 0.$	The zeros of $h(x)$ are the x-coordinates of the **x-intercepts** of the graph of $y = h(x)$.

CONNECTING THE CONCEPTS

COMPARING EXPONENTIAL AND LOGARITHMIC FUNCTIONS

Usually, we use a number that is greater than 1 for the logarithmic base. In the following table, we compare exponential and logarithmic functions with bases a greater than 1. Similar statements would be made for a, where $0 < a < 1$. It is helpful to visualize the differences by carefully observing the graphs.

EXPONENTIAL FUNCTION

$y = a^x$
$f(x) = a^x$
$a > 1$
Continuous
One-to-one
Domain: All real numbers, $(-\infty, \infty)$
Range: All positive real numbers, $(0, \infty)$
Increasing
Horizontal asymptote is x-axis: $(a^x \to 0$ as $x \to -\infty)$
y-intercept: $(0, 1)$
There is no x-intercept.

LOGARITHMIC FUNCTION

$x = a^y$
$f^{-1}(x) = \log_a x$
$a > 1$
Continuous
One-to-one
Domain: All positive real numbers, $(0, \infty)$
Range: All real numbers, $(-\infty, \infty)$
Increasing
Vertical asymptote is y-axis: $(\log_a x \to -\infty$ as $x \to 0^+)$
x-intercept: $(1, 0)$
There is no y-intercept.

Connecting the Concepts

Comprehension is streamlined and retention is maximized when concepts are presented in a visual form. Combining design (usually in outline or chart form) and art, this feature emphasizes the importance of connecting concepts. Its visual aspect invites students to stop and check their understanding of how concepts work together in one or several sections. The increased understanding gained from this feature enhances student performance on homework assignments.

Throughout the exposition, students are directed to investigate new concepts before those concepts are formally developed. This feature invites students to be actively involved with the material in order to identify a mathematical pattern or form an intuitive understanding of a new topic.

Interactive Discovery

Consider the following polynomial function:

$$P(x) = 12x^3 - 5x^2 - 11x + 6.$$

Graph this function using the viewing window $[-1, 2, -2, 10]$. How many zeros does the function appear to have in the interval $[0.5, 1]$?

Now graph the function again using the viewing window $[0.5, 1, -0.1, 0.15]$, with Xscl = 0.1 and Yscl = 0.05. Now how many zeros does the function appear to have in the interval $[0.5, 1]$?

Find $P(0.6)$, $P(0.7)$, and $P(0.8)$ and note any change in sign from one function value to another. Is there a connection between a change in sign and the occurrence of a zero between two function values?

Polynomial functions P are continuous, hence their graphs are unbroken. All polynomial functions have $(-\infty, \infty)$ as the domain. Suppose two function values $P(a)$ and $P(b)$ have opposite signs. Since P is continuous, its graph must be a curve from $(a, P(a))$ to $(b, P(b))$ without a

---

EXAMPLE 3 Solve: $5 + \sqrt{x + 7} = x$.

Algebraic Solution

We first isolate the radical and then use the principle of powers.

$$5 + \sqrt{x + 7} = x$$
$$\sqrt{x + 7} = x - 5 \qquad \text{Subtracting 5 on both sides}$$
$$(\sqrt{x + 7})^2 = (x - 5)^2 \qquad \text{Using the principle of powers; squaring both sides}$$
$$x + 7 = x^2 - 10x + 25$$
$$0 = x^2 - 11x + 18 \qquad \text{Subtracting } x \text{ and 7}$$
$$0 = (x - 9)(x - 2) \qquad \text{Factoring}$$
$$x - 9 = 0 \quad or \quad x - 2 = 0$$
$$x = 9 \quad or \quad x = 2$$

The possible solutions are 9 and 2.

CHECK: For 9:

$$\begin{array}{c} 5 + \sqrt{x + 7} = x \\ \hline 5 + \sqrt{9 + 7} \ ? \ 9 \\ 5 + \sqrt{16} \\ 5 + 4 \\ 9 \ \big| \ 9 \quad \text{TRUE} \end{array}$$

For 2:

$$\begin{array}{c} 5 + \sqrt{x + 7} = x \\ \hline 5 + \sqrt{2 + 7} \ ? \ 2 \\ 5 + \sqrt{9} \\ 5 + 3 \\ 8 \ \big| \ 2 \quad \text{FALSE} \end{array}$$

Since 9 checks but 2 does not, the only solution is 9.

Graphical Solution

We graph $y_1 = 5 + \sqrt{x + 7}$ and $y_2 = x$. Using the INTERSECT feature, we see that the solution is 9.

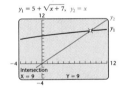

$y_1 = 5 + \sqrt{x + 7}, \ y_2 = x$

We can also use the ZERO feature to get this result. To do so, we first write the equivalent equation

$$5 + \sqrt{x + 7} - x = 0.$$

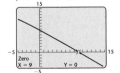

$y = 5 + \sqrt{x + 7} - x$

The zero of the function is 9, so the solution of the original equation is 9. Note that graphs show that the equation has only one solution.

Side-by-Side Algebraic and Graphical Solutions

Many examples in the text are presented in a two-column format that shows simultaneous algebraic and graphical solution methods. This approach allows students to compare the efficiency and appropriateness of each method.

EXAMPLE 6 *Credit Card Volume.* The total credit card volume for Visa, MasterCard, American Express, and Discover has increased dramatically in recent years, as shown in the table on the next page.

a) Use a grapher to fit an exponential function to the data.

b) Graph the function with the scatterplot of the data.

c) Predict the total credit card volume for Visa, MasterCard, American Express, and Discover in 2003.

YEAR, x	CREDIT CARD VOLUME, y (IN BILLIONS)
1988, 0	$261.0
1989, 1	296.3
1990, 2	338.4
1991, 3	361.0
1992, 4	403.1
1993, 5	476.7
1994, 6	584.8
1995, 7	701.2
1996, 8	798.3
1997, 9	885.2

Source: CardWeb Inc.'s CardData

```
ExpReg
y=a*b^x
a=249.7267077
b=1.150881332
r²=.9877691696
r=.9938657704
```

$y = 249.7267077(1.150881332)^x$

Solution

a) We will fit an equation of the type $y = a \cdot b^x$ to the data, where x is the number of years since 1988. Entering the data into the grapher and carrying out the regression procedure, we find that the equation is

$$y = 249.7267077(1.150881332)^x.$$

The correlation coefficient is very close to 1. This gives us a good indication that the exponential function fits the data well.

b) The graph is shown at left.

c) We evaluate the function found in part (a) for $x = 15$ (2003 − 1988 = 15) and estimate that the total credit card volume in 2003 will be about $2056 billion, or $2,056,000,000,000.

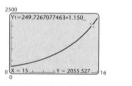

STUDY TIP

It is never too late to begin reviewing for the final examination. The Skill Maintenance exercises found in each exercise set review and reinforce skill taught in earlier sections. Include all of these exercises in your weekly preparation. Answers to both even-numbered and odd-numbered exercises along with section references appear at the back of the text.

Using the results of synthetic division, we can factor further:

$f(x) = (x + 1)\left(x - \frac{2}{3}\right)(3x^2 - 12x + 6)$ Using the results of the last synthetic division

$= (x + 1)\left(x - \frac{2}{3}\right) \cdot 3 \cdot (x^2 - 4x + 2)$.   Removing a factor of 3

The quadratic formula can be used to find the zeros of $x^2 - 4x + 2$:

$$x = \frac{-b \pm \sqrt{b^2 - 4ac}}{2a}$$

$$= \frac{-(-4) \pm \sqrt{(-4)^2 - 4 \cdot 1 \cdot 2}}{2 \cdot 1}$$

$$= \frac{4 \pm \sqrt{8}}{2} = \frac{4 \pm 2\sqrt{2}}{2} = \frac{2(2 \pm \sqrt{2})}{2}$$

$$= 2 \pm \sqrt{2}.$$

The rational zeros are -1 and $\frac{2}{3}$. The other zeros are $2 \pm \sqrt{2}$.

b) The complete factorization of $f(x)$ is

$$f(x) = (x + 1)\left(x - \frac{2}{3}\right) \cdot 3 \cdot [x - (2 - \sqrt{2})][x - (2 + \sqrt{2})]$$

$$= (x + 1)(3x - 2)(x - 2 + \sqrt{2})(x - 2 - \sqrt{2}).$$

Review Icons

These icons prompt students to review topics necessary to understand a concept at hand, and the resulting review enhances students' comprehension. For easy navigation, the section number of the relevant review topic is given in the icon.

Zeros of Linear
Functions
and Models

2.1

• *Find zeros of linear functions and solve linear equations.*
• *Solve applied problems using linear models.*

Zeros of Linear Functions

An input for which a function's output is 0 is called a **zero** of the function.

> **Zeros of Functions**
> An input c of a function f is called a **zero** of the function, if the output for c is 0. That is, $f(c) = 0$.

LINEAR FUNCTIONS
REVIEW SECTION 1.2.

We will restrict our attention to zeros of linear functions in this section. Recall that a linear function is given by $f(x) = mx + b$, where m and b are constants.

For the linear function $f(x) = 2x - 4$, we have $f(2) = 2 \cdot 2 - 4 = 0$, so 2 is a **zero** of the function. In fact, 2 is the *only* zero of this function. In general, a **linear function $f(x) = mx + b$, with $m \neq 0$, has exactly one zero.**

---

38. *Norman Window.* A Norman window is a rectangle with a semicircle on top. Sky Blue Windows is designing a Norman window that will require 24 ft of trim. What dimensions will allow the maximum amount of light to enter a house?

A Norman window

Discussion and Writing

39. Write a problem for a classmate to solve. Design it so that it is a maximum or minimum problem using a quadratic function.

40. Discuss two ways in which we used completing the square in this chapter.

41. Suppose that the graph of $f(x) = ax^2 + bx + c$ has x-intercepts $(x_1, 0)$ and $(x_2, 0)$. What are the x-intercepts of $g(x) = -ax^2 - bx - c$? Explain.

Skill Maintenance

For each function f, construct and simplify the difference quotient

$$\frac{f(x + h) - f(x)}{h}.$$

42. $f(x) = 3x - 7$

43. $f(x) = 2x^2 - x + 4$

A graph of $y = f(x)$ follows. No formula is given for f. Make a hand-drawn graph of each of the following.

44. $g(x) = f(2x)$

45. $g(x) = -2f(x)$

Synthesis

For each equation in Exercises 46–49, under the given condition:

a) *Find k.*
b) *Find a second solution.*

46. $kx^2 - 2x + k = 0$; one solution is -3

47. $kx^2 - 17x + 33 = 0$; one solution is 3

48. $x^2 - (6 + 3i)x + k = 0$; one solution is 3

49. $x^2 - kx + 2 = 0$; one solution is $1 + i$

50. Find b such that
$$f(x) = -4x^2 + bx + 3$$
has a maximum value of 50.

51. Find c such that
$$f(x) = -0.2x^2 - 3x + c$$
has a maximum value of -225.

52. Find a quadratic function with vertex $(4, -5)$ and containing the point $(-3, 1)$.

53. Graph: $f(x) = (|x| - 5)^2 - 3$.

54. *Minimizing Area.* A 24-in. piece of string is cut into two pieces. One piece is used to form a circle while the other is used to form a square. How should the string be cut so that the sum of the areas is a minimum?

Discussion and Writing Exercises

These innovative exercises encourage students both to consider and to write about key mathematical ideas in each section. Many of these exercises are open-ended, making them particularly suitable for use in class discussions or as collaborative activities.

Skill Maintenance Exercises

Skill Maintenance exercises review and reinforce skills previously presented in the text. These exercises provide excellent review for a final examination.

Synthesis Exercises

Synthesis exercises help build critical-thinking skills by requiring students to synthesize or combine learning objectives from the section being studied as well as from preceding sections in the text.

Chapter Summary and Review 4

Important Properties and Formulas

The Composition of Two Functions:	$(f \circ g)(x) = f(g(x))$
One-to-One Function:	$f(a) = f(b) \rightarrow a = b$
Exponential Function:	$f(x) = a^x$
The Number e = 2.7182818284 . . .	
Logarithmic Function:	$f(x) = \log_a x$
A Logarithm is an Exponent:	$\log_a x = y \longleftrightarrow x = a^y$
The Change-of-Base Formula:	$\log_b M = \dfrac{\log_a M}{\log_a b}$
The Product Rule:	$\log_a MN = \log_a M + \log_a N$
The Power Rule:	$\log_a M^p = p \log_a M$
The Quotient Rule:	$\log_a \dfrac{M}{N} = \log_a M - \log_a N$
Other Properties:	$\log_a a = 1, \qquad \log_a 1 = 0,$
	$\log_a a^x = x, \qquad a^{\log_a x} = x$
Base–Exponent Property:	$a^x = a^y \longleftrightarrow x = y,$ for $a > 0$, $a \neq 1$
Exponential Growth Model:	$P(t) = P_0 e^{kt}$
Exponential Decay Model:	$P(t) = P_0 e^{-kt}$
Interest Compounded Continuously:	$P(t) = P_0 e^{kt}$
Limited Growth:	$P(t) = \dfrac{a}{1 + be^{-kt}}$

REVIEW EXERCISES

In Exercises 1 and 2, for the pair of functions:
a) *Find the domain of $f \circ g$ and $g \circ f$.*
b) *Find $(f \circ g)(x)$ and $(g \circ f)(x)$.*

1. $f(x) = \dfrac{4}{x^2}$; $g(x) = 3 - 2x$

2. $f(x) = 3x^2 + 4x$; $g(x) = 2x - 1$

Find $f(x)$ and $g(x)$ such that $h(x) = (f \circ g)x$.

3. $h(x) = \sqrt{5x + 2}$

4. $h(x) = 4(5x - 1)^2 + 9$

5. Find the inverse of the relation
$$\{(1.3, -2.7), (8, -3), (-5, 3), (6, -3), (7, -5)\}.$$

6. Find an equation of the inverse relation.
a) $y = 3x^2 + 2x - 1$
b) $0.8x^3 - 5.4y^2 = 3x$

End-of-Chapter Material

End-of-chapter material includes a summary and review of properties and formulas along with a complete set of Review Exercises. Review Exercises also include Discussion and Writing and Synthesis exercises. The answers to the Review Exercises, which appear at the back of the text, have text section references to further aid students.

College Algebra

GRAPHS & MODELS

SECOND EDITION

Introduction to Graphs and the Graphing Calculator

- *Plot points by hand and using a grapher.*
- *Graph equations by hand and using a grapher.*
- *Find the point(s) of intersection of two graphs.*

Graphing calculators and computers equipped with graphing software are useful tools in applying and understanding mathematical concepts. **All such graphing utilities will be referred to as graphers in this text, although the emphasis will be on graphing calculators.**

When we think of the ways in which a grapher can be used, graphing equations might come to mind first. Although the grapher will be used extensively in this text to graph equations, many of its other uses will also be explored. These include performing calculations, evaluating expressions, solving equations, analyzing the graphs of equations, creating tables of data, finding mathematical models of real data, analyzing those models, and using them to make estimates and predictions.

Keystrokes and features vary among different brands and models of graphers. We will use features that are commonly found on many graphing calculators. **Specific keystrokes and instructions for using these features can be found in the *Graphing Calculator Manual* that accompanies this text.** You can also consult either your instructor or the user's manual for your particular grapher.

The goal of this text is to use the grapher as a tool to enhance the learning and understanding of mathematics and to relieve the tedium of some procedures. *Keep in mind that a grapher cannot be used effectively without a firm mathematical foundation upon which to build.* For example, expressions cannot be entered correctly nor can results be interpreted well if the relevant concepts are not understood and applied correctly.

The Use of the Grapher

A grapher is a tool that can be used in the process of learning and understanding mathematics. It should be used to enhance the understanding of concepts, not to replace the learning of skills.

1

Graphs

Graphs provide a means of displaying, interpreting, and analyzing data in a visual format. It is not uncommon to open a newspaper or magazine and encounter graphs. Shown below are examples of bar, circle, and line graphs.

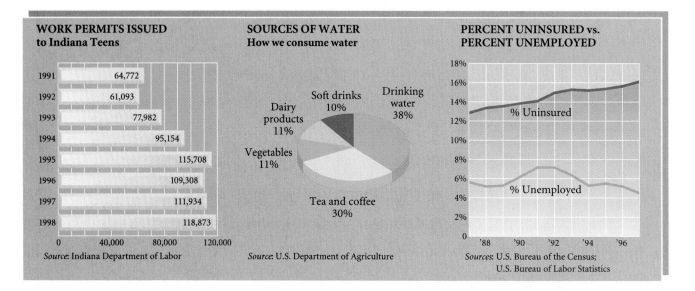

WORK PERMITS ISSUED to Indiana Teens

Year	Permits
1991	64,772
1992	61,093
1993	77,982
1994	95,154
1995	115,708
1996	109,308
1997	111,934
1998	118,873

Source: Indiana Department of Labor

SOURCES OF WATER
How we consume water

Soft drinks 10%
Drinking water 38%
Dairy products 11%
Vegetables 11%
Tea and coffee 30%

Source: U.S. Department of Agriculture

PERCENT UNINSURED vs. PERCENT UNEMPLOYED

% Uninsured
% Unemployed

Sources: U.S. Bureau of the Census;
U.S. Bureau of Labor Statistics

Many real-world situations can be modeled, or described mathematically, using equations in which two variables appear. We use a plane to graph a pair of numbers. To locate points on a plane, we use two perpendicular number lines, called **axes**, which intersect at (0, 0). We call this point the **origin**. The horizontal axis is called the **x-axis**, and the vertical axis is called the **y-axis**. (Other variables, such as a and b, can also be used.) The axes divide the plane into four regions, called **quadrants**, denoted by Roman numerals and numbered counterclockwise from the upper right. Arrows show the positive direction of each axis.

Each point (x, y) in the plane is called an **ordered pair.** The first number, x, indicates the point's horizontal location with respect to the y-axis, and the second number, y, indicates the point's vertical location with respect to the x-axis. We call x the **first coordinate**, **x-coordinate**, or **abscissa**. We call y the **second coordinate, y-coordinate,** or **ordinate**. Such a representation is called the **Cartesian coordinate system** in honor of the great French mathematician and philosopher René Descartes (1596–1650).

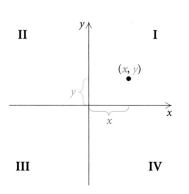

EXAMPLE 1 Graph and label the points $(-3, 5)$, $(4, 3)$, $(3, 4)$, $(-4, -2)$, $(3, -4)$, $(0, 4)$, $(-3, 0)$, and $(0, 0)$.

Solution To graph or **plot** $(-3, 5)$, we note that the x-coordinate tells us to move from the origin 3 units to the left of the y-axis. Then we move

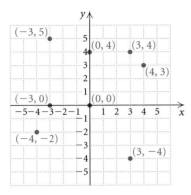

5 units up from the x-axis. To graph the other points, we proceed in a similar manner. (See the graph at left.) Note that the origin, (0, 0), is located at the intersection of the axes and that the point (4, 3) is different from the point (3, 4).

A graph of a set of points like the one in Example 1 is called a **scatterplot**. A grapher can also be used to make a scatterplot. This involves, among other things, determining the portion of the xy-plane that will appear on the grapher's screen. That portion of the plane is called the **viewing window**.

The notation used in this text to denote a window setting consists of four numbers [L, R, B, T], which represent the **L**eft and **R**ight endpoints of the x-axis and the **B**ottom and **T**op endpoints of the y-axis, respectively. The window with the settings $[-10, 10, -10, 10]$ is the **standard viewing window,** shown in the following figure. On some graphers, the standard window can be selected quickly using the ZSTANDARD feature from the ZOOM menu.

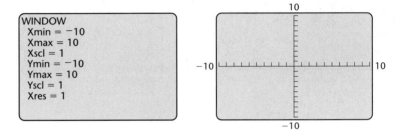

Xmin and Xmax are used to set the left and right endpoints of the x-axis, respectively; Ymin and Ymax are used to set the bottom and top endpoints of the y-axis. The settings Xscl and Yscl give the scales for the axes. For example, Xscl = 1 and Yscl = 1 means that there is 1 unit between tick marks on each of the axes. Generally in this text, Xscl and Yscl are both assumed to be 1 unless different values are given. Some exceptions will occur when the scaling factor is different from 1 but readily apparent.

EXAMPLE 2 Use a grapher to graph the points $(-3, 5)$, $(4, 3)$, $(3, 4)$, $(-4, -2)$, $(3, -4)$, $(0, 4)$, $(-3, 0)$, and $(0, 0)$.

Solution We note that the x-coordinates of the given points range from -4 to 4 and the y-coordinates range from -4 to 5. Thus one good choice for the viewing window is the standard window, because x- and y-values both range from -10 to 10 in this window.

The coordinates of the points are entered into the grapher in lists using the EDIT operation from the STAT menu. Note that the $\boxed{(-)}$ key rather than the $\boxed{-}$ key *must* be used when we are entering negative numbers. We enter the first coordinates in one list, and then the corresponding second coordinates in the same order in a second list. Then we turn on and set up a STAT PLOT to draw a scatterplot using the data from these two lists.

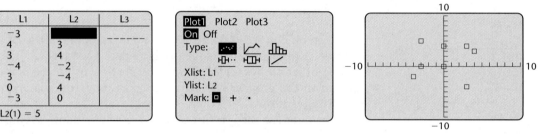

Some graphers have a ZOOMSTAT feature that automatically selects a viewing window that displays all the data points in the lists for a given STAT PLOT. This feature is activated after the data have been entered.

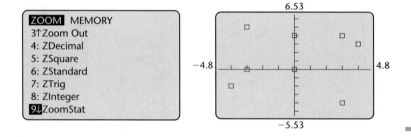

To turn off the plot, we return to the plot screen and select Off.

Solutions of Equations

Equations in two variables, like $2x + 3y = 18$, have solutions (x, y) that are ordered pairs such that when the first coordinate is substituted for x and the second coordinate is substituted for y, the result is a true equation.

EXAMPLE 3   Determine whether each ordered pair is a solution of $2x + 3y = 18$.

a) $(-5, 7)$ **b)** $(3, 4)$

Solution   We substitute the ordered pair into the equation and determine whether the resulting equation is true.

a) $\quad\quad 2x + 3y = 18$

$2(-5) + 3(7)\ ?\ 18 \quad\quad$ We substitute -5 for x and 7 for y (alphabetical order).

$\quad\quad -10 + 21$

$\quad\quad\quad\quad 11\ \big|\ 18 \quad$ FALSE

The equation $11 = 18$ is false, so $(-5, 7)$ is not a solution.

b)
$$2x + 3y = 18$$

$$2(3) + 3(4) \ ? \ 18 \qquad \text{We substitute 3 for } x \text{ and 4 for } y.$$

$$6 + 12 \ \Big|$$

$$18 \ \Big| \ 18 \qquad \text{TRUE}$$

The equation $18 = 18$ is true, so $(3, 4)$ is a solution.

We can also perform these substitutions on a grapher. When we substitute -5 for x and 7 for y, we get 11. Since $11 \neq 18$, $(-5, 7)$ is not a solution of the equation. When we substitute 3 for x and 4 for y, we get 18, so $(3, 4)$ is a solution.

```
2(−5)+3*7
                              11
2*3+3*4
                              18
```

Graphs of Equations

The equation considered in Example 3 actually has an infinite number of solutions. Since we cannot list all the solutions, we will make a drawing, called a **graph**, that represents them.

> ### To Graph an Equation
> To **graph an equation** is to make a drawing that represents the solutions of that equation.

Shown at left are some suggestions for making hand-drawn graphs.

EXAMPLE 4   Graph: $2x + 3y = 18$.

Solution   To find ordered pairs that are solutions of this equation, we can replace either x or y with any number and then solve for the other variable. For instance, if x is replaced with 0, then

$$2 \cdot 0 + 3y = 18$$
$$3y = 18$$
$$y = 6. \qquad \text{Dividing by 3}$$

Thus $(0, 6)$ is a solution. If x is replaced with 5, then

$$2 \cdot 5 + 3y = 18$$
$$10 + 3y = 18$$
$$3y = 8 \qquad \text{Subtracting 10}$$
$$y = \tfrac{8}{3}. \qquad \text{Dividing by 3}$$

Suggestions for Hand-Drawn Graphs

1. Use graph paper.
2. Draw axes and label them with the variables.
3. Use arrows on the axes to indicate positive directions.
4. Scale the axes; that is, mark numbers on the axes.
5. Calculate solutions and list the ordered pairs in a table.
6. Plot the ordered pairs, look for patterns, and complete the graph. Label the graph with the equation being graphed.

Thus $\left(5, \frac{8}{3}\right)$ is a solution. If y is replaced with 0, then

$$2x + 3 \cdot 0 = 18$$
$$2x = 18$$
$$x = 9. \qquad \text{Dividing by 2}$$

Thus (9, 0) is a solution.

We continue choosing values for one variable and finding the corresponding values of the other. We list the solutions in a table, and then plot the points. Note that the points appear to lie on a straight line.

x	y	(x, y)
0	6	(0, 6)
5	$\frac{8}{3}$	$\left(5, \frac{8}{3}\right)$
9	0	(9, 0)
-1	$\frac{20}{3}$	$\left(-1, \frac{20}{3}\right)$

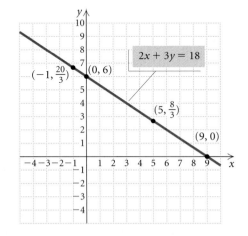

In fact, were we to graph additional solutions of $2x + 3y = 18$, they would be on the same straight line. Thus, to complete the graph, we use a straightedge to draw a line as shown in the figure. That line represents all solutions of the equation.

EXAMPLE 5   Use a grapher to create a table of ordered pairs that are solutions of the equation $y = \frac{1}{2}x + 1$. Then graph the equation by hand.

Solution   We use the TABLE feature on a grapher to create the table of ordered pairs. We must first enter the equation on the *equation-editor*, or "$y =$", screen. Many graphers require an equation to be entered in the form "$y =$" as this one is written. If an equation is not initially given in this form, it must be solved for y before it is entered in the grapher.

```
Plot1   Plot2   Plot3
\Y1 ■(1/2)X+1
\Y2=
\Y3=
\Y4=
\Y5=
\Y6=
\Y7=
```

On a grapher, we enter $y = (1/2)x + 1$. Since $1/2x$ is interpreted as $(1/2)x$ by some graphers and as $1/(2x)$ by others, we use parentheses to ensure that the equation is interpreted correctly. Then we set up a table in AUTO mode by designating a value for TBLSTART and a value for ΔTBL. The grapher will produce a table starting with the value of TBLSTART and continuing by adding ΔTBL to supply succeeding x-values. For the equation $y = \frac{1}{2}x + 1$, we let TBLSTART $= -3$ and ΔTBL $= 1$.

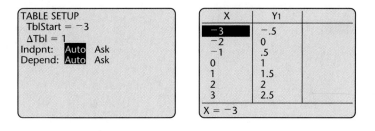

Next, we plot some of the points given in the table and draw the graph.

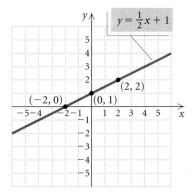

In the equation $y = \frac{1}{2}x + 1$, the value of y *depends* on the value chosen for x, so x is said to be the **independent variable** and y the **dependent variable.**

We can also graph an equation on a grapher. Be sure that the stat plots are turned off, as described on p. 4, before graphing equations like those below. Failure to do this could affect the viewing window and prevent the desired graph from being seen.

EXAMPLE 6 Graph using a grapher: $y = \frac{1}{2}x + 1$.

Solution We enter $y = \frac{1}{2}x + 1$ on the equation-editor screen in the form $y = (1/2)x + 1$, select a viewing window, and draw the graph. The graph is shown in the standard window.

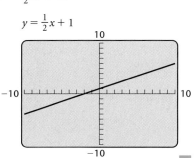

EXAMPLE 7   Graph: $y = x^2 - 5$.

Solution   We select values for x and find the corresponding y-values. Then we plot the points and connect them with a smooth curve.

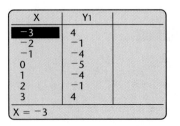

x	y	(x, y)
-3	4	$(-3, 4)$
-2	-1	$(-2, -1)$
-1	-4	$(-1, -4)$
0	-5	$(0, -5)$
1	-4	$(1, -4)$
2	-1	$(2, -1)$
3	4	$(3, 4)$

① Select values for x.

② Compute values for y.

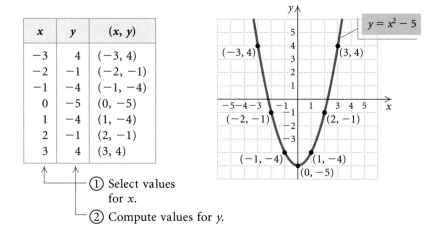

We can also graph this equation using a grapher, as shown below.

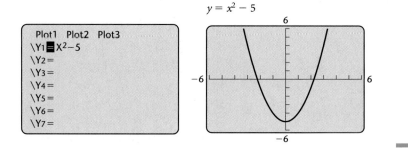

EXAMPLE 8   Graph: $y = x^2 - 9x - 12$.

Solution   We make a table of values, plot enough points to obtain an idea of the shape of the curve, and connect them with a smooth curve.

x	y
-3	24
-1	-2
0	-12
2	-26
4	-32
5	-32
10	-2
12	24

We can also graph the equation using a grapher. We enter $y = x^2 - 9x - 12$ on the equation-editor screen and graph the equation in the standard window.

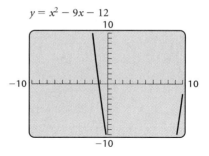

Note that this window does not give a good picture of the graph. Because the graph is cut off to the right of $x = 10$ and below $y = -10$, it appears as though Xmax should be larger than 10 and Ymin should be less than -10. Since the graph rises steeply in the second quadrant, we could also let Xmin be -5 rather than -10. We try $[-5, 15, -15, 10]$, with Yscl $= 2$.

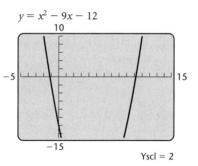

The graph is still cut off below Ymin, or -15. We try other settings for Ymin until we find one that shows the lower portion of the graph. One good window is $[-5, 15, -35, 10]$, with Yscl $= 5$.

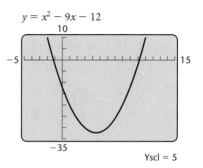

Finding Points of Intersection

There are many situations in which we want to determine the point(s) of intersection of two graphs.

EXAMPLE 9   Use a grapher to find the point of intersection of the graphs of $x - y = -5$ and $y = 4x + 10$.

Solution   Since equations must be entered in the form "$y =$" on many graphers, we solve the first equation for y:

$$
\begin{aligned}
x - y &= -5 \\
x &= y - 5 &&\text{Adding } y \text{ on both sides} \\
x + 5 &= y. &&\text{Adding 5 on both sides}
\end{aligned}
$$

Now we can enter $y_1 = x + 5$ and $y_2 = 4x + 10$ on the equation-editor screen and graph the equations. We begin by using the standard window and see that it is a good choice because it shows the point of intersection of the graphs. Next, we use the INTERSECT feature from the CALC menu to find the coordinates of the point of intersection. (See the *Graphing Calculator Manual* that accompanies this text for the keystrokes.)

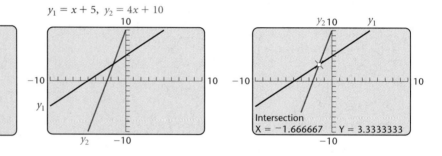

$y_1 = x + 5, \ y_2 = 4x + 10$

Plot1 Plot2 Plot3
\Y1◼X+5
\Y2◼4X+10
\Y3=
\Y4=
\Y5=
\Y6=
\Y7=

Intersection
X = −1.666667   Y = 3.3333333

The graphs intersect at the point $(-1.666667, 3.3333333)$, which is a decimal approximation for the point of intersection. If the coordinates are rational numbers, their exact values can be found using the ▶FRAC feature from the MATH menu. (The keystrokes for doing these conversions are given in the *Graphing Calculator Manual* that accompanies this text.)

RATIONAL NUMBERS
REVIEW SECTION R.1.

X▶Frac
−5/3
Y▶Frac
10/3

The point of intersection is $\left(-\frac{5}{3}, \frac{10}{3}\right)$.

If the coordinates in Example 9 had not been rational numbers, the ▶FRAC operation would have returned the original decimal approximation rather than a fraction.

If a pair of equations has more than one point of intersection, we use the INTERSECT feature repeatedly to find the coordinates of all the points.

Exercise Set

Graph and label the given points by hand.

1. $(4, 0)$, $(-3, -5)$, $(-1, 4)$, $(0, 2)$, $(2, -2)$

2. $(1, 4)$, $(-4, -2)$, $(-5, 0)$, $(2, -4)$, $(4, 0)$

3. $(-5, 1)$, $(5, 1)$, $(2, 3)$, $(2, -1)$, $(0, 1)$

4. $(4, 0)$, $(4, -3)$, $(-5, 2)$, $(-5, 0)$, $(-1, -5)$

Use a grapher to make a scatterplot of the given points.

5. $(-3, 2)$, $(-6, -5)$, $(0, -1)$, $(4, 1)$, $(3, 0)$

6. $(0, 0)$, $(4, -6)$, $(-6, 4)$, $(3, 1)$, $(-5, -3)$

7. $(4.3, -6.7)$, $(-5.1, 3.6)$, $(2.8, 0)$ $(-4.5, -8.4)$, $(7.3, 2.4)$

8. $(-3.2, -4.4)$, $(0, -8.3)$, $(1.6, 7.8)$, $(5.4, -2.9)$, $(-6.7, 1.8)$

Use substitution to determine whether the given ordered pairs are solutions of the given equation.

9. $(1, -1)$, $(0, 3)$; $y = 2x - 3$

10. $(2, 5)$, $(-2, -5)$; $y = 3x - 1$

11. $\left(\frac{2}{3}, \frac{3}{4}\right)$, $\left(1, \frac{3}{2}\right)$; $6x - 4y = 1$

12. $(1.5, 2.6)$, $(-3, 0)$; $x^2 + y^2 = 9$

13. $\left(-\frac{1}{2}, -\frac{4}{5}\right)$, $\left(0, \frac{3}{5}\right)$; $2a + 5b = 3$

14. $\left(0, \frac{3}{2}\right)$, $\left(\frac{2}{3}, 1\right)$; $3m + 4n = 6$

15. $(-0.75, 2.75)$, $(2, -1)$; $x^2 - y^2 = 3$

16. $(2, -4)$, $(4, -5)$; $5x + 2y^2 = 70$

Use a grapher to create a table of values with TBLSTART $= -3$ and ΔTBL $= 1$. Then graph the equation by hand.

17. $y = 3x + 5$ **18.** $y = -2x - 1$

19. $x - y = 3$ **20.** $x + y = 4$

21. $2x + y = 4$ **22.** $3x - y = 6$

23. $x - 4y = 5$ **24.** $x + 3y = 6$

25. $3x - 4y = 12$ **26.** $2x + 3y = -6$

27. $2x + 5y = -10$

28. $4x - 3y = 12$

29. $y = -x^2$

30. $y = x^2$

31. $y = x^2 - 3$

32. $y = 4 - x^2$

33. $y = x^2 - 2x - 3$

34. $y = x^2 + 2x - 1$

In Exercises 35–38, use a grapher to match the equation with one of the graphs (a)–(d), which follow.

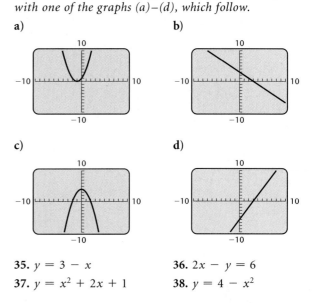

a)

b)

c)

d)

35. $y = 3 - x$

36. $2x - y = 6$

37. $y = x^2 + 2x + 1$

38. $y = 4 - x^2$

Use a grapher to graph the equation in the standard window.

39. $y = 2x + 1$ **40.** $y = 3x - 4$

41. $4x + y = 7$ **42.** $5x + y = -8$

43. $y = \frac{1}{3}x + 2$ **44.** $y = \frac{3}{2}x - 4$

45. $2x + 3y = -5$ **46.** $3x + 4y = 1$

47. $y = x^2 + 6$ **48.** $y = x^2 - 8$

49. $y = 2 - x^2$ **50.** $y = 5 - x^2$

51. $y = x^2 + 4x - 2$ **52.** $y = x^2 - 5x + 3$

Graph the equation in the standard window and in the given window. Determine which window better shows the shape of the graph and where it crosses the x- and y-axes.

53. $y = x - 20$
$[-25, 25, -25, 25]$, with Xscl $= 5$ and Yscl $= 5$

54. $y = -2x + 24$
$[-15, 15, -10, 30]$, with Xscl $= 3$ and Yscl $= 5$

55. $y = 3x^2 - 6$
$[-4, 4, -4, 4]$

56. $y = 8 - x^2$
$[-3, 3, -3, 3]$

57. $y = -\frac{1}{6}x^2 + \frac{1}{12}$
$[-1, 1, -0.3, 0.3]$, with Xscl $= 0.1$ and Yscl $= 0.1$

58. $y = \frac{1}{2}x^2 - \frac{1}{4}x - \frac{1}{8}$
$[-0.8, 1.2, -0.5, 0.5]$, with Xscl $= 0.1$ and Yscl $= 0.1$

Use a grapher to find the point(s) of intersection of the graphs of the given pair of equations.

59. $y = 2x + 3, \ y = x + 4$

60. $y = 3x - 1, \ y = x + 5$

61. $y = 4 - x, \ y = 2x - 1$

62. $y = 5 - 2x, \ y = x + 7$

63. $y = 4x - 7, \ y = 4 - 2x$

64. $y = 5x - 3, \ y = 8 - x$

65. $y = x^2, \ y = 4$

66. $y = -x^2, \ y = -1$

67. $y = x^2 - 3x, \ y = x + 4$

68. $y = x^2 - 5x, \ y = x - 1$

Basic Concepts of Algebra R

The formula

$$M = P\left[\frac{\frac{i}{12}\left(1 + \frac{i}{12}\right)^n}{\left(1 + \frac{i}{12}\right)^n - 1}\right]$$

gives the monthly mortgage payment M on a home loan of P dollars at interest rate i, where n is the total number of payments (12 times the number of years).

Suppose that a house costs $124,000, the down payment is $20,000, the interest rate is $7\frac{3}{4}\%$, and the loan period is 30 yr. What is the monthly mortgage payment?

The monthly mortgage payment is $745.07.

This problem appears as Exercise 68 in Section R.2.

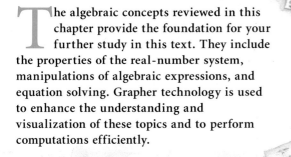

```
124000−20000
                      104000
104000((((.0775/12)(1+(.0775/
12))^360)/((1+(.0775/12))^36
0−1))
                  745.0687354
```

The algebraic concepts reviewed in this chapter provide the foundation for your further study in this text. They include the properties of the real-number system, manipulations of algebraic expressions, and equation solving. Grapher technology is used to enhance the understanding and visualization of these topics and to perform computations efficiently.

The Real-Number System
R.1

- Identify various kinds of real numbers.
- Use interval notation to write a set of numbers.
- Identify the properties of real numbers.
- Find the absolute value of a real number.

Real Numbers

In applications of algebraic concepts, we use real numbers to represent quantities such as distance, time, speed, area, profit, loss, and temperature. Some frequently used sets of real numbers and the relationships among them are shown below.

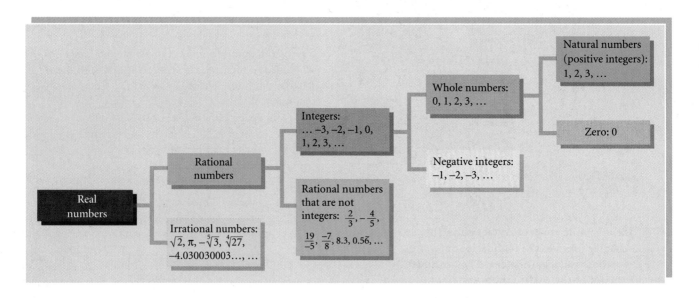

Numbers that can be expressed in the form p/q, where p and q are integers and $q \neq 0$, are **rational numbers.** Decimal notation for rational numbers either *terminates* (ends) or *repeats*. Each of the following is a rational number.

a) 0 $0 = \dfrac{0}{a}$ for any nonzero integer a

b) -7 $-7 = \dfrac{-7}{1}$, or $\dfrac{7}{-1}$, or $-\dfrac{7}{1}$

c) $\dfrac{1}{4} = 0.25$ **Terminating decimal**

d) $-\dfrac{5}{11} = -0.\overline{45}$ **Repeating decimal**

The real numbers that are not rational are **irrational numbers.** Decimal notation for irrational numbers neither terminates nor repeats.

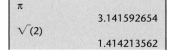

Each of the following is an irrational number.

a) $\pi = 3.1415926535\ldots$ There is no repeating block of digits.

$\left(\frac{22}{7} \text{ and } 3.14 \text{ are rational } approximations \text{ of the irrational number } \pi.\right)$

b) $\sqrt{2} = 1.414213562\ldots$ There is no repeating block of digits.

c) $-6.12122122212222\ldots$ Although there is a pattern, there is no repeating block of digits.

The set of all rational numbers combined with the set of all irrational numbers gives us the set of **real numbers.** The real numbers are *modeled* using a **number line,** as shown below.

Each point on the line represents a real number.

The order of the real numbers can be determined from the number line. If a number a is to the left of a number b, then a **is less than** b ($a < b$). Similarly, a **is greater than** b ($a > b$), if a is to the right of b on the number line. For example, we see from the number line above that $-2.9 < -\frac{3}{5}$, because -2.9 is to the left of $-\frac{3}{5}$. Also, $\frac{17}{4} > \sqrt{3}$, because $\frac{17}{4}$ is to the right of $\sqrt{3}$.

The statement $a \leq b$, read "a is less than or equal to b," is true if either $a < b$ is true or $a = b$ is true.

The symbol $\in$ is used to indicate that a member, or **element**, belongs to a set. Thus if we let $\mathbb{Q}$ represent the set of rational numbers, we can see from the diagram on p. 14 that $0.5\overline{6} \in \mathbb{Q}$. We can also write $\sqrt{2} \notin \mathbb{Q}$ to indicate that $\sqrt{2}$ is *not* an element of the set of rational numbers.

When *all* the elements of one set are elements of a second set, we say that the first set is a **subset** of the second set. The symbol $\subseteq$ is used to denote this. For instance, if we let $\mathbb{R}$ represent the set of real numbers, we can see from the diagram that $\mathbb{Q} \subseteq \mathbb{R}$ (read "$\mathbb{Q}$ is a subset of $\mathbb{R}$").

Interval Notation

Sets of real numbers can be expressed using **interval notation.** For example, for real numbers a and b such that $a < b$, the **open interval** (a, b) is the set of real numbers between, but not including, a and b. That is,

$$(a, b) = \{x \,|\, a < x < b\}.$$

The points a and b are the **endpoints** of the interval. The parentheses indicate that the endpoints are not included in the interval.

Some intervals extend without bound in one or both directions. The interval $[a, \infty)$, for example, begins at a and extends to the right without bound. That is,

$$[a, \infty) = \{x \,|\, x \geq a\}.$$

The bracket indicates that a is included in the interval.

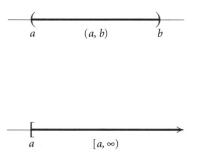

The various types of intervals are listed below.

Intervals: Types, Notation, and Graphs

TYPE	INTERVAL NOTATION	SET NOTATION	GRAPH
Open	(a, b)	$\{x \mid a < x < b\}$	
Closed	$[a, b]$	$\{x \mid a \leq x \leq b\}$	
Half-open	$[a, b)$	$\{x \mid a \leq x < b\}$	
Half-open	$(a, b]$	$\{x \mid a < x \leq b\}$	
Open	(a, ∞)	$\{x \mid x > a\}$	
Half-open	$[a, \infty)$	$\{x \mid x \geq a\}$	
Open	$(-\infty, b)$	$\{x \mid x < b\}$	
Half-open	$(-\infty, b]$	$\{x \mid x \leq b\}$	

The interval $(-\infty, \infty)$ names the set of all real numbers, $\mathbb{R}$.

EXAMPLE 1   Write interval notation for each set.

a) $\{x \mid -4 < x < 5\}$
b) $\{x \mid x \geq 1.7\}$
c) $\{x \mid -10 < x \leq -5\}$
d) $\{x \mid x < \sqrt{5}\}$

Solution

a) $\{x \mid -4 < x < 5\} = (-4, 5)$
b) $\{x \mid x \geq 1.7\} = [1.7, \infty)$
c) $\{x \mid -10 < x \leq -5\} = (-10, -5]$
d) $\{x \mid x < \sqrt{5}\} = (-\infty, \sqrt{5})$

Properties of the Real Numbers

The following properties can be used to manipulate algebraic expressions as well as real numbers.

Properties of the Real Numbers

For any real numbers a, b, and c:

$a + b = b + a$ and $ab = ba$	Commutative properties of addition and multiplication
$a + (b + c) = (a + b) + c$ and $a(bc) = (ab)c$	Associative properties of addition and multiplication
$a + 0 = 0 + a = a$	Additive identity property
$-a + a = a + (-a) = 0$	Additive inverse property
$a \cdot 1 = 1 \cdot a = a$	Multiplicative identity property
$a \cdot \dfrac{1}{a} = 1 \ (a \neq 0)$	Multiplicative inverse property
$a(b + c) = ab + ac$	Distributive property

Note that the distributive property is also true for subtraction since $a(b - c) = a[b + (-c)] = ab + a(-c) = ab - ac$.

EXAMPLE 2   State the property being illustrated in each sentence.

a) $8 \cdot 5 = 5 \cdot 8$ **b)** $5 + (m + n) = (5 + m) + n$

c) $x + 0 = x$ **d)** $14 + (-14) = 0$

e) $2(a - b) = 2a - 2b$

Solution

SENTENCE	PROPERTY
a) $8 \cdot 5 = 5 \cdot 8$	Commutative property of multiplication: $ab = ba$
b) $5 + (m + n) = (5 + m) + n$	Associative property of addition: $a + (b + c) = (a + b) + c$
c) $x + 0 = x$	Additive identity property: $a + 0 = a$
d) $14 + (-14) = 0$	Additive inverse property: $a + (-a) = 0$
e) $2(a - b) = 2a - 2b$	Distributive property: $a(b + c) = ab + ac$

Absolute Value

The number line can be used to provide a geometric interpretation of *absolute value*. The **absolute value** of a number a, denoted $|a|$, is its distance from 0 on the number line. For example, $|-5| = 5$, because the distance of -5 from 0 is 5. Similarly, $\left|\frac{3}{4}\right| = \frac{3}{4}$, because the distance of $\frac{3}{4}$ from 0 is $\frac{3}{4}$.

Absolute Value

For any real number a,

$$|a| = \begin{cases} a, & \text{if } a \geq 0, \\ -a, & \text{if } a < 0. \end{cases}$$

It follows from the definition of absolute value that, for any real number a, $|a| \geq 0$. We can use a grapher to confirm this. We enter $y = |x|$ as $y = \text{abs}(x)$.

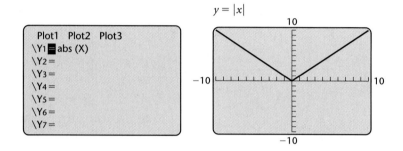

$y = |x|$

We see from the graph that all y-values are nonnegative.

Absolute value can be used to find the distance between two points on the number line.

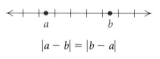

$|a - b| = |b - a|$

Distance Between Two Points on the Number Line

For any real numbers a and b, the **distance between a and b** is $|a - b|$, or, equivalently, $|b - a|$.

EXAMPLE 3   Find the distance between -2 and 3.

Solution

$$|-2 - 3| = |-5| = 5, \quad \text{or, equivalently,}$$
$$|3 - (-2)| = |3 + 2| = |5| = 5.$$

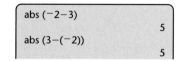

Exercise Set R.1

In Exercises 1–6, consider the numbers -12, $\sqrt{7}$, $5.\overline{3}$, $-\frac{7}{3}$, $\sqrt[3]{8}$, 0, $5.242242224\ldots$, $-\sqrt{14}$, $\sqrt[5]{5}$, -1.96, 9, $4\frac{2}{3}$, $\sqrt{25}$, $\sqrt[3]{4}$, $\frac{5}{7}$.

1. Which are whole numbers? ○
2. Which are integers? -12
3. Which are irrational numbers?
4. Which are natural numbers? 9
5. Which are rational numbers?
6. Which are real numbers? all

Find an example of each of the following. Answers may vary.

7. A rational number that is not an integer
8. A real number that is not a rational number
9. An integer that is not a whole number
10. A real number that is not an irrational number

Write interval notation.

11. $\{x \mid -3 \le x \le 3\}$
12. $\{x \mid -4 < x < 4\}$
13. $\{x \mid -14 \le x < -11\}$
14. $\{x \mid 6 < x \le 20\}$
15. $\{x \mid x \le -4\}$
16. $\{x \mid x > -5\}$
17. $\{x \mid x < 3.8\}$
18. $\{x \mid x \ge \sqrt{3}\}$
19. $\{x \mid x \ne 7\}$
20. $\{x \mid x \ne -3\}$

Write interval notation for the graph.

21.

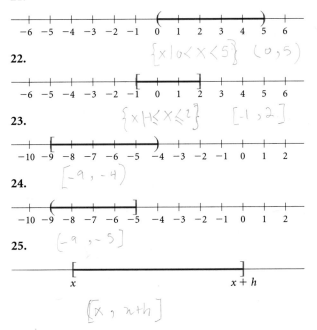

$\{x \mid 0 < x < 5\}$ $(0,5)$

22.

$\{x \mid -1 \le x \le 2\}$ $[-1,2]$

23.

$[-9, -4)$

24.

$(-9, -5]$

25.

$[x, x+h]$

26.

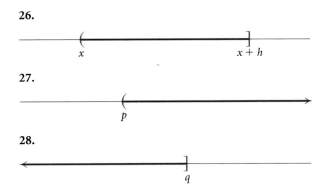

27.

28.

In Exercises 29–46, the following notation is used: $\mathbb{N}$ = the set of natural numbers, $\mathbb{W}$ = the set of whole numbers, $\mathbb{Z}$ = the set of integers, $\mathbb{Q}$ = the set of rational numbers, $\mathbb{I}$ = the set of irrational numbers, and $\mathbb{R}$ = the set of real numbers. Classify the statement as true or false.

29. $6 \in \mathbb{N}$
30. $0 \notin \mathbb{N}$
31. $3.2 \in \mathbb{Z}$
32. $-10.\overline{1} \in \mathbb{R}$
33. $-\dfrac{11}{5} \in \mathbb{Q}$
34. $-\sqrt{6} \in \mathbb{Q}$
35. $\sqrt{11} \notin \mathbb{R}$
36. $-1 \in \mathbb{W}$
37. $24 \notin \mathbb{W}$
38. $1 \in \mathbb{Z}$
39. $1.089 \notin \mathbb{I}$
40. $\mathbb{N} \subseteq \mathbb{W}$
41. $\mathbb{W} \subseteq \mathbb{Z}$
42. $\mathbb{Z} \subseteq \mathbb{N}$
43. $\mathbb{Q} \subseteq \mathbb{R}$
44. $\mathbb{Z} \subseteq \mathbb{Q}$
45. $\mathbb{R} \subseteq \mathbb{Z}$
46. $\mathbb{Q} \subseteq \mathbb{I}$

Name the property illustrated by the sentence.

47. $6x = x6$
48. $3 + (x + y) = (3 + x) + y$
49. $-3 \cdot 1 = -3$
50. $x + 4 = 4 + x$
51. $5(ab) = (5a)b$
52. $4(y - z) = 4y - 4z$
53. $2(a + b) = (a + b)2$
54. $-7 + 7 = 0$
55. $-6(m + n) = -6(n + m)$
56. $t + 0 = t$
57. $8 \cdot \dfrac{1}{8} = 1$
58. $9x + 9y = 9(x + y)$

Simplify.

59. $|-7.1|$
60. $|0|$

61. $\left| \dfrac{5}{4} \right|$

62. $|-\sqrt{3}|$

Find the distance between the given pair of points on the number line.

63. $-5, 6$

64. $0, -2.5$

65. $-2, -8$

66. $\dfrac{15}{8}, \dfrac{23}{12}$

67. $12.1, 6.7$

68. $-3, -14$

Discussion and Writing

To the student and the instructor: The Discussion and Writing exercises are meant to be answered with one or more sentences. These exercises can also be discussed and answered collaboratively by the entire class or by small groups. Because of their open-ended nature, the answers to these exercises do not appear at the back of the book. They are denoted by the words "Discussion and Writing."

69. How would you convince a classmate that division is not associative?

70. Under what circumstances is $\sqrt{a}$ a rational number?

Synthesis

To the student and the instructor: The Synthesis exercises found at the end of every exercise set challenge students to combine concepts or skills studied in that section or in preceding parts of the text.

Between any two (different) real numbers there are many other real numbers. Find each of the following. Answers may vary.

71. An irrational number between 0.124 and 0.125

72. A rational number between $-\sqrt{2.01}$ and $-\sqrt{2}$

73. A rational number between $-\dfrac{1}{101}$ and $-\dfrac{1}{100}$

74. An irrational number between $\sqrt{5.99}$ and $\sqrt{6}$

75. The hypotenuse of an isosceles right triangle with legs of length 1 unit can be used to "measure" a value for $\sqrt{2}$ by using the Pythagorean theorem, as shown.

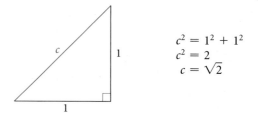

$$c^2 = 1^2 + 1^2$$
$$c^2 = 2$$
$$c = \sqrt{2}$$

Draw a right triangle that could be used to "measure" $\sqrt{10}$ units.

Integer Exponents, Scientific Notation, and Order of Operations

R.2

• *Simplify expressions with integer exponents.*
• *Solve problems using scientific notation.*
• *Use the rules for order of operations.*

Integers as Exponents

When a positive integer is used as an *exponent,* it indicates the number of times a factor appears in a product. For example, 7^3 means $7 \cdot 7 \cdot 7$ and 5^1 means 5.

For any positive integer n,

$$a^n = \underbrace{a \cdot a \cdot a \cdots a}_{n \text{ factors}},$$

where a is the **base** and n is the **exponent**.

Zero and negative-integer exponents are defined as follows.

For any nonzero real number a and any integer n,

$$a^0 = 1 \quad \text{and} \quad a^{-n} = \frac{1}{a^n}.$$

EXAMPLE 1   Simplify each of the following.

a) 6^0 =   **b)** $(-3.4)^0$

Solution

a) $6^0 = 1$ **b)** $(-3.4)^0 = 1$

EXAMPLE 2   Write each of the following with positive exponents.

a) 4^{-5} **b)** $\dfrac{1}{(0.82)^{-7}}$

Solution

a) $4^{-5} = \dfrac{1}{4^5}$

b) $\dfrac{1}{(0.82)^{-7}} = (0.82)^{-(-7)} = (0.82)^7$

The following properties of exponents can be used to simplify expressions.

Properties of Exponents

For any real numbers a and b and any integers m and n, assuming 0 is not raised to a nonpositive power:

$a^m \cdot a^n = a^{m+n}$	Product rule
$\dfrac{a^m}{a^n} = a^{m-n} \ (a \neq 0)$	Quotient rule
$(a^m)^n = a^{mn}$	Power rule
$(ab)^m = a^m b^m$	Raising a product to a power
$\left(\dfrac{a}{b}\right)^m = \dfrac{a^m}{b^m} \ (b \neq 0)$	Raising a quotient to a power

We can use a grapher for partial confirmation of the first three properties. In the case of the product rule, for instance, we can enter

TABLES ON A GRAPHER

REVIEW "INTRODUCTION
TO GRAPHS AND THE
GRAPHING CALCULATOR."

$y_1 = x^3 \cdot x^2$ and $y_2 = x^5$ and compare the values of y_1 and y_2 in a table, as shown below.

$y_1 = x^3 \cdot x^2, \; y_2 = x^5$

X	Y1	Y2
-3	-243	-243
-2	-32	-32
-1	-1	-1
0	0	0
1	1	1
2	32	32
3	243	243

X = -3

Since the values of y_1 and y_2 are the same for each given x-value, it appears that $x^3 \cdot x^2 = x^5$. We can perform similar confirmations of the quotient rule and the power rule.

EXAMPLE 3   Simplify each of the following.

a) $y^{-5} \cdot y^3$

b) $\dfrac{48x^{12}}{16x^4}$

c) $(t^{-3})^5$

d) $(2s^{-2})^5$

e) $\left(\dfrac{45x^{-4}y^2}{9z^{-8}}\right)^{-3}$

Solution

a) $y^{-5} \cdot y^3 = y^{-5+3} = y^{-2}$, or $\dfrac{1}{y^2}$

b) $\dfrac{48x^{12}}{16x^4} = \dfrac{48}{16}x^{12-4} = 3x^8$

c) $(t^{-3})^5 = t^{-3\cdot5} = t^{-15}$, or $\dfrac{1}{t^{15}}$

d) $(2s^{-2})^5 = 2^5(s^{-2})^5 = 32s^{-10}$, or $\dfrac{32}{s^{10}}$

e) $\left(\dfrac{45x^{-4}y^2}{9z^{-8}}\right)^{-3} = \left(\dfrac{5x^{-4}y^2}{z^{-8}}\right)^{-3}$

$= \dfrac{5^{-3}x^{12}y^{-6}}{z^{24}} = \dfrac{x^{12}y^{-6}}{125z^{24}}$, or $\dfrac{x^{12}}{125y^6z^{24}}$

Scientific Notation

We can use scientific notation to name very large and very small positive numbers and to perform computations.

Scientific Notation

Scientific notation for a number is an expression of the type

$$N \times 10^n,$$

where $1 \leq N < 10$, N is in decimal notation, and n is an integer.

Keep in mind that in scientific notation, positive exponents are used for numbers greater than 10 and negative exponents for numbers between 0 and 1.

EXAMPLE 4 *Advertising.* Proctor and Gamble spent $1,703,053,300 on advertising in a recent year (*Source*: Competitive Media Reporting and Publishers Information Bureau). Convert this number to scientific notation.

Solution We want the decimal point to be positioned between the 1 and the 7, so we move it 9 places to the left. Since the number to be converted is greater than 10, the exponent must be positive.

$$1,703,053,300 = 1.7030533 \times 10^9$$

EXAMPLE 5 *Mass of a Neutron.* The mass of a neutron is about 0.000000000000000000000000001167 kg. Convert this number to scientific notation.

Solution We want the decimal point to be positioned between the 1 and the 6, so we move it 27 places to the right. Since the number to be converted is between 0 and 1, the exponent must be negative.

$$0.000000000000000000000000001167 = 1.67 \times 10^{-27}$$

EXAMPLE 6 Convert each of the following to decimal notation.

a) 7.632×10^{-4}

b) 9.4×10^5

Solution

a) The exponent is negative, so the number is between 0 and 1. We move the decimal point 4 places to the left.

$$7.632 \times 10^{-4} = 0.0007632$$

b) The exponent is positive, so the number is greater than 10. We move the decimal point 5 places to the right.

$$9.4 \times 10^5 = 940,000$$

Most calculators make use of scientific notation. For example, a number like 48,000,000,000,000 might be expressed as

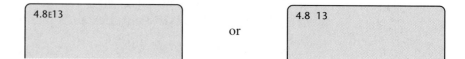

| 4.8ε13 | or | 4.8 13 |

EXAMPLE 7 *Distance to a Star.* Alpha Centauri is about 4.3 light-years from Earth. One **light-year** is the distance that light travels in one year and is about 5.88×10^{12} miles. How many miles is it from Earth to Alpha Centauri? Express your answer in scientific notation.

Solution

4.3∗5.88ε12

         2.5284ε13

$$4.3 \times (5.88 \times 10^{12}) = (4.3 \times 5.88) \times 10^{12}$$
$$= 25.284 \times 10^{12} \qquad \text{This is not}$$
$$\text{scientific notation.}$$
$$= (2.5284 \times 10^{1}) \times 10^{12}$$
$$= 2.5284 \times (10^{1} \times 10^{12})$$
$$= 2.5284 \times 10^{13} \text{ miles} \qquad \text{Writing scientific}$$
$$\text{notation}$$

Order of Operations

Recall that to simplify the expression $3 + 4 \cdot 5$, we multiply 4 and 5 first to get 20 and then add 3 to get 23. Mathematicians have agreed on the following procedure, or rules for order of operations.

Rules for Order of Operations

1. Do all calculations within grouping symbols before operations outside. When nested grouping symbols are present, work from the inside out.

2. Evaluate all exponential expressions.

3. Do all multiplications and divisions in order from left to right.

4. Do all additions and subtractions in order from left to right.

Most graphers use the order of operations above. Thus it is essential to remember these rules when entering a computation.

EXAMPLE 8 Calculate each of the following.

a) $8(5 - 3)^3 - 20$

b) $\dfrac{10 \div (8 - 6) + 9 \cdot 4}{2^5 + 3^2}$

Solution

a) $8(5 - 3)^3 - 20 = 8 \cdot 2^3 - 20$ Doing the calculation within parentheses

$\qquad\qquad\qquad\quad = 8 \cdot 8 - 20$ Evaluating the exponential expression

$\qquad\qquad\qquad\quad = 64 - 20$ Multiplying

$\qquad\qquad\qquad\quad = 44$ Subtracting

On a grapher we would enter this computation using the scientific keys, as follows.

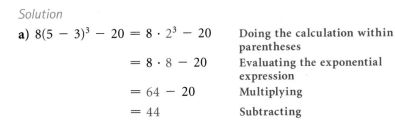

b) $\dfrac{10 \div (8 - 6) + 9 \cdot 4}{2^5 + 3^2} = \dfrac{10 \div 2 + 9 \cdot 4}{32 + 9}$

$\qquad\qquad\qquad\quad = \dfrac{5 + 36}{41} = \dfrac{41}{41} = 1$

Note that fraction bars act as grouping symbols. That is, the given expression is equivalent to $[10 \div (8 - 6) + 9 \cdot 4] \div (2^5 + 3^2)$. On a grapher we would enter the following.

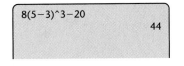

To confirm that parentheses are essential in Example 7(b), enter the computation on a grapher without using parentheses. What is the result?

EXAMPLE 9 *Compound Interest.* If a principal P is invested at an interest rate i, compounded n times per year, in t years it will grow to an amount A given by

$$A = P\left(1 + \frac{i}{n}\right)^{nt}.$$

Suppose that $1250 is invested at 4.6% interest, compounded quarterly. How much is in the account at the end of 8 years?

Solution We have $P = 1250$, $i = 4.6\%$, or 0.046, $n = 4$, and $t = 8$. Substituting, we find that the amount in the account at the end of 8 years is given by

$$A = 1250\left(1 + \frac{0.046}{4}\right)^{4 \cdot 8}.$$

Next, we evaluate this expression:

$$A = 1250\left(1 + \frac{0.046}{4}\right)^{32} \qquad \text{Multiplying in the exponent}$$

$$= 1250(1 + 0.0115)^{32} \qquad \text{Dividing}$$

$$= 1250(1.0115)^{32} \qquad \text{Adding}$$

$$\approx 1250(1.441811175) \qquad \text{Evaluating the exponential expression}$$

$$\approx 1802.263969 \qquad \text{Multiplying}$$

$$\approx 1802.26. \qquad \text{Rounding to the nearest cent}$$

```
1250(1+.046/4)^(4*8)
             1802.263969
```

We can also use the scientific keys on a grapher. Note that the product in the exponent must be enclosed in parentheses.

The amount in the account at the end of 8 years is $1802.26.

Exercise Set R.2

Simplify.

1. 18^0

2. $\left(-\frac{4}{3}\right)^0$

3. $5^8 \cdot 5^{-6}$

4. $6^2 \cdot 6^{-7}$

5. $m^{-5} \cdot m^5$

6. $n^9 \cdot n^{-9}$

7. $7^3 \cdot 7^{-5} \cdot 7$

8. $3^6 \cdot 3^{-5} \cdot 3^4$

9. $2x^3 \cdot 3x^2$

10. $3y^4 \cdot 4y^3$

11. $(5a^2b)(3a^{-3}b^4)$

12. $(4xy^2)(3x^{-4}y^5)$

13. $(2x)^3(3x)^2$

14. $(4y)^2(3y)^3$

15. $\dfrac{b^{40}}{b^{37}}$

16. $\dfrac{a^{39}}{a^{32}}$

17. $\dfrac{x^2y^{-2}}{x^{-1}y}$

18. $\dfrac{x^3y^{-3}}{x^{-1}y^2}$

19. $\dfrac{32x^{-4}y^3}{4x^{-5}y^8}$

20. $\dfrac{20a^5b^{-2}}{5a^7b^{-3}}$

21. $(2ab^2)^3$

22. $(4xy^3)^2$

23. $(-2x^3)^4$

24. $(-3x^2)^4$

25. $(-5c^{-1}d^{-2})^{-2}$

26. $(-4x^{-5}z^{-2})^{-3}$

27. $\left(\dfrac{24a^{10}b^{-8}c^7}{3a^6b^{-3}c^5}\right)^5$

28. $\left(\dfrac{125p^{12}q^{-14}r^{22}}{25p^8q^6r^{-15}}\right)^{-4}$

Convert to scientific notation.

29. 405,000

30. 1,670,000

31. 0.00000039

32. 0.00092

33. One cubic inch is approximately equal to 0.000016 m^3.

34. The United States government received $828,597,000,000 in individual income taxes in a recent year (*Source*: U.S. Department of the Treasury).

Convert to decimal notation.

35. 8.3×10^{-5}

36. 4.1×10^6

37. 2.07×10^7

38. 3.15×10^{-6}

39. In 1997, the revenue of Microsoft Corp. was 1.1358×10^{10} (*Source: Fortune Magazine*).

40. The mass of a proton is about 1.67×10^{-24} g.

Compute. Write the answer using scientific notation.

41. $(3.1 \times 10^5)(4.5 \times 10^{-3})$

42. $(9.1 \times 10^{-17})(8.2 \times 10^3)$

43. $\dfrac{6.4 \times 10^{-7}}{8.0 \times 10^6}$

44. $\dfrac{1.1 \times 10^{-40}}{2.0 \times 10^{-71}}$

Solve. Write the answer using scientific notation.

45. *Chesapeake Bay Bridge-Tunnel.* The 17.6-mile-long Chesapeake Bay Bridge-Tunnel was completed in 1964. Construction costs were $210 million. Find the average cost per mile.

46. *Gambling Revenue.* Casino gambling revenue in Las Vegas and the surrounding county was $6,152,415,000 in 1997. That year 30,500,000 people visited the area. (*Source*: Las Vegas Convention and Visitors Authority) Find the average amount each visitor spent gambling.

47. *Nuclear Disintegration.* One gram of radium produces 37 billion disintegrations per second. How many disintegrations are produced in 1 hr?

48. *Length of Earth's Orbit.* The average distance from the earth to the sun is 93 million mi. About how far does the earth travel in a yearly orbit? (Assume a circular orbit.)

Calculate.

49. $3 \cdot 2 - 4 \cdot 2^2 + 6(3 - 1)$

50. $3[(2 + 4 \cdot 2^2) - 6(3 - 1)]$

51. $16 \div 4 \cdot 4 \div 2 \cdot 256$

52. $2^6 \cdot 2^{-3} \div 2^{10} \div 2^{-8}$

53. $\dfrac{4(8 - 6)^2 - 4 \cdot 3 + 2 \cdot 8}{3^1 + 19^0}$

54. $\dfrac{[4(8 - 6)^2 + 4](3 - 2 \cdot 8)}{2^2(2^3 + 5)}$

Compound Interest. Use the compound interest formula from Example 9 in Exercises 55–58. Round to the nearest cent.

55. Suppose that $2125 is invested at 6.2%, compounded semiannually. How much is in the account at the end of 5 yr?

56. Suppose that $9550 is invested at 5.4%, compounded semiannually. How much is in the account at the end of 7 yr?

57. Suppose that $6700 is invested at 4.5%, compounded quarterly. How much is in the account at the end of 6 yr?

58. Suppose that $4875 is invested at 5.8%, compounded quarterly. How much is in the account at the end of 9 yr?

Discussion and Writing

59. Are parentheses necessary in the expression $4 \cdot 25 \div (10 - 5)$? Why or why not?

60. Is $x^{-2} < x^{-1}$ for any negative value(s) of x? Why or why not?

Synthesis

Simplify. Assume that all exponents are integers, all denominators are nonzero, and zero is not raised to a nonpositive power.

61. $(x^t \cdot x^{3t})^2$

62. $(x^y \cdot x^{-y})^3$

63. $(t^{a+x} \cdot t^{x-a})^4$

64. $(m^{x-b} \cdot n^{x+b})^x (m^b n^{-b})^x$

65. $\left[\dfrac{(3x^a y^b)^3}{(-3x^a y^b)^2} \right]^2$

66. $\left[\left(\dfrac{x^r}{y^t} \right)^2 \left(\dfrac{x^{2r}}{y^{4t}} \right)^{-2} \right]^{-3}$

Mortgage Payments. The formula

$$ M = P \left[\dfrac{\dfrac{i}{12}\left(1 + \dfrac{i}{12}\right)^n}{\left(1 + \dfrac{i}{12}\right)^n - 1} \right] $$

gives the monthly mortgage payment M on a home loan of P dollars at interest rate i, where n is the total number of payments (12 times the number of years). Use this formula in Exercises 67–70.

67. The cost of a house is $98,000. The down payment is $16,000, the interest rate is $8\frac{1}{2}$%, and the loan period is 25 yr. What is the monthly payment?

68. The cost of a house is $124,000. The down payment is $20,000, the interest rate is $7\frac{3}{4}$%, and the loan period is 30 yr. What is the monthly payment?

69. The cost of a house is $135,000. The down payment is $18,000, the interest rate is $7\frac{1}{2}$%, and the loan period is 20 yr. What is the monthly payment?

70. The cost of a house is $151,000. The down payment is $21,000, the interest rate is $8\frac{1}{4}$%, and the loan period is 25 yr. What is the monthly payment?

Use a grapher to graph $y_1 = x^2$, $y_2 = x^3$, and $y_3 = x^4$. Then use the graphs to do Exercises 71 and 72.

71. For what values of x is $x^2 \geq x^3$?

72. For what values of x is $x^4 > x^2$?

Addition, Subtraction, and Multiplication of Polynomials

R.3

- *Identify the terms, coefficients, and degree of a polynomial.*
- *Add, subtract, and multiply polynomials.*

Polynomials

Polynomials are a type of algebraic expression that you will often encounter in your study of algebra. Some examples of polynomials are

$$3x - 4y, \quad 5y^3 - \tfrac{7}{3}y^2 + 3y - 2, \quad 0, \quad -2.3a^4, \quad \text{and} \quad z^6 - \sqrt{5}.$$

All but the first are polynomials in one variable.

Polynomials in One Variable

A **polynomial in one variable** is any expression of the type

$$a_n x^n + a_{n-1} x^{n-1} + \cdots + a_2 x^2 + a_1 x + a_0,$$

where n is a nonnegative integer, $a_n, \ldots, a_0$ are real numbers, called **coefficients**, and $a_n \neq 0$. The parts of a polynomial separated by plus signs are called **terms**. The **degree** of the polynomial is n, the **leading coefficient** is a_n, and the **constant term** is a_0. The polynomial is said to be written in **descending order,** because the exponents decrease from left to right.

EXAMPLE 1   Identify the terms of the polynomial

$$2x^4 - 7.5x^3 + x - 12.$$

Solution   We have

$$2x^4 - 7.5x^3 + x - 12 = 2x^4 + (-7.5x^3) + x + (-12),$$

so the terms are

$$2x^4, \qquad -7.5x^3, \qquad x, \quad \text{and} \quad -12. \qquad \blacksquare$$

A polynomial, like 23, consisting of only a nonzero constant term has degree 0. It is agreed that the polynomial consisting only of 0 has *no* degree.

EXAMPLE 2   Find the degree of each polynomial.

a) $2x^3 - 9$

b) $y^2 - \tfrac{3}{2} + 5y^4$

c) 7

Solution

POLYNOMIAL	DEGREE
a) $2x^3 - 9$	3
b) $y^2 - \frac{3}{2} + 5y^4 = 5y^4 + y^2 - \frac{3}{2}$	4
c) $7 = 7x^0$	0

Algebraic expressions like $3ab^3 - 8$ and $5x^4y^2 - 3x^3y^8 + 7xy^2 + 6$ are **polynomials in several variables.** The **degree of a term** is the sum of the exponents of the variables in that term. The **degree of a polynomial** is the degree of the term of highest degree.

EXAMPLE 3 Find the degree of the polynomial

$$7ab^3 - 11a^2b^4 + 8.$$

Solution The degrees of the terms of $7ab^3 - 11a^2b^4 + 8$ are 4, 6, and 0, respectively, so the degree of the polynomial is 6.

A polynomial with just one term, like $-9y^6$, is a **monomial**. If a polynomial has two terms, like $x^2 + 4$, it is a **binomial**. A polynomial with three terms, like $4x^2 - 4xy + 1$, is a **trinomial**.
Expressions like

$$2x^2 - 5x + \frac{3}{x}, \qquad 9 - \sqrt{x}, \quad \text{and} \quad \frac{x + 1}{x^4 + 5}$$

are not polynomials, because they cannot be written in the form $a_nx^n + a_{n-1}x^{n-1} + \cdots + a_1x + a_0$, where the exponents are all nonnegative integers and the coefficients are all real numbers.

Addition and Subtraction

If two terms of an expression have the same variables raised to the same powers, they are called **like terms,** or **similar terms.** We can **combine**, or **collect like terms** using the distributive property. For example, $3y^2$ and $5y^2$ are like terms and

$$3y^2 + 5y^2 = (3 + 5)y^2$$
$$= 8y^2.$$

We add or subtract polynomials by combining like terms.

EXAMPLE 4 Add or subtract each of the following.

a) $(-5x^3 + 3x^2 - x) + (12x^3 - 7x^2 + 3)$
b) $(6x^2y^3 - 9xy) - (5x^2y^3 - 4xy)$

Solution

a) $(-5x^3 + 3x^2 - x) + (12x^3 - 7x^2 + 3)$

$\qquad = (-5x^3 + 12x^3) + (3x^2 - 7x^2) - x + 3$ 　　Rearranging using the commutative and associative properties

$\qquad = (-5 + 12)x^3 + (3 - 7)x^2 - x + 3$ 　　Using the distributive property

$\qquad = 7x^3 - 4x^2 - x + 3$

b) We can subtract by adding an opposite:

$\qquad (6x^2y^3 - 9xy) - (5x^2y^3 - 4xy)$

$\qquad = (6x^2y^3 - 9xy) + (-5x^2y^3 + 4xy)$ 　　Adding the opposite of $5x^2y^3 - 4xy$

$\qquad = 6x^2y^3 - 9xy - 5x^2y^3 + 4xy$

$\qquad = x^2y^3 - 5xy.$ 　　Combining like terms

Multiplication

Multiplication of polynomials is based on the distributive property. For example,

$$(x + 4)(x + 3) = (x + 4)x + (x + 4)3 \qquad \text{Using the distributive property}$$

$$= x^2 + 4x + 3x + 12 \qquad \text{Using the distributive property two more times}$$

$$= x^2 + 7x + 12. \qquad \text{Combining like terms}$$

We can find the product of two binomials by multiplying the **F**irst terms, then the **O**uter terms, then the **I**nner terms, then the **L**ast terms. Then we collect like terms, if possible. This procedure is sometimes called **FOIL**.

EXAMPLE 5 Multiply: $(2x - 7)(3x + 4)$.

$y_1 = (2x - 7)(3x + 4),$
$y_2 = 6x^2 - 13x - 28$

X	Y₁	Y₂
−3	65	65
−2	22	22
−1	−9	−9
0	−28	−28
1	−35	−35
2	−30	−30
3	−13	−13

X = −3

Solution　We have

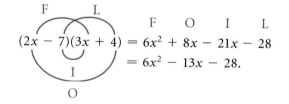

$$(2x - 7)(3x + 4) = 6x^2 + 8x - 21x - 28$$

$$= 6x^2 - 13x - 28.$$

We can use a grapher to do a partial check of this result. We enter $y_1 = (2x - 7)(3x + 4)$ and $y_2 = 6x^2 - 13x - 28$, and construct a table of values. The values of y_1 and y_2 are the same for each given x-value, so the product appears to be correct.

In general, to multiply two polynomials, we multiply each term of one by each term of the other and add the products. One way to do this is to use columns.

EXAMPLE 6   Multiply: $(4x^4y - 7x^2y + 3y)(2y - 3x^2y)$.

Solution   We can use columns to organize our work.

$$
\begin{array}{r}
4x^4y - 7x^2y + 3y \\
2y - 3x^2y \\
\hline
-12x^6y^2 + 21x^4y^2 - 9x^2y^2 \\
8x^4y^2 - 14x^2y^2 + 6y^2 \\
\hline
-12x^6y^2 + 29x^4y^2 - 23x^2y^2 + 6y^2
\end{array}
$$

Multiplying by $-3x^2y$
Multiplying by $2y$

Adding

We can use FOIL to find some special products.

Special Products of Binomials

$(A + B)^2 = A^2 + 2AB + B^2$	Square of a sum
$(A - B)^2 = A^2 - 2AB + B^2$	Square of a difference
$(A + B)(A - B) = A^2 - B^2$	Product of a sum and a difference

---

EXAMPLE 7   Multiply each of the following.

a) $(4x + 1)^2$ **b)** $(3y^2 - 2)^2$ **c)** $(x^2 + 3y)(x^2 - 3y)$

Solution

a) $(4x + 1)^2 = (4x)^2 + 2 \cdot 4x \cdot 1 + 1^2 = 16x^2 + 8x + 1$

b) $(3y^2 - 2)^2 = (3y^2)^2 - 2 \cdot 3y^2 \cdot 2 + 2^2 = 9y^4 - 12y^2 + 4$

c) $(x^2 + 3y)(x^2 - 3y) = (x^2)^2 - (3y)^2 = x^4 - 9y^2$

Exercise Set R.3

Determine the terms and the degree of the polynomial.

1. $-5y^4 + 3y^3 + 7y^2 - y - 4$

2. $2m^3 - m^2 - 4m + 11$

3. $3a^4b - 7a^3b^3 + 5ab - 2$

4. $6p^3q^2 - p^2q^4 - 3pq^2 + 5$

Perform the operations indicated.

5. $(5x^2y - 2xy^2 + 3xy - 5) + (-2x^2y - 3xy^2 + 4xy + 7)$

6. $(6x^2y - 3xy^2 + 5xy - 3) + (-4x^2y - 4xy^2 + 3xy + 8)$

7. $(2x + 3y + z - 7) + (4x - 2y - z + 8) + (-3x + y - 2z - 4)$

8. $(2x^2 + 12xy - 11) + (6x^2 - 2x + 4) + (-x^2 - y - 2)$

9. $(3x^2 - 2x - x^3 + 2) - (5x^2 - 8x - x^3 + 4)$

10. $(5x^2 + 4xy - 3y^2 + 2) - (9x^2 - 4xy + 2y^2 - 1)$

11. $(x^4 - 3x^2 + 4x) - (3x^3 + x^2 - 5x + 3)$

12. $(2x^4 - 3x^2 + 7x) - (5x^3 + 2x^2 - 3x + 5)$

13. $(a - b)(2a^3 - ab + 3b^2)$

14. $(n + 1)(n^2 - 6n - 4)$

15. $(x + 5)(x - 3)$

16. $(y - 4)(y + 1)$

17. $(2a + 3)(a + 5)$

18. $(3b + 1)(b - 2)$

19. Use a grapher to check your answer to Exercise 17.

20. Use a grapher to check your answer to Exercise 18.

21. $(2x + 3y)(2x + y)$ **22.** $(2a - 3b)(2a - b)$

23. $(y + 5)^2$ **24.** $(y + 7)^2$

25. $(5x - 3)^2$ **26.** $(3x - 2)^2$

27. Use a grapher to check your answer to Exercise 25.

28. Use a grapher to check your answer to Exercise 26.

29. $(2x + 3y)^2$ **30.** $(5x + 2y)^2$

31. $(2x^2 - 3y)^2$ **32.** $(4x^2 - 5y)^2$

33. $(a + 3)(a - 3)$ **34.** $(b + 4)(b - 4)$

35. Use a grapher to check your answer to Exercise 33.

36. Use a grapher to check your answer to Exercise 34.

37. $(3x - 2y)(3x + 2y)$

38. $(3x + 5y)(3x - 5y)$

39. $(2x + 3y + 4)(2x + 3y - 4)$

40. $(5x + 2y + 3)(5x + 2y - 3)$

41. $(x + 1)(x - 1)(x^2 + 1)$

42. $(y - 2)(y + 2)(y^2 + 4)$

Discussion and Writing

43. Is the sum of two polynomials of degree n always a polynomial of degree n? Why or why not?

44. Explain how you would convince a classmate that $(A + B)^2 \neq A^2 + B^2$.

Synthesis

Multiply. Assume that all exponents are natural numbers.

45. $(a^n + b^n)(a^n - b^n)$

46. $(t^a + 4)(t^a - 7)$

47. $(a^n + b^n)^2$

48. $(x^{3m} - t^{5n})^2$

49. $(x - 1)(x^2 + x + 1)(x^3 + 1)$

50. $[(2x - 1)^2 - 1]^2$

51. $(x^{a-b})^{a+b}$

52. $(t^{m+n})^{m+n} \cdot (t^{m-n})^{m-n}$

53. $(a + b + c)^2$

R.4
Factoring

- *Factor polynomials by removing a common factor.*
- *Factor special products of polynomials.*

To factor a polynomial, we do the reverse of multiplying; that is, we find an equivalent expression that is written as a product.

Terms with Common Factors

When a polynomial is to be factored, we should always look first to factor out a factor that is common to all the terms using the distributive property. We usually look for the constant common factor with the largest absolute value and for variables with the largest exponent common to all the terms. In this sense, we factor out the "largest" common factor.

EXAMPLE 1 Factor each of the following.

a) $15 + 10x - 5x^2$

b) $12x^2y^2 - 20x^3y$

Solution

a) $15 + 10x - 5x^2 = 5 \cdot 3 + 5 \cdot 2x - 5 \cdot x^2 = 5(3 + 2x - x^2)$

We can check by multiplying: $5(3 + 2x - x^2) = 15 + 10x - 5x^2$.

b) $12x^2y^2 - 20x^3y = 4x^2y(3y - 5x)$

Note that there are several factors common to the terms of $12x^2y^2 - 20x^3y$, but $4x^2y$ is the "largest" of these. ▬

In some polynomials, pairs of terms have a common factor that can be removed in a process called **factoring by grouping.**

EXAMPLE 2 Factor: $x^3 + 3x^2 - 5x - 15$.

Solution We have

$$x^3 + 3x^2 - 5x - 15 = x^2(x + 3) - 5(x + 3)$$
$$= (x + 3)(x^2 - 5).$$

To check using a grapher, we enter $y_1 = x^3 + 3x^2 - 5x - 15$ and $y_2 = (x + 3)(x^2 - 5)$. We can use a table or a graph. To use a graph, we first select SEQUENTIAL mode. Then we select the "line"-graph style for y_1 and the "path"-graph style, denoted by -0, for y_2. The grapher will graph y_1 first. Then y_2 will be graphed as the circular cursor traces the leading edge of the graph, allowing us to determine visually whether the graphs coincide.

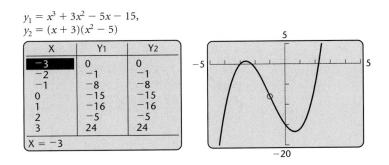

$y_1 = x^3 + 3x^2 - 5x - 15,$
$y_2 = (x + 3)(x^2 - 5)$

X	Y1	Y2
−3	0	0
−2	−1	−1
−1	−8	−8
0	−15	−15
1	−16	−16
2	−5	−5
3	24	24

X = −3

The values of y_1 and y_2 are the same for the given x-values and the graphs coincide, so the factorization appears to be correct. ▬

Trinomials

Some trinomials can be factored into the product of two binomials. To factor a trinomial of the form $x^2 + bx + c$, we look for two numbers with a product of c and a sum of b.

EXAMPLE 3 Factor each of the following.

a) $x^2 + 5x + 6$ **b)** $y^2 - 2y - 24$

Solution

a) We look for two numbers with a product of 6 and a sum of 5. They are 2 and 3. Then

$$x^2 + 5x + 6 = (x + 2)(x + 3).$$ To check, multiply or use a table or a graph.

b) We look for two numbers with a product of -24 and a sum of -2:

$$y^2 - 2y - 24 = (y - 6)(y + 4).$$ Note that $(-6)(4) = -24$ and $-6 + 4 = -2$. ▬

The next example illustrates a method for factoring a trinomial of the form $ax^2 + bx + c, a \neq 1$.

EXAMPLE 4 Factor: $3x^2 - 10x - 8$.

Solution We use FOIL in reverse, looking for two binomial factors of the form ($\boxed{} x +$)($\boxed{} x +$). The product of the First terms is $3x^2$, the product of the Last terms is -8, and the sum of the Inner and Outer products is $-10x$. By trial and error, we determine the factorization:

$$3x^2 - 10x - 8 = (3x + 2)(x - 4).$$ To check, multiply or use a table or a graph. ▬

Special Factorizations

We reverse the equation $(A + B)(A - B) = A^2 - B^2$ to factor a **difference of squares.**

$$A^2 - B^2 = (A + B)(A - B)$$

EXAMPLE 5 Factor each of the following.

a) $x^2 - 16$ **b)** $9a^2 - 25$ **c)** $6x^4 - 6y^4$

Solution

a) $x^2 - 16 = (x + 4)(x - 4)$ To check, multiply or use a table (shown at left) or a graph.

b) $9a^2 - 25 = (3a + 5)(3a - 5)$

c) $6x^4 - 6y^4 = 6(x^4 - y^4)$
$$= 6(x^2 + y^2)(x^2 - y^2)$$
$$= 6(x^2 + y^2)(x + y)(x - y)$$ Because none of these factors can be factored further, we have *factored completely.* ▬

$y_1 = x^2 - 16,$
$y_2 = (x + 4)(x - 4)$

X	Y₁	Y₂
−3	−7	−7
−2	−12	−12
−1	−15	−15
0	−16	−16
1	−15	−15
2	−12	−12
3	−7	−7

X = −3

The rules for squaring binomials can be reversed to factor trinomial squares.

$$A^2 + 2AB + B^2 = (A + B)^2$$
$$A^2 - 2AB + B^2 = (A - B)^2$$

EXAMPLE 6   Factor each of the following.

a) $x^2 + 8x + 16$ **b)** $25y^2 - 30y + 9$

Solution

$y_1 = x^2 + 8x + 16,$
$y_2 = (x + 4)^2$

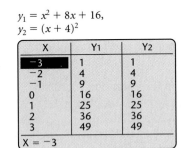

X	Y1	Y2
−3	1	1
−2	4	4
−1	9	9
0	16	16
1	25	25
2	36	36
3	49	49

X = −3

a) $x^2 + 8x + 16 = (x + 4)^2$ To check, multiply or use a table (shown at left) or a graph.

b) $25y^2 - 30y + 9 = (5y - 3)^2$

We can use the following rules to factor a **sum** or a **difference of cubes:**

$$A^3 + B^3 = (A + B)(A^2 - AB + B^2);$$
$$A^3 - B^3 = (A - B)(A^2 + AB + B^2).$$

These rules can be verified by multiplying.

EXAMPLE 7   Factor each of the following.

a) $x^3 + 27$
b) $16y^3 - 250$

Solution

$y_1 = x^3 + 27,$
$y_2 = (x + 3)(x^2 - 3x + 9)$

a) $x^3 + 27 = x^3 + 3^3$
$$= (x + 3)(x^2 - 3x + 9)$$ See the graphical check at left.

b) $16y^3 - 250 = 2(8y^3 - 125)$
$$= 2[(2y)^3 - 5^3]$$
$$= 2(2y - 5)(4y^2 + 10y + 25)$$ To check, multiply or use a table or a graph.

Not all polynomials can be factored into polynomials with integer coefficients. An example is $x^2 - x + 7$. There are no real factors of 7 whose sum is -1. In such a case we say that the polynomial is "not factorable," or **prime**.

A STRATEGY FOR FACTORING

A. Always factor out the largest common factor.

B. Look at the number of terms.

Two terms: Try factoring as a difference of squares first. Next, try factoring as a sum or a difference of cubes. Do *not* try to factor a *sum* of squares.

Three terms: Try factoring as a trinomial square. Next, try trial and error, using FOIL in reverse.

Four or more terms: Try factoring by grouping and factoring out a common binomial factor.

C. Always *factor completely.* If a factor with more than one term can itself be factored further, do so.

Exercise Set R.4

Factor out a common factor.

1. $2x - 10$
2. $7y + 42$

3. $3x^4 - 9x^2$
4. $20y^2 - 5y^5$

5. $4a^2 - 12a + 16$
6. $6n^2 + 24n - 18$

7. $a(b - 2) + c(b - 2)$

8. $a(x^2 - 3) - 2(x^2 - 3)$

Factor by grouping.

9. $x^3 + 3x^2 + 6x + 18$
10. $3x^3 + x^2 - 18x - 6$

11. $y^3 - 3y^2 - 4y + 12$
12. $p^3 - 2p^2 - 9p + 18$

13. Use a grapher to check your answer to Exercise 9.

14. Use a grapher to check your answer to Exercise 10.

Factor the trinomial.

15. $p^2 + 6p + 8$
16. $w^2 - 7w + 10$

17. $2n^2 + 9n - 56$
18. $3y^2 + 7y - 20$

19. $y^4 - 4y^2 - 21$
20. $m^4 - m^2 - 90$

21. Use a grapher to check your answer to Exercise 15.

22. Use a grapher to check your answer to Exercise 16.

Factor the difference of squares.

23. $9x^2 - 25$
24. $16x^2 - 9$

25. $4xy^4 - 4xz^2$
26. $5x^4y - 5yz^4$

Factor the trinomial square.

27. $y^2 - 6y + 9$
28. $x^2 + 8x + 16$

29. $1 - 8x + 16x^2$
30. $1 + 10x + 25x^2$

31. Use a grapher to check your answer to Exercise 29.

32. Use a grapher to check your answer to Exercise 30.

Factor the sum or difference of cubes.

33. $x^3 + 8$
34. $y^3 - 64$

35. $m^3 - 1$
36. $n^3 + 216$

37. Use a grapher to check your answer to Exercise 35.

38. Use a grapher to check your answer to Exercise 36.

Factor completely.

39. $18a^2b - 15ab^2$
40. $4x^2y + 12xy^2$

41. $x^3 - 4x^2 + 5x - 20$
42. $z^3 + 3z^2 - 3z - 9$

43. $8x^2 - 32$
44. $6y^2 - 6$

45. $4y^2 - 5$
46. $16x^2 - 7$

47. Use a grapher to check your answer to Exercise 41.

48. Use a grapher to check your answer to Exercise 42.

49. $x^2 + 9x + 20$

50. $y^2 + y - 6$

51. $y^2 - 6y + 5$

52. $x^2 - 4x - 21$

53. $2a^2 + 9a + 4$

54. $3b^2 - b - 2$

55. $6x^2 + 7x - 3$

56. $8x^2 + 2x - 15$

57. $y^2 - 18y + 81$

58. $n^2 + 2n + 1$

59. $x^2y^2 - 14xy + 49$

60. $x^2y^2 - 16xy + 64$

61. $4ax^2 + 20ax - 56a$

62. $21x^2y + 2xy - 8y$

63. $3z^3 - 24$

64. $4t^3 + 108$

65. $16a^7b + 54ab^7$

66. $24a^2x^4 - 375a^8x$

67. Use a grapher to check your answer to Exercise 55.

68. Use a grapher to check your answer to Exercise 56.

Discussion and Writing

69. Under what circumstances can $A^2 + B^2$ be factored?

70. Explain how the rule for factoring a sum of cubes can be used to factor a difference of cubes.

Synthesis

Factor.

71. $y^4 - 84 + 5y^2$

72. $11x^2 + x^4 - 80$

73. $y^2 - \frac{8}{49} + \frac{2}{7}y$

74. $t^2 - 0.27 + 0.6t$

75. $(x + h)^3 - x^3$

76. $(x + 0.01)^2 - x^2$

77. $(y - 4)^2 + 5(y - 4) - 24$

78. $6(2p + q)^2 - 5(2p + q) - 25$

Factor. Assume that variables in exponents represent natural numbers.

79. $x^{2n} + 5x^n - 24$

80. $4x^{2n} - 4x^n - 3$

81. $x^2 + ax + bx + ab$

82. $bdy^2 + ady + bcy + ac$

83. $25y^{2m} - (x^{2n} - 2x^n + 1)$

84. $x^{6a} - t^{3b}$

85. $(y - 1)^4 - (y - 1)^2$

86. $x^6 - 2x^5 + x^4 - x^2 + 2x - 1$

Rational Expressions
R.5

- *Determine the domain of a rational expression.*
- *Simplify rational expressions.*
- *Multiply, divide, add, or subtract rational expressions.*
- *Simplify complex rational expressions.*

A **rational expression** is the quotient of two polynomials. For example,

$$\frac{3}{5}, \quad \frac{2}{x-3}, \quad \text{and} \quad \frac{x^2 - 4}{x^2 - 4x - 5}$$

are rational expressions.

The Domain of a Rational Expression

The **domain** of an algebraic expression is the set of all real numbers for which the expression is defined. Since division by zero is not defined, any number that makes the denominator zero is not in the domain of a rational expression.

EXAMPLE 1 Find the domain of each of the following.

a) $\frac{2}{x-3}$

b) $\frac{x^2 - 4}{x^2 - 4x - 5}$

Solution

a) Since $x - 3 = 0$ when $x = 3$, the domain of $2/(x - 3)$ is the set of all real numbers except 3. As a partial check, graph $y = 2/(x - 3)$ using DOT mode rather than CONNECTED mode. (See Section 1.4 for a discussion of DOT mode.)

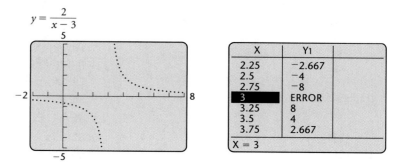

Note that there does not appear to be a point on the graph that has 3 as a first coordinate. Also note that the grapher's TABLE feature produces an error message for $x = 3$, indicating that the expression is not defined for $x = 3$.

b) To determine the domain of $(x^2 - 4)/(x^2 - 4x - 5)$, we first factor the denominator:

$$\frac{x^2 - 4}{x^2 - 4x - 5} = \frac{x^2 - 4}{(x + 1)(x - 5)}.$$

The factor $x + 1$ is 0 when $x = -1$ and the factor $x - 5$ is 0 when $x = 5$. Since $(x + 1)(x - 5) = 0$ when $x = -1$ or $x = 5$, the domain is the set of all real numbers except -1 and 5. The error messages for $x = -1$ and $x = 5$ in a table of values also indicate that -1 and 5 are not in the domain of the expression.

$y_1 = \dfrac{x^2 - 4}{x^2 - 4x - 5}$

X	Y1	
−1	ERROR	
0	.8	
1	.375	
2	0	
3	−.625	
4	−2.4	
5	ERROR	
X = −1		

We can describe the domains found in Example 1 using *set-builder notation*. For example, we write "The set of all real numbers x such that x is not equal to 3" as

$$\{x \mid x \text{ is a real number } and \ x \neq 3\}.$$

Similarly, we write "The set of all real numbers x such that x is not equal to -1 and x is not equal to 5" as

$$\{x \mid x \text{ is a real number } and \ x \neq -1 \ and \ x \neq 5\}.$$

Simplifying, Multiplying, and Dividing Rational Expressions

To simplify rational expressions, we use the fact that

$$\frac{a \cdot c}{b \cdot c} = \frac{a}{b} \cdot \frac{c}{c} = \frac{a}{b} \cdot 1 = \frac{a}{b}.$$

EXAMPLE 2   Simplify: $\dfrac{9x^2 + 6x - 3}{12x^2 - 12}$.

Solution

$$\dfrac{9x^2 + 6x - 3}{12x^2 - 12} = \dfrac{3(x + 1)(3x - 1)}{3 \cdot 4(x + 1)(x - 1)} \qquad \text{Factoring the numerator and the denominator}$$

$$= \dfrac{3(x + 1)}{3(x + 1)} \cdot \dfrac{3x - 1}{4(x - 1)} \qquad \text{Factoring the rational expression}$$

$$= 1 \cdot \dfrac{3x - 1}{4(x - 1)} \qquad \dfrac{3(x + 1)}{3(x + 1)} = 1$$

$$= \dfrac{3x - 1}{4(x - 1)} \qquad \text{Removing a factor of 1}$$

Canceling is a shortcut that is often used to remove a factor of 1.

EXAMPLE 3   Simplify each of the following.

a) $\dfrac{4x^3 + 16x^2}{2x^3 + 6x^2 - 8x}$ **b)** $\dfrac{2 - x}{x^2 + x - 6}$

Solution

a) $\dfrac{4x^3 + 16x^2}{2x^3 + 6x^2 - 8x} = \dfrac{2 \cdot 2 \cdot x \cdot x(x + 4)}{2 \cdot x(x + 4)(x - 1)} \qquad \text{Factoring the numerator and the denominator}$

$$= \dfrac{\cancel{2} \cdot 2 \cdot \cancel{x} \cdot x\cancel{(x + 4)}}{\cancel{2} \cdot \cancel{x}\cancel{(x + 4)}(x - 1)} \qquad \begin{array}{l}\text{Removing a factor of 1:} \\ \dfrac{2x(x + 4)}{2x(x + 4)} = 1\end{array}$$

$$= \dfrac{2x}{x - 1}$$

b) $\dfrac{2 - x}{x^2 + x - 6} = \dfrac{2 - x}{(x + 3)(x - 2)} \qquad \text{Factoring the denominator}$

$$= \dfrac{-1(x - 2)}{(x + 3)(x - 2)} \qquad 2 - x = -1(x - 2)$$

$$= \dfrac{-1\cancel{(x - 2)}}{(x + 3)\cancel{(x - 2)}} \qquad \text{Removing a factor of 1: } \dfrac{x - 2}{x - 2} = 1$$

$$= \dfrac{-1}{x + 3}, \text{ or } -\dfrac{1}{x + 3}$$

In Example 3(b), we saw that

$$\dfrac{2 - x}{x^2 + x - 6} \quad \text{and} \quad \dfrac{-1}{x + 3}$$

are **equivalent expressions.** This means that they have the same value for all numbers that are in *both* domains.

X	Y1	Y2
−3	ERROR	ERROR
−2	−1	−1
−1	−.5	−.5
0	−.3333	−.3333
1	−.25	−.25
2	ERROR	−.2
3	−.1667	−.1667

X = 2

For

$$y_1 = \frac{2 - x}{x^2 + x - 6} \quad \text{and} \quad y_2 = \frac{-1}{x + 3},$$

a table shows that y_1 and y_2 have the same values except for $x = -3$ and $x = 2$. Note that -3 is not in the domain of *either* expression, whereas 2 is in the domain of y_2 but not in the domain of y_1 and thus is not in the domain of *both* expressions.

To multiply rational expressions, we multiply numerators and multiply denominators and, if possible, simplify the result. To divide rational expressions, we multiply the dividend by the reciprocal of the divisor and, if possible, simplify the result—that is,

$$\frac{a}{b} \cdot \frac{c}{d} = \frac{ac}{bd} \quad \text{and} \quad \frac{a}{b} \div \frac{c}{d} = \frac{a}{b} \cdot \frac{d}{c} = \frac{ad}{bc}.$$

EXAMPLE 4   Multiply or divide and simplify each of the following.

a) $\dfrac{x + 4}{x - 3} \cdot \dfrac{x^2 - 9}{x^2 - x - 2}$

b) $\dfrac{y^3 - 1}{y^2 - 1} \div \dfrac{y^2 + y + 1}{y^2 + 2y + 1}$

Solution

a) $\dfrac{x + 4}{x - 3} \cdot \dfrac{x^2 - 9}{x^2 - x - 2} = \dfrac{(x + 4)(x^2 - 9)}{(x - 3)(x^2 - x - 2)}$

Multiplying the numerators and the denominators

$= \dfrac{(x + 4)(x + 3)(x - 3)}{(x - 3)(x - 2)(x + 1)}$ Factoring and removing a factor of 1: $\dfrac{x - 3}{x - 3} = 1$

$= \dfrac{(x + 4)(x + 3)}{(x - 2)(x + 1)}$

b) $\dfrac{y^3 - 1}{y^2 - 1} \div \dfrac{y^2 + y + 1}{y^2 + 2y + 1} = \dfrac{y^3 - 1}{y^2 - 1} \cdot \dfrac{y^2 + 2y + 1}{y^2 + y + 1}$ Multiplying by the reciprocal of the divisor

$= \dfrac{(y^3 - 1)(y^2 + 2y + 1)}{(y^2 - 1)(y^2 + y + 1)}$

$= \dfrac{(y - 1)(y^2 + y + 1)(y + 1)(y + 1)}{(y + 1)(y - 1)(y^2 + y + 1)}$

Factoring and removing a factor of 1

$= y + 1$

Adding and Subtracting Rational Expressions

When rational expressions have the same denominator, we can add or subtract by adding or subtracting the numerators and retaining the common denominator. If the denominators differ, we must find equivalent rational expressions that have a common denominator.

In general, it is most efficient to find the **least common denominator** (**LCD**) of the expressions. To do so, we factor each denominator and form the product that uses each factor the greatest number of times it occurs in any factorization.

EXAMPLE 5 Add or subtract and simplify each of the following.

a) $\dfrac{x^2 - 4x + 4}{2x^2 - 3x + 1} + \dfrac{x + 4}{2x - 2}$

b) $\dfrac{x}{x^2 + 11x + 30} - \dfrac{5}{x^2 + 9x + 20}$

Solution

a) $\dfrac{x^2 - 4x + 4}{2x^2 - 3x + 1} + \dfrac{x + 4}{2x - 2}$

$= \dfrac{x^2 - 4x + 4}{(2x - 1)(x - 1)} + \dfrac{x + 4}{2(x - 1)}$ Factoring the denominators

The LCD is $(2x - 1)(x - 1)(2)$, or $2(2x - 1)(x - 1)$.

$= \dfrac{x^2 - 4x + 4}{(2x - 1)(x - 1)} \cdot \dfrac{2}{2} + \dfrac{x + 4}{2(x - 1)} \cdot \dfrac{2x - 1}{2x - 1}$ Multiplying each term by 1 to get the LCD

$= \dfrac{2x^2 - 8x + 8}{(2x - 1)(x - 1)(2)} + \dfrac{2x^2 + 7x - 4}{2(x - 1)(2x - 1)}$

$= \dfrac{4x^2 - x + 4}{2(2x - 1)(x - 1)}$ Adding the numerators

b) $\dfrac{x}{x^2 + 11x + 30} - \dfrac{5}{x^2 + 9x + 20}$

$= \dfrac{x}{(x + 5)(x + 6)} - \dfrac{5}{(x + 5)(x + 4)}$ Factoring the denominators

The LCD is $(x + 5)(x + 6)(x + 4)$.

$= \dfrac{x}{(x + 5)(x + 6)} \cdot \dfrac{x + 4}{x + 4} - \dfrac{5}{(x + 5)(x + 4)} \cdot \dfrac{x + 6}{x + 6}$

Multiplying each term by 1 to get the LCD

$= \dfrac{x^2 + 4x}{(x + 5)(x + 6)(x + 4)} - \dfrac{5x + 30}{(x + 5)(x + 4)(x + 6)}$

$= \dfrac{x^2 + 4x - 5x - 30}{(x + 5)(x + 6)(x + 4)}$ Be sure to change the sign of *every* term in the numerator of the expression being subtracted: $-(5x + 30) = -5x - 30$

$= \dfrac{x^2 - x - 30}{(x + 5)(x + 6)(x + 4)}$

$= \dfrac{(x + 5)(x - 6)}{(x + 5)(x + 6)(x + 4)}$ Removing a factor of 1: $\dfrac{x + 5}{x + 5} = 1$

$= \dfrac{x - 6}{(x + 6)(x + 4)}$

Complex Rational Expressions

A **complex rational expression** has rational expressions in its numerator or its denominator or both.

To simplify a complex rational expression:

Method 1. Find the LCD of all the denominators within the complex rational expression. Then multiply by 1 using the LCD as the numerator and the denominator of the expression for 1.

Method 2. First add or subtract, if necessary, to get a single rational expression in the numerator and in the denominator. Then divide by multiplying by the reciprocal of the denominator.

EXAMPLE 6   Simplify: $\dfrac{\dfrac{1}{a} + \dfrac{1}{b}}{\dfrac{1}{a^3} + \dfrac{1}{b^3}}$.

Solution

Method 1. The LCD of the four rational expressions in the numerator and the denominator is $a^3 b^3$.

$$\frac{\dfrac{1}{a} + \dfrac{1}{b}}{\dfrac{1}{a^3} + \dfrac{1}{b^3}} = \frac{\dfrac{1}{a} + \dfrac{1}{b}}{\dfrac{1}{a^3} + \dfrac{1}{b^3}} \cdot \frac{a^3 b^3}{a^3 b^3} \qquad \text{Multiplying by 1}\ \text{using } \frac{a^3 b^3}{a^3 b^3}$$

$$= \frac{\left(\dfrac{1}{a} + \dfrac{1}{b}\right)(a^3 b^3)}{\left(\dfrac{1}{a^3} + \dfrac{1}{b^3}\right)(a^3 b^3)}$$

$$= \frac{a^2 b^3 + a^3 b^2}{b^3 + a^3}$$

$$= \frac{a^2 b^2 (b + a)}{(b + a)(b^2 - ba + a^2)} \qquad \text{Factoring and removing a}\ \text{factor of 1: } \frac{b + a}{b + a} = 1$$

$$= \frac{a^2 b^2}{b^2 - ba + a^2}$$

Method 2. We add in the numerator and in the denominator.

$$\frac{\dfrac{1}{a} + \dfrac{1}{b}}{\dfrac{1}{a^3} + \dfrac{1}{b^3}} = \frac{\dfrac{1}{a} \cdot \dfrac{b}{b} + \dfrac{1}{b} \cdot \dfrac{a}{a}}{\dfrac{1}{a^3} \cdot \dfrac{b^3}{b^3} + \dfrac{1}{b^3} \cdot \dfrac{a^3}{a^3}} \quad \longleftarrow \text{ The LCD is } ab.$$
$$\qquad\qquad\qquad\qquad\qquad\qquad \longleftarrow \text{ The LCD is } a^3 b^3.$$

Then

$$= \frac{\dfrac{b}{ab} + \dfrac{a}{ab}}{\dfrac{b^3}{a^3b^3} + \dfrac{a^3}{a^3b^3}}$$

$$= \frac{\dfrac{b+a}{ab}}{\dfrac{b^3 + a^3}{a^3b^3}} \qquad \begin{array}{l}\text{We have a single rational}\\ \text{expression in both the numerator}\\ \text{and the denominator.}\end{array}$$

$$= \frac{b+a}{ab} \cdot \frac{a^3b^3}{b^3 + a^3} \qquad \begin{array}{l}\text{Multiplying by the reciprocal}\\ \text{of the denominator}\end{array}$$

$$= \frac{(b + a)(a)(b)(a^2b^2)}{(a)(b)(b + a)(b^2 - ba + a^2)}$$

$$= \frac{a^2b^2}{b^2 - ba + a^2}.$$

Exercise Set R.5

Find the domain of the rational expression.

1. $-\dfrac{3}{4}$

2. $\dfrac{5}{8 - x}$

3. $\dfrac{3x - 3}{x(x - 1)}$

4. $\dfrac{(x^2 - 4)(x + 1)}{(x + 2)(x^2 - 1)}$

5. $\dfrac{7x^2 - 28x + 28}{(x^2 - 4)(x^2 + 3x - 10)}$

6. $\dfrac{7x^2 + 11x - 6}{x(x^2 - x - 6)}$

7. Use the TABLE feature on a grapher to check your answer to Exercise 3.

8. Use the TABLE feature on a grapher to check your answer to Exercise 4.

Multiply or divide and, if possible, simplify.

9. $\dfrac{x^2 - y^2}{(x - y)^2} \cdot \dfrac{1}{x + y}$

10. $\dfrac{r - s}{r + s} \cdot \dfrac{r^2 - s^2}{(r - s)^2}$

11. $\dfrac{x^2 - 2x - 35}{2x^3 - 3x^2} \cdot \dfrac{4x^3 - 9x}{7x - 49}$

12. $\dfrac{x^2 + 2x - 35}{3x^3 - 2x^2} \cdot \dfrac{9x^3 - 4x}{7x + 49}$

13. $\dfrac{a^2 - a - 6}{a^2 - 7a + 12} \cdot \dfrac{a^2 - 2a - 8}{a^2 - 3a - 10}$

14. $\dfrac{a^2 - a - 12}{a^2 - 6a + 8} \cdot \dfrac{a^2 + a - 6}{a^2 - 2a - 24}$

15. $\dfrac{m^2 - n^2}{r + s} \div \dfrac{m - n}{r + s}$

16. $\dfrac{a^2 - b^2}{x - y} \div \dfrac{a + b}{x - y}$

17. $\dfrac{3x + 12}{2x - 8} \div \dfrac{(x + 4)^2}{(x - 4)^2}$

18. $\dfrac{a^2 - a - 2}{a^2 - a - 6} \div \dfrac{a^2 - 2a}{2a + a^2}$

19. $\dfrac{x^2 - y^2}{x^3 - y^3} \cdot \dfrac{x^2 + xy + y^2}{x^2 + 2xy + y^2}$

20. $\dfrac{c^3 + 8}{c^2 - 4} \div \dfrac{c^2 - 2c + 4}{c^2 - 4c + 4}$

21. $\dfrac{(x - y)^2 - z^2}{(x + y)^2 - z^2} \div \dfrac{x - y + z}{x + y - z}$

22. $\dfrac{(a + b)^2 - 9}{(a - b)^2 - 9} \cdot \dfrac{a - b - 3}{a + b + 3}$

23. Use a grapher to check your answer to Exercise 11.

24. Use a grapher to check your answer to Exercise 12.

Add or subtract and, if possible, simplify.

25. $\dfrac{3}{2a + 3} + \dfrac{2a}{2a + 3}$

26. $\dfrac{a - 3b}{a + b} + \dfrac{a + 5b}{a + b}$

27. $\dfrac{y}{y - 1} + \dfrac{2}{1 - y}$

28. $\dfrac{a}{a - b} + \dfrac{b}{b - a}$

29. $\dfrac{x}{2x - 3y} - \dfrac{y}{3y - 2x}$

30. $\dfrac{3a}{3a - 2b} - \dfrac{2a}{2b - 3a}$

31. $\dfrac{3}{x + 2} + \dfrac{2}{x^2 - 4}$

32. $\dfrac{5}{a - 3} - \dfrac{2}{a^2 - 9}$

33. $\dfrac{y}{y^2 - y - 20} - \dfrac{2}{y + 4}$

34. $\dfrac{6}{y^2 + 6y + 9} - \dfrac{5}{y + 3}$

35. $\dfrac{3}{x + y} + \dfrac{x - 5y}{x^2 - y^2}$

36. $\dfrac{a^2 + 1}{a^2 - 1} - \dfrac{a - 1}{a + 1}$

37. $\dfrac{9x + 2}{3x^2 - 2x - 8} + \dfrac{7}{3x^2 + x - 4}$

38. $\dfrac{3y}{y^2 - 7y + 10} - \dfrac{2y}{y^2 - 8y + 15}$

39. $\dfrac{5a}{a - b} + \dfrac{ab}{a^2 - b^2} + \dfrac{4b}{a + b}$

40. $\dfrac{6a}{a - b} - \dfrac{3b}{b - a} + \dfrac{5}{a^2 - b^2}$

41. $\dfrac{7}{x + 2} - \dfrac{x + 8}{4 - x^2} + \dfrac{3x - 2}{4 - 4x + x^2}$

42. $\dfrac{6}{x + 3} - \dfrac{x + 4}{9 - x^2} + \dfrac{2x - 3}{9 - 6x + x^2}$

43. $\dfrac{1}{x + 1} + \dfrac{x}{2 - x} + \dfrac{x^2 + 2}{x^2 - x - 2}$

44. $\dfrac{x - 1}{x - 2} - \dfrac{x + 1}{x + 2} - \dfrac{x - 6}{4 - x^2}$

45. Use a grapher to check your answer to Exercise 31.

46. Use a grapher to check your answer to Exercise 32.

Simplify.

47. $\dfrac{\dfrac{x^2 - y^2}{xy}}{\dfrac{x - y}{y}}$

48. $\dfrac{\dfrac{a - b}{b}}{\dfrac{a^2 - b^2}{ab}}$

49. $\dfrac{a - a^{-1}}{a + a^{-1}}$

50. $\dfrac{a - \dfrac{a}{b}}{b - \dfrac{b}{a}}$

51. $\dfrac{c + \dfrac{8}{c^2}}{1 + \dfrac{2}{c}}$

52. $\dfrac{x^{-1} + y^{-1}}{x^{-3} + y^{-3}}$

53. $\dfrac{x^2 + xy + y^2}{\dfrac{x^2 - y^2}{y} \cdot x}$

54. $\dfrac{\dfrac{a^2}{b} + \dfrac{b^2}{a}}{a^2 - ab + b^2}$

55. $\dfrac{\dfrac{x}{y} - \dfrac{y}{x}}{\dfrac{1}{y} + \dfrac{1}{x}}$

56. $\dfrac{\dfrac{a}{b} - \dfrac{b}{a}}{\dfrac{1}{a} - \dfrac{1}{b}}$

57. $\dfrac{\dfrac{1}{x - 3} + \dfrac{2}{x + 3}}{\dfrac{3}{x - 1} - \dfrac{4}{x + 2}}$

58. $\dfrac{\dfrac{5}{x + 1} - \dfrac{3}{x - 2}}{\dfrac{1}{x - 5} + \dfrac{2}{x + 2}}$

59. $\dfrac{\dfrac{a}{1 - a} + \dfrac{1 + a}{a}}{\dfrac{1 - a}{a} + \dfrac{a}{1 + a}}$

60. $\dfrac{\dfrac{1 - x}{x} + \dfrac{x}{1 + x}}{\dfrac{1 + x}{x} + \dfrac{x}{1 - x}}$

61. $\dfrac{\dfrac{1}{a^2} + \dfrac{2}{ab} + \dfrac{1}{b^2}}{\dfrac{1}{a^2} - \dfrac{1}{b^2}}$

62. $\dfrac{\dfrac{1}{x^2} - \dfrac{1}{y^2}}{\dfrac{1}{x^2} - \dfrac{2}{xy} + \dfrac{1}{y^2}}$

Discussion and Writing

63. When adding or subtracting rational expressions, we can always find a common denominator by forming the product of all the denominators. Explain why it is usually preferable to find the least common denominator.

64. How would you determine which method to use for simplifying a particular complex rational expression?

Synthesis

Simplify.

65. $\dfrac{(x + h)^2 - x^2}{h}$

66. $\dfrac{\dfrac{1}{x + h} - \dfrac{1}{x}}{h}$

67. $\dfrac{(x + h)^3 - x^3}{h}$

68. $\dfrac{\dfrac{1}{(x + h)^2} - \dfrac{1}{x^2}}{h}$

69. $\left[\dfrac{\dfrac{x + 1}{x - 1} + 1}{\dfrac{x + 1}{x - 1} - 1} \right]^5$

70. $1 + \dfrac{1}{1 + \dfrac{1}{1 + \dfrac{1}{1 + \dfrac{1}{x}}}}$

Perform the indicated operations and, if possible, simplify.

71. $\dfrac{n(n + 1)(n + 2)}{2 \cdot 3} + \dfrac{(n + 1)(n + 2)}{2}$

72. $\dfrac{n(n + 1)(n + 2)(n + 3)}{2 \cdot 3 \cdot 4} + \dfrac{(n + 1)(n + 2)(n + 3)}{2 \cdot 3}$

73. $\dfrac{x^2 - 9}{x^3 + 27} \cdot \dfrac{5x^2 - 15x + 45}{x^2 - 2x - 3} + \dfrac{x^2 + x}{4 + 2x}$

74. $\dfrac{x^2 + 2x - 3}{x^2 - x - 12} \div \dfrac{x^2 - 1}{x^2 - 16} - \dfrac{2x + 1}{x^2 + 2x + 1}$

Radical Notation and Rational Exponents

R.6

- *Simplify radical expressions.*
- *Rationalize denominators or numerators in rational expressions.*
- *Convert between exponential and radical notation.*
- *Simplify expressions with rational exponents.*

A number c is said to be a **square root** of a if $c^2 = a$. Thus, 3 is a square root of 9, because $3^2 = 9$, and -3 is also a square root of 9, because $(-3)^2 = 9$. Similarly, 5 is a third root (called a **cube root**) of 125, because $5^3 = 125$. The number 125 has no other real-number cube root.

---

nth Root

A number c is said to be an **nth root** of a if $c^n = a$.

The symbol $\sqrt{a}$ denotes the nonnegative square root of a, and the symbol $\sqrt[3]{a}$ denotes the real-number cube root of a. The symbol $\sqrt[n]{a}$ denotes the nth root of a, that is, a number whose nth power is a. The symbol $\sqrt[n]{}$ is called a **radical**, and the expression under the radical is called the **radicand**. The number n (which is omitted when it is 2) is called the

index. Examples of roots for $n = 3$, 4, and 2, respectively, are

$$\sqrt[3]{125}, \qquad \sqrt[4]{16}, \quad \text{and} \quad \sqrt{3600}.$$

Any real number has only one real-number odd root. Any positive number has two square roots, one positive and one negative. The same is true for fourth roots or roots with any even index. The positive root is called the **principal root.** When an expression such as $\sqrt{4}$ or $\sqrt[6]{23}$ is used, it is understood to represent the principal (nonnegative) root. To denote a negative root, we use $-\sqrt{4}$, $-\sqrt[6]{23}$, and so on.

EXAMPLE 1   Simplify each of the following.

a) $\sqrt{36}$ **b)** $-\sqrt{36}$ **c)** $\sqrt[5]{\dfrac{32}{243}}$

d) $\sqrt[3]{-8}$ **e)** $\sqrt[4]{-16}$

Solution

a) $\sqrt{36} = 6$, because $6^2 = 36$.

b) $-\sqrt{36} = -6$, because $6^2 = 36$ and $-(\sqrt{36}) = -(6) = -6$.

c) $\sqrt[5]{\dfrac{32}{243}} = \dfrac{2}{3}$, because $\left(\dfrac{2}{3}\right)^5 = \dfrac{2^5}{3^5} = \dfrac{32}{243}$.

d) $\sqrt[3]{-8} = -2$, because $(-2)^3 = -8$.

e) $\sqrt[4]{-16}$ is not a real number, because we cannot find a real number that can be raised to the fourth power to get -16.   ▬

STUDY TIP

The keystrokes for entering $\sqrt[3]{-8}$ and $\sqrt[5]{\frac{32}{243}}$ are found in the *Graphing Calculator Manual* that accompanies this text.

We can find $\sqrt{36}$ and $-\sqrt{36}$ in Example 1 using the square-root feature on the keypad of a grapher, and we can use the cube-root feature to find $\sqrt[3]{-8}$. We can use the xth-root feature to find higher roots.

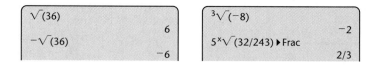

When we try to find $\sqrt[4]{-16}$ on a grapher set in REAL mode, we get an error message indicating that the answer is nonreal.

We can generalize Example 1(e) and say that when a is negative and n is even, $\sqrt[n]{a}$ is not a real number. For example, $\sqrt{-4}$ and $\sqrt[4]{-81}$ are not real numbers.

Simplifying Radical Expressions

Consider the expression $\sqrt{(-3)^2}$. This is equivalent to $\sqrt{9}$, or 3. Similarly, $\sqrt{3^2} = \sqrt{9} = 3$. This illustrates the first of several properties of radicals, listed below.

Properties of Radicals

Let a and b be any real numbers or expressions for which the given roots exist. For any natural numbers m and n ($n \neq 1$):

1. If n is even, $\sqrt[n]{a^n} = |a|$.

2. If n is odd, $\sqrt[n]{a^n} = a$.

3. $\sqrt[n]{a} \cdot \sqrt[n]{b} = \sqrt[n]{ab}$.

4. $\sqrt[n]{\dfrac{a}{b}} = \dfrac{\sqrt[n]{a}}{\sqrt[n]{b}}$ ($b \neq 0$).

5. $\sqrt[n]{a^m} = (\sqrt[n]{a})^m$.

EXAMPLE 2 Simplify each of the following.

a) $\sqrt{(-5)^2}$ b) $\sqrt[3]{(-5)^3}$ c) $\sqrt[4]{4} \cdot \sqrt[4]{5}$ d) $\sqrt{50}$

e) $\dfrac{\sqrt{72}}{\sqrt{6}}$ f) $\sqrt[3]{8^5}$ g) $\sqrt{216x^5y^3}$ h) $\sqrt{\dfrac{x^2}{16}}$

Solution

a) $\sqrt{(-5)^2} = |-5| = 5$ Using Property 1

b) $\sqrt[3]{(-5)^3} = -5$ Using Property 2

c) $\sqrt[4]{4} \cdot \sqrt[4]{5} = \sqrt[4]{4 \cdot 5} = \sqrt[4]{20}$ Using Property 3

d) $\sqrt{50} = \sqrt{25 \cdot 2} = \sqrt{25} \cdot \sqrt{2} = 5\sqrt{2}$ Using Property 3

e) $\dfrac{\sqrt{72}}{\sqrt{6}} = \sqrt{\dfrac{72}{6}}$ Using Property 4

$= \sqrt{12} = \sqrt{4 \cdot 3} = \sqrt{4}\,\sqrt{3}$ Using Property 3

$= 2\sqrt{3}$

f) $\sqrt[3]{8^5} = (\sqrt[3]{8})^5$ Using Property 5

$= 2^5 = 32$

g) $\sqrt{216x^5y^3} = \sqrt{36 \cdot 6 \cdot x^4 \cdot x \cdot y^2 \cdot y}$

$= \sqrt{36x^4y^2}\,\sqrt{6xy}$ Using Property 3

$= |6x^2y|\,\sqrt{6xy}$ Using Property 1

$= 6x^2|y|\,\sqrt{6xy}$ $6x^2$ cannot be negative, so absolute-value signs are not needed for it.

$$y_1 = \sqrt{\frac{x^2}{16}}, \quad y_2 = \frac{|x|}{4}$$

X	Y1	Y2
−3	.75	.75
−2	.5	.5
−1	.25	.25
0	0	0
1	.25	.25
2	.5	.5
3	.75	.75

X = 0

h) $\sqrt{\dfrac{x^2}{16}} = \dfrac{\sqrt{x^2}}{\sqrt{16}}$ **Using Property 4**

$\qquad = \dfrac{|x|}{4}$ **Using Property 1. Check using a table.**

In many situations, radicands are never formed by raising negative quantities to even powers. In such cases, absolute-value notation is not required. For this reason, **we will henceforth assume that no radicands are formed by raising negative quantities to even powers.** For example, we will write $\sqrt{x^2} = x$ and $\sqrt[4]{a^5 b} = a\sqrt[4]{ab}$.

Radical expressions with the same index and the same radicand can be added or subtracted.

EXAMPLE 3   Perform the operations indicated.

a) $3\sqrt{8x^2} - 5\sqrt{2x^2}$

b) $(4\sqrt{3} + \sqrt{2})(\sqrt{3} - 5\sqrt{2})$

Solution

a) $3\sqrt{8x^2} - 5\sqrt{2x^2} = 3\sqrt{4x^2 \cdot 2} - 5\sqrt{x^2 \cdot 2}$

$\qquad\qquad\qquad\quad = 3 \cdot 2x\sqrt{2} - 5x\sqrt{2}$

$\qquad\qquad\qquad\quad = 6x\sqrt{2} - 5x\sqrt{2}$

$\qquad\qquad\qquad\quad = (6x - 5x)\sqrt{2}$ **Using the distributive property**

$\qquad\qquad\qquad\quad = x\sqrt{2}$

b) $(4\sqrt{3} + \sqrt{2})(\sqrt{3} - 5\sqrt{2}) = 4(\sqrt{3})^2 - 20\sqrt{6} + \sqrt{6} - 5(\sqrt{2})^2$

$\qquad\qquad\qquad\qquad\qquad\qquad$ **Multiplying**

$\qquad\qquad\qquad\qquad\quad = 4 \cdot 3 + (-20 + 1)\sqrt{6} - 5 \cdot 2$

$\qquad\qquad\qquad\qquad\quad = 12 - 19\sqrt{6} - 10$

$\qquad\qquad\qquad\qquad\quad = 2 - 19\sqrt{6}$

An Application

The Pythagorean theorem relates the lengths of the sides of a right triangle. The side opposite the triangle's right angle is called the **hypotenuse**. The other sides are the **legs**.

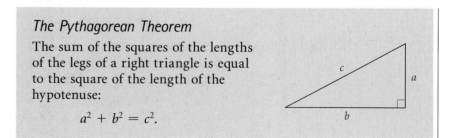

The Pythagorean Theorem

The sum of the squares of the lengths of the legs of a right triangle is equal to the square of the length of the hypotenuse:

$$a^2 + b^2 = c^2.$$

EXAMPLE 4 A surveyor places poles at points A, B, and C in order to measure the distance across a pond. The distances AC and BC are measured as shown. Find the distance AB across the pond.

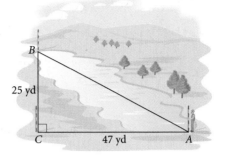

Solution We see that the lengths of the legs of a right triangle are given. Thus we use the Pythagorean theorem to find the length of the hypotenuse:

$$c^2 = a^2 + b^2$$
$$c = \sqrt{a^2 + b^2} \qquad \text{Solving for } c$$
$$= \sqrt{25^2 + 47^2}$$
$$= \sqrt{625 + 2209}$$
$$= \sqrt{2834}$$
$$\approx 53.2.$$

The distance across the pond is about 53.2 yd.

$\sqrt{(25^2+47^2)}$

53.23532662

Rationalizing Denominators or Numerators

There are times when we need to remove the radicals in a denominator or a numerator. This is called **rationalizing the denominator** or **rationalizing the numerator.** It is done by multiplying by 1 in such a way as to obtain a perfect nth power.

EXAMPLE 5 Rationalize the denominator of each of the following.

a) $\sqrt{\dfrac{3}{2}}$ **b)** $\dfrac{\sqrt[3]{7}}{\sqrt[3]{9}}$

Solution

a) $\sqrt{\dfrac{3}{2}} = \sqrt{\dfrac{3}{2} \cdot \dfrac{2}{2}} = \sqrt{\dfrac{6}{4}} = \dfrac{\sqrt{6}}{\sqrt{4}} = \dfrac{\sqrt{6}}{2}$

b) $\dfrac{\sqrt[3]{7}}{\sqrt[3]{9}} = \dfrac{\sqrt[3]{7}}{\sqrt[3]{9}} \cdot \dfrac{\sqrt[3]{3}}{\sqrt[3]{3}} = \dfrac{\sqrt[3]{21}}{\sqrt[3]{27}} = \dfrac{\sqrt[3]{21}}{3}$

Pairs of expressions of the form $a\sqrt{b} + c\sqrt{d}$ and $a\sqrt{b} - c\sqrt{d}$ are called **conjugates**. The product of such a pair contains no radicals and can be used to rationalize a denominator or a numerator.

EXAMPLE 6   Rationalize the numerator: $\dfrac{\sqrt{x} - \sqrt{y}}{5}$.

Solution

$$\frac{\sqrt{x} - \sqrt{y}}{5} = \frac{\sqrt{x} - \sqrt{y}}{5} \cdot \frac{\sqrt{x} + \sqrt{y}}{\sqrt{x} + \sqrt{y}}$$

The conjugate of
$\sqrt{x} - \sqrt{y}$ is
$\sqrt{x} + \sqrt{y}$.

$$= \frac{(\sqrt{x})^2 - (\sqrt{y})^2}{5\sqrt{x} + 5\sqrt{y}}$$

$$= \frac{x - y}{5\sqrt{x} + 5\sqrt{y}}$$

Rational Exponents

We are motivated to define *rational exponents* so that the properties for integer exponents hold for them. For example, we must have

$$a^{1/2} \cdot a^{1/2} = a^{1/2+1/2} = a^1 = a.$$

Thus we are led to define $a^{1/2}$ to mean $\sqrt{a}$. Similarly, $a^{1/n}$ would mean $\sqrt[n]{a}$. Again, if the laws of exponents are to hold, we must have

$$(a^{1/n})^m = (a^m)^{1/n} = a^{m/n}.$$

Thus we are led to define $a^{m/n}$ to mean $(\sqrt[n]{a})^m$, or, equivalently, $\sqrt[n]{a^m}$.

Rational Exponents

For any real number a and any natural numbers m and n for which $\sqrt[n]{a}$ exists,

$$a^{1/n} = \sqrt[n]{a},$$
$$a^{m/n} = \sqrt[n]{a^m} = (\sqrt[n]{a})^m, \quad \text{and}$$

$$a^{-m/n} = \frac{1}{a^{m/n}}.$$

We can use the definition of rational exponents to convert between radical and exponential notation.

EXAMPLE 7   Convert to radical notation and, if possible, simplify each of the following.

a) $7^{3/4}$ **b)** $8^{-5/3}$

c) $m^{1/6}$ **d)** $(-32)^{2/5}$

Solution

a) $7^{3/4} = \sqrt[4]{7^3}$, or $(\sqrt[4]{7})^3$

b) $8^{-5/3} = \dfrac{1}{8^{5/3}} = \dfrac{1}{(\sqrt[3]{8})^5} = \dfrac{1}{2^5} = \dfrac{1}{32}$

c) $m^{1/6} = \sqrt[6]{m}$

d) $(-32)^{2/5} = \sqrt[5]{(-32)^2} = \sqrt[5]{1024} = 4,$ or
$(-32)^{2/5} = (\sqrt[5]{-32})^2 = (-2)^2 = 4$

Some graphers will return an error message when an expression of the form $a^{m/n}$, with $a < 0$ and $m > 1$, such as $(-32)^{2/5}$, is entered. In this case, we enter the expression as $((a)^m)^{1/n}$, or $((a)^{1/n})^m$. For $(-32)^{2/5}$, for example, we enter $((-32)^2)^{1/5}$, or $((-32)^{1/5})^2$.

EXAMPLE 8   Convert to exponential notation and, if possible, simplify each of the following.

a) $(\sqrt[4]{7xy})^5$ **b)** $\sqrt[6]{x^3}$ **c)** $\sqrt[3]{\sqrt{7}}$

Solution

a) $(\sqrt[4]{7xy})^5 = (7xy)^{5/4}$

b) $\sqrt[6]{x^3} = x^{3/6} = x^{1/2}$

c) $\sqrt[3]{\sqrt{7}} = \sqrt[3]{7^{1/2}} = (7^{1/2})^{1/3} = 7^{1/6}$

We can use the laws of exponents to simplify exponential and radical expressions.

EXAMPLE 9   Simplify and then, if appropriate, write radical notation for each of the following.

a) $x^{5/6} \cdot x^{2/3}$ **b)** $(x + 3)^{5/2}(x + 3)^{-1/2}$

Solution

a) $x^{5/6} \cdot x^{2/3} = x^{5/6+2/3} = x^{9/6} = x^{3/2} = \sqrt{x^3} = \sqrt{x^2}\sqrt{x} = x\sqrt{x}$

b) $(x + 3)^{5/2}(x + 3)^{-1/2} = (x + 3)^{5/2-1/2} = (x + 3)^2$

EXAMPLE 10   Write an expression containing a single radical: $a^{1/2}b^{5/6}$.

Solution

$$a^{1/2}b^{5/6} = a^{3/6}b^{5/6} = (a^3b^5)^{1/6} = \sqrt[6]{a^3b^5}$$

```
5/6+2/3 ▶ Frac
                    3/2
```

Exercise Set R.6

Simplify. Assume that variables can represent any real number.

1. $\sqrt{(-11)^2}$

2. $\sqrt{(-1)^2}$

3. $\sqrt{16y^2}$

4. $\sqrt{36t^2}$

5. $\sqrt{(b + 1)^2}$

6. $\sqrt{(2c - 3)^2}$

7. $\sqrt[3]{-27x^3}$

8. $\sqrt[3]{-8y^3}$

9. $\sqrt{x^2 - 4x + 4}$

10. $\sqrt{x^2 + 16x + 64}$

11. $\sqrt[5]{32}$

12. $\sqrt[5]{-32}$

13. $\sqrt{180}$

14. $\sqrt{48}$

15. $\sqrt[3]{54}$

16. $\sqrt[3]{135}$

17. $\sqrt{128c^2d^4}$

18. $\sqrt{162c^4d^6}$

19. Use the TABLE feature on a grapher to check your answer to Exercise 9.

20. Use the TABLE feature on a grapher to check your answer to Exercise 10.

Simplify. Assume that no radicands were formed by raising negative quantities to even powers.

21. $\sqrt{2x^3y}\ \sqrt{12xy}$

22. $\sqrt{3y^4z}\ \sqrt{20z}$

23. $\sqrt[3]{3x^2y}\ \sqrt[3]{36x}$

24. $\sqrt[5]{8x^3y^4}\ \sqrt[5]{4x^4y}$

25. $\sqrt[3]{2(x+4)}\ \sqrt[3]{4(x+4)^4}$

26. $\sqrt[3]{4(x+1)^2}\ \sqrt[3]{18(x+1)^2}$

27. $\sqrt[6]{\dfrac{m^{12}n^{24}}{64}}$

28. $\sqrt[8]{\dfrac{m^{16}n^{24}}{2^8}}$

29. $\dfrac{\sqrt[3]{40m}}{\sqrt[3]{5m}}$

30. $\dfrac{\sqrt{40xy}}{\sqrt{8x}}$

31. $\dfrac{\sqrt[3]{3x^2}}{\sqrt[3]{24x^5}}$

32. $\dfrac{\sqrt{128a^2b^4}}{\sqrt{16ab}}$

33. $\sqrt[3]{\dfrac{64a^4}{27b^3}}$

34. $\sqrt{\dfrac{9x^7}{16y^8}}$

35. $\sqrt{\dfrac{7x^3}{36y^6}}$

36. $\sqrt[3]{\dfrac{2yz}{250z^4}}$

37. $9\sqrt{50}+6\sqrt{2}$

38. $11\sqrt{27}-4\sqrt{3}$

39. $8\sqrt{2x^2}-6\sqrt{20x}-5\sqrt{8x^2}$

40. $2\sqrt[3]{8x^2}+5\sqrt[3]{27x^2}-3\sqrt{x^3}$

41. $(\sqrt{3}-\sqrt{2})(\sqrt{3}+\sqrt{2})$

42. $(\sqrt{8}+2\sqrt{5})(\sqrt{8}-2\sqrt{5})$

43. $(1+\sqrt{3})^2$

44. $(\sqrt{2}-5)^2$

45. Use a grapher to check your answer to Exercise 39.

46. Use a grapher to check your answer to Exercise 40.

47. An airplane is flying at an altitude of 3700 ft. The slanted distance directly to the airport is 14,200 ft. How far horizontally is the airplane from the airport?

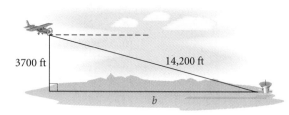

3700 ft   14,200 ft

b

48. During a summer heat wave, a 2-mi bridge expands 2 ft in length. Assuming that the bulge occurs straight up the middle, estimate the height of the bulge. (In reality, bridges are built with expansion joints to control such buckling.)

49. An *equilateral triangle* is shown below.

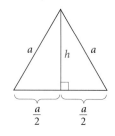

a h a

$\dfrac{a}{2}$ $\dfrac{a}{2}$

a) Find an expression for its height h in terms of a.
b) Find an expression for its area A in terms of a.

50. An isosceles right triangle has legs of length s. Find an expression for the length of the hypotenuse in terms of s.

51. The diagonal of a square has length $8\sqrt{2}$. Find the length of a side of the square.

52. The area of square $PQRS$ is 100 ft^2, and A, B, C, and D are the midpoints of the sides. Find the area of square $ABCD$.

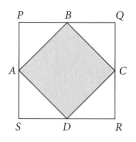

P B Q

A C

S D R

Rationalize the denominator.

53. $\sqrt{\dfrac{2}{3}}$

54. $\sqrt{\dfrac{3}{7}}$

55. $\dfrac{\sqrt[3]{5}}{\sqrt[3]{4}}$

56. $\dfrac{\sqrt[3]{7}}{\sqrt[3]{25}}$

57. $\sqrt[3]{\dfrac{16}{9}}$

58. $\sqrt[3]{\dfrac{3}{5}}$

59. $\dfrac{6}{3 + \sqrt{5}}$

60. $\dfrac{2}{\sqrt{3} - 1}$

61. $\dfrac{6}{\sqrt{m} - \sqrt{n}}$

62. $\dfrac{3}{\sqrt{v} + \sqrt{w}}$

Rationalize the numerator.

63. $\dfrac{\sqrt{12}}{5}$

64. $\dfrac{\sqrt{50}}{3}$

65. $\sqrt[3]{\dfrac{7}{2}}$

66. $\sqrt[3]{\dfrac{2}{5}}$

67. $\dfrac{\sqrt{11}}{\sqrt{3}}$

68. $\dfrac{\sqrt{5}}{\sqrt{2}}$

69. $\dfrac{9 - \sqrt{5}}{3 - \sqrt{3}}$

70. $\dfrac{8 - \sqrt{6}}{5 - \sqrt{2}}$

71. $\dfrac{\sqrt{a} + \sqrt{b}}{3a}$

72. $\dfrac{\sqrt{p} - \sqrt{q}}{1 + \sqrt{q}}$

Convert to radical notation and simplify.

73. $x^{3/4}$

74. $y^{2/5}$

75. $16^{3/4}$

76. $4^{7/2}$

77. $125^{-1/3}$

78. $32^{-4/5}$

79. $a^{5/4}b^{-3/4}$

80. $x^{2/5}y^{-1/5}$

Convert to exponential notation and simplify.

81. $(\sqrt[4]{13})^5$

82. $\sqrt[5]{17^3}$

83. $\sqrt[3]{20^2}$

84. $(\sqrt[5]{12})^4$

85. $\sqrt[3]{\sqrt{11}}$

86. $\sqrt[3]{\sqrt[4]{7}}$

87. $\sqrt{5}\ \sqrt[3]{5}$

88. $\sqrt[3]{2}\ \sqrt{2}$

89. $\sqrt[5]{32^2}$

90. $\sqrt[3]{64^2}$

Simplify and then, if appropriate, write radical notation.

91. $(2a^{3/2})(4a^{1/2})$

92. $(3a^{5/6})(8a^{2/3})$

93. $\left(\dfrac{x^6}{9b^{-4}}\right)^{1/2}$

94. $\left(\dfrac{x^{2/3}}{4y^{-2}}\right)^{1/2}$

95. $\dfrac{x^{2/3}y^{5/6}}{x^{-1/3}y^{1/2}}$

96. $\dfrac{a^{1/2}b^{5/8}}{a^{1/4}b^{3/8}}$

97. Use a grapher to check your answer to Exercise 91.

98. Use a grapher to check your answer to Exercise 92.

Write an expression containing a single radical and simplify.

99. $\sqrt[3]{6}\ \sqrt{2}$

100. $\sqrt{2}\ \sqrt[4]{8}$

101. $\sqrt[4]{xy}\ \sqrt[3]{x^2y}$

102. $\sqrt[3]{ab^2}\ \sqrt{ab}$

103. $\sqrt[3]{a^4\sqrt{a^3}}$

104. $\sqrt{a^3\sqrt[3]{a^2}}$

105. $\dfrac{\sqrt{(a + x)^3}\ \sqrt[3]{(a + x)^2}}{\sqrt[4]{a + x}}$

106. $\dfrac{\sqrt[4]{(x + y)^2}\ \sqrt[3]{x + y}}{\sqrt{(x + y)^3}}$

Discussion and Writing

107. Explain how you would convince a classmate that $\sqrt{a + b}$ is not equivalent to $\sqrt{a} + \sqrt{b}$, for positive real numbers a and b. Give both an algebraic explanation and a graphical explanation.

108. Explain how you would determine whether $10\sqrt{26} - 50$ is positive or negative without carrying out the actual computation.

Synthesis

Simplify.

109. $\sqrt{1 + x^2} + \dfrac{1}{\sqrt{1 + x^2}}$

110. $\sqrt{1 - x^2} - \dfrac{x^2}{2\sqrt{1 - x^2}}$

111. $(\sqrt{a^{\sqrt{a}}})^{\sqrt{a}}$

112. $(2a^3b^{5/4}c^{1/7})^4 \div (54a^{-2}b^{2/3}c^{6/5})^{-1/3}$

113. Use a grapher to determine all values of x for which:
   a) $x^{1/2} = x^{1/3}$;
   b) $x^{1/2} > x^{1/3}$;
   c) $x^{1/2} < x^{1/3}$.

The Basics of Equation Solving

R.7

- *Solve linear and quadratic equations.*
- *Solve a formula for a given variable.*

An **equation** is a statement that two expressions are equal. To **solve** an equation in one variable is to find all the values of the variable that make the equation true. Each of these numbers is a **solution** of the equation. The set of all solutions of an equation is its **solution set**. Equations that have the same solution set are called **equivalent equations**.

Linear and Quadratic Equations

A **linear equation in one variable** is an equation that is equivalent to one of the form $ax + b = 0$, where a and b are real numbers and $a \neq 0$.

A **quadratic equation** is an equation that is equivalent to one of the form $ax^2 + bx + c = 0$, where a, b, and c are real numbers and $a \neq 0$.

The following principles allow us to solve many linear and quadratic equations.

Equation-Solving Principles

For any real numbers a, b, and c,

The Addition Principle: If $a = b$ is true, then $a + c = b + c$ is true.

The Multiplication Principle: If $a = b$ is true, then $ac = bc$ is true.

The Principle of Zero Products: If $ab = 0$ is true, then $a = 0$ or $b = 0$, and if $a = 0$ or $b = 0$, then $ab = 0$.

The Principle of Square Roots: If $x^2 = k$, then $x = \sqrt{k}$ or $x = -\sqrt{k}$.

EXAMPLE 1 Solve: $2(5 - 3x) = 8 - 3(x + 2)$.

Solution We solve this both algebraically and graphically, as shown below.

Algebraic Solution

We have

$2(5 - 3x) = 8 - 3(x + 2)$

$10 - 6x = 8 - 3x - 6$	Using the distributive property
$10 - 6x = 2 - 3x$	Combining like terms
$10 = 2 + 3x$	Using the addition principle to add $6x$ on both sides
$8 = 3x$	Using the addition principle to add -2, or subtract 2, on both sides
$\frac{8}{3} = x.$	Using the multiplication principle to multiply by $\frac{1}{3}$, or divide by 3, on both sides

CHECK:

$$\frac{2(5 - 3x) = 8 - 3(x + 2)}{}$$

$$2\left(5 - 3 \cdot \frac{8}{3}\right) \; ? \; 8 - 3\left(\frac{8}{3} + 2\right) \qquad \text{Substituting } \frac{8}{3} \text{ for } x$$

$$
\begin{array}{c|c}
2(5 - 8) & 8 - 3\left(\frac{14}{3}\right) \\
2(-3) & 8 - 14 \\
-6 & -6 \qquad \text{TRUE}
\end{array}
$$

The solution is $\frac{8}{3}$.

We can also use the TABLE feature on a grapher, set in ASK mode, to check the solution. We let $y_1 = 2(5 - 3x)$ and $y_2 = 8 - 3(x + 2)$. When $\frac{8}{3}$ is entered for x, we see that y_1 and y_2 are both -6. (Note that the grapher converts $\frac{8}{3}$ to decimal notation.)

X	Y₁	Y₂
2.6667	-6	-6
X =		

Graphical Solution

We graph $y_1 = 2(5 - 3x)$ and $y_2 = 8 - 3(x + 2)$. The value of x for which $y_1 = y_2$ is the solution of the original equation. This x-value is the first coordinate of the point of intersection of the graphs of y_1 and y_2. Using the INTERSECT feature (see "Introduction to Graphs and the Graphing Calculator"), we find that the first coordinate of the point of intersection is 2.6666667. This is a decimal approximation of the solution.

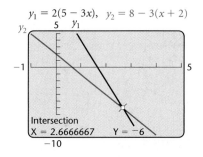

$y_1 = 2(5 - 3x), \; y_2 = 8 - 3(x + 2)$

If the solution is a rational number, we can find fractional notation for the exact solution by using the ▶FRAC feature.

X ▶ Frac

8/3

The solution is $\frac{8}{3}$.

EXAMPLE 2   Solve: $2x^2 - x = 3$.

Algebraic Solution

We have

$$2x^2 - x = 3$$
$$2x^2 - x - 3 = 0 \qquad \text{Subtracting 3 on both sides}$$
$$(x + 1)(2x - 3) = 0 \qquad \text{Factoring}$$
$$x + 1 = 0 \quad or \quad 2x - 3 = 0 \qquad \text{Using the principle of zero products}$$
$$x = -1 \quad or \qquad 2x = 3$$
$$x = -1 \quad or \qquad x = \tfrac{3}{2}.$$

Both numbers check. We can use the TABLE feature on a grapher, set in ASK mode, to confirm this. We let $y = 2x^2 - x$. The y-values in the table should be 3 for both $x = -1$ and $x = \tfrac{3}{2}$.

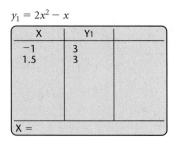

The solutions are -1 and $\tfrac{3}{2}$.

Graphical Solution

We graph $y_1 = 2x^2 - x$ and $y_2 = 3$ and use the INTERSECT feature twice.

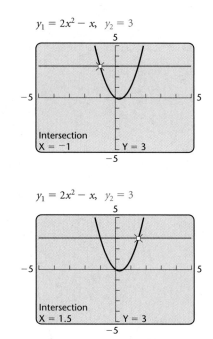

The solutions are -1 and 1.5.

EXAMPLE 3   Solve: $2x^2 - 10 = 0$.

Solution   We solve algebraically:

$$2x^2 - 10 = 0$$
$$2x^2 = 10 \qquad \text{Adding 10 on both sides}$$
$$x^2 = 5 \qquad \text{Dividing by 2 on both sides}$$
$$x = \sqrt{5} \quad or \quad x = -\sqrt{5}. \qquad \text{Using the principle of square roots}$$

Both numbers check. The solutions are $\sqrt{5}$ and $-\sqrt{5}$, or $\pm\sqrt{5}$. We can use the TABLE feature on a grapher, set in ASK mode, to confirm this. Note that the grapher converts $\sqrt{5}$ and $-\sqrt{5}$ to decimal notation.

If we were to use a grapher to solve this equation, we would find that $x \approx 2.236$ or $x \approx -2.236$. Since the exact solutions are not rational numbers, the ▶FRAC operation will not yield exact answers.

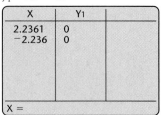

Formulas

A **formula** is an equation that can be used to *model* a situation. For example, the formula $P = 2l + 2w$ gives the perimeter of a rectangle with length l and width w. The equation-solving principles presented earlier can be used to solve a formula for a given variable.

EXAMPLE 4 Solve $P = 2l + 2w$ for l.

Solution

$$P = 2l + 2w$$
$$P - 2w = 2l \qquad \text{Subtracting } 2w \text{ on both sides}$$
$$\frac{P - 2w}{2} = l \qquad \text{Dividing by 2 on both sides}$$

The formula $l = \dfrac{P - 2w}{2}$ can be used to determine a rectangle's length if we are given its perimeter and its width.

On many graphers, an equation must be solved for y before it can be entered. Solving for y is the same as solving a formula for a given variable.

EXAMPLE 5 Solve $3x + 4y = 12$ for y.

Solution We have

$$3x + 4y = 12$$
$$4y = -3x + 12 \qquad \text{Subtracting } 3x \text{ on both sides}$$
$$y = \frac{-3x + 12}{4}. \qquad \text{Dividing by 4 on both sides}$$

If we were to graph the equation $3x + 4y = 12$ on a grapher, we would enter it as $y = (-3x + 12)/4$.

Exercise Set **R.7**

Solve.

1. $4x + 5 = 21$

2. $2y - 1 = 3$

3. $y + 1 = 2y - 7$

4. $5 - 4x = x - 13$

5. $5x - 2 + 3x = 2x + 6 - 4x$

6. $5x - 17 - 2x = 6x - 1 - x$

7. $7(3x + 6) = 11 - (x + 2)$

8. $4(5y + 3) = 3(2y - 5)$

9. $(2x - 3)(3x - 2) = 0$

10. $(5x - 2)(2x + 3) = 0$

11. $3x^2 + x - 2 = 0$

12. $10x^2 - 16x + 6 = 0$

13. $4x^2 - 12 = 0$

14. $6x^2 = 36$

15. $2x^2 = 6x$

16. $18x + 9x^2 = 0$

17. $3y^3 - 5y^2 - 2y = 0$

18. $3t^3 + 2t = 5t^2$

19. $7x^3 + x^2 - 7x - 1 = 0$
(*Hint:* Factor by grouping.)

20. $3x^3 + x^2 - 12x - 4 = 0$
(*Hint:* Factor by grouping.)

Solve.

21. $A = \frac{1}{2}bh$, for b
(Area of a triangle)

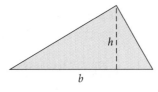

22. $A = \pi r^2$, for π
(Area of a circle)

23. $P = 2l + 2w$, for w
(Perimeter of a rectangle)

24. $A = P + Prt$, for P
(Simple interest)

25. $A = \frac{1}{2}h(b_1 + b_2)$, for h
(Area of a trapezoid)

26. $A = \frac{1}{2}h(b_1 + b_2)$, for b_2

27. $V = \frac{4}{3}\pi r^3$, for π
(Volume of a sphere)

28. $V = \frac{4}{3}\pi r^3$, for r^3

29. $F = \frac{9}{5}C + 32$, for C
(Temperature conversion)

30. $Ax + By = C$, for y
(Standard linear equation)

Discussion and Writing

31. Use a graphical argument to explain why the equation $x^2 - 6x + 9 = 1 - x^4$ has no solution.

32. The formula $A = P + Prt$ gives the amount A in an account when a principal P is invested at interest rate r for t years. Under what circumstances would it be helpful to solve this formula for P?

Synthesis

Solve.

33. $2x - \{x - [3x - (6x + 5)]\} = 4x - 1$

34. $14 - 2[3 + 5(x - 1)] = 3\{x - 4[1 + 6(2 - x)]\}$

35. $(x - 2)^3 = x^3 - 2$

36. $(x + 1)^3 = (x - 1)^3 + 26$

37. $(6x^3 + 7x^2 - 3x)(x^2 - 7) = 0$

38. $\left(x - \frac{1}{5}\right)\left(x^2 - \frac{1}{4}\right) + \left(x - \frac{1}{5}\right)\left(x^2 + \frac{1}{8}\right) = 0$

Chapter Summary and Review R

Important Properties and Formulas

Properties of the Real Numbers

Commutative:	$a + b = b + a$; $ab = ba$
Associative:	$a + (b + c) = (a + b) + c$; $a(bc) = (ab)c$
Additive Identity:	$a + 0 = 0 + a = a$
Additive Inverse:	$-a + a = a + (-a) = 0$
Multiplicative Identity:	$a \cdot 1 = 1 \cdot a = a$
Multiplicative Inverse:	$a \cdot \dfrac{1}{a} = 1 \quad (a \neq 0)$
Distributive:	$a(b + c) = ab + ac$

Absolute Value

For any real number a,

$$|a| = \begin{cases} a, & \text{if } a \geq 0, \\ -a, & \text{if } a < 0. \end{cases}$$

Properties of Exponents

For any real numbers a and b and any integers m and n, assuming 0 is not raised to a nonpositive power:

The Product Rule: $\quad a^m \cdot a^n = a^{m+n}$

The Quotient Rule: $\quad \dfrac{a^m}{a^n} = a^{m-n} \quad (a \neq 0)$

The Power Rule: $\quad (a^m)^n = a^{mn}$

Raising a Product to a Power:
$(ab)^m = a^m b^m$

Raising a Quotient to a Power:
$\left(\dfrac{a}{b}\right)^m = \dfrac{a^m}{b^m} \quad (b \neq 0)$

Compound Interest Formula

$$A = P\left(1 + \frac{i}{n}\right)^{nt}$$

Special Products of Binomials

$(A + B)^2 = A^2 + 2AB + B^2$

$(A - B)^2 = A^2 - 2AB + B^2$

$(A + B)(A - B) = A^2 - B^2$

Sum or Difference of Cubes

$A^3 + B^3 = (A + B)(A^2 - AB + B^2)$

$A^3 - B^3 = (A - B)(A^2 + AB + B^2)$

Properties of Radicals

Let a and b be any real numbers or expressions for which the given roots exist. For any natural numbers m and n ($n \neq 1$):

If n is even, $\sqrt[n]{a^n} = |a|$.

If n is odd, $\sqrt[n]{a^n} = a$.

$\sqrt[n]{a} \cdot \sqrt[n]{b} = \sqrt[n]{ab}$.

$\sqrt[n]{\dfrac{a}{b}} = \dfrac{\sqrt[n]{a}}{\sqrt[n]{b}} \quad (b \neq 0)$.

$\sqrt[n]{a^m} = (\sqrt[n]{a})^m$.

Rational Exponents

For any real number a and any natural numbers m and n for which $\sqrt[n]{a}$ exists,

$a^{1/n} = \sqrt[n]{a}$,

$a^{m/n} = \sqrt[n]{a^m} = (\sqrt[n]{a})^m$, and

$a^{-m/n} = \dfrac{1}{a^{m/n}}$.

(*continued*)

Pythagorean Theorem

$$a^2 + b^2 = c^2$$

Equation-Solving Principles

The Addition Principle: If $a = b$ is true, then $a + c = b + c$ is true.

The Multiplication Principle: If $a = b$ is true, then $ac = bc$ is true.

The Principle of Zero Products: If $ab = 0$ is true, then $a = 0$ or $b = 0$, and if $a = 0$ or $b = 0$, then $ab = 0$.

The Principle of Square Roots: If $x^2 = k$, then $x = \sqrt{k}$ or $x = -\sqrt{k}$.

REVIEW EXERCISES

Consider the numbers -43.89, 12, -3, $-\frac{1}{5}$, $\sqrt{7}$, $\sqrt[3]{10}$, -1, $-\frac{4}{3}$, $7\frac{2}{3}$, -19, 31, 0.

1. Which are integers?

2. Which are natural numbers?

3. Which are rational numbers?

4. Which are real numbers?

5. Which are irrational numbers?

6. Which are whole numbers?

7. Write interval notation for $\{x \mid -3 \le x < 5\}$.

Simplify.

8. $|-3.5|$ 9. $|16|$

10. Find the distance between -7 and 3 on the number line.

Calculate.

11. $5^3 - [2(4^2 - 3^2 - 6)]^3$

12. $\dfrac{3^4 - (6 - 7)^4}{2^3 - 2^4}$

Convert to decimal notation.

13. 3.261×10^6 14. 4.1×10^{-4}

Convert to scientific notation.

15. 0.01432 16. $43{,}210$

Calculate. Write the answer using scientific notation.

17. $\dfrac{2.5 \times 10^{-8}}{3.2 \times 10^{13}}$

18. $(8.4 \times 10^{-17})(6.5 \times 10^{-16})$

Simplify.

19. $(7a^2 b^4)(-2a^{-4} b^3)$ 20. $\dfrac{54x^6 y^{-4} z^2}{9x^{-3} y^2 z^{-4}}$

21. $\sqrt[4]{81}$ 22. $\sqrt[5]{-32}$

23. $\dfrac{b - a^{-1}}{a - b^{-1}}$ 24. $\dfrac{\dfrac{x^2}{y} + \dfrac{y^2}{x}}{y^2 - xy + x^2}$

25. $(\sqrt{3} - \sqrt{7})(\sqrt{3} + \sqrt{7})$

26. $(5x^2 - \sqrt{2})^2$

27. $8\sqrt{5} + \dfrac{25}{\sqrt{5}}$

28. $(x + t)(x^2 - xt + t^2)$

29. $(5a + 4b)(2a - 3b)$

30. $(5xy^4 - 7xy^2 + 4x^2 - 3) - (-3xy^4 + 2xy^2 - 2y + 4)$

Factor.

31. $x^3 + 2x^2 - 3x - 6$

32. $12a^3 - 27ab^4$

33. $24x + 144 + x^2$

34. $9x^3 + 35x^2 - 4x$

35. $8x^3 - 1$

36. $27x^6 + 125y^6$

37. $6x^3 + 48$

38. $4x^3 - 4x^2 - 9x + 9$

39. $9x^2 - 30x + 25$

40. $18x^2 - 3x + 6$

41. $a^2b^2 - ab - 6$

42. Divide and simplify:

$$\frac{3x^2 - 12}{x^2 + 4x + 4} \div \frac{x - 2}{x + 2}.$$

43. Subtract and simplify:

$$\frac{x}{x^2 + 9x + 20} - \frac{4}{x^2 + 7x + 12}.$$

Write an expression containing a single radical.

44. $\sqrt{y^5} \sqrt[3]{y^2}$

45. $\dfrac{\sqrt{(a + b)^3} \sqrt[3]{a + b}}{\sqrt[6]{(a + b)^7}}$

46. Convert to radical notation: $b^{7/5}$.

47. Convert to exponential notation and simplify:

$$\sqrt[8]{\frac{m^{32}n^{16}}{3^8}}.$$

48. Rationalize the denominator:

$$\frac{\sqrt{x} - \sqrt{y}}{\sqrt{x} + \sqrt{y}}.$$

49. How long is a guy wire that reaches from the top of a 17-ft pole to a point on the ground 8 ft from the pole?

Solve.

50. $4y - 5 = 11$

51. $x^2 + 4x - 5 = 0$

52. $3x^2 + 2x = 8$

53. $5x^2 = 15$

54. $x^2 - 10 = 0$

55. Solve $V = lwh$ for h.

56. Solve $M = n + 0.3s$ for s.

Discussion and Writing

57. Explain the difference between equivalent expressions and equivalent equations.

58. As the first step in solving

$$3x - 1 = 8,$$

Juliet multiplies by $\frac{1}{3}$ on both sides. What advice would you give her about the procedure for solving equations?

Synthesis

Multiply. Assume that all exponents are integers.

59. $(x^n + 10)(x^n - 4)$

60. $(t^a + t^{-a})^2$

61. $(y^b - z^c)(y^b + z^c)$

62. $(a^n - b^n)^3$

Factor.

63. $y^{2n} + 16y^n + 64$

64. $x^{2t} - 3x^t - 28$

65. $m^{6n} - m^{3n}$

66. Solve: $\sqrt{\sqrt{\sqrt{x}}} = 2.$

Graphs, Functions, and Models 1

People in the United States create more garbage than people in any other nation. The graph below presents data regarding the total amount of solid waste for certain years (*Source: Parade Magazine*, June 13, 1999). Using linear regression, we can model the data with the linear function

$$y = f(x) = 3.65x - 7069.37,$$

where x is the year and y is the amount of solid waste, in millions of tons. We can use this function to predict amounts of solid waste in future years.

This problem appears as Example 2 in Section 1.3.

T his environmental application is just one example of the widespread use of graphs in today's society. In this chapter, we will focus attention on functions and equations with graphs that can be used to model real-life situations mathematically. Such models are invaluable when doing analyses and making predictions.

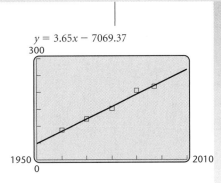

$y = 3.65x - 7069.37$

Functions, Graphs, and Graphers

1.1

- Determine whether a correspondence or a relation is a function.
- Find function values, or outputs, using a formula.
- Find the domain and the range of a function.
- Determine whether a graph is that of a function.
- Solve applied problems using functions.

We now focus attention on a concept that is fundamental to many areas of mathematics—the idea of a *function*.

Functions

We first consider an application.

Thunder Time Related to Lightning Distance. During a thunderstorm, it is possible to calculate how far away, y (in miles), lightning is when the sound of thunder arrives x seconds after the lightning has been sighted. It is known that the distance, in miles, is $\frac{1}{5}$ of the time, in seconds. If we hear the sound of thunder 15 seconds after we've seen the lightning, we know that the lightning is $\frac{1}{5} \cdot 15$, or 3 miles away. Similarly, 5 sec corresponds to 1 mi, 2 sec to $\frac{2}{5}$ mi, and so on. We can express this relationship with a set of ordered pairs, a graph, and an equation.

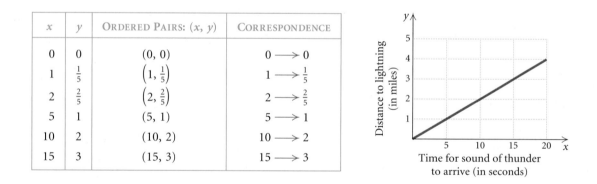

x	y	ORDERED PAIRS: (x, y)	CORRESPONDENCE
0	0	$(0, 0)$	$0 \longrightarrow 0$
1	$\frac{1}{5}$	$\left(1, \frac{1}{5}\right)$	$1 \longrightarrow \frac{1}{5}$
2	$\frac{2}{5}$	$\left(2, \frac{2}{5}\right)$	$2 \longrightarrow \frac{2}{5}$
5	1	$(5, 1)$	$5 \longrightarrow 1$
10	2	$(10, 2)$	$10 \longrightarrow 2$
15	3	$(15, 3)$	$15 \longrightarrow 3$

The ordered pairs express a relationship, or correspondence, between the first and second coordinates. We can see this relationship in the graph as well. The equation that describes the correspondence is

$$y = \frac{1}{5}x.$$

This is an example of a *function*. In this case, distance is a function of time.

Let's consider some other correspondences before giving the definition of a function.

DOMAIN		RANGE
To each registered student	there corresponds	an I. D. number.
To each mountain bike sold	there corresponds	its price.
To each number between −3 and 3	there corresponds	the square of that number.

In each example, the first set is called the **domain** and the second set is called the **range**. For each member, or **element**, in the domain, there is *exactly one* member of the range to which it corresponds. Thus each registered student has exactly *one* I. D. number, each mountain bike has exactly *one* price, and each number between −3 and 3 has exactly *one* square. Each correspondence is a *function*.

Function

A **function** is a correspondence between a first set, called the **domain**, and a second set, called the **range**, such that each member of the domain corresponds to *exactly one* member of the range.

It is important to note that not every correspondence between two sets is a function.

EXAMPLE 1 Determine whether each of the following correspondences is a function.

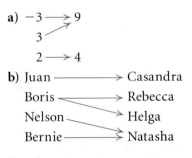

a) −3 ⟶ 9
   3 ⟶
   2 ⟶ 4

b) Juan ⟶ Casandra
  Boris ⟶ Rebecca
  Nelson ⟶ Helga
  Bernie ⟶ Natasha

Solution

a) This correspondence *is* a function because each member of the domain corresponds to exactly one member of the range. Note that the definition of a function does allow more than one member of the domain to correspond to a member of the range.

b) This correspondence *is not* a function because there is a member of the domain (Boris) that is paired with two different members of the range (Rebecca and Helga).

EXAMPLE 2 Determine whether each of the following correspondences is a function.

DOMAIN	CORRESPONDENCE	RANGE
a) Years in which a presidential election occurs	The person elected	A set of presidents
b) The integers	Each integer's cube root	A set of real numbers
c) All states in the United States	A senator from that state	The set of all U.S. senators

Solution

a) This correspondence *is* a function, because in each presidential election *exactly one* president is elected.

b) This correspondence *is* a function, because each integer has *exactly one* cube root.

c) This correspondence *is not* a function, because each state can be paired with *two* different senators.

When a correspondence between two sets is not a function, it is still an example of a **relation**.

Relation

A **relation** is a correspondence between a first set, called the **domain**, and a second set, called the **range**, such that each member of the domain corresponds to *at least one* member of the range.

All the correspondences in Examples 1 and 2 are relations, but, as we have seen, not all are functions. Relations are sometimes written as sets of ordered pairs (as we saw earlier in the example on lightning) in which elements of the domain are the first coordinates of the ordered pairs and elements of the range are the second coordinates. For example, instead of writing $-3 \longrightarrow 9$, as we did in Example 1(a), we could write the ordered pair $(-3, 9)$.

EXAMPLE 3 Determine whether each of the following relations is a function. Identify the domain and the range.

a) $\{(-2, 5), (5, 7), (0, 1), (4, -2)\}$
b) $\{(9, -5), (9, 5), (2, 4)\}$
c) $\{(-5, 3), (0, 3), (6, 3)\}$

Solution

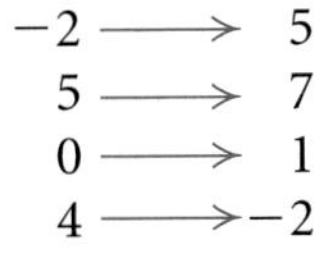

FIGURE 1

a) The relation *is* a function because *no* two ordered pairs have the same first coordinate and different second coordinates (see Fig. 1).

The domain is the set of all first coordinates: $\{-2, 5, 0, 4\}$.

The range is the set of all second coordinates: $\{5, 7, 1, -2\}$.

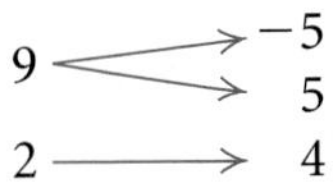

FIGURE 2

b) The relation *is not* a function because the ordered pairs $(9, -5)$ and $(9, 5)$ have the same first coordinate and different second coordinates (see Fig. 2).

The domain is the set of all first coordinates: $\{9, 2\}$.

The range is the set of all second coordinates: $\{-5, 5, 4\}$.

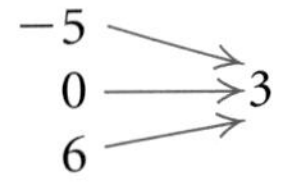

FIGURE 3

c) The relation *is* a function because *no* two ordered pairs have the same first coordinate and different second coordinates (see Fig. 3).

The domain is $\{-5, 0, 6\}$.

The range is $\{3\}$.

Notation for Functions

Functions used in mathematics are often given by equations. They generally require that certain calculations be performed in order to determine which member of the range is paired with each member of the domain. For example, in the "Introduction to Graphs and the Graphing Calculator," we graphed the function $y = x^2 - 5$ by doing calculations like the following:

$$\text{for } x = 3, \ y = 3^2 - 5 = 4,$$
$$\text{for } x = 2, \ y = 2^2 - 5 = -1, \quad \text{and}$$
$$\text{for } x = 1, \ y = 1^2 - 5 = -4.$$

A more concise notation is often used. For $y = x^2 - 5$, the **inputs** (members of the domain) are values of x substituted into the equation. The **outputs** (members of the range) are the resulting values of y. If we call the function f, we can use x to represent an arbitrary *input* and

$f(x)$—read "f of x", or "f at x", or "the value of f at x"—to represent the corresponding *output*. In this notation, the function given by $y = x^2 - 5$ is written as $f(x) = x^2 - 5$ and the above calculations would be

$$f(3) = 3^2 - 5 = 4,$$
$$f(2) = 2^2 - 5 = -1,$$
$$f(1) = 1^2 - 5 = -4.$$

Keep in mind that $f(x)$ *does not* mean $f \cdot x$.

Thus, instead of writing "when $x = 3$, the value of y is 4," we can simply write "$f(3) = 4$," which can also be read as "f of 3 is 4" or "for the input 3, the output of f is 4." The letters g and h are also often used to name functions.

EXAMPLE 4   A function f is given by $f(x) = 2x^2 - x + 3$. Find each of the following.

a) $f(0)$ **b)** $f(-7)$

c) $f(5a)$ **d)** $f(a - 4)$

Solution   We can think of this formula as follows:

$$f(\blacksquare) = 2(\blacksquare)^2 - (\blacksquare) + 3.$$

Then to find an output for a given input we think: "Whatever goes in the blank on the left goes in the blank(s) on the right." This gives us a "recipe" for finding outputs.

a) $f(0) = 2(0)^2 - 0 + 3 = 0 - 0 + 3 = 3$

b) $f(-7) = 2(-7)^2 - (-7) + 3 = 2 \cdot 49 + 7 + 3 = 108$

c) $f(5a) = 2(5a)^2 - 5a + 3 = 2 \cdot 25a^2 - 5a + 3 = 50a^2 - 5a + 3$

d) $f(a - 4) = 2(a - 4)^2 - (a - 4) + 3 = 2(a^2 - 8a + 16) - a + 4 + 3$
$$= 2a^2 - 16a + 32 - a + 4 + 3$$
$$= 2a^2 - 17a + 39$$

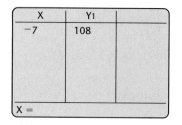

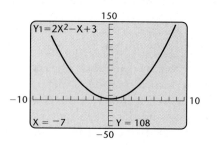

A grapher can be used to find function values. At left, we illustrate finding $f(-7)$ from part (b), first with the TABLE feature set in ASK mode and then with the VALUE feature from the CALC menu. In both screens, we see that $f(-7) = 108$. Consult either the manual for your particular grapher or the *Graphing Calculator Manual* that accompanies this text for further information on finding function values.

Graphs of Functions

We graph functions the same way we graph equations. We find ordered pairs (x, y) or $(x, f(x))$, plot points, and complete the graph.

EXAMPLE 5   Graph each of the following functions using a grapher.

a) $f(x) = x^2 - 5$

b) $f(x) = x^3 - x$

c) $f(x) = \sqrt{x + 4}$

Solution   Most graphers do not use function notation "$f(x) = \ldots$" to enter a function formula. Instead, we must enter the function using "$y = \ldots$". The graphs follow.

a) $f(x) = x^2 - 5$

b) $f(x) = x^3 - x$

c) $f(x) = \sqrt{x + 4}$

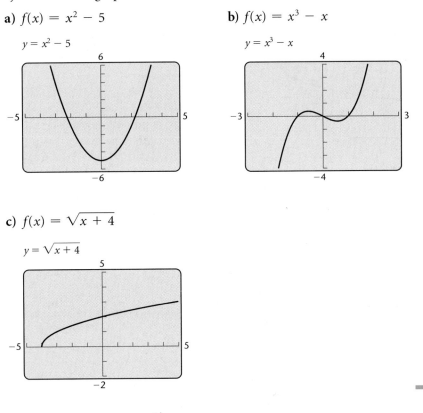

To find a function value, like $f(3)$, from a graph, we locate the input 3 on the horizontal axis, move vertically to the graph of the function, and then move horizontally to find the output on the vertical axis. For the function $f(x) = x^2 - 5$, we see that $f(3) = 4$. We can check our work using a grapher.

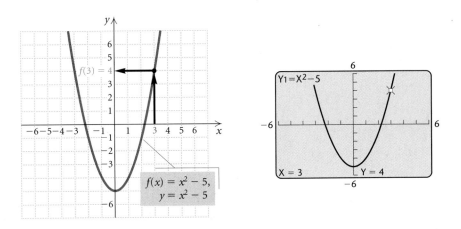

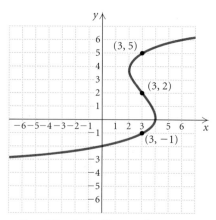

Since 3 is paired with more than one member of the range, the graph does not represent a function.

We know that when one member of the domain is paired with two or more different members of the range, the correspondence *is not* a function. Thus, when a graph contains two or more different points with the same first coordinate, the graph cannot represent a function (see the graph at left). Points sharing a common first coordinate are vertically above or below each other. This leads us to the *vertical-line test*.

The Vertical-Line Test

If it is possible for a vertical line to cross a graph more than once, then the graph is *not* the graph of a function.

To apply the vertical-line test, we try to find a vertical line that crosses the graph more than once. If we succeed, then the graph is not that of a function. If we do not, then the graph is that of a function.

EXAMPLE 6   Which of graphs (a) through (f) (in red) are graphs of functions?

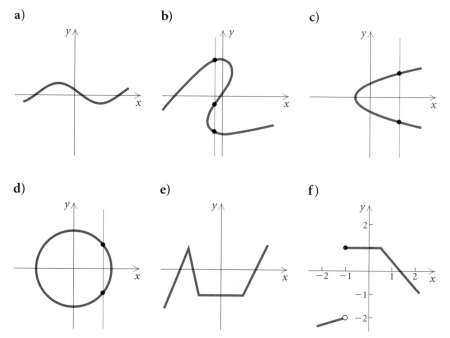

In graph (f), the solid dot shows that $(-1, 1)$ belongs to the graph. The open circle shows that $(-1, -2)$ does *not* belong to the graph.

Solution   Graphs (a), (e), and (f) are graphs of functions because we cannot find a vertical line that crosses any of them more than once. In (b), the vertical line crosses the graph in three points and so it is not that of a function. Also, in (c) and (d), we can find a vertical line that crosses the graph more than once.

Finding Domains of Functions

When a function f, whose inputs and outputs are real numbers, is given by a formula, the *domain* is understood to be the set of all inputs for which the expression is defined as a real number. When a substitution results in an expression that is not defined as a real number, we say that the function value *does not exist* and that the number being substituted *is not* in the domain of the function.

EXAMPLE 7 Find the indicated function values. Simplify, if possible.

a) $f(1)$ and $f(3)$, for $f(x) = \dfrac{1}{x - 3}$

b) $g(16)$ and $g(-7)$, for $g(x) = \sqrt{x} + 5$

Solution

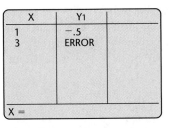

a) $f(1) = \dfrac{1}{1 - 3} = \dfrac{1}{-2} = -\dfrac{1}{2};$

$f(3) = \dfrac{1}{3 - 3} = \dfrac{1}{0}$

Since division by 0 is not defined, the number 3 is not in the domain of f. In a table from a grapher, this is indicated with an ERROR message. Thus, $f(3)$ does not exist.

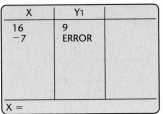

b) $g(16) = \sqrt{16} + 5 = 4 + 5 = 9;$

$g(-7) = \sqrt{-7} + 5$

Since $\sqrt{-7}$ is not defined as a real number, the number -7 is not in the domain of g. Note the ERROR message in the table at left. Thus, $g(-7)$ does not exist.

Inputs that make a denominator 0 or a square-root radicand negative are not in the domain of a function.

EXAMPLE 8 Find the domain of each of the following functions.

a) $f(x) = \dfrac{1}{x - 3}$ **b)** $g(x) = \sqrt{x} + 5$

c) $h(x) = \dfrac{3x^2 - x + 7}{x^2 + 2x - 3}$ **d)** $F(x) = x^3 + |x|$

Solution

a) The input 3 results in a denominator of 0. The domain is $\{x \mid x \neq 3\}$. We can also write the solution using interval notation and the symbol $\cup$ for the **union** or inclusion of both sets: $(-\infty, 3) \cup (3, \infty)$.

INTERVAL NOTATION

REVIEW SECTION R.1.

b) We can substitute any number for which the radicand is nonnegative, that is, for which $x \geq 0$. Thus the domain is $\{x \mid x \geq 0\}$, or the interval $[0, \infty)$.

c) Although we can substitute any real number in the numerator, we must avoid inputs that make the denominator 0. To find those inputs, we solve $x^2 + 2x - 3 = 0$, or $(x + 3)(x - 1) = 0$. Thus the domain consists of the set of all real numbers except -3 and 1, or $\{x \mid x \neq -3 \text{ and } x \neq 1\}$, or $(-\infty, -3) \cup (-3, 1) \cup (1, \infty)$.

d) All substitutions are suitable. The domain is the set of all real numbers, $\mathbb{R}$, or $(-\infty, \infty)$.

Domain and Range Using a Grapher

Keep the following in mind regarding the *graph* of a function:

Domain = the set of a function's inputs, found on the horizontal axis;

Range = the set of a function's outputs, found on the vertical axis.

By carefully examining the graph of a function, we may be able to determine the function's domain as well as its range. Consider the graph of $f(x) = \sqrt{4 - (x - 1)^2}$, shown below. We look for the inputs on the x-axis that correspond to a point on the graph. We see that they extend from -1 to 3, inclusive. Thus the domain is $\{x \mid -1 \leq x \leq 3\}$, or the interval $[-1, 3]$.

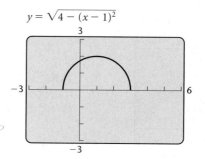

$$y = \sqrt{4 - (x - 1)^2}$$

To find the range, we look for the outputs on the y-axis. We see that they extend from 0 to 2, inclusive. Thus the range of this function is $\{y \mid 0 \leq y \leq 2\}$, or the interval $[0, 2]$. We can confirm our results using the TRACE feature, moving the cursor from left to right along the curve. We can also confirm our results using the TABLE feature.

EXAMPLE 9   Use a grapher to graph each of the following functions. Then estimate the domain and the range of each.

a) $f(x) = \sqrt{x + 4}$ **b)** $f(x) = x^3 - x$

c) $f(x) = \dfrac{12}{x}$ **d)** $f(x) = x^4 - 2x^2 - 3$

Solution

a)

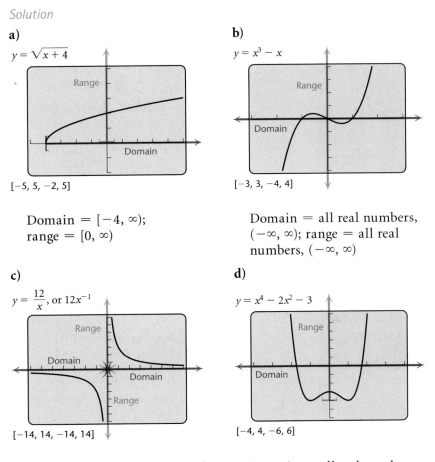

$y = \sqrt{x + 4}$

[−5, 5, −2, 5]

Domain = [−4, ∞);
range = [0, ∞)

b)

$y = x^3 - x$

[−3, 3, −4, 4]

Domain = all real numbers,
(−∞, ∞); range = all real
numbers, (−∞, ∞)

c)

$y = \dfrac{12}{x}$, or $12x^{-1}$

[−14, 14, −14, 14]

Since the graph does not touch or
cross either axis, 0 is excluded
both as an input and as an output.
Domain = (−∞, 0) ∪ (0, ∞);
range = (−∞, 0) ∪ (0, ∞)

d)

$y = x^4 - 2x^2 - 3$

[−4, 4, −6, 6]

Domain = all real numbers,
(−∞, ∞); range = [−4, ∞)

Always consider adding the reasoning of Example 8 to a graphical analysis. Think, "What can I substitute?" to find the domain. Think, "What do I get out?" to find the range. Thus, in Example 9(d), it might not look like the domain is all real numbers because the graph rises steeply, but by reexamining the equation we see that we can indeed substitute any real number for *x*.

Applications of Functions

EXAMPLE 10 *Speed of Sound in Air.* The speed *S* of sound in air is a function of the temperature *t*, in degrees Fahrenheit, and is given by

$$S(t) = 1087.7 \sqrt{\frac{5t + 2457}{2457}},$$

where *S* is in feet per second.

a) Graph the function using the viewing window $[-1000, 2000, -1000, 3000]$, with Xscl = 500 and Yscl = 500.

b) Find the speed of sound in air when the temperature is $0°$, $32°$, $70°$, and $-10°$ Fahrenheit.

Solution

a) The graph is shown below. Note that $S(t)$ must be changed to y and t must be changed to x when the function is entered in a grapher.

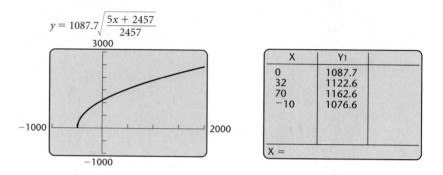

$$y = 1087.7\sqrt{\frac{5x + 2457}{2457}}$$

X	Y1
0	1087.7
32	1122.6
70	1162.6
−10	1076.6

X =

b) We use a grapher with the TABLE feature set in ASK mode to compute the function values. We find that

$$S(0) = 1087.7 \text{ ft/sec},$$
$$S(32) \approx 1122.6 \text{ ft/sec},$$
$$S(70) \approx 1162.6 \text{ ft/sec}, \quad \text{and}$$
$$S(-10) \approx 1076.6 \text{ ft/sec}.$$

CONNECTING THE CONCEPTS

FUNCTION CONCEPTS

Formula for f: $f(x) = 5 + 2x^2 - x^4$.

For every input, there is exactly one output.

$(1, 6)$ is on the graph.

For the input 1, the output is 6.

$f(1) = 6$

Domain: set of all inputs $= (-\infty, \infty)$

Range: set of all outputs $= (-\infty, 6]$

GRAPH

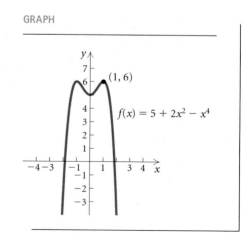

Exercise Set 1.1

In Exercises 1–14, determine whether the correspondence is a function.

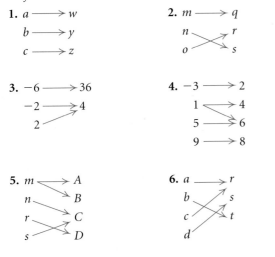

1. $a \longrightarrow w$
$b \longrightarrow y$
$c \longrightarrow z$

2. $m \longrightarrow q$
$n \times r$
$o \quad s$

3. $-6 \longrightarrow 36$
$-2 \longrightarrow 4$
2

4. $-3 \longrightarrow 2$
$1 \longrightarrow 4$
$5 \longrightarrow 6$
$9 \longrightarrow 8$

5. $m \quad A$
$n \quad B$
$r \quad C$
$s \quad D$

6. $a \quad r$
$b \quad s$
$c \quad t$
d

7. WORLD'S TEN LARGEST EARTHQUAKES (1900–1999)
LOCATION AND DATE   MAGNITUDE

Chile (May 22, 1960) ——————————→ 9.5
Alaska (March 28, 1964) ————————→ 9.2
Russia (Nov. 4, 1952) ——————————→ 9.0
Ecuador (Jan. 31, 1906) ———————
Alaska (March 9, 1957) ——————————→ 8.8
Kuril Islands (Nov. 6, 1958) ——————
Alaska (Feb. 4, 1965) ——————————→ 8.7
India (Aug. 15, 1950) ——————————→ 8.6
Argentina (Nov. 11, 1922) ————————
Indonesia (Feb. 1, 1938) ——————————→ 8.5

(*Source*: National Earthquake Information Center, U.S. Geological Survey)

8.   SPEED ON THE GROUND
ANIMAL   (IN MILES PER HOUR)

Cheetah ————————————————→ 70
Lion ————————————————————→ 50
Reindeer —————————————————→ 32
Giraffe ——————
Elephant ————————————————→ 25
Giant tortoise ———————————————→ 0.17

(*Source*: *Time Almanac*, 1999, p. 581)

DOMAIN	CORRESPONDENCE	RANGE
9. A set of cars in a parking lot	Each car's license number	A set of numbers
10. A set of people in a town	A doctor a person uses	A set of doctors
11. A set of members of a family	Each person's eye color	A set of colors
12. A set of members of a rock band	An instrument each person plays	A set of instruments
13. A set of students in a class	A student sitting in a neighboring seat	A set of students
14. A set of bags of chips on a shelf	Each bag's weight	A set of weights

Determine whether the relation is a function. Identify the domain and the range.

15. $\{(2, 10), (3, 15), (4, 20)\}$

16. $\{(3, 1), (5, 1), (7, 1)\}$

17. $\{(-7, 3), (-2, 1), (-2, 4), (0, 7)\}$

18. $\{(1, 3), (1, 5), (1, 7), (1, 9)\}$

19. $\{(-2, 1), (0, 1), (2, 1), (4, 1), (-3, 1)\}$

20. $\{(5, 0), (3, -1), (0, 0), (5, -1), (3, -2)\}$

21. A graph of a function f is shown below. Find $f(-1)$, $f(0)$, and $f(1)$.

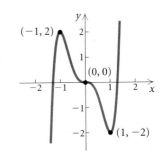

22. A graph of a function g is shown here. Find $g(-2)$, $g(0)$, and $g(2.4)$.

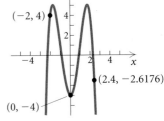

23. Given that $g(x) = 3x^2 - 2x + 1$, find each of the following.

a) $g(0)$ b) $g(-1)$
c) $g(3)$ d) $g(-x)$
e) $g(1 - t)$

24. Given that $f(x) = 5x^2 + 4x$, find each of the following.

a) $f(0)$ b) $f(-1)$
c) $f(3)$ d) $f(t)$
e) $f(t - 1)$

25. Given that $g(x) = x^3$, find each of the following.

a) $g(2)$ b) $g(-2)$
c) $g(-x)$ d) $g(3y)$
e) $g(2 + h)$

26. Given that $f(x) = 2|x| + 3x$, find each of the following.

a) $f(1) = 5$ b) $f(-2)$ -2
c) $f(-x)$ $5x$ d) $f(2y)$ $10y$
e) $f(2 - h)$

27. Given that $g(x) = \dfrac{x - 4}{x + 3}$, find each of the following.

a) $g(5)$ b) $g(4)$
c) $g(-3)$ d) $g(-16.25)$
e) $g(x + h)$

28. Given that $f(x) = \dfrac{x}{2 - x}$, find each of the following.

a) $f(2)$ b) $f(1)$
c) $f(-16)$ d) $f(-x)$
e) $f\left(-\dfrac{2}{3}\right)$

29. Find $g(0)$, $g(-1)$, $g(5)$, and $g\left(\frac{1}{2}\right)$ for

$$g(x) = \dfrac{x}{\sqrt{1 - x^2}}.$$

30. Find $h(0)$, $h(2)$, and $h(-x)$ for

$$h(x) = x + \sqrt{x^2 - 1}.$$

Find the domain of the function. Do not use a grapher.

31. $f(x) = 7x + 4$

32. $f(x) = |3x - 2|$

33. $f(x) = 4 - \dfrac{2}{x}$

34. $f(x) = \dfrac{1}{x^4}$

35. $f(x) = \dfrac{x + 5}{2 - x}$

36. $f(x) = \dfrac{8}{x + 4}$

37. $f(x) = \dfrac{1}{x^2 - 4x - 5}$

38. $f(x) = \dfrac{x^4 - 2x^3 + 7}{3x^2 - 10x - 8}$

Use a grapher to graph the function. Then visually estimate the domain and the range.

39. $f(x) = |x|$ **40.** $f(x) = |x| - 10.3$

41. $f(x) = \sqrt{9 - x^2}$ **42.** $f(x) = -\sqrt{25 - x^2}$

43. $f(x) = (x - 1)^3 + 2$ **44.** $f(x) = (x - 2)^4 + 1$

45. $f(x) = \sqrt{7 - x}$ **46.** $f(x) = \sqrt{x + 8}$

47. $f(x) = -x^2 + 4x - 1$ **48.** $f(x) = 2x^2 - x^4 + 5$

In Exercises 49–56, determine whether the graph is that of a function. An open dot indicates that the point does not belong to the graph.

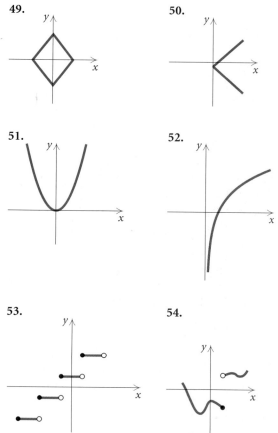

49.

50.

51.

52.

53.

54.

55.   **56.**

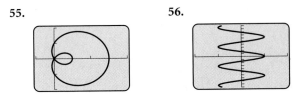

In Exercises 57–60, determine whether the graph is that of a function. Find the domain and the range.

57.   **58.**

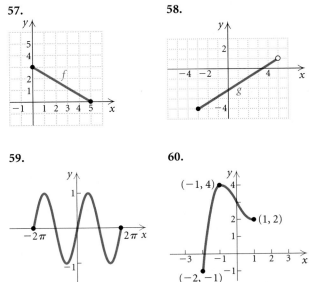

59.   **60.**

61. *Boiling Point and Elevation.*   The elevation E, in meters, above sea level at which the boiling point of water is t degrees Celsius is given by the function

$$E(t) = 1000(100 - t) + 580(100 - t)^2.$$

At what elevation is the boiling point 99.5°? 100°?

62. *Average Price of a Movie Ticket.*   The average price of a movie ticket, in dollars, can be estimated by the function P given by

$$P(x) = 0.1522x - 298.592$$

where x is the year. Thus, $P(2000)$ is the average price of a movie ticket in 2000. The price is lower than what might be expected due to lower prices for matinees, senior citizens' discounts, and so on.

a) Use the function to predict the average price in 2003 and 2010.

b) When will the average price be $8.00?

63. *Territorial Area of an Animal.*   The territorial area of an animal is defined to be its defended, or exclusive region. For example, a lion has a certain region over which it is considered ruler. It has been shown that the territorial area I, in acres, of predatory animals is a function of body weight w, in pounds, and is given by the function

$$T(w) = w^{1.31}.$$

Find the territorial area of animals whose body weights are 0.5 lb, 10 lb, 20 lb, 100 lb, and 200 lb.

Discussion and Writing

To the student and the instructor: The discussion and writing exercises are meant to be answered with one or more sentences. They can be discussed and answered collaboratively by the entire class or by small groups. Because of their open-ended nature, the answers to these exercises do not appear at the back of the book. They are denoted by the words "Discussion and Writing."

64. Explain in your own words what a function is.

65. Explain in your own words the difference between the domain of a function and the range of a function.

Synthesis

To the student and the instructor: The synthesis exercises found at the end of every exercise set challenge students to combine concepts or skills studied in that section or in preceding parts of the text.

66. Make a hand-drawn graph of a function for which the domain is $[-4, 4]$ and the range is $[1, 2] \cup [3, 5]$. Answers may vary.

67. Give an example of two different functions that have the same domain and the same range, but have no pairs in common. Answers may vary.

68. Make a hand-drawn graph of a function for which the domain is $[-3, -1] \cup [1, 5]$ and the range is $\{1, 2, 3, 4\}$. Answers may vary.

69. Suppose that for some function f, $f(x - 1) = 5x$. Find $f(6)$.

Linear Functions, Slope, and Applications

1.2

- Graph linear functions and equations, finding the slope and the y-intercept.
- Determine equations of lines.
- Solve applied problems involving linear functions.

In real-life situations, we often need to make decisions on the basis of limited information. When the given information is used to formulate an equation or inequality that at least approximates the situation mathematically, we have created a **model**. One of the most frequently used mathematical models is *linear*—the graph of a linear model is a straight line.

Linear Functions

Let's begin to examine the connections among equations, functions, and graphs that are straight lines.

Interactive Discovery

Graph each of the following equations using a grapher. (Some graphers are not able to graph equations of the type $x = a$. Consult your manual.) Which of these equations have graphs that are lines? Which of the equations whose graphs are lines are also functions? Look for patterns.

$$y = x, \qquad\qquad y = x^2,$$
$$4x + 5y = 20, \qquad y = 5 - x^2,$$
$$y = \frac{x}{2} + 12, \qquad y = 1.5x - 0.05,$$
$$y = -2x - 5, \qquad y = \frac{1}{x},$$
$$x = -3, \qquad\qquad y = 6.2,$$
$$y = \sqrt{x}, \qquad\qquad x = 4.9.$$

We have the following results and related terminology.

Linear Functions

A function f is a **linear function** if it can be written as

$$f(x) = mx + b,$$

where m and b are constants.

(If $m = 0$, the function is the **constant function** $f(x) = b$. If $m = 1$ and $b = 0$, the function is the **identity function** $f(x) = x$.)

Horizontal and Vertical Lines

Horizontal lines are given by equations of the type $y = b$ or $f(x) = b$. (They are functions.)

Vertical lines are given by equations of the type $x = a$. (They are *not* functions.)

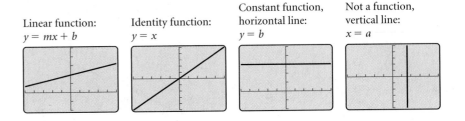

Linear function:
$y = mx + b$

Identity function:
$y = x$

Constant function, horizontal line:
$y = b$

Not a function, vertical line:
$x = a$

The Linear Function $f(x) = mx + b$ and Slope

To attach meaning to the constant m in the equation $f(x) = mx + b$, we first consider an application. FaxMax is an office machine business that currently has two stores in locations A and B in the same city. Their total costs for the same time period are given by two functions shown in the tables and graphs that follow. The variable x represents time, in months. The variable y represents total costs, in thousands of dollars, over that amount of time. Look for a pattern.

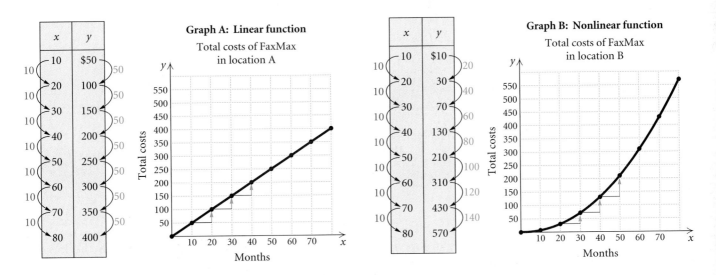

x	y
10	$50
20	100
30	150
40	200
50	250
60	300
70	350
80	400

Graph A: Linear function

Total costs of FaxMax in location A

x	y
10	$10
20	30
30	70
40	130
50	210
60	310
70	430
80	570

Graph B: Nonlinear function

Total costs of FaxMax in location B

We see in graph A that *every* change of 10 months results in a $50 thousand change in total costs. But in graph B, changes of 10 months do *not* result in constant changes in total costs. This is a way to distinguish linear from nonlinear functions. The rate at which a linear function changes, or the steepness of its graph, is constant.

Mathematically, we define a line's steepness, or **slope**, as the ratio of its vertical change (rise) to the corresponding horizontal change (run).

Slope

The **slope m** of a line containing points (x_1, y_1) and (x_2, y_2) is given by

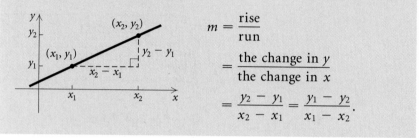

$$m = \frac{\text{rise}}{\text{run}}$$

$$= \frac{\text{the change in } y}{\text{the change in } x}$$

$$= \frac{y_2 - y_1}{x_2 - x_1} = \frac{y_1 - y_2}{x_1 - x_2}.$$

EXAMPLE 1 Make a hand-drawn graph of the function $f(x) = -\frac{2}{3}x + 1$ and determine its slope.

Solution Since the equation for f is in the form $f(x) = mx + b$, we know it is a linear function. We can graph it by connecting two points on the graph with a straight line. We calculate two ordered pairs, plot the points, graph the function, and determine the slope:

$$f(3) = -\frac{2}{3} \cdot 3 + 1 = -1;$$

$$f(9) = -\frac{2}{3} \cdot 9 + 1 = -5;$$

Pairs: $(3, -1)$, $(9, -5)$;

$$\text{Slope} = m = \frac{y_2 - y_1}{x_2 - x_1}$$

$$= \frac{-5 - (-1)}{9 - 3}$$

$$= \frac{-4}{6} = -\frac{2}{3}.$$

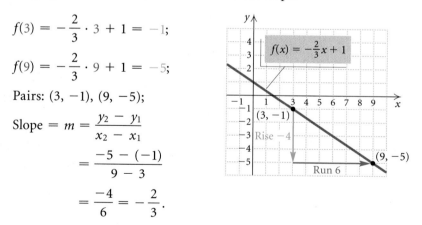

The slope is the same for any two points on a line. Thus, to check our work, note that $f(6) = -\frac{2}{3} \cdot 6 + 1 = -3$. Using the points $(6, -3)$ and $(3, -1)$, we have

$$m = \frac{-1 - (-3)}{3 - 6} = \frac{2}{-3} = -\frac{2}{3}.$$

We can also use the points in the opposite order when computing slope, so long as we are consistent:

$$m = \frac{-3 - (-1)}{6 - 3} = \frac{-2}{3} = -\frac{2}{3}.$$

Note also that the slope of the line is the number m in the equation for the function $f(x) = -\frac{2}{3}x + 1$. ▬

The slope of the line given by $y = mx + b$ is m.

Let's explore the effect of the slope m in linear equations of the type $f(x) = mx$.

Interactive Discovery

Graph the following equations using a grapher:

$$y_1 = x, \qquad y_2 = 2x, \qquad y_3 = 5x, \quad \text{and} \quad y_4 = 10x.$$

Try entering these equations as $y_1 = \{1, 2, 5, 10\}x$. What do you think the graph of $y = 128x$ will look like?

Clear the screen and graph the following equations:

$$y_1 = x, \qquad y_2 = 0.75x, \qquad y_3 = 0.48x, \quad \text{and} \quad y_4 = 0.12x.$$

What do you think the graph of $y = 0.000029x$ will look like?

Again clear the screen and graph each set of equations:

$$y_1 = -x, \qquad y_2 = -2x, \qquad y_3 = -4x, \quad \text{and} \quad y_4 = -10x$$

and $\quad y_1 = -x, \qquad y_2 = -\frac{2}{3}x, \qquad y_3 = -\frac{7}{20}x, \quad \text{and} \quad y_4 = -\frac{1}{10}x.$

From your observations, what do you think the graphs of $y = -200x$ and $y = -\frac{17}{100,000}x$ will look like?

If a line slants up from left to right, the change in x and the change in y have the same sign, so the line has a positive slope. The larger the slope is, the steeper the line. If a line slants down from left to right, the change in x and the change in y are of opposite signs, so the line has a negative slope. The larger the absolute value of the slope, the steeper the line. When $m = 0$, $y = 0x$, or $y = 0$. The graph of this equation is shown on the right below. Note that this horizontal line is the x-axis.

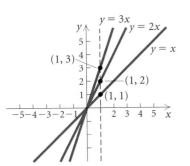

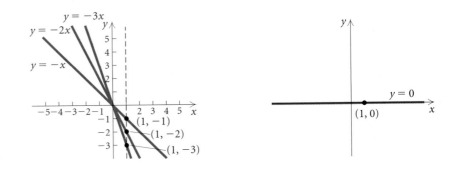

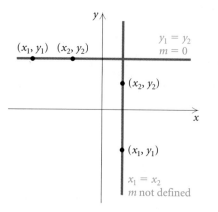

Horizontal and Vertical Lines

If a line is horizontal, the change in y for any two points is 0 and the change in x is nonzero. Thus a horizontal line has slope 0.

If a line is vertical, the change in y for any two points is nonzero and the change in x is 0. Thus the slope is *not defined* because we cannot divide by 0.

Note that zero slope and an undefined slope are two very different concepts.

EXAMPLE 2 Graph each linear equation and determine its slope.

a) $x = -2$

b) $y = \dfrac{5}{2}$

Solution

a) Since y is missing in $x = -2$, any value for y will do.

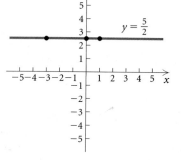

FIGURE 1

x	y
-2	0
-2	3
-2	-4

Choose any number for y; x must be -2.

The graph (see Fig. 1) is a *vertical line* 2 units to the left of the y-axis. The slope is not defined. The graph is not the graph of a function.

b) Since x is missing in $y = \frac{5}{2}$, any value for x will do.

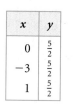

FIGURE 2

x	y
0	$\frac{5}{2}$
-3	$\frac{5}{2}$
1	$\frac{5}{2}$

Choose any number for x; y must be $\frac{5}{2}$.

The graph (see Fig. 2) is a *horizontal line* $\frac{5}{2}$, or $2\frac{1}{2}$, units above the x-axis. The slope is 0. The graph is the graph of a function.

Applications of Slope

Slope has many real-world applications. Numbers like 2%, 4%, and 7% are often used to represent the **grade** of a road. Such a number is meant to tell how steep a road is on a hill or mountain. For example, a 4% grade means that the road rises 4 ft for every horizontal distance of 100 ft.

The concept of grade is also used with a treadmill. During a treadmill test, a cardiologist might change the slope, or grade, of the treadmill to measure its effect on heart rate. Another example occurs in hydrology. The strength or force of a river depends on how far the river falls vertically compared to how far it flows horizontally.

EXAMPLE 3 *Ramp for the Handicapped.* Construction laws regarding access ramps for the handicapped state that every vertical rise of 1 ft requires a horizontal run of 12 ft. What is the grade, or slope, of such a ramp?

Solution The grade, or slope, is given by

$$m = \tfrac{1}{12} \approx 0.083 \approx 8.3\%.$$

Slope can also be considered as an **average rate of change.** To find the average rate of change between any two data points on a graph, we determine the slope of the line that passes through the two points.

EXAMPLE 4   *Newspaper Circulation.* On September 30, 1999, the last edition of *The Indianapolis News*, an afternoon daily paper, rolled off the press. Circulation of the paper had dropped approximately 70% in the 10 preceding years, as shown in the following graph. Find the average rate of change in the circulation from 1988 to 1998.

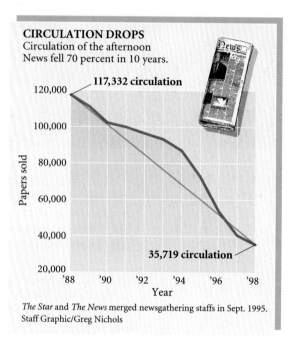

CIRCULATION DROPS
Circulation of the afternoon
News fell 70 percent in 10 years.

The Star and *The News* merged newsgathering staffs in Sept. 1995.
Staff Graphic/Greg Nichols

Solution   We determine the coordinates of two points on the graph. In this case, we will use (1988, 117,332) and (1998, 35,719). Then we compute the slope, or average rate of change, as follows:

$$\text{Slope} = \text{Average rate of change} = \frac{\text{change in } y}{\text{change in } x}$$

$$= \frac{35{,}719 - 117{,}332}{1998 - 1988} = \frac{-81{,}613}{10} = -8161.3.$$

This result tells us that, on average, each year the circulation of the paper decreased by approximately 8161 copies. The *average rate of change* over the 10-year period was -8161 copies per year.

Slope–Intercept Equations of Lines

Let's explore the effect of the constant b in linear equations of the type $f(x) = mx + b$.

Interactive Discovery

Begin with the graph of $y = x$. Now graph the lines $y = x + 3$ and $y = x - 4$ in the same viewing window. Try entering these equations as $y = x + \{0, 3, -4\}$ and compare the graphs. How do the lines differ from $y = x$? What do you think the line $y = x - 6$ will look like?

Try graphing $y = -0.5x$, $y = -0.5x - 4$, and $y = -0.5x + 3$ in the same viewing window. Describe what happens to the graph of $y = -0.5x$ when a number b is added.

Compare the graphs of the equations

$$y = 3x$$

and

$$y = 3x - 2.$$

Note that the graph of $y = 3x - 2$ is a shift down of the graph of $y = 3x$, and that $y = 3x - 2$ has y-intercept $(0, -2)$. That is, the graph is parallel to $y = 3x$ and it crosses the y-axis at $(0, -2)$. The point $(0, -2)$ is called the **y-intercept** of the graph.

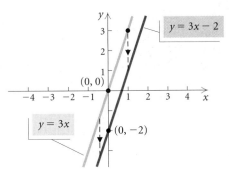

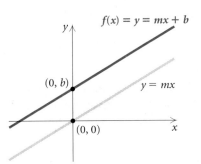

The Slope–Intercept Equation

The linear function f given by

$$f(x) = mx + b$$

has a graph that is a straight line parallel to $y = mx$. The constant m is called the slope, and the y-intercept is $(0, b)$.

We know that a nonvertical line has slope m and y-intercept $(0, b)$ and that its equation can be written as $y = mx + b$. The advantage of writing the equation in the form $y = mx + b$ is that we can read the slope m and the y-intercept $(0, b)$ directly from the equation.

EXAMPLE 5 Find the slope and the y-intercept of the line with equation $y = -0.25x - 3.8$.

$y = -0.25x - 3.8$

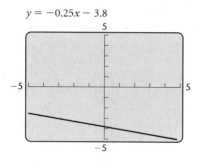

Solution

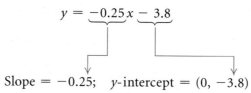

Slope $= -0.25$; y-intercept $= (0, -3.8)$

Any equation whose graph is a straight line is a **linear equation.** To find the slope and the y-intercept of the graph of a linear equation, we can solve for y, and then read the information from the equation.

EXAMPLE 6   Find the slope and the y-intercept of the line with equation $3x - 6y - 7 = 0$.

Solution   We solve for y:

$$3x - 6y - 7 = 0$$
$$-6y = -3x + 7 \qquad \text{Adding } -3x \text{ and 7 on both sides}$$
$$-\tfrac{1}{6}(-6y) = -\tfrac{1}{6}(-3x + 7) \qquad \text{Multiplying by } -\tfrac{1}{6}$$
$$y = \tfrac{1}{2}x - \tfrac{7}{6}.$$

Thus the slope is $\tfrac{1}{2}$, and the y-intercept is $\left(0, -\tfrac{7}{6}\right)$.

$y = \tfrac{1}{2}x - \tfrac{7}{6}$

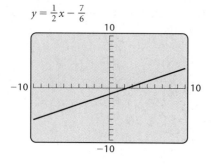

EXAMPLE 7   A line has slope $-\tfrac{7}{9}$ and y-intercept $(0, 16)$. Find an equation of the line.

Solution   We use the slope–intercept equation and substitute $-\tfrac{7}{9}$ for m and 16 for b:

$$y = mx + b$$
$$y = -\tfrac{7}{9}x + 16.$$

Point–Slope Equations of Lines

Suppose that we have a nonvertical line and that the coordinates of point P_1 are (x_1, y_1). We can think of P_1 as fixed and imagine a movable point P on the line with coordinates (x, y). Thus the slope is given by

$$\frac{y - y_1}{x - x_1} = m.$$

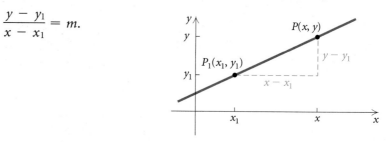

Multiplying on both sides by $x - x_1$, we get the *point–slope equation* of the line:

$$(x - x_1) \cdot \frac{y - y_1}{x - x_1} = m \cdot (x - x_1)$$
$$y - y_1 = m(x - x_1).$$

Point–Slope Equation

The **point–slope equation** of the line with slope m passing through (x_1, y_1) is

$$y - y_1 = m(x - x_1).$$

Thus if we know the slope of a line and the coordinates of one point on the line, we can find an equation of the line.

EXAMPLE 8 Find an equation of the line containing the point $\left(\frac{1}{2}, -1\right)$ and with slope 5.

Solution If we substitute in $y - y_1 = m(x - x_1)$, we get

$$y - (-1) = 5\left(x - \tfrac{1}{2}\right),$$

which simplifies as follows:

$$y + 1 = 5x - \tfrac{5}{2}$$
$$y = 5x - \tfrac{5}{2} - 1$$
$$y = 5x - \tfrac{7}{2}. \qquad \text{Slope–intercept equation}$$

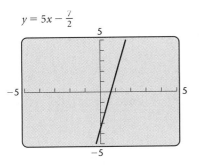

$y = 5x - \frac{7}{2}$

EXAMPLE 9 Find an equation of the line containing the points $(2, 3)$ and $(1, -4)$.

Solution We first determine the slope:

$$m = \frac{-4 - 3}{1 - 2}$$
$$= \frac{-7}{-1}$$
$$= 7.$$

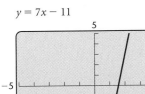

$y = 7x - 11$

Then we substitute 7 for m and either of the points $(2, 3)$ or $(1, -4)$ for (x_1, y_1) in the point–slope equation. In this case, we use $(2, 3)$. We get

$$y - 3 = 7(x - 2),$$

which simplifies to

$$y - 3 = 7x - 14$$
$$y = 7x - 11.$$

Parallel Lines

How can we examine a pair of equations to determine whether their graphs are parallel—that is, they do not intersect? We can explore this using a grapher.

Interactive Discovery

Graph each of the following pairs of lines. Try to determine whether they are parallel. You will probably need to change viewing windows and either zoom in or zoom out to decide for sure. Complete the table and look for a pattern.

PAIRS OF EQUATIONS	SLOPES		PARALLEL?
	m_1	m_2	
$y = -0.38x + 4.2,\ y = -0.37x - 5.1$			Yes
$y = -0.38x + 4.2,\ 50y + 19x = 21$			
$2x + 1 = y,\ y + 3x = 4$			
$y = 8.02x - 11.3,\ y = 7.98x - 11.3$			
$y = -3,\ y = 4.2$			
$x = -5,\ x = 2$ (Use the DRAW menu to graph $x = -5$.)			

Can you find a way to determine whether two lines are parallel without graphing?

If two different lines are vertical, then they are parallel. Thus two equations such as $x = a_1$, $x = a_2$, where $a_1 \neq a_2$, have graphs that are *parallel lines*. Two nonvertical lines such as $y = mx + b_1$, $y = mx + b_2$, where the slopes are the same and $b_1 \neq b_2$, also have graphs that are *parallel lines*.

> *Parallel Lines*
>
> Vertical lines are **parallel**. Nonvertical lines are **parallel** if and only if they have the same slope and different y-intercepts.

Perpendicular Lines

How can we examine a pair of equations to determine whether they are perpendicular? Let's explore this using a grapher.

*Interactive
Discovery*

Graph each of the following pairs of lines using the window $[-12, 12, -8, 8]$. This is an example of a **square** window that will be discussed in more detail in Section 1.7. Try to determine whether the lines are perpendicular. *You must use square viewing windows so that you can see whether angles between lines seem to be 90°.* Complete the table and look for a pattern.

PAIRS OF EQUATIONS	SLOPES m_1	m_2	PRODUCT OF SLOPES $m_1 \cdot m_2$	PERPENDICULAR?
$y = -\frac{2}{5}x + 3, \ y = \frac{5}{2}x + 3$				
$y = 0.1875x - 2, \ y = -\frac{16}{3}x - 5.1$				
$3y + 4x = -21, \ 4y - 3x = 8$				
$y = -3x + 4, \ y = 2x - 7$				
$x = -5, \ y = 2$				

Can you find a way to determine whether two lines are perpendicular without graphing?

Perpendicular Lines

Two lines with slopes m_1 and m_2 are **perpendicular** if and only if the product of their slopes is -1:

$$m_1 m_2 = -1.$$

Lines are also **perpendicular** if one is vertical ($x = a$) and the other is horizontal ($y = b$).

If a line has slope m_1, the slope m_2 of a line perpendicular to it is $-1/m_1$ (the slope of one line is the opposite of the reciprocal of the other).

EXAMPLE 10 Determine whether each of the following pairs of lines is parallel, perpendicular, or neither.

a) $y + 2 = 5x, \ 5y + x = -15$
b) $2y + 4x = 8, \ 5 + 2x = -y$
c) $2x + 1 = y, \ y + 3x = 4$

Solution   We use the slopes of the lines to determine whether the lines are parallel or perpendicular. We can create graphs on a grapher as a check.

a) We solve each equation for y:

$$y = 5x - 2, \qquad y = -\tfrac{1}{5}x - 3.$$

The slopes are 5 and $-\tfrac{1}{5}$. Their product is -1, so the lines are perpendicular (see Fig. 1).

b) Solving each equation for y, we get

$$y = -2x + 4, \qquad y = -2x - 5.$$

We see that $m_1 = -2$ and $m_2 = -2$. Since the slopes are the same and the y-intercepts, 4 and -5, are different, the lines are parallel (see Fig. 2).

c) Solving each equation for y, we get

$$y = 2x + 1, \qquad y = -3x + 4.$$

Since $m_1 = 2$ and $m_2 = -3$, we know that $m_1 \neq m_2$ and that $m_1 m_2 \neq -1$. It follows that the lines are neither parallel nor perpendicular (see Fig. 3).

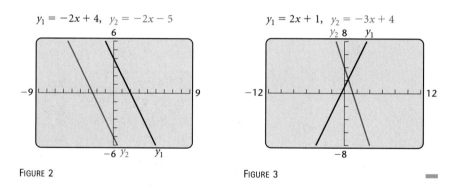

$y_1 = 5x - 2, \; y_2 = -\tfrac{1}{5}x - 3$

FIGURE 1

$y_1 = -2x + 4, \; y_2 = -2x - 5$

FIGURE 2

$y_1 = 2x + 1, \; y_2 = -3x + 4$

FIGURE 3

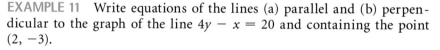

EXAMPLE 11   Write equations of the lines (a) parallel and (b) perpendicular to the graph of the line $4y - x = 20$ and containing the point $(2, -3)$.

Solution   We first solve $4y - x = 20$ for y to get $y = \tfrac{1}{4}x + 5$. Thus the slope is $\tfrac{1}{4}$.

a) The line parallel to the given line will have slope $\tfrac{1}{4}$. We use the point–slope equation for a line with slope $\tfrac{1}{4}$ and containing the point $(2, -3)$:

$$y - y_1 = m(x - x_1)$$
$$y - (-3) = \tfrac{1}{4}(x - 2)$$
$$y + 3 = \tfrac{1}{4}x - \tfrac{1}{2}$$
$$y = \tfrac{1}{4}x - \tfrac{7}{2}.$$

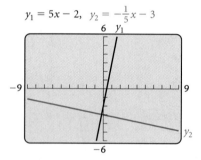

$y_1 = \tfrac{1}{4}x + 5,$

$y_2 = \tfrac{1}{4}x - \tfrac{7}{2},$

$y_3 = -4x + 5$

b) The slope of the perpendicular line is -4. Now we use the point–slope equation to write an equation for a line with slope -4 and containing the point $(2, -3)$:

$$y - y_1 = m(x - x_1)$$
$$y - (-3) = -4(x - 2)$$
$$y + 3 = -4x + 8$$
$$y = -4x + 5.$$

We can visualize our work using a grapher.

Summary of Terminology About Lines

TERMINOLOGY	MATHEMATICAL INTERPRETATION
Slope	$m = \dfrac{y_2 - y_1}{x_2 - x_1}$, or $\dfrac{y_1 - y_2}{x_1 - x_2}$
Slope–intercept equation	$y = mx + b$
Point–slope equation	$y - y_1 = m(x - x_1)$
Horizontal lines	$y = b$
Vertical lines	$x = a$
Parallel lines	$m_1 = m_2,\ b_1 \neq b_2$
Perpendicular lines	$m_1 m_2 = -1$

Applications of Linear Functions

We now consider an application of linear functions.

Humerus

EXAMPLE 12 *Height Estimates.* An anthropologist can use linear functions to estimate the height of a male or a female, given the length of the humerus, the bone from the elbow to the shoulder. The height, in centimeters, of an adult male with a humerus of length x, in centimeters, is given by the function

$$M(x) = 2.89x + 70.64.$$

The height, in centimeters, of an adult female with a humerus of length x is given by the function

$$F(x) = 2.75x + 71.48.$$

A 26-cm humerus was uncovered in a ruins. Assuming it was from a female, how tall was she?

Algebraic Solution

We substitute into the function:

$$F(26) = 2.75(26) + 71.48$$
$$= 142.98.$$

Thus the female was 142.98 cm tall.

Graphical Solution

We graph $F(x) = 2.75x + 71.48$ and use the VALUE feature to find $F(26)$.

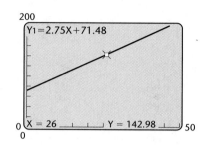

The female was 142.98 cm tall.

Theoretically, the domain of the function is the set of all real numbers. However, the context of the problem dictates a different domain. One could not find a bone with a length of 0 or less. Thus the domain consists of positive real numbers, that is, the interval $(0, \infty)$. (A more realistic domain might be 20 cm to 60 cm, the interval $[20, 60]$.) ▬

Exercise Set 1.2

In Exercises 1–4, the table of data contains input–output values for a function. Answer the following questions for each table.

a) Is the change in the inputs x the same?
b) Is the change in the outputs y the same?
c) Is the function linear?

1.

x	y
-3	7
-2	10
-1	13
0	16
1	19
2	22
3	25

2.

x	y
20	12.4
30	24.8
40	49.6
50	99.2
60	198.4
70	396.8
80	793.6

3.

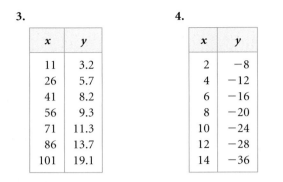

x	y
11	3.2
26	5.7
41	8.2
56	9.3
71	11.3
86	13.7
101	19.1

4.

x	y
2	−8
4	−12
6	−16
8	−20
10	−24
12	−28
14	−36

Find the slope of the line containing the given points.

5.

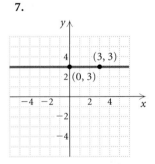

6.

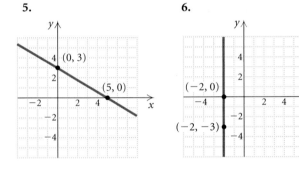

7.

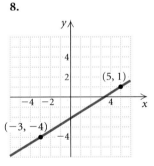

8.

9. $(9, 4)$ and $(-1, 2)$

10. $(-3, 7)$ and $(5, -1)$

11. $(4, -9)$ and $(-5, 6)$

12. $(-6, -1)$ and $(2, -13)$

13. $(\pi, -3)$ and $(\pi, 2)$

14. $(\sqrt{2}, -4)$ and $(0.56, -4)$

15. (a, a^2) and $(a + h, (a + h)^2)$

16. $(a, 3a + 1)$ and $(a + h, 3(a + h) + 1)$

17. *Road Grade.* Using the following figure, find the road grade and an equation giving the height y as a function of the horizontal distance x.

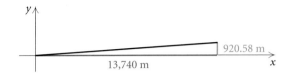

18. *Grade of Treadmills.* A treadmill is 5 ft long and is set at an 8% grade. How high is the end of the treadmill?

19. Find the rate of change of the tuition and fees at public two-year colleges.

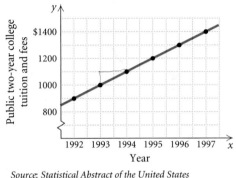

Source: Statistical Abstract of the United States

20. Find the rate of change of the cost of a formal wedding.

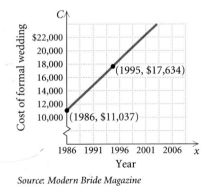

Source: Modern Bride Magazine

21. *Running Rate.* An Olympic marathoner passes the 10-km point of a race after 50 min and arrives at the 25-km point $1\frac{1}{2}$ hr later. Find the speed (average rate of change) of the runner.

22. *Work Rate.* As a typist resumes work on a project, $\frac{1}{6}$ of the paper has already been typed. Six hours later, the paper is $\frac{3}{4}$ done. Calculate the worker's typing rate.

Find the slope and the y-intercept of the equation.

23. $y = \frac{3}{5}x - 7$

24. $f(x) = -2x + 3$

25. $f(x) = 5 - \frac{1}{2}x$

26. $y = 2 + \frac{3}{7}x$

27. $3x + 2y = 10$

28. $2x - 3y = 12$

29. $4x - 3f(x) - 15 = 6$

30. $9 = 3 + 5x - 2f(x)$

Write a slope–intercept equation for a line with the given characteristics.

31. $m = \frac{2}{9}$, *y*-intercept $(0, 4)$

32. $m = -\frac{3}{8}$, *y*-intercept $(0, 5)$

33. $m = -4$, *y*-intercept $(0, -7)$

34. $m = \frac{2}{7}$, *y*-intercept $(0, -6)$

35. $m = -4.2$, *y*-intercept $\left(0, \frac{3}{4}\right)$

36. $m = -4$, *y*-intercept $\left(0, -\frac{3}{2}\right)$

37. $m = \frac{2}{9}$, passes through $(3, 7)$

38. $m = -\frac{3}{8}$, passes through $(5, 6)$

39. $m = 3$, passes through $(1, -2)$

40. $m = -2$, passes through $(-5, 1)$

41. $m = -\frac{3}{5}$, passes through $(-4, -1)$

42. $m = \frac{2}{3}$, passes through $(-4, -5)$

43. Passes through $(-1, 5)$ and $(2, -4)$

44. Passes through $(2, -1)$ and $(7, -11)$

45. Passes through $(7, 0)$ and $(-1, 4)$

46. Passes through $(-3, 7)$ and $(-1, -5)$

Determine whether the pair of lines is parallel, perpendicular, or neither.

47. $x + 2y = 5$,
$2x + 4y = 8$

48. $2x - 5y = -3$,
$2x + 5y = 4$

49. $y = 4x - 5$,
$4y = 8 - x$

50. $y = 7 - x$,
$y = x + 3$

Write a slope–intercept equation for a line passing through the given point that is parallel to the given line. Then write a second equation for a line passing through the given point that is perpendicular to the given line.

51. $(3, 5)$, $y = \frac{2}{7}x + 1$

52. $(-1, 6)$, $f(x) = 2x + 9$

53. $(-7, 0)$, $y = -0.3x + 4.3$

54. $(-4, -5)$, $2x + y = -4$

55. $(3, -2)$, $3x + 4y = 5$

56. $(8, -2)$, $y = 4.2(x - 3) + 1$

57. $(3, -3)$, $x = -1$

58. $(4, -5)$, $y = -1$

59. *Ideal Weight.* One way to estimate the ideal weight of a woman is to multiply her height by 3.5 and subtract 110. Let $W =$ the ideal weight, in pounds, and $h =$ height, in inches.

a) Express W as a linear function of h.

b) Graph W.

c) Find the ideal weight of a woman whose height is 62 in.

d) Find the domain of the function.

60. *Pressure at Sea Depth.* The function P, given by
$$P(d) = \frac{1}{33}d + 1,$$
gives the pressure, in atmospheres (atm), at a depth d, in feet, under the sea.

a) Graph P.

b) Find $P(0)$, $P(5)$, $P(10)$, $P(33)$, and $P(200)$.

c) Find the domain of the function.

61. *Stopping Distance on Glare Ice.* The stopping distance (at some fixed speed) of regular tires on glare ice is a function of the air temperature F, in degrees Fahrenheit. This function is estimated by
$$D(F) = 2F + 115,$$
where $D(F)$ is the stopping distance, in feet, when the air temperature is F, in degrees Fahrenheit.

a) Graph D.

b) Find $D(0°)$, $D(-20°)$, $D(10°)$, and $D(32°)$.

c) Explain why the domain should be restricted to $[-57.5°, 32°]$.

62. *Anthropology Estimates.* Consider Example 12 and the function
$$M(x) = 2.89x + 70.64$$
for estimating the height of a male.

a) Find the height of a male if a 26-cm humerus is found in an archeological dig.

b) What is the domain of M?

63. *Reaction Time.* Suppose that while driving a car, you suddenly see a school crossing guard standing in the road. Your brain registers the information and sends a signal to your foot to hit the brake. The car travels a distance D, in feet, during this time, where D is a function of the speed r, in miles per hour, that the car is traveling when you see the crossing guard. That reaction distance is a linear function given by

$$D(r) = \frac{11r + 5}{10}.$$

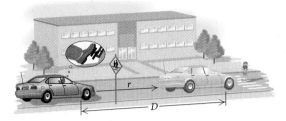

a) Find the slope of this line and interpret its meaning in this application.
b) Graph D.
c) Find $D(5)$, $D(10)$, $D(20)$, $D(50)$, and $D(65)$.
d) What is the domain of this function? Explain.

64. *Straight-line Depreciation.* A company buys a new color printer for $5200 to print banners for a sales campaign. The printer is purchased on January 1 and is expected to last 8 yr, at the end of which time its *trade-in*, or *salvage value*, will be $1100. If the company figures the decline or depreciation in value to be the same each year, then the salvage value V, after t years, is given by the linear function

$$V(t) = \$5200 - \$512.50t, \quad \text{for } 0 \le t \le 8.$$

a) Graph V.
b) Find $V(0)$, $V(1)$, $V(2)$, $V(3)$, and $V(8)$.
c) Find the domain and the range of this function.

In Exercises 65 and 66, express the slope as a ratio of two quantities.

65.

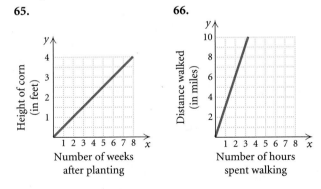

Number of weeks after planting

66.

Number of hours spent walking

67. *Total Cost.* The Cellular Connection charges $60 for a phone and $40 per month under its economy plan. Write and graph an equation that can be used to determine the total cost, $C(t)$, of operating a

Cellular Connection phone for t months. Then find the total cost for 6 months.

68. *Total Cost.* Twin Cities Cable Television charges a $35 installation fee and $30 per month for "deluxe" service. Write and graph an equation that can be used to determine the total cost, $C(t)$, for t months of deluxe cable television service. Then find the total cost for 8 months of service.

*In Exercises 69 and 70, the term **fixed costs** refers to the start-up costs of operating a business. This includes machinery and building costs. The term **variable costs** refers to what it costs a business to produce or service one item.*

69. Kara's Custom Tees experienced fixed costs of $800 and variable costs of $3 per shirt. Write an equation that can be used to determine the total expenses encountered by Kara's Custom Tees. Then graph the equation and determine the total cost of producing 75 shirts.

70. It's My Racquet experienced fixed costs of $500 and variable costs of $2 for each tennis racquet that is restrung. Write an equation that can be used to determine the total expenses encountered by It's My Racquet. Then graph the equation and determine the total cost of restringing 150 tennis racquets.

Discussion and Writing

71. Discuss why the graph of a vertical line $x = a$ cannot represent y as a function of x.

72. Explain as you would to a fellow student how the numerical value of slope can be used to describe the slant and the steepness of a line.

Skill Maintenance

To the student and the instructor: The skill maintenance exercises review skills covered previously in the text. You can expect such exercises in every exercise set. They provide excellent review for a final examination. Answers to all skill maintenance exercises appear, along with section references, in the answer section at the back of the book.

If $f(x) = x^2 - 3x$, find each of the following.

73. $f(-5)$

74. $f(5)$

75. $f(-a)$

76. $f(a + h)$

Synthesis

77. Find k so that the line containing the points $(-3, k)$ and $(4, 8)$ is parallel to the line containing the points $(5, 3)$ and $(1, -6)$.

78. Find an equation of the line passing through the point $(4, 5)$ perpendicular to the line passing through the points $(-1, 3)$ and $(2, 9)$.

79. *Fahrenheit and Celsius Temperatures.* Fahrenheit temperature F is a linear function of Celsius temperature C. When C is 0, F is 32; and when C is 100, F is 212. Use these data to express F as a function of C and to express C as a function of F.

Suppose that f is a linear function. Then $f(x) = mx + b$. Determine whether each of the following is true or false.

80. $f(c + d) = f(c) + f(d)$

81. $f(cd) = f(c)f(d)$

82. $f(kx) = kf(x)$

83. $f(c - d) = f(c) - f(d)$

Let $f(x) = mx + b$. Find a formula for $f(x)$ given each of the following.

84. $f(3x) = 3f(x)$

85. $f(x + 2) = f(x) + 2$

Modeling: Data Analysis, Curve Fitting, and Linear Regression

1.3

• *Analyze a set of data to determine whether it can be modeled by a linear function.*
• *Fit a regression line to a set of data; then use the linear model to make predictions.*

Mathematical Models

When a real-world problem can be described in mathematical language, we have a **mathematical model.** For example, the natural numbers constitute a mathematical model for situations in which counting is essential. Situations in which algebra can be brought to bear often require the use of functions as models.

Mathematical models are abstracted from real-world situations. Procedures within the mathematical model give results that allow one to predict what will happen in that real-world situation. If the predictions are inaccurate or the results of experimentation do not conform to the model, the model needs to be changed or discarded.

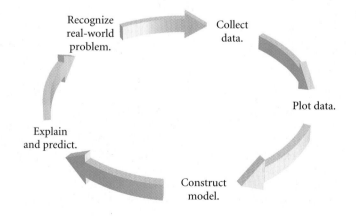

Mathematical modeling can be an ongoing process. For example, finding a mathematical model that will enable an accurate prediction of population growth is not a simple problem. Any population model that one might devise would need to be reshaped as further information is acquired.

Curve Fitting

We will develop and use many kinds of mathematical models in this text. In this chapter, we have seen some types of functions that can be used as models. Let's look more closely at three of them. The quadratic and cubic functions are nonlinear. Modeling with these functions will be discussed in Chapters 2 and 3.

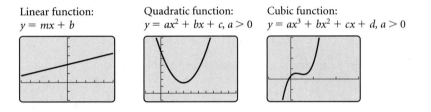

Linear function: $y = mx + b$

Quadratic function: $y = ax^2 + bx + c, a > 0$

Cubic function: $y = ax^3 + bx^2 + cx + d, a > 0$

In general, we try to find a function that fits, as well as possible, observations (data), theoretical reasoning, and common sense. We call this **curve fitting;** it is one aspect of mathematical modeling.

Let's first look at some data and related graphs or scatterplots and determine whether any of the above functions, linear or nonlinear, seem to fit either set of data.

Number of Apartment Households in the United States

YEAR, x	NUMBER OF APARTMENT HOUSEHOLDS, y (IN MILLIONS)	SCATTERPLOT	
1970, 0	8.5	Apartment households	It appears that the data points can be represented or fitted by a straight line.
1975, 5	9.9		
1980, 10	10.8		
1985, 15	12.9		
1990, 20	14.2		The graph is linear.
1997, 27	14.5		

Sources: U.S. Bureau of the Census; National Multi-Housing Council

Number of Cellular Phone Subscribers

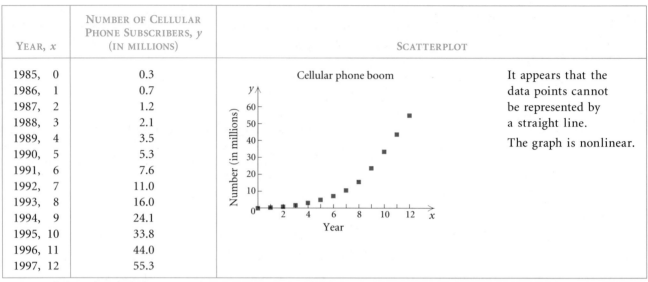

Year, *x*	Number of Cellular Phone Subscribers, *y* (in millions)	Scatterplot	
1985, 0	0.3		It appears that the
1986, 1	0.7		data points cannot
1987, 2	1.2		be represented by
1988, 3	2.1		a straight line.
1989, 4	3.5		
1990, 5	5.3		The graph is nonlinear.
1991, 6	7.6		
1992, 7	11.0		
1993, 8	16.0		
1994, 9	24.1		
1995, 10	33.8		
1996, 11	44.0		
1997, 12	55.3		

Source: Cellular Telecommunications Industry Association

Looking at the scatterplots, we see that the apartment households data seem to be rising in a manner to suggest that a linear function might fit, although a "perfect" straight line cannot be drawn through the data points. A linear function does not seem to fit the cellular phone data.

The Regression Line

We now consider **linear regression,** a procedure that can be used to model a set of data using a linear function. Although discussion leading to a complete understanding of this method belongs in a statistics course, we present the procedure here because we can carry it out easily using technology. The grapher gives us the powerful capability to find linear models and to make predictions using them.

EXAMPLE 1 *Number of Apartment Households in the United States.* Consider the data just presented on the number of apartment households in the United States.

a) Fit a regression line to the data using the linear regression feature on a grapher.

b) Use the linear model to predict the number of apartment households in 2003.

Solution

a) First, we enter the data in lists on a grapher. We enter the values of the independent variable *x* in list L1 and the corresponding values of the

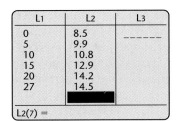

FIGURE 1

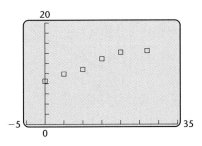

FIGURE 2

dependent variable y in L2 (see Fig. 1). The grapher can then create a scatterplot of the data, as shown in Fig. 2.

Consider the data points (0, 8.5), (5, 9.9), (10, 10.8), (15, 12.9), (20, 14.2), and (27, 14.5) on the graph. We want to fit a **regression line,**

$$y = mx + b,$$

to these data points. We use the LINEAR REGRESSION feature to find the regression line.

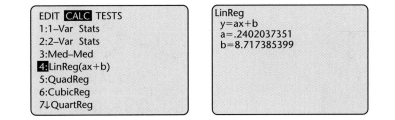

The result is

$$y = 0.2402037351x + 8.717385399. \quad \textbf{Regression line}$$

We can then graph the regression line on the same graph as the scatterplot.

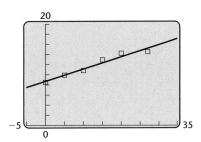

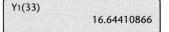

b) To predict the number of apartment households in 2003, we evaluate the function found in part (a) for $x = 33$. (In 1970, $x = 0$; in 1990, $x = 20$. Thus in 2003, $x = 33$.)

We see from the grapher that the number of apartment households in 2003 is predicted to be about 16.6 million.

The Correlation Coefficient

On some graphers with DIAGNOSTICS turned on, a constant r between -1 and 1, called the **coefficient of linear correlation,** appears with the equation of the regression line. Though we cannot develop a formula for calculating r in this text, keep in mind that it is used to describe the strength of the linear relationship between x and y. The closer $|r|$ is to 1, the better the correlation. A positive value of r also indicates that the regression line has a positive slope and a negative value of r indicates that the regression line has a negative slope. For the apartment household data just discussed, $r = 0.9755068364$, which indicates a very good linear correlation. The following scatterplots summarize the interpretation of a correlation coefficient.

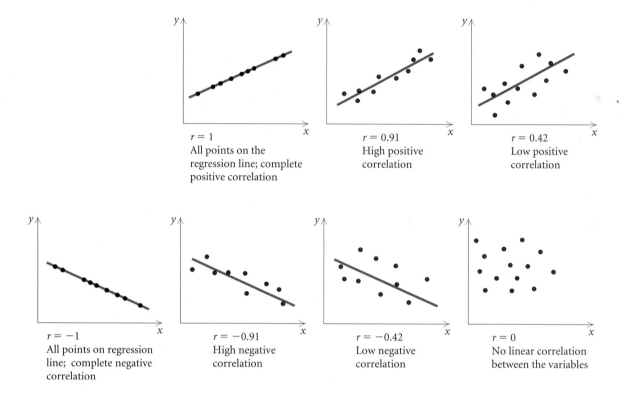

$r = 1$
All points on the regression line; complete positive correlation

$r = 0.91$
High positive correlation

$r = 0.42$
Low positive correlation

$r = -1$
All points on regression line; complete negative correlation

$r = -0.91$
High negative correlation

$r = -0.42$
Low negative correlation

$r = 0$
No linear correlation between the variables

Keep in mind that a high linear correlation coefficient does not necessarily indicate a "cause-and-effect" connection between the variables. We might be able to calculate a high positive correlation between stock prices and rainfall in India, but before we place our life savings in the market, further analysis and common sense should be applied!

EXAMPLE 2 *Garbage Production.* America creates more garbage than any other nation. According to Denis Hayes, president of Seattle's non-profit Bullitt Foundation and one of the founders of Earth Day in 1970, "we need to be an Heirloom Society instead of a Throw-Away Society." The Environmental Protection Agency estimates that, on average, we each produce 4.4 lb of garbage a day (*Source*: "Take Out the Trash, and Put It—Where?", by Bernard Bavzer, *Parade Magazine*, June 13, 1999). The following graph illustrates the annual production of municipal solid waste in the United States for certain years.

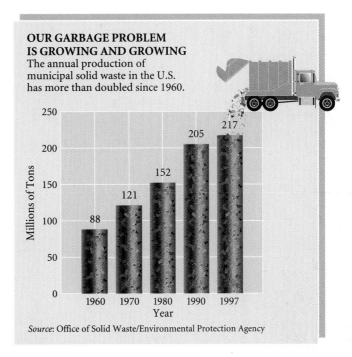

**OUR GARBAGE PROBLEM
IS GROWING AND GROWING**
The annual production of
municipal solid waste in the U.S.
has more than doubled since 1960.

Source: Office of Solid Waste/Environmental Protection Agency

a) Fit a regression line to the data and use it to predict the amount of municipal solid waste in the United States in 2005. Enter the data as (1960, 88), (1970, 121), and so on.

b) What is the correlation coefficient for the regression line? How close a fit is the regression line?

Solution

a) Using a grapher, we get the following.

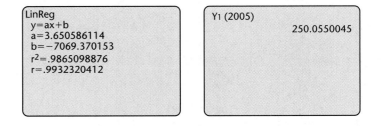

The regression line is

$$y = 3.650586114x - 7069.370153.$$

To predict the amount of municipal solid waste in 2005, we find the value of the regression function when $x = 2005$. Thus the amount of solid waste will be about 250 million tons in 2005.

b) The correlation coefficient for the regression line is about 0.9932. Since the closer $|r|$ is to 1 the better the correlation, this equation fits the data closely.

Exercise Set 1.3

In Exercises 1–4, determine whether the graph might be modeled by a linear function.

1.

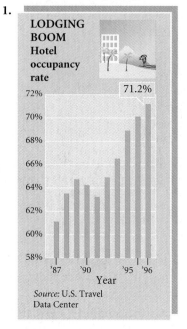

LODGING BOOM
Hotel occupancy rate

71.2%

72%
70%
68%
66%
64%
62%
60%
58%

'87   '90   '95 '96
Year

Source: U.S. Travel Data Center

2.

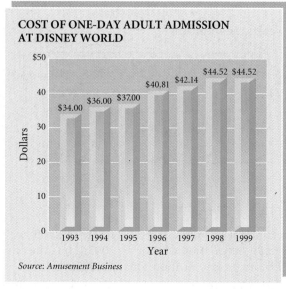

COST OF ONE-DAY ADULT ADMISSION AT DISNEY WORLD

$50

40

30

Dollars

20

10

0

$34.00 $36.00 $37.00 $40.81 $42.14 $44.52 $44.52

1993 1994 1995 1996 1997 1998 1999
Year

Source: Amusement Business

3.

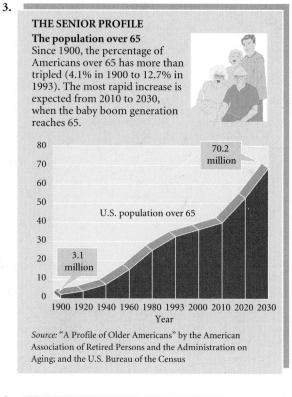

THE SENIOR PROFILE

The population over 65
Since 1900, the percentage of Americans over 65 has more than tripled (4.1% in 1900 to 12.7% in 1993). The most rapid increase is expected from 2010 to 2030, when the baby boom generation reaches 65.

80
70 70.2 million
60
50
40 U.S. population over 65
30
20 3.1 million
10
0

1900 1920 1940 1960 1980 1993 2000 2010 2020 2030
Year

Source: "A Profile of Older Americans" by the American Association of Retired Persons and the Administration on Aging; and the U.S. Bureau of the Census

4.

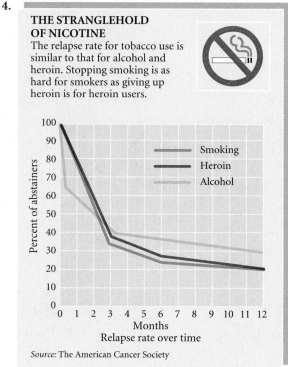

THE STRANGLEHOLD OF NICOTINE
The relapse rate for tobacco use is similar to that for alcohol and heroin. Stopping smoking is as hard for smokers as giving up heroin is for heroin users.

100
90
80
70
60

Percent of abstainers

50
40
30
20
10
0

— Smoking
— Heroin
— Alcohol

0 1 2 3 4 5 6 7 8 9 10 11 12
Months
Relapse rate over time

Source: The American Cancer Society

In Exercises 5–8, determine whether the scatterplot might be fit by a linear model.

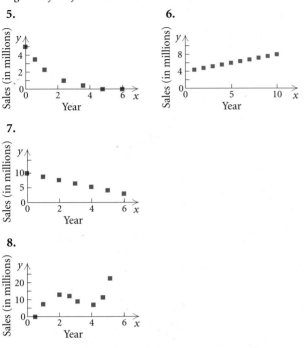

5.

6.

7.

8.

Exercises 9–14 involve using a grapher and linear regression to model data.

9.

Number of Shopping Centers

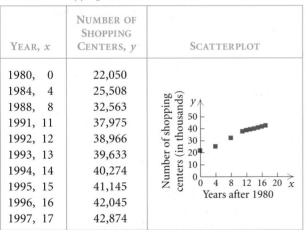

YEAR, x	NUMBER OF SHOPPING CENTERS, y	SCATTERPLOT
1980, 0	22,050	
1984, 4	25,508	
1988, 8	32,563	
1991, 11	37,975	
1992, 12	38,966	
1993, 13	39,633	
1994, 14	40,274	
1995, 15	41,145	
1996, 16	42,045	
1997, 17	42,874	

Source: International Council of Shopping Centers

a) Use regression to model the data with a linear function and use it to predict the number of shopping centers in 2005.
b) What is the correlation coefficient for the regression line? How close a fit is the regression line?

10.

Cases of Skin Cancers in the United States

YEAR, x	NUMBER OF CASES, y (IN THOUSANDS)	SCATTERPLOT
1988, 0	5800	Skin cancer on the rise
1989, 1	6000	
1990, 2	6300	
1991, 3	6500	
1992, 4	6700	

Source: American Cancer Society

a) Use regression to model the data with a linear function and use it to estimate the number of cases of skin cancer in the United States in 1998 and 2000. Also use the function to predict the number of cases in 2010.
b) What is the correlation coefficient for the regression line? How close a fit is the regression line?

11. *The Cost of Tuition at State Universities*

COLLEGE YEAR, x	ESTIMATED TUITION, y
1997–1998, 0	$ 9,838
1998–1999, 1	10,424
1999–2000, 2	11,054
2000–2001, 3	11,717
2001–2002, 4	12,420

Source: College Board, Senate Labor Committee

a) Use regression to model the data with a linear function and use it to predict the cost of college tuition in 2004–2005, 2006–2007, and 2010–2011.
b) What is the correlation coefficient for the regression line? How close a fit is the regression line?
c) A college president asserts that in the college year 2006–2007, the cost of tuition will be $16,619. How well does your estimate compare?

12.   *Book Buying Growth in the United States*

YEAR, x	BOOK SALES, y (IN BILLIONS)
1992, 0	$21
1993, 1	23
1994, 2	24
1995, 3	25
1996, 4	26

Source: Book Industry Trends 1995

a) Use regression to model the data with a linear function and use it to predict book sales in 2000, 2005, and 2010.

b) What is the correlation coefficient for the regression line? How close a fit is the regression line?

c) A bookstore owner asserts that in 2000, book sales will be $45 billion. How well does your estimate stack up to this?

13. *Maximum Heart Rate.*   A person who is exercising should not exceed his or her maximum heart rate, which is determined on the basis of that person's sex, age, and resting heart rate. The following table relates resting heart rate and maximum heart rate for a 20-yr-old man.

RESTING HEART RATE, H (IN BEATS PER MINUTE)	MAXIMUM HEART RATE, M (IN BEATS PER MINUTE)
50	166
60	168
70	170
80	172

Source: American Heart Association

a) Use regression to model the data with a linear function.

b) Predict a maximum heart rate if the resting heart rate is 40, 65, 76, and 84.

c) What is the correlation coefficient? How confident are you about using the regression line as a predictor?

14. *Study Time versus Grades.*   A math instructor asked her students to keep track of how much time each spent studying a chapter on functions in her algebra–trigonometry course. She collected the information together with test scores from that chapter's test. The data are listed in the table at the top of the next column.

STUDY TIME, x (IN HOURS)	TEST GRADE, y (IN PERCENT)
23	81
15	85
17	80
9	75
21	86
13	80
16	85
11	93

a) Use regression to model the data with a linear function.

b) Predict a student's score if he or she studies 24 hr, 6 hr, and 18 hr.

c) What is the correlation coefficient? How confident are you about using the regression line as a predictor?

Discussion and Writing

15. Why does a correlation coefficient of 1 not guarantee the existence of a cause-and-effect relationship?

16. Conduct your own experiment. Use some kind of measuring device and gather data relating body height to arm length. Fit a regression line to the data and find the correlation coefficient. How confident would you be about using these data to predict Shaquille O'Neal's arm length, knowing that his height is 7'0"?

Skill Maintenance

Find the slope of the line containing the given points.

17. (2, −8) and (−5, −1)   **18.** (5, 7) and (5, −7)

Write a slope–intercept equation for a line with the given characteristics.

19. $m = \frac{3}{4}$; passes through (0, −5)

20. Passes through (2, −10) and (−2, −10)

Synthesis

21. *Ted Williams and the War Years.*   Ted Williams played for the Boston Red Sox from 1939–1960. Many credit him with being the greatest hitter of all time. Unfortunately, his career totals are somewhat less impressive than others because his career was interrupted from 1943–1945 by World War II and from 1952–1953 by the Korean War. Some assert that if he had been playing during the war years, he would have broken Hank Aaron's home run record of 755 and runs batted in (RBI) record of 2297. Below are Williams' statistics.

YEAR, x	NUMBER OF HOME RUNS, H	NUMBER OF RBIS, R
1939	31	145
1940	23	113
1941	37	120
1942	36	137
1943	0*	0*
1944	0*	0*
1945	0*	0*
1946	38	123
1947	32	114
1948	25	127
1949	43	159
1950	28	97
1951	30	126
1952	1†	3†
1953	13†	34†
1954	29	89
1955	28	83
1956	24	82
1957	38	87
1958	26	85
1959	10	43
1960	29	72

*World War II
†Korean War

a) Excluding all data from the war years, fit a linear regression equation $H = mx + b$ to the data regarding the number of home runs. Then use the equation to predict how many home runs Williams would have hit in each of the war years.

What is the correlation coefficient? How confident are you of your prediction?
b) Would Williams have broken Aaron's home run record?
c) Excluding all data from the war years, fit a linear regression equation $R = mx + b$ to the data regarding the number of RBIs. Then use the equation to predict how many RBIs Williams would have had in each of the war years. What is the correlation coefficient? How confident are you of your prediction?
d) Would Williams have broken Aaron's RBI record?

More on Functions

1.4

- Graph functions, looking for intervals on which the function is increasing, decreasing, or constant, and determine relative maxima and minima.
- Given an application, find a function formula that models the application; find the domain of the function and function values, and then graph the function.
- Graph functions defined piecewise.
- Find the sum, the difference, the product, and the quotient of two functions, and determine their domains.

Because functions occur in so many real-world situations, it is important to be able to analyze them carefully.

Increasing, Decreasing, and Constant Functions

On a given interval, if the graph of a function rises from left to right, it is said to be **increasing** on that interval. If the graph drops from left to right, it is said to be **decreasing**. If the function stays the same from left to right, it is said to be **constant**.

Interactive Discovery

Graph each of the following functions on a grapher using the given viewing window. Then use the TRACE feature, moving from left to right along each graph. As you move the cursor from left to right, note that the x-coordinate always increases. Observe what happens to the y-coordinate. Where does it increase, decrease, or remain constant? Confirm the results with a table.

$$y = 5 - (x + 2)^2, \quad [-2, 4, -10, 6];$$
$$y = 4, \quad [-20, 20, -3, 6];$$
$$y = |x + 1| + |x - 2| - 5, \quad [-6, 6, -6, 6].$$

We are led to the following definitions.

Increasing, Decreasing, and Constant Functions

A function f is said to be **increasing** on an open interval I, if for all a and b in that interval, $a < b$ implies $f(a) < f(b)$.

A function f is said to be **decreasing** on an open interval I, if for all a and b in that interval, $a < b$ implies $f(a) > f(b)$.

A function f is said to be **constant** on an open interval I, if for all a and b in that interval, $f(a) = f(b)$.

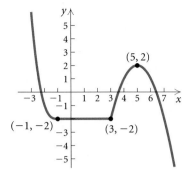

EXAMPLE 1 Determine the intervals on which the function in the figure at left is **(a)** increasing; **(b)** decreasing; and **(c)** constant.

Solution

a) The function is increasing on the interval $(3, 5)$.

b) The function is decreasing on the intervals $(-\infty, -1)$ and $(5, \infty)$.

c) The function is constant on the interval $(-1, 3)$.　　　　　▬

Relative Maximum and Minimum Values

Consider the graph shown below. Note the "peaks" and "valleys" at the points c_1, c_2, and c_3. The function value $f(c_2)$ is called a **relative maximum** (plural, **maxima**). Each of the function values $f(c_1)$ and $f(c_3)$ is called a **relative minimum** (plural, **minima**).

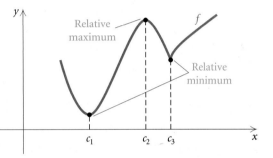

Relative Maxima and Minima

Suppose that f is a function for which $f(c)$ exists for some c in the domain of f. Then:

$f(c)$ is a **relative maximum** if there exists an open interval I containing c such that $f(c) > f(x)$, for all x in I where $x \neq c$; and

$f(c)$ is a **relative minimum** if there exists an open interval I containing c such that $f(c) < f(x)$, for all x in I where $x \neq c$.

Simply stated, $f(c)$ is a relative maximum if $f(c)$ is the highest point in some open interval, and $f(c)$ is a relative minimum if $f(c)$ is the lowest point in some open interval.

We can use the MAXIMUM or MINIMUM features on a grapher to approximate relative maxima or minima. If you take a calculus course, you will learn an algebraic method for determining exact values of relative maxima and minima.

EXAMPLE 2 Use a grapher to determine any relative maxima or minima of the function $f(x) = 0.1x^3 - 0.6x^2 - 0.1x + 2$ and to determine intervals on which the function is increasing or decreasing.

Solution We first graph the function, experimenting with the window as needed. The curvature is seen fairly well with $[-4, 6, -3, 3]$. Using the MAXIMUM and MINIMUM features, we determine the relative maximum value and the relative minimum value of the function.

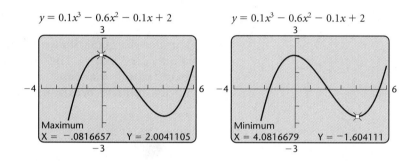

We see that the relative maximum value of the function is about 2.004. It occurs when $x \approx -0.082$. We also approximate the relative minimum: -1.604 at $x \approx 4.082$.

We note that the graph starts rising, or increasing, from the left and stops increasing at the relative maximum. From this point, the graph decreases to the relative minimum and then begins to rise again. The function is *increasing* on the intervals

$$(-\infty, -0.082) \quad \text{and} \quad (4.082, \infty)$$

and *decreasing* on the interval

$$(-0.082, 4.082).$$

Applications of Functions

Many real-world situations can be modeled by functions.

EXAMPLE 3 *Car Distance.* Jamie and Ruben drive away from a restaurant at right angles to each other. Jamie's speed is 65 mph and Ruben's is 55 mph.

a) Express the distance between the cars as a function of time.

b) Find the domain of the function.

c) Graph the function.

Solution

a) Suppose 1 hr goes by. At that time, Jamie has traveled 65 mi and Ruben has traveled 55 mi. We can use the Pythagorean theorem then to find the distance between them. This distance would be the length of the hypotenuse of a triangle with legs measuring 65 mi and 55 mi. After

2 hr, the triangle's legs would measure 130 mi and 110 mi. Observing that the distances will always be changing, we make a drawing and let t = the time, in hours, that Jamie and Ruben have been driving since leaving the restaurant.

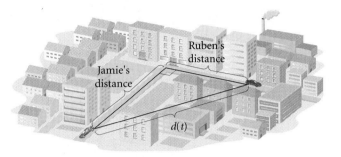

After t hours, Jamie has traveled $65t$ miles and Ruben $55t$ miles. We can use the Pythagorean theorem:

$$[d(t)]^2 = (65t)^2 + (55t)^2.$$

Because distance must be nonnegative, we need only consider the positive square root when solving for $d(t)$:

$$
\begin{aligned}
d(t) &= \sqrt{(65t)^2 + (55t)^2} \\
&= \sqrt{4225t^2 + 3025t^2} \\
&= \sqrt{7250t^2} \\
&= \sqrt{7250}\sqrt{t^2} \\
&\approx 85.15|t| \qquad \text{Approximating the root to two decimal places} \\
&\approx 85.15t. \qquad \text{Since } t \geq 0, |t| = t.
\end{aligned}
$$

Thus, $d(t) = 85.15t,\ t \geq 0$.

b) Since the time traveled must be nonnegative, the domain is the set of nonnegative real numbers $[0, \infty)$.

c) Because of the ease of using a grapher, we can almost always visualize a problem by making a graph. Here we observe that the function is increasing.

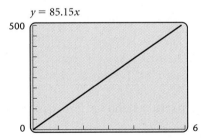

$y = 85.15x$

EXAMPLE 4 *Storage Area.* The Sound Shop has 20 ft of dividers with which to set off a rectangular area for the storage of overstock. If a corner of the store is used, the partition need only form two sides of a rectangle.

a) Express the floor area of the storage space as a function of the length of the partition.

b) Find the domain of the function.

c) Graph the function.

d) Find the dimensions that maximize the floor area.

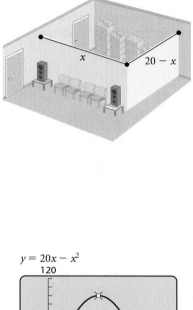

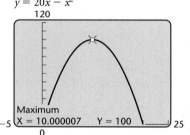

$y = 20x - x^2$

Solution

a) Note that the dividers will form two sides of a rectangle. If, for example, 14 ft of dividers are used for the length of the rectangle, that would leave $20 - 14$, or 6 ft of dividers for the width. Thus if $x =$ the length, in feet, of the rectangle, then $20 - x =$ the width. We represent this information in a sketch, as shown at left.

The area, $A(x)$, is given by

$$A(x) = x(20 - x) \qquad \text{Area} = \text{length} \cdot \text{width}$$
$$= 20x - x^2.$$

The function $A(x) = 20x - x^2$ can be used to express the rectangle's area as a function of the length.

b) Because the rectangle's length must be positive and less than 20 ft (why?), we restrict the domain of A to $\{x \mid 0 < x < 20\}$, that is, the interval $(0, 20)$.

c) The graph is shown at left in the viewing window $[-5, 25, 0, 120]$.

d) We use the MAXIMUM feature as in Example 2. The maximum value of the area function on $(0, 20)$ is 100 when $x = 10$. Thus the dimensions that maximize the area are

$$\text{Length} = x = 10 \text{ ft} \quad \text{and}$$
$$\text{Width} = 20 - x = 20 - 10 = 10 \text{ ft.}$$

Functions Defined Piecewise

Sometimes functions are defined **piecewise** using different output formulas for different parts of the domain.

EXAMPLE 5 Make a hand-drawn graph of the function defined as

$$f(x) = \begin{cases} 4, & \text{for } x \leq 0, \\ 4 - x^2, & \text{for } 0 < x \leq 2, \\ 2x - 6, & \text{for } x > 2. \end{cases}$$

Solution We create the graph in three parts, as shown and described below.

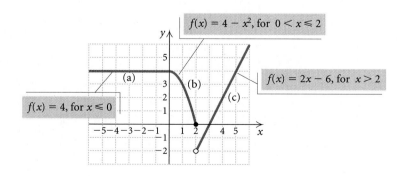

a) We graph $f(x) = 4$ *only* for inputs x less than or equal to 0 (that is, on the interval $(-\infty, 0]$).

b) We graph $f(x) = 4 - x^2$ *only* for inputs x greater than 0 and less than or equal to 2 (that is, on the interval $(0, 2]$).

c) We graph $f(x) = 2x - 6$ *only* for inputs x greater than 2 (that is, on the interval $(2, \infty)$).

To graph a function defined piecewise using a grapher, consult your user's manual or the *Graphing Calculator Manual* that accompanies this text. You might try the following way to enter the function formula for Example 5. It incorporates parenthetic descriptions of the intervals. With this method, you need to graph in DOT mode.

$$f(x) = 4 \;(x \leq 0) \; + \; (4 - x^2) \;(x > 0)(x \leq 2)\; + \;(2x - 6)\;(x > 2)$$

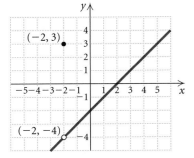

EXAMPLE 6 Make a hand-drawn graph of the function defined as

$$f(x) = \begin{cases} \dfrac{x^2 - 4}{x + 2}, & \text{for } x \neq -2, \\ 3, & \text{for } x = -2. \end{cases}$$

Solution When $x \neq -2$, the denominator of $(x^2 - 4)/(x + 2)$ is nonzero, so we can simplify:

$$\frac{x^2 - 4}{x + 2} = \frac{(x + 2)(x - 2)}{x + 2} = x - 2.$$

Thus,

$$f(x) = x - 2, \quad \text{for } x \neq -2.$$

The graph of this part of the function consists of a line with a "hole" at the point $(-2, -4)$, indicated by the open dot. At $x = -2$, we have $f(-2) = 3$, so the point $(-2, 3)$ is plotted above the open dot.

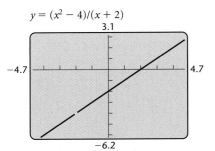

$y = (x^2 - 4)/(x + 2)$

The hand-drawn graph of $f(x) = (x^2 - 4)/(x + 2)$ for $x \neq -2$, in Example 6, can be checked using a grapher. When $y = (x^2 - 4)/(x + 2)$ is graphed, the hole may or may not be visible, depending on the window dimensions chosen. First, graph using the dimensions of your choosing. Then use the ZDECIMAL feature from the ZOOM menu. Note the hole will appear, as shown in the graph at left. (See the *Graphing Calculator Manual* that accompanies this text for further details on selecting window dimensions.) If we use the TABLE feature set in ASK mode, as shown in the table at left, the ERROR message indicates that -2 is *not* in the domain of the function g given by $g(x) = (x^2 - 4)/(x + 2)$. However, -2 *is* in the domain of the piecewise function f because $f(-2)$ is defined to be 3.

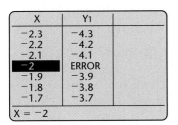

A function with importance in calculus and computer programming is the *greatest integer function*, f, denoted as $f(x) = [\![x]\!]$, or $\text{int}(x)$.

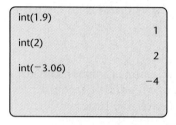

> **Greatest Integer Function**
>
> $f(x) = [\![x]\!] = $ the greatest integer less than or equal to x.

The greatest integer function pairs each input with the greatest integer less than or equal to that input. Thus, 1, 1.2, and 1.9 are all paired with the y-value 1. Similarly, 2, 2.3, and 2.8 are all paired with 2, and -3.4, -3.06, and -3.99027 are all paired with -4. These values can be checked with a grapher using the int(feature from the NUM submenu in the MATH menu.

EXAMPLE 7   Graph $f(x) = [\![x]\!]$ and determine its domain and range.

Solution   The greatest integer function can also be defined by a piece-wise function with an infinite number of statements. When plotting points by hand, it can be helpful to use the TABLE feature on a grapher.

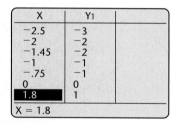

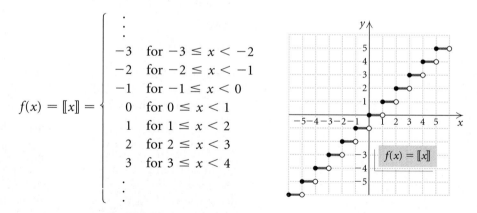

$$f(x) = [\![x]\!] = \begin{cases} \vdots & \\ -3 & \text{for } -3 \le x < -2 \\ -2 & \text{for } -2 \le x < -1 \\ -1 & \text{for } -1 \le x < 0 \\ 0 & \text{for } 0 \le x < 1 \\ 1 & \text{for } 1 \le x < 2 \\ 2 & \text{for } 2 \le x < 3 \\ 3 & \text{for } 3 \le x < 4 \\ \vdots & \end{cases}$$

We see that the domain of $f(x)$ is the set of all real numbers, $(-\infty, \infty)$, and the range is the set of all integers, $\{\ldots, -3, -2, -1, 0, 1, 2, 3, \ldots\}$.

If we used a grapher set in CONNECTED mode for Example 7, we would see the graph on the left below. CONNECTED mode connects points (or rectangles called **pixels** on a grapher) with line segments. We would see the graph on the right if we used DOT mode. The DOT mode is preferable, though even it does not show the open dots at the endpoints of the segments.

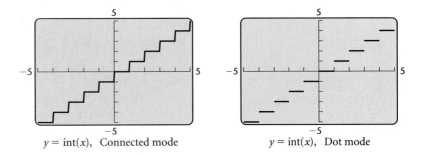

$y = \text{int}(x)$, Connected mode $y = \text{int}(x)$, Dot mode

Thus we must be careful when analyzing graphs of functions on graphers. The use of technology must be combined with an understanding of the basic concepts of functions. This is why we balance technology with a discussion of hand-drawn graphs.

The Algebra of Functions: Sums, Differences, Products, and Quotients

We now use addition, subtraction, multiplication, and division to combine functions and obtain new functions.

Consider the following two functions f and g:

$$f(x) = x + 2 \quad \text{and} \quad g(x) = x^2 + 1.$$

Since $f(3) = 3 + 2 = 5$ and $g(3) = 3^2 + 1 = 10$, we have

$$f(3) + g(3) = 5 + 10 = 15,$$
$$f(3) - g(3) = 5 - 10 = -5,$$
$$f(3) \cdot g(3) = 5 \cdot 10 = 50, \quad \text{and} \quad \frac{f(3)}{g(3)} = \frac{5}{10} = \frac{1}{2}.$$

In fact, so long as x is in the domain of both f and g, we can easily compute $f(x) + g(x)$, $f(x) - g(x)$, $f(x) \cdot g(x)$, and, assuming $g(x) \neq 0$, $f(x)/g(x)$. Notation has been developed to facilitate this work.

Sums, Differences, Products, and Quotients of Functions

If f and g are functions and x is in the domain of each function, then

$(f + g)(x) = f(x) + g(x), \qquad (f - g)(x) = f(x) - g(x),$

$(fg)(x) = f(x) \cdot g(x), \qquad (f/g)(x) = f(x)/g(x),$

provided $g(x) \neq 0.$

EXAMPLE 8   Given that $f(x) = x + 1$ and $g(x) = \sqrt{x + 3}$, find each of the following.

a) $(f + g)(x)$ **b)** $(f + g)(6)$ **c)** $(f + g)(-4)$

Solution

a) $(f + g)(x) = f(x) + g(x)$
$$= x + 1 + \sqrt{x + 3} \qquad \text{This cannot be simplified.}$$

b) We can find $(f + g)(6)$ provided 6 is in the domain of each function. The domain of f is all real numbers. The domain of g is all real numbers x for which $x + 3 \geq 0$, or $x \geq -3$. This is the interval $[-3, \infty)$. Thus, 6 is in both domains, so we have

$$f(6) = 6 + 1 = 7, \qquad g(6) = \sqrt{6 + 3} = \sqrt{9} = 3,$$
$$(f + g)(6) = f(6) + g(6) = 7 + 3 = 10.$$

Another method is to use the formula found in part (a):

$$(f + g)(6) = 6 + 1 + \sqrt{6 + 3} = 7 + \sqrt{9} = 7 + 3 = 10.$$

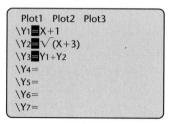

We can check our work using a grapher by entering $y_1 = x + 1$, $y_2 = \sqrt{x + 3}$, and $y_3 = y_1 + y_2$ in the $y =$ screen. Then on the home screen, we find Y3(6).

c) To find $(f + g)(-4)$, we must first determine whether -4 is in the domain of each function. We note that -4 is not in the domain of g, $[-3, \infty)$. That is, $\sqrt{-4 + 3}$ is not a real number. Thus, $(f + g)(-4)$ does not exist.

It is useful to view the concept of the sum of two functions graphically. In the graph below, we see the graphs of two functions f and g and their sum, $f + g$. Consider finding $(f + g)(4) = f(4) + g(4)$. We can locate $g(4)$ on the graph of g and use a compass to measure it. Then we move that length on top of $f(4)$ and add. The sum gives us $(f + g)(4)$.

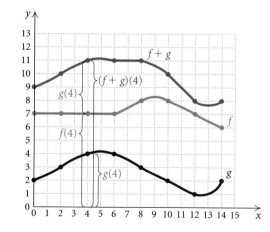

With this in mind, let's view Example 8 from a graphical perspec-

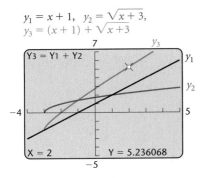

$y_1 = x + 1$, $y_2 = \sqrt{x + 3}$,
$y_3 = (x + 1) + \sqrt{x + 3}$

tive. Let's use a grapher to graph

$$y_1 = x + 1, \qquad y_2 = \sqrt{x + 3}, \quad \text{and} \quad y_3 = x + 1 + \sqrt{x + 3}.$$

See the graph at left. Note that because the domain of y_2 is $[-3, \infty)$, the graph of the sum exists only on $[-3, \infty)$. By using the VALUE feature and jumping the cursor vertically among the three graphs, we can confirm that the y-coordinates of the graph of $(f + g)(x)$ are the sums of the corresponding y-coordinates of the graphs of $f(x)$ and $g(x)$.

EXAMPLE 9   Given that $f(x) = x^2 - 4$ and $g(x) = x + 2$, find each of the following.

a) $(f + g)(x)$ b) $(f - g)(x)$
c) $(fg)(x)$ d) $(f/g)(x)$
e) $(gg)(x)$
f) The domain of $f + g$, $f - g$, fg, and f/g

Solution

a) $(f + g)(x) = f(x) + g(x) = (x^2 - 4) + (x + 2) = x^2 + x - 2$
b) $(f - g)(x) = f(x) - g(x) = (x^2 - 4) - (x + 2) = x^2 - x - 6$
c) $(fg)(x) = f(x) \cdot g(x) = (x^2 - 4)(x + 2) = x^3 + 2x^2 - 4x - 8$

d) $(f/g)(x) = \dfrac{f(x)}{g(x)}$

$\qquad = \dfrac{x^2 - 4}{x + 2}$ Note that $(f/g)(x)$ is undefined when $x = -2$.

$\qquad = \dfrac{(x + 2)(x - 2)}{x + 2}$ Factoring

$\qquad = x - 2$ Removing a factor of 1: $\dfrac{x + 2}{x + 2} = 1$

Thus, $(f/g)(x) = x - 2$ with the added stipulation that $x \neq -2$.

e) $(gg)(x) = g(x) \cdot g(x) = [g(x)]^2 = (x + 2)^2 = x^2 + 4x + 4$

f) The domain of f is the set of all real numbers. The domain of g is also the set of all real numbers. The domain of $f + g$, $f - g$, and fg is the set of numbers in the intersection of the domains—that is, the set of numbers in both domains, which is again the set of real numbers. For f/g, we must exclude -2, since $g(-2) = 0$. Thus the domain of f/g is the set of real numbers excluding -2, or $(-\infty, -2) \cup (-2, \infty)$. ▬

Difference Quotients

It is important for students preparing for calculus to be able to simplify rational expressions like

$$\frac{f(x + h) - f(x)}{h}.$$

In calculus, this process will be referred to as finding the **difference quotient.**

EXAMPLE 10   For the function f given by $f(x) = 2x^2 - x - 3$, find the difference quotient

$$\frac{f(x + h) - f(x)}{h}.$$

SOLUTION   We first find $f(x + h)$:

$$
\begin{aligned}
f(x + h) &= 2(x + h)^2 - (x + h) - 3 \\
&= 2[x^2 + 2xh + h^2] - x - h - 3 \\
&= 2x^2 + 4xh + 2h^2 - x - h - 3.
\end{aligned}
$$

Then

$$
\begin{aligned}
\frac{f(x + h) - f(x)}{h} \\
&= \frac{[2x^2 + 4xh + 2h^2 - x - h - 3] - [2x^2 - x - 3]}{h} \\
&= \frac{2x^2 + 4xh + 2h^2 - x - h - 3 - 2x^2 + x + 3}{h} \\
&= \frac{4xh + 2h^2 - h}{h} \\
&= \frac{h(4x + 2h - 1)}{h \cdot 1} = \frac{h}{h} \cdot \frac{4x + 2h - 1}{1} = 4x + 2h - 1.
\end{aligned}
$$

Exercise Set 1.4

Determine intervals on which the function is (a) increasing, (b) decreasing, and (c) constant.

1.

2.

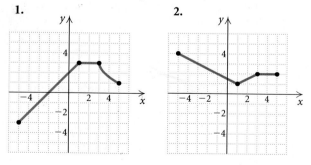

3.

4.

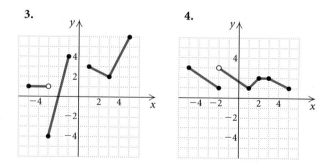

Graph the function using the given viewing window. Find the intervals on which the function is increasing or decreasing and find any relative maxima or minima. Change viewing windows if it seems appropriate for further analysis.

5. $f(x) = x^2$,
   $[-4, 4, -1, 10]$

6. $f(x) = 4 - x^2$,
   $[-5, 5, -5, 5]$

7. $f(x) = 5 - |x|$,
$[-10, 10, -10, 10]$

8. $f(x) = |x + 3| - 5$,
$[-10, 10, -10, 10]$

9. $f(x) = x^2 - 6x + 10$,
$[-4, 10, -1, 20]$

10. $f(x) = -x^2 - 8x - 9$,
$[-10, 2, -10, 10]$

11. $f(x) = -x^3 + 6x^2 - 9x - 4$,
$[-3, 7, -20, 15]$

12. $f(x) = 0.2x^3 - 0.2x^2 - 5x - 4$,
$[-10, 10, -30, 20]$

13. $f(x) = 1.1x^4 - 5.3x^2 + 4.07$,
$[-4, 4, -4, 8]$

14. $f(x) = 1.2(x + 3)^4 + 10.3(x + 3)^2 + 9.78$,
$[-9, 3, -40, 100]$

15. *Temperature During an Illness.* The temperature of a patient during an illness is given by the function

$$T(t) = -0.1t^2 + 1.2t + 98.6, \quad 0 \le t \le 12,$$

where T is the temperature, in degrees Fahrenheit, at time t, in days, after the onset of the illness.

a) Graph the function using a grapher.
b) At what time was the patient's temperature the highest? What was the highest temperature?

16. *Advertising Effect.* A software firm estimates that it will sell N units of a new CD-ROM video game after spending a dollars on advertising, where

$$N(a) = -a^2 + 300a + 6, \quad 0 \le a \le 300,$$

and a is measured in thousands of dollars.

a) Graph the function using a grapher.
b) Find the relative maximum.
c) For what advertising expenditure will the greatest number of games be sold? How many games will be sold for that amount?

Use a grapher. Find the intervals on which each function is increasing or decreasing. Consider the entire set of real numbers if no domain is given.

17. $f(x) = \dfrac{8x}{x^2 + 1}$ **18.** $f(x) = \dfrac{-4}{x^2 + 1}$

19. $f(x) = x\sqrt{4 - x^2}$, for $-2 \le x \le 2$
20. $f(x) = -0.8x\sqrt{9 - x^2}$, for $-3 \le x \le 3$

21. *Rising Balloon.* A hot-air balloon rises straight up from the ground at a rate of 120 ft/min. The balloon is tracked from a rangefinder on the ground at point P, which is 400 ft from the release point Q of the balloon. Let $d =$ the distance from the balloon to the rangefinder at point P and $t =$ the

time, in minutes, since the balloon was released. Express d as a function of t.

22. *Airplane Distance.* An airplane is flying at an altitude of 3700 ft. The slanted distance directly to the airport is d feet. Express the horizontal distance h as a function of d.

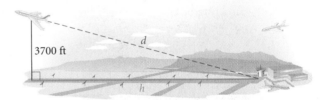

23. *Garden Area.* Yardbird Landscaping has 48 m of fencing with which to enclose a rectangular garden. If the garden is x meters long, express the garden's area as a function of the length.

24. *Triangular Flag.* A scout troop is designing a triangular flag so that the length of its base, in inches, is 7 less than twice the height, h. Express the area of the flag as a function of the height.

25. *Inscribed Rhombus.* A rhombus is inscribed in a rectangle that is w meters wide with a perimeter of 40 m. Each vertex of the rhombus is a midpoint of a side of the rectangle. Express the area of the rhombus as a function of the rectangle's width. (*Hint*: Consider the area of the rectangle.)

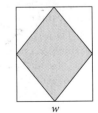

26. *Tablecloth Area.*   A tailor uses 16 ft of lace to trim the edges of a rectangular tablecloth. If the tablecloth is *w* feet wide, express its area as a function of the width.

27. *Golf Distance Finder.*   A device used in golf to estimate the distance *d*, in yards, to a hole measures the size *s*, in inches, that the 7-ft pin appears to be in a viewfinder. Express the distance *d* as a function of *s*.

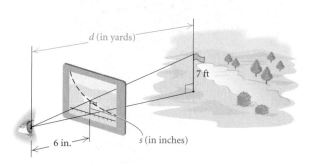

28. *Gas Tank Volume.*   A gas tank has ends that are hemispheres of radius *r* feet. The cylindrical midsection is 6 ft long. Express the volume of the tank as a function of *r*.

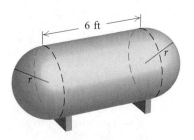

29. *Play Space.*   A daycare center has 30 ft of dividers with which to enclose a rectangular play space in a corner of a large room. The sides against the wall require no partition. Suppose the play space is *x* feet long.

a) Express the area of the play space as a function of *x*.
b) Find the domain of the function.
c) Graph the function.
d) What dimensions yield the maximum area?

30. *Corral Design.*   A rancher has 360 yd of fencing with which to enclose two adjacent rectangular corrals, one for horses and one for cattle. A river forms one side of the corrals. Suppose the width of each corral is *x* yards.

a) Express the total area of the two corrals as a function of *x*.
b) Find the domain of the function.
c) Graph the function.
d) What dimensions yield the maximum area?

31. *Volume of a Box.*   From a 12-cm by 12-cm piece of cardboard, square corners are cut out so that the sides can be folded up to make a box.

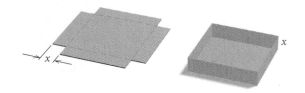

a) Express the volume of the box as a function of the length *x*, in centimeters, of a cut-out square.
b) Find the domain of the function.
c) Graph the function.
d) What dimensions yield the maximum volume?

32. *Molding Plastics.*   Perfect Plastics plans to produce a one-component vertical file by bending the long side of an 8-in. by 14-in. sheet of plastic along two lines to form a U shape.

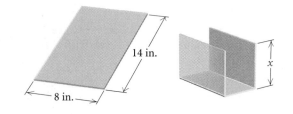

a) Express the volume of the file as a function of the height *x*, in inches, of the file.
b) Find the domain of the function.
c) Graph the function.
d) How tall should the file be in order to maximize the volume that the file can hold?

33. *Area of an Inscribed Rectangle.* A rectangle that is x feet wide is inscribed in a circle of radius 8 ft.

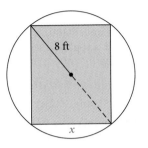

8 ft

x

a) Express the area of the rectangle as a function of x.
b) Find the domain of the function.
c) Graph the function.
d) What dimensions maximize the area of the rectangle?

34. *Cost of Material.* A rectangular box with volume 320 ft³ is built with a square base and top. The cost is \$1.50/ft² for the bottom, \$2.50/ft² for the sides, and \$1/ft² for the top. Let x = the length of the base, in feet.

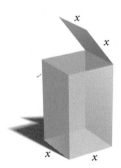

x

x

x x

a) Express the cost of the box as a function of x.
b) Find the domain of the function.
c) Graph the function.
d) What dimensions minimize the cost of the box?

Make a hand-drawn graph of each of the following. Check your results using a grapher.

35. $f(x) = \begin{cases} \frac{1}{2}x, & \text{for } x < 0, \\ x + 3, & \text{for } x \geq 0 \end{cases}$

36. $f(x) = \begin{cases} -\frac{1}{3}x + 2, & \text{for } x \leq 0, \\ x - 5, & \text{for } x > 0 \end{cases}$

37. $f(x) = \begin{cases} -\frac{3}{4}x + 2, & \text{for } x < 4, \\ -1, & \text{for } x \geq 4 \end{cases}$

38. $f(x) = \begin{cases} 4, & \text{for } x \leq -2, \\ x + 1, & \text{for } -2 < x < 3, \\ -x, & \text{for } x \geq 3 \end{cases}$

39. $f(x) = \begin{cases} x + 1, & \text{for } x \leq -3, \\ -1, & \text{for } -3 < x < 4, \\ \frac{1}{2}x, & \text{for } x \geq 4 \end{cases}$

40. $f(x) = \begin{cases} \dfrac{x^2 - 9}{x + 3}, & \text{for } x \neq -3, \\ 5, & \text{for } x = -3 \end{cases}$

41. $f(x) = \begin{cases} 2, & \text{for } x = 5, \\ \dfrac{x^2 - 25}{x - 5}, & \text{for } x \neq 5 \end{cases}$

42. $f(x) = \begin{cases} \dfrac{x^2 + 3x + 2}{x + 1}, & \text{for } x \neq -1, \\ 7, & \text{for } x = -1 \end{cases}$

43. $f(x) = [\![x]\!]$
44. $f(x) = 2[\![x]\!]$
45. $g(x) = 1 + [\![x]\!]$
46. $h(x) = \frac{1}{2}[\![x]\!] - 2$

Graph using a grapher.

47. $f(x) = \begin{cases} \sqrt[3]{x}, & \text{for } x \leq -1, \\ x^2 - 3x, & \text{for } -1 < x < 4, \\ \sqrt{x - 4}, & \text{for } x \geq 4 \end{cases}$

48. $f(x) = \begin{cases} \left|\dfrac{x}{3} + 2\right|, & \text{for } x < 5, \\ \sqrt[3]{x - 8}, & \text{for } x \geq 5 \end{cases}$

Given that $f(x) = x^2 - 3$ and $g(x) = 2x + 1$, find each of the following, if it exists.

49. $(f + g)(5)$
50. $(f - g)(3)$
51. $(f - g)(-1)$
52. $(fg)(0)$
53. $(fg)(2)$
54. $(fg)(-2)$
55. $(f/g)(-1)$
56. $(f/g)\left(-\frac{1}{2}\right)$
57. $(fg)\left(-\frac{1}{2}\right)$
58. $(f/g)(-\sqrt{3})$
59. $(f/g)(\sqrt{3})$
60. $(f - g)(0)$

For each pair of functions in Exercises 61–64:

a) Find the domain of f, g, $f + g$, $f - g$, fg, ff, f/g, and g/f.

b) Find $(f + g)(x)$, $(f - g)(x)$, $(fg)(x)$, $(ff)(x)$, $(f/g)(x)$, and $(g/f)(x)$.

61. $f(x) = x - 3$, $g(x) = \sqrt{x + 4}$

62. $f(x) = x^2 - 1$, $g(x) = 2x + 5$

63. $f(x) = x^3$, $g(x) = 2x^2 + 5x - 3$

64. $f(x) = x^2$, $g(x) = \sqrt{x}$

For each pair of functions in Exercises 65–68, find
$(f + g)(x)$, $(f - g)(x)$, $(fg)(x)$, and $(f/g)(x)$.

65. $f(x) = x^2 - 4$, $g(x) = x^2 + 2$

66. $f(x) = x^2 + 2$, $g(x) = 4x - 7$

67. $f(x) = \sqrt{x - 7}$, $g(x) = x^2 - 25$

68. $f(x) = \sqrt{x - 1}$, $g(x) = \sqrt{3x - 8}$

In Exercises 69–74, consider the functions F and G as shown in the following graph.

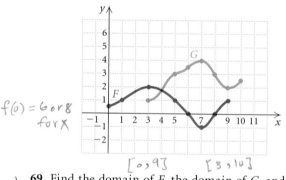

$f(0) = 6 \text{ or } 8$
$\text{for } X$

$[0, 9]$ $[3, 10]$

69. Find the domain of F, the domain of G, and the domain of $F + G$.

70. Find the domain of $F - G$, FG, and F/G.

71. Find the domain of G/F. $[3, 6) \cup (6, 8) \cup (8, 9]$

72. Graph $F + G$.

73. Graph $G - F$.

74. Graph $F - G$.

75. *Total Cost, Revenue, and Profit.* In economics, functions that involve revenue, cost, and profit are used. For example, suppose that $R(x)$ and $C(x)$ denote the total revenue and the total cost, respectively, of producing a new kind of tool for King Hardware Stores. Then the difference

$$P(x) = R(x) - C(x)$$

represents the total profit for producing x tools.

Given

$$R(x) = 60x - 0.4x^2 \quad \text{and} \quad C(x) = 3x + 13,$$

find each of the following.

a) $P(x)$

b) $R(100)$, $C(100)$, and $P(100)$

c) Using a grapher, graph the three functions in the viewing window $[0, 160, 0, 3000]$.

76. *Total Cost, Revenue, and Profit.* Given that

$$R(x) = 200x - x^2 \quad \text{and} \quad C(x) = 5000 + 8x,$$

for a new radio produced by Clear Communication, find each of the following. (See Exercise 75.)

a) $P(x)$

b) $R(175)$, $C(175)$, and $P(175)$

c) Using a grapher, graph the three functions in the viewing window $[0, 200, 0, 10,000]$.

For each function f, construct and simplify the difference quotient

$$\frac{f(x + h) - f(x)}{h}.$$

77. $f(x) = 3x^2 - 2x + 1$

78. $f(x) = 5x^2 + 4x$

79. $f(x) = x^3$

80. $f(x) = 2|x| + 3x$

81. $f(x) = \dfrac{x - 4}{x + 3}$

82. $f(x) = \dfrac{x}{2 - x}$

Discussion and Writing

83. Describe a real-world situation that could be modeled by a function that is, in turn, increasing, then constant, and finally decreasing.

84. Simply stated, a *continuous function* is a function whose graph can be drawn without lifting the pencil from the paper. Examine several functions in this exercise set to see if they are continuous. Then explore the continuous functions to find the relative maxima and minima. For continuous functions, how can you connect the ideas of increasing and decreasing on an interval to relative maxima and minima?

Skill Maintenance

Find the domain of the function.

85. $f(x) = -2x - 3$

86. $g(x) = \dfrac{1}{(x - 3)^2}$

87. $h(x) = \dfrac{x}{|x + 5|}$

88. $f(x) = -5$

Synthesis

Using a grapher, estimate the interval on which each function is increasing or decreasing and any relative maxima or minima.

89. $f(x) = x^4 + 4x^3 - 36x^2 - 160x + 400$

90. $f(x) = 3.22x^5 - 5.208x^3 - 11$

91. *Parking Costs.* A parking garage charges $2 for up to (but not including) 1 hr of parking, $4 for up to 2 hr of parking, $6 for up to 3 hr of parking, and so on. Let $C(t)$ = the cost of parking for t hours.

a) Graph the function.

b) Write an equation for $C(t)$ using the greatest integer notation $[\![t]\!]$.

92. If $[\![x + 2]\!] = -3$, what are the possible inputs for x?

93. If $([\![x]\!])^2 = 25$, what are the possible inputs for x?

94. *Minimizing Power Line Costs.* A power line is constructed from a power station at point A to an island at point C, which is 1 mi directly out in the water from a point B on the shore. Point B is 4 mi downshore from the power station at A. It costs

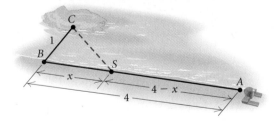

$5000 per mile to lay the power line under water and $3000 per mile to lay the power line under ground. The line comes to the shore at point S downshore from A. Let x = the distance from B to S.

a) Express the cost C of laying the line as a function of x.

b) At what distance x should the line come to shore in order to minimize cost?

95. *Volume of an Inscribed Cylinder.* A right circular cylinder of height h and radius r is inscribed in a right circular cone with a height of 10 ft and a base with radius 6 ft.

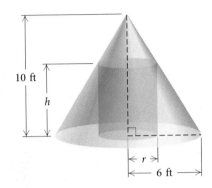

a) Express the height h of the cylinder as a function of r.

b) Express the volume V of the cylinder as a function of r.

c) Express the volume V of the cylinder as a function of h.

Symmetry and Transformations

1.5

• *Determine whether a graph is symmetric with respect to the x-axis, the y-axis, and the origin.*
• *Determine whether a function is even, odd, or neither even nor odd.*
• *Given the graph of a function, graph its transformation under translations, reflections, stretchings, and shrinkings.*

Symmetry   قرينه وتَقارن

Symmetry occurs often in nature and in art. For example, when viewed from the front, the bodies of most animals are at least approximately symmetric. This means that each eye is the same distance from the center of the bridge of the nose, each shoulder is the same distance from the center of the chest, and so on. Architects have used symmetry for thousands of years to enhance the beauty of buildings.

A knowledge of symmetry in mathematics helps us graph and analyze equations and functions.

Consider the points (4, 2) and (4, −2) that appear in the graph of $x = y^2$ (see Fig. 1 at the top of the following page). Points like these have the same x-value but opposite y-values and are **reflections** of each other across the x-axis. If for any point (x, y) on a graph the point $(x, -y)$ is also on the graph, then the graph is said to be **symmetric with respect to the x-axis.** If we fold the graph on the x-axis, the parts above and below the x-axis will coincide. منطبق شدن

متقارن

Consider the points (3, 4) and (−3, 4) that appear in the graph of $y = x^2 - 5$ (see Fig. 2 at the top of the following page). Points like these have the same y-value but opposite x-values and are **reflections** of each other across the y-axis. If for any point (x, y) on a graph the point $(-x, y)$ is also on the graph, then the graph is said to be **symmetric**

with respect to the *y*-axis. If we fold the graph on the *y*-axis, the parts to the left and right of the *y*-axis will coincide.

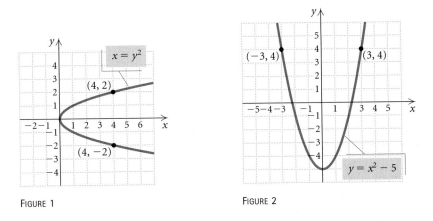

FIGURE 1   FIGURE 2

Consider the points $(3, \sqrt{7})$ and $(-3, -\sqrt{7})$ that appear in the graph of $x^2 = y^2 + 2$ (see Fig. 3). Note that if we take the opposites of the coordinates of one pair, we get the other pair. If for any point (x, y) on a graph the point $(-x, -y)$ is also on the graph, then the graph is said to be **symmetric with respect to the origin.** Visually, if we rotate the graph $180°$ about the origin, the resulting figure coincides with the original.

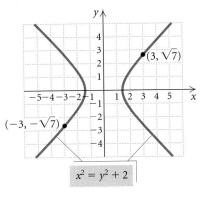

FIGURE 3

Algebraic Tests of Symmetry

x-axis: If replacing *y* with $-y$ produces an equivalent equation, then the graph is *symmetric with respect to the x-axis.*

y-axis: If replacing *x* with $-x$ produces an equivalent equation, then the graph is *symmetric with respect to the y-axis.*

Origin: If replacing *x* with $-x$ and *y* with $-y$ produces an equivalent equation, then the graph is *symmetric with respect to the origin.*

EXAMPLE 1   Test $y = x^2 + 2$ for symmetry with respect to the x-axis, the y-axis, and the origin.

Algebraic Solution

x-AXIS

We replace y with $-y$:

$$y = x^2 + 2$$
$$\downarrow$$
$$-y = x^2 + 2$$
$$y = -x^2 - 2. \qquad \text{Multiplying by } -1 \text{ on both sides}$$

The resulting equation is not equivalent to the original equation, so the graph *is not* symmetric with respect to the x-axis.

y-AXIS

We replace x with $-x$:

$$y = x^2 + 2$$
$$\searrow$$
$$y = (-x)^2 + 2$$
$$y = x^2 + 2. \qquad \text{Simplifying}$$

The resulting equation is equivalent to the original equation, so the graph *is* symmetric with respect to the y-axis.

ORIGIN

We replace x with $-x$ and y with $-y$:

$$y = x^2 + 2$$
$$\downarrow \qquad \searrow$$
$$-y = (-x)^2 + 2$$
$$-y = x^2 + 2 \qquad \text{Simplifying}$$
$$y = -x^2 - 2.$$

The resulting equation is not equivalent to the original equation, so the graph *is not* symmetric with respect to the origin.

Graphical Solution

We use a grapher to graph the equation.

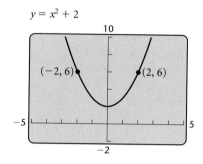

$y = x^2 + 2$

Note that if the graph were folded on the x-axis, the parts above and below the x-axis would not coincide. If it were folded on the y-axis, the parts to the left and right of the y-axis would coincide. If we rotated it 180°, the resulting graph would not coincide with the original graph.

Thus we see that the graph *is not* symmetric with respect to the x-axis or the origin. The graph *is* symmetric with respect to the y-axis.

The algebraic method is often easier to apply than the graphical, especially with equations that we may not be able to graph easily. It is also often more precise.

EXAMPLE 2 Test $x^2 + y^4 = 5$ for symmetry with respect to the x-axis, the y-axis, and the origin.

Algebraic Solution

x-AXIS

We replace y with $-y$:

$$x^2 + y^4 = 5$$

$$x^2 + (-y)^4 = 5$$

$$x^2 + y^4 = 5.$$

The resulting equation is equivalent to the original equation. Thus the graph is symmetric with respect to the x-axis.

    We leave it to the student to verify algebraically that the graph is also symmetric with respect to the y-axis and the origin.

Graphical Solution

To graph $x^2 + y^4 = 5$ using a grapher, we first solve the equation for y:

$$y = \pm\sqrt[4]{5 - x^2}.$$

Then we enter on the $y=$ screen the equations

$$y_1 = \sqrt[4]{5 - x^2} \quad \text{and} \quad y_2 = -\sqrt[4]{5 - x^2}.$$

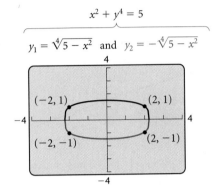

$$x^2 + y^4 = 5$$

$$y_1 = \sqrt[4]{5 - x^2} \quad \text{and} \quad y_2 = -\sqrt[4]{5 - x^2}$$

We see symmetry with respect to both axes and with respect to the origin.

$$\sqrt[3]{5-x^2} \; , \; -\sqrt[3]{5-x^2}$$

Even and Odd Functions

Now we relate symmetry to graphs of functions.

n and $-n$ *[-0;*

Even and Odd Functions

If the graph of a function f is symmetric with respect to the y-axis, we say that it is an **even function.** That is, for each x in the domain of f, $f(x) = f(-x)$.

If the graph of a function f is symmetric with respect to the origin, we say that it is an **odd function.** That is, for each x in the domain of f, $f(-x) = -f(x)$.

Algebraic Procedure for Determining Even and Odd Functions
Given the function $f(x)$:

1. Find $f(-x)$ and simplify. If $f(x) = f(-x)$, then f is even.
2. Find $-f(x)$, simplify, and compare with $f(-x)$ from step (1).
   If $f(-x) = -f(x)$, then f is odd.

Except for the function $f(x) = 0$, a function cannot be *both* even and odd. Thus if we see in step (1) that $f(x) = f(-x)$ (that is, f is even), we need not continue.

EXAMPLE 3 Determine whether each of the following functions is even, odd, or neither.

a) $f(x) = 5x^7 - 6x^3 - 2x$ **b)** $h(x) = 5x^6 - 3x^2 - 7$

a)

Algebraic Solution

$f(x) = 5x^7 - 6x^3 - 2x$

1. $f(-x) = 5(-x)^7 - 6(-x)^3 - 2(-x)$
$\qquad = 5(-x^7) - 6(-x^3) + 2x$
$\qquad\qquad\qquad (-x)^7 = (-1 \cdot x)^7 = (-1)^7 x^7 = -x^7$
$\qquad = -5x^7 + 6x^3 + 2x$

We see that $f(x) \neq f(-x)$. Thus, f is *not* even.

2. $-f(x) = -(5x^7 - 6x^3 - 2x)$
$\qquad = -5x^7 + 6x^3 + 2x$

We see that $f(-x) = -f(x)$. Thus, f is odd.

Graphical Solution

$y = 5x^7 - 6x^3 - 2x$

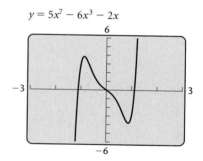

We see that the graph appears to be symmetric with respect to the origin. The function is odd.

b)

Algebraic Solution

$h(x) = 5x^6 - 3x^2 - 7$

1. $h(-x) = 5(-x)^6 - 3(-x)^2 - 7$
$\qquad = 5x^6 - 3x^2 - 7$

We see that $h(x) = h(-x)$. Thus the function is even.

Graphical Solution

$y = 5x^6 - 3x^2 - 7$

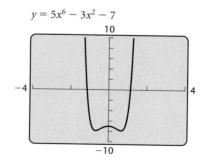

We see that the graph appears to be symmetric with respect to the y-axis. The function is even.

Transformations of Functions

The graphs of some basic functions are shown below. Others can be seen on the inside back cover.

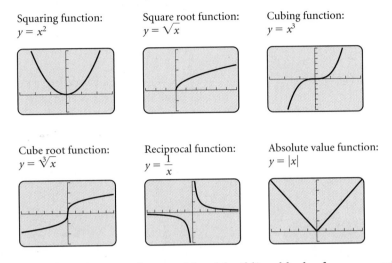

Squaring function:
$y = x^2$

Square root function:
$y = \sqrt{x}$

Cubing function:
$y = x^3$

Cube root function:
$y = \sqrt[3]{x}$

Reciprocal function:
$y = \dfrac{1}{x}$

Absolute value function:
$y = |x|$

These functions can be considered building blocks for many other functions. We can create graphs of new functions by shifting them horizontally or vertically, stretching or shrinking them, and reflecting them across an axis. We now consider these *transformations*.

Vertical and Horizontal Translations

Suppose that we have a function given by $y_1 = f(x)$. Let's explore the graphs of new functions $y_2 = f(x) + b$ and $y_3 = f(x) - b$, for $b > 0$.

Interactive Discovery

Consider the function $y_1 = \frac{1}{5}x^4$. Graph it and the functions $y_2 = \frac{1}{5}x^4 + 5$ and $y_3 = \frac{1}{5}x^4 - 3$ using different graph styles and the same viewing window, $[-5, 5, -8, 10]$. What pattern do you see? Test it with some other graphs.

The effect of adding a constant to or subtracting a constant from $f(x)$ in $y = f(x)$ is a shift of $f(x)$ up or down. Such a shift is called a **vertical translation.**

Vertical Translation

For $b > 0$,

the graph of $y = f(x) + b$ is the graph of $y = f(x)$ shifted *up* b units;

the graph of $y = f(x) - b$ is the graph of $y = f(x)$ shifted *down* b units.

Suppose that we have a function given by $y_1 = f(x)$. Let's explore the graphs of the new functions $y_2 = f(x + d)$ and $y_3 = f(x - d)$, for $d > 0$.

Interactive Discovery

Consider the function $y_1 = \frac{1}{5}x^4$. Graph it and the functions $y_2 = \frac{1}{5}(x - 3)^4$ and $y_3 = \frac{1}{5}(x + 7)^4$ using the viewing window $[-10, 10, -2, 10]$. What pattern do you see? Test it with some other graphs.

The effect of adding a constant to the x-value or subtracting a constant from the x-value in $y = f(x)$ is a shift of $f(x)$ to the right or left. Such a shift is called a **horizontal translation**.

Horizontal Translation

For $d > 0$:

the graph of $y = f(x - d)$ is the graph of $y = f(x)$ shifted *right* d units;

the graph of $y = f(x + d)$ is the graph of $y = f(x)$ shifted *left* d units.

EXAMPLE 4 Graph each of the following. Before doing so, describe how each graph can be obtained from one of the basic graphs shown on the preceding page.

a) $g(x) = x^2 - 6$

b) $g(x) = \sqrt{x + 2}$

c) $g(x) = |x - 4|$

d) $h(x) = \sqrt{x + 2} - 3$

Solution

a) To graph $g(x) = x^2 - 6$, think of the graph of $f(x) = x^2$. Since $g(x) = f(x) - 6$, the graph of $g(x) = x^2 - 6$ is the graph of $f(x) = x^2$, shifted, or translated, *down* 6 units (see Fig. 1).

b) To graph $g(x) = \sqrt{x + 2}$, think of the graph of $f(x) = \sqrt{x}$. Since $g(x) = f(x + 2)$, the graph of $g(x) = \sqrt{x + 2}$ is the graph of $f(x) = \sqrt{x}$, shifted *left* 2 units (see Fig. 2).

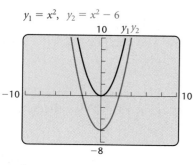

$y_1 = x^2,\ \ y_2 = x^2 - 6$

FIGURE 1

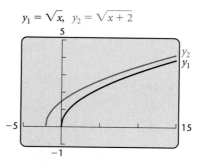

$y_1 = \sqrt{x},\ \ y_2 = \sqrt{x + 2}$

FIGURE 2

c) To graph $g(x) = |x - 4|$, think of the graph of $f(x) = |x|$. Since $g(x) = f(x - 4)$, the graph of $g(x) = |x - 4|$ is the graph of $f(x) = |x|$ shifted *right* 4 units (see Fig. 3).

$$y_1 = |x|, \quad y_2 = |x - 4|$$

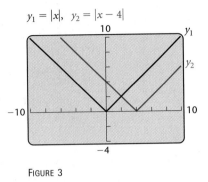

FIGURE 3

d) To graph $h(x) = \sqrt{x + 2} - 3$, think of the graph of $f(x) = \sqrt{x}$. In part (b), we found that the graph of $g(x) = \sqrt{x + 2}$ is the graph of $f(x) = \sqrt{x}$ shifted <u>left 2</u> units. Since $h(x) = g(x) - 3$, we shift the graph of $g(x) = \sqrt{x + 2}$ *down* 3 units. Together, the graph of $f(x) = \sqrt{x}$ is shifted *left* 2 units and *down* 3 units (see Fig. 4).

$$y_1 = \sqrt{x}, \quad y_2 = \sqrt{x + 2}, \quad y_3 = \sqrt{x + 2} - 3$$

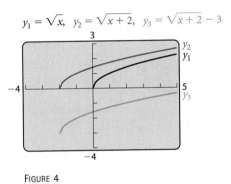

FIGURE 4

Reflections

Suppose that we have a function given by $y = f(x)$. Let's explore the graphs of the new functions $g(x) = -f(x)$ and $g(x) = f(-x)$.

Interactive Discovery

Consider the functions $y_1 = \frac{1}{5}x^4$ and $y_2 = -\frac{1}{5}x^4$. (Here $y_1 = f(x)$ and $y_2 = -f(x)$.) Graph y_1 using the viewing window $[-10, 10, -10, 10]$. Then graph y_2 using the same viewing window. What do you observe? Test your observation with some other functions in which $y_2 = -y_1$.

Consider the functions $y_1 = 2x^3 - x^4 + 5$ and $y_2 = 2(-x)^3 - (-x)^4 + 5$. (Here $y_1 = f(x)$ and $y_2 = f(-x)$.) Graph y_1 using the viewing window $[-4, 4, -10, 10]$. Then graph y_2 in the same viewing window. What do you observe? Test your observation with some other functions in which x is replaced with $-x$.

Given the graph of $y_1 = f(x)$, we can reflect each point across the x-axis to obtain the graph of $y_2 = -f(x)$. We can reflect each point of y_1 across the y-axis to obtain the graph of $y_2 = f(-x)$. The new graphs are called **reflections** of $f(x)$.

Reflections

The graph of $y = -f(x)$ is the **reflection** of the graph of $y = f(x)$ across the x-axis.

The graph of $y = f(-x)$ is the **reflection** of the graph of $y = f(x)$ across the y-axis.

EXAMPLE 5   Graph each of the following. Before doing so, describe how each graph can be obtained from the graph of $f(x) = x^3 - 4x^2$.

a) $g(x) = (-x)^3 - 4(-x)^2$ **b)** $h(x) = 4x^2 - x^3$

Solution

a) We first note that

$$f(-x) = (-x)^3 - 4(-x)^2 = g(x).$$

Thus the graph of g is a reflection of the graph of f across the y-axis (see Fig. 1).

b) We first note that

$$-f(x) = -(x^3 - 4x^2)$$
$$= -x^3 + 4x^2$$
$$= 4x^2 - x^3$$
$$= h(x).$$

Thus the graph of h is a reflection of the graph of f across the x-axis (see Fig. 2).

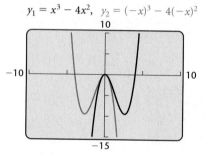

$y_1 = x^3 - 4x^2, \quad y_2 = (-x)^3 - 4(-x)^2$

FIGURE 1

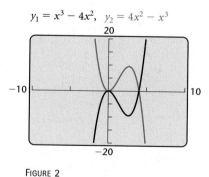

$y_1 = x^3 - 4x^2, \quad y_2 = 4x^2 - x^3$

FIGURE 2

Vertical and Horizontal Stretchings and Shrinkings

Suppose that we have a function given by $y_1 = f(x)$. Let's explore the graphs of the new functions $y_2 = af(x)$ and $y_3 = f(cx)$.

Interactive Discovery

Graph the function $y_1 = x^3 - x$ using the viewing window $[-3, 3, -1, 1]$. Then graph $y_2 = \frac{1}{10}(x^3 - x)$ using the same viewing window. Turn off y_2 and graph $y_3 = 2(x^3 - x)$. Then turn off y_3 and graph $y_4 = -2(x^3 - x)$. What pattern do you see? Test it with some other graphs.

Now graph y_1 and $y_2 = (2x)^3 - (2x)$ using the same viewing window. Turn off y_2 and graph $y_3 = \left(\frac{1}{2}x\right)^3 - \left(\frac{1}{2}x\right)$. Then turn off y_3 and graph $y_4 = \left(-\frac{1}{2}x\right)^3 - \left(-\frac{1}{2}x\right)$. What pattern do you see? Test it with some other graphs.

Consider any function f given by $y = f(x)$. Multiplying on the right by any constant a, where $|a| > 1$, to obtain $g(x) = af(x)$ will *stretch* the graph vertically away from the x-axis. If $0 < |a| < 1$, then the graph will be flattened or *shrunk* vertically toward the x-axis. If $a < 0$, the graph is also reflected across the x-axis.

Vertical Stretching and Shrinking

The graph of $y = af(x)$ can be obtained from the graph of $y = f(x)$ by

> stretching vertically for $|a| > 1$, or
>
> shrinking vertically for $0 < |a| < 1$.

For $a < 0$, the graph is also reflected across the x-axis.

The constant c in the equation $g(x) = f(cx)$ will *stretch* the graph of $y = f(x)$ horizontally away from the y-axis if $0 < |c| < 1$. If $|c| > 1$, the graph will be *shrunk* horizontally toward the y-axis. If $c < 0$, the graph is also reflected across the y-axis.

Horizontal Stretching and Shrinking

The graph of $y = f(cx)$ can be obtained from the graph of $y = f(x)$ by

> shrinking horizontally for $|c| > 1$, or
>
> stretching horizontally for $0 < |c| < 1$.

For $c < 0$, the graph is also reflected across the y-axis.

It is instructive to use these concepts now to create hand-drawn graphs from a given graph.

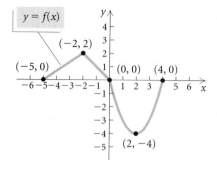

EXAMPLE 6   Shown at left is a graph of $y = f(x)$ for some function f. No formula for f is given. Make a hand-drawn graph of each of the following.

a) $g(x) = f(2x)$ **b)** $h(x) = f\left(\frac{1}{2}x\right)$ **c)** $t(x) = f\left(-\frac{1}{2}x\right)$

Solution

a) Since $|2| > 1$, the graph of $g(x) = f(2x)$ is a horizontal shrinking of the graph of $y = f(x)$. We can consider the key points $(-5, 0)$, $(-2, 2)$, $(0, 0)$, $(2, -4)$, and $(4, 0)$ on the graph of $y = f(x)$. The transformation divides each x-coordinate by 2 to obtain the key points $(-2.5, 0)$, $(-1, 2)$, $(0, 0)$, $(1, -4)$, and $(2, 0)$ of the graph of $g(x) = f(2x)$. The graph is shown below.

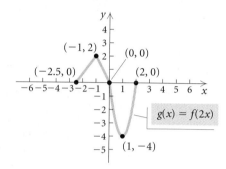

b) Since $\left|\frac{1}{2}\right| < 1$, the graph of $h(x) = f\left(\frac{1}{2}x\right)$ is a horizontal stretching of the graph of $y = f(x)$. We again consider the key points $(-5, 0)$, $(-2, 2)$, $(0, 0)$, $(2, -4)$, and $(4, 0)$ on the graph of $y = f(x)$. The transformation divides each x-coordinate by $\frac{1}{2}$ (which is the same as multiplying by 2) to obtain the key points $(-10, 0)$, $(-4, 2)$, $(0, 0)$, $(4, -4)$, and $(8, 0)$ of the graph of $h(x) = f\left(\frac{1}{2}x\right)$. The graph is shown below.

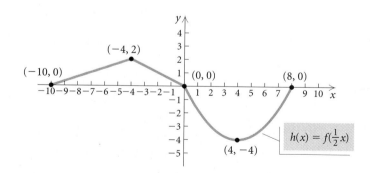

c) The graph of $t(x) = f\left(-\frac{1}{2}x\right)$ can be obtained by reflecting the graph in part (b) across the y-axis. It is shown at the top of the following page.

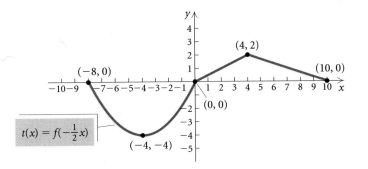

$$t(x) = f\left(-\tfrac{1}{2}x\right)$$

EXAMPLE 7 Use the graph of $y = f(x)$ given in Example 6. Make a hand-drawn graph of

$$y = -2f(x - 3) + 1.$$

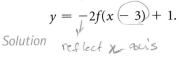

reflect x-axis

Solution

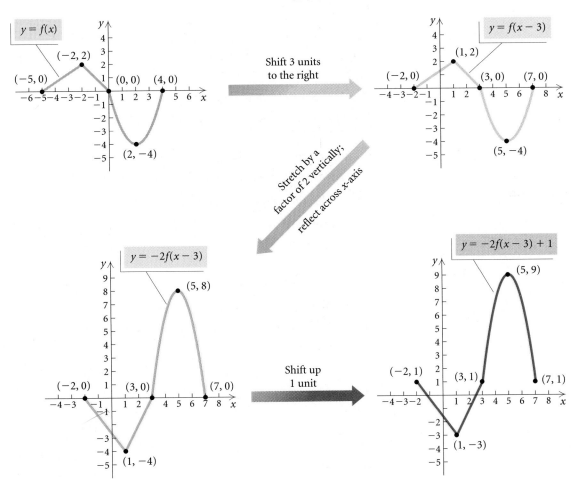

Summary of Transformations of $y = f(x)$

Vertical Translation: $y = f(x) \pm b$

For $b > 0$,

> the graph of $y = f(x) + b$ is the graph of $y = f(x)$ shifted *up b* units;
>
> the graph of $y = f(x) - b$ is the graph of $y = f(x)$ shifted *down b* units.

Horizontal Translation: $y = f(x \mp d)$

For $d > 0$,

> the graph of $y = f(x - d)$ is the graph of $y = f(x)$ shifted *right d* units;
>
> the graph of $y = f(x + d)$ is the graph of $y = f(x)$ shifted *left d* units.

Reflections

Across the x-axis: The graph of $y = -f(x)$ is the reflection of the graph of $y = f(x)$ across the x-axis.

Across the y-axis: The graph of $y = f(-x)$ is the reflection of the graph of $y = f(x)$ across the y-axis.

Vertical Stretching or Shrinking: $y = af(x)$

The graph of $y = af(x)$ can be obtained from the graph of $y = f(x)$ by

> stretching vertically for $|a| > 1$, or
>
> shrinking vertically for $0 < |a| < 1$.

For $a < 0$, the graph is also reflected across the x-axis.

Horizontal Stretching or Shrinking: $y = f(cx)$

The graph of $y = f(cx)$ can be obtained from the graph of $y = f(x)$ by

> shrinking horizontally for $|c| > 1$, or
>
> stretching horizontally for $0 < |c| < 1$.

For $c < 0$, the graph is also reflected across the y-axis.

Exercise Set 1.5

21. $2x^4 + 3 = y^2$ **22.** $2y^2 = 5x^2 + 12$

23. $3y^3 = 4x^3 + 2$ **24.** $3x = |y|$

25. $xy = 12$ **26.** $xy - x^2 = 3$

Determine visually whether the function is even, odd, or neither even nor odd.

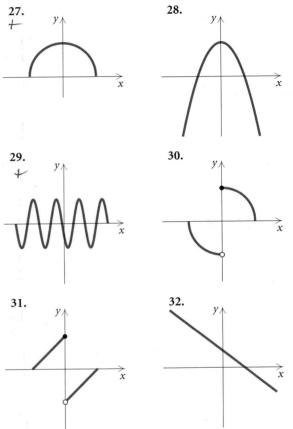

27.
28.
29.
30.
31.
32.

Determine visually whether the graph is symmetric with respect to the x-axis, the y-axis, and the origin.

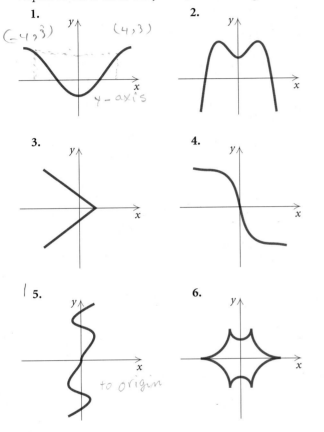

1. (-4,3)   (4,3)   y-axis
2.
3.
4.
5.   to origin
6.

First, graph the equation and determine visually whether it is symmetric with respect to the x-axis, the y-axis, and the origin. Then verify your assertion algebraically.

7. $y = |x| - 2$ **8.** $y = |x + 5|$

9. $5y = 4x + 5$ **10.** $2x - 5 = 3y$

11. $5y = 2x^2 - 3$ **12.** $x^2 + 4 = 3y$

13. $y = \dfrac{1}{x}$ **14.** $y = -\dfrac{4}{x}$

Test algebraically whether the graph is symmetric with respect to the x-axis, the y-axis, and the origin. Then check your work graphically, if possible, using a grapher.

15. $5x - 5y = 0$ **16.** $6x + 7y = 0$

17. $3x^2 - 2y^2 = 3$ **18.** $5y = 7x^2 - 2x$

19. $y = |2x|$ **20.** $y^3 = 2x^2$

Test algebraically whether the function is even, odd, or neither even nor odd. Then check your work graphically, where possible, using a grapher.

33. $f(x) = -3x^3 + 2x$ **34.** $f(x) = 7x^3 + 4x - 2$

35. $f(x) = 5x^2 + 2x^4 - 1$ **36.** $f(x) = x + \dfrac{1}{x}$

37. $f(x) = x^{17}$ **38.** $f(x) = \sqrt[3]{x}$

39. $f(x) = \dfrac{1}{x^2}$ **40.** $f(x) = x - |x|$

41. $f(x) = 8$ **42.** $f(x) = \sqrt{x^2 + 1}$

Graph each of the following using a grapher. Before doing so, describe how the graph can be obtained from one of the basic graphs at the beginning of this section.

43. $f(x) = x^2 + 1$ **44.** $g(x) = |3x|$

45. $g(x) = (x + 5)^3$ **46.** $f(x) = \frac{1}{2}\sqrt[3]{x}$

47. $f(x) = -x^2$

48. $f(x) = |x - 3| - 4$

49. $f(x) = 3\sqrt{x} - 5$

50. $f(x) = 5 - \dfrac{1}{x}$

51. $g(x) = \left|\frac{1}{3}x\right| - 4$

52. $f(x) = \frac{2}{3}x^3 - 4$

53. $f(x) = (x + 5)^2 - 4$

54. $f(x) = (-x)^3 - 5$

55. $f(x) = -\frac{1}{4}(x - 5)^2$

56. $g(x) = \sqrt{-x} + 5$

57. $f(x) = \dfrac{1}{x + 3} + 2$

58. $f(x) = 3(x + 4)^2 - 3$

Write an equation for a function that has a graph with the given characteristics. Check your answer using a grapher.

59. The shape of $y = x^2$, but upside-down and shifted right 8 units

60. The shape of $y = \sqrt{x}$, but shifted left 6 units and down 5 units

61. The shape of $y = |x|$, but shifted left 7 units and up 2 units

62. The shape of $y = x^3$, but upside-down and shifted right 5 units

63. The shape of $y = 1/x$, but shrunk vertically by a factor of $\frac{1}{2}$ and shifted down 3 units

64. The shape of $y = x^2$, but shifted right 6 units and up 2 units

65. The shape of $y = x^2$, but upside-down and shifted right 3 units and up 4 units

66. The shape of $y = |x|$, but stretched horizontally by a factor of 2 and shifted down 5 units

67. The shape of $y = \sqrt{x}$, but reflected across the y-axis and shifted left 2 units and down 1 unit

68. The shape of $y = 1/x$, but reflected across the x-axis and shifted up 1 unit

69. The shape of $y = x^3$, but shifted left 4 units and shrunk vertically by a factor of 0.83

A graph of $y = f(x)$ follows. No formula for f is given. In Exercises 70–75, make a hand-drawn graph of the equation.

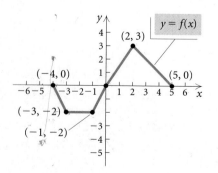

70. $g(x) = \frac{1}{2}f(x)$

71. $g(x) = -2f(x)$

72. $g(x) = f(2x)$

73. $g(x) = f\left(-\frac{1}{2}x\right)$

74. $g(x) = -3f(x + 1) - 4$

75. $g(x) = -\frac{1}{2}f(x - 1) + 3$

The graph of the function f is shown in figure (a). In Exercises 76–83, match the function g with one of the graphs (a)–(h), which follow. Some graphs may be used more than once.

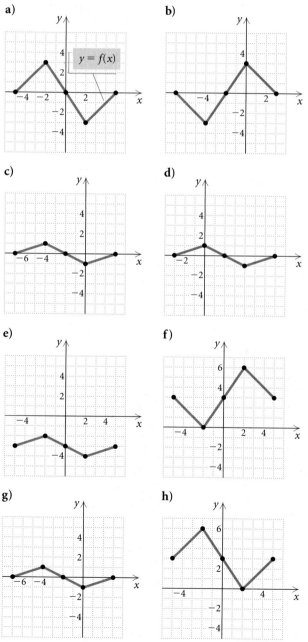

76. $g(x) = f(x) + 3$ **77.** $g(x) = f(-x) + 3$

78. $g(x) = -f(-x)$ **79.** $g(x) = -f(x) + 3$

80. $g(x) = \frac{1}{3}f(x) - 3$ **81.** $g(x) = \frac{1}{3}f(x - 2)$

82. $g(x) = -f(x + 2)$ **83.** $g(x) = \frac{1}{3}f(x + 2)$

For each pair of functions, determine algebraically if $g(x) = f(-x)$. Then, using the TABLE *feature on a grapher, check your answers.*

84. $f(x) = \frac{1}{4}x^4 + \frac{1}{5}x^3 - 81x^2 - 17$,
$g(x) = \frac{1}{4}x^4 + \frac{1}{5}x^3 + 81x^2 - 17$

85. $f(x) = 2x^4 - 35x^3 + 3x - 5$,
$g(x) = 2x^4 + 35x^3 - 3x - 5$

A graph of the function $f(x) = x^3 - 3x^2$ is shown below. Exercises 86–89 show graphs of functions transformed from this one. Find a formula for each function.

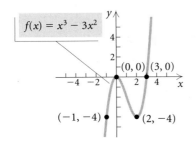

86.

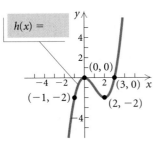

87.

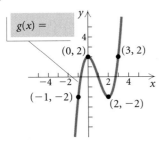

88.

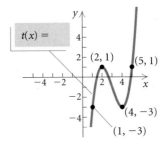

89.

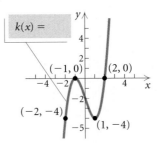

Discussion and Writing

90. Describe conditions under which you would know whether a polynomial function $f(x) = a_n x^n + a_{n-1} x^{n-1} + \cdots + a_2 x^2 + a_1 x + a_0$ is even or odd without using an algebraic or graphical procedure. Explain.

91. Consider the constant function $f(x) = 0$. Determine whether the graph of this function is symmetric with respect to the x-axis, the y-axis, and the origin. Determine whether this function is even or odd. In general, can a function be symmetric with respect to the x-axis? Explain.

92. Without drawing the graph, describe what the graph of $f(x) = |x^2 - 9|$ looks like.

93. Explain in your own words why the graph of $y = f(-x)$ is a reflection of the graph of $y = f(x)$ across the y-axis.

Skill Maintenance

94. Given $f(x) = 4x^3 - 5x$, find each of the following.
 a) $f(2)$ **b)** $f(-2)$ **c)** $f(a)$ **d)** $f(-a)$

95. Given $f(x) = 5x^2 - 7$, find each of the following.
 a) $f(-3)$ **b)** $f(3)$ **c)** $f(a)$ **d)** $f(-a)$

96. Find the slope and the y-intercept of the line with equation $2x - 9y + 1 = 0$.

97. Write an equation of the line perpendicular to the graph of the line $8x - y = 10$ and containing the point $(-1, 1)$.

Synthesis

Determine whether the function is even, odd, or neither even nor odd. Use any method.

98. $f(x) = \dfrac{x^2 + 1}{x^3 - 1}$

99. $f(x) = x\sqrt{10 - x^2}$

Determine whether the graph is symmetric with respect to the x-axis, the y-axis, and the origin. Use any method.

100.

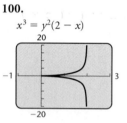

$x^3 = y^2(2 - x)$

101.

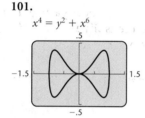

$x^4 = y^2 + x^6$

102.

$(x^2 + y^2)^2 = 2xy$

103.

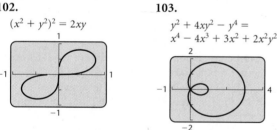

$y^2 + 4xy^2 - y^4 = x^4 - 4x^3 + 3x^2 + 2x^2y^2$

104. The graph of $f(x) = |x|$ passes through the points $(-3, 3)$, $(0, 0)$, and $(3, 3)$. Transform this function to one whose graph passes through the points $(5, 1)$, $(8, 4)$, and $(11, 1)$. Check your answer using a grapher.

Graph each of the following using a grapher. Before doing so, describe how the graph can be obtained from a more basic graph. Give the domain and the range of the function.

105. $f(x) = \left\lbrack\!\left\lbrack x - \tfrac{1}{2} \right\rbrack\!\right\rbrack$

106. $f(x) = |\sqrt{x} - 1|$

107. If $(3, 4)$ is a point on the graph of $y = f(x)$, what point do you know is on the graph of $y = 2f(x)$? of $y = 2 + f(x)$? of $y = f(2x)$?

108. Find the zeros of $f(x) = 3x^5 - 20x^3$. Then without using a grapher, state the zeros of $f(x - 3)$ and $f(x + 8)$.

109. If $(-1, 5)$ is a point on the graph of $y = f(x)$, find b such that $(2, b)$ is on the graph of $y = f(x - 3)$.

State whether each of the following is true or false.

110. The product of two odd functions is odd.

111. The sum of two even functions is even.

112. The product of an even function and an odd function is odd.

113. Show that if f is *any* function, then the function E defined by

$$E(x) = \frac{f(x) + f(-x)}{2}$$

is even.

114. Show that if f is *any* function, then the function O defined by

$$O(x) = \frac{f(x) - f(-x)}{2}$$

is odd.

115. Consider the functions E and O of Exercises 113 and 114.

a) Show that $f(x) = E(x) + O(x)$. This means that every function can be expressed as the sum of an even and an odd function.

b) Let $f(x) = 4x^3 - 11x^2 + \sqrt{x} - 10$. Express f as a sum of an even function and an odd function.

Variation and Applications

1.6

- *Find equations of direct, inverse, and combined variation given values of the variables.*
- *Solve applied problems involving variation.*

We now extend our study of formulas and functions by considering applications involving variation.

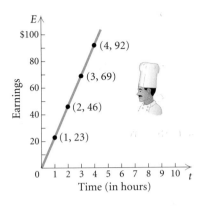

Time (in hours)

Direct Variation

An executive chef earns \$23 per hour. In 1 hr, \$23 is earned; in 2 hr, \$46 is earned; in 3 hr, \$69 is earned; and so on. This gives rise to a set of ordered pairs of numbers:

$$(1, 23), \qquad (2, 46), \qquad (3, 69), \qquad (4, 92),$$

and so on. Note that the ratio of the second coordinate to the first is the same number for each pair:

$$\frac{23}{1} = 23, \qquad \frac{46}{2} = 23, \qquad \frac{69}{3} = 23, \qquad \frac{92}{4} = 23, \quad \text{and so on.}$$

Whenever a situation produces pairs of numbers in which the *ratio is constant,* we say that there is **direct variation.** Here the amount earned varies directly as the time worked:

$$\frac{E}{t} = 23 \text{ (a constant),} \quad \text{or} \quad E = 23t,$$

or, using function notation, $E(t) = 23t$. The equation is an equation of **direct variation.** The coefficient, 23 in the situation above, is called the **variation constant.** In this case, it is the rate of change of earnings with respect to time.

The graph of $y = kx$, $k > 0$, always goes through the origin and rises from left to right. Note that as x increases, y increases; that is, the function is increasing on the interval $(0, \infty)$. The constant k is also the slope of the line.

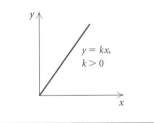

$y = kx,$
$k > 0$

Direct Variation

If a situation gives rise to a linear function $f(x) = kx$, or $y = kx$, where k is a positive constant, we say that we have **direct variation,** or that **y varies directly as x,** or that **y is directly proportional to x.** The number k is called the **variation constant,** or **constant of proportionality.**

EXAMPLE 1   Find the variation constant and an equation of variation in which y varies directly as x, and $y = 32$ when $x = 2$.

Solution   We know that $(2, 32)$ is a solution of $y = kx$. Thus,

$$y = kx$$
$$32 = k \cdot 2 \qquad \textbf{Substituting}$$
$$\frac{32}{2} = k, \text{ or } k = 16. \qquad \textbf{Solving for } k$$

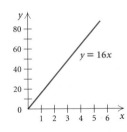

$y = 16x$

The variation constant, 16, is the rate of change of y with respect to x. The equation of variation is $y = 16x$.

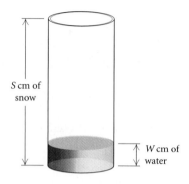

S cm of snow

W cm of water

EXAMPLE 2 *Water from Melting Snow.* The number of centimeters *W* of water produced from melting snow varies directly as *S*, the number of centimeters of snow. Meteorologists have found that 150 cm of snow will melt to 16.8 cm of water. To how many centimeters of water will 200 cm of snow melt?

Solution We can express the amount of water as a function of the amount of snow. Thus, $W(S) = kS$, where k is the variation constant. We first find k using the given data and then find an equation of variation:

$W(S) = kS$	W varies directly as S.
$W(150) = k \cdot 150$	Substituting 150 for S
$16.8 = k \cdot 150$	Replacing $W(150)$ with 16.8
$\dfrac{16.8}{150} = k$	Solving for k
$0.112 = k.$	This is the variation constant.

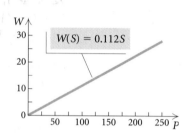

$W(S) = 0.112S$

The equation of variation is $W(S) = 0.112S$.

Next, we use the equation to find how many centimeters of water will result from melting 200 cm of snow:

$$W(S) = 0.112S$$
$$W(200) = 0.112(200) \qquad \text{Substituting}$$
$$= 22.4.$$

Thus, 200 cm of snow will melt to 22.4 cm of water. ▬

Inverse Variation

A bus is traveling a distance of 20 mi. At a speed of 5 mph, the trip will take 4 hr; at 10 mph, it will take 2 hr; at 20 mph, it will take 1 hr; at 40 mph, it will take $\frac{1}{2}$ hr; and so on. We plot this information on a graph, using speed as the first coordinate and time as the second coordinate to determine a set of ordered pairs:

$$(5, 4), \qquad (10, 2), \qquad (20, 1), \qquad \left(40, \tfrac{1}{2}\right), \quad \text{and so on.}$$

Note that the products of the coordinates are all the same number:

$$5 \cdot 4 = 20, \qquad 10 \cdot 2 = 20, \qquad 20 \cdot 1 = 20, \qquad 40 \cdot \tfrac{1}{2} = 20, \quad \text{and so on.}$$

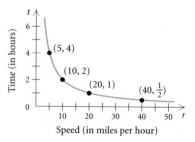

Time (in hours)

(5, 4)

(10, 2)

(20, 1)

$(40, \frac{1}{2})$

Speed (in miles per hour)

Whenever a situation produces pairs of numbers in which the *product is constant,* we say that there is **inverse variation.** Here the time varies inversely as the speed:

$$rt = 20 \text{ (a constant)}, \quad \text{or} \quad t = \frac{20}{r},$$

or, using function notation, $t(r) = 20/r$. The equation is an equation of **inverse variation.** The coefficient, 20 in the situation above, is called the **variation constant.** Note that as the first number increases, the second number decreases.

> It is helpful to look at the graph of $y = k/x$, $k > 0$. The graph is like the one shown below for positive values of x. Note that as x increases, y decreases; that is, the function is decreasing on the interval $(0, \infty)$.
>
>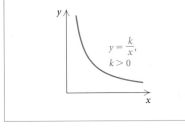

Inverse Variation

If a situation gives rise to a function $f(x) = k/x$, or $y = k/x$, where k is a positive constant, we say that we have **inverse variation,** or that **y varies inversely as x,** or that **y is inversely proportional to x.** The number k is called the **variation constant,** or **constant of proportionality.**

EXAMPLE 3 Find the variation constant and an equation of variation in which y varies inversely as x, and $y = 16$ when $x = 0.3$.

Solution We know that $(0.3, 16)$ is a solution of $y = k/x$. We substitute:

$$y = \frac{k}{x}$$

$$16 = \frac{k}{0.3} \qquad \text{Substituting}$$

$$(0.3)16 = k \qquad \text{Solving for } k$$

$$4.8 = k.$$

$$y = \frac{4.8}{x}$$

The variation constant is 4.8. The equation of variation is $y = 4.8/x$.

There are many problems that translate to an equation of inverse variation.

EXAMPLE 4 *Building a House.* The time t required to do a job varies inversely as the number of people P who work on the job (assuming that all work at the same rate). If it takes 32 hr for 30 people to build a house, how long will it take 20 people to complete the same job?

Solution We can express the amount of time required, in hours, as a function of the number of people working. Thus we have $t(P) = k/P$. We first find k using the given information and then find an equation of variation:

$$t(P) = \frac{k}{P} \qquad t \text{ varies inversely as } P.$$

$$t(30) = \frac{k}{30} \qquad \text{Substituting 30 for } P$$

$$32 = \frac{k}{30} \qquad \text{Replacing } t(30) \text{ with 32}$$

$$30 \cdot 32 = k \qquad \text{Solving for } k$$

$$960 = k. \qquad \text{This is the variation constant.}$$

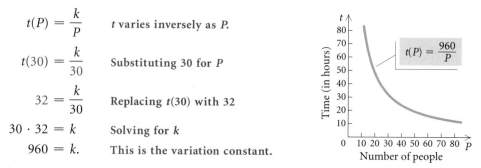

The equation of variation is $t(P) = 960/P$.

Next, we use the equation to find the time that it would take 20 people to do the job. We compute $t(20)$:

$$t(P) = \frac{960}{P} \qquad t \text{ varies inversely as } P.$$

$$t(20) = \frac{960}{20} \qquad \text{Substituting}$$

$$t = 48.$$

Thus it would take 48 hr for 20 people to complete the job.

Combined Variation

We now look at other kinds of variation.

---

y varies **directly as the nth power of x** if there is some positive constant k such that

$$y = kx^n.$$

y varies **inversely as the nth power of x** if there is some positive constant k such that

$$y = \frac{k}{x^n}.$$

y varies **jointly as x and z** if there is some positive constant k such that

$$y = kxz.$$

There are other types of combined variation as well. Consider the formula $V = \pi r^2 h$, in which V, r, and h are variables and π is a constant. We say that V varies jointly as h and the square of r.

EXAMPLE 5 Find an equation of variation in which y varies directly as the square of x, and $y = 12$ when $x = 2$.

Solution We write an equation of variation and find k:

$$y = kx^2$$
$$12 = k \cdot 2^2 \qquad \text{Substituting}$$
$$12 = k \cdot 4$$
$$3 = k.$$

Thus, $y = 3x^2$.

EXAMPLE 6 Find an equation of variation in which y varies jointly as x and z, and $y = 42$ when $x = 2$ and $z = 3$.

Solution We have

$$y = kxz$$
$$42 = k \cdot 2 \cdot 3 \qquad \text{Substituting}$$
$$42 = k \cdot 6$$
$$7 = k.$$

Thus, $y = 7xz$.

EXAMPLE 7 Find an equation of variation in which y varies jointly as x and z and inversely as the square of w, and $y = 105$ when $x = 3$, $z = 20$, and $w = 2$.

Solution We have

$$y = k \cdot \frac{xz}{w^2}$$
$$105 = k \cdot \frac{3 \cdot 20}{2^2} \qquad \text{Substituting}$$
$$105 = k \cdot 15$$
$$7 = k.$$

Thus, $y = 7\dfrac{xz}{w^2}$.

Many applied problems can be modeled using equations of combined variation.

EXAMPLE 8 *Volume of a Tree.* The volume of wood V in a tree varies jointly as the height h and the square of the girth g (girth is distance around). If the volume of a redwood tree is 216 m³ when the height is 30 m and the girth is 1.5 m, what is the height of a tree whose volume is 960 m³ and girth is 2 m?

Solution   We first find k using the first set of data. Then we solve for h using the second set of data.

$$V = khg^2$$
$$216 = k \cdot 30 \cdot 1.5^2$$
$$3.2 = k$$

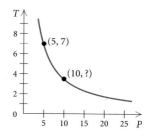

Then the equation of variation is $V = 3.2hg^2$. We substitute the second set of data into the equation:

$$960 = 3.2 \cdot h \cdot 2^2$$
$$75 = h.$$

Thus the height of the tree is 75 m.

Exercise Set 1.6

Find the variation constant and an equation for the given situation.

1. y varies directly as x, and $y = 54$ when $x = 12$

2. y varies directly as x, and $y = 0.1$ when $x = 0.2$

3. y varies inversely as x, and $y = 3$ when $x = 12$

4. y varies inversely as x, and $y = 12$ when $x = 5$

5. y varies directly as x, and $y = 1$ when $x = \frac{1}{4}$

6. y varies inversely as x, and $y = 0.1$ when $x = 0.5$

7. y varies inversely as x, and $y = 32$ when $x = \frac{1}{8}$

8. y varies directly as x, and $y = 3$ when $x = 33$

9. y varies directly as x, and $y = \frac{3}{4}$ when $x = 2$

10. y varies inversely as x, and $y = \frac{1}{5}$ when $x = 35$

11. y varies inversely as x, and $y = 1.8$ when $x = 0.3$

12. y varies directly as x, and $y = 0.9$ when $x = 0.4$

13. *Work Rate.*   The time T required to do a job varies inversely as the number of people P working. It takes 5 hr for 7 bricklayers to build a park wall (see the graph at the top of the next column). How long will it take 10 bricklayers to complete the job?

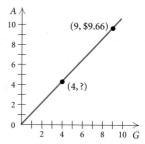

14. *Weekly Allowance.*   According to Fidelity Investments *Investment Vision Magazine*, the average weekly allowance A of children varies directly as their grade level G. It is known that the average allowance of a 9th-grade student is $9.66 per week. What then is the average allowance of a 4th-grade student?

15. *Fat Intake.* The maximum number of grams of fat that should be in a diet varies directly as a person's weight. A person weighing 120 lb should have no more than 60 g of fat per day. What is the maximum daily fat intake for a person weighing 180 lb?

16. *Rate of Travel.* The time t required to drive a fixed distance varies inversely as the speed r. It takes 5 hr at a speed of 80 km/h to drive a fixed distance. How long will it take to drive the same distance at a speed of 70 km/h?

17. *Beam Weight.* The weight W that a horizontal beam can support varies inversely as the length L of the beam. Suppose an 8-m beam can support 1200 kg. How many kilograms can a 14-m beam support?

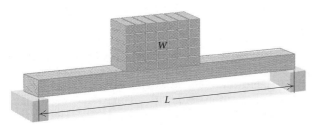

18. *House of Representatives.* The number of representatives N that each state has varies directly as the number of people P living in the state. If New York, with 18,137,226 residents, has 31 representatives, how many representatives does Colorado, with a population of 3,892,644, have?

19. *Weight on Mars.* The weight M of an object on Mars varies directly as its weight E on Earth. A person who weighs 95 lb on Earth weighs 38 lb on Mars. How much would a 100-lb person weigh on Mars?

20. *Pumping Rate.* The time t required to empty a tank varies inversely as the rate r of pumping. If a pump can empty a tank in 45 min at the rate of 600 kL/min, how long will it take the pump to empty the same tank at the rate of 1000 kL/min?

21. *Hooke's Law.* Hooke's law states that the distance d that a spring will stretch varies directly as the mass m of an object hanging from the spring. If a

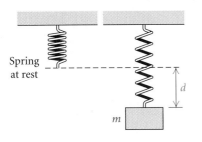

3-kg mass stretches a spring 40 cm, how far will a 5-kg mass stretch the spring?

22. *Relative Aperture.* The relative aperture, or f-stop, of a 23.5-mm diameter lens is directly proportional to the focal length F of the lens. If a 150-mm focal length has an f-stop of 6.3, find the f-stop of a 23.5-mm diameter lens with a focal length of 80 mm.

23. *Musical Pitch.* The pitch P of a musical tone varies inversely as its wavelength W. One tone has a pitch of 330 vibrations per second and a wavelength of 3.2 ft. Find the wavelength of another tone that has a pitch of 550 vibrations per second.

24. *Lead Pollution.* The average U.S. community of population 12,500 released about 385 tons of lead into the environment in a recent year (*Source*: *Conservation Matters,* Autumn 1995 issue. Boston: Conservation Law Foundation, p. 30). How many tons were released nationally? Use 250,000,000 as the U.S. population.

Find an equation of variation for the given situation.

25. y varies inversely as the square of x, and $y = 0.15$ when $x = 0.1$

26. y varies inversely as the square of x, and $y = 6$ when $x = 3$

27. y varies directly as the square of x, and $y = 0.15$ when $x = 0.1$

28. y varies directly as the square of x, and $y = 6$ when $x = 3$

29. y varies jointly as x and z, and $y = 56$ when $x = 7$ and $z = 8$

30. y varies directly as x and inversely as z, and $y = 4$ when $x = 12$ and $z = 15$

31. y varies jointly as x and the square of z, and $y = 105$ when $x = 14$ and $z = 5$

32. y varies jointly as x and z and inversely as w, and $y = \frac{3}{2}$ when $x = 2$, $z = 3$, and $w = 4$

33. y varies jointly as x and z and inversely as the product of w and p, and $y = \frac{3}{28}$ when $x = 3$, $z = 10$, $w = 7$, and $p = 8$

34. y varies jointly as x and z and inversely as the square of w, and $y = \frac{12}{5}$ when $x = 16$, $z = 3$, and $w = 5$

35. *Intensity of Light.* The intensity I of light from a light bulb varies inversely as the square of the distance d from the bulb. Suppose that I is 90 W/m² (watts per square meter) when the distance is 5 m. How much *further* would it be to a point where the intensity is 40 W/m²?

36. *Atmospheric Drag.* Wind resistance, or atmospheric drag, tends to slow down moving objects. Atmospheric drag varies jointly as an object's surface area A and velocity v. If a car traveling at a speed of 40 mph with a surface area of 37.8 ft^2 experiences a drag of 222 N (Newtons), how fast must a car with 51 ft^2 of surface area travel in order to experience a drag force of 430 N?

37. *Stopping Distance of a Car.* The stopping distance d of a car after the brakes have been applied varies directly as the square of the speed r. If a car traveling 60 mph can stop in 200 ft, how fast can a car travel and still stop in 72 ft?

38. *Weight of an Astronaut.* The weight W of an object varies inversely as the square of the distance d from the center of the earth. At sea level (3978 mi from the center of the earth), an astronaut weighs 220 lb. Find his weight when he is 200 mi above the surface of the earth and the spacecraft is not in motion.

39. *Earned-Run Average.* A pitcher's earned-run average E varies directly as the number R of earned runs allowed and inversely as the number I of innings pitched. In a recent year, Shawn Estes of the San Francisco Giants had an earned-run average of 3.18. He gave up 71 earned runs in 201 innings. How many earned runs would he have given up had he pitched 300 innings with the same average? Round to the nearest whole number.

40. *Boyle's Law.* The volume V of a given mass of a gas varies directly as the temperature T and inversely as the pressure P. If $V = 231$ cm^3 when $T = 42°$ and $P = 20$ kg/cm^2, what is the volume when $T = 30°$ and $P = 15$ kg/cm^2?

Discussion and Writing

41. If y varies directly as x^2, explain why doubling x would not cause y to be doubled as well.

42. If y varies directly as x and x varies inversely as z, how does y vary with regard to z? Why?

Skill Maintenance

43. Graph: $f(x) = \begin{cases} x - 2, & \text{for } x \le -1, \\ 3, & \text{for } -1 < x \le 2, \\ x, & \text{for } x > 2. \end{cases}$

Determine algebraically whether the graph is symmetric with respect to the x-axis, the y-axis, and the origin.

44. $y = 3x^4 - 3$

45. $y^2 = x$

46. $2x - 5y = 0$

Synthesis

47. *Volume and Cost.* A peanut butter jar in the shape of a right circular cylinder is 4 in. high and 3 in. in diameter and sells for $1.20. If we assume that cost is directly proportional to volume, how much should a jar 6 in. high and 6 in. in diameter cost?

48. In each of the following equations, state whether y varies directly as x, inversely as x, or neither directly nor inversely as x.

a) $7xy = 14$
b) $x - 2y = 12$
c) $-2x + 3y = 0$
d) $x = \dfrac{3}{4}\, y$
e) $\dfrac{x}{y} = 2$

49. *Area of a Circle.* The area of a circle varies directly as the square of the length of a diameter. What is the variation constant?

50. Describe in words the variation given by the equation

$$Q = \frac{kp^2}{q^3}.$$

Distance, Midpoints, and Circles

1.7

- *Find the distance between two points in the plane and find the midpoint of a segment.*
- *Find an equation of a circle with a given center and radius, and given an equation of a circle, find the center and the radius.*
- *Graph equations of circles.*

In carpentry, surveying, engineering, and other fields, it is often necessary to determine distances and midpoints and to produce accurately drawn circles.

The Distance Formula

Suppose that a conservationist needs to determine the distance across an irregularly shaped pond. One way in which the conservationist might proceed is to measure two legs of a right triangle that is situated as shown below. The Pythagorean theorem, $a^2 + b^2 = c^2$, can then be used to find the length of the hypotenuse, which is the distance across the pond.

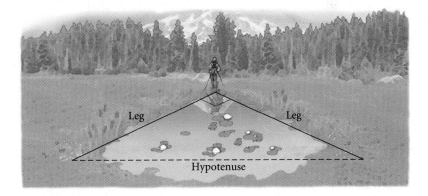

A similar strategy is used to find the distance between two points in a plane. For two points (x_1, y_1) and (x_2, y_2), we can draw a right triangle in which the legs have lengths $|x_2 - x_1|$ and $|y_2 - y_1|$.

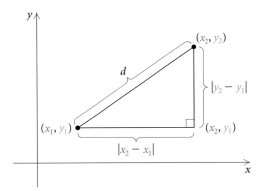

Using the Pythagorean theorem, we have

$$d^2 = |x_2 - x_1|^2 + |y_2 - y_1|^2.$$

Because we are squaring, parentheses can replace the absolute-value symbols:

$$d^2 = (x_2 - x_1)^2 + (y_2 - y_1)^2.$$

Taking the principal square root, we obtain the distance formula.

The Distance Formula

The **distance d** between any two points (x_1, y_1) and (x_2, y_2) is given by

$$d = \sqrt{(x_2 - x_1)^2 + (y_2 - y_1)^2}.$$

The subtraction of the x-coordinates can be done in any order, as can the subtraction of the y-coordinates. Although we derived the distance formula by considering two points not on a horizontal or a vertical line, the distance formula holds for *any* two points.

EXAMPLE 1 The point $(-2, 5)$ is on a circle that has $(3, -1)$ as its center. Find the length of the radius of the circle.

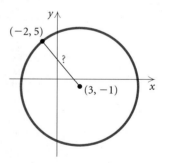

Solution Since the length of the radius is the distance from the center to a point on the circle, we substitute into the distance formula:

$$\sqrt{(3-(-2))^2+(-1-5)^2}$$
$$7.810249676$$

$$r = \sqrt{[3 - (-2)]^2 + [-1 - 5]^2} \qquad \text{Either point can serve as } (x_1, y_1).$$
$$= \sqrt{5^2 + (-6)^2} = \sqrt{61} \approx 7.8. \qquad \text{Rounded to the nearest thousandth}$$

The radius of the circle is approximately 7.8. ▬

Midpoints of Segments

The distance formula can be used to develop a way of determining the *midpoint* of a segment when the endpoints are known. We state the formula and leave its proof to the exercises.

> ### The Midpoint Formula
>
> If the endpoints of a segment are (x_1, y_1) and (x_2, y_2), then the coordinates of the **midpoint** are
>
> $$\left(\frac{x_1 + x_2}{2}, \frac{y_1 + y_2}{2}\right).$$

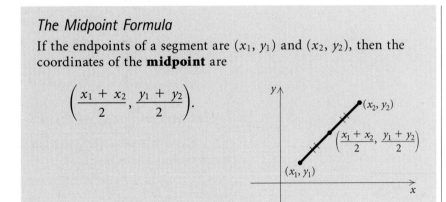

Note that we obtain the coordinates of the midpoint by averaging the coordinates of the endpoints. This is an easy way to remember the midpoint formula.

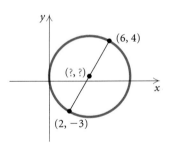

EXAMPLE 2   The diameter of a circle connects two points $(2, -3)$ and $(6, 4)$ on the circle. Find the coordinates of the center of the circle.

Solution   Using the midpoint formula, we obtain

$$\left(\frac{2 + 6}{2}, \frac{-3 + 4}{2}\right), \quad \text{or} \quad \left(\frac{8}{2}, \frac{1}{2}\right), \quad \text{or} \quad \left(4, \frac{1}{2}\right).$$

The coordinates of the center are $\left(4, \frac{1}{2}\right)$.

Circles

A **circle** is the set of all points in a plane that are a fixed distance r from a *center* (h, k). Thus if a point (x, y) is to be r units from the center, we must have

$$r = \sqrt{(x - h)^2 + (y - k)^2}. \qquad \text{Using the distance formula}$$

Squaring both sides gives an equation of a circle. The distance r is the length of a *radius* of the circle.

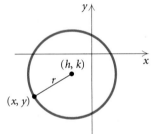

> ### The Equation of a Circle
>
> The equation, in standard form, of a circle with center (h, k) and radius r is
>
> $$(x - h)^2 + (y - k)^2 = r^2.$$

EXAMPLE 3   Find an equation of the circle having radius 5 and center $(3, -7)$.

Solution   Using the standard form, we have

$$[x - 3]^2 + [y - (-7)]^2 = 5^2 \qquad \text{Substituting}$$
$$(x - 3)^2 + (y + 7)^2 = 25.$$

EXAMPLE 4   Graph the circle $(x + 5)^2 + (y - 2)^2 = 16$.

Solution   We write the equation in standard form to determine the center and the radius:

$$[x - (-5)]^2 + [y - 2]^2 = 4^2.$$

The center is $(-5, 2)$ and the radius is 4. We locate the center and draw the circle using a compass.

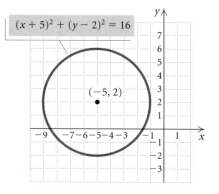

Circles can also be graphed using a grapher. We show one method of doing so here. Another method will be discussed in Section 6.2.

When we graph a circle, we select a viewing window in which the distance between units is visually the same on both axes. This procedure is called **squaring the viewing window.** We do this so that the graph will not be distorted. A graph of the circle $x^2 + y^2 = 36$ in a nonsquared window is shown in Fig. 1.

On many graphers, the distance between units on the y-axis is about $\frac{2}{3}$ the distance between units on the x-axis. When we choose a window in which the length of the y-axis is $\frac{2}{3}$ the length of the x-axis, the window will be squared. The windows with dimensions $[-6, 6, -4, 4]$, $[-9, 9, -6, 6]$, and $[-12, 12, -8, 8]$ are examples of squared windows. A graph of the circle $x^2 + y^2 = 36$ in a squared window is shown in Fig. 2.

Many graphers have an option on the ZOOM menu that squares the window automatically.

EXAMPLE 5   Graph the circle $(x - 2)^2 + (y + 1)^2 = 16$.

Solution   The circle $(x - 2)^2 + (y + 1)^2 = 16$ has center $(2, -1)$ and radius 4, so the viewing window $[-9, 9, -6, 6]$ is a good choice in this

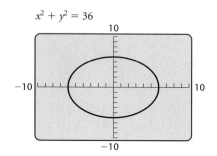

FIGURE 1

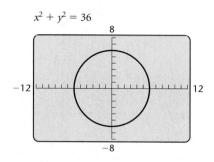

FIGURE 2

case. After entering the window dimensions, we select the CIRCLE feature from the DRAW menu and enter the coordinates of the center and the radius to draw the graph of the circle.

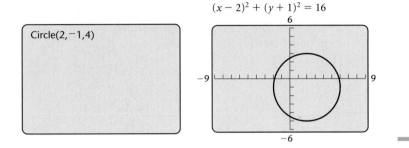

$(x - 2)^2 + (y + 1)^2 = 16$

Exercise Set 1.7

Find the distance between the pair of points. Give an exact answer and, where appropriate, an approximation to three decimal places.

1. (4, 6) and (5, 9)
2. (−3, 7) and (2, 11)
3. (6, −1) and (9, 5)
4. (−4, −7) and (−1, 3)
5. $(\sqrt{3}, -\sqrt{5})$ and $(-\sqrt{6}, 0)$
6. $(-\sqrt{2}, 1)$ and $(0, \sqrt{7})$

7. The points (−3, −1) and (9, 4) are the endpoints of the diameter of a circle. Find the length of the radius of the circle.

8. The point (0, 1) is on a circle that has center (−3, 5). Find the length of the diameter of the circle.

Use the distance formula and the Pythagorean theorem to determine whether the set of points could be vertices of a right triangle.

9. (−4, 5), (6, 1), and (−8, −5)
10. (−3, 1), (2, −1), and (6, 9)

11. The points (−3, 4), (0, 5), and (3, −4) are all on the circle $x^2 + y^2 = 25$. Show that these three points are vertices of a right triangle.

12. The points (−3, 4), (2, −1), (5, 2), and (0, 7) are vertices of a quadrilateral. Show that the quadrilateral is a rectangle. (*Hint*: Show that the quadrilateral's opposite sides are the same length and that the two diagonals are the same length.)

13.–18. Find the midpoint of each segment having the endpoints given in Exercises 1–6, respectively.

19. Graph the rectangle described in Exercise 12. Then determine the coordinates of the midpoint for each of the four sides. Are the midpoints vertices of a rectangle?

20. Graph the square with vertices (−5, −1), (7, −6), (12, 6), and (0, 11). Then determine the midpoint for each of the four sides. Are the midpoints vertices of a square?

21. The points $(\sqrt{7}, -4)$ and $(\sqrt{2}, 3)$ are endpoints of the diameter of a circle. Determine the center of the circle.

22. The points $(-3, \sqrt{5})$ and $(1, \sqrt{2})$ are endpoints of the diagonal of a square. Determine the center of the square.

In Exercises 23 and 24, how would you change the window so that the circle is not distorted? Answers may vary.

23.   **24.**

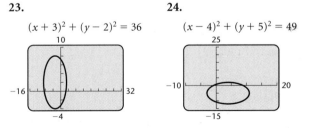

$(x + 3)^2 + (y - 2)^2 = 36$ $(x - 4)^2 + (y + 5)^2 = 49$

Find an equation for a circle satisfying the given conditions.

25. Center (2, 3), radius of length $\frac{5}{3}$
26. Center (4, 5), diameter of length 8.2
27. Center (−1, 4), passes through (3, 7)
28. Center (6, −5), passes through (1, 7)
29. The points (7, 13) and (−3, −11) are at either end of a diameter.

30. The points $(-9, 4)$, $(-2, 5)$, $(-8, -3)$, and $(-1, -2)$ are vertices of an inscribed square.

31. Center $(-2, 3)$, tangent (touching at one point) to the y-axis

32. Center $(4, -5)$, tangent to the x-axis

Find the center and the radius of the circle. Then graph the circle using a square viewing window.

33. $x^2 + y^2 = 4$

34. $x^2 + y^2 = 81$

35. $x^2 + (y - 3)^2 = 16$

36. $(x + 2)^2 + y^2 = 100$

37. $(x - 1)^2 + (y - 5)^2 = 36$

38. $(x - 7)^2 + (y + 2)^2 = 25$

39. $(x + 4)^2 + (y + 5)^2 = 9$

40. $(x + 1)^2 + (y - 2)^2 = 64$

Discussion and Writing

41. Explain how the Pythagorean theorem is used to develop the equation of a circle in standard form.

42. Explain how you could find the coordinates of a point $\frac{7}{8}$ of the way from point A to point B.

Skill Maintenance

Find the slope and the y-intercept of the equation.

43. $y = -\frac{3}{5}x + 4$

44. $2x - 3y = 15$

Write a slope–intercept equation for a line with the given characteristics.

45. $m = 3$, y-intercept $(0, -1)$

46. Passes through $(-2, 3)$ and $(4, 1)$

Synthesis

Find the distance between the pair of points and find the midpoint of the segment having the given points as endpoints.

47. $(a, \sqrt{a})$ and $(a + h, \sqrt{a + h})$

48. $\left(a, \dfrac{1}{a}\right)$ and $\left(a + h, \dfrac{1}{a + h}\right)$

Find an equation of a circle satisfying the given conditions.

49. Center $(2, -7)$ with an area of 36π square units

50. Center $(-5, 8)$ with a circumference of 10π units

51. *Swimming Pool.* A swimming pool is being constructed in the corner of a yard, as shown. Before installation, the contractor needs to know measurements a_1 and a_2. Find them.

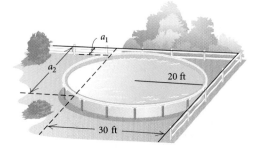

52. *An Arch of a Circle in Carpentry.* Ace Carpentry needs to cut an arch for the top of an entranceway. The arch needs to be 8 ft wide and 2 ft high. To draw the arch, the carpenters will use a stretched string with chalk attached at an end as a compass.

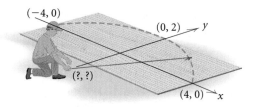

a) Using a coordinate system, locate the center of the circle.

b) What radius should the carpenters use to draw the arch?

*Determine whether each of the following points lies on the **unit circle**, $x^2 + y^2 = 1$.*

53. $\left(\dfrac{\sqrt{3}}{2}, -\dfrac{1}{2}\right)$

54. $(0, -1)$

55. $\left(-\dfrac{\sqrt{2}}{2}, \dfrac{\sqrt{2}}{2}\right)$

56. $\left(\dfrac{1}{2}, -\dfrac{\sqrt{3}}{2}\right)$

57. Find the point on the y-axis that is equidistant from the points $(-2, 0)$ and $(4, 6)$.

58. Consider any right triangle with base b and height h, situated as shown. Show that the midpoint of the hypotenuse P is equidistant from the three vertices of the triangle.

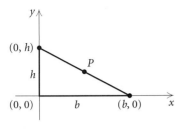

59. Prove the midpoint formula by showing that:

a) $\left(\dfrac{x_1 + x_2}{2}, \dfrac{y_1 + y_2}{2}\right)$ is equidistant from the points (x_1, y_1) and (x_2, y_2); and

b) the distance from (x_1, y_1) to the midpoint plus the distance from (x_2, y_2) to the midpoint equals the distance from (x_1, y_1) to (x_2, y_2).

Chapter Summary and Review   1

Important Properties and Formulas

Terminology about Lines

Slope: $m = \dfrac{y_2 - y_1}{x_2 - x_1}$

The Slope–Intercept Equation:
$$y = mx + b$$

The Point–Slope Equation:
$$y - y_1 = m(x - x_1)$$

Horizontal Lines: $y = b$

Vertical Lines: $x = a$

Parallel Lines: $m_1 = m_2,\ b_1 \neq b_2$

Perpendicular Lines:
$$m_1 m_2 = -1, \text{ or}$$
$$x = a,\ y = b$$

The Algebra of Functions

The Sum of Two Functions:
$(f + g)(x) = f(x) + g(x)$

The Difference of Two Functions:
$(f - g)(x) = f(x) - g(x)$

The Product of Two Functions:
$(fg)(x) = f(x) \cdot g(x)$

The Quotient of Two Functions:
$(f/g)(x) = f(x)/g(x),\ g(x) \neq 0$

Tests for Symmetry

x-axis: If replacing y with $-y$ produces an equivalent equation, then the graph is symmetric with respect to the x-axis.

y-axis: If replacing x with $-x$ produces an equivalent equation, then the graph is symmetric with respect to the y-axis.

Origin: If replacing x with $-x$ and y with $-y$ produces an equivalent equation, then the graph is symmetric with respect to the origin.

Even Function: $f(-x) = f(x)$ y-axis

Odd Function: $f(-x) = -f(x)$

Transformations

Vertical Translation: $y = f(x) \pm b$

Horizontal Translation: $y = f(x \pm d)$

Reflection across the x-axis: $y = -f(x)$

Reflection across the y-axis: $y = f(-x)$

Vertical Stretching or Shrinking:
$$y = af(x)$$

Horizontal Stretching or Shrinking:
$$y = f(cx)$$

(continued)

Variation

Direct: $y = kx$

Inverse: $y = \dfrac{k}{x}$

Joint: $y = kxz$

The Distance Formula

$$d = \sqrt{(x_2 - x_1)^2 + (y_2 - y_1)^2}$$

The Midpoint Formula

$$\left(\dfrac{x_1 + x_2}{2}, \dfrac{y_1 + y_2}{2}\right)$$

Equation of a Circle

$$(x - h)^2 + (y - k)^2 = r^2$$

REVIEW EXERCISES

Determine whether the relation is a function. Identify the domain and the range.

1. $\{(3, 1), (5, 3), (7, 7), (3, 5)\}$

2. $\{(2, 7), (-2, -7), (7, -2), (0, 2), (1, -4)\}$

Determine whether the graph is that of a function.

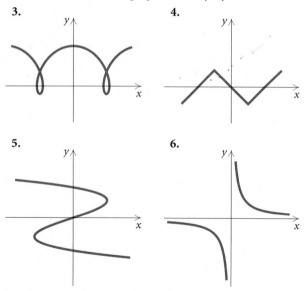

3.

4.

5.

6.

Find the domain of the function. Do not use a grapher.

7. $f(x) = 4 - 5x + x^2$ **8.** $f(x) = \dfrac{3}{x} + 2$

9. $f(x) = \dfrac{1}{x^2 - 6x + 5}$ **10.** $f(x) = \dfrac{-5x}{|16 - x^2|}$

Use a grapher to graph the function. Then visually estimate the domain and the range.

11. $f(x) = \sqrt{16 - (x + 1)^2}$

12. $f(x) = |x - 3| + |x + 4| - 12$

13. $f(x) = x^3 - 5x^2 + 2x - 7$

14. $f(x) = x^4 - 8x^2 - 3$

15. Given that $f(x) = x^2 - x - 3$, find each of the following.

a) $f(0)$ b) $f(-3)$

c) $f(a - 1)$ d) $\dfrac{f(x + h) - f(x)}{h}$

In Exercises 16 and 17, the table of data contains input–output values for a function. Answer the following questions.

a) *Is the change in the inputs, x, the same?*
b) *Is the change in the outputs, y, the same?*
c) *Is the function linear?*

16.

x	y
-3	8
-2	11
-1	14
0	17
1	20
2	22
3	26

17.

x	y
20	11.8
30	24.2
40	36.6
50	49.0
60	61.4
70	73.8
80	86.2

18. Find the slope and the y-intercept of the graph of $-2x - y = 7$.

Write a point–slope equation for a line with the following characteristics.

19. $m = -\frac{2}{3}$, y-intercept $(0, -4)$

20. $m = 3$, passes through $(-2, -1)$

21. Passes through $(4, 1)$ and $(-2, -1)$

Determine whether the lines are parallel, perpendicular, or neither.

22. $3x - 2y = 8$,
$6x - 4y = 2$

23. $y - 2x = 4$,
$2y - 3x = -7$

24. $y = \frac{3}{2}x + 7$,
$y = -\frac{2}{3}x - 4$

Given the point $(1, -1)$ and the line $2x + 3y = 4$:

25. Find an equation of the line containing the given point and parallel to the given line.

26. Find an equation of the line containing the given point and perpendicular to the given line.

27. *Total Cost.* Clear County Cable Television charges a \$25 installation fee and \$20 per month for "basic" service. Write and graph an equation that can be used to determine the total cost, $C(t)$, of t months of basic cable television service. Find the total cost of 6 months of service.

28. *Temperature and Depth of the Earth.* The function T given by $T(d) = 10d + 20$ can be used to determine the temperature T, in degrees Celsius, at a depth d, in kilometers, inside the earth.

a) Find $T(5)$, $T(20)$, and $T(1000)$.
b) Graph T.
c) The radius of the earth is about 5600 km. Use this fact to determine the domain of the function.

29. *Motor Vehicle Production.* The data in the following table shows the increase in world motor vehicle production since 1950.

YEAR, x	NUMBER OF VEHICLES PRODUCED IN THE WORLD, y (IN MILLIONS)
1950, 0	10.6
1960, 10	16.5
1970, 20	29.4
1980, 30	38.6
1990, 40	48.6
1995, 45	50.0

Source: The Wall Street Journal Almanac, 1999

a) Fit a regression line to the data and use it to estimate the number of vehicles produced in 2000.
b) What is the correlation coefficient for the regression line? How close a fit is the regression line?

Make a hand-drawn graph of each of the following. Check your results using a grapher.

30. $f(x) = \begin{cases} x^3, & \text{for } x < -2, \\ |x|, & \text{for } -2 \le x \le 2, \\ \sqrt{x-1}, & \text{for } x > 2 \end{cases}$

31. $f(x) = \begin{cases} \dfrac{x^2 - 1}{x + 1}, & \text{for } x \ne -1, \\ 3, & \text{for } x = -1 \end{cases}$

32. $f(x) = [\![x]\!]$

33. $f(x) = [\![x - 3]\!]$

In Exercises 34–36, use a grapher to estimate the following for each function.

a) *The intervals on which the function is increasing or decreasing*
b) *Any relative maxima or minima*

34. $f(x) = 1.1x^3 - 9.04x^2 - 18x + 702$

35. $f(x) = \dfrac{20}{(x + 4)^2 + 1} - 5$

36. $f(x) = 0.8x\sqrt{16 - (x + 1)^2}$

37. *Inscribed Rectangle.* A rectangle is inscribed in a semicircle of radius 2, as shown. The variable $x = $ half the length of the rectangle. Express the area of the rectangle as a function of x.

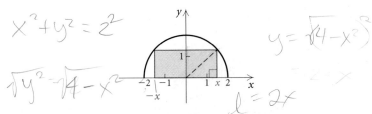

38. *Minimizing Surface Area.* A container firm is designing an open-top rectangular box, with a square base, that will hold 108 in³. Let $x = $ the length of a side of the base.

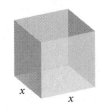

a) Express the surface area as a function of x.
b) Find the domain of the function.
c) Graph the function.
d) What dimensions minimize the surface area of the box?

First, graph the equation and determine visually whether it is symmetric with respect to the x-axis, the y-axis, and the origin. Then verify your assertion algebraically.

39. $x^2 + y^2 = 4$ **40.** $y^2 = x^2 + 3$

41. $x + y = 3$ **42.** $y = x^2$

43. $y = x^3$ **44.** $y = x^4 - x^2$

Determine visually whether the function is even, odd, or neither even nor odd.

45.   **46.**

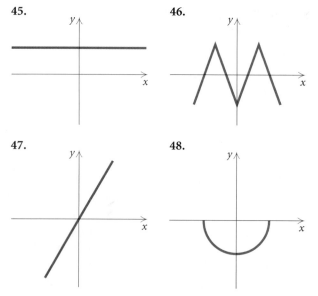

47.   **48.**

Test algebraically whether the function is even, odd, or neither even nor odd. Then check your work graphically, where possible, using a grapher.

49. $f(x) = 9 - x^2$ **50.** $f(x) = x^3 - 2x + 4$

51. $f(x) = x^7 - x^5$ **52.** $f(x) = |x|$

53. $f(x) = \sqrt{16 - x^2}$ **54.** $f(x) = \dfrac{10x}{x^2 + 1}$

Write an equation for a function that has a graph with the given characteristics. Check your answer using a grapher.

55. The shape of $y = x^2$, but shifted left 3 units

56. The shape of $y = \sqrt{x}$, but upside down and shifted right 3 units and up 4 units

57. The shape of $y = |x|$, but stretched vertically by a factor of 2 and shifted right 3 units

A graph of $y = f(x)$ is shown below. No formula for f is given. Make a hand-drawn graph of each of the following.

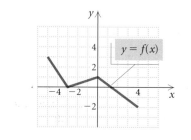

58. $y = f(x - 1)$ **59.** $y = f(2x)$

60. $y = -2f(x)$ **61.** $y = 3 + f(x)$

For each pair of functions in Exercises 62 and 63:
a) *Find the domain of f, g, f + g, f − g, fg, and f/g.*
b) *Find $(f + g)(x)$, $(f - g)(x)$, $fg(x)$, and $(f/g)(x)$.*

62. $f(x) = \dfrac{4}{x^2}$; $g(x) = 3 - 2x$

63. $f(x) = 3x^2 + 4x$; $g(x) = 2x - 1$

64. Given the total-revenue and total-cost functions
$$R(x) = 120x - 0.5x^2 \quad \text{and} \quad C(x) = 15x + 6,$$
find total profit $P(x)$.

65. Find an equation of variation in which y varies directly as x, and $y = 100$ when $x = 25$.

66. Find an equation of variation in which y varies inversely as x, and $y = 100$ when $x = 25$.

67. *Pumping Time.* The time t required to empty a tank varies inversely as the rate r of pumping. If a pump can empty a tank in 35 min at the rate of 800 kL per minute, how long will it take the pump to empty the same tank at the rate of 1400 kL per minute?

68. *Power of Electric Current.* The power P expended by heat in an electric circuit of fixed resistance varies directly as the square of the current C in the circuit. A circuit expends 180 watts when a current of 6 amperes is flowing. What is the amount of heat expended when the current is 10 amperes?

69. *Test Score.* The score N on a test varies directly as the number of correct responses a. Ellen answers 28 questions correctly and earns a score of 87. What would Ellen's score have been if she had answered 25 questions correctly?

70. Find the distance between $(3, 7)$ and $(-2, 4)$.

71. Find the midpoint of the segment with endpoints $(3, 7)$ and $(-2, 4)$.

72. Find an equation of the circle with center $(-2, 6)$ and radius $\sqrt{13}$.

73. Find the center and the radius of the circle
$$(x + 1)^2 + (y - 3)^2 = 16.$$

74. Find an equation of the circle having a diameter with endpoints $(-3, 5)$ and $(7, 3)$.

Discussion and Writing

75. Given that $f(x) = 4x^3 - 2x + 7$, find each of the following. Then discuss how each expression differs from the other.

a) $f(x) + 2$ **b)** $f(x + 2)$ **c)** $f(x) + f(2)$

76. Given the graph of $y = f(x)$, explain and contrast the effect of the constant c on the graphs of $y = f(cx)$ and $y = cf(x)$.

77. a) Graph several functions of the type $y_1 = f(x)$ and $y_2 = |f(x)|$ using a grapher. Describe a procedure, involving transformations, for creating hand-drawn graphs of y_2 from y_1.

b) Describe a procedure, involving transformations, for creating hand-drawn graphs of $y_2 = f(|x|)$ from $y_1 = f(x)$.

Synthesis

Find the domain.

78. $f(x) = \dfrac{\sqrt{1 - x}}{x - |x|}$

79. $f(x) = (x - 9x^{-1})^{-1}$

80. Prove that the sum of two odd functions is odd.

81. Describe how the graph of $y = -f(-x)$ is obtained from the graph of $y = f(x)$.

Functions and Equations: Zeros and Solutions 2

The median number of leisure hours that Americans had each week decreased in the 1970s and 1980s, but this number is now on the rise. The quadratic function

$$y = f(x) = 0.04x^2 - 1.21x + 26.03,$$

where x is the number of years since 1973, models this situation (*Source*: Louis Harris and Associates). We can use this function to estimate that Americans had about 21.0 leisure hours per week in 1978 ($x = 5$) and about 17.0 hours in 1990 ($x = 17$) and to predict that they will have about 28.4 leisure hours per week in 2005 ($x = 32$).

This problem appears as Example 1 in Section 2.5.

I n this chapter, we study the zeros of linear and quadratic functions and their relation to the solutions of equations and to the x-intercepts of graphs.

We also use these types of functions to solve applied problems and to model real data and make predictions.

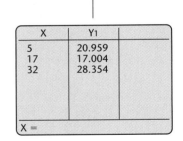

X	Y₁	
5	20.959	
17	17.004	
32	28.354	
X =		

Zeros of Linear Functions and Models

2.1

- • *Find zeros of linear functions and solve linear equations.*
- • *Solve applied problems using linear models.*

Zeros of Linear Functions

An input for which a function's output is 0 is called a **zero** of the function.

> ### Zeros of Functions
>
> An input c of a function f is called a **zero** of the function, if the output for c is 0. That is, $f(c) = 0$.

LINEAR FUNCTIONS

REVIEW SECTION 1.2.

We will restrict our attention to zeros of linear functions in this section. Recall that a linear function is given by $f(x) = mx + b$, where m and b are constants.

For the linear function $f(x) = 2x - 4$, we have $f(2) = 2 \cdot 2 - 4 = 0$, so 2 is a **zero** of the function. In fact, 2 is the *only* zero of this function. In general, a **linear function $f(x) = mx + b$, with $m \neq 0$, has exactly one zero.**

Consider the graph of $f(x) = 2x - 4$, shown here.

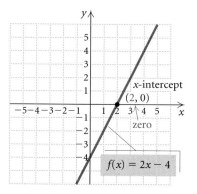

We see from the graph that the zero, 2, is the first coordinate of the point at which the graph crosses the x-axis. This point, $(2, 0)$, is the **x-intercept** of the graph. Thus when we find the zero of a linear function, we are also finding the first coordinate of the x-intercept of the graph of the function.

For every linear function $f(x) = mx + b$, there is an associated linear equation $mx + b = 0$. When we find the zero of a function $f(x) = mx + b$, we are also finding the solution of the equation $mx + b = 0$.

EXAMPLE 1   Find the zero of $f(x) = 5x - 9$.

Algebraic Solution

We find the value of x for which $f(x) = 0$:

$5x - 9 = 0$ Setting $f(x) = 0$

$5x - 9 + 9 = 0 + 9$ Adding 9 on both sides

$5x = 9$

$\dfrac{5x}{5} = \dfrac{9}{5}$ Dividing by 5 on both sides

$x = \dfrac{9}{5}$, or 1.8.

Using a table, set in ASK mode, we can check the solution. We enter $y = 5x - 9$ on the equation-editor screen and then enter the value $x = 9/5$, or 1.8, in the table. Note that if 9/5 is entered, the grapher will express it as 1.8 in the table.

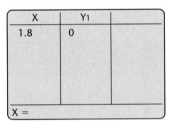

We see that $y = 0$ when $x = 1.8$, so the number 1.8 checks. The zero is 9/5, or 1.8. Note that the *zero* of the function $f(x) = 5x - 9$ is the *solution* of the equation $5x - 9 = 0$.

Graphical Solution

We graph $y = 5x - 9$ in the standard window and use the ZERO feature from the CALC menu to find the zero of $f(x) = 5x - 9$. We call this the **Zero method.** Note that the x-intercept must appear in the window when the ZERO feature is used.

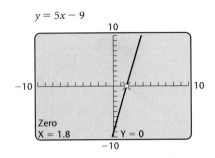

We can check algebraically by substituting 1.8 for x:

$$f(1.8) = 5(1.8) - 9 = 9 - 9 = 0.$$

The zero is 1.8, or 9/5.

EXAMPLE 2   Solve: $-3x + 33 = 0$.

Algebraic Solution

We solve algebraically as follows:

$-3x + 33 = 0$

$-3x + 33 - 33 = 0 - 33$ Subtracting 33 on both sides

$-3x = -33$

$\dfrac{-3x}{-3} = \dfrac{-33}{-3}$ Dividing by -3 on both sides

$x = 11.$

We can use function notation on the home screen to check the solution. We enter $y_1 = -3x + 33$ on the equation-editor screen and then enter $y_1(11)$ on the home screen. This is equivalent to substituting 11 for x in $-3x + 33$.

The solution is 11.

Graphical Solution

We will use the Zero method. We graph $y = -3x + 33$ in the standard window. Since this window, shown on the left below, does not show the x-intercept, we choose a different window. It appears that the x-intercept will be to the right of the point $(10, 0)$. We try the dimensions $[-5, 15, -10, 10]$, shown on the right below.

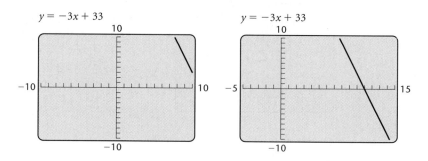

Now we see the x-intercept displayed. We use the ZERO feature from the CALC menu to find the zero of $f(x) = -3x + 33$. This is the solution of $-3x + 33 = 0$.

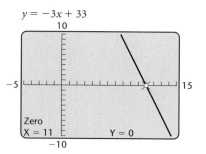

The solution is 11.

We can also use the Zero method to solve an equation that is not written with 0 on one side.

FINDING POINTS OF INTERSECTION

REVIEW "INTRODUCTION TO GRAPHS AND THE GRAPHING CALCULATOR."

EXAMPLE 3   Solve $3x - 7 = 5x + 1$ using the Zero method.

Solution   First, we subtract $5x$ and 1 on both sides to get 0 on one side of the equation:

$$3x - 7 = 5x + 1$$
$$3x - 7 - 5x - 1 = 0.$$

The expression on the left side of the equation can be entered in the grapher without further simplification. We graph $y = 3x - 7 - 5x - 1$ and use the ZERO feature.

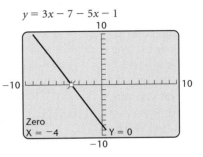

We see that -4 is the zero of the function $f(x) = 3x - 7 - 5x - 1$ and thus it is the solution of the equation $3x - 7 - 5x - 1 = 0$. Since $3x - 7 - 5x - 1 = 0$ is equivalent to the original equation, $3x - 7 = 5x + 1$, we know that -4 is the solution of this equation. We can substitute -4 in the *original* equation to check the solution.

We can also solve equations graphically using the INTERSECT feature from the CALC menu. We call this the **Intersect method.** Let's reconsider the equation $3x - 7 = 5x + 1$ that we solved using the Zero method in Example 3.

EXAMPLE 4   Solve $3x - 7 = 5x + 1$ using the Intersect method.

Solution   We enter $y_1 = 3x - 7$ and $y_2 = 5x + 1$ in the grapher and use the INTERSECT feature to find the first coordinate of the point of intersection of the graphs. Trial and error reveals that one good viewing window is $[-10, 10, -30, 5]$, with Yscl $= 5$.

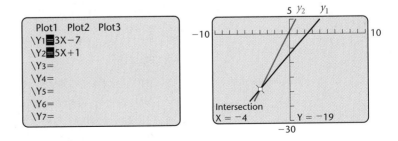

The solution of the equation is the first coordinate of the point of intersection, -4. In Example 3, we saw that this is also the zero of the function $f(x) = 3x - 7 - 5x - 1$.

CONNECTING THE CONCEPTS

THE INTERSECT AND ZERO METHODS

An equation $f(x) = g(x)$ can be solved using the Intersect method by graphing $y_1 = f(x)$ and $y_2 = g(x)$ and using the INTERSECT feature to find the first coordinate of the point of intersection of the graphs. The equation can also be solved using the Zero method by rewriting it with 0 on one side of the equals sign, if necessary, and then using the ZERO feature.

Solve: $x - 1 = 2x - 6$.

The Intersect Method

Graph

$$f(x) = y_1 = x - 1$$

and

$$g(x) = y_2 = 2x - 6.$$

The Zero Method

Graph

$$y = x - 1 - 2x + 6.$$

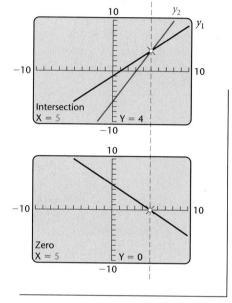

Applications Using Linear Models

Mathematical techniques are used to answer questions arising from real-world situations. Functions and equations of the form $f(x) = mx + b$ and $g(x) = h(x)$, where g and h are linear functions, *model* many of these situations.

The following strategy is of great assistance in problem solving.

Five Steps for Problem Solving

1. **Familiarize** yourself with the problem situation. If the problem is presented in words, then, of course, this means to read carefully. Some or all of the following can also be helpful.

   a) Make a drawing, if it makes sense to do so.

   b) Make a written list of the known facts and a list of what you wish to find out.

   c) Assign variables to represent unknown quantities.

   d) Organize the information in a chart or a table.

   (continued)

e) Find further information. Look up a formula, consult a reference book or an expert in the field, or do research on the Internet.

f) Guess or estimate the answer and check your guess or estimate.

2. **Translate** the problem situation to mathematical language or symbolism. For most of the problems you will encounter in algebra, this means to write one or more equations, but sometimes an inequality or some other mathematical symbolism may be appropriate.

3. **Carry out** some type of mathematical manipulation. Use your mathematical skills to find a possible solution. In algebra, this usually means to solve an equation, an inequality, or a system of equations.

4. **Check** to see whether your possible solution actually fits the problem situation and is thus really a solution of the problem. You might be able to solve an equation, but the solution(s) of the equation might or might not be solution(s) of the original problem.

5. **State** the answer clearly using a complete sentence.

EXAMPLE 5 *Library Acquisitions.* In 1998, college and university libraries in the United States spent $1.9 billion on books, periodicals, and other materials for their collections. This amount is 5.6% more than the amount spent in 1997. (*Source*: *Book Industry Trends 1998*, Book Industry Study Group, Inc.) How much was spent in 1997?

Solution

1. **Familiarize.** Let's estimate that $1.5 billion was spent in 1997. Then the amount spent in 1998, in billions of dollars, would have been

$$1.5 + 5.6\% \text{ of } 1.5 = 1(1.5) + 0.056(1.5)$$
$$= 1.056(1.5) = \$1.584 \text{ billion.}$$

Since $1.9 billion was the actual amount spent in 1998, the estimate is too low. Nevertheless, the calculations performed in making the estimate indicate how we can translate this problem to an equation. We let $a =$ the amount that libraries spent, in billions of dollars, adding to their collections in 1997. Then $a + 5.6\%a$, or $1.056a$, is the amount spent in 1998.

2. **Translate.** We translate to an equation.

$$\underbrace{\text{The amount spent in 1998}}_{1.056a} \quad \underset{=}{\overset{\text{is}}{\downarrow}} \quad \underset{1.9}{\overset{\$1.9 \text{ billion.}}{\downarrow}}$$

3. **Carry out.** We carry out the solution as follows.

Algebraic Solution

We solve the equation:

$$1.056a = 1.9$$

$$\frac{1.056a}{1.056} = \frac{1.9}{1.056} \qquad \text{Dividing by}$$
$$\qquad\qquad\qquad\qquad \text{1.056 on}$$
$$\qquad\qquad\qquad\qquad \text{both sides}$$

$$a \approx 1.8.$$

Graphical Solution

To use the Intersect method, we replace a with x and graph $y_1 = 1.056x$ and $y_2 = 1.9$. We begin with the standard window and find that it is a good choice because it displays the point of intersection of the graphs. We use the INTERSECT feature to find the first coordinate of the point of intersection, as shown on the left below.

To use the Zero method, we first subtract 1.9 on both sides of the equation to obtain $1.056a - 1.9 = 0$, an equation with 0 on one side. Then we graph $y = 1.056x - 1.9$ and use the ZERO feature to find the zero of this function, as shown on the right below.

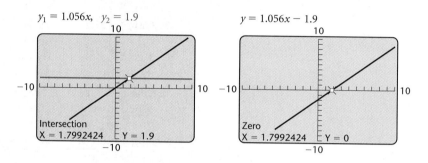

4. **Check.** We see that 5.6% of 1.8 is 0.1008 and $1.8 + 0.1008 = 1.9008 \approx 1.9$, so the answer checks. (Remember that we rounded the value of a.)

5. **State.** Colleges and universities in the United States spent about $1.8 billion adding books, periodicals, and other materials to their collections in 1997.

In some applications, we need to use a formula that describes the relationships among variables. When a situation involves distance, speed, and time, for example, we need to recall the **motion formula:**

$$d = rt, \quad \text{where } d = \text{distance}, \; r = \text{rate (or speed), and } t = \text{time.}$$

EXAMPLE 6 *Airplane Speed.* Continental Airlines' fleet includes 727s with a cruising speed of 545 mph and B1900Ds with a cruising speed of 285 mph (*Source:* Continental Airlines). Suppose that a B1900D takes off and travels at its cruising speed. One hour later a 727 takes off and follows the same route, traveling at its cruising speed. How long will it take the 727 to overtake the B1900D?

Solution

1. **Familiarize.** We make a drawing showing both the known and the unknown information. We let $t =$ the time, in hours, that the 727 travels before it overtakes the B1900D. Since the B1900D takes off 1 hr before the 727, it will travel for $t + 1$ hr before being overtaken. The planes will have traveled the same distance, d, when one overtakes the other.

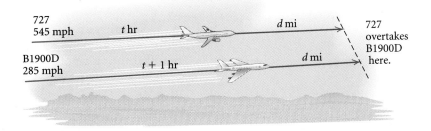

We can also organize the information in a table, as follows.

$$d \quad = \quad r \quad \cdot \quad t$$

	DISTANCE	RATE	TIME	
727	d	545	t	$\longrightarrow d = 545t$
B1900D	d	285	$t + 1$	$\longrightarrow d = 285(t + 1)$

2. **Translate.** Using the formula $d = rt$ in each row of the table, we get two expressions for d:

$$d = 545t \quad \text{and} \quad d = 285(t + 1).$$

Substituting $545t$ for d in the second equation, we have

$$545t = 285(t + 1).$$

3. **Carry out.** We carry out the solution as follows.

Algebraic Solution

We solve the equation:

$$545t = 285(t + 1)$$

$545t = 285t + 285$	Using the distributive property
$260t = 285$	Subtracting $285t$ on both sides
$t \approx 1.1.$	Dividing by 260 on both sides

Graphical Solution

Using the Intersect method, we replace t with x, and graph $y_1 = 545x$ and $y_2 = 285(x + 1)$. We must choose a window that contains the point of intersection. Experimentation with several windows reveals that a window of $[0, 3, 0, 1000]$, with Yscl = 100, is a good choice. We use the INTERSECT feature to find the first coordinate of the point of intersection of the graphs.

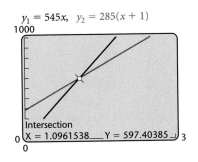

$y_1 = 545x, \quad y_2 = 285(x + 1)$

Intersection
X = 1.0961538   Y = 597.40385

The solution is about 1.1.

4. **Check.**   If the 727 travels for about 1.1 hr, then the B1900D travels for about $1.1 + 1$, or 2.1 hr. In 2.1 hr, the B1900D travels $285(2.1)$, or 598.5 mi; and in 1.1 hr, the 727 travels $545(1.1)$, or 599.5 mi. Since 598.5 mi $\approx$ 599.5 mi, the answer checks. (Remember that we rounded the value of t.)

5. **State.**   About 1.1 hr after the 727 has taken off, it will overtake the B1900D.

CONNECTING THE CONCEPTS

ZEROS, SOLUTIONS, AND INTERCEPTS

The zero of a linear function $f(x) = mx + b$, with $m \neq 0$, is the solution of the linear equation $mx + b = 0$ and is the first coordinate of the x-intercept of the graph of $f(x) = mx + b$. To find the zero of $f(x) = mx + b$ algebraically, we solve $f(x) = 0$, or $mx + b = 0$. To find the zero graphically, we graph $y = mx + b$ and use the ZERO feature from the CALC menu of a grapher.

FUNCTION	ZERO OF THE FUNCTION; SOLUTION OF THE EQUATION	X–INTERCEPT OF THE GRAPH

Linear Function

$f(x) = 2x - 4$, or

$\quad y = 2x - 4$

To find the **zero** of $f(x)$, we solve $f(x) = 0$:

$$2x - 4 = 0$$
$$2x = 4$$
$$x = 2.$$

The **solution** of $2x - 4 = 0$ is 2. This is the zero of the function $f(x) = 2x - 4$. That is, $f(2) = 0$.

The zero of $f(x)$ is the first coordinate of the **x-intercept** of the graph of $y = f(x)$.

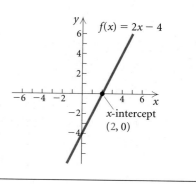

x-intercept $(2, 0)$

Exercise Set 2.1

In Exercises 1 and 2, use the given graph to find each of the following:

a) the x-intercept;

b) the zero of the function.

1.

2.

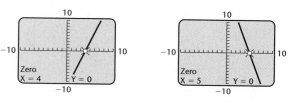

Find the zero of the linear function algebraically.

3. $f(x) = x + 5$ **4.** $f(x) = 5x + 20$

5. $f(x) = -x + 18$ **6.** $f(x) = 8 + x$

7. $f(x) = 16 - x$ **8.** $f(x) = -2x + 7$

Find the zero of the linear function graphically.

9. $f(x) = x + 12$ **10.** $f(x) = 8x + 2$

11. $f(x) = -x + 6$ **12.** $f(x) = 4 + x$

13. $f(x) = 20 - x$ **14.** $f(x) = -3x + 13$

Find the zero of the linear function.

15. $f(x) = x - 6$ **16.** $f(x) = 3x - 9$

17. $f(x) = -x + 15$ **18.** $f(x) = 4 - x$

Solve algebraically.

19. $4x + 3 = 0$ **20.** $3x - 16 = 0$

21. $2x + 7 = x + 3$ **22.** $5x - 4 = 2x + 5$

23. $3(x + 1) = 5 - 2(3x + 4)$

24. $4(3x + 2) - 7 = 3(x - 2)$

Solve graphically.

25. $5x - 8 = 0$ **26.** $6x + 5 = 0$

27. $3x - 5 = 2x + 1$ **28.** $4x + 3 = 2x - 7$

29. $2(x - 4) = 3 - 5(2x + 1)$

30. $3(2x - 5) + 4 = 2(4x + 3)$

Solve.

31. $8x + 25 = 0$

32. $7 - 9x = 0$

33. $6x - 5 = 3x + 10$

34. $2(3x - 1) = 3 - 4(x + 2)$

35. *Tourism.* France was the leading destination of foreign tourists in 1998, with 70 million arrivals. This exceeded the number of foreign-tourist arrivals in the United States by 22.9 million. (*Source*: World Tourism Organization) How many foreign tourists visited the United States in 1998?

36. *Express Delivery.* DHL Worldwide Express has determined that its couriers in Cairo walk an average of 11 mi on a single route when delivering packages. This is about six times the average distance walked by its couriers in Amsterdam. (*Source*: DHL Worldwide Express) How far, on average, does a courier in Amsterdam walk on a route?

37. *Direct Marketing.* In a recent year, 8000 catalog companies mailed 14 billion catalogs to the 112 million households in the United States (*Source*: W. A. Dean and Associates). On average, how many catalogs were received in each household?

38. *Traffic Delays.* In a recent study of 70 metropolitan areas, it was determined that drivers in Indianapolis are delayed in traffic 32 hr annually. This is eight times the annual delay in 1982. (*Source*: Texas Transportation Institute Urban Mobility Study) What was the annual delay in 1982?

39. *Braille Reading Material.* With an increasing reliance on tape recorders, letter magnifiers, and computer voice translators, the percentage of legally blind students learning Braille, a system that translates language into a code of raised dots, has dropped from 53% in 1963 to 10% in 1997. Nevertheless, an increasing number of Braille materials are being published, because those who do read Braille continue to seek reading material. In 1997, the Braille Press published 5.8 million pages. This was about three times the number of pages published in 1977. (*Source*: The American Foundation for the Blind) How many Braille pages were published by the Braille Press in 1977?

40. *Corporate Fax Expense.* The average *Fortune 500* company spends $13 million annually for local and domestic long-distance fax service. The amount spent for local faxes is about one third the amount spent for domestic long-distance faxes. (*Source*: Gallup survey of *Fortune 500* companies for Pitney Bowes) Find the amount spent for each type of fax service.

41. *Test Plot Dimensions.* A flower seed company has a rectangular test plot with a perimeter of 322 m. The length is 25 m more than the width. Find the dimensions of the plot.

w + 25

42. *Garden Dimensions.* The children at Tiny Tots Day Care plant a rectangular vegetable garden with a perimeter of 39 m. The length is twice the width. Find the dimensions of the garden.

43. *Angle Measure.* In triangle *ABC*, angle *B* is five times as large as angle *A*. The measure of angle *C* is 2° less than that of angle *A*. Find the measures of the angles. (*Hint*: The sum of the angle measures is 180°.)

44. *Angle Measure.* In triangle *ABC*, angle *B* is twice as large as angle *A*. Angle *C* measures 20° more than angle *A*. Find the measures of the angles.

45. *HMO Enrollment.* The number of enrollees in health maintenance organizations (HMOs) in Kentucky in a recent year was 416,300. This was a 48.4% increase over the previous year. (*Source*: Marion Merrell Dow Managed Care Department) What was the former number of enrollees?

46. *Moving Costs.* The average cost of moving a distance of 1255 mi is $1035 if you handle the move yourself. This is about 52% of the cost of hiring a professional moving firm. (*Source*: American Movers Conference) What is the cost of the professionally handled move?

47. *Train Speeds.* The speed of an Amtrak passenger train is 14 mph faster than the speed of a Central Railway freight train. The passenger train travels 400 mi in the same time it takes the freight train to travel 330 mi. Find the speed of each train.

48. *Distance Traveled.* A private airplane leaves Midway airport and flies due east at a speed of 180 km/h. Two hours later, a jet leaves Midway and flies due east at a speed of 900 km/h. How far from the airport will the jet overtake the private plane?

49. *Water Weight.* Water accounts for 50% of a woman's weight (*Source*: National Institute for Fitness and Sport). Jocelyn weighs 135 lb. How much of her body weight is water?

50. *Water Weight.* Water accounts for 60% of a man's weight (*Source*: National Institute for Fitness and Sport). Reggie weighs 186 lb. How much of his body weight is water?

51. *Cardiovascular Procedures.* About 6 million Americans had cardiovascular procedures performed in 1999. This was about 9% more than the number who had such procedures performed in 1996. (*Source*: *Parade Magazine*, July 4, 1999) How many Americans had cardiovascular procedures performed in 1996?

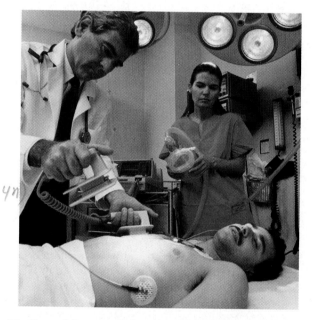

52. *Cost of Commercials.* In the fall of 1998, the average cost of a 30-sec commercial on NBC television was $168,000. This was a 10.2% decrease from the cost a year earlier. (*Source*: *Advertising Age*) What was the cost of a 30-sec commercial on NBC in the fall of 1997?

53. *Programming Cost.* Rights to the medical drama *ER* cost NBC television $13 million per episode. NBC charges advertisers $565,000 per 30-sec commercial on *ER*. (*Source*: *Indianapolis Star*, October 1, 1998) How many 30-sec commercials must the network sell in order to break even on the show?

54. *Book Buyers.* 15% of book buyers are 40–44 years old. This is two and one-half times the number who are 18–24 years old. (*Source*: Consumer Research Study on Book Purchasing) What percent of book buyers are 18–24 years old?

55. *Erosion.* Because of erosion, Horseshoe Falls, one of the two falls that make up Niagara Falls, is migrating upstream at a rate of 2 ft per year (*Source*: *Indianapolis Star,* February 14, 1999). At this rate, how long will it take the falls to move one-fourth mile?

56. *Volcanic Activity.* A volcano that is currently about one-half mile below the surface of the Pacific Ocean near the Big Island of Hawaii will eventually become a new Hawaiian island, Loihi. The volcano will break the surface of the ocean in about 50,000 yr. (*Source*: U.S. Geological Survey) On average, how many inches does the volcano rise in a year?

Discussion and Writing

57. Explain in your own words why a linear function $f(x) = mx + b$, with $m \neq 0$, has exactly one zero.

58. In the statement "A linear function $f(x) = mx + b$, with $m \neq 0$, has exactly one zero," explain why it is necessary to include the restriction $m \neq 0$.

Skill Maintenance

59. Write a slope–intercept equation for the line containing the point $(-1, 4)$ and parallel to the line $3x + 4y = 7$.

60. Fit a regression line to the data in the following table and use it to predict the number of households with personal computers in 2005.

Households with Personal Computers

YEAR, x	NUMBER OF HOUSEHOLDS (IN MILLIONS)
1996, 0	38.7
1997, 1	44.0
1998, 2	47.8
1999, 3	51.9
2000, 4	55.1

Source: Jupiter Communications

Given that $f(x) = 2x - 1$ and $g(x) = x^2 - 5$, find each of the following.

61. $(f + g)(x)$

62. $(fg)(-1)$

Synthesis

State whether each of the following is a linear function.

63. $f(x) = 7 - \dfrac{3}{2}x$

64. $f(x) = \dfrac{3}{2x} + 5$

65. $f(x) = x^2 + 1$

66. $f(x) = \dfrac{3}{4}x - (2.4)^2$

The Complex Numbers

2.2

- *Perform computations involving complex numbers.*

Some functions have zeros that are not real numbers. In order to find the zeros of such functions, we must consider the **complex-number system.**

The Complex-Number System

Let's begin by considering the function $f(x) = x^2 - 5$. To find the zeros of this function, we solve the associated equation $x^2 - 5 = 0$.

EXAMPLE 1   Solve: $x^2 - 5 = 0$.

Algebraic Solution

We use the principle of square roots:

$$x^2 - 5 = 0$$
$$x^2 = 5 \qquad \text{Adding 5}$$
$$x = \pm\sqrt{5}. \qquad \text{Using the principle of square roots}$$

The solutions of the equation are the real numbers $-\sqrt{5}$ and $\sqrt{5}$, or approximately -2.236 and 2.236.

Graphical Solution

Using the Zero method, we graph the function $f(x) = x^2 - 5$ and use the ZERO feature to find its zeros.

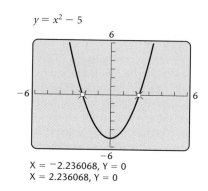

$y = x^2 - 5$

X = −2.236068, Y = 0
X = 2.236068, Y = 0

The zeros of the function $f(x) = x^2 - 5$ are the solutions of the equation $x^2 - 5 = 0$. They are approximately -2.236 and 2.236.

Note that the real solutions found with many graphers are approximations. Now let's consider an equation that has no real-number solutions.

EXAMPLE 2   Solve: $x^2 + 1 = 0$.

Algebraic Solution

We have

$$x^2 + 1 = 0$$
$$x^2 = -1 \qquad \text{Subtracting 1}$$
$$x = \pm\sqrt{-1}. \qquad \text{Using the principle of square roots}$$

There are no real-number square roots of negative numbers. Thus the equation $x^2 + 1 = 0$ has no real-number solutions.

Graphical Solution

We graph the function $f(x) = x^2 + 1$.

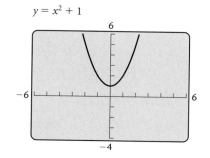

$y = x^2 + 1$

We see that since the graph does not cross the x-axis, it has no x-intercepts. This means that the function $f(x) = x^2 + 1$ has no *real-number* zeros. Thus there are no *real-number* solutions of the equation $x^2 + 1 = 0$.

We can define a number that is a solution of the equation in Example 2.

The Number i

The number i is defined such that

$$i = \sqrt{-1} \quad \text{and} \quad i^2 = -1.$$

To express roots of negative numbers in terms of i, we can use the fact that

$$\sqrt{-p} = \sqrt{-1 \cdot p} = \sqrt{-1}\,\sqrt{p} = i\sqrt{p}$$

when p is a positive real number.

EXAMPLE 3   Express each number in terms of i.

a) $\sqrt{-7}$ b) $\sqrt{-16}$

c) $-\sqrt{-13}$ d) $-\sqrt{-64}$

e) $\sqrt{-48}$

Solution

a) $\sqrt{-7} = \sqrt{-1 \cdot 7} = \sqrt{-1} \cdot \sqrt{7} = i\sqrt{7}$, or $\sqrt{7}i$

b) $\sqrt{-16} = \sqrt{-1 \cdot 16} = \sqrt{-1} \cdot \sqrt{16} = i \cdot 4 = 4i$

c) $-\sqrt{-13} = -\sqrt{-1 \cdot 13} = -\sqrt{-1} \cdot \sqrt{13} = -i\sqrt{13}$, or $-\sqrt{13}i$

d) $-\sqrt{-64} = -\sqrt{-1 \cdot 64} = -\sqrt{-1} \cdot \sqrt{64} = -i \cdot 8 = -8i$

e) $\sqrt{-48} = \sqrt{-1 \cdot 48} = \sqrt{-1} \cdot \sqrt{48}$
$= i\sqrt{16 \cdot 3} = i \cdot 4\sqrt{3} = 4\sqrt{3}i$, or $4i\sqrt{3}$ ▬

> i is *not* under the radical.

The complex numbers are formed by adding real numbers and multiples of i.

Complex Numbers

A **complex number** is a number of the form $a + bi$, where a and b are real numbers. The number a is said to be the **real part** of $a + bi$ and the number b is said to be the **imaginary part** of $a + bi$.*

Note that either a or b or both can be 0. When $b = 0$, $a + bi = a + 0i = a$, so every real number is a complex number. A complex number like $3 + 4i$ or $17i$, for which $b \neq 0$, is called an **imaginary number.** A complex number like $17i$ or $-4i$, in which $a = 0$ and $b \neq 0$, is some-

---

*Sometimes bi is considered to be the imaginary part.

times called a **pure imaginary number.** The relationships among various types of complex numbers are shown in the figure below.

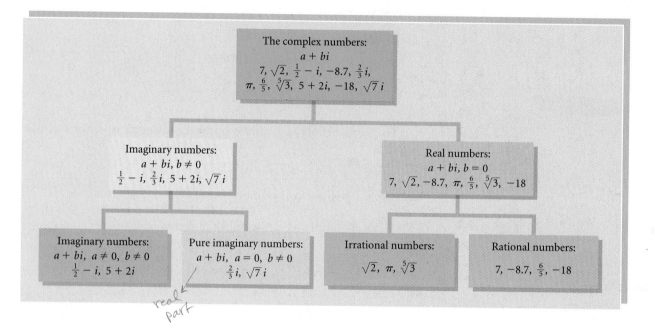

Addition and Subtraction

The complex numbers obey the commutative, associative, and distributive laws. Thus we can add and subtract them as we do binomials.

EXAMPLE 4   Add or subtract and simplify each of the following.

a) $(8 + 6i) + (3 + 2i)$ **b)** $(4 + 5i) - (6 - 3i)$

Solution

a) $(8 + 6i) + (3 + 2i) = (8 + 3) + (6i + 2i)$

                                  Collecting the real parts and the imaginary parts

             $= 11 + (6 + 2)i = 11 + 8i$

b) $(4 + 5i) - (6 - 3i) = (4 - 6) + [5i - (-3i)]$

                                  Note that 6 and $-3i$ are both being subtracted.

             $= -2 + 8i$

Most graphers perform operations on complex numbers as seen in the window below. Some graphers will express a complex number in the form (a, b) rather than $a + bi$.

```
(8+6i)+(3+2i)
                          11+8i
(4+5i)−(6−3i)
                          −2+8i
```

Multiplication

When $\sqrt{a}$ and $\sqrt{b}$ are real numbers, $\sqrt{a} \cdot \sqrt{b} = \sqrt{ab}$, but this is not true when $\sqrt{a}$ and $\sqrt{b}$ are not real numbers. Thus,

$$\sqrt{-2} \cdot \sqrt{-5} = \sqrt{-1} \cdot \sqrt{2} \cdot \sqrt{-1} \cdot \sqrt{5}$$
$$= i\sqrt{2} \cdot i\sqrt{5}$$
$$= i^2\sqrt{10} = -1\sqrt{10} = -\sqrt{10} \quad \text{is correct!}$$

But

$$\sqrt{-2} \cdot \sqrt{-5} = \sqrt{(-2)(-5)} = \sqrt{10} \quad \text{is wrong!}$$

Keeping this and the fact that $i^2 = -1$ in mind, we multiply in much the same way that we do with real numbers.

EXAMPLE 5 Multiply and simplify each of the following.

a) $\sqrt{-16} \cdot \sqrt{-25}$ **b)** $(1 + 2i)(1 + 3i)$ **c)** $(3 - 7i)^2$

Solution

a) $\sqrt{-16} \cdot \sqrt{-25} = \sqrt{-1} \cdot \sqrt{16} \cdot \sqrt{-1} \cdot \sqrt{25}$
$$= i \cdot 4 \cdot i \cdot 5$$
$$= i^2 \cdot 20$$
$$= -1 \cdot 20 \qquad i^2 = -1$$
$$= -20$$

$\sqrt{(-16)}\sqrt{(-25)}$	
	-20
$(1+2i)(1+3i)$	
	$-5+5i$
$(3-7i)^2$	
	$-40-42i$

b) $(1 + 2i)(1 + 3i) = 1 + 3i + 2i + 6i^2$ Multiplying each term of one number by every term of the other (FOIL)
$$= 1 + 3i + 2i - 6 \qquad i^2 = -1$$
$$= -5 + 5i \qquad\qquad \text{Collecting like terms}$$

c) $(3 - 7i)^2 = 3^2 - 2 \cdot 3 \cdot 7i + (7i)^2$ Recall that $(A - B)^2 = A^2 - 2AB + B^2$
$$= 9 - 42i + 49i^2$$
$$= 9 - 42i - 49 \qquad i^2 = -1$$
$$= -40 - 42i$$

Recall that -1 raised to an *even* power is 1, and -1 raised to an *odd* power is -1. Simplifying powers of i can then be done by using the fact that $i^2 = -1$ and expressing the given power of i in terms of i^2. Consider the following:

$$i = \sqrt{-1}, \qquad\qquad i^5 = i^4 \cdot i = (i^2)^2 \cdot i = (-1)^2 \cdot i = i,$$
$$i^2 = -1, \qquad\qquad i^6 = (i^2)^3 = (-1)^3 = -1,$$
$$i^3 = i^2 \cdot i = (-1)i = -i, \qquad i^7 = (i^2)^3 \cdot i = (-1)^3 \cdot i = -i,$$
$$i^4 = (i^2)^2 = (-1)^2 = 1, \qquad i^8 = (i^2)^4 = (-1)^4 = 1.$$

Note that the powers of i cycle through the values i, -1, $-i$, and 1.

EXAMPLE 6 Simplify each of the following.

a) i^{37} 　　　　　　　　　　　　　**b)** i^{58}

c) i^{75} 　　　　　　　　　　　　　**d)** i^{80}

Solution

a) $i^{37} = i^{36} \cdot i = (i^2)^{18} \cdot i = (-1)^{18} \cdot i = 1 \cdot i = i$

b) $i^{58} = (i^2)^{29} = (-1)^{29} = -1$

c) $i^{75} = i^{74} \cdot i = (i^2)^{37} \cdot i = (-1)^{37} \cdot i = -1 \cdot i = -i$

d) $i^{80} = (i^2)^{40} = (-1)^{40} = 1$

Conjugates and Divison

Conjugates of complex numbers are defined as follows.

Conjugate of a Complex Number

The **conjugate** of a complex number $a + bi$ is $a - bi$. The numbers $a + bi$ and $a - bi$ are **complex conjugates.**

Each of the following pairs of numbers are complex conjugates: $-3 + 7i$ and $-3 - 7i$; $14 - 5i$ and $14 + 5i$; and $8i$ and $-8i$.

The product of a complex number and its conjugate is a real number.

EXAMPLE 7 Multiply each of the following.

a) $(5 + 7i)(5 - 7i)$

b) $(8i)(-8i)$

Solution

```
(5+7i)(5-7i)
                74
(8i)(-8i)
                64
```

a) $(5 + 7i)(5 - 7i) = 5^2 - (7i)^2$ 　　　Using $(A + B)(A - B) = A^2 - B^2$

$$= 25 - 49i^2$$
$$= 25 - 49(-1)$$
$$= 74$$

b) $(8i)(-8i) = -64i^2$

$$= -64(-1)$$
$$= 64$$

When we are dividing complex numbers, conjugates are used.

EXAMPLE 8 Divide $2 - 5i$ by $1 - 6i$.

Solution We write fractional notation and then multiply by 1, using the conjugate of the denominator to form the symbol for 1.

```
(2−5i)/(1−6i) ▶ Frac
              32/37+7/37i
```

$$\frac{2 - 5i}{1 - 6i} = \frac{2 - 5i}{1 - 6i} \cdot \frac{1 + 6i}{1 + 6i}$$

Note that $1 + 6i$ is the conjugate of the divisor.

$$= \frac{(2 - 5i)(1 + 6i)}{(1 - 6i)(1 + 6i)}$$

$$= \frac{2 + 7i - 30i^2}{1 - 36i^2}$$

$$= \frac{2 + 7i + 30}{1 + 36} \qquad i^2 = -1$$

$$= \frac{32 + 7i}{37}$$

$$= \frac{32}{37} + \frac{7}{37}i.$$

STUDY TIP

Prepare yourself for your homework assignment by reading the explanations of concepts and following the step-by-step solutions of examples in the text. The time you spend preparing yourself will save you valuable time when you do your assignment.

Exercise Set 2.2

Simplify. Write answers in the form $a + bi$, where a and b are real numbers.

1. $(-5 + 3i) + (7 + 8i)$ $= 2 + 11i$
2. $(-6 - 5i) + (9 + 2i)$ $= 3 - 3i$
3. $(4 - 9i) + (1 - 3i)$ $= 5 - 12i$
4. $(7 - 2i) + (4 - 5i)$ $= 11 - 7i$
5. $(3 + \sqrt{-16}) + (2 + \sqrt{-25})$ $= 5 + 9i$
6. $(7 - \sqrt{-36}) + (2 + \sqrt{-9})$ $= 9 - 9i$
7. $(10 + 7i) - (5 + 3i)$
8. $(-3 - 4i) - (8 - i)$
9. $(13 + 9i) - (8 + 2i)$
10. $(-7 + 12i) - (3 - 6i)$
11. $(1 + 3i)(1 - 4i)$
12. $(1 - 2i)(1 + 3i)$
13. $(2 + 3i)(2 + 5i)$
14. $(3 - 5i)(8 - 2i)$
15. $7i(2 - 5i)$
16. $3i(6 + 4i)$
17. $(3 + \sqrt{-16})(2 + \sqrt{-25})$
18. $(7 - \sqrt{-16})(2 + \sqrt{-9})$
19. $(5 - 4i)(5 + 4i)$
20. $(5 + 9i)(5 - 9i)$

21. $(4 + 2i)^2$
22. $(5 - 4i)^2$
23. $(-2 + 7i)^2$
24. $(-3 + 2i)^2$
25. $\dfrac{2 + \sqrt{3}i}{5 - 4i}$
26. $\dfrac{\sqrt{5} + 3i}{1 - i}$
27. $\dfrac{4 + i}{-3 - 2i}$
28. $\dfrac{5 - i}{-7 + 2i}$
29. $\dfrac{3}{5 - 11i}$
30. $\dfrac{i}{2 + i}$
31. $\dfrac{1 + i}{(1 - i)^2}$
32. $\dfrac{1 - i}{(1 + i)^2}$
33. $\dfrac{4 - 2i}{1 + i} + \dfrac{2 - 5i}{1 + i}$
34. $\dfrac{3 + 2i}{1 - i} + \dfrac{6 + 2i}{1 - i}$

Simplify.

35. i^{11} $= -i$
36. i^7
37. i^{35}
38. i^{24}
39. i^{64}
40. i^{42}
41. $(-i)^{71}$
42. $(-i)^6$
43. $(5i)^4$
44. $(2i)^5$

Discussion and Writing

45. Is the sum of two imaginary numbers always an imaginary number? Explain your answer.
46. Is the product of two imaginary numbers always an imaginary number? Explain your answer.

Skill Maintenance

47. Write a slope–intercept equation for the line containing the point $(3, -5)$ and perpendicular to the line $3x - 6y = 7$.

Given that $f(x) = x^2 + 4$ and $g(x) = 3x + 5$, find each of the following.

48. $(f - g)(x)$

49. $(f/g)(2)$

50. For the function $f(x) = x^2 - 3x + 4$, construct and simplify the difference quotient

$$\frac{f(x + h) - f(x)}{h}.$$

Synthesis

Determine whether each of the following is true or false.

51. The sum of two numbers that are conjugates of each other is always a real number.

52. The conjugate of a sum is the sum of the conjugates of the individual complex numbers.

53. The conjugate of a product is the product of the conjugates of the individual complex numbers.

Let $z = a + bi$ and $\bar{z} = a - bi$.

54. Find a general expression for $1/z$.

55. Find a general expression for $z\bar{z}$.

56. Solve $z + 6\bar{z} = 7$ for z.

Zeros of Quadratic Functions and Models

2.3

- *Find zeros of quadratic functions and solve quadratic equations by completing the square and by using the quadratic formula.*
- *Solve equations that are reducible to quadratic.*
- *Solve applied problems.*

توابع   معادلات

Quadratic Functions and Quadratic Equations

In this section, we will explore the relationship between the zeros of quadratic functions and the solutions of quadratic equations. We define quadratic functions and equations as follows.

Quadratic Functions

A **quadratic function** f is a second-degree polynomial function

$$f(x) = ax^2 + bx + c, \quad a \neq 0,$$

where a, b, and c are real numbers.

Quadratic Equations

A **quadratic equation** is an equation equivalent to

$$ax^2 + bx + c = 0, \quad a \neq 0,$$

where a, b, and c are real numbers.

ZEROS OF A FUNCTION
REVIEW SECTION 2.1.

The zeros of a quadratic function $f(x) = ax^2 + bx + c$ are the solutions of the associated quadratic equation $ax^2 + bx + c = 0$. (These solutions are sometimes called *roots* of the equation.) Quadratic functions can have real-number or imaginary-number zeros and quadratic equations can have real-number or imaginary-number solutions. If the zeros

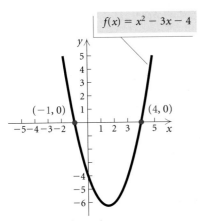

Two real-number zeros
Two x-intercepts

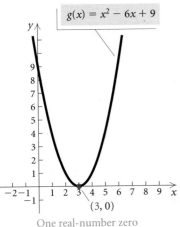

One real-number zero
One x-intercept

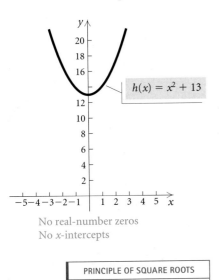

No real-number zeros
No x-intercepts

| PRINCIPLE OF SQUARE ROOTS |
| REVIEW SECTION R.7. |

or solutions are real numbers, they are also the first coordinates of the x-intercepts of the graph of the quadratic function or equation.

We now develop methods for finding the exact solutions of *any* quadratic equation.

Completing the Square

Quadratic equations like $x^2 - 3x - 4 = 0$ and $x^2 - 6x + 9 = 0$ can be solved by factoring and using the principle of zero products:

$$x^2 - 3x - 4 = 0$$
$$(x + 1)(x - 4) = 0 \qquad \text{Factoring}$$
$$x + 1 = 0 \quad or \quad x - 4 = 0 \qquad \text{Using the principle of zero products}$$
$$x = -1 \quad or \qquad x = 4.$$

The equation $x^2 - 3x - 4 = 0$ has *two real-number* solutions, -1 and 4. These are the zeros of the associated quadratic function $f(x) = x^2 - 3x - 4$ and the first coordinates of the x-intercepts of the graph of this function.

Next, we have

$$x^2 - 6x + 9 = 0$$
$$(x - 3)(x - 3) = 0 \qquad \text{Factoring}$$
$$x - 3 = 0 \quad or \quad x - 3 = 0 \qquad \text{Using the principle of zero products}$$
$$x = 3 \quad or \qquad x = 3.$$

The equation $x^2 - 6x + 9 = 0$ has *one real-number* solution, 3. It is the zero of the associated quadratic function $g(x) = x^2 - 6x + 9$ and the first coordinate of the x-intercept of the graph of this function.

The principle of square roots can be used to solve quadratic equations like $x^2 + 13 = 0$:

$$x^2 + 13 = 0$$
$$x^2 = -13$$
$$x = \pm\sqrt{-13} \qquad \text{Using the principle of square roots}$$
$$x = \pm\sqrt{13}i.$$

The equation has *two imaginary-number* solutions, $-\sqrt{13}i$ and $\sqrt{13}i$. These are the zeros of the associated quadratic function $h(x) = x^2 + 13$. Since the zeros are not real numbers, the graph of the function has no x-intercepts.

We can find the solutions of $x^2 - 3x - 4 = 0$ and $x^2 - 6x + 9 = 0$ graphically, but a graphical approach would not yield the solutions of $x^2 + 13 = 0$. Some graphers can find complex-number solutions of quadratic equations, but the procedures used are not graphical. Neither the principle of zero products nor the principle of square roots would yield the *exact* zeros of a function like $f(x) = x^2 - 6x - 10$ or the *exact* solutions of the associated equation $x^2 - 6x - 10 = 0$. If we wish to do that, we can use a procedure called **completing the square** and then use the principle of square roots.

EXAMPLE 1   Find the zeros of $f(x) = x^2 - 6x - 10$ by completing the square.

Solution   We find the values of x for which $f(x) = 0$. That is, we solve the associated equation $x^2 - 6x - 10 = 0$. Our goal is to find an equivalent equation of the form $x^2 + bx + c = d$ in which $x^2 + bx + c$ is a perfect square. Since

$$x^2 + bx + \left(\frac{b}{2}\right)^2 = \left(x + \frac{b}{2}\right)^2,$$

the number c is found by taking half the coefficient of the x-term and squaring it. Then for the equation $x^2 - 6x - 10 = 0$, we have

$$x^2 - 6x - 10 = 0$$
$$x^2 - 6x \qquad = 10 \qquad \text{Adding 10}$$
$$x^2 - 6x + 9 = 10 + 9 \qquad \text{Adding 9 to complete the square:}$$
$$\left(\frac{b}{2}\right)^2 = \left(\frac{-6}{2}\right)^2 = 9$$
$$x^2 - 6x + 9 = 19.$$

Because $x^2 - 6x + 9$ is a perfect square, we are able to write it as the square of a binomial. We can then use the principle of square roots to finish the solution:

$$(x - 3)^2 = 19 \qquad \text{Factoring}$$
$$x - 3 = \pm\sqrt{19} \qquad \text{Using the principle of square roots}$$
$$x = 3 \pm \sqrt{19}. \qquad \text{Adding 3}$$

Thus the solutions of the equation are $3 + \sqrt{19}$ and $3 - \sqrt{19}$, or simply $3 \pm \sqrt{19}$. The zeros of $f(x) = x^2 - 6x - 10$ are also $3 + \sqrt{19}$ and $3 - \sqrt{19}$, or $3 \pm \sqrt{19}$.

Decimal approximations for $3 \pm \sqrt{19}$ can be found using a grapher:

$$3 + \sqrt{19} \approx 7.359 \quad \text{and} \quad 3 - \sqrt{19} \approx -1.359.$$

Approximations for the zeros can also be found by using the Zero method to find the zeros of the quadratic function $f(x) = x^2 - 6x - 10$, as shown below.

3+√(19)	
	7.358898944
3−√(19)	
	−1.358898944

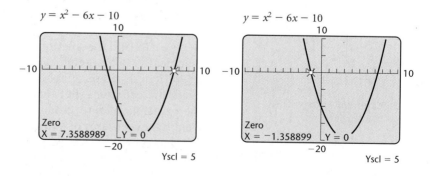

The zeros are approximately 7.359 and -1.359.

EXAMPLE 2 Solve: $2x^2 - 1 = 3x$.

Solution We have

$$2x^2 - 1 = 3x$$

$$2x^2 - 3x - 1 = 0 \qquad \text{Subtracting } 3x. \text{ We are unable to factor the result.}$$

$$2x^2 - 3x = 1 \qquad \text{Adding 1}$$

$$x^2 - \frac{3}{2}x = \frac{1}{2} \qquad \text{Dividing by 2 to make the } x^2\text{-coefficient 1}$$

$$x^2 - \frac{3}{2}x + \frac{9}{16} = \frac{1}{2} + \frac{9}{16} \qquad \text{Completing the square: } \frac{1}{2}\left(-\frac{3}{2}\right) = -\frac{3}{4} \text{ and } \left(-\frac{3}{4}\right)^2 = \frac{9}{16}; \text{ adding } \frac{9}{16}$$

$$\left(x - \frac{3}{4}\right)^2 = \frac{17}{16} \qquad \text{Factoring and simplifying}$$

$$x - \frac{3}{4} = \pm\frac{\sqrt{17}}{4} \qquad \text{Using the principle of square roots and the quotient rule for radicals}$$

$$x = \frac{3 \pm \sqrt{17}}{4}. \qquad \text{Adding } \frac{3}{4}$$

The solutions are

$$\frac{3 + \sqrt{17}}{4} \quad \text{and} \quad \frac{3 - \sqrt{17}}{4}, \quad \text{or} \quad \frac{3 \pm \sqrt{17}}{4}.$$

We can calculate approximate solutions using a grapher and check the solutions graphically by graphing the associated function $f(x) = 2x^2 - 3x - 1$.

To solve a quadratic equation by completing the square:

1. Isolate the terms with variables on one side of the equation and arrange them in descending order.
2. Divide by the coefficient of the squared term if that coefficient is not 1.
3. Complete the square by taking half the coefficient of the first-degree term and adding its square on both sides of the equation.
4. Express one side of the equation as the square of a binomial.
5. Use the principle of square roots.
6. Solve for the variable.

Using the Quadratic Formula

Because completing the square works for *any* quadratic equation, it can be used to solve the general quadratic equation $ax^2 + bx + c = 0$ for x. The result will be a formula that can be used to solve any quadratic equation quickly.

Consider any quadratic equation in standard form:

$$ax^2 + bx + c = 0, \quad a \neq 0.$$

For now, we assume that $a > 0$ and solve by completing the square. As the steps are carried out, compare them with those of Example 2.

$$ax^2 + bx + c = 0 \qquad \text{Standard form}$$

$$ax^2 + bx = -c \qquad \text{Adding } -c$$

$$x^2 + \frac{b}{a}x = -\frac{c}{a} \qquad \text{Dividing by } a$$

Half of $\dfrac{b}{a}$ is $\dfrac{b}{2a}$ and $\left(\dfrac{b}{2a}\right)^2 = \dfrac{b^2}{4a^2}$. Thus we add $\dfrac{b^2}{4a^2}$:

$$x^2 + \frac{b}{a}x + \frac{b^2}{4a^2} = -\frac{c}{a} + \frac{b^2}{4a^2} \qquad \text{Adding } \dfrac{b^2}{4a^2} \text{ to complete the square}$$

$$\left(x + \frac{b}{2a}\right)^2 = -\frac{4ac}{4a^2} + \frac{b^2}{4a^2} \qquad \begin{array}{l}\text{Factoring and finding a common}\\ \text{denominator:} \\ -\dfrac{c}{a} = -\dfrac{4a}{4a} \cdot \dfrac{c}{a} = -\dfrac{4ac}{4a^2}\end{array}$$

$$\left(x + \frac{b}{2a}\right)^2 = \frac{b^2 - 4ac}{4a^2}$$

$$x + \frac{b}{2a} = \pm\frac{\sqrt{b^2 - 4ac}}{2a} \qquad \begin{array}{l}\text{Using the principle of square roots}\\ \text{and the quotient rule for radicals.}\\ \text{Since } a > 0, \sqrt{4a^2} = 2a.\end{array}$$

$$x = \frac{-b \pm \sqrt{b^2 - 4ac}}{2a}. \qquad \text{Adding } -\dfrac{b}{2a}$$

It can also be shown that this result holds if $a < 0$.

The Quadratic Formula

The solutions of $ax^2 + bx + c = 0$, $a \neq 0$, are given by

$$x = \frac{-b \pm \sqrt{b^2 - 4ac}}{2a}.$$

EXAMPLE 3   Solve $3x^2 + 2x = 7$. Find exact solutions and approximate solutions rounded to the nearest thousandth.

Solution   We show both algebraic and graphical solutions.

Algebraic Solution

After finding standard form, we are unable to factor, so we identify a, b, and c in order to use the quadratic formula:

$$3x^2 + 2x - 7 = 0;$$
$$a = 3, \quad b = 2, \quad c = -7.$$

We then use the quadratic formula:

$$x = \frac{-b \pm \sqrt{b^2 - 4ac}}{2a}$$

$$= \frac{-2 \pm \sqrt{2^2 - 4(3)(-7)}}{2(3)} \qquad \textbf{Substituting}$$

$$= \frac{-2 \pm \sqrt{4 + 84}}{6} = \frac{-2 \pm \sqrt{88}}{6}$$

$$= \frac{-2 \pm \sqrt{4 \cdot 22}}{6} = \frac{-2 \pm 2\sqrt{22}}{6}$$

$$= \frac{2}{2} \cdot \frac{-1 \pm \sqrt{22}}{3} = \frac{-1 \pm \sqrt{22}}{3}.$$

The exact solutions are

$$\frac{-1 - \sqrt{22}}{3} \quad \text{and} \quad \frac{-1 + \sqrt{22}}{3}.$$

Using a grapher, we approximate the solutions to be -1.897 and 1.230.

A table set in ASK mode can be used to check the solutions. If the exact solutions are entered, the grapher converts them to decimal approximations in the table.

X	Y1	
−1.897	0	
1.2301	0	
X =		

A graph can also provide a visual check of the algebraic solution.

Graphical Solution

Using the Intersect method, we graph $y_1 = 3x^2 + 2x$ and $y_2 = 7$ and use the INTERSECT feature to find the coordinates of the points of intersection. The first coordinates of these points are the solutions of the equation $y_1 = y_2$, or $3x^2 + 2x = 7$.

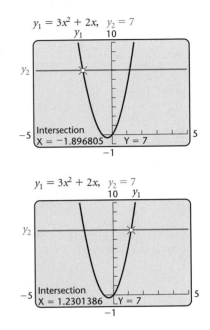

The solutions are approximately -1.897 and 1.230. We could also write the equation in standard form, $3x^2 + 2x - 7 = 0$, and use the Zero method.

Not all quadratic equations can be solved graphically.

EXAMPLE 4   Solve: $x^2 + 5x + 8 = 0$.

Algebraic Solution

To find the solutions, we use the quadratic formula. Here

$$a = 1, \quad b = 5, \quad c = 8;$$

$$x = \frac{-b \pm \sqrt{b^2 - 4ac}}{2a}$$

$$= \frac{-5 \pm \sqrt{5^2 - 4(1)(8)}}{2 \cdot 1} \qquad \text{Substituting}$$

$$= \frac{-5 \pm \sqrt{-7}}{2} \qquad \text{Simplifying}$$

$$= \frac{-5 \pm \sqrt{7}i}{2}.$$

The solutions are $-\dfrac{5}{2} - \dfrac{\sqrt{7}}{2}i$ and $-\dfrac{5}{2} + \dfrac{\sqrt{7}}{2}i$.

Graphical Solution

The graph of the function $f(x) = x^2 + 5x + 8$ shows no x-intercepts.

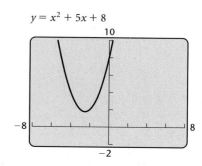

$y = x^2 + 5x + 8$

Thus the function has no zeros and there are no real-number solutions of the associated equation $x^2 + 5x + 8 = 0$. This is a quadratic equation that cannot be solved graphically.

The Discriminant

From the quadratic formula, we know that the solutions x_1 and x_2 of a quadratic equation are given by

$$x_1 = \frac{-b + \sqrt{b^2 - 4ac}}{2a} \quad \text{and} \quad x_2 = \frac{-b - \sqrt{b^2 - 4ac}}{2a}.$$

The expression $b^2 - 4ac$ shows the nature of the solutions. This expression is called the **discriminant**. If it is 0, then it makes no difference whether we choose the plus or the minus sign in the formula. There is just one solution. In this case, we sometimes say that there is one repeated real solution. If the discriminant is positive, there will be two real solutions. If it is negative, we will be taking the square root of a negative number; hence there will be two imaginary-number solutions, and they will be complex conjugates.

Discriminant

For $ax^2 + bx + c = 0$:

$b^2 - 4ac = 0 \longrightarrow$ One real-number solution;

$b^2 - 4ac > 0 \longrightarrow$ Two different real-number solutions;

$b^2 - 4ac < 0 \longrightarrow$ Two different imaginary-number solutions, complex conjugates.

In Example 3, the discriminant, 88, is positive, indicating that there are two different real-number solutions. If the discriminant is negative, as it is in Example 4, we know that there are two different imaginary-number solutions.

Equations Reducible to Quadratic

Some equations can be treated as quadratic, provided that we make a suitable substitution. For example, consider the following:

$$x^4 - 5x^2 + 4 = 0$$
$$(x^2)^2 - 5x^2 + 4 = 0 \qquad x^4 = (x^2)^2$$
$$u^2 - 5u + 4 = 0. \qquad \text{Substituting } u \text{ for } x^2$$

The equation $u^2 - 5u + 4 = 0$ can be solved for u by factoring or using the quadratic formula. Then we can reverse the substitution, replacing u with x^2, and solve for x. Equations like the one above are said to be **reducible to quadratic,** or **quadratic in form.** No substitution is required if this equation is solved graphically.

EXAMPLE 5 Solve: $x^4 - 5x^2 + 4 = 0$.

Algebraic Solution

We let $u = x^2$ and substitute:

$$u^2 - 5u + 4 = 0 \qquad \text{Substituting } u \text{ for } x^2$$
$$(u - 1)(u - 4) = 0 \qquad \text{Factoring}$$
$$u - 1 = 0 \quad or \quad u - 4 = 0 \qquad \begin{array}{l}\text{Using the} \\ \text{principle of} \\ \text{zero products}\end{array}$$
$$u = 1 \quad or \quad u = 4.$$

Don't stop here! We must solve for the original variable. We substitute x^2 for u and solve for x:

$$x^2 = 1 \qquad or \quad x^2 = 4$$
$$x = \pm 1 \quad or \quad x = \pm 2. \qquad \begin{array}{l}\text{Using the principle of} \\ \text{square roots}\end{array}$$

The solutions are -1, 1, -2, and 2.

Graphical Solution

Using the Zero method, we graph the function $y = x^4 - 5x^2 + 4$ and use the ZERO feature to find the zeros.

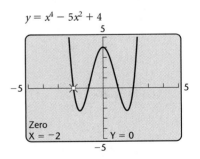

The leftmost zero is -2. Using the Zero feature three more times, we find that the other zeros are -1, 1, and 2.

Applications

Some applied problems can be translated to quadratic equations.

EXAMPLE 6 *Time of a Free Fall.* The World Trade Center in New York City is 1350 ft tall. How long would it take an object falling freely from the top to reach the ground?

Solution

1. **Familiarize.** The formula $s = 16t^2$ is used to approximate the distance s, in feet, that an object falls freely from rest in t seconds. In this case, the distance is 1350 ft.

2. **Translate.** We substitute 1350 for s in the formula:

$$1350 = 16t^2.$$

3. **Carry out.** We carry out the solution as follows.

Algebraic Solution

We use the principle of square roots:

$$1350 = 16t^2$$

$$\frac{1350}{16} = t^2 \qquad \text{Dividing by 16}$$

$$\sqrt{\frac{1350}{16}} = t \qquad \begin{array}{l}\text{Taking the positive} \\ \text{square root. Time} \\ \text{cannot be negative in} \\ \text{this application.}\end{array}$$

$$9.186 \approx t.$$

Graphical Solution

Using the Intersect method, we replace t with x, graph $y_1 = 1350$ and $y_2 = 16x^2$, and find the first coordinates of the points of intersection. Time cannot be negative in this application, so we need to find only the positive first coordinate. Since $y_1 = 1350$, we must choose a viewing window with Ymax greater than this value. Trial and error shows that a good choice is $[-15, 15, 0, 1500]$, with Xscl = 3 and Yscl = 100.

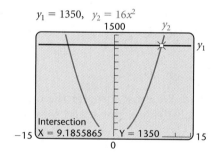

$$y_1 = 1350, \quad y_2 = 16x^2$$

Intersection
X = 9.1855865   Y = 1350

We see that $x \approx 9.186$.

4. **Check.** In 9.186 sec, an object falling freely from rest would travel a distance of $16(9.186)^2$, or about 1350 ft. The answer checks.

5. **State.** It would take about 9.186 sec for an object falling freely from the top of the World Trade Center to reach the ground. ▬

EXAMPLE 7 *Bicycling Speed.* Castulo and Linette leave a campsite, Castulo biking due north and Linette biking due east. Castulo bikes 7 km/h slower than Linette. After 4 hr, they are 68 km apart. Find the speed of each bicyclist.

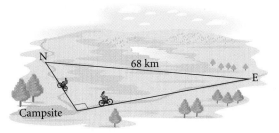

Solution

1. **Familiarize.** We let r = Linette's speed, in kilometers per hour. Then $r - 7$ = Castulo's speed, in kilometers per hour. We will use the motion formula $d = rt$, where d is the distance, r is the rate (or speed), and t is the time. Then, after 4 hr, Linette has traveled $4r$ km and Castulo has traveled $4(r - 7)$ km. We add these distances to the drawing, as shown below.

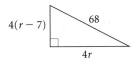

2. **Translate.** We use the Pythagorean theorem, $a^2 + b^2 = c^2$, where a and b are the lengths of the legs of a right triangle and c is the length of the hypotenuse:

$$(4r)^2 + [4(r - 7)]^2 = 68^2.$$

3. **Carry out.** We carry out the solution as follows.

Algebraic Solution

We have

$$(4r)^2 + [4(r - 7)]^2 = 68^2$$
$$16r^2 + 16(r^2 - 14r + 49) = 4624$$
$$16r^2 + 16r^2 - 224r + 784 = 4624$$

$$32r^2 - 224r - 3840 = 0 \qquad \text{Subtracting 4624}$$
$$r^2 - 7r - 120 = 0 \qquad \text{Dividing by 32}$$
$$(r + 8)(r - 15) = 0 \qquad \text{Factoring}$$
$$r + 8 = 0 \quad or \quad r - 15 = 0 \qquad \text{Principle of zero products}$$
$$r = -8 \quad or \qquad\quad r = 15.$$

Graphical Solution

Using the Intersect method, we enter the functions

$$y_1 = (4x)^2 + (4(x - 7))^2 \quad \text{and} \quad y_2 = 68^2$$

and find the points of intersection. It will probably be necessary to experiment with several viewing windows before an appropriate one is found. Note that $y_2 = 68^2$, or 4624, so Ymax must be greater than 4624. We try $[-10, 10, 0, 6000]$, with Yscl = 500.

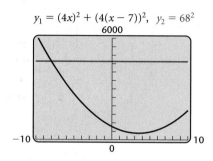

We note that we must increase Xmax in order to see the second point of intersection. The window $[-10, 20, 0, 6000]$, with Xscl = 5 and Yscl = 500 works well. The first coordinates of the points of intersection are the solutions of the equation $y_1 = y_2$, or $(4x)^2 + [4(x - 7)]^2 = 68^2$. Since the speed cannot be negative, we need to find only the positive first coordinate.

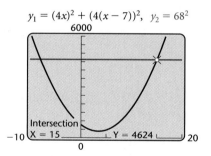

4. **Check.** Since speed cannot be negative, we need to check only 15. Note that this is the possible solution we considered in the graphical method. If Linette's speed is 15 km/h, then Castulo's speed is $15 - 7$, or 8 km/h. In 4 hr, Linette travels $4 \cdot 15$, or 60 km, and Castulo travels $4 \cdot 8$, or 32 km. Then they are $60^2 + 32^2$, or 4624 km apart. Since $4624 = 68^2$, the answer checks.

5. **State.** Linette's speed is 15 km/h and Castulo's speed is 8 km/h.

CONNECTING THE CONCEPTS

ZEROS, SOLUTIONS, AND INTERCEPTS

The zeros of a function $y = f(x)$ are also the solutions of the equation $f(x) = 0$ and the first coordinates of the x-intercepts of the graph of the function.

FUNCTION	ZEROS OF THE FUNCTION; SOLUTIONS OF THE EQUATION	X-INTERCEPTS OF THE GRAPH

Linear Function

$f(x) = 2x - 4$, or

$\quad y = 2x - 4$

To find the **zero** of $f(x)$, we solve $f(x) = 0$:

$$2x - 4 = 0$$
$$2x = 4$$
$$x = 2.$$

The **solution** of $2x - 4 = 0$ is 2. This is the zero of the function $f(x) = 2x - 4$. That is, $f(2) = 0$.

The zero of $f(x)$ is the first coordinate of the **x-intercept** of the graph of $y = f(x)$.

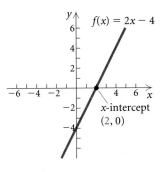

Quadratic Function

$g(x) = x^2 - 3x - 4$, or

$\quad y = x^2 - 3x - 4$

To find the **zeros** of $g(x)$, we solve $g(x) = 0$:

$$x^2 - 3x - 4 = 0$$
$$(x + 1)(x - 4) = 0$$
$$x + 1 = 0 \quad or \quad x - 4 = 0$$
$$x = -1 \quad or \quad x = 4.$$

The **solutions** of $x^2 - 3x - 4 = 0$ are -1 and 4. They are the zeros of the function $g(x)$. That is, $g(-1) = 0$ and $g(4) = 0$.

The zeros of $g(x)$ are the first coordinates of the **x-intercepts** of the graph of $y = g(x)$.

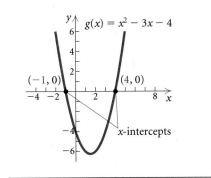

Exercise Set  2.3

In Exercises 1 and 2, use the given graph to find each of the following:

a) the x-intercepts and $(-4,0), (2,0)$
b) the zeros of the function. $-4, 2$

1.   **2.**

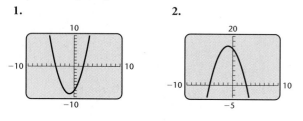

Solve by completing the square to obtain exact solutions. Check your work using a grapher, if appropriate.

3. $x^2 + 6x = 7$ **4.** $x^2 + 8x = -15$

5. $x^2 = 8x - 9$ **6.** $x^2 = 22 + 10x$

7. $x^2 + 8x + 25 = 0$ **8.** $x^2 + 6x + 13 = 0$

9. $3x^2 + 5x - 2 = 0$ **10.** $2x^2 - 5x - 3 = 0$

Use the quadratic formula to find exact solutions. Check your work using a grapher, if appropriate.

11. $x^2 - 2x = 15$ **12.** $x^2 + 4x = 5$

13. $5m^2 + 3m = 2$ **14.** $2y^2 - 3y - 2 = 0$

15. $3x^2 + 6 = 10x$ **16.** $3t^2 + 8t + 3 = 0$

17. $x^2 + x + 2 = 0$ **18.** $x^2 + 1 = x$

19. $5t^2 - 8t = 3$ **20.** $5x^2 + 2 = x$

21. $3x^2 + 4 = 5x$ **22.** $2t^2 - 5t = 1$

For each of the following, consider only $b^2 - 4ac$, the discriminant of the quadratic formula, to determine whether imaginary solutions exist.

23. $4x^2 = 8x + 5$ **24.** $4x^2 - 12x + 9 = 0$

25. $x^2 + 3x + 4 = 0$ **26.** $x^2 - 2x + 4 = 0$

27. $5t^2 - 7t = 0$ **28.** $5t^2 - 4t = 11$

Solve graphically. Round solutions to three decimal places, where appropriate.

29. $x^2 - 8x + 12 = 0$ **30.** $5x^2 + 42x + 16 = 0$

31. $7x^2 - 43x + 6 = 0$ **32.** $10x^2 - 23x + 12 = 0$

33. $6x + 1 = 4x^2$ **34.** $3x^2 + 5x = 3$

35. $2x^2 - 4 = 5x$ **36.** $4x^2 - 2 = 3x$

Find the zeros of the function algebraically. Give exact answers.

37. $f(x) = x^2 + 6x + 5$ **38.** $f(x) = x^2 - x - 2$

39. $f(x) = x^2 - 3x - 3$ **40.** $f(x) = 3x^2 + 8x + 2$

41. $f(x) = x^2 - 5x + 1$ **42.** $f(x) = x^2 - 3x - 7$

Use a grapher to find the zeros of the function. Round to three decimal places.

43. $f(x) = x^2 + 2x - 5$ **44.** $f(x) = x^2 - x - 4$

45. $f(x) = 3x^2 + 2x - 4$ **46.** $f(x) = 9x^2 - 8x - 7$

47. $f(x) = 5.02x^2 - 4.19x - 2.057$

48. $f(x) = 1.21x^2 - 2.34x - 5.63$

Solve.

49. $x^4 - 3x^2 + 2 = 0$

50. $x^4 + 3 = 4x^2$

51. $x - 3\sqrt{x} - 4 = 0$
    (Hint: Let $u = \sqrt{x}$.)

52. $2x - 9\sqrt{x} + 4 = 0$

53. $m^{2/3} - 2m^{1/3} - 8 = 0$
    (Hint: Let $u = m^{1/3}$.)

54. $t^{2/3} + t^{1/3} - 6 = 0$

55. $(2x - 3)^2 - 5(2x - 3) + 6 = 0$
    (Hint: Let $u = 2x - 3$.)

56. $(3x + 2)^2 + 7(3x + 2) - 8 = 0$

57. $(2t^2 + t)^2 - 4(2t^2 + t) + 3 = 0$

58. $12 = (m^2 - 5m)^2 + (m^2 - 5m)$

Time of a Free Fall.   The formula $s = 16t^2$ is used to approximate the distance s, in feet, that an object falls freely from rest in t seconds.

59. The Warszawa Radio Mast in Poland, at 2120 ft, is the world's tallest structure (*Source: The Cambridge Fact Finder*). How long would it take an object falling freely from the top to reach the ground?

60. The tallest structure in the United States, at 2063 ft, is the KTHI-TV tower in North Dakota (*Source: The Cambridge Fact Finder*). How long would it take an object falling freely from the top to reach the ground?

61. The length of a rectangular rug is 1 ft more than the width and a diagonal of the rug is 5 ft. Find the length and the width.

62. One leg of a right triangle is 7 cm less than the length of the other leg. The length of the hypotenuse is 13 cm. Find the lengths of the legs.

63. One number is 5 greater than another. The product of the numbers is 36. Find the numbers.

64. One number is 6 less than another. The product of the numbers is 72. Find the numbers.

65. *Box Construction.* An open box is made from a 10-cm by 20-cm piece of tin by cutting a square from each corner and folding up the edges. The area of the resulting base is 96 cm². What is the length of the sides of the squares?

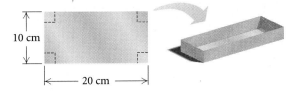

66. *Picture Frame Dimensions.* The frame of a picture is 28 cm by 32 cm outside and is of uniform width. What is the width of the frame if 192 cm² of the picture shows?

State whether the function is linear or quadratic.

67. $f(x) = 4 - 5x$ **68.** $f(x) = 4 - 5x^2$

69. $f(x) = 7x^2$ **70.** $f(x) = 23x + 6$

71. $f(x) = 1.2x - (3.6)^2$ **72.** $f(x) = 2 - x - x^2$

linear

Discussion and Writing

73. Is it possible for a quadratic function to have one real zero and one imaginary zero? Why or why not?

74. The graph of a quadratic function can have 0, 1, or 2 x-intercepts. How can you predict the number of x-intercepts without drawing the graph or (completely) solving an equation?

Skill Maintenance

Test algebraically whether the graph is symmetric with respect to the x-axis, the y-axis, and the origin. Check your work using a grapher.

75. $3x^2 + 4y^2 = 5$ **76.** $y^3 = 6x^2$

Test algebraically whether the function is even, odd, or neither even nor odd. Check your work using a grapher.

77. $f(x) = 2x^3 - x$ **78.** $f(x) = 4x^2 + 2x - 3$

Synthesis

Solve.

79. $x^2 + x - \sqrt{2} = 0$

80. $x^2 + \sqrt{5}x - \sqrt{3} = 0$

81. $2t^2 + (t - 4)^2 = 5t(t - 4) + 24$

82. $9t(t + 2) - 3t(t - 2) = 2(t + 4)(t + 6)$

83. $\sqrt{x - 3} - \sqrt[4]{x - 3} = 2$

84. $x^6 - 28x^3 + 27 = 0$

85. $\left(y + \dfrac{2}{y}\right)^2 + 3y + \dfrac{6}{y} = 4$

86. $x^2 + 3x + 1 - \sqrt{x^2 + 3x + 1} = 8$

87. Solve $\frac{1}{2}at^2 + v_0t + x_0 = 0$ for t.

• *Find the vertex, the line of symmetry, and the maximum or minimum value of a quadratic function using the method of completing the square.*
• *Graph quadratic functions.*
• *Solve applied problems involving maximum and minimum function values.*

Graphing Quadratic Functions of the Type $f(x) = a(x - h)^2 + k$

The graph of a quadratic function is called a **parabola**. We will see that the graph of every parabola evolves from the graph of the squaring function $f(x) = x^2$ using transformations.

Interactive Discovery

Think of transformations and look for patterns. Consider the following functions:

$$y_1 = x^2, \qquad y_2 = -0.4x^2,$$
$$y_3 = -0.4(x - 2)^2, \qquad y_4 = -0.4(x - 2)^2 + 3.$$

Graph y_1 and y_2. How do you get from the graph of y_1 to y_2?
Graph y_2 and y_3. How do you get from the graph of y_2 to y_3?
Graph y_3 and y_4. How do you get from the graph of y_3 to y_4?
  Consider the following functions:

$$y_1 = x^2, \qquad y_2 = 2x^2,$$
$$y_3 = 2(x + 3)^2, \qquad y_4 = 2(x + 3)^2 - 5.$$

Graph y_1 and y_2. How do you get from the graph of y_1 to y_2?
Graph y_2 and y_3. How do you get from the graph of y_2 to y_3?
Graph y_3 and y_4. How do you get from the graph of y_3 to y_4?

TRANSFORMATIONS

REVIEW SECTION 1.5.

We get the graph of $f(x) = a(x - h)^2 + k$ from the graph of $f(x) = x^2$ as follows:

$$f(x) = x^2$$
$$\downarrow$$
$$f(x) = ax^2 \qquad \textbf{Vertical stretching or shrinking with a reflection across the } x\textbf{-axis if } a < 0$$
$$\downarrow$$
$$f(x) = a(x - h)^2 \qquad \textbf{Horizontal translation}$$
$$\downarrow$$
$$f(x) = a(x - h)^2 + k. \qquad \textbf{Vertical translation}$$

Consider the following graphs of the form $f(x) = a(x - h)^2 + k$. The point (h, k) at which the graph turns is called the **vertex**. The maximum or minimum value of $f(x)$ occurs at the vertex. Each graph has a line $x = h$ that is called the **line of symmetry.**

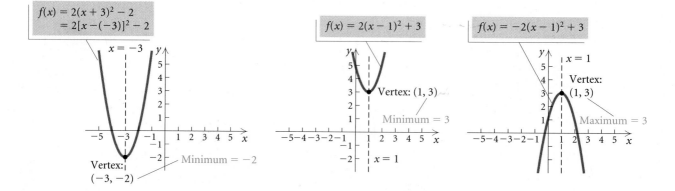

CONNECTING THE CONCEPTS

GRAPHING QUADRATIC FUNCTIONS

The graph of $f(x) = a(x - h)^2 + k$ is a parabola that:

opens up if $a > 0$ and down if $a < 0$;

has (h, k) as the vertex;

has $x = h$ as the line of symmetry;

has k as a minimum value (output) if $a > 0$ and has k as a maximum value if $a < 0$.

If the equation is in the form $f(x) = a(x - h)^2 + k$, we can learn a great deal about the graph without graphing.

FUNCTION	$f(x) = 3\left(x - \frac{1}{4}\right)^2 - 2$ $= 3\left(x - \frac{1}{4}\right)^2 + (-2)$	$g(x) = -3(x + 5)^2 + 7$ $= -3[x - (-5)]^2 + 7$
VERTEX	$\left(\frac{1}{4}, -2\right)$	$(-5, 7)$
LINE OF SYMMETRY	$x = \frac{1}{4}$	$x = -5$
MAXIMUM	No: $3 > 0$, graph opens up.	Yes, 7 $(-3 < 0$, graph opens down.)
MINIMUM	Yes, -2 $(3 > 0$, graph opens up.)	No: $-3 < 0$ graph opens down.

Note that the vertex (h, k) is used to find the maximum or the minimum value of the function. The maximum or minimum value is the number k, *not* the ordered pair (h, k).

Graphing Quadratic Functions of the Type $f(x) = ax^2 + bx + c, a \neq 0$

We now use a modification of the method of completing the square as an aid in graphing and analyzing quadratic functions of the form $f(x) = ax^2 + bx + c, a \neq 0$.

EXAMPLE 1 Complete the square to find the vertex, the line of symmetry, and the maximum or minimum value of $f(x) = x^2 + 10x + 23$. Then graph the function.

Solution To express $f(x) = x^2 + 10x + 23$ in the form $f(x) = a(x - h)^2 + k$, we construct a trinomial square. To do so, we take half the coefficient of x and square it, obtaining $(10/2)^2$, or 25. We now add and subtract that number on the right-hand side. We can think of this as adding $25 - 25$, which is 0.

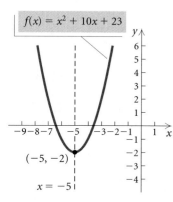

$f(x) = x^2 + 10x + 23$

$f(x) = x^2 + 10x + 23$		Note that 25 completes the square for $x^2 + 10x$.
$= x^2 + 10x + 25 - 25 + 23$		Adding $25 - 25$, or 0, to the right side
$= (x^2 + 10x + 25) - 25 + 23$		
$= (x + 5)^2 - 2$		Factoring and simplifying
$= [x - (-5)]^2 + (-2)$		Writing in the form $f(x) = a(x - h)^2 + k$

From this form of the function, we know the following:

Vertex: $(-5, -2)$;

Line of symmetry: $x = -5$;

Minimum value of the function $= -2$.

The graph of f, shown at left, is a shift of the graph of $y = x^2$ left 5 units and down 2 units.

Keep in mind that the line of symmetry is not part of the graph; it is a characteristic of the graph. If you fold the graph on its line of symmetry, the two halves of the graph will coincide.

EXAMPLE 2 Complete the square to find the vertex, the line of symmetry, and the maximum or minimum value of $g(x) = x^2/2 - 4x + 8$. Then graph the function.

Solution To complete the square, we factor $\frac{1}{2}$ out of the first two terms. This makes the coefficient of x^2 within the parentheses 1:

$$g(x) = \frac{x^2}{2} - 4x + 8$$

$$= \frac{1}{2}(x^2 - 8x) + 8. \qquad \text{Factoring } \tfrac{1}{2} \text{ out of the first two terms}$$

Now we complete the square inside the parentheses: Half of -8 is -4, and $(-4)^2 = 16$. We add and subtract 16 inside the parentheses:

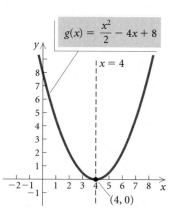

$g(x) = \dfrac{x^2}{2} - 4x + 8$

$g(x) = \frac{1}{2}(x^2 - 8x + 16 - 16) + 8$	
$= \frac{1}{2}(x^2 - 8x + 16) - \frac{1}{2} \cdot 16 + 8$	Using the distributive law to remove -16 from within the parentheses
$= \frac{1}{2}(x - 4)^2 + 0$, or $\frac{1}{2}(x - 4)^2$.	Factoring and simplifying

We know the following:

Vertex: $(4, 0)$;

Line of symmetry: $x = 4$;

Minimum value of the function $= 0$.

The graph of g is a vertical shrinking of the graph of $y = x^2$ followed by a shift 4 units to the right.

EXAMPLE 3   Complete the square to find the vertex, the line of symmetry, and the maximum or minimum value of $f(x) = -2x^2 + 10x - \frac{23}{2}$. Then graph the function.

Solution   We have

$$f(x) = -2x^2 + 10x - \frac{23}{2}$$

$$= -2(x^2 - 5x) - \frac{23}{2} \qquad \text{Factoring } -2 \text{ out of the first two terms}$$

$$= -2\left(x^2 - 5x + \frac{25}{4} - \frac{25}{4}\right) - \frac{23}{2} \qquad \text{Completing the square inside the parentheses}$$

$$= -2\left(x^2 - 5x + \frac{25}{4}\right) - 2\left(-\frac{25}{4}\right) - \frac{23}{2} \qquad \text{Removing } -\frac{25}{4} \text{ from within the parentheses}$$

$$= -2\left(x - \frac{5}{2}\right)^2 + \frac{25}{2} - \frac{23}{2}$$

$$= -2\left(x - \frac{5}{2}\right)^2 + 1.$$

This form of the function yields the following:

Vertex: $\left(\frac{5}{2}, 1\right)$;

Line of symmetry: $x = \frac{5}{2}$;

Maximum value of the function $= 1$.

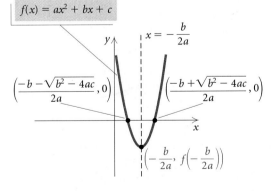

The graph is found by vertically stretching the graph of $f(x) = x^2$, reflecting the graph across the x-axis, shifting the graph to the right $\frac{5}{2}$ units, and shifting that curve up 1 unit.

In many situations, we want to find the coordinates of the vertex directly from the equation $f(x) = ax^2 + bx + c$ using a formula. One way to develop such a formula is to observe that the x-coordinate of the vertex is centered between the x-intercepts, or zeros, of the function. By averaging the two solutions of $ax^2 + bx + c = 0$, we find a formula for the x-coordinate of the vertex:

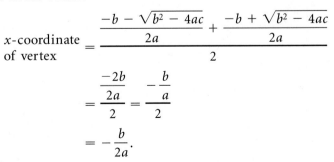

$$\begin{aligned}
x\text{-coordinate} \atop \text{of vertex} &= \frac{\dfrac{-b - \sqrt{b^2 - 4ac}}{2a} + \dfrac{-b + \sqrt{b^2 - 4ac}}{2a}}{2} \\[2mm]
&= \frac{\dfrac{-2b}{2a}}{2} = \frac{-\dfrac{b}{a}}{2} \\[2mm]
&= -\frac{b}{2a}.
\end{aligned}$$

We use this value of x to find the y-coordinate of the vertex,

$$f\left(-\frac{b}{2a}\right).$$

The Vertex of a Parabola

The **vertex** of the graph of $f(x) = ax^2 + bx + c$ is

$$\left(-\frac{b}{2a}, f\left(-\frac{b}{2a}\right)\right).$$

We calculate the   We substitute to
x-coordinate.   find the y-coordinate.

EXAMPLE 4   For the function $f(x) = -x^2 + 14x - 47$:

y-coordinate

x-coordinate

a) Find the vertex.

b) Determine whether there is a maximum or minimum value and find that value.

c) Find the range.

d) On what intervals is the function increasing? decreasing?

Solution   There is no need to graph the function.

a) The x-coordinate of the vertex is

$$-\frac{b}{2a} = -\frac{14}{2(-1)}, \text{ or } 7.$$

Since

$$f(7) = -7^2 + 14 \cdot 7 - 47 = 2,$$

the vertex is $(7, 2)$.

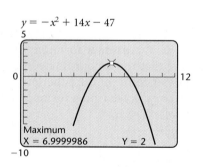

$y = -x^2 + 14x - 47$

Maximum
X = 6.9999986   Y = 2

b) Since a is negative ($a = -1$), the graph opens down so the second co-ordinate of the vertex, 2, is the maximum value of the function.

c) The range is $(-\infty, 2]$.

d) Since the graph opens down, function values increase to the left of the vertex and decrease to the right of the vertex. Thus the function is increasing on the interval $(-\infty, 7)$ and decreasing on $(7, \infty)$.

You should check these results visually using a grapher.

Applications

Many real-world situations involve finding the maximum or minimum value of a quadratic function.

EXAMPLE 5   *Maximizing Area.* A stone mason has enough stones to enclose a rectangular patio with 60 ft of stone wall. If the house forms one side of the rectangle, what is the maximum area that the mason can enclose? What should the dimensions of the patio be in order to yield this area?

PROBLEM-SOLVING STRATEGY

REVIEW SECTION 2.1.

Solution We will use the five-step problem-solving strategy.

1. **Familiarize.** We make a drawing of the situation, using w to represent the width of the patio, in feet. Then $(60 - 2w)$ feet of stone is available for the length. Suppose the patio were 10 ft wide. It would then be $60 - 2 \cdot 10 = 40$ ft long. The area would be $(10\text{ ft})(40\text{ ft}) = 400\text{ ft}^2$. If the patio were 12 ft wide, it would be $60 - 2 \cdot 12 = 36$ ft long. The area would be $(12\text{ ft})(36\text{ ft}) = 432\text{ ft}^2$. If it were 16 ft wide, it would be $60 - 2 \cdot 16 = 28$ ft long and the area would be $(16\text{ ft})(28\text{ ft}) = 448\text{ ft}^2$. There are more combinations of length and width than we could possibly try. Instead we will find a function that represents the area and then determine the maximum value of the function.

2. **Translate.** Since the area of a rectangle is given by length times width, we have

$$A(w) = (60 - 2w)w \qquad A = lw; \ l = 60 - 2w$$
$$= -2w^2 + 60w,$$

where $A(w)$ is the area of the patio, in square feet, as a function of the width, w.

3. **Carry out.** To solve this problem, we need to determine the maximum value of $A(w)$ and find the dimensions for which that maximum occurs. Since A is a quadratic function and w^2 has a negative coefficient, we know that the function has a maximum value that occurs at the vertex of the graph of the function. The first coordinate of the vertex, $(w, A(w))$, is

$$w = -\frac{b}{2a} = -\frac{60}{2(-2)} = 15 \text{ ft.}$$

Thus, if $w = 15$ ft, then the length $l = 60 - 2 \cdot 15 = 30$ ft; and the area is $15 \cdot 30$, or 450 ft^2.

4. **Check.** As a partial check, we note that $450\text{ ft}^2 > 448\text{ ft}^2$, which is the largest area we found in a guess in the *Familiarize* step. A more complete check, assuming that the function $A(w)$ is correct, would examine a table of values for $A(w) = (60 - 2w)w$ and/or examine its graph.

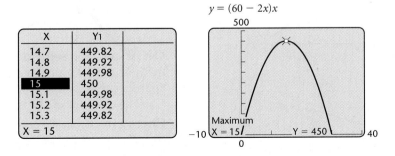

X	Y1	
14.7	449.82	
14.8	449.92	
14.9	449.98	
15	**450**	
15.1	449.98	
15.2	449.92	
15.3	449.82	
X = 15		

$$y = (60 - 2x)x$$

5. **State.** The maximum possible area is 450 ft^2 when the patio is 15 ft wide and 30 ft long.

EXAMPLE 6   *Finding the Depth of a Well.* Two seconds after a chlorine tablet has been dropped into a well, a splash is heard. The speed of sound is 1100 ft/sec. How far is the top of the well from the water?

Solution

1. **Familiarize.** We first make a drawing and label it with known and unknown information. We let s = the depth of the well, in feet, t_1 = the time, in seconds, that it takes for the tablet to hit the water, and t_2 = the time, in seconds, that it takes for the sound to reach the top of the well. This gives us the equation

$$t_1 + t_2 = 2. \tag{1}$$

2. **Translate.** Can we find any relationship between the two times and the distance s? Often in problem solving you may need to look up related formulas in a physics book, another mathematics book, or on the Internet. We find that the formula

$$s = 16t^2$$

gives the distance, in feet, that a dropped object falls in t seconds. The time t_1 that it takes the tablet to hit the water can be found as follows:

$$s = 16t_1^2, \quad \text{or} \quad t_1 = \frac{\sqrt{s}}{4}. \tag{2}$$

To find an expression for t_2, the time it takes the sound to travel to the top of the well, recall that *Distance = Rate · Time.* Thus,

$$s = 1100t_2, \quad \text{or} \quad t_2 = \frac{s}{1100}. \tag{3}$$

We now have expressions for t_1 and t_2, both in terms of s. Substituting into equation (1), we obtain

$$t_1 + t_2 = 2, \quad \text{or} \quad \frac{\sqrt{s}}{4} + \frac{s}{1100} = 2. \tag{4}$$

3. **Carry out.** We solve the equation graphically using the Intersect method. It will probably require some trial and error to determine an appropriate window.

    The solution of the equation is approximately 60.5.

4. **Check.** To check, we can substitute 60.5 for s in equation (4) and see that $t_1 + t_2 \approx 2$. We leave the mathematics for the student.

5. **State.** The top of the well is about 60.5 ft above the water.

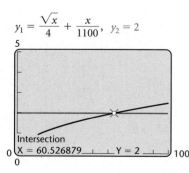

$y_1 = \dfrac{\sqrt{x}}{4} + \dfrac{x}{1100}, \; y_2 = 2$

Intersection
X = 60.526879   Y = 2

Exercise Set 2.4

In Exercises 1 and 2, use the given graph to find each of the following:

a) the vertex;
b) the line of symmetry; and
c) the maximum or minimum value of the function.

1.

2.

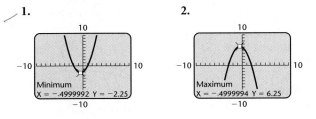

In Exercises 3–10, complete the square to:

a) find the vertex;
b) find the line of symmetry; and
c) determine whether there is a maximum or minimum value and find that value.

Then check your answers using a grapher.

3. $f(x) = x^2 - 8x + 12$

4. $g(x) = x^2 + 7x - 8$

5. $f(x) = x^2 - 7x + 12$

6. $g(x) = x^2 - 5x + 6$

7. $g(x) = 2x^2 + 6x + 8$

8. $f(x) = 2x^2 - 10x + 14$

9. $g(x) = -2x^2 + 2x + 1$

10. $f(x) = -3x^2 - 3x + 1$

In Exercises 11–18, match the equation with one of the figures (a)–(h), which follow. Use a grapher only as a check.

a)

b)

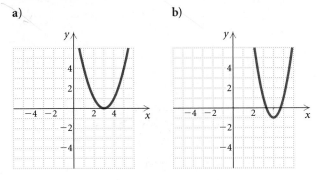

c)

d)

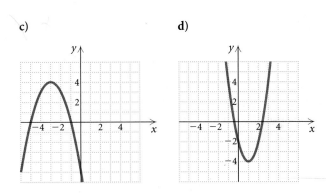

e)

f)

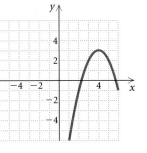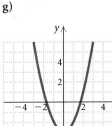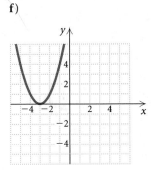

g)

h)

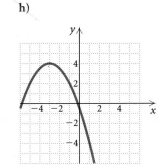

11. $y = (x + 3)^2$ f

12. $y = -(x - 4)^2 + 3$ e (4,3)

13. $y = 2(x - 4)^2 - 1$ b

14. $y = x^2 - 3$ g

15. $y = -\frac{1}{2}(x + 3)^2 + 4$ h

16. $y = (x - 3)^2$ a

17. $y = -(x + 3)^2 + 4$ c

18. $y = 2(x - 1)^2 - 4$ (1,-4) d

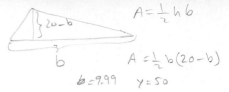

In Exercises 19–26:

a) *Find the vertex.*

b) *Determine whether there is a maximum or minimum value and find that value.*

c) *Find the range.*

d) *Find the intervals on which the function is increasing and the intervals on which the function is decreasing.*

Then check your answers using a grapher.

19. $f(x) = x^2 - 6x + 5$

20. $f(x) = x^2 + 4x - 5$

21. $f(x) = 2x^2 + 4x - 16$

22. $f(x) = \frac{1}{2}x^2 - 3x + \frac{5}{2}$

23. $f(x) = -\frac{1}{2}x^2 + 5x - 8$

24. $f(x) = -2x^2 - 24x - 64$

25. $f(x) = 3x^2 + 6x + 5$

26. $f(x) = -3x^2 + 24x - 49$

27. *Maximizing Volume.* Plasco Manufacturing plans to produce a one-compartment vertical file by bending the long side of an 10-in. by 18-in. sheet of plastic along two lines to form a U-shape. How tall should the file be in order to maximize the volume that the file can hold?

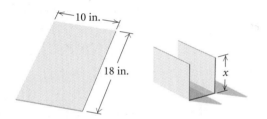

28. *Maximizing Area.* A fourth-grade class decides to enclose a rectangular garden, using the side of the school as one side of the rectangle. What is the maximum area that the class can enclose with 32 ft of fence? What should the dimensions of the garden be in order to yield this area?

29. *Maximizing Area.* The sum of the base and the height of a triangle is 20 cm. Find the dimensions for which the area is a maximum.

30. *Maximizing Area.* The sum of the base and the height of a parallelogram is 69 cm. Find the dimensions for which the area is a maximum.

31. *Finding the Height of an Elevator Shaft.* Jenelle dropped a screwdriver from the top of an elevator shaft. Exactly 5 sec later, she hears the sound of the screwdriver hitting the bottom of the shaft. How tall is the elevator shaft? (*Hint:* See Example 6.)

32. *Finding the Height of a Cliff.* A water balloon is dropped from a cliff. Exactly 3 sec later, the sound of the balloon hitting the ground reaches the top of the cliff. How high is the cliff? (*Hint:* See Example 6.)

33. *Minimizing Cost.* Aki's Bicycle Designs has determined that when x hundred bicycles are built, the average cost per bicycle is given by

$$C(x) = 0.1x^2 - 0.7x + 2.425,$$

where $C(x)$ is in hundreds of dollars. How many bicycles should be built in order to minimize the average cost per bicycle?

Maximizing Profit. *In business, profit is the difference between revenue and cost, that is,*

$$\text{Total profit} = \text{Total revenue} - \text{Total cost,}$$
$$P(x) = R(x) - C(x),$$

where x is the number of units sold. Find the maximum profit and the number of units that must be sold in order to yield the maximum profit for each of the following.

34. $R(x) = 5x, \ C(x) = 0.001x^2 + 1.2x + 60$

35. $R(x) = 50x - 0.5x^2, \ C(x) = 10x + 3$

36. $R(x) = 20x - 0.1x^2, \ C(x) = 4x + 2$

37. *Maximizing Area.* A rancher needs to enclose two adjacent rectangular corrals, one for sheep and one for cattle. If a river forms one side of the corrals and 240 yd of fencing is available, what is the largest total area that can be enclosed?

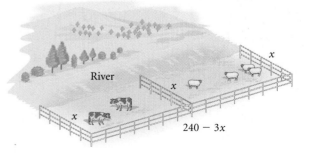

38. *Norman Window.*   A Norman window is a
rectangle with a semicircle on top. Sky Blue
Windows is designing a Norman window that
will require 24 ft of trim. What dimensions
will allow the maximum amount of light to
enter a house?

A Norman
window

Discussion and Writing

39. Write a problem for a classmate to solve. Design it
so that it is a maximum or minimum problem
using a quadratic function.

40. Discuss two ways in which we used completing the
square in this chapter.

41. Suppose that the graph of $f(x) = ax^2 + bx + c$ has
x-intercepts $(x_1, 0)$ and $(x_2, 0)$. What are the
x-intercepts of $g(x) = -ax^2 - bx - c$? Explain.

Skill Maintenance

*For each function f, construct and simplify the difference
quotient*

$$\frac{f(x + h) - f(x)}{h}.$$

42. $f(x) = 3x - 7$

43. $f(x) = 2x^2 - x + 4$

*A graph of $y = f(x)$ follows. No formula is given for f.
Make a hand-drawn graph of each of the following.*

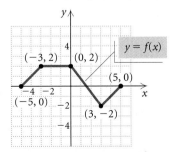

44. $g(x) = f(2x)$

45. $g(x) = -2f(x)$

Synthesis

*For each equation in Exercises 46–49, under the given
condition:*

a) *Find k.*

b) *Find a second solution.*

46. $kx^2 - 2x + k = 0$; one solution is -3

47. $kx^2 - 17x + 33 = 0$; one solution is 3

48. $x^2 - (6 + 3i)x + k = 0$; one solution is 3

49. $x^2 - kx + 2 = 0$; one solution is $1 + i$

50. Find b such that
$$f(x) = -4x^2 + bx + 3$$
has a maximum value of 50.

51. Find c such that
$$f(x) = -0.2x^2 - 3x + c$$
has a maximum value of -225.

52. Find a quadratic function with vertex $(4, -5)$ and
containing the point $(-3, 1)$.

53. Graph: $f(x) = (|x| - 5)^2 - 3$.

54. *Minimizing Area.*   A 24-in. piece of string is cut
into two pieces. One piece is used to form a circle
while the other is used to form a square. How
should the string be cut so that the sum of the areas
is a minimum?

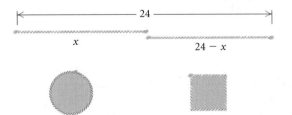

Modeling: Data Analysis, Curve Fitting, and Quadratic Regression

2.5

 Fit quadratic functions to data and make estimates and predictions.

As we proceed through this text, we are building a "library" of functions that can be used to model data. Thus far, our library consists of the following functions.

Linear Function:
$y_1 = mx + b$

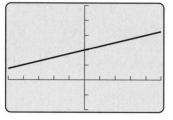

Quadratic Function:
$y_2 = ax^2 + bx + c, a > 0$

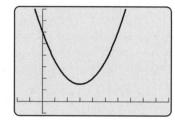

Quadratic Function:
$y_3 = ax^2 + bx + c, a < 0$

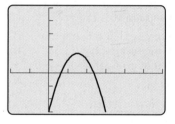

Given a set of data, how can we determine if it can be modeled with one of these functions? One way is to graph the data. For example, consider the scatterplots shown below.

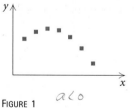

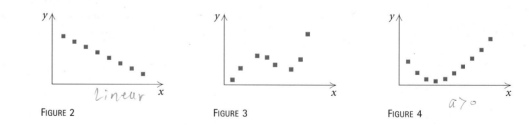

FIGURE 1 $a < 0$

FIGURE 2   Linear

FIGURE 3

FIGURE 4 $a > 0$

The data points in Fig. 1 rise and then fall, suggesting that a quadratic function $f(x) = ax^2 + bx + c$, $a < 0$, might fit the data. In Fig. 2, the data points fall at a steady rate, so a linear function $f(x) = mx + b$ could be used. The data points in Fig. 3 rise, then fall, and then rise again. Thus neither a linear function nor a quadratic function would fit

these data. And, finally, the data points in Fig. 4 fall and then rise, suggesting that a quadratic function $f(x) = ax^2 + bx + c$, $a > 0$, might fit the data.

EXAMPLE 1 *Leisure Time.* The following table shows the median number of hours of leisure time that Americans had each week in various years.

YEAR, x	MEDIAN NUMBER OF LEISURE HOURS PER WEEK
1973, 0	26.2
1980, 7	19.2
1987, 14	16.6
1993, 20	18.8
1997, 24	19.5

Source: Louis Harris and Associates

a) Make a scatterplot of the data, letting x represent the number of years since 1973, and determine whether a linear function, a quadratic function, or neither seems to fit the data.

b) Use a grapher to fit the type of function determined in part (a) to the data.

c) Graph the equation with the scatterplot.

d) Use the function found in part (b) to estimate the number of leisure hours per week in 1978, 1990, and 2005.

Solution

a) We enter the data in a grapher, letting x represent the number of years since 1973. That is, 1973 corresponds to $x = 0$. Then, since $1980 - 1973 = 7$, we know that 1980 corresponds to $x = 7$. Similarly, 1987 corresponds to $x = 14$, 1993 to $x = 20$, and 1997 to $x = 24$. Some graphers have a ZOOM option that automatically selects a window that includes all the data points. We use this option to obtain the window and the scatterplot shown below.

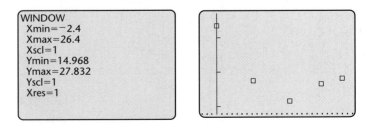

The data points fall and then rise, so a quadratic function $f(x) = ax^2 + bx + c$, $a > 0$, could be used to model the data.

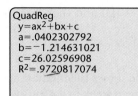

```
QuadReg
y=ax²+bx+c
a=.0402302792
b=−1.214631021
c=26.02596908
R²=.9720817074
```

b) Using the quadratic regression feature on a grapher, we find a quadratic function:

$$f(x) = 0.0402302792x^2 - 1.214631021x + 26.02596908.$$

c) We can store this equation on the equation-editor screen at the same time that it is found or immediately after it is found. Then we can graph the regression equation with the scatterplot.

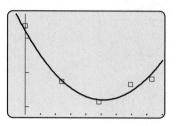

X	Y₁
5	20.959
17	17.004
32	28.354

X =

d) We use a table set in ASK mode to find the desired function values. Note that since 1978 − 1973 = 5, the year 1978 corresponds to $x = 5$. Similarly, 1990 corresponds to $x = 17$ and 2005 to $x = 32$.

We estimate that Americans had about 21.0 leisure hours per week in 1978 and about 17.0 hours in 1990, and we predict that they will have about 28.4 leisure hours per week in 2005.

Exercise Set 2.5

In Exercises 1–6, determine which of the following types of functions might be used to model the data shown in the scatterplot.

a) Linear, $f(x) = mx + b$
b) Quadratic, $f(x) = ax^2 + bx + c,\ a > 0$
c) Quadratic, $f(x) = ax^2 + bx + c,\ a < 0$
d) Neither a linear nor a quadratic function

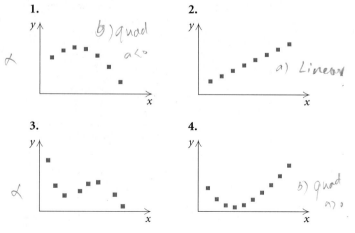

1.

2.

3.

4.

5.

6.

7. *Morning Newspapers.* The number of morning newspapers in the United States in various years is shown in the following table.

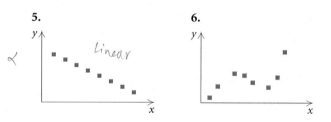

YEAR	NUMBER OF MORNING NEWSPAPERS
1920	437
1940	380
1960	312
1980	387
1990	559
1997	705

Source: Editor & Publisher

a) Make a scatterplot of the data, letting x represent the number of years since 1920.

b) Determine whether a linear function, a quadratic function, or neither seems to fit the data.

c) Use a grapher to fit the type of function determined in part (b) to the data.

d) Graph the equation with the scatterplot.

e) Use the function found in part (c) to predict the number of morning newspapers in 2004 and 2007.

8. *Home Education.* The number of children who were home-educated in the United States in various years is shown in the following table.

YEAR	NUMBER OF HOME-EDUCATED CHILDREN (IN THOUSANDS)
1983	92.5
1988	225
1992	703
1995	1060
1997	1347

Source: National Home Education Research Institute

a) Make a scatterplot of the data, letting *x* represent the number of years since 1983.

b) Determine whether a linear function, a quadratic function, or neither seems to fit the data.

c) Use a grapher to fit the type of function determined in part (b) to the data.

d) Graph the equation with the scatterplot.

e) Use the function found in part (c) to predict the number of children who will be home-educated in 2005 and 2010.

9. *Medicare Assets.* The end-of-year assets of the U.S. Hospital Insurance program, also known as Medicare Part A, for several recent years are shown in the following table.

YEAR	ASSETS (IN MILLIONS)
1990	$ 95,631
1993	126,131
1995	129,520
1998	110,584
2000	100,852

Source: U.S. Health Care Financing Administration

a) Make a scatterplot of the data, letting *x* represent the number of years since 1990.

b) Determine whether a linear function, a quadratic function, or neither seems to fit the data.

c) Use a grapher to fit the type of function determined in part (b) to the data.

d) Graph the equation with the scatterplot.

e) Use the function found in part (c) to predict the assets of the Hospital Insurance program in 2003 and 2005.

10. *Social Security Assets.* The assets of the U.S. Social Security Trust Fund for several recent years are shown in the following table.

YEAR	ASSETS (IN BILLIONS)
1995	$339.8
1996	353.6
1997	369.1
1998	382.9
1999	396.3
2000	413.4

Source: Social Security Administration

a) Make a scatterplot of the data, letting *x* represent the number of years since 1995.

b) Determine whether a linear function, a quadratic function, or neither seems to fit the data.

c) Use a grapher to fit the type of function determined in part (b) to the data.

d) Graph the equation with the scatterplot.

e) Use the function found in part (c) to predict the assets of the Social Security Trust Fund in 2004 and 2010.

11. *Mothers in the Labor Force.* The following table shows the percentage of mothers in the labor force who have children under 6 yr old.

YEAR	PERCENTAGE OF MOTHERS IN THE LABOR FORCE
1975	39.0%
1985	53.5
1997	65.0

Source: U.S. Bureau of Labor Statistics

a) Make a scatterplot of the data, letting x represent the number of years since 1975.

b) Determine whether a linear function, a quadratic function, or neither seems to fit the data.

c) Use a grapher to fit the type of function determined in part (b) to the data.

d) Graph the equation with the scatterplot.

e) Use the function found in part (c) to predict the percentage of mothers with children under 6 yr old who will be in the labor force in 2003 and 2010.

12. *Senior Citizens.* The following table shows the number of Americans age 65 and over in various years.

YEAR	AMERICANS AGE 65 AND OVER (IN MILLIONS)
1900	3.1
1920	4.9
1940	9.0
1960	16.7
1980	25.5
2000	35.3

Source: U.S. Bureau of the Census

a) Make a scatterplot of the data, letting x represent the number of years since 1900.

b) Determine whether a linear function, a quadratic function, or neither seems to fit the data.

c) Use a grapher to fit the type of function determined in part (b) to the data.

d) Graph the equation with the scatterplot.

e) Use the function found in part (c) to predict the number of Americans age 65 and over in 2006 and 2011.

13. *Hours of Sleep versus Death Rate.* The following table shows the relationship between the average number of hours that a group of males slept per day and their death rates.

AVERAGE NUMBER OF HOURS OF SLEEP, x	DEATH RATE PER 100,000 MALES, y
5	1121
6	805
7	626
8	813
9	967

Source: "Hiding in the Hammond Report," *Hospital Practice*, by Harold J. Morowitz

a) Make a scatterplot of the data, letting x represent the average number of hours of sleep.

b) Determine whether a linear function, a quadratic function, or neither seems to fit the data.

c) Use a grapher to fit the type of function determined in part (b) to the data.

d) Graph the equation with the scatterplot.

e) Use the function found in part (c) to estimate the death rate of males who slept for an average of 7.5 hr per day and 10 hr per day.

14. *Federal Research Outlays.* Federal outlays for basic science research, in terms of 1992 dollars, are shown in the following table.

YEAR	RESEARCH OUTLAY (IN BILLIONS)
1970	$3.4
1980	2.4
1985	2.6
1990	3.1
1995	3.7
2000	4.8

Source: Office of Management and Budget

a) Make a scatterplot of the data, letting x represent the number of years since 1970.

b) Determine whether a linear function, a quadratic function, or neither seems to fit the data.

c) Use a grapher to fit the type of function determined in part (b) to the data.

d) Graph the equation with the scatterplot.

e) Use the function found in part (c) to predict the federal outlay for basic science research, in terms of 1992 dollars, in 2004 and 2008.

Shoe Size and Life Expectancy.   *The data in the following table are the result of a Swedish study relating one's life expectancy to the size of one's feet. Use these data in Exercises 15 and 16.*

SHOE SIZE	LIFE EXPECTANCY FOR WOMEN (IN YEARS)	SHOE SIZE	LIFE EXPECTANCY FOR MEN (IN YEARS)
4	69	5	66
5	76	6	69
6	82	7	69
7	84	8	72
8	75	9	75
9	72	10	77
10	70	11	82
11	69	12	79
		13	72
		14	69

Source: Orthopedic Quarterly

15. a) Use a grapher to fit a quadratic function to the data for women.

   b) Estimate the life expectancy of women with shoe sizes of $5\frac{1}{2}$ and $8\frac{1}{2}$.

16. a) Use a grapher to fit a quadratic function to the data for men.

   b) Estimate the life expectancy of men with shoe sizes of $10\frac{1}{2}$ and $15\frac{1}{2}$.

Discussion and Writing

17. Discuss the limitations of the function found in Exercise 13 to estimate death rates. For example, consider the meaning of values such as $f(0)$, $f(24)$, and $f(25)$.

18. Discuss some ways in which you could test the accuracy of the function found in Exercise 12.

Skill Maintenance

Given that

$$g(x) = \frac{x + 5}{x - 1},$$

find each of the following.

19. $g(-2)$ **20.** $g(1)$

21. $g(-5)$ **22.** The domain of g

Synthesis

23. *The Effect of Advertising on Movie Revenue.*   The following table shows data about the box office revenues and advertising budgets of several movies.

MOVIE	ADVERTISING BUDGET (IN MILLIONS)	BOX OFFICE REVENUE (IN MILLIONS)
The Lion King	$23.3	$300.4
Forrest Gump	25.0	298.5
True Lies	20.5	146.3
The Santa Clause	19.4	137.8
The Flintstones	10.6	130.6
Clear & Present Danger	17.9	121.8
Speed	17.2	121.2
The Mask	11.2	118.8
Maverick	13.8	101.6
Interview with a Vampire	15.4	100.7

Source: The Hollywood Reporter

a) Make a scatterplot of the data, letting x represent advertising budget and y represent box office revenue.

b) Determine whether a linear function, a quadratic function, or neither seems to fit the data.

c) Use a grapher to fit the type of function determined in part (b) to the data.

d) Graph the equation with the scatterplot.

e) Suppose a producer wants to make a movie that will have a box office revenue of $180,000,000. What amount should be allotted for the advertising budget of the movie?

Zeros and More Equation Solving

2.6
_____

- *Solve rational and radical equations and equations with absolute value and find zeros of their associated functions.*

Rational Equations

Equations containing rational expressions are called **rational equations.** Solving such equations algebraically involves multiplying on both sides by the least common denominator (LCD).

EXAMPLE 1   Solve: $\dfrac{x-8}{3} + \dfrac{x-3}{2} = 0$.

Algebraic Solution

We have

$$\frac{x-8}{3} + \frac{x-3}{2} = 0$$ The LCD is $3 \cdot 2$, or 6.

$$6\left(\frac{x-8}{3} + \frac{x-3}{2}\right) = 6 \cdot 0$$ Multiplying by the LCD on both sides

$$6 \cdot \frac{x-8}{3} + 6 \cdot \frac{x-3}{2} = 0$$

$$2(x-8) + 3(x-3) = 0$$

$$2x - 16 + 3x - 9 = 0$$

$$5x - 25 = 0$$

$$5x = 25$$

$$x = 5.$$

The possible solution is 5. We check using a table in ASK mode.

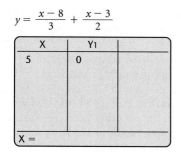

$$y = \frac{x-8}{3} + \frac{x-3}{2}$$

X	Y1	
5	0	

X =

Since the value of

$$\frac{x-8}{3} + \frac{x-3}{2}$$

is 0 when $x = 5$, the number 5 is the solution.

Graphical Solution

We use the Zero method. The solutions of the given equation are the zeros of the function

$$f(x) = \frac{x-8}{3} + \frac{x-3}{2}.$$

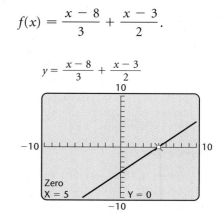

$$y = \frac{x-8}{3} + \frac{x-3}{2}$$

The solution is 5.

When we use the multiplication principle to multiply (or divide) both sides of an equation by an expression with a variable, we might not obtain an equivalent equation. We must check possible solutions obtained in this manner by substituting in the original equation. The next example illustrates this.

EXAMPLE 2   Solve: $\dfrac{x^2}{x-3} = \dfrac{9}{x-3}$.

Solution   The LCD is $x - 3$.

$$(x-3) \cdot \frac{x^2}{x-3} = (x-3) \cdot \frac{9}{x-3}$$

$$x^2 = 9$$

$$x = -3 \quad or \quad x = 3 \qquad \text{Using the principle of square roots}$$

The possible solutions are -3 and 3. We check.

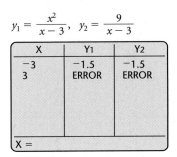

$$y_1 = \frac{x^2}{x-3}, \ y_2 = \frac{9}{x-3}$$

X	Y1	Y2
-3	-1.5	-1.5
3	ERROR	ERROR

X =

CHECK: For -3:

$$\frac{x^2}{x-3} = \frac{9}{x-3}$$

$$\frac{(-3)^2}{-3-3} \ \overset{?}{\mid} \ \frac{9}{-3-3}$$

$$\frac{9}{-6} \ \Big| \ \frac{9}{-6} \qquad \text{TRUE}$$

For 3:

$$\frac{x^2}{x-3} = \frac{9}{x-3}$$

$$\frac{3^2}{3-3} \ \overset{?}{\mid} \ \frac{9}{3-3}$$

$$\frac{9}{0} \ \Big| \ \frac{9}{0} \qquad \text{UNDEFINED}$$

The number -3 checks, so it is a solution. Since division by 0 is undefined, 3 is not a solution. Note that 3 is not in the domain of $x^2/(x-3)$ or $9/(x-3)$. (See the table at left.)

Radical Equations

A **radical equation** is an equation in which variables appear in one or more radicands. For example,

$$\sqrt{2x-5} - \sqrt{x-3} = 1$$

is a radical equation. The following principle is used to solve such equations.

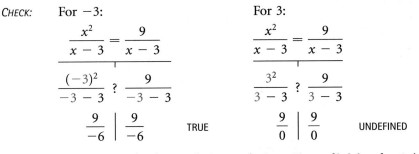

The Principle of Powers

For any positive integer n:

If $a = b$ is true, then $a^n = b^n$ is true.

EXAMPLE 3 Solve: $5 + \sqrt{x + 7} = x$.

Algebraic Solution

We first isolate the radical and then use the principle of powers.

$$5 + \sqrt{x + 7} = x$$

$$\sqrt{x + 7} = x - 5 \qquad \text{Subtracting 5 on both sides}$$

$$(\sqrt{x + 7})^2 = (x - 5)^2 \qquad \text{Using the principle of powers; squaring both sides}$$

$$x + 7 = x^2 - 10x + 25$$

$$0 = x^2 - 11x + 18 \qquad \text{Subtracting } x \text{ and 7}$$

$$0 = (x - 9)(x - 2) \qquad \text{Factoring}$$

$$x - 9 = 0 \quad \text{or} \quad x - 2 = 0$$

$$x = 9 \quad \text{or} \quad x = 2$$

The possible solutions are 9 and 2.

CHECK: For 9:

$$\frac{5 + \sqrt{x + 7} = x}{}$$

$$5 + \sqrt{9 + 7} \;?\; 9$$

$$5 + \sqrt{16}$$

$$5 + 4$$

$$9 \;\Big|\; 9 \quad \text{TRUE}$$

For 2:

$$\frac{5 + \sqrt{x + 7} = x}{}$$

$$5 + \sqrt{2 + 7} \;?\; 2$$

$$5 + \sqrt{9}$$

$$5 + 3$$

$$8 \;\Big|\; 2 \quad \text{FALSE}$$

Since 9 checks but 2 does not, the only solution is 9.

Graphical Solution

We graph $y_1 = 5 + \sqrt{x + 7}$ and $y_2 = x$. Using the INTERSECT feature, we see that the solution is 9.

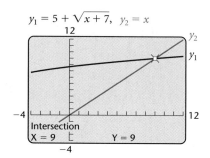

We can also use the ZERO feature to get this result. To do so, we first write the equivalent equation

$$5 + \sqrt{x + 7} - x = 0.$$

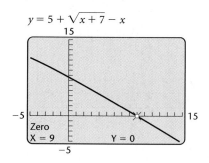

The zero of the function is 9, so the solution of the original equation is 9. Note that graphs show that the equation has only one solution.

When we raise both sides of an equation to an even power, the resulting equation can have solutions that the original equation does not. This is because the converse of the principle of powers is not necessarily true. That is, if $a^n = b^n$ is true, we do not know that $a = b$ is true. For example, $(-2)^2 = 2^2$, but $-2 \neq 2$. Thus, as we see in Example 3, it is necessary to check the possible solutions in the original equation when the principle of powers is used to raise both sides of an equation to an even power.

When a radical equation has two radical terms on one side, we isolate one of them and then use the principle of powers. If, after doing so, a radical term remains, we repeat these steps.

EXAMPLE 4   Solve: $\sqrt{x-3} + \sqrt{x+5} = 4$.

Solution   We have

$$\sqrt{x-3} = 4 - \sqrt{x+5}$$ Isolating one radical

$$(\sqrt{x-3})^2 = (4 - \sqrt{x+5})^2$$ Using the principle of powers; squaring both sides

$$x - 3 = 16 - 8\sqrt{x+5} + (x+5)$$

$$x - 3 = 21 - 8\sqrt{x+5} + x$$ Combining like terms

$$-24 = -8\sqrt{x+5}$$ Isolating the remaining radical; subtracting x and 21 on both sides

$$3 = \sqrt{x+5}$$ Dividing by -8 on both sides

$$3^2 = (\sqrt{x+5})^2$$ Using the principle of powers; squaring both sides

$$9 = x + 5$$

$$4 = x.$$

The number 4 checks and is the solution. Graphs or the TABLE feature on a grapher can be used to confirm this.

Equations with Absolute Value

Recall that the absolute value of a number is its distance from 0 on the number line. We use this concept to solve equations with absolute value.

> For $a > 0$:
>
> $|X| = a$ is equivalent to $X = -a$ or $X = a$.

EXAMPLE 5   Solve each of the following.

a) $|x| = 5$
b) $|x - 3| = 2$

Solution

a) We solve $|x| = 5$ both algebraically and graphically.

Algebraic Solution

We have

$$|x| = 5$$
$$x = -5 \quad or \quad x = 5.$$
Writing an equivalent statement

The solutions are -5 and 5.

We can use the TABLE feature on a grapher to confirm this.

$y = |x|$

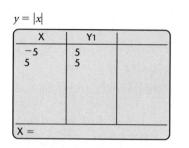

Graphical Solution

Using the Intersect method, we graph $y_1 = |x|$ and $y_2 = 5$ and find the first coordinates of the points of intersection.

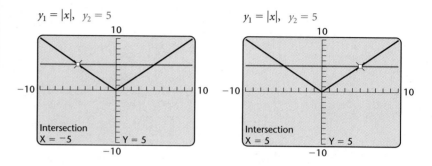

The solutions are -5 and 5.

We could also have used the Zero method to get this result, graphing $y = |x| - 5$ and using the ZERO feature twice.

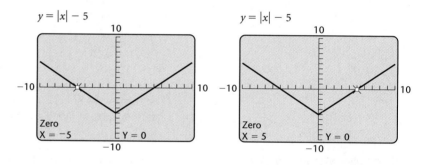

The zeros of $f(x) = |x| - 5$ are -5 and 5, so the solutions of the original equation are -5 and 5.

b) We solve $|x - 3| = 2$ algebraically:

$$|x - 3| = 2$$
$$x - 3 = -2 \quad or \quad x - 3 = 2 \qquad \text{Writing an equivalent statement}$$
$$x = 1 \quad or \qquad x = 5. \qquad \text{Adding 3}$$

The solutions are 1 and 5.

$y_1 = |x|, \quad y_2 = -3$

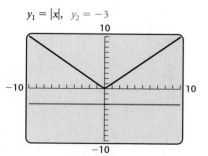

When $a = 0$, $|X| = a$ is equivalent to $x = 0$. Note that for $a < 0$, $|X| = a$ has no solution, because the absolute value of an expression is never negative. We can use a graph to illustrate the last statement for a specific value of a. For example, we let $a = -3$ and graph $y_1 = |x|$ and $y_2 = -3$. The graphs do not intersect, as shown at left. Thus the equation $|x| = -3$ has no solution. The solution set is the **empty set,** denoted $\varnothing$.

Exercise Set 2.6

Solve.

1. $\dfrac{1}{4} + \dfrac{1}{5} = \dfrac{1}{t}$

2. $\dfrac{1}{3} - \dfrac{5}{6} = \dfrac{1}{x}$

3. $\dfrac{x + 2}{4} - \dfrac{x - 1}{5} = 15$

4. $\dfrac{t + 1}{3} - \dfrac{t - 1}{2} = 1$

5. $\dfrac{1}{2} + \dfrac{2}{x} = \dfrac{1}{3} + \dfrac{3}{x}$

6. $\dfrac{1}{t} + \dfrac{1}{2t} + \dfrac{1}{3t} = 5$

7. $\dfrac{3x}{x + 2} + \dfrac{6}{x} = \dfrac{12}{x^2 + 2x}$

8. $\dfrac{5x}{x - 4} - \dfrac{20}{x} = \dfrac{80}{x^2 - 4x}$

9. $\dfrac{4}{x^2 - 1} - \dfrac{2}{x - 1} = \dfrac{3}{x + 1}$

10. $\dfrac{3y + 5}{y^2 + 5y} + \dfrac{y + 4}{y + 5} = \dfrac{y + 1}{y}$

11. $\dfrac{490}{x^2 - 49} = \dfrac{5x}{x - 7} - \dfrac{35}{x + 7}$

12. $\dfrac{3}{m + 2} + \dfrac{2}{m} = \dfrac{4m - 4}{m^2 - 4}$

13. $\dfrac{1}{x - 6} - \dfrac{1}{x} = \dfrac{6}{x^2 - 6x}$

14. $\dfrac{8}{x^2 - 4} = \dfrac{x}{x - 2} - \dfrac{2}{x + 2}$

15. $\dfrac{8}{x^2 - 2x + 4} = \dfrac{x}{x + 2} + \dfrac{24}{x^3 + 8}$

16. $\dfrac{18}{x^2 - 3x + 9} - \dfrac{x}{x + 3} = \dfrac{81}{x^3 + 27}$

17. $\sqrt{3x - 4} = 1$

18. $\sqrt[3]{2x + 1} = -5$

19. $\sqrt[4]{x^2 - 1} = 1$

20. $\sqrt{m + 1} - 5 = 8$

21. $\sqrt{y - 1} + 4 = 0$

22. $\sqrt[5]{3x + 4} = 2$

23. $\sqrt[3]{6x + 9} + 8 = 5$

24. $\sqrt{6x + 7} = x + 2$

25. $\sqrt{x - 3} + \sqrt{x + 2} = 5$

26. $\sqrt{x} - \sqrt{x - 5} = 1$

27. $\sqrt{3x - 5} + \sqrt{2x + 3} + 1 = 0$

28. $\sqrt{2m - 3} = \sqrt{m + 7} - 2$

29. $\sqrt{x} - \sqrt{3x - 3} = 1$

30. $\sqrt{2x + 1} - \sqrt{x} = 1$

31. $\sqrt{2y - 5} - \sqrt{y - 3} = 1$

32. $\sqrt{4p + 5} + \sqrt{p + 5} = 3$

33. $x^{1/3} = -2$

34. $t^{1/5} = 2$

35. $t^{1/4} = 3$

36. $m^{1/2} = -7$

Solve.

37. $|x| = 7$

38. $|x| = 4.5$

39. $|x| = -10.7$

40. $|x| = -\dfrac{3}{5}$

41. $|x - 1| = 4$

42. $|x - 7| = 5$

43. $|3x| = 1$

44. $|5x| = 4$

45. $|x| = 0$

46. $|6x| = 0$

47. $|3x + 2| = 1$

48. $|7x - 4| = 8$

49. $\left|\dfrac{1}{2}x - 5\right| = 17$

50. $\left|\dfrac{1}{3}x - 4\right| = 13$

51. $|x - 1| + 3 = 6$

52. $|x + 2| - 5 = 9$

Solve.

53. $\dfrac{P_1 V_1}{T_1} = \dfrac{P_2 V_2}{T_2}$, for T_1
(A chemistry formula for gases)

54. $\dfrac{1}{F} = \dfrac{1}{m} + \dfrac{1}{p}$, for F
(A formula from optics)

55. $\dfrac{1}{R} = \dfrac{1}{R_1} + \dfrac{1}{R_2}$, for R_2
(Resistance)

56. $A = P(1 + i)^2$, for i
(Compound interest)

57. $\dfrac{1}{F} = \dfrac{1}{m} + \dfrac{1}{p}$, for p
(A formula from optics)

Discussion and Writing

58. Explain why it is necessary to check the possible solutions of a rational equation.

59. Explain in your own words why it is necessary to check the possible solutions when the principle of powers is used to solve an equation.

Skill Maintenance

Use a grapher to find the zero of the function.

60. $f(x) = -3x + 9$ **61.** $f(x) = 15 - 2x$

62. In 1991, the U.S. per-capita consumption of bananas was 25.1 lb. In 1997, this number rose to 27.7 lb. (*Source*: U.S. Department of Agriculture) Find the percent of increase.

63. In 1998, 506,000 adults earned high school equivalency diplomas by passing the General Educational Development (GED) test. This was 25,000 more than the number who passed the test in 1997. (*Source*: *Indianapolis Star*, August 7, 1999) How many adults passed the GED in 1997?

Synthesis

Solve.

64. $\dfrac{x+3}{x+2} - \dfrac{x+4}{x+3} = \dfrac{x+5}{x+4} - \dfrac{x+6}{x+5}$

65. $(x-3)^{2/3} = 2$

66. $\sqrt{15 + \sqrt{2x+80}} = 5$

67. $\sqrt{x+5} + 1 = \dfrac{6}{\sqrt{x+5}}$

68. $x^{2/3} = x + 1$

Solving Inequalities

2.7

- *Solve linear inequalities, using interval notation to express solution sets.*
- *Solve compound inequalities.*
- *Solve inequalities with absolute value.*

An **inequality** is a sentence with $<$, $>$, $\le$, or $\ge$ as its verb. An example is $3x - 5 < 6 - 2x$. To **solve** an inequality is to find all values of the variable that make the inequality true. Each of these numbers is a **solution** of the inequality, and the set of all such solutions is its **solution set**. Inequalities that have the same solution set are called **equivalent inequalities.**

Linear Inequalities

The principles for solving inequalities are similar to those for solving equations.

Principles for Solving Inequalities

For any real numbers a, b, and c:

The Addition Principle for Inequalities: If $a < b$ is true, then $a + c < b + c$ is true.

(continued)

> ***The Multiplication Principle for Inequalities:***   If $a < b$ and $c > 0$ are true, then $ac < bc$ is true. If $a < b$ and $c < 0$ are true, then $ac > bc$ is true.
>
> Similar statements hold for $a \le b$.

> When both sides of an inequality are multiplied by a negative number, we must reverse the inequality sign.

First-degree inequalities with one variable, like those in Example 1 below, are **linear inequalities.**

EXAMPLE 1   Solve each of the following.

a) $3x - 5 < 6 - 2x$ b) $13 - 7x \ge 10x - 4$

Solution

a) $3x - 5 < 6 - 2x$

$\quad\quad 5x - 5 < 6$ Using the addition principle for inequalities; adding $2x$

$\quad\quad\quad 5x < 11$ Using the addition principle for inequalities; adding 5

$\quad\quad\quad\quad x < \frac{11}{5}$ Using the multiplication principle for inequalities; multiplying by $\frac{1}{5}$ or dividing by 5

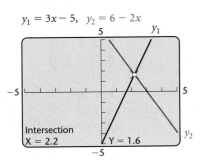

$y_1 = 3x - 5, \quad y_2 = 6 - 2x$

Intersection
X = 2.2   Y = 1.6

Any number less than $\frac{11}{5}$ is a solution. To check graphically, we graph $y_1 = 3x - 5$ and $y_2 = 6 - 2x$. The graph (shown at left) confirms that for $x < 2.2$, or $x < \frac{11}{5}$, we have $y_1 < y_2$. The solution set is $\left\{ x \mid x < \frac{11}{5} \right\}$, or $\left(-\infty, \frac{11}{5} \right)$.

b) $13 - 7x \ge 10x - 4$

$\quad\quad 13 - 17x \ge -4$ Subtracting $10x$

$\quad\quad\quad -17x \ge -17$ Subtracting 13

$\quad\quad\quad\quad\quad x \le 1$ Dividing by -17 and reversing the inequality sign

The solution set is $\{ x \mid x \le 1 \}$, or $(-\infty, 1]$.

Compound Inequalities

When two inequalities are joined by the word *and* or the word *or*, a **compound inequality** is formed. A compound inequality like $-3 < 2x + 5$ and $2x + 5 \le 7$ is called a **conjunction**, because it uses the word *and*. The sentence $-3 < 2x + 5 \le 7$ is an abbreviation for the preceding conjunction.

Compound inequalities can be solved using the addition and multiplication principles for inequalities.

EXAMPLE 2   Solve $-3 < 2x + 5 \leq 7$. Then graph the solution set.

Solution   We have

$$-3 < 2x + 5 \leq 7$$
$$-8 < 2x \leq 2 \qquad \text{Subtracting 5}$$
$$-4 < x \leq 1. \qquad \text{Dividing by 2}$$

The solution set is $\{x \mid -4 < x \leq 1\}$, or $(-4, 1]$. The graph of the solution set is as shown below.

We can perform a partial check of the solution graphically using operations from the TEST menu. We graph $y_1 = (-3 < 2x + 5)$ *and* $(2x + 5 \leq 7)$ in DOT mode. The calculator graphs a segment 1 unit above the x-axis for the values of x for which this expression for y is true. Here the number 1 corresponds to "true."

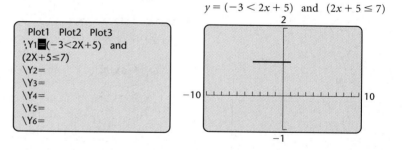

The segment extends from -4 to 1, confirming that all x-values from -4 to 1 are in the solution set. The algebraic solution indicates that the endpoint 1 is also in the solution set.   ▬

A compound inequality like $2x - 5 \leq -7$ *or* $2x - 5 > 1$ is called a **disjunction**, because it contains the word *or*. Unlike some conjunctions, it cannot be abbreviated; that is, it cannot be written without the word *or*.

EXAMPLE 3   Solve $2x - 5 \leq -7$ *or* $2x - 5 > 1$. Then graph the solution set.

Solution   We have

$$2x - 5 \leq -7 \quad or \quad 2x - 5 > 1$$
$$2x \leq -2 \quad or \qquad 2x > 6 \qquad \text{Adding 5}$$
$$x \leq -1 \quad or \qquad x > 3. \qquad \text{Dividing by 2}$$

The solution set is $\{x \mid x \leq -1 \text{ or } x > 3\}$. We can also write the solution using interval notation and the symbol $\cup$ for the **union** or inclusion

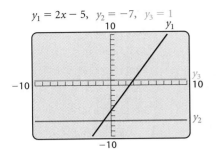

of both sets: $(-\infty, -1] \cup (3, \infty)$. The graph of the solution set is shown below.

To check, we graph $y_1 = 2x - 5$, $y_2 = -7$, and $y_3 = 1$. Note that for $\{x \mid x \le -1 \text{ or } x > 3\}$, $y_1 < y_2 \text{ or } y_1 > y_3$. ▬

Inequalities with Absolute Value

Inequalities sometimes contain absolute-value notation. The following properties are used to solve them.

For $a > 0$:

$\quad |X| < a$ is equivalent to $-a < X < a$.

$\quad |X| > a$ is equivalent to $X < -a \text{ or } X > a$.

Similar statements hold for $|X| \le a$ and $|X| \ge a$.

EXAMPLE 4   Solve each of the following.

a) $|3x + 2| < 5$

b) $|5 - 2x| \ge 1$

Solution

a) $|3x + 2| < 5$

$\quad\quad -5 < 3x + 2 < 5$ **Writing an equivalent inequality**

$\quad\quad\quad -7 < 3x < 3$ **Subtracting 2**

$\quad\quad\quad -\frac{7}{3} < x < 1$ **Dividing by 3**

The solution is $\left\{x \mid -\frac{7}{3} < x < 1\right\}$, or $\left(-\frac{7}{3}, 1\right)$.

To perform a partial check with a grapher, we graph $y = |3x + 2| < 5$ in DOT mode. The calculator graphs a segment 1 unit above the x-axis for the values of x for which this expression for y is true. The graph shows that the solution is probably correct.

b) $|5 - 2x| \ge 1$

$\quad 5 - 2x \le -1 \quad or \quad 5 - 2x \ge 1$ **Writing an equivalent inequality**

$\quad\quad -2x \le -6 \quad or \quad\quad -2x \ge -4$ **Subtracting 5**

$\quad\quad\quad x \ge 3 \quad or \quad\quad\quad x \le 2$ **Dividing by -2 and reversing the inequality signs**

The solution is $\{x \mid x \le 2 \text{ or } x \ge 3\}$, or $(-\infty, 2] \cup [3, \infty)$. ▬

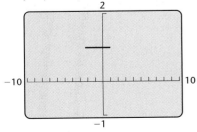

Handwritten at top:
$x - 5 > 0.1$ or $x - 5 < 0.1$
$x > 5.1$ or $x < 4.9$
$(-\infty, 4.9) \cup (5.1, \infty)$

Exercise Set 2.7

Solve.

1. $x + 6 < 5x - 6$
2. $3 - x < 4x + 7$
3. $3x - 3 + 2x \geq 1 - 7x - 9$
4. $5y - 5 + y \leq 2 - 6y - 8$
5. $14 - 5y \leq 8y - 8$
6. $8x - 7 < 6x + 3$
7. $-\frac{3}{4}x \geq -\frac{5}{8} + \frac{2}{3}x$
8. $-\frac{5}{6}x \leq \frac{3}{4} + \frac{8}{3}x$
9. $4x(x - 2) < 2(2x - 1)(x - 3)$
10. $(x + 1)(x + 2) > x(x + 1)$

11. Use a grapher to check your answer to Exercise 9.
12. Use a grapher to check your answer to Exercise 10.

Solve. Write interval notation.

13. $-2 \leq x + 1 < 4$
14. $-3 < x + 2 \leq 5$
15. $5 \leq x - 3 \leq 7$
16. $-1 < x - 4 < 7$
17. $-3 \leq x + 4 \leq 3$
18. $-5 < x + 2 < 15$
19. $-2 < 2x + 1 < 5$
20. $-3 \leq 5x + 1 \leq 3$
21. $-4 \leq 6 - 2x < 4$
22. $-3 < 1 - 2x \leq 3$
23. $-5 < \frac{1}{2}(3x + 1) \leq 7$
24. $\frac{2}{3} \leq -\frac{4}{5}(x - 3) < 1$
25. $3x \leq -6$ or $x - 1 > 0$
26. $2x < 8$ or $x + 3 \geq 10$
27. $2x + 3 \leq -4$ or $2x + 3 \geq 4$
28. $3x - 1 < -5$ or $3x - 1 > 5$
29. $2x - 20 < -0.8$ or $2x - 20 > 0.8$
30. $5x + 11 \leq -4$ or $5x + 11 \geq 4$
31. $x + 14 \leq -\frac{1}{4}$ or $x + 14 \geq \frac{1}{4}$
32. $x - 9 < -\frac{1}{2}$ or $x - 9 > \frac{1}{2}$

33. Use a grapher to check your answer to Exercise 21.
34. Use a grapher to check your answer to Exercise 22.

Solve and graph the solution set.

35. $|x| < 7$
36. $|x| \leq 4.5$
37. $|x| \geq 4.5$
38. $|x| > 7$

Solve.

39. $|x + 8| < 9$
40. $|x + 6| \leq 10$
41. $|x + 8| \geq 9$
42. $|x + 6| > 10$

43. $\left|x - \frac{1}{4}\right| < \frac{1}{2}$
44. $|x - 0.5| \leq 0.2$
45. $|3x| < 1$
46. $|5x| \leq 4$
47. $|2x + 3| \leq 9$
48. $|3x + 4| < 13$
49. $|x - 5| > 0.1$
50. $|x - 7| \geq 0.4$
51. $|6 - 4x| \leq 8$
52. $|5 - 2x| > 10$
53. $\left|x + \frac{2}{3}\right| \leq \frac{5}{3}$
54. $\left|x + \frac{3}{4}\right| < \frac{1}{4}$
55. $\left|\dfrac{2x + 1}{3}\right| > 5$
56. $\left|\dfrac{2x - 1}{3}\right| \geq \dfrac{5}{6}$
57. $|2x - 4| < -5$
58. $|3x + 5| < 0$

59. Use a grapher to check your answer to Exercise 41.
60. Use a grapher to check your answer to Exercise 42.

Discussion and Writing

61. Explain why $|x| < p$ has no solution for $p \leq 0$.
62. Explain why all real numbers are solutions of $|x| > p$, for $p < 0$.
63. Give a graphical explanation why
$$-\frac{1}{2}x + 1 > |x - 5|$$
has no solution.

Skill Maintenance

Simplify. Write the answer in the form $a + bi$, where a and b are real numbers.

64. $(-1 + 2i) + (3 - i)$
65. $(6 - 4i) - (-4 + 3i)$
66. $(2 - 3i)(5 + i)$
67. $\dfrac{5 + 2i}{3 - 4i}$

Synthesis

Solve.

68. $x \leq 3x - 2 \leq 2 - x$
69. $2x \leq 5 - 7x < 7 + x$
70. $|x + 2| \leq |x - 5|$
71. $|3x - 1| > 5x - 2$
72. $|x| + |x + 1| < 10$
73. $|p - 4| + |p + 4| < 8$
74. $|x - 3| + |2x + 5| > 6$

Handwritten at bottom:
$-x < 5 > 0.1$ or $x - 5 < 0.1$

$|x-5| = 0.1$
$n-5 = \pm 0.1$
$x = -0.1 + 5 = 4.9$
$x = 0.1 + 5 = 5.1$

Chapter Summary and Review 2

Important Properties and Formulas

Zero of a Function: An input c of a function f is a zero of f if $f(c) = 0$.

Complex Number: $a + bi$, a, b real, $i^2 = -1$

Imaginary Number: $a + bi$, $b \neq 0$

Complex Conjugates: $a + bi$, $a - bi$

Quadratic Equation:

$$ax^2 + bx + c = 0, \ a \neq 0, \ a, b, c \text{ real}$$

Quadratic Function:

$$f(x) = ax^2 + bx + c, \ a \neq 0, \ a, b, c \text{ real}$$

Quadratic Formula:

For $ax^2 + bx + c = 0$, $a \neq 0$,

$$x = \frac{-b \pm \sqrt{b^2 - 4ac}}{2a}.$$

Five Steps for Problem Solving

1. Familiarize.
2. Translate.
3. Carry out.
4. Check.
5. State.

Principles for Solving Equations

The Principle of Square Roots:

If $x^2 = k$, then $x = \sqrt{k}$ or $x = -\sqrt{k}$.

The Principle of Powers:

For any positive integer n, if $a = b$ is true, then $a^n = b^n$ is true.

Principles for Solving Inequalities

The Addition Principle for Inequalities:

If $a < b$ is true,
then $a + c < b + c$ is true.

The Multiplication Principle for Inequalities:

If $a < b$ and $c > 0$ are true,
then $ac < bc$ is true.

If $a < b$ and $c < 0$ are true,
then $ac > bc$ is true.

Similar statements hold for $\leq$.

Equations and Inequalities with Absolute Value

For $a > 0$,

$$|X| = a \longrightarrow X = -a \ \text{ or } \ X = a$$
$$|X| < a \longrightarrow -a < X < a$$
$$|X| > a \longrightarrow X < -a \ \text{ or } \ X > a$$

REVIEW EXERCISES

Find the zero(s) of the function.

1. $f(x) = 6x - 18$

2. $f(x) = x - 4$

3. $f(x) = 2 - 10x$

4. $f(x) = 8 - 2x$

5. $f(x) = x^2 - 2x + 1$

6. $f(x) = x^2 + 2x - 15$

7. $f(x) = 2x^2 - x - 5$

8. $f(x) = 3x^2 + 2x - 3$

Solve.

9. $5(3x + 1) = 2(x - 4)$

10. $(2y + 5)(3y - 1) = 0$

11. $\dfrac{5}{2x + 3} + \dfrac{1}{x - 6} = 0$

12. $\dfrac{3}{8x + 1} + \dfrac{8}{2x + 5} = 1$

13. $\sqrt{5x + 1} - 1 = \sqrt{3x}$

14. $\sqrt{x - 1} - \sqrt{x - 4} = 1$

15. $|x - 4| = 3$

16. $|2y + 7| = 9$

17. $-3 \le 3x + 1 < 5$

18. $-2 < 5x - 4 \le 6$

19. $2x < -1 \text{ or } x + 3 > 0$

20. $3x + 7 \le 2 \text{ or } 2x + 3 \ge 5$

21. $|6x - 1| < 5$

22. $|x + 4| \ge 2$

23. Solve $v = \sqrt{2gh}$ for h.

24. Solve $\dfrac{1}{a} + \dfrac{1}{b} = \dfrac{1}{t}$ for t.

Express in terms of i.

25. $-\sqrt{-40}$

26. $\sqrt{-12} \cdot \sqrt{-20}$

27. $\dfrac{\sqrt{-49}}{-\sqrt{-64}}$

Simplify each of the following. Write the answer in the form $a + bi$, where a and b are real numbers.

28. $(6 + 2i)(-4 - 3i)$

29. $\dfrac{2 - 3i}{1 - 3i}$

30. $(3 - 5i) - (2 - i)$

31. $(6 + 2i) + (-4 - 3i)$

32. i^{23}

33. $(-3i)^{28}$

Solve by completing the square to obtain exact solutions. Show your work. Check your work using a grapher.

34. $x^2 - 3x = 18$

35. $3x^2 - 12x - 6 = 0$

Solve. Use any method, but obtain exact solutions. Check your work using a grapher, if possible.

36. $3x^2 + 2x = 8$

37. $r^2 - 2r + 10 = 0$

38. $x^2 = 18 + 3x$

39. $x = 2\sqrt{x} - 1$

40. $y^4 - 3y^2 + 1 = 0$

41. $(x^2 - 1)^2 - (x^2 - 1) - 2 = 0$

42. $(p - 3)(3p + 2)(p + 2) = 0$

43. $x^3 + 5x^2 - 4x - 20 = 0$

In Exercises 44 and 45, complete the square to:

a) *find the vertex;*

b) *find the line of symmetry; and*

c) *determine whether there is a maximum or minimum value and find that value.*

Then check your answers using a grapher.

44. $f(x) = -4x^2 + 3x - 1$

45. $f(x) = 5x^2 - 10x + 3$

In Exercises 46–49, match the equation with one of the figures (a)–(d), which follow. Do not use a grapher except as a check.

a)

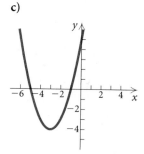

b)

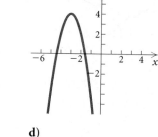

c)

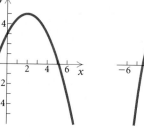

d)

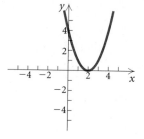

46. $y = (x - 2)^2$

47. $y = (x + 3)^2 - 4$

48. $y = -2(x + 3)^2 + 4$

49. $y = -\frac{1}{2}(x - 2)^2 + 5$

50. *Legs of a Right Triangle.* The hypotenuse of a right triangle is 50 ft. One leg is 10 ft longer than the other. What are the lengths of the legs?

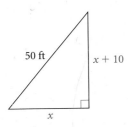

51. *Motion.* A Riverboat Cruise Line boat travels 8 mi upstream and 8 mi downstream. The total time for both parts of the trip is 3 hr. The speed of the stream is 2 mph. What is the speed of the boat in still water?

52. *Motion.* Two freight trains leave the same city at right angles. The first train travels at a speed of 60 km/h. In 1 hr, the trains are 100 km apart. How fast is the second train traveling?

53. *Sidewalk Width.* A 60-ft by 80-ft parking lot is torn up to install a sidewalk of uniform width around its perimeter. The new area of the parking lot is two thirds of the old area. How wide is the sidewalk?

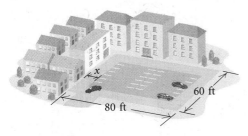

54. *Maximizing Volume.* The Berniers have 24 ft of flexible fencing with which to build a rectangular "toy corral." If the fencing is 2 ft high, what dimensions should the corral have in order to maximize its volume?

55. *Dimensions of a Box.* An open box is made from a 10-cm by 20-cm piece of aluminum by cutting a square from each corner and folding up the edges.

The area of the resulting base is 90 cm². What is the length of the sides of the squares?

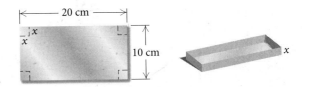

56. *Maximizing Box Dimensions.* An open box is to be made from a 10-cm by 20-cm piece of aluminum by cutting a square from each corner and folding up the edges. What should the length of the sides of the squares be in order to maximize the volume?

57. *Cholesterol Level and the Risk of Heart Attack.* The data in the following table show the relationship of cholesterol level in men to the risk of a heart attack.

CHOLESTEROL LEVEL	NUMBER OF MEN, PER 10,000, WHO SUFFER A HEART ATTACK
100	30
200	65
250	100
275	130

Source: Nutrition Action Healthletter

a) Use a grapher to fit a quadratic function to the data.

b) Use the function found in part (a) to estimate the heart attack rate for men with a cholesterol level of 350 and 400.

Discussion and Writing

58. Explain how the quadratic formula could be used to factor the expression $ax^2 + bx + c$.

59. If the graphs of

$$f(x) = a_1(x - h_1)^2 + k_1$$

and

$$g(x) = a_2(x - h_2)^2 + k_2$$

have the same shape, what, if anything, can you conclude about the a's, the h's, and the k's? Explain your answer.

Synthesis

Solve.

60. $(x - 1)^{2/3} = 4$

61. $(t - 4)^{4/5} = 3$

62. $\sqrt{x + 2} + \sqrt[4]{x + 2} - 2 = 0$

63. $(2y - 2)^2 + y - 1 = 5$

64. Find b such that $f(x) = -3x^2 + bx - 1$ has a maximum value of 2.

65. At the beginning of the year, $3500 was deposited in a savings account. One year later, $4000 was deposited in another account. The interest rate was the same for both accounts. At the end of the second year, there was a total of $8518.35 in the accounts. What was the annual interest rate?

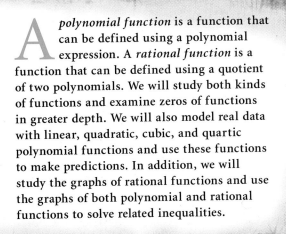

Polynomial and Rational Functions 3

become smaller

A *polynomial function* is a function that can be defined using a polynomial expression. A *rational function* is a function that can be defined using a quotient of two polynomials. We will study both kinds of functions and examine zeros of functions in greater depth. We will also model real data with linear, quadratic, cubic, and quartic polynomial functions and use these functions to make predictions. In addition, we will study the graphs of rational functions and use the graphs of both polynomial and rational functions to solve related inequalities.

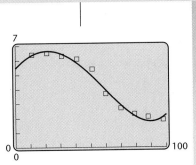

223

Polynomial Functions and Modeling

3.1

- *Use a grapher to graph a polynomial function and find its real-number zeros, relative maximum and minimum values, and domain and range.*
- *Solve applied problems using polynomial models.*
- *Fit linear, quadratic, power, cubic, and quartic polynomial functions to data and make predictions.*

There are many different kinds of functions. The constant, linear, and quadratic functions that we studied in Chapters 1 and 2 are part of a larger group of functions called *polynomial functions*.

Polynomial Function

A **polynomial function** P is given by

$$P(x) = a_n x^n + a_{n-1} x^{n-1} + a_{n-2} x^{n-2} + \cdots + a_1 x + a_0,$$

where the coefficients a_n, a_{n-1}, ..., a_1, a_0 are real numbers and the exponents are whole numbers.

The term $a_n x^n$ is called the **leading term.** The first nonzero coefficient, a_n, is called the **leading coefficient.** The **degree** of the polynomial function is n. Some examples of types of polynomial functions and their names are as follows.

POLYNOMIAL FUNCTION	DEGREE	EXAMPLE
Constant	0	$f(x) = 3$
Linear	1	$f(x) = \frac{2}{3}x + 1$
Quadratic	2	$f(x) = 2x^2 - x + 3$
Cubic	3	$f(x) = x^3 + 2x^2 + x - 5$
Quartic	4	$f(x) = -x^4 - 1.1x^3 + 0.3x^2 - 2.8x - 1.7$

The function $f(x) = 0$ can be described in many ways: $f(x) = 0 = 0x^2 = 0x^{15} = 0x^{48}$, and so on. For this reason, we say that the constant function $f(x) = 0$ has no degree.

From our study of functions in Chapters 1 and 2, we know how to use a grapher to find or at least estimate many characteristics of a polynomial function. Let's consider an example for review.

Function:	$f(x) = x^4 - 6x^3 - 4x^2 + 53.2x - 42.6$
Zeros:	-2.975, 0.950, 3, 5.025
Relative maximum:	15.921 at $x = 1.914$
Relative minima:	-106.907 at $x = -1.643$, -23.102 at $x = 4.229$
Domain:	All real numbers, $(-\infty, \infty)$
Range:	$[-106.907, \infty)$

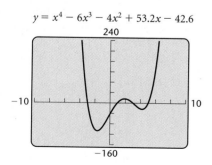

$y = x^4 - 6x^3 - 4x^2 + 53.2x - 42.6$

The graphs of polynomial functions have specific characteristics.

The following table lists some polynomial functions and some nonpolynomial functions. Graph all eight functions and compare the graphs.

POLYNOMIAL FUNCTIONS	NONPOLYNOMIAL FUNCTIONS
$f(x) = x^2 - 5x + 2$	$f(x) = |x + 2|$
$f(x) = -3$	$f(x) = \sqrt{x + 1} - 1$
$f(x) = 3x^3 - 2x^2 + x - 4$	$f(x) = x^{4/5}$
$f(x) = x^4 + 2x^2$	$f(x) = \dfrac{x - 1}{x + 5}$

Describe some characteristics of the graphs of polynomial functions that you observe. How do the graphs of polynomial functions differ from the graphs of nonpolynomial functions?

You probably noted that the graph of a polynomial function is continuous, that is, it has no holes or breaks. It is also smooth; there are no sharp corners. Furthermore, the *domain* of a polynomial function is the set of all real numbers, $(-\infty, \infty)$.

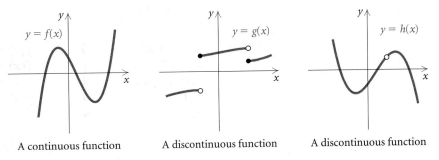

A continuous function   A discontinuous function   A discontinuous function

The Leading-Term Test

It would seem reasonable that as x becomes very large $(x \to \infty)$ or as x becomes very small $(x \to -\infty)$, the behavior of the graph of a polynomial function would be determined by the *leading term*. Let's see if we can discover some general patterns.

Graph each of the following groups of functions.

Degree even:
$y = x^2$, $y = -x^4 - 2x^3 + x - 1$, $y = \frac{1}{2}x^6 + 3$, and
$y = 1 - x - x^{10}$

Degree odd:

$y = x^3$, $y = -3x^3 - 5x + 2$, $y = 0.9x^5 - 2x^3 + x^2$, and
$y = -x^7 - 2x^4$

How are the shapes of the graphs in each group similar? Can we predict the behavior of the graph of a polynomial function as $x \to \infty$ and as $x \to -\infty$ by observing the leading term?

We can summarize our observations as follows.

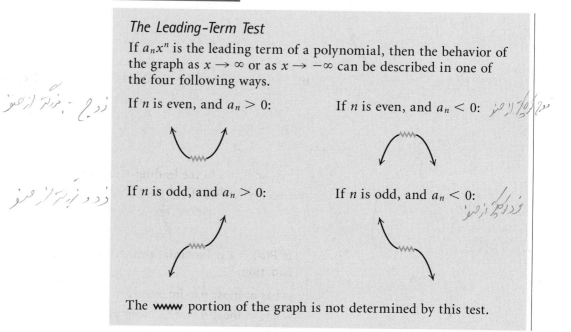

The Leading-Term Test

If $a_n x^n$ is the leading term of a polynomial, then the behavior of the graph as $x \to \infty$ or as $x \to -\infty$ can be described in one of the four following ways.

If n is even, and $a_n > 0$:   If n is even, and $a_n < 0$:

If n is odd, and $a_n > 0$:   If n is odd, and $a_n < 0$:

The ⌇⌇⌇ portion of the graph is not determined by this test.

EXAMPLE 1 Using the leading-term test, match each of the following functions with one of the graphs A–D, which follow.

a) $f(x) = 3x^4 - 2x^3 + 3$

b) $f(x) = -5x^3 - x^2 + 4x + 2$

c) $f(x) = x^5 + \frac{1}{4}x + 1$

d) $f(x) = -x^6 + x^5 - 4x^3$

A.

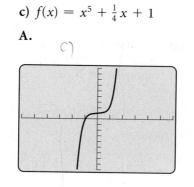

B.

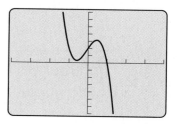

C. *d)*

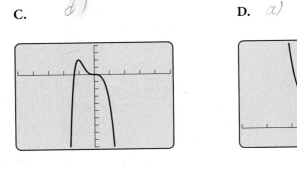

D. *a)*

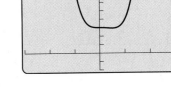

Solution

	LEADING TERM	DEGREE OF LEADING TERM	SIGN OF LEADING COEFFICIENT	GRAPH
a)	$3x^4$	Even	Positive	D
b)	$-5x^3$	Odd	Negative	B
c)	x^5	Odd	Positive	A
d)	$-x^6$	Even	Negative	C

In addition to the leading-term test, it is helpful to consider the following facts when graphing polynomial functions.

If $P(x)$ is a polynomial function of degree n, the graph of the function:

- has at most n x-intercepts, and thus at most n real zeros;
- has at most $n - 1$ turning points.

(Turning points on a graph occur when the function changes from decreasing to increasing or from increasing to decreasing.)

ZEROS OF FUNCTIONS

REVIEW SECTIONS 2.1 AND 2.3.

Finding Zeros on a Grapher

An input for which a function's output is 0 is called a **zero** of the function.

Zeros of Functions

An input c of a function f is called a **zero** of the function, if the output for c is 0. That is, $f(c) = 0$.

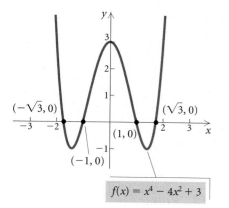

$f(x) = x^4 - 4x^2 + 3$

For example, the function given by $f(x) = x^4 - 4x^2 + 3$ (see the graph at left) has the zeros $-\sqrt{3}$, -1, 1, and $\sqrt{3}$. These numbers are the first coordinates of the x-intercepts, $(-\sqrt{3}, 0)$, $(-1, 0)$, $(1, 0)$, and $(\sqrt{3}, 0)$.

Finding exact values of the zeros of a function can be difficult. We can find approximations using a grapher.

EXAMPLE 2   Find the zeros of the function f given by

$$f(x) = 0.1x^3 - 0.6x^2 - 0.1x + 2.$$

Approximate the zeros to three decimal places.

Solution   We use a grapher, trying to create a graph that clearly shows the curvature. Then we look for points where the graph crosses the x-axis. It appears that there are three zeros, one near -2, one near 2, and one near 6. We use the ZERO feature to find them.

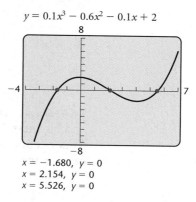

$y = 0.1x^3 - 0.6x^2 - 0.1x + 2$

$x = -1.680, \; y = 0$
$x = 2.154, \; y = 0$
$x = 5.526, \; y = 0$

The zeros are approximately -1.680, 2.154, and 5.526.

Interactive Discovery

Consider the following polynomial function:

$$P(x) = 12x^3 - 5x^2 - 11x + 6.$$

Graph this function using the viewing window $[-1, 2, -2, 10]$. How many zeros does the function appear to have in the interval $[0.5, 1]$?

Now graph the function again using the viewing window $[0.5, 1, -0.1, 0.15]$, with Xscl = 0.1 and Yscl = 0.05. Now how many zeros does the function appear to have in the interval $[0.5, 1]$?

Find $P(0.6)$, $P(0.7)$, and $P(0.8)$ and note any change in sign from one function value to another. Is there a connection between a change in sign and the occurrence of a zero between two function values?

Polynomial functions P are continuous, hence their graphs are unbroken. All polynomial functions have $(-\infty, \infty)$ as the domain. Suppose two function values $P(a)$ and $P(b)$ have opposite signs. Since P is continuous, its graph must be a curve from $(a, P(a))$ to $(b, P(b))$ without a break. Then it follows that the curve must cross the x-axis somewhere between a and b—that is, the function has a zero between a and b.

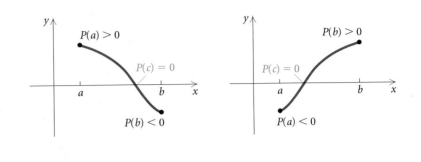

The Intermediate Value Theorem

For any polynomial function $P(x)$ with real coefficients, suppose that for $a \neq b$, $P(a)$ and $P(b)$ are of opposite signs. Then the function has a real zero between a and b.

Polynomial Models

Polynomial functions have many uses as models in science, engineering, and business. The simplest use of polynomial functions in applied problems occurs when we merely evaluate a polynomial function. In such cases, a model has already been developed.

EXAMPLE 3 *Ibuprofen in the Bloodstream.* Ibuprofen is a pain relief medication. The polynomial function

$$M(t) = 0.5t^4 + 3.45t^3 - 96.65t^2 + 347.7t$$

can be used to estimate the number of milligrams of ibuprofen in the bloodstream t hours after 400 mg of the medication has been taken.

a) Graph the function using the viewing windows $[-24, 15, -14{,}400, 8000]$ with Xscl = 3 and Yscl = 800 and $[0, 6, -200, 500]$ with Xscl = 1 and Yscl = 100. Discuss the limitations of the model.

b) Find the domain, the relative maximum and where it occurs, and the range.

c) Find the number of milligrams in the bloodstream at $t = 0, 0.5, 1, 1.5$, and so on, up to 3 hr.

Solution

a) The graphs using the two viewing windows are shown at left. The domain of a polynomial function, unless restricted by a statement of the function, is $(-\infty, \infty)$. The first graph on the left gives a fairly complete picture of the curvature of the polynomial M. The implications of the application restrict the domain of the function, however. If we assume

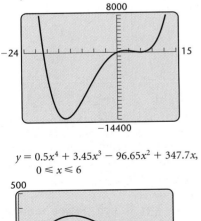

$y = 0.5x^4 + 3.45x^3 - 96.65x^2 + 347.7x$

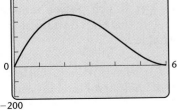

$y = 0.5x^4 + 3.45x^3 - 96.65x^2 + 347.7x,$
$0 \leq x \leq 6$

that a patient had not taken any of the medication before, it seems reasonable that $M(0) = 0$; that is, at time 0, there is 0 mg of the medication in the bloodstream. After the medication has been taken, $M(t)$ will be positive for a period of time and eventually decrease back to 0 and certainly not increase again (unless another dose is taken). Thus the second graph suggests that a better model includes a restriction on t,

$$M(t) = 0.5t^4 + 3.45t^3 - 96.65t^2 + 347.7t, \quad 0 \le t \le 6.$$

b) Given the restriction, the domain is [0, 6]. To find the range, we need to find the relative maximum value of the function using the MAXI-MUM feature on a grapher.

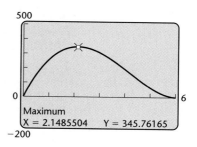

The maximum is about 345.76. It occurs approximately 2.15 hr, or 2 hr 9 min, after the initial dose has been taken. The range is about [0, 345.76].

c) We can evaluate the function in a number of ways. With the TABLE feature, we can start at 0 and use a step-value of 0.5.

X	Y₁
0	0
.5	150.15
1	255
1.5	318.26
2	344.4
2.5	338.63
3	306.9

X = 0

LINEAR AND
QUADRATIC REGRESSION

REVIEW SECTIONS 1.3 AND 2.5.

So far in this text, we have used regression to model data with two types of polynomial functions: linear and quadratic. We now expand that procedure to include power, cubic, and quartic models.

POWER MODELS   There are many situations in which **power models** $y = ax^b$ can be fit to data using regression. Note that the constants to be determined are a and b. The base is the variable x.

EXAMPLE 4   *Cholesterol Level and the Risk of Heart Attack.* The data in the following table show the relationship of cholesterol level in men to the risk of a heart attack.

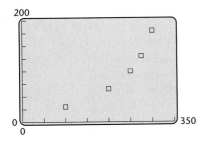

CHOLESTEROL LEVEL, x	MEN, PER 10,000, WHO SUFFER A HEART ATTACK, y
100	30
200	65
250	100
275	130
300	180

Source: Nutrition Action Healthletter

a) Use a grapher to fit a power function to the data.

b) Graph the function with the scatterplot of the data.

c) Use the answer to part (a) to estimate the heart attack rate for men with cholesterol levels of 150, 350, and 400.

Solution

a) We are fitting an equation of the type $y = ax^b$ to the data. Entering the data into the grapher with DIAGNOSTIC turned on and carrying out the regression procedure, we find that

$$a = 0.0241789574,$$
$$b = 1.527457172,$$
$$r = 0.9739361336.$$

This tells us that the equation is about $y = 0.024x^{1.527}$. The *correlation coefficient, r,* of about 0.974 is very close to 1. This indicates that the data fit a power function fairly well.

b) The graph is shown at left.

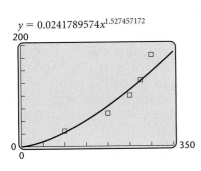

$y = 0.0241789574x^{1.527457172}$

c) To estimate the heart attack rate for men with cholesterol levels of 150, 350, and 400, we evaluate the function found in part (a) for $x = 150$, $x = 350$, and $x = 400$.

Y₁(150)	
	50.97107227
Y₁(350)	
	185.9484444
Y₁(400)	
	228.0199073

We see that the heart attack rate is about 51 out of 10,000 for men with a cholesterol level of 150, about 186 out of 10,000 for men with a cholesterol level of 350, and about 228 out of 10,000 for men with a cholesterol level of 400.

CUBIC AND QUARTIC MODELS   We continue our discussion of curve fitting with examples of cubic and quartic regression.

EXAMPLE 5 *Declining Number of Farms in the United States.* Today U.S. farm acreage is about the same as it was in the early part of the twentieth century, but the number of farms has shrunk.

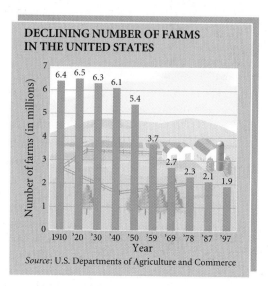

Looking at the graph above, we note that the data could be modeled with a cubic or a quartic function.

a) Model the data with both cubic and quartic functions. Let the first coordinate of each data point be the number of years after 1900. That is, enter the data as (10, 6.4), (20, 6.5), and so on. Then using R^2, the **coefficient of determination,** decide which function is the better fit. The R^2-value gives an indication of how well the function fits the data. The closer R^2 is to 1, the better the fit.

b) Graph the function with the scatterplot of the data.

c) Use the answer to part (a) to estimate the number of farms in 1900, 1975, and 2003.

Solution

a) Using the REGRESSION feature with DIAGNOSTIC turned on, we get the following.

```
CubicReg
y=ax³+bx²+cx+d
a=2.9566515E−5
b=−.0049844076
c=.1742288914
d=4.983981559
R²=.9828373098
```

```
QuarticReg
y=ax⁴+bx³+...+e
a=1.5229297E−7
b=−3.092184E−6
c=−.0026500648
d=.1117312855
↓e=5.463775105
```

```
QuarticReg
y=ax⁴+bx³+...+e
↑b=−3.092184E−6
c=−.0026500648
d=.1117312855
e=5.463775105
R²=.9836750763
```

Since the R^2-value for the quartic function is closer to 1 than that for the cubic function, the quartic function is the better fit. Note that

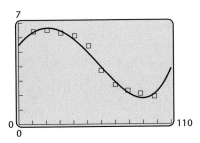

a and *b* are given in scientific notation on the grapher, but we convert to decimal notation when we write the function.

$$f(x) = 0.00000015229297x^4 - 0.000003092184x^3$$
$$- 0.0026500648x^2 + 0.1117312855x + 5.463775105$$

b) The graph is shown at left.

c) We evaluate the function found in part (a).

Y₁(0)	5.463775105
Y₁(75)	2.451136695
Y₁(103)	2.61935514

Thus we estimate that there were about 5.5 million farms in 1900 and about 2.5 million in 1975. Looking at the bar graph shown above, we see that these estimates appear to be fairly accurate.

Since the number of farms will probably continue to decrease, the estimate of 2.6 million for 2003 is probably not a good prediction. Both the cubic and the quartic models have high correlation coefficients over the domain of the data, but this correlation does not reflect the degree of accuracy for extended values. It is always important when using regression to evaluate predictions with common sense and knowledge of current trends.

As useful as regression models might be in certain situations, we must keep in mind that almost every model has its limitations.

Exercise Set 3.1

Classify the polynomial as constant, linear, quadratic, cubic, or quartic and determine the leading term, the leading coefficient, and the degree of the polynomial.

1. $g(x) = \frac{1}{2}x^3 - 10x + 8$

2. $f(x) = 15x^2 - 10 + 0.11x^4 - 7x^3$

3. $h(x) = 0.9x - 0.13$

4. $f(x) = -6$

5. $g(x) = 305x^4 + 4021$

6. $h(x) = 2.4x^3 + 5x^2 - x + \frac{7}{8}$

In Exercises 7–12, use the leading-term test and your knowledge of y-intercepts to match the function with one of graphs (a)–(f), which follow.

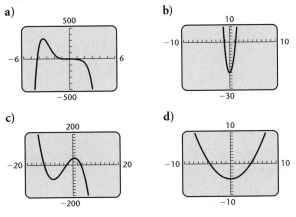

a)

b)

c)

d)

e)

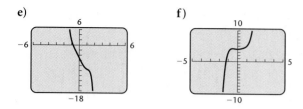

f)

7. $f(x) = \frac{1}{4}x^2 - 5$

8. $f(x) = -0.5x^6 - x^5 + 4x^4 - 5x^3 - 7x^2 + x - 3$

9. $f(x) = x^5 - x^4 + x^2 + 4$

10. $f(x) = -\frac{1}{3}x^3 - 4x^2 + 6x + 42$

11. $f(x) = x^4 - 2x^3 + 12x^2 + x - 20$

12. $f(x) = -0.3x^7 + 0.11x^6 - 0.25x^5 + x^4 + x^3 - 6x - 5$

Using a grapher, find the zeros of the function.

13. $f(x) = x^3 - 3x - 1$

14. $f(x) = x^3 + 3x^2 - 9x - 13$

15. $f(x) = x^4 - 2x^2$

16. $f(x) = x^4 - 2x^3 - 5.6$

17. $f(x) = x^3 - x$

18. $f(x) = 2x^3 - x^2 - 14x - 10$

19. $f(x) = x^8 + 8x^7 - 28x^6 - 56x^5 + 70x^4$
$\qquad + 56x^3 - 28x^2 - 8x + 1$

20. $f(x) = x^6 - 10x^5 + 13x^3 - 4x^2 - 5$

Using a grapher, estimate the real zeros, the relative maxima and minima, and the range of the polynomial function.

21. $g(x) = x^3 - 1.2x + 1$

22. $g(x) = 5x - 14$

23. $h(x) = -\frac{1}{2}x^4 + 3x^3 - 5x^2 + 3x + 6$

24. $f(x) = -3.5x^3 - x^2 + 24.5x - 4$

25. $f(x) = x^6 - 3.8$

26. $f(x) = x^5 - 3x^4 + 6x + 6$

27. $g(x) = 12 - 3.14x$

28. $h(x) = 2x^3 - x^4 + 20$

29. $f(x) = x^2 + 10x - x^5$

30. $g(x) = 2x^2 - 5.8x + 1$

31. $h(x) = -x^5 + 4x^3 - x$

32. $f(x) = 2x^4 - 5.6x^2 + 10$

Using the intermediate value theorem, show that the function f has a zero between a and b.

33. $f(x) = x^3 + 3x^2 - 9x - 13;\ a = -5,\ b = -4$

34. $f(x) = x^3 + 3x^2 - 9x - 13;\ a = 2,\ b = 3$

35. *Vertical Leap.* A formula relating an athlete's vertical leap V, in inches, to hang time T, in seconds, is

$$V = 48T^2.$$

Anfernee Hardaway of the Phoenix Suns has a vertical leap of 36 in. (*Source:* National Basketball Association). What is his hang time?

36. *Projectile Motion.* A stone thrown downward with an initial velocity of 34.3 m/sec will travel a distance of s meters, where

$$s(t) = 4.9t^2 + 34.3t$$

and t is in seconds. If a stone is thrown downward at 34.3 m/sec from a height of 294 m, how long will it take the stone to hit the ground?

37. *Games in a Sports League.* If there are x teams in a sports league and all the teams play each other twice, a total of $N(x)$ games are played, where

$$N(x) = x^2 - x.$$

A softball league has 9 teams, each of which plays the others twice. If the league pays $45 per game for the field and umpires, how much will it cost to play the entire schedule?

38. *Windmill Power.* Under certain conditions, the power P, in watts per hour, generated by a windmill with winds blowing v miles per hour is given by

$$P(v) = 0.015v^3.$$

a) Find the power generated by 15-mph winds.

b) How fast must the wind blow in order to generate 120 watts of power in 1 hr?

$$3610 = 2560(1+i)^2$$
$$0 = 2560(1+i)^2 - 3610$$

39. *Interest Compounded Annually.* When P dollars is invested at interest rate i, compounded annually, for t years, the investment grows to A dollars, where

$$A = P(1 + i)^t.$$

a) Find the interest rate i if $2560 grows to $3610 in 2 yr.

b) Find the interest rate i if $10,000 grows to $13,310 in 3 yr.

40. *Threshold Weight.* In a study performed by Alvin Shemesh, it was found that the **threshold weight** W, defined as the weight above which the risk of death rises dramatically, is given by

$$W(h) = \left(\frac{h}{12.3}\right)^3,$$

where W is in pounds and h is a person's height, in inches. Find the threshold weight of a person who is 5 ft, 7 in. tall.

For the scatterplots and graphs in Exercises 41–48, determine which, if any, of the following functions might be used as a model for the data.

a) *Linear,* $f(x) = mx + b$
b) *Quadratic,* $f(x) = ax^2 + bx + c, \ a > 0$
c) *Quadratic,* $f(x) = ax^2 + bx + c, \ a < 0$
d) *Polynomial, not quadratic or linear*

41.   **42.**

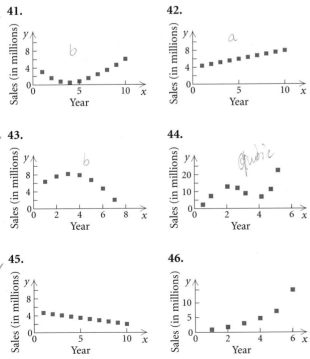

43.   **44.**

45.   **46.**

47.

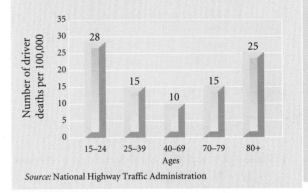

DRIVER FATALITIES BY AGE
Number of licensed drivers per 100,000 who died in motor vehicle accidents in 1990. The fatality rates for both the 70–79 group and 80+ age group were lower than for the 15- to 24-year-olds.

Source: National Highway Traffic Administration

48.

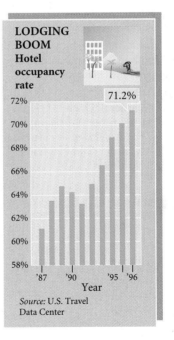

LODGING BOOM Hotel occupancy rate

71.2%

Source: U.S. Travel Data Center

49. *U.S. Farms.* As the number of farms has decreased in the United States, the average size of the remaining farms has grown larger, as shown in the table on the next page. (See Example 5 in this section.)

YEAR	AVERAGE ACREAGE PER FARM
1910	139
1920	149
1930	157
1940	175
1950	216
1959	303
1969	390
1978	449
1987	462
1997	487

Source: U.S. Departments of Agriculture and of Commerce

a) Make a scatterplot of the data, letting x represent the number of years since 1900.
b) Use a grapher to fit linear, quadratic, cubic, and power functions to the data. By comparing the values of R^2, determine the function that best fits the data.
c) Graph the function of best fit with the scatterplot of the data.
d) With each function found in part (b), predict the average acreage in 2000 and 2010 and determine which function gives the most realistic predictions.

50. *Funeral Costs.*   Since the Federal Trade Commission began regulating funeral directors in 1984, the average cost for the funeral of an adult has greatly increased, as shown in the following table.

YEAR	AVERAGE COST OF FUNERAL
1980	$1809
1985	2737
1991	3742
1995	4624
1996	4782
1998	5020

Source: National Funeral Directors Association

a) Make a scatterplot of the data, letting x represent the number of years since 1980.
b) Use a grapher to fit quadratic, cubic, and quartic functions to the data. By comparing the values of R^2, determine the function that best fits the data.

c) Graph the function of best fit with the scatterplot.
d) With each function found in part (b), estimate the cost of a funeral in 2000 and 2010 and determine which function gives the most realistic estimates.

51. *Damage to the Ozone Layer.*   The concentration of chlorine compounds in the stratosphere serves as an indicator of damage to the ozone layer. The data in the following table show the relationship of estimated chlorine concentration in the atmosphere, in parts per billion (ppb), to the year.

YEAR	CHLORINE CONCENTRATION (IN PARTS PER BILLION)
1985	2.5
1995	3.3
2010	3.9
2035	3.7

Source: Adapted by U.S. EPA by NRDC Earth Action Guide, "Saving the Ozone"

a) Make a scatterplot of the data, letting x represent the number of years since 1985.
b) Use a grapher to fit linear, quadratic, and cubic functions to the data. By comparing the values of R^2, determine the function that best fits the data.
c) Graph the function of best fit with the scatterplot.
d) Use the functions found in part (b) to predict the chlorine concentration in 2060. Compare the predictions and determine which model gives the most realistic prediction.

52. *Consumer Debt.*   Nonmortgage consumer debt is mounting in the United States, as shown in the table below.

YEAR	NONMORTGAGE DEBT (IN BILLIONS)
1989	$ 762
1990	789
1991	783
1992	775
1993	804
1994	902
1995	1038
1996	1161
1997	1216
1998	1266

Source: Federal Reserve

a) Make a scatterplot of the data, letting x represent the number of years since 1989.

b) Use a grapher to fit linear, quadratic, cubic, and quartic functions to the data. By comparing the values of R^2, determine the function that best fits.

c) Graph the function of best fit with the scatterplot.

d) Use the functions found in part (b) to predict the debt in 2003. Compare the predictions and determine which model gives the most realistic prediction.

Discussion and Writing

53. Find a cubic function that fits the data shown in the graph, noting that $R^2 = 1$. Then, using this model, estimate the number of gunfire deaths in 2000 and discuss the limitations of the model.

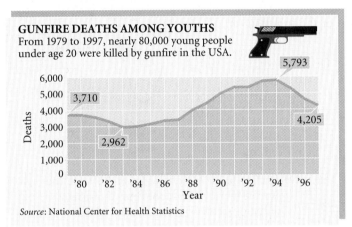

GUNFIRE DEATHS AMONG YOUTHS
From 1979 to 1997, nearly 80,000 young people under age 20 were killed by gunfire in the USA.

Source: National Center for Health Statistics

54. The value of R^2 is used in regression to describe the strength of the relationship between the independent and dependent variables over the domain of the given data. Does it also reflect the accuracy of the fit beyond the limited domain of the data? Discuss its connection to the reality of the predicted values.

55. How is the range of a polynomial function related to the degree of the polynomial?

Skill Maintenance

Find the distance between the pair of points.

56. $(4, 2)$ and $(-2, -4)$

57. $(3, -5)$ and $(0, -1)$

58. The diameter of a circle connects two points $(-6, 5)$ and $(-2, 1)$ on the circle. Find the coordinates of the center of the circle and the length of the radius.

59. Find the center and the radius of the circle

$$(x - 3)^2 + (y + 5)^2 = 49.$$

Synthesis

60. In early 1995, $2000 was deposited at a certain interest rate compounded annually. One year later, $1200 was deposited in another account at the same rate. At the end of that year, there was a total of $3573.80 in both accounts. What is the annual interest rate?

Polynomial Division; The Remainder and Factor Theorems

3.2

- *Perform long division with polynomials and determine whether one polynomial is a factor of another.*
- *Use synthetic division to divide a polynomial by $x - c$.*
- *Use the remainder theorem to find a function value $f(c)$.*
- *Use the factor theorem to determine whether $x - c$ is a factor of $f(x)$.*

In general, finding exact zeros of many polynomial functions is neither easy nor straightforward. In this section and the one that follows, we develop concepts that help us find exact zeros of certain polynomial functions with degree 3 or greater.

Let's review the meaning of the real zeros of a function and their connection to the x-intercepts of the function's graph.

CONNECTING THE CONCEPTS

ZEROS, SOLUTIONS, AND INTERCEPTS

| FUNCTION | REAL ZEROS OF THE FUNCTION; SOLUTIONS OF THE EQUATION | x-INTERCEPTS OF THE GRAPH |

Quadratic Polynomial

$g(x) = x^2 - 2x - 8$
$\quad = (x + 2)(x - 4),$

or

$\quad y = (x + 2)(x - 4)$

To find the **zeros** of $g(x)$, we solve $g(x) = 0$:

$x^2 - 2x - 8 = 0$
$(x + 2)(x - 4) = 0$
$x + 2 = 0 \quad or \quad x - 4 = 0$
$x = -2 \quad or \quad x = 4.$

The **solutions** of $x^2 - 2x - 8 = 0$ are -2 and 4. They are the zeros of the function $g(x)$. That is,

$g(-2) = 0 \quad and \quad g(4) = 0.$

The zeros of $g(x)$ are the x-coordinates of the **x-intercepts** of the graph of $y = g(x)$.

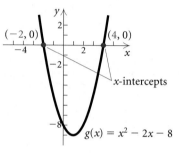

Cubic Polynomial

$h(x)$
$\quad = x^3 + 2x^2 - 5x - 6$
$\quad = (x + 3)(x + 1)(x - 2),$

or

$y = (x + 3)(x + 1)(x - 2)$

To find the **zeros** of $h(x)$, we solve $h(x) = 0$:

$x^3 + 2x^2 - 5x - 6 = 0$
$(x + 3)(x + 1)(x - 2) = 0$
$x + 3 = 0 \quad or \quad x + 1 = 0 \quad or \quad x - 2 = 0$
$x = -3 \quad or \quad x = -1 \quad or \quad x = 2.$

The **solutions** of $x^3 + 2x^2 - 5x - 6 = 0$ are -3, -1, and 2. They are the zeros of the function $h(x)$. That is,

$h(-3) = 0,$
$h(-1) = 0, \quad and$
$h(2) = 0.$

The zeros of $h(x)$ are the x-coordinates of the **x-intercepts** of the graph of $y = h(x)$.

The connection between the zeros of a function and the x-intercepts of the graph of the function is easily seen in the examples above. If c is a real zero of a function (that is, $f(c) = 0$), then $(c, 0)$ is an x-intercept of the graph of the function.

EXAMPLE 1   Consider $P(x) = x^3 + x^2 - 17x + 15$. Determine whether each of the numbers 2 and -5 is a zero of $P(x)$.

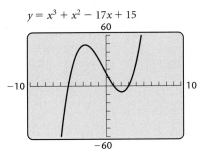

$y = x^3 + x^2 - 17x + 15$

Solution   We have

$$P(2) = (2)^3 + (2)^2 - 17(2) + 15 = -7.$$ **Substituting 2 into the polynomial**

Since $P(2) \neq 0$, we know that 2 is *not* a zero of the polynomial.
    We also have

$$P(-5) = (-5)^3 + (-5)^2 - 17(-5) + 15 = 0.$$ **Substituting -5 into the polynomial**

Since $P(-5) = 0$, we know that -5 is a zero of $P(x)$.   ▬

    Let's take a closer look at the polynomial

$$h(x) = x^3 + 2x^2 - 5x - 6$$

(see Connecting the Concepts on p. 238). Is there a connection between the factors of the polynomial and the zeros of the function? For $h(x)$, the factors are

$$x + 3, \qquad x + 1, \quad \text{and} \quad x - 2,$$

and the zeros are

$$-3, \qquad -1, \quad \text{and} \quad 2.$$

We note that when the polynomial is expressed in factored form, each factor determines a zero of the function. Thus if we know the factors of a polynomial, we can easily find the zeros. We now show how this idea can be "reversed" so that if we know the zeros of a polynomial function, we can find the factors of the polynomial.

Division and Factors

When we divide one polynomial by another, we obtain a quotient and a remainder. If the remainder is 0, then the divisor is a factor of the dividend.

EXAMPLE 2   Divide to determine whether $x + 1$ and $x - 3$ are factors of

$$x^3 + 2x^2 - 5x - 6.$$

Solution   We have

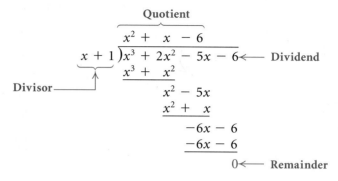

Since the remainder is 0, we know that $x + 1$ is a factor of $x^3 + 2x^2 - 5x - 6$.

We also have

$$
\begin{array}{r}
x^2 + 5x + 10 \\
x - 3 \overline{)\smash{x^3 + 2x^2 -\ \ 5x -\ \ 6}} \\
\underline{x^3 - 3x^2} \\
5x^2 -\ \ 5x \\
\underline{5x^2 - 15x} \\
10x -\ \ 6 \\
\underline{10x - 30} \\
24 \longleftarrow \text{Remainder}
\end{array}
$$

Since the remainder is not 0, we know that $x - 3$ is *not* a factor of $x^3 + 2x^2 - 5x - 6$.

EXAMPLE 3   Divide to determine whether $x^2 + 3x - 1$ is a factor of $x^4 - 81$.

Solution   We have

$$
\begin{array}{r}
x^2 - 3x + 10 \\
x^2 + 3x - 1 \overline{)\smash{x^4 + 0x^3 +\ \ 0x^2 +\ \ 0x - 81}} \\
\underline{x^4 + 3x^3 -\ \ \ \ x^2} \\
-3x^3 +\ \ \ x^2 +\ \ 0x \\
\underline{-3x^3 -\ \ 9x^2 +\ \ 3x} \\
10x^2 -\ \ 3x - 81 \\
\underline{10x^2 + 30x - 10} \\
-33x - 71
\end{array}
$$

Note that terms in which the coefficient is 0 have been inserted for the missing terms. Spaces can also be used to indicate missing terms.

Since the remainder is not 0, we know that $x^2 + 3x - 1$ is *not* a factor of $x^4 - 81$.

When we divide a polynomial $P(x)$ by a divisor $d(x)$, a polynomial $Q(x)$ is the quotient and a polynomial $R(x)$ is the remainder. The remainder must either be 0 or have degree less than that of $d(x)$.

As in arithmetic, to check a division, we multiply the quotient by the divisor and add the remainder, to see if we get the dividend. Thus these polynomials are related as follows:

$$P(x) = d(x) \cdot Q(x) + R(x)$$

Dividend   Divisor   Quotient   Remainder

For example, if $P(x) = x^3 + 2x^2 - 5x - 6$ and $d(x) = x + 1$, as in Example 2, then $Q(x) = x^2 + x - 6$ and $R(x) = 0$, and

$$\underbrace{x^3 + 2x^2 - 5x - 6}_{P(x)} = \underbrace{(x + 1)}_{d(x)} \cdot \underbrace{(x^2 + x - 6)}_{Q(x)} + \underbrace{0}_{R(x).}$$

If $P(x) = x^4 - 81$ and $d(x) = x^2 + 3x - 1$, as in Example 3, then $Q(x) = x^2 - 3x + 10$ and $R(x) = -33x - 71$, and

$$\underbrace{x^4 - 81}_{P(x)} = \underbrace{(x^2 + 3x - 1)}_{d(x)} \cdot \underbrace{(x^2 - 3x + 10)}_{Q(x)} + \underbrace{(-33x - 71)}_{R(x).}$$

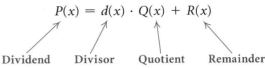

The Remainder Theorem and Synthetic Division

Interactive Discovery

Consider

$$h(x) = x^3 + 2x^2 - 5x - 6.$$

Find $h(-1)$ and $h(3)$ and compare your answers to the remainders found in Example 2 when the polynomial $x^3 + 2x^2 - 5x - 6$ was divided by the polynomials $x + 1$ and $x - 3$. Also, divide the polynomial by $x + 3$ and $x - 2$ and then find $h(-3)$ and $h(2)$. How are the remainders and the function values related?

The Interactive Discovery suggests the following theorem.

> ### The Remainder Theorem
>
> If a number c is substituted for x in the polynomial $f(x)$, then the result $f(c)$ is the remainder that would be obtained by dividing $f(x)$ by $x - c$. That is, if $f(x) = (x - c) \cdot Q(x) + R$, then $f(c) = R$.

PROOF (OPTIONAL). The equation $f(x) = d(x) \cdot Q(x) + R(x)$, where $d(x) = x - c$, is the basis of this proof. If we divide $f(x)$ by $x - c$, we obtain a quotient $Q(x)$ and a remainder $R(x)$ related as follows:

$$f(x) = (x - c) \cdot Q(x) + R(x).$$

The remainder $R(x)$ must either be 0 or have degree less than $x - c$. Thus, $R(x)$ must be a constant. Let's call this constant R. The equation above is true for any replacement of x, so we replace x with c. We get

$$
\begin{aligned}
f(c) &= (c - c) \cdot Q(c) + R \\
&= 0 \cdot Q(c) + R \\
&= R.
\end{aligned}
$$

Thus the function value $f(c)$ is the remainder obtained when we divide $f(x)$ by $x - c$.

The remainder theorem motivates us to find a rapid way of dividing by $x - c$, in order to find function values. To streamline division, we can arrange the work so that duplicate and unnecessary writing is avoided. Consider the following.

A.
$$
\begin{array}{r}
4x^2 + 5x + 11 \\
x - 2 \overline{)4x^3 - 3x^2 + x + 7} \\
\underline{4x^3 - 8x^2} \\
5x^2 + x \\
\underline{5x^2 - 10x} \\
11x + 7 \\
\underline{11x - 22} \\
29
\end{array}
$$

B.
$$
\begin{array}{r}
4 \quad 5 \quad 11 \\
1 - 2 \overline{)4 - 3 + 1 + 7} \\
\underline{4 - 8} \\
5 + 1 \\
\underline{5 - 10} \\
11 + 7 \\
\underline{11 - 22} \\
29
\end{array}
$$

The division in (B) is the same as that in (A), but we wrote only the coefficients. The color numerals are duplicated, so we look for an arrangement in which they are not duplicated. In place of the divisor in the form $x - c$, we can simply use c and then add rather than subtract. When the procedure is "collapsed," we have the algorithm known as **synthetic division.**

C. *Synthetic Division*

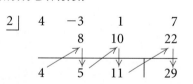

We "bring down" the 4. Then we multiply it by the 2 to get 8 and add to get 5. We then multiply 5 by 2 to get 10, add, and so on. The last number, 29, is the remainder. The others, 4, 5, and 11, are the coefficients of the quotient.

When using synthetic division, we write a 0 for a missing term in the dividend.

EXAMPLE 4   Use synthetic division to find the quotient and the remainder:

$$(2x^3 + 7x^2 - 5) \div (x + 3).$$

Solution First, we note that $x + 3 = x - (-3)$.

$$
\begin{array}{r|rrrr}
-3 & 2 & 7 & 0 & -5 \\
   &   & -6 & -3 & 9 \\
\hline
   & 2 & 1 & -3 & \enspace 4
\end{array}
$$

Note: We must write a 0 for the missing term.

The quotient is $2x^2 + x - 3$. The remainder is 4. (Note that the degree of the quotient is 1 less than the degree of the dividend when the degree of the divisor is 1.)

We can now use synthetic division to find polynomial function values.

EXAMPLE 5 Given that $f(x) = 2x^5 - 3x^4 + x^3 - 2x^2 + x - 8$, find $f(10)$.

Solution By the remainder theorem, $f(10)$ is the remainder when $f(x)$ is divided by $x - 10$. We use synthetic division to find that remainder.

$$
\begin{array}{r|rrrrrr}
10 & 2 & -3 & 1 & -2 & 1 & -8 \\
   &   & 20 & 170 & 1710 & 17{,}080 & 170{,}810 \\
\hline
   & 2 & 17 & 171 & 1708 & 17{,}081 & \enspace 170{,}802
\end{array}
$$

Thus, $f(10) = 170{,}802$.

Compare the computations in Example 5 with those in a direct substitution:

$$
\begin{aligned}
f(10) &= 2(10)^5 - 3(10)^4 + (10)^3 - 2(10)^2 + 10 - 8 \\
      &= 170{,}802.
\end{aligned}
$$

The computations in synthetic division are less complicated than those involved in substituting. The easiest way to find $f(10)$ is by using a grapher, as shown below.

$$y = 2x^5 - 3x^4 + x^3 - 2x^2 + x - 8$$

```
Y1(10)
                      170802
```

EXAMPLE 6 Determine whether -4 is a zero of $f(x)$, where
$$f(x) = x^3 + 8x^2 + 8x - 32.$$

Solution We use synthetic division and the remainder theorem to find $f(-4)$.

$$
\begin{array}{r|rrrr}
-4 & 1 & 8 & 8 & -32 \\
   &   & -4 & -16 & 32 \\
\hline
   & 1 & 4 & -8 & \enspace 0
\end{array}
$$

Since $f(-4) = 0$, the number -4 is a zero of $f(x)$.

Finding Factors of Polynomials

We now consider a useful result that follows from the remainder theorem.

The Factor Theorem

For a polynomial $f(x)$, if $f(c) = 0$, then $x - c$ is a factor of $f(x)$.

PROOF (OPTIONAL).   If we divide $f(x)$ by $x - c$, we obtain a quotient and a remainder, related as follows:

$$f(x) = (x - c) \cdot Q(x) + f(c).$$

Then if $f(c) = 0$, we have

$$f(x) = (x - c) \cdot Q(x),$$

so $x - c$ is a factor of $f(x)$.

The factor theorem is very useful in factoring polynomials, and hence in solving polynomial equations.

EXAMPLE 7   Let $f(x) = x^3 + 2x^2 - 5x - 6$. Factor $f(x)$ and solve the equation $f(x) = 0$.

Solution   We look for linear factors of the form $x - c$. Let's try $x - 1$. (In the next section, we will learn a method for choosing the numbers to try for c.) We use synthetic division to determine whether $f(1) = 0$.

$$
\begin{array}{r|rrrr}
1 & 1 & 2 & -5 & -6 \\
  &   & 1 & 3 & -2 \\
\hline
  & 1 & 3 & -2 & -8
\end{array}
$$

Since $f(1) \neq 0$, we know that $x - 1$ is *not a factor* of $f(x)$. We try $x + 1$ or $x - (-1)$.

$$
\begin{array}{r|rrrr}
-1 & 1 & 2 & -5 & -6 \\
   &   & -1 & -1 & 6 \\
\hline
   & 1 & 1 & -6 & 0
\end{array}
$$

Since $f(-1) = 0$, we know that $x + 1$ *is one factor* and the quotient, $x^2 + x - 6$, is another. Thus,

$$f(x) = (x + 1)(x^2 + x - 6).$$

The trinomial is easily factored in this case, so we have

$$f(x) = (x + 1)(x + 3)(x - 2).$$

Our goal is to solve the equation $f(x) = 0$. To do so, we use the principle of zero products. The solutions are -1, -3, and 2. A check of these solutions on a grapher is shown at left.

$y = x^3 + 2x^2 - 5x - 6$

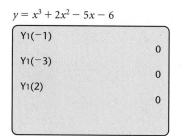

Exercise Set 3.2

1. Use substitution to determine whether 4, 5, and -2 are zeros of
$$f(x) = x^3 - 9x^2 + 14x + 24.$$

2. Use substitution to determine whether 2, 3, and -1 are zeros of
$$f(x) = 2x^3 - 3x^2 + x - 1.$$

3. For $f(x)$ in Exercise 1, use long division to determine which of the following are factors of $f(x)$.
   a) $x - 4$ **b)** $x - 5$ **c)** $x + 2$

4. For $f(x)$ in Exercise 2, use long division to determine which of the following are factors of $f(x)$.
   a) $x - 2$ **b)** $x - 3$ **c)** $x + 1$

In each of the following, a polynomial $P(x)$ and a divisor $d(x)$ are given. Use long division to find the quotient $Q(x)$ and the remainder $R(x)$ when $P(x)$ is divided by $d(x)$, and express $P(x)$ in the form $d(x) \cdot Q(x) + R(x)$.

5. $P(x) = x^3 - 8,$
   $d(x) = x + 2$

6. $P(x) = 2x^3 - 3x^2 + x - 1,$
   $d(x) = x - 3$

7. $P(x) = x^4 + 9x^2 + 20,$
   $d(x) = x^2 + 4$

8. $P(x) = x^4 + x^2 + 2,$
   $d(x) = x^2 + x + 1$

Use synthetic division to find the quotient and the remainder.

9. $(2x^4 + 7x^3 + x - 12) \div (x + 3)$

10. $(x^3 - 7x^2 + 13x + 3) \div (x - 2)$

11. $(x^3 - 2x^2 - 8) \div (x + 2)$

12. $(x^3 - 3x + 10) \div (x - 2)$

13. $(x^4 - 1) \div (x - 1)$

14. $(x^5 + 32) \div (x + 2)$

15. $(2x^4 + 3x^2 - 1) \div \left(x - \frac{1}{2}\right)$

16. $(3x^4 - 2x^2 + 2) \div \left(x - \frac{1}{4}\right)$

17. $(x^4 - y^4) \div (x - y)$

18. $(x^3 + 3ix^2 - 4ix - 2) \div (x + i)$

Use synthetic division to find the function values. Then check your work using a grapher.

19. $f(x) = x^3 - 6x^2 + 11x - 6$; find $f(1)$, $f(-2)$, and $f(3)$.

20. $f(x) = x^3 + 7x^2 - 12x - 3$; find $f(-3)$, $f(-2)$, and $f(1)$.

21. $f(x) = 2x^5 - 3x^4 + 2x^3 - x + 8$; find $f(20)$ and $f(-3)$.

22. $f(x) = x^5 - 10x^4 + 20x^3 - 5x - 100$; find $f(-10)$ and $f(5)$.

23. $f(x) = x^4 - 16$; find $f(2)$, $f(-2)$, $f(3)$, and $f(1 - \sqrt{2})$.

24. $f(x) = x^5 + 32$; find $f(2)$, $f(-2)$, $f(3)$, and $f(2 + 3i)$.

Using synthetic division, determine whether the numbers are zeros of the polynomials.

25. $-3, 2$; $f(x) = 3x^3 + 5x^2 - 6x + 18$

26. $-4, 2$; $f(x) = 3x^3 + 11x^2 - 2x + 8$

27. $-3, \frac{1}{2}$; $f(x) = x^3 - \frac{7}{2}x^2 + x - \frac{3}{2}$

28. $i, -i, -2$; $f(x) = x^3 + 2x^2 + x + 2$

Factor the polynomial $f(x)$. Then solve the equation $f(x) = 0$.

29. $f(x) = x^3 + 4x^2 + x - 6$ $(+)(1)$

30. $f(x) = x^3 + 5x^2 - 2x - 24$

31. $f(x) = x^3 - 6x^2 + 3x + 10$ $(+)(-1)$

32. $f(x) = x^3 + 2x^2 - 13x + 10$

33. $f(x) = x^3 - x^2 - 14x + 24$

34. $f(x) = x^3 - 3x^2 - 10x + 24$

35. $f(x) = x^4 - x^3 - 19x^2 + 49x - 30$

36. $f(x) = x^4 + 11x^3 + 41x^2 + 61x + 30$

Discussion and Writing

37. Is synthetic division always the fastest way to evaluate a polynomial function? What about using a grapher? Why or why not?

38. Can an nth-degree polynomial function have more than n zeros? Why or why not?

Skill Maintenance

Find exact solutions.

39. $2x - 7 = 5x + 8$

40. $2x^2 + 12 = 5x$

41. $7x^2 + 4x = 3$

42. The sum of the base and the height of a triangle is 30 in. Find the dimensions for which the area is a maximum.

Synthesis

In Exercises 43 and 44, a graph of a polynomial function is given. On the basis of the graph:

a) *Find as many factors as you can of the polynomial.*
b) *Construct a polynomial function with the zeros shown in the graph.*
c) *Can you find any other polynomial functions with the given zeros?*
d) *Can you find any other polynomial functions with the given zeros and the same graph?*

43. $(-3, 0)$ **44.** $(-3, 0)$

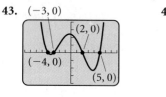

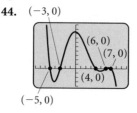

45. Find k such that $x + 2$ is a factor of $x^3 - kx^2 + 3x + 7k$.

46. For what values of k will the remainder be the same when $x^2 + kx + 4$ is divided by $x - 1$ or $x + 1$?

47. *Beam Deflection.*   A beam rests at two points A and B and has a concentrated load applied to its center. Let $y = $ the deflection, in feet, of the beam at a distance of x feet from A. Under certain conditions, this deflection is given by

$$y = \frac{1}{13}x^3 - \frac{1}{14}x.$$

Find the zeros of the polynomial in the interval $[0, 2]$.

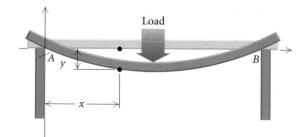

Solve.

48. $\dfrac{6x^2}{x^2 + 11} + \dfrac{60}{x^3 - 7x^2 + 11x - 77} = \dfrac{1}{x - 7}$

49. $\dfrac{2x^2}{x^2 - 1} + \dfrac{4}{x + 3} = \dfrac{32}{x^3 + 3x^2 - x - 3}$

50. Use a grapher to graph $f(x) = x^3 + kx^2 - 19x - 35$ for $k = 15, 19, 20,$ and 24. Compare the graphs. Do the zeros change when you change a coefficient?

51. Find a 15th-degree polynomial for which $x - 1$ is a factor. Answers may vary.

Use synthetic division to divide.

52. $(x^2 - 4x - 2) \div [x - (3 + 2i)]$

53. $(x^2 - 3x + 7) \div (x - i)$

Theorems about Zeros of Polynomial Functions

3.3

- *Factor polynomial functions and find the zeros and their multiplicities.*
- *Find a polynomial with specified zeros.*
- *For a polynomial function with integer coefficients, find the rational zeros and the other zeros, if possible.*

We will now allow the coefficients of a polynomial to be complex numbers. In certain cases, we will restrict the coefficients to be real numbers, rational numbers, or integers, as shown in the following examples.

POLYNOMIAL	TYPE OF COEFFICIENT
$5x^3 - 3x^2 + (2 + 4i)x + i$	Complex
$5x^3 - 3x^2 + \sqrt{2}x - \pi$	Real
$5x^3 - 3x^2 + \frac{2}{3}x - \frac{7}{4}$	Rational
$5x^3 - 3x^2 + 8x - 11$	Integer

The Fundamental Theorem of Algebra

A linear, or first-degree, polynomial function $f(x) = mx + b$ (where $m \neq 0$) has just one zero, $-b/m$. It can be shown that any quadratic polynomial function with complex numbers for coefficients has at least one, and at most two, complex zeros. The following theorem is a generalization. No proof is given.

The Fundamental Theorem of Algebra
Every polynomial function of degree n, with $n \geq 1$, has at least one zero in the system of complex numbers.

Note that although the fundamental theorem of algebra guarantees that a zero exists, it does not tell how to find it. Recall that the zeros of a polynomial function $f(x)$ are the solutions of the polynomial equation $f(x) = 0$. We now develop some concepts that can help in finding zeros. First, we consider one of the results of the fundamental theorem of algebra.

Every polynomial function f of degree n, with $n \geq 1$, can be factored into n linear factors (not necessarily unique); that is, $f(x) = a_n(x - c_1)(x - c_2) \cdots (x - c_n)$.

Finding Zeros of Factored Polynomial Functions

When a polynomial function is factored into a product of linear factors, it is easy to find the zeros by solving the equation $f(x) = 0$ using the principle of zero products.

EXAMPLE 1   Find the zeros of

$$f(x) = 5(x - 2)(x - 2)(x - 2)(x + 1).$$

Solution   To solve the equation $f(x) = 0$, we use the principle of zero products. The zeros of $f(x)$ are 2 and -1.   ▬

In Example 1, the factor $x - 2$ occurs three times. In a case like this, we sometimes say that the zero we obtain from this factor, 2, has a **multiplicity** of 3. If we multiply out the right side, we obtain

$$f(x) = 5x^4 - 25x^3 + 30x^2 + 20x - 40.$$

Had we started with this form of the function, we might have had trouble finding the zeros. Some polynomials, however, can be factored using techniques we already know, such as factoring by grouping.

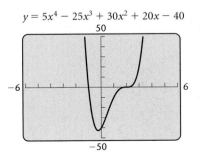
$y = 5x^4 - 25x^3 + 30x^2 + 20x - 40$

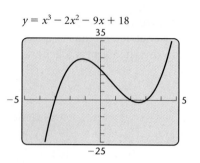

$y = x^3 - 2x^2 - 9x + 18$

EXAMPLE 2 Find the zeros of

$$f(x) = x^3 - 2x^2 - 9x + 18.$$

Solution We factor by grouping, as follows:

$$\begin{aligned} f(x) &= x^3 - 2x^2 - 9x + 18 \\ &= x^2(x - 2) - 9(x - 2) \\ &= (x^2 - 9)(x - 2) \\ &= (x + 3)(x - 3)(x - 2). \end{aligned}$$

Then, by the principle of zero products, the solutions of the equation $f(x) = 0$ are -3, 3, and 2. These are the zeros of $f(x)$.

Other factoring techniques can also be used.

EXAMPLE 3 Find the zeros of

$$f(x) = x^4 + 4x^2 - 45.$$

Solution We factor as follows:

$$\begin{aligned} f(x) &= x^4 + 4x^2 - 45 \\ &= (x^2 - 5)(x^2 + 9). \end{aligned}$$

We now solve the equation $f(x) = 0$ to determine the zeros. We use the principle of zero products:

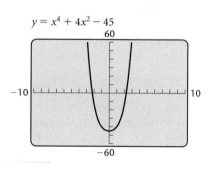

$y = x^4 + 4x^2 - 45$

$$(x^2 - 5)(x^2 + 9) = 0$$

$$x^2 - 5 = 0 \quad or \quad x^2 + 9 = 0$$

$$x^2 = 5 \quad or \quad x^2 = -9$$

$$x = \pm\sqrt{5} \quad or \quad x = \pm\sqrt{-9} = \pm 3i.$$

The solutions are $\pm\sqrt{5}$ and $\pm 3i$. These are the zeros of $f(x)$.

If we tried to use a grapher to find the zeros of f in Example 3, as shown at left, the nonreal solutions $3i$ and $-3i$ could not be seen.

Every polynomial function of degree n, with $n \geq 1$, has at least one zero and at most n zeros.

This is often stated as follows: "Every polynomial function of degree n, with $n \geq 1$, has *exactly* n zeros." This statement is not incompatible with the preceding statement, if one takes multiplicities into account.

Finding Polynomials with Given Zeros

Graph each of the following:

$$y_1 = (x + 2)(x - 1)\left(x - \tfrac{5}{2}\right),$$
$$y_2 = 2(x + 2)(x - 1)\left(x - \tfrac{5}{2}\right),$$
$$y_3 = -\tfrac{1}{2}(x + 2)(x - 1)\left(x - \tfrac{5}{2}\right).$$

How are the polynomials different? How do the graphs compare? How do the zeros compare?

Given several numbers, we can find a polynomial function with those numbers as its zeros.

EXAMPLE 4 Find a polynomial function of degree 3, having the zeros -2, 1, and $3i$.

Solution Such a polynomial has factors $x + 2$, $x - 1$, and $x - 3i$, so we have

$$f(x) = a_n(x + 2)(x - 1)(x - 3i).$$

The number a_n can be any nonzero number. The simplest polynomial will be obtained if we let it be 1. If we then multiply the factors, we obtain

$$f(x) = x^3 + (1 - 3i)x^2 + (-2 - 3i)x + 6i. \quad \blacksquare$$

$y = x^5 - x^4 - 9x^3 - 11x^2 - 4x$

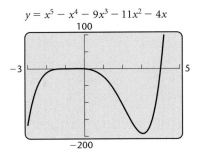

EXAMPLE 5 Find a polynomial function of degree 5 with -1 as a zero of multiplicity 3, 4 as a zero of multiplicity 1, and 0 as a zero of multiplicity 1.

Solution Proceeding as in Example 4, letting $a_n = 1$, we obtain

$$f(x) = (x + 1)^3(x - 4)(x - 0)$$
$$= x^5 - x^4 - 9x^3 - 11x^2 - 4x. \quad \blacksquare$$

Zeros of Polynomial Functions with Real Coefficients

Consider the quadratic equation $x^2 - 2x + 2 = 0$, with real coefficients. Its solutions are $1 + i$ and $1 - i$. Note that they are complex conjugates. This generalizes to any polynomial with real coefficients.

> If a complex number $a + bi$, $b \neq 0$, is a zero of a polynomial function $f(x)$ with *real* coefficients, then its conjugate, $a - bi$, is also a zero. (Nonreal zeros occur in conjugate pairs.)

For the preceding to be true, it is essential that the coefficients be real numbers. We see this in Example 4, where some of the coefficients of the polynomial are not real and the root $3i$ occurs, but its conjugate does not.

Rational Coefficients

When a polynomial has rational numbers for coefficients, certain irrational zeros also occur in pairs, as described in the following theorem.

> If $a + c\sqrt{b}$, a and c rational, b not a square, is a zero of a polynomial function $f(x)$ with *rational* coefficients, then $a - c\sqrt{b}$ is also a zero.

EXAMPLE 6 Suppose that a polynomial function of degree 6 with rational coefficients has $-2 + 5i$, $-2i$, and $1 - \sqrt{3}$ as three of its zeros. Find the other zeros.

Solution The other zeros are $-2 - 5i$, $2i$, and $1 + \sqrt{3}$. There are no other zeros because a polynomial of degree 6 can have at most 6 zeros. ▬

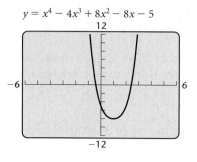

$y = x^4 - 4x^3 + 8x^2 - 8x - 5$

EXAMPLE 7 Find a polynomial function of lowest degree with rational coefficients that has $1 - \sqrt{2}$ and $1 + 2i$ as two of its zeros.

Solution The function must also have the zeros $1 + \sqrt{2}$ and $1 - 2i$. Because we want to find the polynomial function of lowest degree with the given zeros, we will not include additional zeros. That is, we will write a polynomial function of degree 4. Thus the polynomial function is

$$f(x) = [x - (1 - \sqrt{2})][x - (1 + \sqrt{2})][x - (1 + 2i)][x - (1 - 2i)]$$
$$= (x^2 - 2x - 1)(x^2 - 2x + 5)$$
$$= x^4 - 4x^3 + 8x^2 - 8x - 5.$$

▬

Integer Coefficients and the Rational Zeros Theorem

It is not always easy to find the zeros of a polynomial function. However, if a polynomial function has integer coefficients, there is a procedure that will yield all the rational zeros.

> ### The Rational Zeros Theorem
> Let
> $$P(x) = a_n x^n + a_{n-1} x^{n-1} + \cdots + a_1 x + a_0,$$
> where all the coefficients are integers. Consider a rational number denoted by p/q, where p and q are relatively prime (having no common factor besides -1 and 1). If p/q is a zero of $P(x)$, then p is a factor of a_0 and q is a factor of a_n.

EXAMPLE 8   Given $f(x) = 2x^5 - x^4 - 4x^3 + 2x^2 - 30x + 15$:

a) Find the rational zeros, and then the other zeros; that is, solve $f(x) = 0$.

b) Factor $f(x)$ into linear factors.

Solution

a) Because the degree of $f(x)$ is 5, there are at most 5 distinct zeros. According to the rational zeros theorem, any rational zero of f must be of the form p/q, where p is a factor of 15 and q is a factor of 2. The possibilities are

$$\frac{\textit{Possibilities for } p}{\textit{Possibilities for } q} : \frac{\pm 1, \pm 3, \pm 5, \pm 15}{\pm 1, \pm 2};$$

$$\textit{Possibilities for } p/q: \quad 1, -1, 3, -3, 5, -5, 15, -15, \tfrac{1}{2}, -\tfrac{1}{2}, \tfrac{3}{2},$$
$$-\tfrac{3}{2}, \tfrac{5}{2}, -\tfrac{5}{2}, \tfrac{15}{2}, -\tfrac{15}{2}.$$

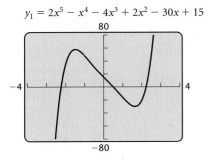

$$y_1 = 2x^5 - x^4 - 4x^3 + 2x^2 - 30x + 15$$

Rather than use synthetic division to check each of these possibilities, we graph $y = 2x^5 - x^4 - 4x^3 + 2x^2 - 30x + 15$ (see the graph at left). We can then inspect the graph for zeros that appear to be near any of the possible rational zeros.

From the graph, we see that of the possibilities in the list, only the numbers $-\tfrac{5}{2}, \tfrac{1}{2}$, and $\tfrac{5}{2}$ might be rational zeros. By synthetic division, we see that only $\tfrac{1}{2}$ is actually a rational zero.

$$\begin{array}{r|rrrrrr} \tfrac{1}{2} & 2 & -1 & -4 & 2 & -30 & 15 \\ & & 1 & 0 & -2 & 0 & -15 \\ \hline & 2 & 0 & -4 & 0 & -30 & \big|\ \ 0 \end{array}$$

This means that $x - \tfrac{1}{2}$ is a factor of $f(x)$. We write the factorization and try to factor further:

$$f(x) = \left(x - \tfrac{1}{2}\right)(2x^4 - 4x^2 - 30)$$
$$= \left(x - \tfrac{1}{2}\right) \cdot 2 \cdot (x^4 - 2x^2 - 15) \qquad \text{Factoring out the 2}$$
$$= \left(x - \tfrac{1}{2}\right) \cdot 2 \cdot (x^2 - 5)(x^2 + 3). \qquad \text{Factoring the trinomial}$$

We now solve the equation $f(x) = 0$ to determine the zeros. We use the principle of zero products:

$$\left(x - \tfrac{1}{2}\right) \cdot 2 \cdot (x^2 - 5)(x^2 + 3) = 0$$
$$x - \tfrac{1}{2} = 0 \quad \textit{or} \quad x^2 - 5 = 0 \quad \textit{or} \quad x^2 + 3 = 0$$
$$x = \tfrac{1}{2} \quad \textit{or} \quad x^2 = 5 \quad \textit{or} \quad x^2 = -3$$
$$x = \tfrac{1}{2} \quad \textit{or} \quad x = \pm\sqrt{5} \quad \textit{or} \quad x = \pm\sqrt{3}\,i.$$

There is only one rational zero, $\tfrac{1}{2}$. The other zeros are $\pm\sqrt{5}$ and $\pm\sqrt{3}i$.

b) The factorization into linear factors is

$$f(x) = 2\left(x - \tfrac{1}{2}\right)(x + \sqrt{5})(x - \sqrt{5})(x + \sqrt{3}\,i)(x - \sqrt{3}\,i). \qquad \blacksquare$$

EXAMPLE 9   Given $f(x) = 3x^4 - 11x^3 + 10x - 4$:

a) Find the rational zeros, and then the other zeros; that is, solve $f(x) = 0$.

b) Factor $f(x)$ into linear factors.

Solution

a) Because the degree of $f(x)$ is 4, there are at most 4 distinct zeros. The rational zeros theorem says that if p/q is a zero of $f(x)$, then p must be a factor of -4 and q must be a factor of 3. Thus the possibilities for p/q are

$$\frac{\text{Possibilities for } p}{\text{Possibilities for } q} : \frac{\pm 1, \pm 2, \pm 4}{\pm 1, \pm 3}.$$

Possibilities for p/q: $1, -1, 2, -2, 4, -4, \frac{1}{3}, -\frac{1}{3}, \frac{2}{3}, -\frac{2}{3}, \frac{4}{3}, -\frac{4}{3}$.

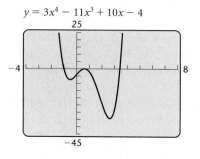

$y = 3x^4 - 11x^3 + 10x - 4$

We could use the TABLE feature or some other method to find function values. However, if we use synthetic division, the quotient polynomial becomes a beneficial by-product if a zero is found. Rather than use synthetic division to check *each* of these possibilities, we graph the function and inspect the graph for zeros that appear to be near any of the possible rational zeros (see the graph at left).

From the graph, we see that of the possibilities in the list, only the numbers -1, $\frac{1}{3}$, and $\frac{2}{3}$ might be rational zeros.

We try -1.

$$
\begin{array}{r|rrrrr}
-1 & 3 & -11 & 0 & 10 & -4 \\
   &   & -3 & 14 & -14 & 4 \\
\hline
   & 3 & -14 & 14 & -4 & 0
\end{array}
$$

We see that -1 is a zero. Thus, $x + 1$ is a factor. Using the results of the synthetic division, we can express $f(x)$ as follows:

$$f(x) = (x + 1)(3x^3 - 14x^2 + 14x - 4).$$

We now use $3x^3 - 14x^2 + 14x - 4$ and check the other possible zeros. We try $\frac{1}{3}$.

$$
\begin{array}{r|rrrr}
1/3 & 3 & -14 & 14 & -4 \\
    &   & 1 & -\frac{13}{3} & \frac{29}{9} \\
\hline
    & 3 & -13 & \frac{29}{3} & -\frac{7}{9}
\end{array}
$$

Since $f\left(\frac{1}{3}\right) \neq 0$, we know that $\frac{1}{3}$ is not a zero.

Let's now try $\frac{2}{3}$.

$$
\begin{array}{r|rrrr}
2/3 & 3 & -14 & 14 & -4 \\
    &   & 2 & -8 & 4 \\
\hline
    & 3 & -12 & 6 & 0
\end{array}
$$

Since $f\left(\frac{2}{3}\right) = 0$, we know that $\frac{2}{3}$ is also a zero.

Using the results of synthetic division, we can factor further:

$$f(x) = (x + 1)\left(x - \tfrac{2}{3}\right)(3x^2 - 12x + 6)$$ **Using the results of the last synthetic division**

$$= (x + 1)\left(x - \tfrac{2}{3}\right) \cdot 3 \cdot (x^2 - 4x + 2).$$ **Removing a factor of 3**

The quadratic formula can be used to find the zeros of $x^2 - 4x + 2$:

$$x = \frac{-b \pm \sqrt{b^2 - 4ac}}{2a}$$

$$= \frac{-(-4) \pm \sqrt{(-4)^2 - 4 \cdot 1 \cdot 2}}{2 \cdot 1}$$

$$= \frac{4 \pm \sqrt{8}}{2} = \frac{4 \pm 2\sqrt{2}}{2} = \frac{2(2 \pm \sqrt{2})}{2}$$

$$= 2 \pm \sqrt{2}.$$

The rational zeros are -1 and $\tfrac{2}{3}$. The other zeros are $2 \pm \sqrt{2}$.

b) The complete factorization of $f(x)$ is

$$f(x) = (x + 1)\left(x - \tfrac{2}{3}\right) \cdot 3 \cdot [x - (2 - \sqrt{2})][x - (2 + \sqrt{2})]$$

$$= (x + 1)(3x - 2)(x - 2 + \sqrt{2})(x - 2 - \sqrt{2}).$$ ▬

Exercise Set 3.3

Find the zeros of the polynomial function and state the multiplicity of each.

1. $f(x) = (x + 3)^2(x - 1)$

2. $f(x) = -8(x - 3)^2(x + 4)^3 x^4$

3. $f(x) = x^3(x - 1)^2(x + 4)$

4. $f(x) = (x^2 - 5x + 6)^2$

5. $f(x) = x^4 - 4x^2 + 3$

6. $f(x) = x^4 - 10x^2 + 9$

7. $f(x) = x^3 + 3x^2 - x - 3$

8. $f(x) = x^3 - x^2 - 2x + 2$

Find a polynomial function of degree 3 with the given numbers as zeros.

9. $-2, 3, 5$

10. $2, i, -i$

11. $-3, 2i, -2i$

12. $1 + 4i, 1 - 4i, -1$

13. $\sqrt{2}, -\sqrt{2}, \sqrt{3}$

14. $-3, 0, \tfrac{1}{2}$

15. Find a polynomial function of degree 5 with -1 as a zero of multiplicity 3, 0 as a zero of multiplicity 1, and 1 as a zero of multiplicity 1.

16. Find a polynomial function of degree 4 with -2 as a zero of multiplicity 1, 3 as a zero of multiplicity 2, and -1 as a zero of multiplicity 1.

Suppose that a polynomial function of degree 5 with rational coefficients has the given numbers as zeros. Find the other zeros.

17. $6, -3 + 4i, 4 - \sqrt{5}$

18. $-2, 3, 4, 1 - i$

Find a polynomial function of lowest degree with rational coefficients that has the given numbers as some of its zeros.

19. $1 + i, 2$ *also have to have conjugate pairs....*

20. $2 - i, -1$

21. $-4i, 5$ *pairs....*

22. $2 - \sqrt{3}, 1 + i$

23. $\sqrt{5}, -3i$

24. $-\sqrt{2}, 4i$

Given that the polynomial function has the given zero, find the other zeros.

25. $f(x) = x^4 - 5x^3 + 7x^2 - 5x + 6;$ $-i$

26. $f(x) = x^4 - 16;$ $2i$

27. $f(x) = x^3 - 6x^2 + 13x - 20;$ 4

28. $f(x) = x^3 - 8;$ 2

List all possible rational zeros.

29. $f(x) = x^5 - 3x^2 + 1$

30. $f(x) = x^7 + 37x^5 - 6x^2 + 12$

31. $f(x) = 15x^6 + 47x^2 + 2$

32. $f(x) = 10x^{25} + 3x^{17} - 35x + 6$

For each polynomial function:

a) *Find the rational zeros, and then the other zeros; that is, solve $f(x) = 0$.*

b) *Factor $f(x)$ into linear factors.*

33. $f(x) = x^3 + 3x^2 - 2x - 6$

34. $f(x) = x^3 - x^2 - 3x + 3$

35. $f(x) = x^3 - 3x + 2$

36. $f(x) = x^3 - 2x + 4$

37. $f(x) = x^3 - 5x^2 + 11x + 17$

38. $f(x) = 2x^3 + 7x^2 + 2x - 8$

39. $f(x) = 5x^4 - 4x^3 + 19x^2 - 16x - 4$

40. $f(x) = 3x^4 - 4x^3 + x^2 + 6x - 2$

41. $f(x) = x^4 - 3x^3 - 20x^2 - 24x - 8$

42. $f(x) = x^4 + 5x^3 - 27x^2 + 31x - 10$

43. $f(x) = x^3 - 4x^2 + 2x + 4$

44. $f(x) = x^3 - 8x^2 + 17x - 4$

45. $f(x) = x^3 + 8$

46. $f(x) = x^3 - 8$

47. $f(x) = \frac{1}{3}x^3 - \frac{1}{2}x^2 - \frac{1}{6}x + \frac{1}{6}$

48. $f(x) = \frac{2}{3}x^3 - \frac{1}{2}x^2 + \frac{2}{3}x - \frac{1}{2}$

Find only the rational zeros.

49. $f(x) = x^4 + 32$

50. $f(x) = x^6 + 8$

51. $f(x) = x^4 + 5x^3 - 13x^2 - 35x + 42$

52. $f(x) = 2x^3 + 3x^2 + 2x + 3$

53. $f(x) = x^4 + 2x^3 + 2x^2 - 4x - 8$

54. $f(x) = x^4 + 6x^3 + 17x^2 + 36x + 66$

55. $f(x) = x^5 - 5x^4 + 5x^3 + 15x^2 - 36x + 20$

56. $f(x) = x^5 - 3x^4 - 3x^3 + 9x^2 - 4x + 12$

Discussion and Writing

57. Is it possible for a third-degree polynomial with rational coefficients to have no real zeros? Why or why not?

58. If $Q(x) = -P(x)$, do $P(x)$ and $Q(x)$ have the same zeros? Why or why not?

Skill Maintenance

For Exercises 59 and 60, complete the square to:

a) *find the vertex;*

b) *find the line of symmetry; and*

c) *determine whether there is a maximum or minimum value and find that value.*

59. $f(x) = x^2 - 8x + 10$

60. $f(x) = 3x^2 - 6x - 1$

Find the zeros of the function.

61. $f(x) = -\frac{4}{5}x + 8$

62. $g(x) = x^2 - 8x - 33$

Synthesis

63. Consider $f(x) = 2x^3 - 5x^2 - 4x + 3$. Find the solutions of each equation.

a) $f(x) = 0$ **b)** $f(x - 1) = 0$

c) $f(x + 2) = 0$ **d)** $f(2x) = 0$

64. Use the rational zeros theorem and the equation $x^4 - 12 = 0$ to show that $\sqrt[4]{12}$ is irrational.

65. Use the rational zeros theorem and the equation $x^2 - 5 = 0$ to show that $\sqrt{5}$ is not a rational number.

Find the rational zeros.

66. $P(x) = x^6 - 6x^5 - 72x^4 - 81x^2 + 486x + 5832$

67. $P(x) = 2x^5 - 33x^4 - 84x^3 + 2203x^2 - 3348x - 10,080$

Rational Functions

3.4

• *Graph a rational function, identifying all asymptotes.*
• *Solve applied problems involving rational functions.*

The sum, difference, or product of two polynomials is a polynomial. Now we turn our attention to functions that represent the quotient of two polynomials. In general, the quotient of two polynomials is *not* itself a polynomial.

A *rational number* can be expressed as the quotient of two integers, p/q, where $q \neq 0$. A *rational function* is formed by the quotient of two polynomials, $p(x)/q(x)$, where $q(x) \neq 0$. Here are some examples of rational functions and their graphs.

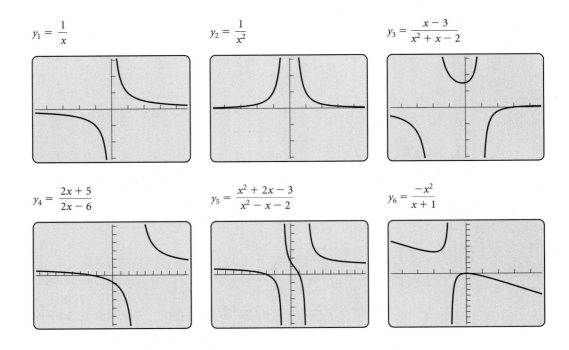

$$y_1 = \frac{1}{x} \qquad y_2 = \frac{1}{x^2} \qquad y_3 = \frac{x - 3}{x^2 + x - 2}$$

$$y_4 = \frac{2x + 5}{2x - 6} \qquad y_5 = \frac{x^2 + 2x - 3}{x^2 - x - 2} \qquad y_6 = \frac{-x^2}{x + 1}$$

Rational Function

A **rational function** is a function f that is a quotient of two polynomials, that is,

$$f(x) = \frac{p(x)}{q(x)},$$

where $p(x)$ and $q(x)$ are polynomials and where $q(x)$ is not the zero polynomial. The domain of f consists of all inputs x for which $q(x) \neq 0$.

The Domain of a Rational Function

EXAMPLE 1   Consider

$$f(x) = \frac{1}{x - 3}.$$

Find the domain and graph f.

Solution   The only input that results in a denominator of 0 is 3. Thus the domain is

$$\{x \mid x \neq 3\}, \text{ or } (-\infty, 3) \cup (3, \infty).$$

The graph of this function is the graph of $y = 1/x$ translated right 3 units. Two versions of the graph on a grapher are shown below.

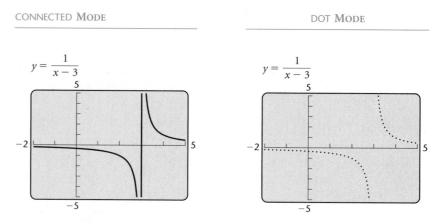

CONNECTED MODE   DOT MODE

Using CONNECTED mode can lead to an incorrect graph. In CONNECTED mode, a grapher connects plotted points with line segments. In DOT mode, it simply plots unconnected points. In the graph on the left above, the grapher has connected the points plotted on either side of the x-value 3 with a line that appears to be the vertical line $x = 3$. (It is not actually vertical since it connects the last point to the left of $x = 3$ with the first point to the right of $x = 3$.) Since 3 is not in the domain of the function, the vertical line, $x = 3$, cannot be part of the graph. We will see later in this section that vertical lines like $x = 3$, although not part of the graph, are important in the construction of graphs. If you have a choice when graphing rational functions, use DOT mode. With some graphers set in CONNECTED mode, you can choose window dimensions for which these extra lines do not appear. See the window below for an example.

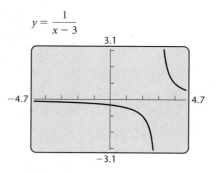

DOMAINS OF FUNCTIONS
REVIEW SECTION 1.1.

EXAMPLE 2   Determine the domain of each of the functions y_1 through y_6 illustrated at the beginning of this section.

Solution   The domain of each rational function will be the set of all real numbers except those values that make the denominator 0. To determine those exceptions, we set the denominator equal to 0 and solve for *x*.

FUNCTION	DOMAIN
$y_1 = \dfrac{1}{x}$	$\{x \mid x \neq 0\}$, or $(-\infty, 0) \cup (0, \infty)$
$y_2 = \dfrac{1}{x^2}$	$\{x \mid x \neq 0\}$, or $(-\infty, 0) \cup (0, \infty)$
$y_3 = \dfrac{x - 3}{x^2 + x - 2} = \dfrac{x - 3}{(x + 2)(x - 1)}$	$\{x \mid x \neq -2 \text{ and } x \neq 1\}$, or $(-\infty, -2) \cup (-2, 1) \cup (1, \infty)$
$y_4 = \dfrac{2x + 5}{2x - 6}$	$\{x \mid x \neq 3\}$, or $(-\infty, 3) \cup (3, \infty)$
$y_5 = \dfrac{x^2 + 2x - 3}{x^2 - x - 2} = \dfrac{x^2 + 2x - 3}{(x + 1)(x - 2)}$	$\{x \mid x \neq -1 \text{ and } x \neq 2\}$, or $(-\infty, -1) \cup (-1, 2) \cup (2, \infty)$
$y_6 = \dfrac{-x^2}{x + 1}$	$\{x \mid x \neq -1\}$, or $(-\infty, -1) \cup (-1, \infty)$

As a partial check of the domains, we can observe the discontinuities (breaks) in the graphs of these functions (see p. 255).

Asymptotes

Look at the graph of $f(x) = 1/(x - 3)$ in Example 1. Let's explore what happens as *x*-values get closer and closer to 3 from the left. We then explore what happens as *x*-values get closer and closer to 3 from the right.

Interactive Discovery

Consider $y = 1/(x - 3)$. Complete the following input–output tables and look for a pattern. What happens to the *y*-values when the *x*-values approach 3 from the left? What happens to the *y*-values when the *x*-values approach 3 from the right?

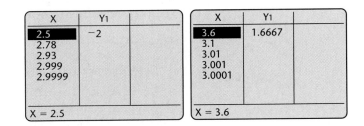

Complete:

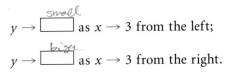

$y \rightarrow \boxed{}$ as $x \rightarrow 3$ from the left;

$y \rightarrow \boxed{}$ as $x \rightarrow 3$ from the right.

We see that as x-values get closer and closer to 3 from the left, the function values (y-values) decrease without bound. Similarly, as the x-values approach 3 from the right, the function values increase without bound. We write this as

$$f(x) \rightarrow -\infty \text{ as } x \rightarrow 3^- \quad \text{and} \quad f(x) \rightarrow \infty \text{ as } x \rightarrow 3^+.$$

We read "$f(x) \rightarrow -\infty$ as $x \rightarrow 3^-$" as "$f(x)$ decreases without bound as x approaches 3 from the left." We read "$f(x) \rightarrow \infty$ as $x \rightarrow 3^+$" as "$f(x)$ increases without bound as x approaches 3 from the right." The vertical line $x = 3$ is said to be a **vertical asymptote** for this curve.

Vertical Asymptote

The line $x = a$ is a **vertical asymptote** for the graph of f if any of the following is true:

$$f(x) \rightarrow \infty \text{ as } x \rightarrow a^- \quad \text{or} \quad f(x) \rightarrow -\infty \text{ as } x \rightarrow a^-, \quad \text{or}$$
$$f(x) \rightarrow \infty \text{ as } x \rightarrow a^+ \quad \text{or} \quad f(x) \rightarrow -\infty \text{ as } x \rightarrow a^+.$$

The following figures show the four ways in which a vertical asymptote can occur.

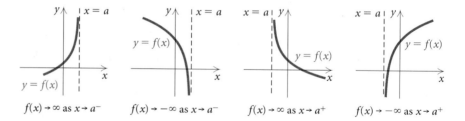

The vertical asymptotes of a rational function $f(x) = p(x)/q(x)$ are found by determining the zeros of $q(x)$ that are not also zeros of $p(x)$. If $p(x)$ and $q(x)$ are polynomials with no common factors other than constants, we need determine only the zeros of the denominator $q(x)$.

Determining Vertical Asymptotes

For a rational function $f(x) = p(x)/q(x)$, where $p(x)$ and $q(x)$ are polynomials with no common factors other than constants, if a is a zero of the denominator, then the line $x = a$ is a vertical asymptote for the graph of the function.

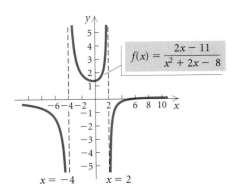

FIGURE 1

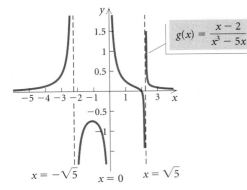

FIGURE 2

EXAMPLE 3 Determine the vertical asymptotes of each of the following functions.

a) $f(x) = \dfrac{2x - 11}{x^2 + 2x - 8}$

b) $g(x) = \dfrac{x - 2}{x^3 - 5x}$

Solution

a) We factor to find the zeros of the denominator:

$$x^2 + 2x - 8 = (x + 4)(x - 2).$$

The zeros of the denominator are -4 and 2. Thus the vertical asymptotes are the lines $x = -4$ and $x = 2$ (see Fig. 1).

b) We factor to find the zeros of the denominator:

$$x^3 - 5x = x(x^2 - 5).$$

Setting $x(x^2 - 5) = 0$ and solving for x, we get

$$
\begin{aligned}
x = 0 \quad &or \quad x^2 - 5 = 0 \\
x = 0 \quad &or \quad x^2 = 5 \\
x = 0 \quad &or \quad x = \pm\sqrt{5}.
\end{aligned}
$$

The zeros of the denominator are 0, $\sqrt{5}$, and $-\sqrt{5}$. Thus the vertical asymptotes are the lines $x = 0$, $x = \sqrt{5}$, and $x = -\sqrt{5}$ (see Fig. 2).

Referring again to Example 1, let's explore what happens to $f(x) = 1/(x - 3)$ as x increases without bound (for example, as $x = 100{,}000$, $750{,}000$, $1{,}000{,}000$, and beyond) and as x decreases without bound (for example, as $x = -200{,}000$, $-695{,}000$, $-1{,}200{,}000$, and beyond).

Interactive Discovery

Consider $y = 1/(x - 3)$. Complete the following tables and look for a pattern. What happens to the y-values as the x-values become large without bound? become small without bound?

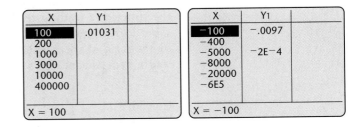

We see from the table on the left that "as x increases without bound, y goes to 0." We write this as $y \to 0$ as $x \to \infty$. Using the table on the right, complete:

$$y \to \boxed{} \text{ as } x \to -\infty.$$

We see that

$$\frac{1}{x-3} \to 0 \text{ as } x \to \infty \quad \text{and} \quad \frac{1}{x-3} \to 0 \text{ as } x \to -\infty.$$

Since $y = 0$ is the equation of the x-axis, we say that the curve approaches the x-axis asymptotically and that the x-axis is a horizontal asymptote for the curve.

Horizontal Asymptote

The line $y = b$ is a **horizontal asymptote** for the graph of f if either or both of the following are true:

$$f(x) \to b \text{ as } x \to \infty \quad \text{or} \quad f(x) \to b \text{ as } x \to -\infty.$$

The following figures illustrate four ways in which horizontal asymptotes can occur. In each case, the curve gets close to the line $y = b$ either as $x \to \infty$ or as $x \to -\infty$. Keep in mind that the symbols ∞ and $-\infty$ convey the idea of increasing without bound and decreasing without bound, respectively.

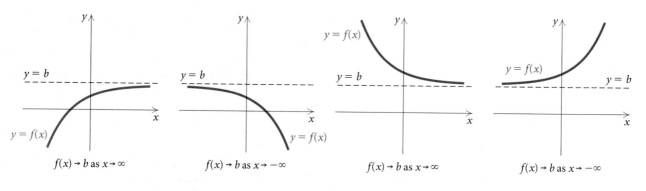

How can we determine a horizontal asymptote? As x gets very large or very small, the value of the polynomial function $p(x)$ is dominated by the function's leading term. Because of this, if $p(x)$ and $q(x)$ have the *same* degree, the value of $p(x)/q(x)$ as $x \to \infty$ or as $x \to -\infty$ is dominated by the ratio of the numerator's leading coefficient to the denominator's leading coefficient. Let's explore this further.

Interactive Discovery

Consider

$$f(x) = \frac{3x^2 + 2x - 4}{2x^2 - x + 1}.$$

Let $y = (3x^2 + 2x - 4)/(2x^2 - x + 1)$. Complete the following tables and look for a pattern. What happens to the y-values as the x-values become large without bound? become small without bound?

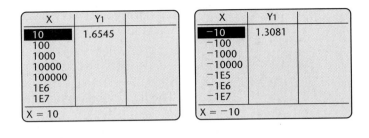

Complete:

$$y \to \boxed{} \text{ as } x \to \infty;$$

$$y \to \boxed{} \text{ as } x \to -\infty.$$

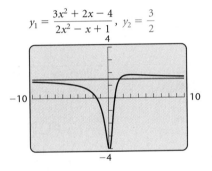

$$y_1 = \frac{3x^2 + 2x - 4}{2x^2 - x + 1}, \ y_2 = \frac{3}{2}$$

For $f(x) = (3x^2 + 2x - 4)/(2x^2 - x + 1)$, we see that

$$\frac{3x^2 + 2x - 4}{2x^2 - x + 1} \to 1.5, \text{ or } \frac{3}{2}, \text{ as } x \to \infty, \text{ and}$$

$$\frac{3x^2 + 2x - 4}{2x^2 - x + 1} \to 1.5, \text{ or } \frac{3}{2}, \text{ as } x \to -\infty.$$

We say that the curve approaches the horizontal line $y = \frac{3}{2}$ asymptotically and that $y = \frac{3}{2}$ is a horizontal asymptote for the curve. As a partial check, we can graph both the function and its horizontal asymptote in the same window.

It follows that when the numerator and the denominator of a rational function have the same degree, the line $y = a/b$ is the horizontal asymptote, where a and b are the leading coefficients of the numerator and the denominator, respectively.

EXAMPLE 4 Find the horizontal asymptote: $f(x) = \dfrac{-7x^4 - 10x^2 + 1}{11x^4 + x - 2}.$

Solution The numerator and the denominator have the same degree, so the line $y = -\frac{7}{11}$, or $0.\overline{63}$, is a horizontal asymptote. ▬

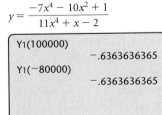

$$y = \frac{-7x^4 - 10x^2 + 1}{11x^4 + x - 2}$$

To check Example 4, we could examine a table of values or evaluate the function for a very large and a very small value of x. (See the window at left.) A third check, one that is useful in calculus, is to multiply by 1, using $(1/x^4)/(1/x^4)$:

$$f(x) = \frac{-7x^4 - 10x^2 + 1}{11x^4 + x - 2} \cdot \frac{\dfrac{1}{x^4}}{\dfrac{1}{x^4}}$$

$$= \frac{-7 - \dfrac{10}{x^2} + \dfrac{1}{x^4}}{11 + \dfrac{1}{x^3} - \dfrac{2}{x^4}}.$$

As $|x|$ becomes very large, each expression with x in the denominator tends toward 0. Specifically, as $x \to \infty$ or as $x \to -\infty$, we have

$$f(x) \to \frac{-7 - 0 + 0}{11 + 0 - 0}, \quad \text{or} \quad f(x) \to -\frac{7}{11}.$$

The horizontal asymptote is $y = -\frac{7}{11}$.

We now investigate the occurrence of a horizontal asymptote when the degree of the numerator is less than the degree of the denominator. We saw in Example 1 that $y = 0$, the x-axis, is the horizontal asymptote of $f(x) = 1/(x - 3)$.

EXAMPLE 5 Find the horizontal asymptote: $f(x) = \dfrac{2x + 3}{x^3 - 2x^2 + 4}$.

Solution We let $y_1 = 2x + 3$, $y_2 = x^3 - 2x^2 + 4$, and $y_3 = y_1/y_2$, and form a table of values, as follows.

X	Y₁	Y₂	Y₃
	23	804	.02861
100	203	980004	2.1E−4
1000	2003	9.98E8	2E−6
10000	20003	1E12	2E−8

X = 10

X	Y₁	Y₂	Y₃
	−17	−1196	.01421
−100	−197	−1E6	1.9E−4
−1000	−1997	−1E9	2E−6
−10000	−19997	−1E12	2E−8

X = −10

Note that as x grows large, the value of y_2 grows much faster than the value of y_1. Because of this, the ratio of y_1/y_2 shrinks toward 0. As $x \to -\infty$, the ratio y_1/y_2 behaves in a similar manner. The horizontal asymptote is $y = 0$, the x-axis.

The following statements describe the two ways in which a horizontal asymptote occurs.

Determining a Horizontal Asymptote

- When the numerator and the denominator of a rational function have the same degree, the line $y = a/b$ is the horizontal asymptote, where a and b are the leading coefficients of the numerator and the denominator, respectively.
- When the degree of the numerator of a rational function is less than the degree of the denominator, the x-axis, or $y = 0$, is the horizontal asymptote.
- When the degree of the numerator of a rational function is greater than the degree of the denominator, there is no horizontal asymptote.

The following statements are also true.

> The graph of a rational function never crosses a vertical asymptote.

> The graph of a rational function may or may not cross a horizontal asymptote.

EXAMPLE 6 Make a hand-drawn graph of

$$g(x) = \frac{2x^2 + 1}{x^2}.$$

Include and label all asymptotes.

Solution Since 0 is the zero of the denominator, the y-axis, $x = 0$, is the vertical asymptote. Note also that the degree of the numerator is the same as the degree of the denominator. Thus, $y = 2/1$, or 2, is the horizontal asymptote.

To draw the graph, we first draw the asymptotes with dashed lines. Then we compute some ordered pairs and draw the two branches of the curve.

HAND-DRAWN GRAPH   CHECK ON GRAPHER

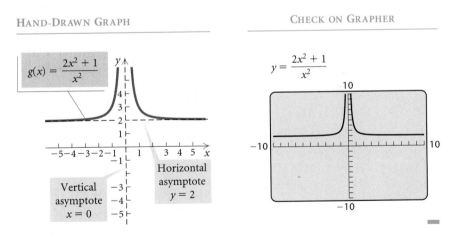

Sometimes a line that is neither horizontal nor vertical is an asymptote. Such a line is called an **oblique asymptote,** or a **slant asymptote.**

EXAMPLE 7 Find all the asymptotes of

$$f(x) = \frac{2x^2 - 3x - 1}{x - 2}.$$

Solution The line $x = 2$ is the vertical asymptote because 2 is the zero of the denominator. There is no horizontal asymptote because the degree of the numerator is greater than the degree of the denominator. When the

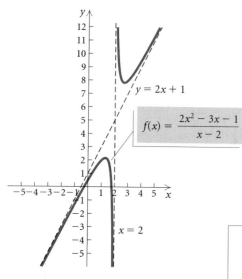

$$f(x) = \frac{2x^2 - 3x - 1}{x - 2}$$

degree of the numerator is 1 greater than the degree of the denominator, we divide to find an equivalent expression:

$$\frac{2x^2 - 3x - 1}{x - 2} = (2x + 1) + \frac{1}{x - 2}.$$

$$\begin{array}{r} 2x + 1 \\ x - 2 \overline{\smash{)}2x^2 - 3x - 1} \\ \underline{2x^2 - 4x} \\ x - 1 \\ \underline{x - 2} \\ 1 \end{array}$$

Now we see that when $x \to \infty$ or $x \to -\infty$, $1/(x - 2) \to 0$ and the value of $f(x) \to 2x + 1$. This means that as $|x|$ becomes very large, the graph of $f(x)$ gets very close to the graph of $y = 2x + 1$. Thus the line $y = 2x + 1$ is the oblique asymptote.

Occurrence of Lines as Asymptotes

Vertical asymptotes of a rational function $p(x)/q(x)$, where $p(x)$ and $q(x)$ have no common factors other than constants, occur at any x-values that make the denominator 0.

The x-axis is the horizontal asymptote when the degree of the numerator is less than the degree of the denominator.

A horizontal asymptote other than the x-axis occurs when the numerator and the denominator have the same degree.

An oblique asymptote occurs when the degree of the numerator is 1 greater than the degree of the denominator.

There can be only one horizontal asymptote or one oblique asymptote and never both.

An asymptote is *not* part of the graph of the function.

The following is an outline of a procedure that we can follow to create accurate hand-drawn graphs of rational functions.

To graph a rational function $f(x) = p(x)/q(x)$, where $p(x)$ and $q(x)$ have no common factor:

1. Find the real zeros of the denominator. Determine the domain of the function and sketch the vertical asymptotes.
2. Find the horizontal or the oblique asymptote, if there is one, and sketch it.
3. Find the zeros of the function. These are the first coordinates of the x-intercepts of the function. The zeros are found by determining the zeros of the numerator.
4. Find $f(0)$. This gives the y-intercept of the function.
5. Find other function values to determine the general shape. Then draw the graph.

EXAMPLE 8   Graph: $f(x) = \dfrac{2x + 3}{3x^2 + 7x - 6}$.

Solution

1. We find the zeros of the denominator by solving $3x^2 + 7x - 6 = 0$. Since

$$3x^2 + 7x - 6 = (3x - 2)(x + 3),$$

the zeros are $\frac{2}{3}$ and -3. Thus the domain excludes $\frac{2}{3}$ and -3 and is

$$(-\infty, -3) \cup \left(-3, \tfrac{2}{3}\right) \cup \left(\tfrac{2}{3}, \infty\right).$$

The graph has vertical asymptotes $x = -3$ and $x = \frac{2}{3}$. We sketch these as dashed lines.

2. Because the degree of the numerator is less than the degree of the denominator, the x-axis is the horizontal asymptote.

3. To find the zeros of the numerator, we solve $2x + 3 = 0$ and get $x = -\frac{3}{2}$. Thus, $-\frac{3}{2}$ is the zero of the function, and the pair $\left(-\frac{3}{2}, 0\right)$ is the x-intercept.

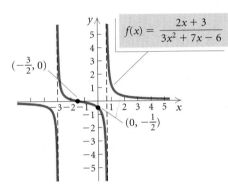

4. We find $f(0)$:

$$f(0) = \frac{2 \cdot 0 + 3}{3 \cdot 0^2 + 7 \cdot 0 - 6}$$

$$= \frac{3}{-6} = -\frac{1}{2}.$$

Thus, $\left(0, -\frac{1}{2}\right)$ is the y-intercept.

5. We find other function values to determine the general shape and then draw the graph. Note that the graph of this function crosses its horizontal asymptote at $x = -1.5$. ▬

EXAMPLE 9   Graph: $g(x) = \dfrac{x^2 - 1}{x^2 + x - 6}$.

Solution

1. We find the zeros of the denominator by solving $x^2 + x - 6 = 0$. Since

$$x^2 + x - 6 = (x + 3)(x - 2),$$

the zeros are -3 and 2. Thus the domain excludes the x-values -3 and 2 and is $(-\infty, -3) \cup (-3, 2) \cup (2, \infty)$. The graph has vertical asymptotes $x = -3$ and $x = 2$. We sketch these as dashed lines.

2. The numerator and the denominator have the same degree, so the horizontal asymptote is determined by the ratio of the leading coefficients: $1/1$, or 1. Thus, $y = 1$ is the horizontal asymptote. We sketch it with a dashed line.

3. To find the zeros of the numerator, we solve $x^2 - 1 = 0$. The solutions are -1 and 1. Thus, -1 and 1 are the zeros of the function and the pairs $(-1, 0)$ and $(1, 0)$ are the x-intercepts.

4. We find $g(0)$:

$$g(0) = \frac{0^2 - 1}{0^2 + 0 - 6} = \frac{-1}{-6} = \frac{1}{6}.$$

Thus, $\left(0, \frac{1}{6}\right)$ is the y-intercept.

5. We find other function values to determine the general shape and then draw the graph.

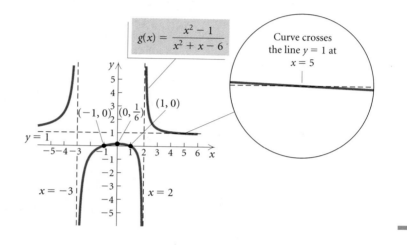

$$y_1 = \frac{x^2 - 1}{x^2 + x - 6}, \ y_2 = 1$$

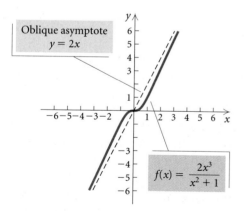

The magnified portion of the graph in Example 9 shows another situation in which a graph can cross its horizontal asymptote. Using a grapher, let's observe the behavior of the curve after it crosses the horizontal asymptote at $x = 5$. (See the window at left.) It continues to decrease for a short interval and then begins to increase, getting closer and closer to $y = 1$ as $x \to \infty$.

Graphs of rational functions can also cross an oblique asymptote. The graph of

$$f(x) = \frac{2x^3}{x^2 + 1}$$

shown at left crosses its oblique asymptote $y = 2x$. Remember, graphs can cross horizontal or oblique asymptotes, but they cannot cross vertical asymptotes.

Let's now graph a rational function $p(x)/q(x)$, where $p(x)$ and $q(x)$ have a common factor.

EXAMPLE 10   Graph: $g(x) = \dfrac{x - 2}{x^2 - x - 2}.$

Solution   We first express the denominator in factored form:

$$g(x) = \frac{x - 2}{x^2 - x - 2} = \frac{x - 2}{(x + 1)(x - 2)}.$$

The domain of the function is $\{x \mid x \neq -1 \ and \ x \neq 2\}$, or $(-\infty, -1) \cup (-1, 2) \cup (2, \infty)$. The zeros of the denominator are

X	Y1	
-3	-.5	
-2	-1	
-1	ERROR	
0	1	
1	.5	
2	ERROR	
3	.25	

X = 3

−1 and 2, and the zero of the numerator is 2. Thus the graph of the function has $x = -1$ as its only vertical asymptote, since −1 is the only zero of the denominator that is not a zero of the numerator. The degree of the numerator is less than the degree of the denominator, so $y = 0$ is the horizontal asymptote. There are no zeros of the function and thus no x-intercepts, because 2 is the only zero of the numerator and 2 is not in the domain of the function. Since $g(0) = 1$, $(0, 1)$ is the y-intercept. We draw the graph indicating the "hole" when $x = 2$ with an open circle.

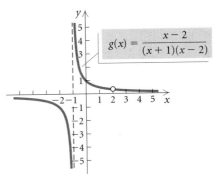

$$g(x) = \frac{x - 2}{(x + 1)(x - 2)}$$

The rational expression $(x - 2)/((x + 1)(x - 2))$ can be simplified. Thus,

$$g(x) = \frac{x - 2}{(x + 1)(x - 2)} = \frac{1}{x + 1}, \quad \text{where } x \neq -1 \text{ and } x \neq 2.$$

The graph of $g(x)$ is the graph of $y = 1/(x + 1)$ with the point $\left(2, \frac{1}{3}\right)$ missing. With certain window dimensions, the "hole" is visible on a grapher.

$$y = \frac{x - 2}{x^2 - x - 2}$$

Applications

EXAMPLE 11 *Temperature During an Illness.* The temperature T, in degrees Fahrenheit, of a person during an illness is given by the function

$$T(t) = \frac{4t}{t^2 + 1} + 98.6,$$

where time t is given in hours since the onset of the illness.

a) Graph the function on the interval $[0, 48]$.
b) Find the temperature at $t = 0, 1, 2, 5, 12,$ and 24.
c) Find the horizontal asymptote. Complete:

$$T(t) \rightarrow \boxed{} \text{ as } t \rightarrow \infty.$$

d) Give the meaning of the answer to part (c) in terms of the application.
e) Find the maximum temperature during the illness.

Solution

a) The graph is shown at left.

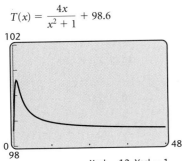

$$T(x) = \frac{4x}{x^2 + 1} + 98.6$$

Xscl = 12, Yscl = 1

b) We have

$$T(0) = 98.6, \qquad T(1) = 100.6, \qquad T(2) = 100.2,$$
$$T(5) = 99.369, \qquad T(12) = 98.931, \quad \text{and} \quad T(24) = 98.766.$$

c) Since

$$T(t) = \frac{4t}{t^2 + 1} + 98.6$$

$$= \frac{98.6t^2 + 4t + 98.6}{t^2 + 1},$$

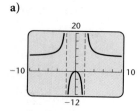

the horizontal asymptote is $y = 98.6/1$, or 98.6. Then it follows that $T(t) \to 98.6$ as $t \to \infty$. We can check this using a table of values.

d) As time goes on, the temperature returns to "normal," which is $98.6°$.

e) Using the MAXIMUM feature on a grapher, we find the maximum temperature to be $100.6°$ at $t = 1$ hr.

Maximum
X = 1   . Y = 100.6

Exercise Set 3.4

In Exercises 1–6, use your knowledge of asymptotes and intercepts to match the equation with one of the graphs (a)–(f), which follow. List all asymptotes. Check your work using a grapher.

1. $f(x) = \dfrac{8}{x^2 - 4}$

2. $f(x) = \dfrac{8}{x^2 + 4}$

3. $f(x) = \dfrac{8x}{x^2 - 4}$

4. $f(x) = \dfrac{8x^2}{x^2 - 4}$

5. $f(x) = \dfrac{8x^3}{x^2 - 4}$

6. $f(x) = \dfrac{8x^3}{x^2 + 4}$

a)

b)

c)

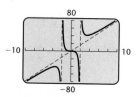

d)

e)

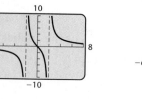

f)

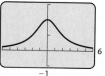

Make a hand-drawn graph for each of the following. Be sure to label all the asymptotes. List the x- and y-intercepts. Check your work using a grapher.

7. $f(x) = \dfrac{1}{x + 3}$

8. $f(x) = \dfrac{1}{x - 5}$

9. $f(x) = \dfrac{-2}{x - 5}$

10. $f(x) = \dfrac{3}{3 - x}$

11. $f(x) = \dfrac{2x + 1}{x}$

12. $f(x) = \dfrac{3x - 1}{x}$

13. $f(x) = \dfrac{1}{(x - 2)^2}$

14. $f(x) = \dfrac{-2}{(x - 3)^2}$

15. $f(x) = \dfrac{1}{x^2}$

16. $f(x) = \dfrac{1}{3x^2}$

17. $f(x) = \dfrac{1}{x^2 + 3}$

18. $f(x) = \dfrac{-1}{x^2 + 2}$

19. $f(x) = \dfrac{x^2 - 4}{x - 2}$

20. $f(x) = \dfrac{x^2 - 9}{x + 3}$

21. $f(x) = \dfrac{x - 1}{x + 2}$

22. $f(x) = \dfrac{x - 2}{x + 1}$

23. $f(x) = \dfrac{x + 3}{2x^2 - 5x - 3}$

24. $f(x) = \dfrac{3x}{x^2 + 5x + 4}$

25. $f(x) = \dfrac{x^2 - 9}{x + 1}$

26. $f(x) = \dfrac{x^2 - 4}{x - 1}$

27. $f(x) = \dfrac{x^2 + x - 2}{2x^2 + 1}$

28. $f(x) = \dfrac{x^2 - 2x - 3}{3x^2 + 2}$

29. $g(x) = \dfrac{3x^2 - x - 2}{x - 1}$

30. $f(x) = \dfrac{2x + 1}{2x^2 - 5x - 3}$

31. $f(x) = \dfrac{x - 1}{x^2 - 2x - 3}$

32. $f(x) = \dfrac{x + 2}{x^2 + 2x - 15}$

33. $f(x) = \dfrac{x - 3}{(x + 1)^3}$

34. $f(x) = \dfrac{x + 2}{(x - 1)^3}$

35. $f(x) = \dfrac{x^3 + 1}{x}$

36. $f(x) = \dfrac{x^3 - 1}{x}$

37. $f(x) = \dfrac{x^3 + 2x^2 - 15x}{x^2 - 5x - 14}$

38. $f(x) = \dfrac{x^3 + 2x^2 - 3x}{x^2 - 25}$

39. $f(x) = \dfrac{5x^4}{x^4 + 1}$

40. $f(x) = \dfrac{x + 1}{x^2 + x - 6}$

41. $f(x) = \dfrac{x^2}{x^2 - x - 2}$

42. $f(x) = \dfrac{x^2 - x - 2}{x + 2}$

Find a rational function that satisfies the given conditions for each of the following. Answers may vary, but try to give the simplest answer possible.

43. Vertical asymptotes $x = -4$, $x = 5$

44. Vertical asymptotes $x = -4$, $x = 5$; x-intercept $(-2, 0)$

45. Vertical asymptotes $x = -4$, $x = 5$; horizontal asymptote $y = \frac{3}{2}$; x-intercept $(-2, 0)$

46. Oblique asymptote $y = x - 1$

47. *Medical Dosage.* The function

$$N(t) = \dfrac{0.8t + 1000}{5t + 4}, \quad t \geq 15$$

gives the body concentration $N(t)$, in parts per million, of a certain dosage of medication after time t, in hours.

a) Graph the function on the interval $[15, \infty)$ and complete the following:

$$N(t) \to \boxed{} \text{ as } t \to \infty.$$

b) Explain the meaning of the answer to p(a) in terms of the application.

48. *Average Cost.* The average cost per tape, in dollars, for a company to produce x videotapes on exercising is given by the function

$$A(x) = \dfrac{13x + 100}{x}.$$

a) Graph the function on the interval $(0, \infty)$ and complete the following:

$$A(x) \to \boxed{} \text{ as } x \to \infty.$$

b) Explain the meaning of the answer to part (a) in terms of the application.

49. *Population Growth.* The population P, in thousands, of Lordsburg is given by

$$P(t) = \dfrac{500t}{2t^2 + 9},$$

where t is the time, in months.

a) Graph the function on the interval $[0, \infty)$.

b) Find the population at $t = 0$, 1, 3, and 8 months.

c) Find the horizontal asymptote. Complete:

$$P(t) \to \boxed{} \text{ as } t \to \infty.$$

d) Give the meaning of the answer to p(c).

e) Find the maximum population.

50. *Minimizing Surface Area.* The Hold-It Container Co. is designing an open-top rectangular box, with a square base, that will hold 108 cubic centimeters (cc).

a) Express the surface area S as a function of the length x of a side of the base.

b) Graph the function on the interval $(0, \infty)$.

c) Estimate the minimum surface area and the value of x that will yield it.

Discussion and Writing

51. Under what circumstances will a rational function have a domain consisting of all real numbers?

52. Explain why the graph of a rational function cannot have both a horizontal and an oblique asymptote.

53. Graph

$$y_1 = \dfrac{x^3 + 4}{x} \quad \text{and} \quad y_2 = x^2$$

using the same viewing window. Explain how the parabola $y_2 = x^2$ can be thought of as a nonlinear asymptote for y_1.

Skill Maintenance

Find the equation of variation for the given situation.

54. y varies directly as x, and $y = 22$ when $x = \frac{1}{2}$

55. y varies inversely as x, and $y = \frac{2}{5}$ when $x = 10$

56. y varies jointly as x and the square of z and inversely as w, and $y = 0.1$ when $x = 10{,}000$, $z = 10$, and $w = 5$

57. The time T required to do a job varies inversely as the number of people P working. Opinions Count is a company that collects data through phone surveys. If it takes 8 employees 6 hr to make 600 calls, how long will it take 12 employees to make the same number of calls?

Synthesis

Find the nonlinear asymptotes of the function.

58. $f(x) = \dfrac{x^4 + 3x^2}{x^2 + 1}$

59. $f(x) = \dfrac{x^5 + 2x^3 + 4x^2}{x^2 + 2}$

Graph the function. Check using a grapher.

60. $f(x) = \dfrac{x^3 + 4x^2 + x - 6}{x^2 - x - 2}$

61. $f(x) = \dfrac{2x^3 + x^2 - 8x - 4}{x^3 + x^2 - 9x - 9}$

62. $f(x) = \dfrac{x^4 + 3x^3 + 21x^2 - 50x + 80}{x^4 + 8x^3 - x^2 + 20x - 10}$

Find the domain.

63. $f(x) = \sqrt{\dfrac{72}{x^2 - 4x - 21}}$

64. $f(x) = \sqrt{x^2 - 4x - 21}$

Polynomial and Rational Inequalities

3.5

• *Solve polynomial and rational inequalities.*

We will use a combination of algebraic and graphical methods to solve polynomial and rational inequalities.

Polynomial Inequalities

Just as a quadratic equation can be written in the form $ax^2 + bx + c = 0$, a **quadratic inequality** can be written in the form $ax^2 + bx + c \,\blacksquare\, 0$, where $\blacksquare$ is $<$, $>$, $\leq$, or $\geq$. Here are some examples of quadratic inequalities:

$$3x^2 - 2x - 5 > 0, \qquad -\tfrac{1}{2}x^2 + 4x - 7 \leq 0.$$

Quadratic inequalities are one type of **polynomial inequality.** Other examples of polynomial inequalities are

$$-2x^4 + x^2 - 3 < 7, \qquad \tfrac{2}{3}x + 4 \geq 0, \quad \text{and} \quad 4x^3 - 2x^2 > 5x + 7.$$

When the inequality symbol in a polynomial inequality is replaced with an equals sign, a **related equation** is formed. Polynomial inequalities can be easily solved once the related equation has been solved.

EXAMPLE 1   Solve: $x^3 - x > 0$.

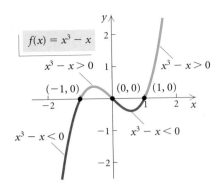

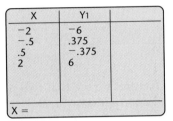

To solve a polynomial inequality:

1. Find an equivalent inequality with 0 on one side.

2. Solve the related polynomial equation.

3. Use the solutions to divide the x-axis into intervals. Then select a test value from each interval and determine the polynomial's sign on the interval.

4. Determine the intervals for which the inequality is satisfied and write interval notation or set-builder notation for the solution set. Include the endpoints of the intervals in the solution set if the inequality symbol is $\leq$ or $\geq$.

Solution   We are asked to find all x-values for which $x^3 - x > 0$. To locate these values, we graph $f(x) = x^3 - x$. Then we note that whenever the function changes sign, its graph passes through an x-intercept. Thus to solve $x^3 - x > 0$, we first solve the related equation $x^3 - x = 0$ to find all zeros of the function:

$$x^3 - x = 0$$
$$x(x^2 - 1) = 0$$
$$x(x + 1)(x - 1) = 0.$$

The zeros are -1, 0, and 1. Thus the x-intercepts of the graph are $(-1, 0)$, $(0, 0)$, and $(1, 0)$, as shown in the figure at left. The zeros divide the x-axis into four intervals:

$$(-\infty, -1), \quad (-1, 0), \quad (0, 1), \quad \text{and} \quad (1, \infty).$$

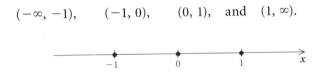

For all x-values within a given interval, the sign of $x^3 - x$ must be either *positive* or *negative*. To determine which, we choose a test value for x from each interval and find $f(x)$. We can use the TABLE feature set in ASK mode to determine the sign of $f(x)$ in each interval (see the table at left). We can also determine the sign of $f(x)$ in each interval by simply looking at the graph of the function.

INTERVAL	TEST VALUE	SIGN OF $f(x)$
$(-\infty, -1)$	$f(-2) = -6$	Negative
$(-1, 0)$	$f(-0.5) = 0.375$	Positive
$(0, 1)$	$f(0.5) = -0.375$	Negative
$(1, \infty)$	$f(2) = 6$	Positive

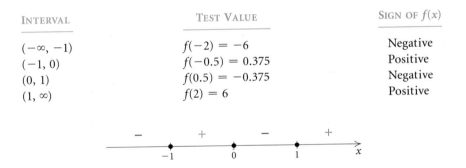

Since we are solving $x^3 - x > 0$, the solution set consists of only two of the four intervals, those in which the sign of $f(x)$ is *positive*. We see that the solution set is $(-1, 0) \cup (1, \infty)$, or $\{x \mid -1 < x < 0 \ or \ x > 1\}$.

Shown in the box at left is a method for solving polynomial inequalities.

EXAMPLE 2   Solve: $3x^4 + 10x \leq 11x^3 + 4$.

By subtracting $11x^3 + 4$, we form the equivalent inequality

$$3x^4 - 11x^3 + 10x - 4 \leq 0.$$

Algebraic Solution

To solve the related equation

$$3x^4 - 11x^3 + 10x - 4 = 0,$$

we need to use a grapher and/or the theorems of Section 3.3 (see Example 9 in Section 3.3). The solutions are

$$-1, \quad 2 - \sqrt{2}, \quad \tfrac{2}{3}, \quad \text{and} \quad 2 + \sqrt{2},$$

or approximately

$$-1, \quad 0.586, \quad 0.667, \quad \text{and} \quad 3.414.$$

These numbers divide the x-axis into five intervals: $(-\infty, -1)$, $(-1, 2 - \sqrt{2})$, $\left(2 - \sqrt{2}, \tfrac{2}{3}\right)$, $\left(\tfrac{2}{3}, 2 + \sqrt{2}\right)$, and $(2 + \sqrt{2}, \infty)$.

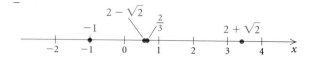

We then let $f(x) = 3x^4 - 11x^3 + 10x - 4$ and, using test values for $f(x)$, determine the sign of $f(x)$ in each interval:

X	Y₁	
-2	112	
0	-4	
.6	.0128	
1	-2	
4	100	
X =		

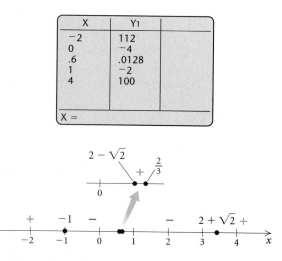

Function values are negative in the intervals $(-1, 2 - \sqrt{2})$ and $\left(\tfrac{2}{3}, 2 + \sqrt{2}\right)$. Since the inequality sign is $\leq$, we include the endpoints of the intervals in the solution set. The solution set is

$$\left[-1, 2 - \sqrt{2}\right] \cup \left[\tfrac{2}{3}, 2 + \sqrt{2}\right], \quad \text{or}$$
$$\left\{x \mid -1 \leq x \leq 2 - \sqrt{2} \text{ or } \tfrac{2}{3} \leq x \leq 2 + \sqrt{2}\right\}.$$

Graphical Solution

We graph $y = 3x^4 - 11x^3 + 10x - 4$ using a viewing window that reveals the curvature of the graph.

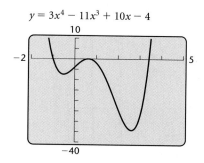

Using the ZERO feature, we see that two of the zeros are -1 and approximately 3.414. However, this window leaves us uncertain about the number of zeros of the function in the interval $[0, 1]$. The following window shows another view of the zeros in the interval $[0, 1]$. Those zeros are about 0.586 and 0.667.

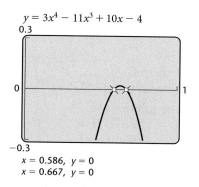

The intervals to be considered are $(-\infty, -1)$, $(-1, 0.586)$, $(0.586, 0.667)$, $(0.667, 3.414)$, and $(3.414, \infty)$. We note on the graph where the function is negative. Then including appropriate endpoints, we find that the solution set is approximately

$$[-1, 0.586] \cup [0.667, 3.414], \quad \text{or}$$
$$\{x \mid -1 \leq x \leq 0.586 \text{ or } 0.667 \leq x \leq 3.414\}.$$

Rational Inequalities

Some inequalities involve rational expressions and functions. These are called **rational inequalities.** To solve rational inequalities, we need to make some adjustments to the preceding method.

EXAMPLE 3   Solve: $\dfrac{x-3}{x+4} \geq \dfrac{x+2}{x-5}$.

Solution   We first subtract $(x+2)/(x-5)$ in order to find an equivalent inequality with 0 on one side:

$$\frac{x-3}{x+4} - \frac{x+2}{x-5} \geq 0.$$

Algebraic Solution

We look for all values of x for which the related function

$$f(x) = \frac{x-3}{x+4} - \frac{x+2}{x-5}$$

is not defined or is 0. These are called **critical values.**
A look at the denominators shows that $f(x)$ is not defined for $x = -4$ and $x = 5$. Next, we solve $f(x) = 0$:

$$\frac{x-3}{x+4} - \frac{x+2}{x-5} = 0$$

$$(x+4)(x-5)\left(\frac{x-3}{x+4} - \frac{x+2}{x-5}\right) = (x+4)(x-5) \cdot 0$$

$$(x-5)(x-3) - (x+4)(x+2) = 0$$

$$(x^2 - 8x + 15) - (x^2 + 6x + 8) = 0$$

$$-14x + 7 = 0$$

$$x = \tfrac{1}{2}.$$

The critical values are $-4, \tfrac{1}{2}$, and 5. These values divide the x-axis into four intervals:

$$(-\infty, -4), \qquad \left(-4, \tfrac{1}{2}\right), \qquad \left(\tfrac{1}{2}, 5\right), \quad \text{and} \quad (5, \infty).$$

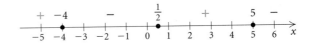

We then use a test value to determine the sign of $f(x)$ in each interval.

X	Y₁	
−5	7.7	
−2	−2.5	
3	2.5	
6	−7.7	

X = −5

Function values are positive in the intervals $(-\infty, -4)$ and $\left(\frac{1}{2}, 5\right)$. Since $f\left(\frac{1}{2}\right) = 0$ and the inequality symbol is $\geq$, we know that $\frac{1}{2}$ must be in the solution set. Note that since neither -4 nor 5 is in the domain of f, they cannot be part of the solution set.

The solution set is $(-\infty, -4) \cup \left[\frac{1}{2}, 5\right)$.

Graphical Solution

We graph

$$y = \frac{x - 3}{x + 4} - \frac{x + 2}{x - 5}$$

in the standard window, which shows its curvature.

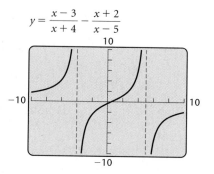

By using the ZERO feature, we find that 0.5 is a zero.

We then look for values where the function is not defined. By examining the denominators $x + 4$ and $x - 5$, we see that $f(x)$ is not defined for $x = -4$ and $x = 5$.

The **critical values,** where y is either not defined or 0, are -4, 0.5, and 5.

The graph shows where y is positive and where it is negative. Note that -4 and 5 cannot be in the solution set since y is not defined for these values. We do include 0.5, however, since the inequality symbol is $\geq$ and $f(0.5) = 0$. The solution set is

$$(-\infty, -4) \cup [0.5, 5).$$

The following is a method for solving rational inequalities.

To solve a rational inequality:

1. Find an equivalent inequality with 0 on one side.
2. Change the inequality symbol to an equals sign and solve the related equation.
3. Find values of the variable for which the related rational function is not defined.
4. The numbers found in steps (2) and (3) are called critical values. Use the critical values to divide the x-axis into intervals. Then test an x-value from each interval to determine the function's sign in that interval.
5. Select the intervals for which the inequality is satisfied and write interval notation or set-builder notation for the solution set. If the inequality symbol is $\leq$ or $\geq$, then the solutions to step (2) should be included in the solution set. The x-values found in step (3) are never included in the solution set.

It works well to use a combination of algebraic and graphical methods to solve polynomial and rational inequalities. The algebraic methods give exact numbers for the critical values, and the graphical methods allow us to see easily what intervals satisfy the inequality.

Exercise Set **3.5**

In Exercises 1–4, a related function is graphed. Solve the given inequality.

1. $x^3 + 6x^2 < x + 30$

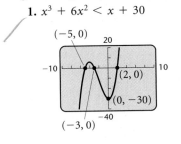

2. $x^4 - 27x^2 - 14x + 120 \geq 0$

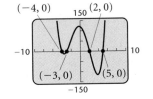

3. $\dfrac{8x}{x^2 - 4} \geq 0$

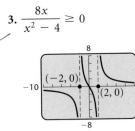

4. $\dfrac{8}{x^2 - 4} < 0$

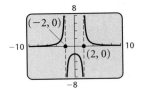

Solve.

5. $(x - 1)(x + 4) < 0$

6. $(x + 3)(x - 5) < 0$

7. $(x - 4)(x + 2) \geq 0$

8. $(x - 2)(x + 1) \geq 0$

9. $x^2 + x - 2 > 0$

10. $x^2 - x - 6 > 0$

11. $x^2 > 25$

12. $x^2 \leq 1$

13. $4 - x^2 \leq 0$

14. $11 - x^2 \geq 0$

15. $6x - 9 - x^2 < 0$

16. $x^2 + 2x + 1 \leq 0$

17. $x^2 + 12 < 4x$

18. $x^2 - 8 > 6x$

19. $4x^3 - 7x^2 \leq 15x$

20. $2x^3 - x^2 < 5$

21. $x^3 + 3x^2 - x - 3 \geq 0$

22. $x^3 + x^2 - 4x - 4 \geq 0$

23. $x^3 - 2x^2 < 5x - 6$

24. $x^3 + x \leq 6 - 4x^2$

25. $x^5 + x^2 \geq 2x^3 + 2$

26. $x^5 + 24 > 3x^3 + 8x^2$

27. $2x^3 + 6 \leq 5x^2 + x$

28. $2x^3 + x^2 < 10 + 11x$

29. $x^3 + 5x^2 - 25x \leq 125$

30. $x^3 - 9x + 27 \geq 3x^2$

31. $0.1x^3 - 0.6x^2 - 0.1x + 2 < 0$

32. $19.2x^3 + 12.8x^2 + 144 \geq 172.8x + 3.2x^4$

33. $\dfrac{1}{x + 4} > 0$

34. $\dfrac{1}{x - 3} \leq 0$

35. $\dfrac{-4}{2x + 5} < 0$

36. $\dfrac{-2}{5 - x} \geq 0$

37. $\dfrac{x - 4}{x + 3} - \dfrac{x + 2}{x - 1} \leq 0$

38. $\dfrac{x + 1}{x - 2} + \dfrac{x - 3}{x - 1} < 0$

39. $\dfrac{2x - 1}{x + 3} \geq \dfrac{x + 1}{3x + 1}$

40. $\dfrac{x + 5}{x - 4} > \dfrac{3x + 2}{2x + 1}$

41. $\dfrac{x + 1}{x - 2} \geq 3$

42. $\dfrac{x}{x - 5} < 2$

43. $x - 2 > \dfrac{1}{x}$

44. $4 \geq \dfrac{4}{x} + x$

45. $\dfrac{2}{x^2 - 4x + 3} \leq \dfrac{5}{x^2 - 9}$

46. $\dfrac{3}{x^2 - 4} \leq \dfrac{5}{x^2 + 7x + 10}$

47. $\dfrac{3}{x^2 + 1} \geq \dfrac{6}{5x^2 + 2}$

48. $\dfrac{4}{x^2 - 9} < \dfrac{3}{x^2 - 25}$

49. $\dfrac{5}{x^2 + 3x} < \dfrac{3}{2x + 1}$

50. $\dfrac{2}{x^2 + 3} > \dfrac{3}{5 + 4x^2}$

51. $\dfrac{5x}{7x - 2} > \dfrac{x}{x + 1}$

52. $\dfrac{x^2 - x - 2}{x^2 + 5x + 6} < 0$

53. $\dfrac{x}{x^2 + 4x - 5} + \dfrac{3}{x^2 - 25} \leq \dfrac{2x}{x^2 - 6x + 5}$

54. $\dfrac{2x}{x^2 - 9} + \dfrac{x}{x^2 + x - 12} \geq \dfrac{3x}{x^2 + 7x + 12}$

55. *Temperature During an Illness.* The temperature T, in degrees Fahrenheit, of a person during an illness is given by the function

$$T(t) = \frac{4t}{t^2 + 1} + 98.6,$$

where t is the time, in hours. Find the interval on which the temperature was over $100°$. (See Example 11 in Section 3.4.)

56. *Population Growth.* The population P, in thousands, of Lordsburg is given by

$$P(t) = \frac{500t}{2t^2 + 9},$$

where t is the time, in months. Find the interval on which the population was 40 thousand or greater. (See Exercise 49 in Exercise Set 3.4.)

57. *Total Profit.* Flexl, Inc., determines that its total profit is given by the function

$$P(x) = -3x^2 + 630x - 6000.$$

a) Flexl makes a profit for those nonnegative values of x for which $P(x) > 0$. Find the values of x for which Flexl makes a profit.

b) Flexl loses money for those nonnegative values of x for which $P(x) < 0$. Find the values of x for which Flexl loses money.

58. *Height of a Thrown Object.* The function

$$S(t) = -16t^2 + 32t + 1920$$

gives the height S, in feet, of an object thrown from a cliff that is 1920 ft high. Here t is the time, in seconds, that the object is in the air.

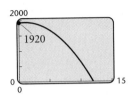

a) For what times is the height greater than 1920 ft?
b) For what times is the height less than 640 ft?

59. *Number of Diagonals.* A polygon with n sides has D diagonals, where D is given by the function

$$D(n) = \frac{n(n-3)}{2}.$$

Find the number of sides n if

$$27 \leq D \leq 230.$$

60. *Number of Handshakes.* There are n people in a room. The number N of possible handshakes by all the people in the room is given by the function

$$N(n) = \frac{n(n-1)}{2}.$$

For what number n of people is

$$66 \leq N \leq 300?$$

Discussion and Writing

61. Under what circumstances would a quadratic inequality have a solution set that is a closed interval? Under what circumstances would a quadratic inequality have an empty solution set?

62. Why, when solving rational inequalities, do we need to find values for which the function is undefined as well as zeros of the function?

Skill Maintenance

Find an equation for a circle satisfying the given conditions.

63. Center: $(-2, 4)$; radius of length 3

64. Center: $(0, -3)$; diameter of length $\frac{7}{2}$

In Exercises 65 and 66:

a) Find the vertex.
b) Determine whether there is a maximum or minimum value and find that value.
c) Find the range.

65. $h(x) = -2x^2 + 3x - 8$

66. $g(x) = x^2 - 10x + 2$

Synthesis

Solve.

67. $x^2 + 9 \leq 6x$

68. $x^4 - 6x^2 + 5 > 0$

69. $x^4 + 3x^2 > 4x - 15$

70. $\left| \dfrac{x+3}{x-4} \right| < 2$

71. $|x^2 - 5| = 5 - x^2$

72. $(7 - x)^{-2} < 0$

73. $2|x|^2 - |x| + 2 \leq 5$

74. $|x|^2 - 4|x| + 4 \geq 9$

75. $\left| 1 + \dfrac{1}{x} \right| < 3$

76. $\left| 2 - \dfrac{1}{x} \right| \leq 2 + \left| \dfrac{1}{x} \right|$

77. $|x^2 + 3x - 1| < 3$

78. $|1 + 5x - x^2| \geq 5$

79. Write a quadratic inequality for which the solution set is $(-4, 3)$.

80. Write a polynomial inequality for which the solution set is $[-4, 3] \cup [7, \infty)$.

Chapter Summary and Review 3

Important Properties and Formulas

Polynomial Function:

$$P(x) = a_n x^n + a_{n-1} x^{n-1} + a_{n-2} x^{n-2} + \cdots + a_1 x + a_0$$

The Leading-Term Test: If $a_n x^n$ is the leading term of a polynomial, then the behavior of the graph as $x \to \infty$ and as $x \to -\infty$ can be described in one of the four following ways.

If n is even, and $a_n > 0$:	If n is even, and $a_n < 0$:

If n is odd, and $a_n > 0$:	If n is odd, and $a_n < 0$:

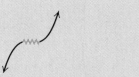

The Intermediate Value Theorem: For any polynomial function $P(x)$ with real coefficients, suppose that $P(a) \neq P(b)$ for $a \neq b$ and that $P(a)$ and $P(b)$ are of opposite signs. Then the polynomial has a real zero between a and b.

Polynomial Division:

$$P(x) = D(x) \cdot Q(x) + R(x)$$

Dividend   Divisor   Quotient   Remainder

The Remainder Theorem: The remainder found by dividing $P(x)$ by $x - c$ is $P(c)$.

The Factor Theorem: $P(c) = 0 \leftrightarrow x - c$ is a factor of $P(x)$.

The Fundamental Theorem of Algebra: Every polynomial of degree n, $n \geq 1$, with complex coefficients has at least one complex-number zero.

The Rational Zeros Theorem: Consider the polynomial equation

$$a_n x^n + a_{n-1} x^{n-1} + a_{n-2} x^{n-2} + \cdots + a_1 x + a_0 = 0,$$

where all the coefficients are integers and $n \geq 1$. Also, consider a rational number p/q, where p and q have no common factor other than -1 and 1. If p/q is a solution of the polynomial equation, then p is a factor of a_0 and q is a factor of a_n.

Rational Function:

$$f(x) = \frac{p(x)}{q(x)},$$

where $p(x)$ and $q(x)$ are polynomials and where $q(x)$ is not the zero polynomial, and the domain of $f(x)$ consists of all x for which $q(x) \neq 0$.

Occurrence of Lines as Asymptotes

For a rational function $p(x)/q(x)$, where $p(x)$ and $q(x)$ have no common factors other than constants:

Vertical asymptotes occur at any x-values that make the denominator 0.

The x-axis is the horizontal asymptote when the degree of the numerator is less than the degree of the denominator.

A horizontal asymptote other than the x-axis occurs when the numerator and the denominator have the same degree.

An oblique asymptote occurs when the degree of the numerator is 1 greater than the degree of the denominator.

REVIEW EXERCISES

Use a grapher to graph the polynomial function. Then estimate the function's (a) zeros; (b) relative maxima; (c) relative minima; (d) domain and range.

1. $f(x) = -2x^2 - 3x + 6$

2. $f(x) = x^3 + 3x^2 - 2x - 6$

3. $f(x) = x^4 - 3x^3 + 2x^2$

4. *Interest Compounded Annually.* When P dollars is invested at interest rate i, compounded annually, for t years, the investment grows to A dollars, where

$$A = P(1 + i)^t.$$

a) Find the interest rate i if $6250 grows to $6760 in 2 yr.

b) Find the interest rate i if $1,000,000 grows to $1,215,506.25 in 4 yr.

5. *Cholesterol Level and the Risk of Heart Attack.* The data in the following table show the relationship of cholesterol level in men to the risk of a heart attack.

CHOLESTEROL LEVEL	NUMBER OF MEN, PER 10,000, WHO SUFFER A HEART ATTACK
100	30
200	65
250	100
275	130

Source: Nutrition Action Healthletter

a) Use regression on a grapher to fit linear, quadratic, and cubic functions to the data.

b) It is also known that 180 of 10,000 men have a heart attack with a cholesterol level of 300. Which function in part (a) would best make this prediction?

c) Use the answer to part (b) to predict the heart attack rate for men with cholesterol levels of 350 and of 400.

Do the long division. Check by multiplying or by using a grapher.

6. $(20x^3 - 6x^2 - 9x + 10) \div (5x + 2)$

7. $(2x^5 - 3x^3 + x^2 + 4) \div (x^2 - 2)$

In Exercises 8 and 9, a polynomial equation is given along with information about factors. Use the information to solve the equation.

8. $x^3 + 2x^2 - 13x + 10 = 0$; $x + 5$ is a factor of $x^3 + 2x^2 - 13x + 10$

9. $x^4 + 5x^3 - 23x^2 - 87x + 140 = 0$; $x^2 + x - 20$ is a factor of $x^4 + 5x^3 - 23x^2 - 87x + 140$

Use synthetic division to find the quotient and the remainder.

10. $(x^3 + 2x^2 - 13x + 10) \div (x - 5)$

11. $(x^4 + 3x^3 + 3x^2 + 3x + 2) \div (x + 2)$

Use synthetic division to find the indicated function value.

12. $P(x) = x^3 + 2x^2 - 13x + 10$; $P(5)$

13. $P(x) = x^4 - 16$; $P(-2)$

Determine whether the given values are zeros of $P(x)$. Then factor and find any other zeros that exist.

14. $P(x) = x^3 + 7x^2 + 7x - 15$; $-1, 2$, and -3

15. $P(x) = x^4 + 9x^3 + 19x^2 + 9x - 2$; $1, -4$, and -2

Find a polynomial equation of lowest degree with integer coefficients and the following as some of its zeros. Use a grapher as a partial check.

16. $4, -1, -2$

17. $-4, -\sqrt{3}, \sqrt{3}, 1$

18. $3, 2 - 3i, 4 + \sqrt{5}, -2$

Use the rational zeros theorem and/or factoring to solve the equation.

19. $x^3 - 4x^2 + x + 6 = 0$

20. $x^3 + 7x^2 + 7x = 15$

21. $x^4 + 36x + 63 = 4x^3 + 16x^2$

In Exercises 22 and 23, for the polynomial function:

a) *Solve $P(x) = 0$.*

b) *Graph $y = P(x)$.*

c) *Express $P(x)$ as a product of linear factors.*

22. $P(x) = x^3 + 8x^2 - 19x + 10$

23. $P(x) = x^6 + x^5 - 28x^4 - 16x^3 + 192x^2$

Make a hand-drawn graph for each of the following. Be sure to label all the asymptotes.

24. $f(x) = \dfrac{x^2 - 5}{x + 2}$

25. $f(x) = \dfrac{5}{(x - 2)^2}$

26. $f(x) = \dfrac{x^2 + x - 6}{x^2 - x - 20}$ **27.** $f(x) = \dfrac{x - 2}{x^2 - 2x - 15}$

In Exercises 28 and 29, find a rational function that satisfies the given conditions. Answers may vary, but try to give the simplest answer possible.

28. Vertical asymptotes $x = -2$, $x = 3$

29. Vertical asymptotes $x = -2$, $x = 3$; horizontal asymptote $y = 4$; x-intercept $(-3, 0)$

30. *Medical Dosage.* The function

$$N(t) = \frac{0.7t + 2000}{8t + 9}, \quad t \geq 5$$

gives the body concentration $N(t)$, in parts per million, of a certain dosage of medication after time t, in hours.

a) Graph the function on the interval $[5, \infty)$ and complete the following:

$$N(t) \rightarrow \boxed{} \text{ as } t \rightarrow \infty.$$

b) Explain the meaning of the answer to part (a) in terms of the application.

Solve.

31. $x^2 - 9 < 0$ **32.** $2x^2 > 3x + 2$

33. $(1 - x)(x + 4)(x - 2) \leq 0$

34. $\dfrac{x - 2}{x + 3} < 4$

35. $\dfrac{x - 3}{x^2 + x - 20} \geq \dfrac{4}{x^2 - 4}$

36. *Height of a Launched Rocket.* The function

$$S(t) = -16t^2 + 80t + 224$$

gives the height S, in feet, of a model rocket launched from a hill that is 224 ft high with a velocity of 80 ft/sec, where t is the time, in seconds.

a) Find the maximum height of the rocket and when that height is attained.

b) Determine when the rocket reaches the ground.

c) On what interval is the height greater than 320 ft?

37. *Population Growth.* The population P, in thousands, of Novi is given by

$$P(t) = \frac{8000t}{4t^2 + 10},$$

where t is the time, in months. Find the interval on which the population was 400,000 or greater.

Discussion and Writing

38. Explain the difference between a polynomial function and a rational function.

39. Explain and contrast the three types of asymptotes considered for rational functions.

Synthesis

40. *Interest Rate.* In early 2000, $3500 was deposited at a certain interest rate. One year later, $4000 was deposited in another account at the same rate. At the end of that year, there was a total of $8518.35 in both accounts. What is the annual interest rate?

Solve.

41. $x^2 \geq 5 - 2x$ **42.** $\left| 1 - \dfrac{1}{x^2} \right| < 3$

43. $x^4 - 2x^3 + 3x^2 - 2x + 2 = 0$

44. $(x - 2)^{-3} < 0$

45. Express $x^3 - 1$ as a product of linear factors.

46. Find k such that $x + 3$ is a factor of $x^3 + kx^2 + kx - 15$.

47. When $x^2 - 4x + 3k$ is divided by $x + 5$, the remainder is 33. Find the value of k.

Find the domain of the function.

48. $f(x) = \sqrt{x^2 + 3x - 10}$

49. $f(x) = \sqrt{x^2 - 3.1x + 2.2} + 1.75$

50. $f(x) = \dfrac{1}{\sqrt{5 - |7x + 2|}}$

Exponential and Logarithmic Functions 4

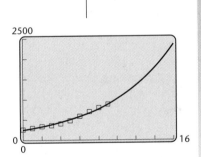

I n this chapter, we will consider two kinds of closely related functions. The first type, *exponential functions*, has a variable in the exponent. Such functions have many applications to the growth of populations, commodities, and investments.

Recall that a function takes an input to an output. Suppose that we can reverse the process and take the output back to an input. That process produces the *inverse* of the original function. Functions that are inverses of each other are closely related. The inverses of exponential functions, called *logarithmic functions*, or *logarithm functions*, are also important in many applications such as earthquake magnitude, sound level, and chemical pH.

Composite and Inverse Functions

4.1

- *Find the composition of two functions and the domain of the composition; decompose a function as a composition of two functions.*
- *Determine whether a function is one-to-one, and if it is, find a formula for its inverse.*
- *Simplify expressions of the type $(f \circ f^{-1})(x)$ and $(f^{-1} \circ f)(x)$.*

The Composition of Functions

In the real world, it is not uncommon for a function's output to depend on some input that is itself an output of another function. For instance, the amount a person pays as state income tax usually depends on the amount of adjusted gross income on the person's federal tax return, which, in turn, depends on his or her annual earnings. Such functions are called **composite functions.**

To illustrate how composite functions work, suppose a chemistry student needs a formula to convert Fahrenheit temperatures to Kelvin units. The formula $c(t) = \frac{5}{9}(t - 32)$ gives the Celsius temperature $c(t)$ that corresponds to the Fahrenheit temperature t. The formula $k(c) = c + 273$ gives the Kelvin temperature $k(c)$ that corresponds to the Celsius temperature c. Thus, 50° Fahrenheit corresponds to

$$c(50) = \tfrac{5}{9}(50 - 32)$$
$$= \tfrac{5}{9}(18) = 10° \text{ Celsius}$$

and 10° Celsius corresponds to

$$k(10) = 10 + 273$$
$$= 283° \text{ Kelvin.}$$

We see that 50° Fahrenheit is the same as 283° Kelvin. This two-step procedure can be used to convert any Fahrenheit temperature to Kelvin units.

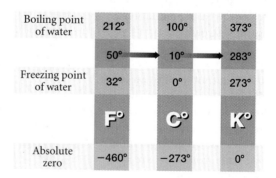

$y_1 = \frac{5}{9}(x - 32), \quad y_2 = y_1 + 273$

X	Y1	Y2
50	10	283
59	15	288
68	20	293
77	25	298
86	30	303
95	35	308
104	40	313
X = 50		

In the table shown at left, we use a grapher to convert Fahrenheit temperatures, x, to Celsius temperatures, y_1, using $y_1 = \frac{5}{9}(x - 32)$. We also convert Celsius temperatures to Kelvin units, y_2, using $y_2 = y_1 + 273$.

A student making numerous conversions might look for a formula that converts directly from Fahrenheit to Kelvin. Such a formula can be found by substitution:

$$y_2 = y_1 + 273$$

$$= \left[\frac{5}{9}(x - 32)\right] + 273 \qquad \text{Substituting}$$

$$= \frac{5}{9}x - \frac{160}{9} + 273 = \frac{5}{9}x - \frac{160}{9} + \frac{2457}{9}$$

$$= \frac{5x + 2297}{9}. \qquad \text{Simplifying}$$

We can show on a grapher that the same values that appear in the table for y_2 will appear when y_2 is entered as

$$y_2 = \frac{5x + 2297}{9}.$$

In the more commonly used function notation, we have

$$k(c(t)) = c(t) + 273$$

$$= \frac{5}{9}(t - 32) + 273 \qquad \text{Substituting}$$

$$= \frac{5t + 2297}{9}. \qquad \text{Simplifying as above}$$

Since the last equation expresses the Kelvin temperature as a new function, K, of the Fahrenheit temperature, t, we can write

$$K(t) = \frac{5t + 2297}{9},$$

where $K(t)$ is the Kelvin temperature corresponding to the Fahrenheit temperature, t. Here we have $K(t) = k(c(t))$. The new function K is called the **composition** of k and c and can be denoted $k \circ c$ (read "k composed with c," "the composition of k and c," or "k circle c").

Composition of Functions

The **composite function** $f \circ g$, the **composition** of f and g, is defined as

$$(f \circ g)(x) = f(g(x)),$$

where x is in the domain of g and $g(x)$ is in the domain of f.

EXAMPLE 1 Given that $f(x) = 2x - 5$ and $g(x) = x^2 - 3x + 8$, find each of the following.

a) $(f \circ g)(7)$ and $(g \circ f)(7)$ **b)** $(f \circ g)(x)$ and $(g \circ f)(x)$

Solution   Consider each function separately:

$f(x) = 2x - 5$ **This function multiplies each input by 2 and subtracts 5.**

and

$g(x) = x^2 - 3x + 8.$ **This function squares an input, subtracts 3 times the input from the result, and then adds 8.**

a) To find $(f \circ g)(7)$, we first find $g(7)$. Then we use $g(7)$ as an input for f:

$$(f \circ g)(7) = f(g(7)) = f(7^2 - 3 \cdot 7 + 8)$$
$$= f(36) = 2 \cdot 36 - 5$$
$$= 67.$$

To find $(g \circ f)(7)$, we first find $f(7)$. Then we use $f(7)$ as an input for g:

$$(g \circ f)(7) = g(f(7)) = g(2 \cdot 7 - 5)$$
$$= g(9) = 9^2 - 3 \cdot 9 + 8$$
$$= 62.$$

We can check our work using a grapher by entering on the equation-editor screen

$$y_1 = f(x) = 2x - 5$$

and

$$y_2 = g(x) = x^2 - 3x + 8.$$

Then, on the home screen, we find $(f \circ g)(7)$ and $(g \circ f)(7)$ using the function notation Y1(Y2(7)) and Y2(Y1(7)), respectively.

b) To find $(f \circ g)(x)$, we substitute $g(x)$ for x in the equation for $f(x)$:

$$(f \circ g)(x) = f(g(x)) = f(x^2 - 3x + 8)$$
$$\text{Substituting } x^2 - 3x + 8 \text{ for } g(x)$$
$$= 2(x^2 - 3x + 8) - 5$$
$$= 2x^2 - 6x + 16 - 5$$
$$= 2x^2 - 6x + 11.$$

To find $(g \circ f)(x)$, we substitute $f(x)$ for x in the equation for $g(x)$:

$$(g \circ f)(x) = g(f(x)) = g(2x - 5)$$ **Substituting $2x - 5$ for $f(x)$**
$$= (2x - 5)^2 - 3(2x - 5) + 8$$
$$= 4x^2 - 20x + 25 - 6x + 15 + 8$$
$$= 4x^2 - 26x + 48.$$

Note in Example 1 that, as a rule, $(f \circ g)(x) \neq (g \circ f)(x)$. We can check this graphically, as shown in the graphs at left.

EXAMPLE 2   Given that $f(x) = \sqrt{x}$ and $g(x) = x - 3$:

a) Find $h(x)$ and $k(x)$ if $h = f \circ g$ and $k = g \circ f$.

b) Find the domains of h and k.

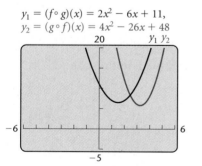

$y_1 = 2x - 5, \ y_2 = x^2 - 3x + 8$

Y1(Y2(7))
                            67
Y2(Y1(7))
                            62

$y_1 = (f \circ g)(x) = 2x^2 - 6x + 11,$
$y_2 = (g \circ f)(x) = 4x^2 - 26x + 48$

$y = h(x) = \sqrt{x - 3}$

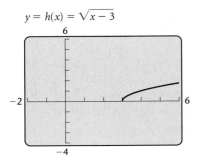

FIGURE 1

$y = k(x) = \sqrt{x} - 3$

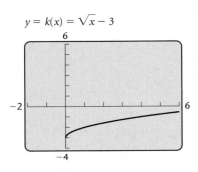

FIGURE 2

Solution

a) $h(x) = (f \circ g)(x) = f(g(x)) = f(x - 3) = \sqrt{x - 3}$

$k(x) = (g \circ f)(x) = g(f(x)) = g(\sqrt{x}) = \sqrt{x} - 3$

b) The domain of f is $\{x \mid x \geq 0\}$, or the interval $[0, \infty)$. The domain of g is $(-\infty, \infty)$.

To find the domain of $h = f \circ g$, we must consider that the outputs for g will serve as inputs for f. Since the inputs for f cannot be negative, we must have

$$g(x) = x - 3, \quad \text{where } x - 3 \geq 0, \text{ or } x \geq 3.$$

Thus the domain of $h = \{x \mid x \geq 3\}$, or the interval $[3, \infty)$, as the graph in Fig. 1 confirms.

To find the domain of $k = g \circ f$, we must consider that the outputs for f will serve as inputs for g. Since g can accept *any* real number as an input, any output from f is acceptable. Thus the domain of $k = \{x \mid x \geq 0\}$, as the graph in Fig. 2 confirms. ▬

Decomposing a Function as a Composition

In calculus, one often needs to recognize how a function can be expressed as the composition of two functions. In this way, we are "decomposing" the function.

EXAMPLE 3 If $h(x) = (2x - 3)^5$, find $f(x)$ and $g(x)$ such that $h(x) = (f \circ g)(x)$.

Solution The function $h(x)$ raises $(2x - 3)$ to the 5th power. Two functions that can be used for the composition are

$$f(x) = x^5 \quad \text{and} \quad g(x) = 2x - 3.$$

We can check by forming the composition:

$$h(x) = (f \circ g)(x) = f(g(x)) = f(2x - 3) = (2x - 3)^5.$$

This is the most "obvious" answer to the question. There can be other less obvious answers. For example, if

$$f(x) = (x + 7)^5 \quad \text{and} \quad g(x) = 2x - 10,$$

then

$$\begin{aligned} h(x) &= (f \circ g)(x) = f(g(x)) \\ &= f(2x - 10) \\ &= [(2x - 10) + 7]^5 = (2x - 3)^5. \end{aligned}$$ ▬

Inverses

When we go from an output of a function back to its input or inputs, we get an inverse relation. When that relation is a function, we have an inverse function. We now study inverse functions and how to find their formulas when the original function has a formula.

Consider the relation h given as follows:

$$h = \{(-8, 5), (4, -2), (-7, 1), (3.8, 6.2)\}.$$

Suppose we *interchange* the first and second coordinates. The relation we obtain is called the **inverse** of the relation h and is given as follows:

$$\text{Inverse of } h = \{(5, -8), (-2, 4), (1, -7), (6.2, 3.8)\}.$$

Inverse Relation

Interchanging the first and second coordinates of each ordered pair in a relation produces the **inverse relation.**

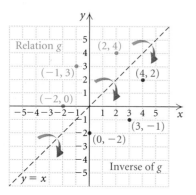

EXAMPLE 4   Consider the relation g given by

$$g = \{(2, 4), (-1, 3), (-2, 0)\}.$$

Graph the relation in blue. Find the inverse and graph it in red.

Solution   The relation g is shown in blue in the figure at left. The inverse of the relation is

$$\{(4, 2), (3, -1), (0, -2)\}$$

and is shown in red. The pairs in the inverse are reflections across the line $y = x$.   —

Inverse Relation

If a relation is defined by an equation, interchanging the variables produces an equation of the **inverse relation.**

EXAMPLE 5   Find an equation for the inverse of the relation:

$$y = x^2 - 5x.$$

Solution   We interchange x and y and obtain an equation of the inverse:

$$x = y^2 - 5y.$$ —

---

*Interactive
Discovery*

Graph each of the following relations. Then find the inverse of each and graph it using the same set of axes. What pattern do you see? Use a square viewing window for a true geometric perspective.

$$y = 3x + 2, \qquad y = x, \qquad xy = 2,$$
$$x^2 + 3y^2 = 4, \qquad y^2 = 4x - 5.$$

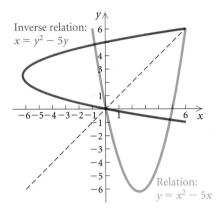

Inverse relation:
$x = y^2 - 5y$

Relation:
$y = x^2 - 5x$

If a relation is given by an equation, then the solutions of the inverse can be found from those of the original equation by interchanging the first and second coordinates of each ordered pair. Thus the graphs of a relation and its inverse are always reflections of each other across the line $y = x$. This is illustrated with the equations of Example 5 in the graph at left. We will explore inverses and their graphs later in this section.

Inverses and One-to-One Functions

Let's consider the following two functions.

YEAR (DOMAIN)	FIRST-CLASS POSTAGE COST, IN CENTS (RANGE)
1978 ⟶	15
1983 ⟶	20
1984 ⟶	
1989 ⟶	25
1991 ⟶	29
1995 ⟶	32
1999 ⟶	33

Source: U.S. Postal Service

NUMBER (DOMAIN)	CUBING (RANGE)
−3 ⟶	−27
−2 ⟶	−8
−1 ⟶	−1
0 ⟶	0
1 ⟶	1
2 ⟶	8
3 ⟶	27

Suppose we reverse the arrows. Are these inverse relations functions?

YEAR (RANGE)	FIRST-CLASS POSTAGE COST, IN CENTS (DOMAIN)
1978 ⟵	15
1983 ⟵	20
1984 ⟵	
1989 ⟵	25
1991 ⟵	29
1995 ⟵	32
1999 ⟵	33

NUMBER (RANGE)	CUBING (DOMAIN)
−3 ⟵	−27
−2 ⟵	−8
−1 ⟵	−1
0 ⟵	0
1 ⟵	1
2 ⟵	8
3 ⟵	27

We see that the inverse of the postage function is not a function. Like all functions, each input in the postage function has exactly one output. However, the outputs for both 1983 and 1984 are the same, 20. Thus in the inverse of the postage function, the input 20 has *two* outputs, 1983 and 1984. When the same output of a function comes from two or more different inputs, the inverse cannot be a function. In the cubing function, each output corresponds to exactly one input, so its inverse is also a function. The cubing function is an example of a **one-to-one function.** If the inverse of a function f is also a function, it is named f^{-1} (read "f-inverse").

$$\text{The } -1 \text{ in } f^{-1} \text{ is } \textit{not} \text{ an exponent!}$$

One-to-One Functions and Inverses

A function f is **one-to-one** if different inputs have different outputs—that is,

$$\text{if} \quad a \neq b, \quad \text{then} \quad f(a) \neq f(b). \quad \text{Or,}$$

A function f is **one-to-one** if when the outputs are the same, the inputs are the same—that is,

$$\text{if} \quad f(a) = f(b), \quad \text{then} \quad a = b.$$

If a function is one-to-one, then its inverse is a function.

The domain of a one-to-one function f is the range of the inverse f^{-1}.

The range of a one-to-one function f is the domain of the inverse f^{-1}.

EXAMPLE 6 Given the function f described by $f(x) = 2x - 3$, prove that f is one-to-one (that is, it has an inverse that is a function).

Solution To show that f is one-to-one, we show that if $f(a) = f(b)$, then $a = b$. Assume that $f(a) = f(b)$ for any numbers a and b in the domain of f. Since $f(a) = 2a - 3$ and $f(b) = 2b - 3$, we have

$$2a - 3 = 2b - 3$$
$$2a = 2b \qquad \text{Adding 3}$$
$$a = b. \qquad \text{Dividing by 2}$$

Thus, if $f(a) = f(b)$, then $a = b$. This shows that f is one-to-one. ▬

EXAMPLE 7 Given the function g described by $g(x) = x^2$, prove that g is not one-to-one.

Solution To prove that g is not one-to-one, we need to find two numbers a and b for which $a \neq b$ and $g(a) = g(b)$. Two such numbers are -3 and 3, because $-3 \neq 3$ and $g(-3) = g(3) = 9$. Thus g is not one-to-one. ▬

The graph at the top of the next page shows a function, in blue, and its inverse, in red. To determine whether the inverse is a function, we can apply the vertical-line test to its graph. By reflecting each such vertical line back across the line $y = x$, we obtain an equivalent **horizontal-line test** for the original function.

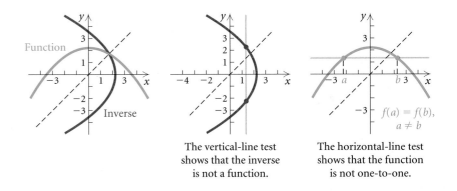

The vertical-line test
shows that the inverse
is not a function.

The horizontal-line test
shows that the function
is not one-to-one.

Horizontal-Line Test

If it is possible for a horizontal line to intersect the graph of a function more than once, then the function is *not* one-to-one and its inverse is *not* a function.

EXAMPLE 8 Graph each of the following functions. Then determine whether each is one-to-one and thus has an inverse that is a function.

a) $f(x) = 4 - x$ **b)** $f(x) = x^2$

c) $f(x) = \sqrt[3]{x + 2} + 3$ **d)** $f(x) = 3x^5 - 20x^3$

Solution We graph each function using a grapher. Then we apply the horizontal-line test.

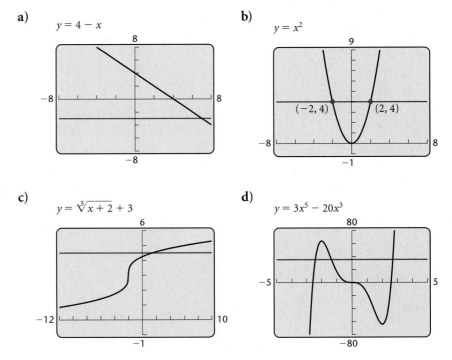

RESULT	REASON
a) One-to-one; inverse is a function	No horizontal line intersects the graph more than once.
b) Not one-to-one; inverse is not a function	There are many horizontal lines that intersect the graph more than once. Note that where the line $y = 4$ intersects the graph, the first coordinates are -2 and 2. Although these are different inputs, they have the same output, 4.
c) One-to-one; inverse is a function	No horizontal line intersects the graph more than once.
d) Not one-to-one; inverse is not a function	There are many horizontal lines that intersect the graph more than once.

Finding Formulas for Inverses

Suppose that a function is described by a formula. If it has an inverse that is a function, we proceed as follows to find a formula for f^{-1}.

Obtaining a Formula for an Inverse

If a function f is one-to-one, a formula for its inverse can generally be found as follows:

1. Replace $f(x)$ with y.

2. Interchange x and y.

3. Solve for y.

4. Replace y with $f^{-1}(x)$.

EXAMPLE 9 Determine whether the function $f(x) = 2x - 3$ is one-to-one, and if it is, find a formula for $f^{-1}(x)$.

Solution The graph of f is shown at left. It passes the horizontal-line test. Thus it is one-to-one and its inverse is a function.

We also proved that f is one-to-one in Example 6.

$y = 2x - 3$

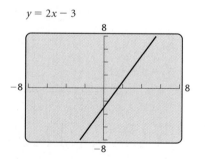

1. Replace $f(x)$ with y: $\qquad\qquad y = 2x - 3$

2. Interchange x and y: $\qquad\qquad x = 2y - 3$

3. Solve for y: $\qquad\qquad\quad x + 3 = 2y$

$$\frac{x + 3}{2} = y$$

4. Replace y with $f^{-1}(x)$: $\quad f^{-1}(x) = \dfrac{x + 3}{2}.$

Consider

$$f(x) = 2x - 3 \quad \text{and} \quad f^{-1}(x) = \frac{x + 3}{2}$$

from Example 9. For the input 5, we have

$$f(5) = 2 \cdot 5 - 3 = 10 - 3 = 7.$$

The output is 7. Now we use 7 for the input in the inverse:

$$f^{-1}(7) = \frac{7 + 3}{2}$$
$$= \frac{10}{2} = 5.$$

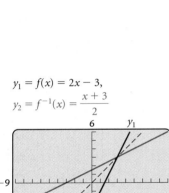

The function f takes the number 5 to 7. The inverse function f^{-1} takes the number 7 back to 5.

EXAMPLE 10   Graph

$$f(x) = 2x - 3 \quad \text{and} \quad f^{-1}(x) = \frac{x + 3}{2}$$

using the same set of axes. Then compare the two graphs.

Solution   The graphs of f and f^{-1} are shown at left. They are graphed in a square window. Note that the graph of f^{-1} can be drawn by reflecting the graph of f across the line $y = x$. That is, if we were to graph $f(x) = 2x - 3$ in wet ink and fold along the line $y = x$, the graph of $f^{-1}(x) = (x + 3)/2$ would be formed by the ink transferred from f.

When we interchange x and y in finding a formula for the inverse of $f(x) = 2x - 3$, we are in effect reflecting the graph of that function across the line $y = x$. For example, when the coordinates of the y-intercept, $(0, -3)$, of the graph of f are reversed, we get the x-intercept, $(-3, 0)$, of the graph of f^{-1}.

On some graphers, we can graph the inverse of a function after graphing the function itself by accessing a drawing feature. Consult your user's manual or the *Graphing Calculator Manual* that accompanies this text for the procedure.

> The graph of f^{-1} is a reflection of the graph of f across the line $y = x$.

EXAMPLE 11   Consider $g(x) = x^3 + 2$.

a) Determine whether the function is one-to-one.

b) If it is one-to-one, find a formula for its inverse.

c) Graph the function and its inverse.

Solution

a) The graph of $g(x) = x^3 + 2$ is shown below. It passes the horizontal-line test and thus has an inverse that is a function.

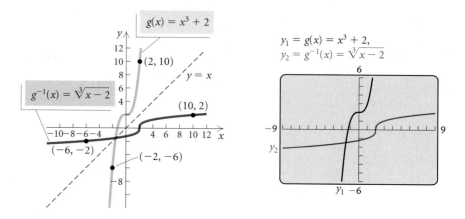

b) We follow the procedure for finding an inverse.

1. Replace $g(x)$ with y: $y = x^3 + 2$
2. Interchange x and y: $x = y^3 + 2$
3. Solve for y: $x - 2 = y^3$
   $$\sqrt[3]{x - 2} = y$$
4. Replace y with $g^{-1}(x)$: $g^{-1}(x) = \sqrt[3]{x - 2}$.

c) To find the graph, we reflect the graph of $g(x) = x^3 + 2$ across the line $y = x$. This can be done by plotting points or by using a grapher, as shown in part (a) above.

Inverse Functions and Composition

Suppose that we were to use some input a for a one-to-one function f and find its output, $f(a)$. The function f^{-1} would then take that output back to a. Similarly, if we began with an input b for the function f^{-1} and found its output, $f^{-1}(b)$, the original function f would then take that output back to b. This is summarized as follows.

If a function f is one-to-one, then f^{-1} is the unique function such that each of the following holds:

$$(f^{-1} \circ f)(x) = f^{-1}(f(x)) = x, \quad \text{for each } x \text{ in the domain of } f, \text{ and}$$

$$(f \circ f^{-1})(x) = f(f^{-1}(x)) = x, \quad \text{for each } x \text{ in the domain of } f^{-1}.$$

EXAMPLE 12   Given that $f(x) = 5x + 8$, use composition of functions to show that $f^{-1}(x) = (x - 8)/5$.

Solution   We find $(f^{-1} \circ f)(x)$ and $(f \circ f^{-1})(x)$ and check to see that each is x:

$$(f^{-1} \circ f)(x) = f^{-1}(f(x))$$

$$= f^{-1}(5x + 8) = \frac{(5x + 8) - 8}{5}$$

$$= \frac{5x}{5} = x;$$

$$(f \circ f^{-1})(x) = f(f^{-1}(x))$$

$$= f\left(\frac{x - 8}{5}\right) = 5\left(\frac{x - 8}{5}\right) + 8$$

$$= x - 8 + 8 = x.$$

Restricting a Domain

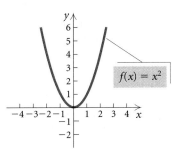

$f(x) = x^2$

In the case in which the inverse of a function is not a function, the domain of the function can be restricted to allow the inverse to be a function. We saw in Example 8 that $f(x) = x^2$ is not one-to-one. The graph is shown at left.

Suppose that we had tried to find a formula for the inverse as follows:

$$y = x^2 \qquad \text{Replacing } f(x) \text{ with } y$$
$$x = y^2 \qquad \text{Interchanging } x \text{ and } y$$
$$\pm\sqrt{x} = y. \qquad \text{Solving for } y$$

This is not the equation of a function. An input of, say, 4 would yield two outputs, -2 and 2. In such cases, it is convenient to consider "part" of the function by restricting the domain of $f(x)$. For example, if we restrict the domain of $f(x) = x^2$ to nonnegative numbers, then its inverse is a function as shown with the graphs of $f(x) = x^2$, $x \geq 0$, and $f^{-1}(x) = \sqrt{x}$, $x \geq 0$ below.

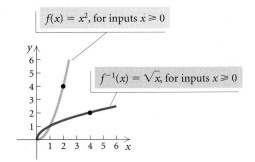

$f(x) = x^2$, for inputs $x \geq 0$

$f^{-1}(x) = \sqrt{x}$, for inputs $x \geq 0$

Exercise Set 4.1

Find $(f \circ g)(x)$ and $(g \circ f)(x)$.

1. $f(x) = x + 3, \ g(x) = x - 3$

2. $f(x) = \frac{4}{5}x, \ g(x) = \frac{5}{4}x$

3. $f(x) = 3x - 7, \ g(x) = \dfrac{x + 7}{3}$

4. $f(x) = \frac{2}{3}x - \frac{4}{5}, \ g(x) = 1.5x + 1.2$

5. $f(x) = 20, \ g(x) = 0.05$

6. $f(x) = x^4, \ g(x) = \sqrt[4]{x}$

7. $f(x) = \sqrt{x + 5}, \ g(x) = x^2 - 5$

8. $f(x) = x^5 - 2, \ g(x) = \sqrt[5]{x + 2}$

9. $f(x) = \dfrac{1 - x}{x}, \ g(x) = \dfrac{1}{1 + x}$

10. $f(x) = \dfrac{x^2 - 1}{x^2 + 1}, \ g(x) = \dfrac{3x - 4}{5x - 2}$

11. $f(x) = x^3 - 5x^2 + 3x + 7, \ g(x) = x + 1$

12. $f(x) = x - 1, \ g(x) = x^3 + 2x^2 - 3x - 9$

Find $f(x)$ and $g(x)$ such that $h(x) = (f \circ g)(x)$. Answers may vary, but try to select the most obvious answer.

13. $h(x) = (4 + 3x)^5$

14. $h(x) = \sqrt[3]{x^2 - 8}$

15. $h(x) = \dfrac{1}{(x - 2)^4}$

16. $h(x) = \dfrac{1}{\sqrt{3x + 7}}$

17. $h(x) = \dfrac{x^3 - 1}{x^3 + 1}$

18. $h(x) = |9x^2 - 4|$

19. $h(x) = \left(\dfrac{2 + x^3}{2 - x^3}\right)^6$

20. $h(x) = (\sqrt{x} - 3)^4$

21. $h(x) = \sqrt{\dfrac{x - 5}{x + 2}}$

22. $h(x) = \sqrt{1 + \sqrt{1 + x}}$

23. $h(x) = (x + 2)^3 - 5(x + 2)^2 + 3(x + 2) - 1$

24. $h(x) = 2(x - 1)^{5/3} + 5(x - 1)^{2/3}$

25. *Dress Sizes.* A dress that is size x in France is size $s(x)$ in the United States, where $s(x) = x - 32$. A

dress that is size x in the United States is size $y(x)$ in Italy, where $y(x) = 2(x + 12)$. Find a function that will convert French dress sizes to Italian dress sizes.

26. *Ripple Spread.* A stone is thrown into a pond. A circular ripple is spreading over the pond in such a way that the radius is increasing at the rate of 3 ft/sec.

a) Find a function $r(t)$ for the radius in terms of t.
b) Find a function $A(r)$ for the area of the ripple in terms of the radius r.
c) Find $(A \circ r)(t)$. Explain the meaning of this function.

Find the inverse of the relation.

27. $\{(7, 8), (-2, 8), (3, -4), (8, -8)\}$

28. $\{(0, 1), (5, 6), (-2, -4)\}$

29. $\{(-1, -1), (-3, 4)\}$

30. $\{(-1, 3), (2, 5), (-3, 5), (2, 0)\}$

Find an equation of the inverse relation.

31. $y = 4x - 5$ **32.** $2x^2 + 5y^2 = 4$

33. $x^3y = -5$ **34.** $y = 3x^2 - 5x + 9$

Graph the equation by substituting and plotting points. Then reflect the graph across the line $y = x$ to obtain the graph of its inverse.

35. $x = y^2 - 3$ **36.** $y = x^2 + 1$

37. $y = |x|$ **38.** $x = |y|$

Using the horizontal-line test, determine whether the function is one-to-one.

39. $f(x) = 2.7^x$ **40.** $f(x) = 2^{-x}$

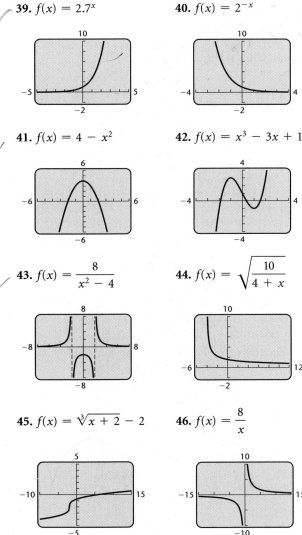

41. $f(x) = 4 - x^2$ **42.** $f(x) = x^3 - 3x + 1$

43. $f(x) = \dfrac{8}{x^2 - 4}$ **44.** $f(x) = \sqrt{\dfrac{10}{4 + x}}$

45. $f(x) = \sqrt[3]{x + 2} - 2$ **46.** $f(x) = \dfrac{8}{x}$

Determine whether the function is one-to-one. Use a grapher and the horizontal-line test or use an algebraic procedure.

47. $f(x) = 5x - 8$ **48.** $f(x) = 3 + 4x$

49. $f(x) = 1 - x^2$ **50.** $f(x) = |x| - 2$

51. $f(x) = |x + 2|$ **52.** $f(x) = -0.8$

53. $f(x) = -\dfrac{4}{x}$ **54.** $f(x) = \dfrac{2}{x + 3}$

Graph the function and its inverse using a grapher. Use an inverse drawing feature, if available. Find the domain and the range of f. Find the domain and the range of the inverse f^{-1}.

55. $f(x) = 0.8x + 1.7$ **56.** $f(x) = 2.7 - 1.08x$

57. $f(x) = \frac{1}{2}x - 4$ **58.** $f(x) = x^3 - 1$

59. $f(x) = \sqrt{x - 3}$ **60.** $f(x) = -\dfrac{2}{x}$

61. $f(x) = x^2 - 4, \ x \geq 0$

62. $f(x) = 3 - x^2, \ x \geq 0$

63. $f(x) = (3x - 9)^3$

64. $f(x) = \sqrt[3]{\dfrac{x - 3.2}{1.4}}$

In Exercises 65–80, for each function:

a) *Determine whether it is one-to-one, using a grapher if desired.*

b) *If it is one-to-one, find a formula for the inverse.*

65. $f(x) = x + 4$ **66.** $f(x) = 7 - x$

67. $f(x) = 2x - 1$ **68.** $f(x) = 5x + 8$

69. $f(x) = \dfrac{4}{x + 7}$ **70.** $f(x) = -\dfrac{3}{x}$

71. $f(x) = \dfrac{x + 4}{x - 3}$ **72.** $f(x) = \dfrac{5x - 3}{2x + 1}$

73. $f(x) = x^3 - 1$ **74.** $f(x) = (x + 5)^3$

75. $f(x) = x\sqrt{4 - x^2}$

76. $f(x) = 4x^5 - 20x^3 + 2x^2 - 5x + 1$

77. $f(x) = 5x^2 - 2, \ x \geq 0$

78. $f(x) = 4x^2 + 3, \ x \geq 0$

79. $f(x) = \sqrt{x + 1}$

80. $f(x) = \sqrt[3]{x - 8}$

Find the inverse by thinking about the operations of the function and then reversing, or undoing, them. Check your work algebraically.

FUNCTION	INVERSE
81. $f(x) = 3x$	$f^{-1}(x) = $ 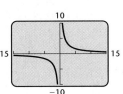
82. $f(x) = \frac{1}{4}x + 7$	$f^{-1}(x) = $
83. $f(x) = -x$	$f^{-1}(x) = $
84. $f(x) = \sqrt[3]{x} - 5$	$f^{-1}(x) = $

85. $f(x) = \sqrt[3]{x - 5}$ $\qquad$ $f^{-1}(x) =$ ▨▨

86. $f(x) = x^{-1}$ $\qquad$ $f^{-1}(x) =$ ▨▨

For the function f, use composition of functions to show that f^{-1} is as given.

87. $f(x) = \dfrac{7}{8}x,\ f^{-1}(x) = \dfrac{8}{7}x$

88. $f(x) = \dfrac{(x + 5)}{4},\ f^{-1}(x) = 4x - 5$

89. $f(x) = \dfrac{(1 - x)}{x},\ f^{-1}(x) = \dfrac{1}{x + 1}$

90. $f(x) = \sqrt[3]{x + 4},\ f^{-1}(x) = x^3 - 4$

91. Find $f(f^{-1}(5))$ and $f^{-1}(f(a))$:
$$f(x) = x^3 - 4.$$

92. Find $f^{-1}(f(p))$ and $f(f^{-1}(1253))$:
$$f(x) = \sqrt[5]{\dfrac{2x - 7}{3x + 4}}.$$

93. *Dress Sizes in the United States and Italy.* A function that will convert dress sizes in the United States to those in Italy is
$$g(x) = 2(x + 12).$$

a) Find the dress sizes in Italy that correspond to sizes 6, 8, 10, 14, and 18 in the United States.
b) Find a formula for the inverse of the function.
c) Use the inverse function to find the dress sizes in the United States that correspond to 36, 40, 44, 52, and 60 in Italy.

94. *Bus Chartering.* An organization determines that the cost per person of chartering a bus is given by the formula
$$C(x) = \dfrac{100 + 5x}{x},$$
where x is the number of people in the group and $C(x)$ is in dollars. Determine $C^{-1}(x)$ and explain what it represents.

95. *Reaction Distance.* You are driving a car when a deer suddenly darts across the road in front of you. Your brain registers the emergency and sends a signal to your foot to hit the brake. The car travels a distance D, in feet, during this time, where D is a function of the speed r, in miles per hour, that the car is traveling when you see the deer. That reaction distance D is a linear function given by
$$D(r) = \dfrac{11r + 5}{10}.$$

a) Find $D(0)$, $D(10)$, $D(20)$, $D(50)$, and $D(65)$.
b) Graph $D(r)$.
c) Find $D^{-1}(r)$ and explain what it represents.
d) Graph the inverse.

96. *Bread Consumption.* The number N of 1-lb loaves of bread consumed per person per year t years after 1995 is given by the function
$$N(t) = 0.6514t + 53.1599$$
(*Source*: U.S. Department of Agriculture).

a) Find the consumption of bread per person in 1998 and 2000.
b) Use a grapher to graph the function and its inverse.
c) Explain what the inverse represents.

Discussion and Writing

97. Suppose that you have graphed a function using a grapher and you see that it is one-to-one. How could you then use the TRACE feature to make a hand-drawn graph of the inverse?

98. The following formulas for the conversion between Fahrenheit and Celsius temperatures have been considered several times in this text:
$$C = \tfrac{5}{9}(F - 32)$$
and
$$F = \tfrac{9}{5}C + 32.$$
Discuss these formulas from the standpoint of inverses.

Skill Maintenance

Graph.

99. $y = \dfrac{3}{2}x - 4$

100. $(x - 1)^2 + y^2 = 9$

101. $y = \dfrac{x + 3}{x^2 - x - 2}$

102. $y = -x^2 + x - 4$

Synthesis

Using only a grapher, determine whether the functions are inverses of each other.

103. $f(x) = \sqrt[3]{\dfrac{x - 3.2}{1.4}}, \quad g(x) = 1.4x^3 + 3.2$

104. $f(x) = \dfrac{2x - 5}{4x + 7}, \quad g(x) = \dfrac{7x - 4}{5x + 2}$

105. $f(x) = \dfrac{2}{3}, \quad g(x) = \dfrac{3}{2}$

106. $f(x) = x^4, \ x \geq 0; \ g(x) = \sqrt[4]{x}$

107. The function $f(x) = x^2 - 3$ is not one-to-one. Restrict the domain of f so that its inverse is a function. Find the inverse and state the restriction on the domain of the inverse.

108. Consider the function f given by

$$f(x) = \begin{cases} x^3 + 2, & x \leq -1, \\ x^2, & -1 < x < 1, \\ x + 1, & x \geq 1. \end{cases}$$

Does f have an inverse that is a function? Why or why not?

109. Find three examples of functions that are their own inverses, that is, $f = f^{-1}$.

Exponential
Functions
and Graphs

4.2

- *Graph exponential equations and functions.*
- *Solve applied problems involving exponential functions and their graphs.*

We now turn our attention to the study of a set of functions that are very rich in application. Consider the following graphs. Each one illustrates an *exponential function*. In this section, we consider such functions, their inverses, and some important applications of exponential functions.

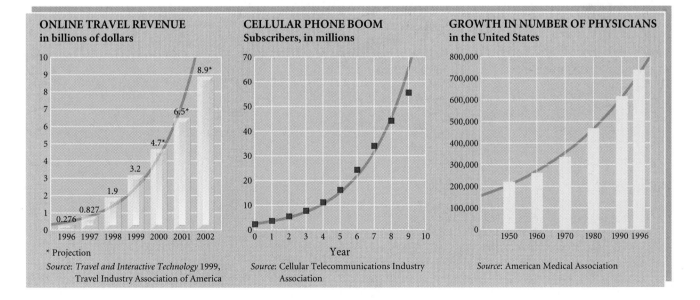

ONLINE TRAVEL REVENUE
in billions of dollars

8.9*
6.5*
4.7*
3.2
1.9
0.827
0.276

1996 1997 1998 1999 2000 2001 2002

* Projection

Source: Travel and Interactive Technology 1999, Travel Industry Association of America

CELLULAR PHONE BOOM
Subscribers, in millions

0 1 2 3 4 5 6 7 8 9 10
Year

Source: Cellular Telecommunications Industry Association

GROWTH IN NUMBER OF PHYSICIANS
in the United States

1950 1960 1970 1980 1990 1996

Source: American Medical Association

Graphing Exponential Functions

EXPONENTS

REVIEW SECTIONS R.2 AND R.6.

We now define exponential functions. We assume that a^x has meaning for any real number x and any positive real number a and that the laws of exponents still hold, though we will not prove them here.

Exponential Function

The function $f(x) = a^x$, where x is a real number, $a > 0$ and $a \neq 1$, is called the **exponential function,** base a.

We require the **base** to be positive in order to avoid the complex numbers that would occur by taking even roots of negative numbers—an example is $(-1)^{1/2}$, the square root of -1, which is not a real number. The restriction $a \neq 1$ is made to exclude the constant function $f(x) = 1^x = 1$, which does not have an inverse because it is not one-to-one.

The following are examples of exponential functions:

$$f(x) = 2^x, \qquad f(x) = \left(\frac{1}{2}\right)^x, \qquad f(x) = (3.57)^x.$$

Note that, in contrast to functions like $f(x) = x^5$ and $f(x) = x^{1/2}$ in which the variable is the base of an exponential expression, the variable in an exponential function is *in the exponent*. Let's now consider graphs of exponential functions.

EXAMPLE 1   Graph the exponential function

$$y = f(x) = 2^x.$$

Solution   We compute some function values and list the results in a table (at left).

$$f(0) = 2^0 = 1;$$

$$f(1) = 2^1 = 2;$$

$$f(2) = 2^2 = 4;$$

$$f(3) = 2^3 = 8;$$

$$f(-1) = 2^{-1} = \frac{1}{2^1} = \frac{1}{2};$$

$$f(-2) = 2^{-2} = \frac{1}{2^2} = \frac{1}{4};$$

$$f(-3) = 2^{-3} = \frac{1}{2^3} = \frac{1}{8}$$

x	$y = f(x) = 2^x$	(x, y)
0	1	$(0, 1)$
1	2	$(1, 2)$
2	4	$(2, 4)$
3	8	$(3, 8)$
-1	$\frac{1}{2}$	$\left(-1, \frac{1}{2}\right)$
-2	$\frac{1}{4}$	$\left(-2, \frac{1}{4}\right)$
-3	$\frac{1}{8}$	$\left(-3, \frac{1}{8}\right)$

Next, we plot these points and connect them with a smooth curve. Be sure to plot enough points to determine how steeply the curve rises.

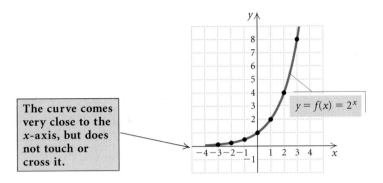

The curve comes very close to the x-axis, but does not touch or cross it.

$$y = f(x) = 2^x$$

Note that as x increases, the function values increase without bound. As x decreases, the function values decrease, getting close to 0. That is, as $x \to -\infty$, $y \to 0$. The x-axis, or the line $y = 0$, is a horizontal asymptote. As the x-inputs decrease, the curve gets closer and closer to this line, but does not cross it.

Check the graph of $y = f(x) = 2^x$ using a grapher. Then graph $y = f(x) = 3^x$. Use the TRACE and TABLE features to confirm that the graph never crosses the x-axis.

EXAMPLE 2   Graph the exponential function $y = f(x) = \left(\dfrac{1}{2}\right)^x$.

Solution   We compute some function values and list the results in a table (at left). Before we plot these points and draw the curve, note that

$$y = f(x) = \left(\frac{1}{2}\right)^x = (2^{-1})^x = 2^{-x}.$$

This tells us, before we begin graphing, that this graph is a reflection of the graph of $y = 2^x$ across the y-axis.

$$f(0) = 2^{-0} = 1;$$

$$f(1) = 2^{-1} = \frac{1}{2^1} = \frac{1}{2};$$

$$f(2) = 2^{-2} = \frac{1}{2^2} = \frac{1}{4};$$

$$f(3) = 2^{-3} = \frac{1}{2^3} = \frac{1}{8};$$

$$f(-1) = 2^{-(-1)} = 2^1 = 2;$$

$$f(-2) = 2^{-(-2)} = 2^2 = 4;$$

$$f(-3) = 2^{-(-3)} = 2^3 = 8$$

	y	
x	$y = f(x) = 2^{-x}$	(x, y)
0	1	$(0, 1)$
1	$\frac{1}{2}$	$\left(1, \frac{1}{2}\right)$
2	$\frac{1}{4}$	$\left(2, \frac{1}{4}\right)$
3	$\frac{1}{8}$	$\left(3, \frac{1}{8}\right)$
-1	2	$(-1, 2)$
-2	4	$(-2, 4)$
-3	8	$(-3, 8)$

Next, we plot these points and connect them with a smooth curve.

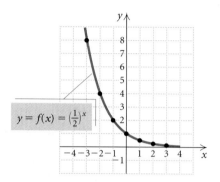

$$y = f(x) = \left(\tfrac{1}{2}\right)^x$$

Graph each of the following functions using the same set of axes and a viewing window of $[-10, 10, -10, 10]$. Try entering these functions with one equation: $y_1 = \{1.1, 2.7, 4\} \wedge x$. Look for patterns in the graphs.

$$f(x) = 1.1^x, \qquad f(x) = 2.7^x, \qquad f(x) = 4^x$$

Next, clear the screen and graph each of the following functions using the same set of axes and viewing window. Again, look for patterns in the graphs.

$$f(x) = 0.1^x, \qquad f(x) = 0.6^x, \qquad f(x) = 0.23^x$$

What relationship do you see between the base a and the shape of the resulting graph of a^x? What do all the graphs have in common? How do they differ?

CONNECTING THE CONCEPTS

PROPERTIES OF EXPONENTIAL FUNCTIONS

Let's list and compare some characteristics of exponential functions, keeping in mind that the definition of an exponential function, $f(x) = a^x$, requires that a be positive and different from 1.

$f(x) = a^x$ for $0 < a < 1$ or
$f(x) = a^{-x}$, $a > 1$:

Continuous

One-to-one

Domain: $(-\infty, \infty)$

Range: $(0, \infty)$

Decreasing

Horizontal asymptote is x-axis

y-intercept: $(0, 1)$

$y = \left(\tfrac{1}{4}\right)^x$, or 4^{-x}

$y = \left(\tfrac{1}{2}\right)^x$, or 2^{-x}

$y = 4^x$

$y = 2^x$

$f(x) = a^x$, $a > 1$:

Continuous

One-to-one

Domain: $(-\infty, \infty)$

Range: $(0, \infty)$

Increasing

Horizontal asymptote is x-axis

y-intercept: $(0, 1)$

TRANSFORMATIONS
OF FUNCTIONS

REVIEW SECTION 1.5.

To graph other types of exponential functions, keep in mind the ideas of translation, stretching, and reflection. All these concepts allow us to visualize the graph before drawing it.

EXAMPLE 3 Graph each of the following. Before doing so, describe how each graph can be obtained from $f(x) = 2^x$.

a) $f(x) = 2^{x-2}$

b) $f(x) = 2^x - 4$

c) $f(x) = 5 - 2^{-x}$

Solution

a) The graph of this function is the graph of $y = 2^x$ shifted *right* 2 units.

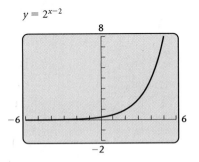

$y = 2^{x-2}$

b) The graph is the graph of $y = 2^x$ shifted *down* 4 units (see the graph at left).

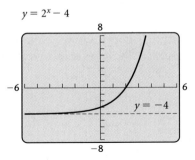

$y = 2^x - 4$

c) The graph is a reflection of the graph of $y = 2^x$ across the y-axis, followed by a reflection across the x-axis and then a shift *up* 5 units.

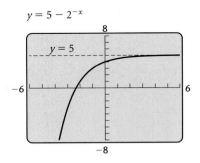

$y = 5 - 2^{-x}$

Graphs of Inverses of Exponential Functions

We have noted that every exponential function (with $a > 0$ and $a \neq 1$) is one-to-one. Thus each such function has an inverse that is a function. In the next section, we will name these inverse functions and use them in applications. For now, we draw their graphs by interchanging x and y.

EXAMPLE 4 Graph: $x = 2^y$.

Solution   Note that x is alone on one side of the equation. We can find ordered pairs that are solutions by choosing values for y and then computing the corresponding x-values.

For $y = 0$, $x = 2^0 = 1$.
For $y = 1$, $x = 2^1 = 2$.
For $y = 2$, $x = 2^2 = 4$.
For $y = 3$, $x = 2^3 = 8$.
For $y = -1$, $x = 2^{-1} = \dfrac{1}{2^1} = \dfrac{1}{2}$.
For $y = -2$, $x = 2^{-2} = \dfrac{1}{2^2} = \dfrac{1}{4}$.
For $y = -3$, $x = 2^{-3} = \dfrac{1}{2^3} = \dfrac{1}{8}$.

| x | | |
$x = 2^y$	y	(x, y)
1	0	$(1, 0)$
2	1	$(2, 1)$
4	2	$(4, 2)$
8	3	$(8, 3)$
$\dfrac{1}{2}$	-1	$\left(\dfrac{1}{2}, -1\right)$
$\dfrac{1}{4}$	-2	$\left(\dfrac{1}{4}, -2\right)$
$\dfrac{1}{8}$	-3	$\left(\dfrac{1}{8}, -3\right)$

(1) Choose values for y.
(2) Compute values for x.

We plot the points and connect them with a smooth curve. Note that the curve does not touch or cross the y-axis. The y-axis is a vertical asymptote.

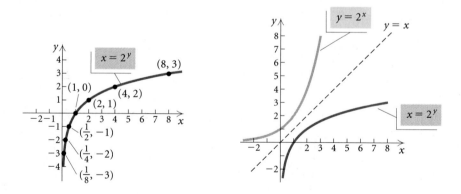

Note too that this curve looks just like the graph of $y = 2^x$, except that it is reflected across the line $y = x$, as we would expect for an inverse. The inverse of $y = 2^x$ is $x = 2^y$.

Applications

Graphers are especially helpful when working with exponential functions. They not only facilitate computations but they also allow us to visualize the functions.

EXAMPLE 5 *Compound Interest.* The amount of money A that a principal P will be worth after t years at interest rate i, compounded n times per year, is given by the formula

$$A = P\left(1 + \frac{i}{n}\right)^{nt}.$$

Suppose that $100,000 is invested at 8% interest, compounded semi-annually.

a) Find a function for the amount of money after t years.

b) Graph the function.

c) Find the amount of money in the account at $t = 0$, 4, 8, and 10 yr.

d) When will the amount of money in the account reach $400,000?

Solution

a) Since $P = \$100{,}000$, $i = 8\% = 0.08$, and $n = 2$, we can substitute these values and write the following function:

$$A(t) = 100{,}000\left(1 + \frac{0.08}{2}\right)^{2 \cdot t} = 100{,}000(1.04)^{2t}.$$

b) For the graph, we use the viewing window [0, 20, 0, 500,000] because of the large numbers and the fact that negative time values and amounts of money have no meaning in this application.

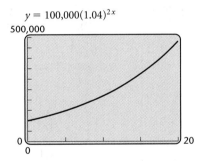

$y = 100{,}000(1.04)^{2x}$

c) We can compute function values using function notation on the home screen of a grapher. (See the window at left.) We can also calculate the values directly on a grapher by substituting in the expression for $A(t)$:

$$A(0) = 100{,}000(1.04)^{2 \cdot 0} = 100{,}000;$$
$$A(4) = 100{,}000(1.04)^{2 \cdot 4} \approx 136{,}856.91;$$
$$A(8) = 100{,}000(1.04)^{2 \cdot 8} \approx 187{,}298.12;$$
$$A(10) = 100{,}000(1.04)^{2 \cdot 10} \approx 219{,}112.31.$$

d) To find the amount of time it takes for the account to grow to $400,000, we set

$$100{,}000(1.04)^{2t} = 400{,}000$$

and solve. One way we can do this is by graphing the equations

$$y_1 = 100{,}000(1.04)^{2x} \quad \text{and} \quad y_2 = 400{,}000.$$

$y = 100{,}000(1.04)^{2x}$

Y₁(0)	
	100000
Y₁(4)	
	136856.905
Y₁(8)	
	187298.1246

| INTERSECT METHOD |
| REVIEW SECTION 2.1. |

Then we can use the Intersect method to estimate the first coordinate of the point of intersection. (See Fig. 1 below.)

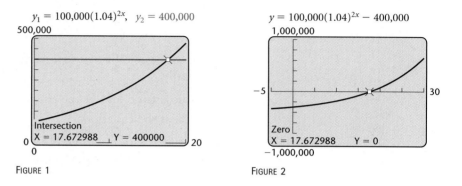

$y_1 = 100,000(1.04)^{2x}, \quad y_2 = 400,000$

FIGURE 1

$y = 100,000(1.04)^{2x} - 400,000$

FIGURE 2

| ZERO METHOD |
| REVIEW SECTION 2.1. |

We can also use the Zero method to estimate the zero of the function $y = 100,000(1.04)^{2x} - 400,000$. (See Fig. 2 above.)

Regardless of the method we use, we see that the account grows to $400,000 after about 17.67 yr, or about 17 yr, 8 mo, and 2 days. ▬

Interactive Discovery

Take a sheet of $8\frac{1}{2}$-in. by 11-in. paper and cut it into two equal pieces. Then cut each of these in half. Next, cut each of the four pieces in half again. Continue this process.

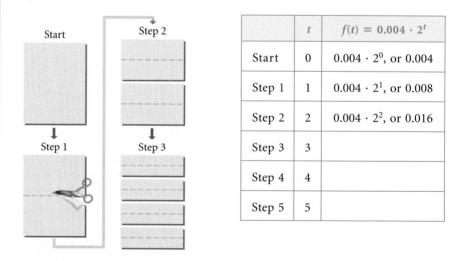

	t	$f(t) = 0.004 \cdot 2^t$
Start	0	$0.004 \cdot 2^0$, or 0.004
Step 1	1	$0.004 \cdot 2^1$, or 0.008
Step 2	2	$0.004 \cdot 2^2$, or 0.016
Step 3	3	
Step 4	4	
Step 5	5	

a) Place all the pieces in a stack and measure the thickness with a micrometer or other measuring device such as a dial caliper.

b) A piece of paper is typically 0.004 in. thick. Check the measurement in part (a) by completing the table shown here.

c) Graph the function $f(t) = 0.004(2)^t$.

d) Compute the thickness of the paper (in miles) after 25 steps.

The Number e

We now consider a very special number in mathematics. In 1741, Leonhard Euler named this number e. Though you may not have encountered it before, you will see here and in future mathematics that it has many important applications. To explain this number, we use the compound interest formula $A = P(1 + i/n)^{nt}$ in Example 5. Suppose that $1 is an initial investment at 100% interest for 1 yr. (No bank would pay this.) The formula above becomes a function A defined in terms of the number of compounding periods n:

$$A(n) = \left(1 + \frac{1}{n}\right)^n.$$

Let's visualize the function

$$A(n) = (1 + 1/n)^n$$

using a grapher and explore the values of $A(n)$ as $n \to \infty$.

Interactive Discovery

Graph the function $y = (1 + 1/x)^x$. Consider the graph for larger and larger values of x. Does this function have a horizontal asymptote? Explore with a grapher to determine the asymptote, if it exists. You might try to use the TABLE feature.

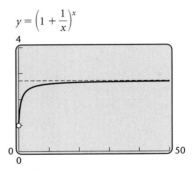

$$y = \left(1 + \frac{1}{x}\right)^x$$

Let's find some function values using the TABLE feature on a grapher.

n	$A(n) = \left(1 + \dfrac{1}{n}\right)^n$
1 (compounded annually)	$2.00
2 (compounded semiannually)	$2.25
3	$2.3704
4 (compounded quarterly)	$2.4414
5	$2.4883
100	$2.7048
365 (compounded daily)	$2.7146
8760 (compounded hourly)	$2.7181

As the values of n get larger and larger, the function values get closer and closer to the number Euler named e. Its decimal representation does not terminate or repeat; it is irrational.

$$e = 2.7182818284\ldots$$

EXAMPLE 6   Find each value of e^x, to four decimal places, using the $\boxed{e^x}$ key on a grapher.

a) e^3 **b)** $e^{-0.23}$

c) e^0 **d)** e^1

Solution

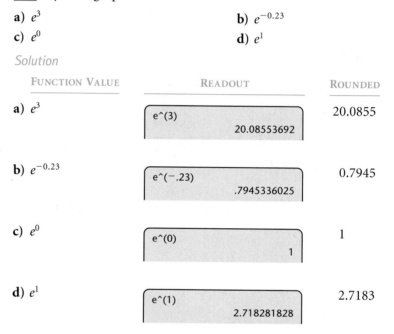

FUNCTION VALUE	READOUT	ROUNDED
a) e^3	e^(3) 20.08553692	20.0855
b) $e^{-0.23}$	e^(−.23) .7945336025	0.7945
c) e^0	e^(0) 1	1
d) e^1	e^(1) 2.718281828	2.7183

Graphs of Exponential Functions, Base e

We demonstrate ways in which to graph exponential functions.

EXAMPLE 7   Graph $f(x) = e^x$ and $g(x) = e^{-x}$.

Solution   We can compute points for each equation using the $\boxed{e^x}$ key on a grapher. Then we plot these points and draw the graphs of the functions.

$y_1 = e^x,\ \ y_2 = e^{-x}$

X	Y1	Y2
−3	.04979	20.086
−2	.13534	7.3891
−1	.36788	2.7183
0	1	1
1	2.7183	.36788
2	7.3891	.13534
3	20.086	.04979

X = 3

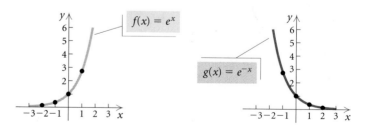

Note that the graph of g is a reflection of the graph of f across the y-axis.

EXAMPLE 8 Graph each of the following using a grapher. Briefly describe how each graph can be obtained from $y = e^x$.

a) $f(x) = e^{-0.5x}$ **b)** $f(x) = 1 - e^{-2x}$ **c)** $f(x) = e^{x+3}$

Solution

a) We note that the graph of this function is a horizontal stretching of the graph of $y = e^x$ followed by a reflection across the y-axis. (See Fig. 1.)

b) The graph is a horizontal shrinking of the graph of $y = e^x$, followed by a reflection across the y-axis, then across the x-axis, followed by a translation up 1 unit. (See Fig. 2.)

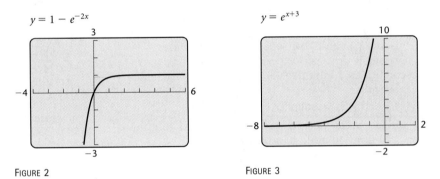

$y = e^{-0.5x}$

FIGURE 1

$y = 1 - e^{-2x}$

FIGURE 2

$y = e^{x+3}$

FIGURE 3

c) The graph is a translation of the graph of $y = e^x$ to the left 3 units. (See Fig. 3.)

Exercise Set 4.2

Find each of the following, to four decimal places, using a grapher.

1. e^4 **2.** e^{10}

3. $e^{-2.458}$ **4.** $\left(\dfrac{1}{e^3}\right)^2$

Make a hand-drawn graph of the function. Then check your work using a grapher, if possible.

5. $f(x) = 3^x$ **6.** $f(x) = 5^x$

7. $f(x) = 6^x$ **8.** $f(x) = 3^{-x}$

9. $f(x) = \left(\dfrac{1}{4}\right)^x$ **10.** $f(x) = \left(\dfrac{2}{3}\right)^x$

11. $x = 3^y$ **12.** $x = 4^y$

13. $x = \left(\dfrac{1}{2}\right)^y$ **14.** $x = \left(\dfrac{4}{3}\right)^y$

15. $y = \dfrac{1}{4}e^x$ **16.** $y = 2e^{-x}$

17. $f(x) = 1 - e^{-x}$ **18.** $f(x) = e^x - 2$

Graph the function using a grapher. Describe how each graph can be obtained from a basic exponential function.

19. $f(x) = 2^{x+1}$ **20.** $f(x) = 2^{x-1}$

21. $f(x) = 2^x - 3$ **22.** $f(x) = 2^x + 1$

23. $f(x) = 4 - 3^{-x}$ **24.** $f(x) = 2^{x-1} - 3$

25. $f(x) = \left(\dfrac{3}{2}\right)^{x-1}$ **26.** $f(x) = 3^{4-x}$

27. $f(x) = 2^{x+3} - 5$ **28.** $f(x) = -3^{x-2}$

29. $f(x) = e^{2x}$ **30.** $f(x) = e^{-0.2x}$

31. $y = e^{-x+1}$ **32.** $y = e^{2x} + 1$

33. $f(x) = 2(1 - e^{-x})$ **34.** $f(x) = 1 - e^{-0.01x}$

35. *Growth of AIDS.* The total number of AIDS cases reported in the United States is approximated by the exponential function

$$N(t) = 274,390(1.195)^t,$$

where $t = 0$ corresponds to 1992 (*Source*: U.S. Centers for Disease Control and Prevention, HIV/AID Surveillance Reports).

a) According to this function, how many AIDS cases had been reported by 1998?

b) Predict the number of AIDS cases that will be reported by 2001.

c) Graph the function.

d) Using a grapher and either the Intersect or the Zero method (see Example 5), estimate the length of time it will take for the number of AIDS cases to reach 3 million.

36. *Growth of Bacteria* Escherichia coli. The bacteria *Escherichia coli* are commonly found in the human bladder. Suppose that 3000 of the bacteria are present at time $t = 0$. Then under certain conditions, t minutes later, the number of bacteria present is

$$N(t) = 3000(2)^{t/20}.$$

a) How many bacteria will be present after 10 min? 20 min? 30 min? 40 min? 60 min?

b) Graph the function.

c) This bacteria can cause bladder infections in humans when the number of bacteria reaches 100,000,000. Using a grapher and either the Intersect or the Zero method (see Example 5), find the length of time it takes for a bladder infection to be possible.

37. *Recycling Aluminum Cans.* It is estimated that two thirds of all aluminum cans distributed will be recycled each year (*Source*: Alcoa Corporation). A beverage company distributes 350,000 cans. The number still in use after time t, in years, is given by the exponential function

$$N(t) = 350,000\left(\tfrac{2}{3}\right)^t.$$

a) How many cans are still in use after 0 yr? 1 yr? 4 yr? 10 yr?

b) Graph the function.

c) After how long will 2000 of the cans still be in use?

38. *Interest in a College Trust Fund.* Following the birth of a child, Juan deposits $10,000 in a college trust fund where interest is 6.4%, compounded semiannually.

a) Find a function for the amount in the account after t years.

b) Find the amount of money in the account at $t = 0, 4, 8, 10,$ and 18 yr.

c) Graph the function.

d) After how long will the account contain $100,000?

39. *Salvage Value.* A top-quality fax–copying machine is purchased for $5800. Its value each year is about 80% of the value of the preceding year. After t years, its value, in dollars, is given by the exponential function

$$V(t) = 5800(0.8)^t.$$

a) Graph the function.

b) Find the value of the machine after 0 yr, 1 yr, 2 yr, 5 yr, and 10 yr.

c) The company decides to replace the machine when its value has declined to $500. After how long will the machine be replaced?

40. *Cellular Phone Boom.* The number of people using cellular phones has grown exponentially in recent years. The total number of subscribers C, in millions, is given by

$$C(t) = 0.4703477987(1.523963441)^t,$$

where t is the number of years since 1985.

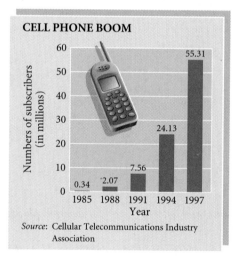

CELL PHONE BOOM

Source: Cellular Telecommunications Industry Association

a) Find the total number of subscribers in 1995, 1999, 2000, and 2004.

b) Graph the function with a scatterplot of the data.

c) After how long will the number of subscribers be 500,000,000?

41. *Timber Demand.* World demand for timber is increasing exponentially. The demand N, in billions of cubic feet, purchased is given by

$$N(t) = 46.6(1.018)^t,$$

where t is the number of years since 1981 (*Sources:* U.N. Food and Agricultural Organization, American Forest and Paper Association).

a) Graph the function.
b) Find the demand for timber in 2000 and 2010.
c) After how many years will the demand for timber be 93.4 billion cubic feet?

42. *Typing Speed.* Sarah is taking keyboarding at a community college. After she practices for t hours, her speed, in words per minute, is given by the function

$$S(t) = 200[1 - (0.86)^t].$$

a) Graph the function.
b) What is Sarah's speed after practicing for 10 hr? 20 hr? 40 hr? 100 hr?
c) How much time passes before Sarah's speed is 100 words per minute?
d) Does this graph have an asymptote? If so, what is it, and what is its significance to Sarah's learning?

43. *Advertising.* A company begins a radio advertising campaign in New York City to market a new CD-ROM video game. The percentage of the target market that buys a game is generally a function of the length of the advertising campaign. The estimated percentage is given by

$$f(t) = 100(1 - e^{-0.04t}),$$

where t is the number of days of the campaign.

a) Graph the function.
b) Find $f(25)$, the percentage of the target market that has bought the product after a 25-day advertising campaign.
c) After how long will 90% of the target market have bought the product?

44. *Growth of a Stock.* The value of a stock is given by the function

$$V(t) = 58(1 - e^{-1.1t}) + 20,$$

where V is the value of the stock after time t, in months.

a) Graph the function.
b) Find $V(1)$, $V(2)$, $V(4)$, $V(6)$, and $V(12)$.
c) After how long will the value of the stock be $75?

In Exercises 45–58, use a grapher to match the equation with one of figures (a)–(n), which follow.

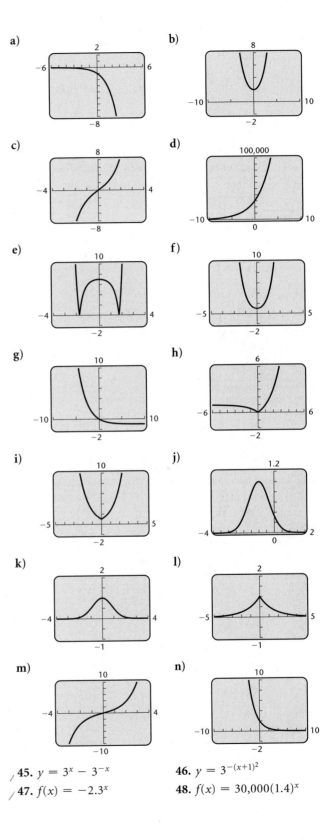

a)

b)

c)

d)

e)

f)

g)

h)

i)

j)

k)

l)

m)

n)

45. $y = 3^x - 3^{-x}$

46. $y = 3^{-(x+1)^2}$

47. $f(x) = -2.3^x$

48. $f(x) = 30,000(1.4)^x$

49. $y = 2^{-|x|}$

50. $y = 2^{-(x-1)}$

51. $f(x) = (0.58)^x - 1$

52. $y = 2^x + 2^{-x}$

53. $g(x) = e^{|x|}$

54. $f(x) = |2^x - 1|$

55. $y = 2^{-x^2}$

56. $y = |2^{x^2} - 8|$

57. $g(x) = \dfrac{e^x - e^{-x}}{2}$

58. $f(x) = \dfrac{e^x + e^{-x}}{2}$

Use a grapher to find the point(s) of intersection of the graphs of each of the following pairs of equations.

59. $y = |1 - 3^x|$,
     $y = 4 + 3^{-x^2}$

60. $y = 4^x + 4^{-x}$,
     $y = 8 - 2x - x^2$

61. $y = 2e^x - 3, \ y = \dfrac{e^x}{x}$

62. $y = \dfrac{1}{e^x + 1}, \ y = 0.3x + \dfrac{7}{9}$

Use a grapher. Solve graphically.

63. $5.3^x - 4.2^x = 1073$

64. $e^x = x^3$

65. $2^x > 1$

66. $3^x \le 1$

67. $2^x + 3^x = x^2 + x^3$

68. $31{,}245e^{-3x} = 523{,}467$

Discussion and Writing

69. Describe the differences between the graphs of $f(x) = x^3$ and $g(x) = 3^x$.

70. Suppose that \$10,000 is invested for 8 yr at 6.4%

interest, compounded annually. In what year will the most interest be earned? Why?

71. Graph each pair of equations using the same set of axes. Then compare the results in parts (a) and (b).

   a) $y = 3^x, \ x = 3^y$ **b)** $y = 1^x, \ x = 1^y$

Skill Maintenance

Find the zeros of the function using a grapher.

72. $g(x) = x^3 - 2x + 4$ **73.** $h(x) = x^4 - 5x + 3$

Solve.

74. $3x^2 - 6 = 5x$ **75.** $x^3 + 6x^2 - 16x = 0$

Synthesis

76. Which is larger, 7^π or π^7? 70^{80} or 80^{70}?

77. Graph $f(x) = x^{1/(x-1)}$ for $x > 0$. Use a grapher and the TABLE feature to identify the horizontal asymptote.

In Exercises 78 and 79:

a) *Graph using a grapher.*
b) *Approximate the zeros.*
c) *Approximate the relative maximum and minimum values. If your grapher has a MAX–MIN feature, use it.*

78. $f(x) = x^2 e^{-x}$ **79.** $f(x) = e^{-x^2}$

Logarithmic Functions and Graphs

4.3

- *Graph logarithmic functions.*
- *Convert between exponential and logarithmic equations.*
- *Find common and natural logarithms using a grapher.*

We now consider *logarithmic*, or *logarithm*, *functions*. These functions are inverses of exponential functions and have many applications.

Logarithmic Functions

Consider the function $f(x) = 2^x$ graphed at right. We see from the graph that this function passes the horizontal-line test and is therefore one-to-one. Thus, f has an inverse that is a function.

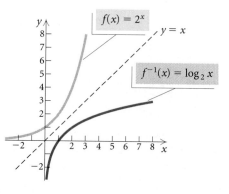

To find a formula for f^{-1} when $f(x) = 2^x$, we try to use the method of Section 4.1:

1. Replace $f(x)$ with y:
$$f(x) = 2^x$$
$$y = 2^x$$

2. Interchange x and y:
$$x = 2^y$$

3. Solve for y:
$$y = \text{the power to which we raise 2 to get } x$$

4. Replace y with $f^{-1}(x)$:
$$f^{-1}(x) = \text{the power to which we raise 2 to get } x.$$

Mathematicians have defined a new symbol to replace the words "the power to which we raise 2 to get x." That symbol is "$\log_2 x$," read "the logarithm, base 2, of x."

Logarithmic Function, Base 2

"$\log_2 x$," read "the logarithm, base 2, of x," means "the power to which we raise 2 to get x."

Thus if $f(x) = 2^x$, then $f^{-1}(x) = \log_2 x$. For example,

$$f^{-1}(8) = \log_2 8 = 3,$$

because

3 is the power to which we raise 2 to get 8.

Similarly, $\log_2 13$ is the power to which we raise 2 to get 13. As yet, we have no simpler way to say this other than

"$\log_2 13$ is the power to which we raise 2 to get 13."

Later, however, we will learn how to approximate this expression using a grapher.

For any exponential function $f(x) = a^x$, its inverse is called a **logarithmic function, base a.** The graph of the inverse can be obtained by reflecting the graph of $y = a^x$ across the line $y = x$, to obtain $x = a^y$. Then $x = a^y$ is equivalent to $y = \log_a x$. We read $\log_a x$ as "the logarithm, base a, of x."

The inverse of $f(x) = a^x$ is given by $f^{-1}(x) = \log_a x$.

Logarithmic Function, Base a

We define $y = \log_a x$ as that number y such that $x = a^y$, where $x > 0$ and a is a positive constant other than 1.

Let's look at the graphs of $f(x) = a^x$ and $f^{-1}(x) = \log_a x$ for $a > 1$ and $0 < a < 1$.

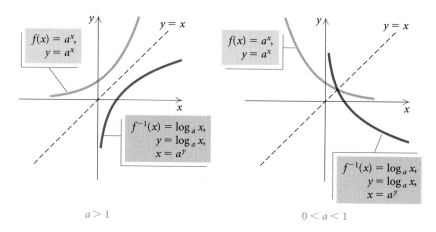

$f(x) = a^x,$
$y = a^x$

$y = x$

$f^{-1}(x) = \log_a x,$
$y = \log_a x,$
$x = a^y$

$a > 1$

$f(x) = a^x,$
$y = a^x$

$y = x$

$f^{-1}(x) = \log_a x,$
$y = \log_a x,$
$x = a^y$

$0 < a < 1$

CONNECTING THE CONCEPTS

COMPARING EXPONENTIAL AND LOGARITHMIC FUNCTIONS

Usually, we use a number that is greater than 1 for the logarithmic base. In the following table, we compare exponential and logarithmic functions with bases a greater than 1. Similar statements would be made for a, where $0 < a < 1$. It is helpful to visualize the differences by carefully observing the graphs.

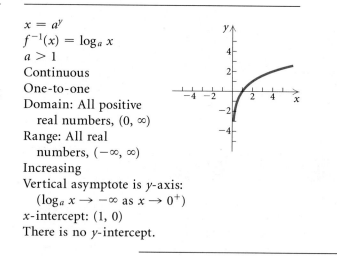

EXPONENTIAL FUNCTION

$y = a^x$
$f(x) = a^x$
$a > 1$
Continuous
One-to-one
Domain: All real
   numbers, $(-\infty, \infty)$
Range: All positive
   real numbers, $(0, \infty)$
Increasing
Horizontal asymptote is x-axis:
   $(a^x \to 0$ as $x \to -\infty)$
y-intercept: $(0, 1)$
There is no x-intercept.

LOGARITHMIC FUNCTION

$x = a^y$
$f^{-1}(x) = \log_a x$
$a > 1$
Continuous
One-to-one
Domain: All positive
   real numbers, $(0, \infty)$
Range: All real
   numbers, $(-\infty, \infty)$
Increasing
Vertical asymptote is y-axis:
   $(\log_a x \to -\infty$ as $x \to 0^+)$
x-intercept: $(1, 0)$
There is no y-intercept.

Converting Between Exponential and Logarithmic Equations

It is helpful in dealing with logarithmic functions to remember that a logarithm of a number is an *exponent*. It is the exponent y in $x = a^y$. You might think to yourself, "the logarithm, base a, of a number x is the power to which a must be raised to get x."

We are led to the following. (The symbol $\longleftrightarrow$ means that the two statements are equivalent; that is, when one is true, the other is true. The words "if and only if" can be used in place of $\longleftrightarrow$.)

$$\log_a x = y \longleftrightarrow x = a^y \qquad \text{A logarithm is an exponent!}$$

EXAMPLE 1   Convert each of the following to a logarithmic equation.

a) $16 = 2^x$ **b)** $10^{-3} = 0.001$ **c)** $e^t = 70$

Solution

The exponent is the logarithm.

a) $16 = 2^x \qquad \log_2 16 = x$

The base remains the same.

b) $10^{-3} = 0.001 \rightarrow \log_{10} 0.001 = -3$

c) $e^t = 70 \rightarrow \log_e 70 = t$

EXAMPLE 2   Convert each of the following to an exponential equation.

a) $\log_2 32 = 5$ **b)** $\log_a Q = 8$ **c)** $x = \log_t M$

Solution

The logarithm is the exponent.

a) $\log_2 32 = 5 \qquad 2^5 = 32$

The base remains the same.

b) $\log_a Q = 8 \rightarrow a^8 = Q$

c) $x = \log_t M \rightarrow t^x = M$

Finding Certain Logarithms

Let's use the definition of logarithms to find some logarithmic values.

EXAMPLE 3   Find each of the following logarithms.

a) $\log_{10} 10{,}000$ **b)** $\log_{10} 0.01$
c) $\log_2 8$ **d)** $\log_9 3$
e) $\log_6 1$ **f)** $\log_8 8$

Solution

RESULT	REASON
a) $\log_{10} 10,000 = 4$	Think of the meaning of $\log_{10} 10,000$. It is the exponent to which we raise 10 to get 10,000. That exponent is 4.
b) $\log_{10} 0.01 = -2$	We have $0.01 = \dfrac{1}{100} = \dfrac{1}{10^2} = 10^{-2}$. The exponent to which we raise 10 to get 0.01 is -2.
c) $\log_2 8 = 3$	$8 = 2^3$. The exponent to which we raise 2 to get 8 is 3.
d) $\log_9 3 = \frac{1}{2}$	$3 = \sqrt{9} = 9^{1/2}$. The exponent to which we raise 9 to get 3 is $\frac{1}{2}$.
e) $\log_6 1 = 0$	$1 = 6^0$. The exponent to which we raise 6 to get 1 is 0.
f) $\log_8 8 = 1$	$8 = 8^1$. The exponent to which we raise 8 to get 8 is 1.

Examples 3(e) and 3(f) illustrate two important properties of logarithms. The property $\log_a 1 = 0$ follows from the fact that $a^0 = 1$. Thus, $\log_5 1 = 0$, $\log_{10} 1 = 0$, and so on. The property $\log_a a = 1$ follows from the fact that $a^1 = a$. Thus, $\log_5 5 = 1$, $\log_{10} 10 = 1$, and so on.

$$\log_a 1 = 0 \quad \text{and} \quad \log_a a = 1, \quad \text{for any logarithmic base } a.$$

Finding Logarithms on a Grapher

Before calculators became so widely available, base-10, or **common logarithms,** were used extensively to simplify complicated calculations. In fact, that is why logarithms were invented. The abbreviation **log**, with no base written, is used to represent the common logarithms, or base-10 logarithms. Thus,

$$\log 29 \quad \text{means} \quad \log_{10} 29.$$

Let's compare log 29 with log 10 and log 100:

$$\left.\begin{array}{l} \log 10 = \log_{10} 10 = 1 \\ \quad\quad \log 29 = ? \\ \log 100 = \log_{10} 100 = 2 \end{array}\right\}$$ Since 29 is between 10 and 100, it seems reasonable that log 29 is between 1 and 2.

On a grapher, the scientific key for common logarithms is generally marked ⌐ LOG ⌐. Using that key, we find that

$$\log 29 \approx 1.462397998 \approx 1.4624$$

rounded to four decimal places. Since $1 < 1.4624 < 2$, our answer seems reasonable. This also tells us that $10^{1.4624} \approx 29$.

Interactive Discovery

Find log 1000 and log 10,000 without using a grapher. Between what two whole numbers is log 9874? Now approximate log 9874, rounded to four decimal places, on a grapher.

EXAMPLE 4 Find each of the following common logarithms on a grapher set in REAL mode. Round to four decimal places.

a) log 645,778
b) log 0.0000239
c) log (−3)

Solution

FUNCTION VALUE	READOUT	ROUNDED
a) log 645,778	log(645778) 5.810083246	5.8101
b) log 0.0000239	log(0.0000239) −4.621602099	−4.6216
c) log (−3)	ERR:NONREAL ANS *	Does not exist

A check for part (a) is shown below. Since 5.810083246 is the power to which we raise 10 to get 645,778, we can check by finding $10^{5.810083246}$. In part (c), log (−3) does not exist as a real number because there is no real-number power to which we can raise 10 to get −3. The number 10 raised to any real-number power is positive. The common logarithm of a negative number does not exist as a real number.

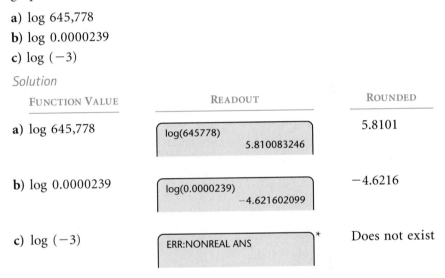

```
log(645778)
              5.810083246
10^(Ans)
                    645778
```

*If the grapher is set in $a + bi$ mode, the readout is $.4771212547 + 1.364376354i$.

Natural Logarithms

Logarithms, base e, are called **natural logarithms.** The abbreviation "ln" is generally used for natural logarithms. Thus,

$$\ln 53 \quad \text{means} \quad \log_e 53.$$

On a grapher, the scientific key for natural logarithms is generally marked $\boxed{\text{LN}}$. Using that key, we find that

$$\ln 53 \approx 3.970291914$$
$$\approx 3.9703$$

rounded to four decimal places.

EXAMPLE 5   Find each of the following natural logarithms on a grapher set in REAL mode. Round to four decimal places.

a) $\ln 645{,}778$ b) $\ln 0.0000239$

c) $\ln(-5)$ d) $\ln e$

e) $\ln 1$

Solution

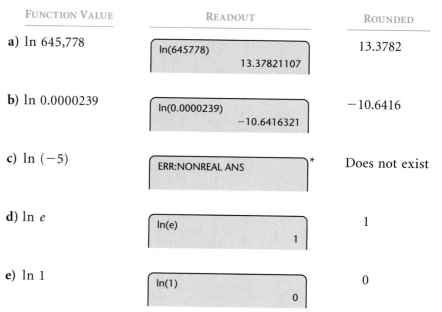

FUNCTION VALUE	READOUT	ROUNDED
a) $\ln 645{,}778$	ln(645778) 13.37821107	13.3782
b) $\ln 0.0000239$	ln(0.0000239) −10.6416321	−10.6416
c) $\ln(-5)$	ERR:NONREAL ANS *	Does not exist
d) $\ln e$	ln(e) 1	1
e) $\ln 1$	ln(1) 0	0

ln(0.0000239)
          −10.6416321
e^(Ans)
          2.39E−5

Note that $\ln e = \log_e e = 1$ and $\ln 1 = \log_e 1 = 0$. A check for part (b) is shown at left.

---

*If the grapher is set in $a + bi$ mode, the readout is $1.609437912 + 3.141592654i$.

Changing Logarithmic Bases

Most graphers give the values of both common logarithms and natural logarithms. To find a logarithm with a base other than 10 or e, we can use the following conversion formula.

The Change-of-Base Formula

For any logarithmic bases a and b, and any positive number M,

$$\log_b M = \frac{\log_a M}{\log_a b}.$$

We will prove this result in the next section.

EXAMPLE 6   Find $\log_5 8$ using common logarithms.

Solution   First, we let $a = 10$, $b = 5$, and $M = 8$. Then we substitute into the change-of-base formula:

$$\log_5 8 = \frac{\log_{10} 8}{\log_{10} 5} \qquad \text{Substituting}$$

$$\approx 1.2920. \qquad \text{Using a grapher}$$

The check is shown in Fig. 1.

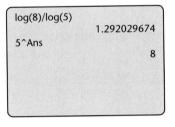

FIGURE 1

We can also use base e for a conversion.

EXAMPLE 7   Find $\log_5 8$ using natural logarithms.

Solution   Substituting e for a, 5 for b, and 8 for M, we have

$$\log_5 8 = \frac{\log_e 8}{\log_e 5}$$

$$= \frac{\ln 8}{\ln 5} \approx 1.2920.$$

The check is shown in Fig. 2.

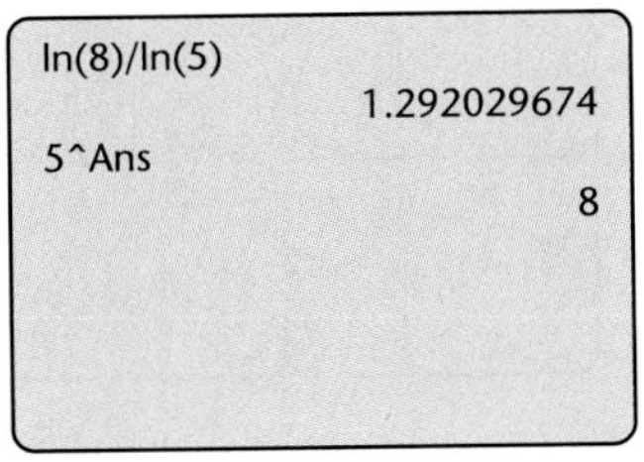

FIGURE 2

Graphs of Logarithmic Functions

We demonstrate several ways to graph logarithmic functions.

EXAMPLE 8   Graph: $y = f(x) = \log_5 x$.

Solution

Method 1. The equation $y = \log_5 x$ is equivalent to $x = 5^y$. We can find ordered pairs that are solutions by choosing values for y and computing the x-values. We then plot points, remembering that x still is the first coordinate.

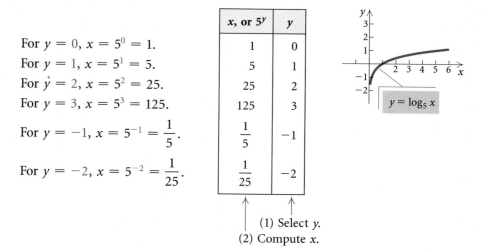

For $y = 0$, $x = 5^0 = 1$.
For $y = 1$, $x = 5^1 = 5$.
For $y = 2$, $x = 5^2 = 25$.
For $y = 3$, $x = 5^3 = 125$.
For $y = -1$, $x = 5^{-1} = \dfrac{1}{5}$.
For $y = -2$, $x = 5^{-2} = \dfrac{1}{25}$.

x, or 5^y	y
1	0
5	1
25	2
125	3
$\dfrac{1}{5}$	-1
$\dfrac{1}{25}$	-2

(1) Select y.
(2) Compute x.

Method 2. To use a grapher, we must first change the base. Here we change from base 5 to base e:

$$y = \log_5 x = \frac{\ln x}{\ln 5}. \qquad \text{Using } \log_b M = \frac{\log_a M}{\log_a b}$$

The graph is as shown in Fig. 3 below.

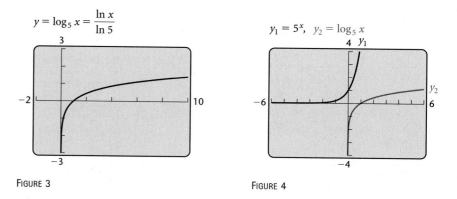

FIGURE 3

FIGURE 4

Method 3. Some graphers have a feature that graphs inverses automatically. If we begin with $y_1 = 5^x$, the graphs of both y_1 and its inverse $y_2 = \log_5 x$ will be drawn. (See Fig. 4 above.)

EXAMPLE 9   Graph: $g(x) = \ln x$.

Solution   To graph $y = g(x) = \ln x$, we first write its equivalent exponential equation, $x = e^y$. Next, we select values for y and use a grapher to find the corresponding values of e^y. We then plot points and draw the curve.

x, or e^y	y
1	0
2.7	1
7.4	2
20.1	3
0.1	-2
0.4	-1

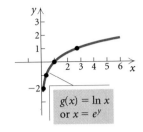

$g(x) = \ln x$
or $x = e^y$

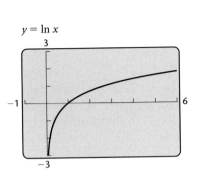

$y = \ln x$

We could also use the ⎡LN⎤ key on a grapher to find function values. The easiest way is to graph the function $y = \ln x$ using a grapher. (See the window at left.)

Recall that the graph of $f(x) = \log_a x$, for any base a, has the x-intercept $(1, 0)$. The domain is the set of positive real numbers. The range is the set of all real numbers.

EXAMPLE 10   Graph each of the following using a grapher. Describe how each graph can be obtained from $y = \ln x$. Give the domain of each function and discuss the vertical asymptotes.

a) $f(x) = \ln (x + 3)$
b) $f(x) = 3 - \frac{1}{2} \ln x$
c) $f(x) = |\ln (x - 1)|$

Solution
a) The graph is a shift of the graph of $y = \ln x$ left 3 units (see Fig. 5). The domain is the set of all real numbers greater than -3, $(-3, \infty)$. The line $x = -3$ is the vertical asymptote.

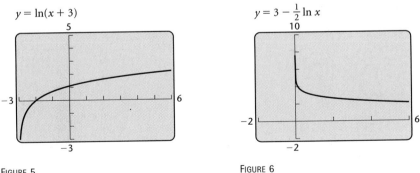

FIGURE 5

$y = \ln(x + 3)$

$y = 3 - \frac{1}{2} \ln x$

FIGURE 6

b) The graph is a shrinking of the graph of $y = \ln x$ in the y-direction, followed by a reflection across the x-axis, and then a translation up 3 units (see Fig. 6). The domain is the set of all positive real numbers, $(0, \infty)$. The y-axis is the vertical asymptote.

$y = |\ln(x - 1)|$

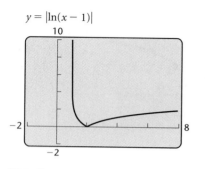

FIGURE 7

c) The graph is a translation of the graph of $y = \ln x$, 1 unit to the right. Then the absolute value has the effect of reflecting negative outputs across the x-axis (see Fig. 7). The domain is the set of all real numbers greater than 1, $(1, \infty)$. The line $x = 1$ is the vertical asymptote.

Applications

EXAMPLE 11 *Walking Speed.* In a study by psychologists Bornstein and Bornstein, it was found that the average walking speed w, in feet per second, of a person living in a city of population P, in thousands, is given by the function

$$w(P) = 0.37 \ln P + 0.05$$

(*Source: International Journal of Psychology*).

a) The population of San Diego, California, is 1,171,000. Find the average walking speed of people living in San Diego.

b) Graph the function.

c) A sociologist computes the average walking speed in a city to be approximately 2.0 ft/sec. Use this information to estimate the population of the city.

Solution

a) We substitute 1171 for P, since P is in thousands:

$$w(1171) = 0.37 \ln 1171 + 0.05 \qquad \text{Substituting}$$
$$\approx 2.7 \text{ ft/sec.} \qquad \text{Finding the natural logarithm and simplifying}$$

The average walking speed of people living in San Diego is about 2.7 ft/sec.

b) We graph with a viewing window of [0, 600, 0, 4] because inputs are very large and outputs are very small by comparison.

$y = 0.37 \ln x + 0.05$

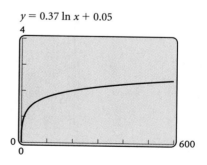

c) To find the population for which the average walking speed is 2.0 ft/sec, we set

$$2.0 = 0.37 \ln P + 0.05$$

and solve.

$y_1 = 0.37 \ln x + 0.05$, $y_2 = 2$

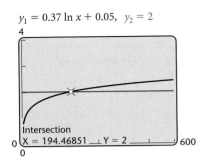

We will use the Intersect method. We graph the equations $y_1 = 0.37 \ln x + 0.05$ and $y_2 = 2$ and use the INTERSECT feature to approximate the point of intersection (see the window at left). We see that in a city in which the average walking speed is 2.0 ft/sec, the population is about 194.5 thousand, or 194,500.

EXAMPLE 12 *Earthquake Magnitude.* The magnitude R, measured on the Richter scale, of an earthquake of intensity I is defined as

$$R = \log \frac{I}{I_0},$$

where I_0 is a minimum intensity used for comparison. We can think of I_0 as a threshold intensity that is the weakest earthquake that can be recorded on a seismograph. If one earthquake is 10 times as intense as another, its magnitude on the Richter scale is 1 greater than that of the other. If one earthquake is 100 times as intense as another, its magnitude on the Richter scale is 2 higher, and so on. Thus an earthquake whose magnitude is 7 on the Richter scale is 10 times as intense as an earthquake whose magnitude is 6. Earthquakes can be interpreted as multiples of the minimum intensity I_0.

The earthquake in western Turkey on August 17, 1999, had an intensity of $10^{7.4} \cdot I_0$. What was its magnitude on the Richter scale?

Solution   We substitute into the formula:

$$R = \log \frac{I}{I_0} = \log \frac{10^{7.4} I_0}{I_0}$$
$$= \log 10^{7.4} = 7.4.$$

The magnitude of the earthquake was 7.4 on the Richter scale.

Exercise Set 4.3

Make a hand-drawn graph of each of the following. Then check your work using a grapher.

1. $y = \log_3 x$ **2.** $y = \log_4 x$

3. $f(x) = \log x$ **4.** $f(x) = \ln x$

Find each of the following. Do not use a grapher.

5. $\log_2 16$ **6.** $\log_3 9$

7. $\log_5 125$ **8.** $\log_2 64$

9. $\log 0.001$ **10.** $\log 100$

11. $\log_2 \frac{1}{4}$ **12.** $\log_8 2$

13. $\ln 1$ **14.** $\ln e$

15. $\log 10$ **16.** $\log 1$

Convert to a logarithmic equation.

17. $10^3 = 1000$ **18.** $5^{-3} = \frac{1}{125}$

19. $8^{1/3} = 2$ **20.** $10^{0.3010} = 2$

21. $e^3 = t$ **22.** $Q^t = x$

23. $e^2 = 7.3891$ **24.** $e^{-1} = 0.3679$

25. $p^k = 3$ **26.** $e^{-t} = 4000$

Convert to an exponential equation.

27. $\log_5 5 = 1$ **28.** $t = \log_4 7$

29. $\log 0.01 = -2$ **30.** $\log 7 = 0.845$

31. $\ln 30 = 3.4012$ **32.** $\ln 0.38 = -0.9676$

33. $\log_a M = -x$

34. $\log_t Q = k$

35. $\log_a T^3 = x$

36. $\ln W^5 = t$

Find each of the following using a grapher. Round to four decimal places.

37. log 3   **38.** log 8

39. log 532   **40.** log 93,100

41. log 0.57   **42.** log 0.082

43. log (-2)   **44.** ln 50

45. ln 2   **46.** ln (-4)

47. ln 809.3   **48.** ln 0.00037

49. ln (-1.32)   **50.** ln 0

Find the logarithm using the change-of-base formula.

51. $\log_4 100$ **52.** $\log_3 20$

53. $\log_{100} 0.3$ **54.** $\log_\pi 100$

55. $\log_{200} 50$ **56.** $\log_{5.3} 1700$

For each of the following functions, briefly describe how the graph can be obtained from a basic logarithmic function. Then graph the function using a grapher. Give the domain of the function and discuss the vertical asymptotes.

57. $f(x) = \log_2 (x + 3)$ **58.** $f(x) = \log_3 (x - 2)$

59. $y = \log_3 x - 1$ **60.** $y = 3 + \log_2 x$

61. $f(x) = 4 \ln x$ **62.** $f(x) = \frac{1}{2} \ln x$

63. $y = 2 - \ln x$ **64.** $y = \ln (x + 1)$

Graph the function and its inverse using the same set of axes. Use any method.

65. $f(x) = 3^x$, $f^{-1}(x) = \log_3 x$

66. $f(x) = \log_4 x$, $f^{-1}(x) = 4^x$

67. $f(x) = \log x$, $f^{-1}(x) = 10^x$

68. $f(x) = e^x$, $f^{-1}(x) = \ln x$

69. *Walking Speed.* Refer to Example 11. Various cities and their populations are given below. Find the average walking speed in each city.

a) Albuquerque, New Mexico: 419,681
b) Chicago, Illinois: 2,721,547
c) Pittsburgh, Pennsylvania: 350,363
d) Durham, North Carolina: 149,799
e) Green Bay, Wisconsin: 94,466

70. *Earthquake Magnitude.* Refer to Example 12. Various locations of earthquakes and their intensities are given below. What was the magnitude on the Richter scale?

a) Mexico City, 1978: $10^{7.85} \cdot I_0$
b) San Francisco, 1906: $10^{8.25} \cdot I_0$
c) Chile, 1960: $10^{9.6} \cdot I_0$
d) Italy, 1980: $10^{7.85} \cdot I_0$
e) San Francisco, 1989: $10^{6.9} \cdot I_0$

71. *Forgetting.*   Students in an accounting class took a final exam. They took equivalent forms of the exam in monthly intervals thereafter. The average score $S(t)$, in percent, after t months was found to be given by the function

$$S(t) = 78 - 15 \log (t + 1), \quad t \geq 0.$$

a) What was the average score when they initially took the test, $t = 0$?
b) What was the average score after 4 months? 24 months?
c) Graph the function.
d) After what time t was the average score 50?

72. *pH of Substances in Chemistry.*   In chemistry, the pH of a substance is defined as

$$\text{pH} = -\log [H^+],$$

where H^+ is the hydrogen ion concentration, in moles per liter. Find the pH of each substance.

Litmus paper is used to test pH.

SUBSTANCE	HYDROGEN ION CONCENTRATION
a) Pineapple juice	1.6×10^{-4}
b) Hair rinse	0.0013
c) Mouthwash	6.3×10^{-7}
d) Eggs	1.6×10^{-8}
e) Tomatoes	6.3×10^{-5}

73. Find the hydrogen ion concentration of each substance, given the pH (see Exercise 72). Express the answer in scientific notation.

SUBSTANCE	pH
a) Tap water	7
b) Rainwater	5.4
c) Orange juice	3.2
d) Wine	4.8

74. *Advertising.*   A model for advertising response is given by the function

$$N(a) = 1000 + 200 \ln a, \quad a \geq 1,$$

where $N(a)$ is the number of units sold and a is the amount spent on advertising, in thousands of dollars.

a) How many units were sold after spending \$1000 ($a = 1$) on advertising?
b) How many units were sold after spending \$5000?
c) Graph the function.
d) How much would have to be spent in order to sell 2000 units?

75. *Loudness of Sound.*   The **loudness L**, in bels (after Alexander Graham Bell), of a sound of intensity I is defined to be

$$L = \log \frac{I}{I_0},$$

where I_0 is the minimum intensity detectable by the human ear (such as the tick of a watch at 20 ft under quiet conditions). If a sound is 10 times as intense as another, its loudness is 1 bel greater than that of the other. If a sound is 100 times as intense as another, its loudness is 2 bels greater, and so on. The bel is a large unit, so a subunit, the **decibel**, is generally used. For L, in decibels, the formula is

$$L = 10 \log \frac{I}{I_0}.$$

Find the loudness, in decibels, of each sound with the given intensity.

SOUND	INTENSITY
a) Library	$2510 \cdot I_0$
b) Dishwasher	$2{,}500{,}000 \cdot I_0$
c) Conversational speech	$10^6 \cdot I_0$
d) Heavy truck	$10^9 \cdot I_0$

Discussion and Writing

76. Explain how the graph of $f(x) = \ln x$ can be used to obtain the graph of $g(x) = e^{x-2}$.

77. If $\log b < 0$, what can you say about b?

Skill Maintenance

Use synthetic division to find the function values.

78. $f(x) = x^4 - 2x^3 + x - 6$; find $f(-1)$

79. $g(x) = x^3 - 6x^2 + 3x + 10$; find $f(-5)$

Find a polynomial function of degree 3 with the given numbers as zeros.

80. $4i, -4i, 1$

81. $\sqrt{7}, -\sqrt{7}, 0$

Synthesis

Simplify.

82. $\dfrac{\log_3 64}{\log_3 16}$

83. $\dfrac{\log_5 8}{\log_5 2}$

Find the domain of the function.

84. $f(x) = \log_4 x^2$

85. $f(x) = \log_5 x^3$

86. $f(x) = \log (3x - 4)$

87. $f(x) = \ln |x|$

Solve.

88. $\log_2 (x - 3) \geq 4$

89. $\log_2 (2x + 5) < 0$

90. Using a grapher, find the point(s) of intersection of the graphs.

$$y = 4 \ln x, \qquad y = \frac{4}{e^x + 1}$$

In Exercises 91–94, match the equation with one of figures (a)–(d), which follow. If needed, use a grapher.

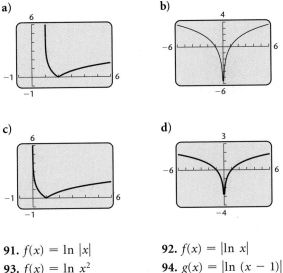

91. $f(x) = \ln |x|$

92. $f(x) = |\ln x|$

93. $f(x) = \ln x^2$

94. $g(x) = |\ln (x - 1)|$

For Exercises 95–98:
a) *Graph the function.*
b) *Estimate the zeros.*
c) *Estimate the relative maximum and the minimum values.*

95. $f(x) = x \ln x$

96. $f(x) = x^2 \ln x$

97. $f(x) = \dfrac{\ln x}{x^2}$

98. $f(x) = e^{-x} \ln x$

Properties of Logarithmic Functions

4.4

- Convert from logarithms of products, powers, and quotients to expressions in terms of individual logarithms, and conversely.
- Simplify expressions of the type $\log_a a^x$ and $a^{\log_a x}$.

We now establish some properties of logarithmic functions. These properties are based on corresponding rules for exponents.

Logarithms of Products

Interactive Discovery

Graph each of the following functions. Use the graphs to discover a pair of equivalent equations.

$$y_1 = \log (4x^2 \cdot x^3), \qquad y_2 = \log (4x^2) + \log (x^3),$$
$$y_3 = \log (4x^2 + x^3), \qquad y_4 = [\log (4x^2)][\log (x^3)]$$

Test the result by using the TABLE feature to complete the following tables and compare the y-values.

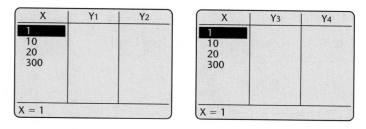

The first property of logarithms corresponds to the rule of exponents: $a^m \cdot a^n = a^{m+n}$.

The Product Rule

For any positive numbers M and N and any logarithmic base a,

$$\log_a MN = \log_a M + \log_a N.$$

(The logarithm of a product is the sum of the logarithms of the factors.)

EXAMPLE 1 Express as a sum of logarithms: $\log_3 (9 \cdot 27)$.

Solution We have

$$\log_3 (9 \cdot 27) = \log_3 9 + \log_3 27. \qquad \textbf{Using the product rule}$$

As a check, note that

$$\log_3 (9 \cdot 27) = \log_3 243 = 5$$

and $\log_3 9 + \log_3 27 = 2 + 3 = 5.$

EXAMPLE 2 Express as a single logarithm: $\log_2 p^3 + \log_2 q$.

Solution We have

$$\log_2 p^3 + \log_2 q = \log_2 (p^3 q).$$

A PROOF OF THE PRODUCT RULE: Let $\log_a M = x$ and $\log_a N = y$. Converting to exponential equations, we have $a^x = M$ and $a^y = N$. Then

$$MN = a^x \cdot a^y = a^{x+y}.$$

Converting back to a logarithmic equation, we get

$$\log_a MN = x + y.$$

Remembering what x and y represent, we know it follows that

$$\log_a MN = \log_a M + \log_a N.$$

Logarithms of Powers

Interactive Discovery

Graph each of the following functions. Use the graphs to discover a pair of equivalent equations.

$$y_1 = \log x^5, \qquad y_2 = \log x + \log 5,$$
$$y_3 = 5 \log x, \qquad y_4 = \log (x + 5)$$

Test the result by using the TABLE feature to complete the following tables and compare the y-values.

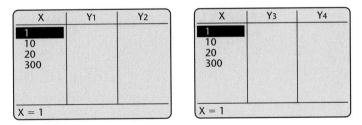

The second property of logarithms corresponds to the rule of exponents: $(a^m)^n = a^{mn}$.

The Power Rule

For any positive number M, any logarithmic base a, and any real number p,

$$\log_a M^p = p \log_a M.$$

(The logarithm of a power of M is the exponent times the logarithm of M.)

EXAMPLE 3   Express each of the following as a product.

a) $\log_a 11^{-3}$

b) $\log_a \sqrt[4]{7}$

Solution

a) $\log_a 11^{-3} = -3\log_a 11$ Using the power rule

b) $\log_a \sqrt[4]{7} = \log_a 7^{1/4}$ Writing exponential notation

  $= \frac{1}{4}\log_a 7$ Using the power rule

A PROOF OF THE POWER RULE: Let $x = \log_a M$. The equivalent exponential equation is $a^x = M$. Raising both sides to the power p, we obtain

$$(a^x)^p = M^p, \quad \text{or} \quad a^{xp} = M^p.$$

Converting back to a logarithmic equation, we get

$$\log_a M^p = xp.$$

But $x = \log_a M$, so substituting gives us

$$\log_a M^p = (\log_a M)p$$
$$= p\log_a M.$$

Logarithms of Quotients

Interactive Discovery

Graph each of the following functions. Use the graphs to discover a pair of equivalent equations.

$$y_1 = \log\left(\frac{4x^5}{x^2}\right),$$

$$y_2 = \log(4x^5) + \log(x^2),$$

$$y_3 = \log(4x^5) - \log(x^2),$$

$$y_4 = \log(4x + 5x^2)$$

Test the result by using the TABLE feature to complete the following tables and compare the y-values.

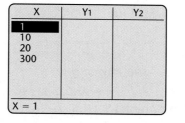

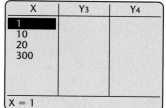

The third property of logarithms corresponds to the rule of exponents: $a^m/a^n = a^{m-n}$.

The Quotient Rule

For any positive numbers M and N, and any logarithmic base a,

$$\log_a \frac{M}{N} = \log_a M - \log_a N.$$

(The logarithm of a quotient is the logarithm of the numerator minus the logarithm of the denominator.)

EXAMPLE 4   Express as a difference of logarithms: $\log_t \dfrac{8}{w}$.

Solution

$$\log_t \frac{8}{w} = \log_t 8 - \log_t w \qquad \text{Using the quotient rule}$$

EXAMPLE 5   Express as a single logarithm: $\log_b 64 - \log_b 16$.

Solution

$$\log_b 64 - \log_b 16 = \log_b \frac{64}{16} = \log_b 4$$

A PROOF OF THE QUOTIENT RULE: The proof follows from both the product and the power rules:

$$\log_a \frac{M}{N} = \log_a MN^{-1}$$

$$= \log_a M + \log_a N^{-1} \qquad \text{Using the product rule}$$

$$= \log_a M + (-1) \log_a N \qquad \text{Using the power rule}$$

$$= \log_a M - \log_a N.$$

Note the following.

Common Errors

$\log_a MN \neq (\log_a M)(\log_a N)$	The logarithm of a product is *not* the product of the logarithms.
$\log_a (M + N) \neq \log_a M + \log_a N$	The logarithm of a sum is *not* the sum of the logarithms.
$\log_a \dfrac{M}{N} \neq \dfrac{\log_a M}{\log_a N}$	The logarithm of a quotient is *not* the quotient of the logarithms.

Using the Properties Together

EXAMPLE 6   Express each of the following in terms of sums and differences of logarithms.

a) $\log_a \dfrac{x^2 y^5}{z^4}$
b) $\log_a \sqrt[3]{\dfrac{a^2 b}{c^5}}$
c) $\log_b \dfrac{a y^5}{m^3 n^4}$

Solution

a) $\log_a \dfrac{x^2 y^5}{z^4} = \log_a (x^2 y^5) - \log_a z^4$ Using the quotient rule

$\qquad\qquad = \log_a x^2 + \log_a y^5 - \log_a z^4$ Using the product rule

$\qquad\qquad = 2 \log_a x + 5 \log_a y - 4 \log_a z$ Using the power rule

b) $\log_a \sqrt[3]{\dfrac{a^2 b}{c^5}} = \log_a \left(\dfrac{a^2 b}{c^5} \right)^{1/3}$ Writing exponential notation

$\qquad\qquad = \dfrac{1}{3} \log_a \dfrac{a^2 b}{c^5}$ Using the power rule

$\qquad\qquad = \dfrac{1}{3} (\log_a a^2 b - \log_a c^5)$ Using the quotient rule

$\qquad\qquad = \dfrac{1}{3} (2 \log_a a + \log_a b - 5 \log_a c)$ Using the product and power rules. The parentheses are important.

$\qquad\qquad = \dfrac{1}{3} (2 + \log_a b - 5 \log_a c)$ $\log_a a = 1$

$\qquad\qquad = \dfrac{2}{3} + \dfrac{1}{3} \log_a b - \dfrac{5}{3} \log_a c$ Multiplying to remove parentheses

c) $\log_b \dfrac{a y^5}{m^3 n^4} = \log_b a y^5 - \log_b m^3 n^4$ Using the quotient rule

$\qquad\qquad = (\log_b a + \log_b y^5) - (\log_b m^3 + \log_b n^4)$ Using the product rule

$\qquad\qquad = \log_b a + \log_b y^5 - \log_b m^3 - \log_b n^4$ Removing parentheses

$\qquad\qquad = \log_b a + 5 \log_b y - 3 \log_b m - 4 \log_b n$ Using the power rule

EXAMPLE 7   Express as a single logarithm:

$$5 \log_b x - \log_b y + \dfrac{1}{4} \log_b z.$$

Solution

$$5 \log_b x - \log_b y + \frac{1}{4} \log_b z = \log_b x^5 - \log_b y + \log_b z^{1/4}$$

Using the power rule

$$= \log_b \frac{x^5}{y} + \log_b z^{1/4}$$

Using the quotient rule

$$= \log_b \frac{x^5 z^{1/4}}{y}, \text{ or } \log_b \frac{x^5 \sqrt[4]{z}}{y}$$

Using the product rule

EXAMPLE 8 Given that $\log_a 2 \approx 0.301$ and $\log_a 3 \approx 0.477$, find each of the following.

a) $\log_a 6$ 　　　　　　**b)** $\log_a \dfrac{2}{3}$ 　　　　　　**c)** $\log_a 81$

d) $\log_a \sqrt{a}$ 　　　　　**e)** $\log_a 5$ 　　　　　　**f)** $\dfrac{\log_a 3}{\log_a 2}$

Solution

a) $\log_a 6 = \log_a (2 \cdot 3) = \log_a 2 + \log_a 3$ 　　Using the product rule

$$\approx 0.301 + 0.477$$

$$\approx 0.778$$

b) $\log_a \frac{2}{3} = \log_a 2 - \log_a 3$ 　　Using the quotient rule

$$\approx 0.301 - 0.477 \approx -0.176$$

c) $\log_a 81 = \log_a 3^4 = 4 \log_a 3$ 　　Using the power rule

$$\approx 4(0.477) \approx 1.908$$

d) $\log_a \sqrt{a} = \log_a a^{1/2} = \frac{1}{2} \log_a a$ 　　Using the power rule

$$= \frac{1}{2} \cdot 1 = \frac{1}{2}$$

e) $\log_a 5$ *cannot be found using these properties and the given information.*

$$(\log_a 5 \neq \log_a 2 + \log_a 3)$$

f) $\dfrac{\log_a 3}{\log_a 2} \approx \dfrac{0.477}{0.301} \approx 1.585$ 　　We simply divided, not using any of the properties.

   In Example 8, a is actually 10 so we have common logarithms. Check as many results as possible using a grapher.

Simplifying Expressions of the Type $\log_a a^x$ and $a^{\log_a x}$

We have two final properties to consider. The first follows from the product rule: Since $\log_a a^x = x \log_a a = x \cdot 1 = x$, we have $\log_a a^x = x$. This property also follows from the definition of a logarithm: x is the power to which we raise a in order to get a^x.

The Logarithm of a Base to a Power

For any base a and any real number x,

$$\log_a a^x = x.$$

(The logarithm, base a, of a to a power is the power.)

EXAMPLE 9   Simplify each of the following.

a) $\log_a a^8$ **b)** $\ln e^{-t}$ **c)** $\log 10^{3k}$

Solution

a) $\log_a a^8 = 8$ 8 is the power to which we raise a in order to get a^8.

b) $\ln e^{-t} = \log_e e^{-t} = -t$

c) $\log 10^{3k} = \log_{10} 10^{3k} = 3k$

    Let $M = \log_a x$. Then $a^M = x$. Substituting $\log_a x$ for M, we obtain $a^{\log_a x} = x$. This also follows from the definition of a logarithm: $\log_a x$ is the power to which a is raised in order to get x.

A Base to a Logarithmic Power

For any base a and any positive real number x,

$$a^{\log_a x} = x.$$

(The number a raised to the power $\log_a x$ is x.)

EXAMPLE 10   Simplify each of the following.

a) $4^{\log_4 k}$ **b)** $e^{\ln 5}$ **c)** $10^{\log 7t}$

Solution

a) $4^{\log_4 k} = k$

b) $e^{\ln 5} = e^{\log_e 5} = 5$

c) $10^{\log 7t} = 10^{\log_{10} 7t} = 7t$

A PROOF OF THE CHANGE-OF-BASE FORMULA: We close this section by proving the change-of-base formula and summarizing the properties of logarithms considered thus far in this chapter. In Section 4.3, we used the change-of-base formula,

$$\log_b M = \frac{\log_a M}{\log_a b},$$

to make base conversions in order to graph logarithmic functions on a grapher. Let $x = \log_b M$. Then

$$b^x = M \qquad \text{Definition of logarithm}$$
$$\log_a b^x = \log_a M \qquad \text{Taking the log on both sides}$$
$$x \log_a b = \log_a M \qquad \text{Using the power rule}$$
$$x = \frac{\log_a M}{\log_a b}, \qquad \text{Dividing by } \log_a b$$

so

$$x = \log_b M = \frac{\log_a M}{\log_a b}.$$

Following is a summary of the properties of logarithms.

Summary of the Properties of Logarithms

The Product Rule:	$\log_a MN = \log_a M + \log_a N$
The Power Rule:	$\log_a M^p = p \log_a M$
The Quotient Rule:	$\log_a \dfrac{M}{N} = \log_a M - \log_a N$
The Change-of-Base Formula:	$\log_b M = \dfrac{\log_a M}{\log_a b}$
Other Properties:	$\log_a a = 1, \qquad \log_a 1 = 0,$
	$\log_a a^x = x, \qquad a^{\log_a x} = x$

Exercise Set **4.4**

Express as a sum of logarithms.

1. $\log_3 (81 \cdot 27)$

2. $\log_2 (8 \cdot 64)$

3. $\log_5 (5 \cdot 125)$

4. $\log_4 (64 \cdot 32)$

5. $\log_t 8Y$

6. $\log_e Qx$

Express as a product.

7. $\log_b t^3$

8. $\log_a x^4$

9. $\log y^8$

10. $\ln y^5$

11. $\log_c K^{-6}$

12. $\log_b Q^{-8}$

Express as a difference of logarithms.

13. $\log_t \dfrac{M}{8}$

14. $\log_a \dfrac{76}{13}$

15. $\log_a \dfrac{x}{y}$

16. $\log_b \dfrac{3}{w}$

Express in terms of sums and differences of logarithms.

17. $\log_a 6xy^5z^4$

18. $\log_a x^3y^2z$

19. $\log_b \dfrac{p^2q^5}{m^4b^9}$

20. $\log_b \dfrac{x^2y}{b^3}$

21. $\log_a \sqrt{\dfrac{x^6}{p^5q^8}}$

22. $\log_c \sqrt[3]{\dfrac{y^3z^2}{x^4}}$

23. $\log_a \sqrt[4]{\dfrac{m^8n^{12}}{a^3b^5}}$

24. $\log_a \sqrt{\dfrac{a^6b^8}{a^2b^5}}$

Express as a single logarithm and, if possible, simplify.

25. $\log_a 75 + \log_a 2$

26. $\log 0.01 + \log 1000$

27. $\log 10,000 - \log 100$

28. $\ln 54 - \ln 6$

29. $\frac{1}{2} \log_a x + 4 \log_a y - 3 \log_a x$

30. $\frac{2}{5} \log_a x - \frac{1}{3} \log_a y$

31. $\ln x^2 - 2 \ln \sqrt{x}$

32. $\ln 2x + 3(\ln x - \ln y)$

33. $\ln (x^2 - 4) - \ln (x + 2)$

34. $\log_a \dfrac{a}{\sqrt{x}} - \log_a \sqrt{ax}$

35. $\ln x - 3[\ln (x - 5) + \ln (x + 5)]$

36. $\frac{2}{3}[\ln (x^2 - 9) - \ln (x + 3)] + \ln (x + y)$

37. $\frac{3}{2} \ln 4x^6 - \frac{4}{5} \ln 2y^{10}$

38. $120(\ln \sqrt[5]{x^3} + \ln \sqrt[3]{y^2} - \ln \sqrt[4]{16z^5})$

Given that $\log_b 3 = 1.0986$ *and* $\log_b 5 = 1.6094$, *find each of the following.*

39. $\log_b \frac{3}{5}$

40. $\log_b 15$

41. $\log_b \frac{1}{5}$

42. $\log_b \frac{5}{3}$

43. $\log_b \sqrt{b}$

44. $\log_b \sqrt{b^3}$

45. $\log_b 5b$

46. $\log_b 9$

47. $\log_b 75$

48. $\log_b \dfrac{1}{b}$

Simplify.

49. $\log_p p^3$

50. $\log_t t^{2713}$

51. $\log_e e^{|x-4|}$

52. $\log_q q^{\sqrt{3}}$

53. $3^{\log_3 4x}$

54. $5^{\log_5 (4x-3)}$

55. $10^{\log w}$

56. $e^{\ln x^3}$

57. $\ln e^{8t}$

58. $\log 10^{-k}$

Discussion and Writing

59. Given that $f(x) = a^x$ and $g(x) = \log_a x$, find $(f \circ g)(x)$ and $(g \circ f)(x)$. These results are alternative proofs of what properties of logarithms already proven in this section? Explain.

60. Explain the errors, if any, in the following:
$$\log_a ab^3 = (\log_a a)(\log_a b^3) = 3 \log_a b.$$

Skill Maintenance

Simplify.

61. $(1 - 4i)(7 + 6i)$

62. $\dfrac{2 - i}{3 + i}$

Find the x-intercepts and the zeros of the function.

63. $f(x) = 2x^2 - 13x - 7$

64. $h(x) = x^3 - 3x^2 + 3x - 1$

Synthesis

Solve for x. Do an algebraic solution.

65. $5^{\log_5 8} = 2x$

66. $\ln e^{3x-5} = -8$

Express as a single logarithm and, if possible, simplify.

67. $\log_a (x^2 + xy + y^2) + \log_a (x - y)$

68. $\log_a (a^{10} - b^{10}) - \log_a (a + b)$

Express as a sum or a difference of logarithms.

69. $\log_a \dfrac{x - y}{\sqrt{x^2 - y^2}}$

70. $\log_a \sqrt{9 - x^2}$

71. Given that $\log_a x = 2$, $\log_a y = 3$, and $\log_a z = 4$, find
$$\log_a \frac{\sqrt[4]{y^2 z^5}}{\sqrt[4]{x^3 z^{-2}}}.$$

Determine whether each of the following is true. Assume that a, x, M, and N are positive.

72. $\log_a M + \log_a N = \log_a (M + N)$

73. $\log_a M - \log_a N = \log_a \dfrac{M}{N}$

74. $\dfrac{\log_a M}{\log_a N} = \log_a M - \log_a N$

75. $\dfrac{\log_a M}{x} = \log_a M^{1/x}$

76. $\log_a x^3 = 3 \log_a x$

77. $\log_a 8x = \log_a x + \log_a 8$

78. $\log_N (MN)^x = x \log_N M + x$

Suppose that $\log_a x = 2$. *Find each of the following.*

79. $\log_a \left(\dfrac{1}{x}\right)$

80. $\log_{1/a} x$

81. Simplify:
$$\log_{10} 11 \cdot \log_{11} 12 \cdot \log_{12} 13 \cdots \log_{998} 999 \cdot \log_{999} 1000.$$

Prove each of the following for any base a and any positive number x.

82. $\log_a \left(\dfrac{1}{x}\right) = -\log_a x = \log_{1/a} x$

83. $\log_a \left(\dfrac{x + \sqrt{x^2 - 5}}{5}\right) = -\log_a (x - \sqrt{x^2 - 5})$

Solving Exponential and Logarithmic Equations

4.5

• *Solve exponential and logarithmic equations.*

Solving Exponential Equations

Equations with variables in the exponents, such as

$$3^x = 20 \quad \text{and} \quad 2^{5x} = 64,$$

are called **exponential equations.** We now consider solving exponential equations.

Sometimes, as is the case with the equation $2^{5x} = 64$, we can write each side as a power of the same number:

$$2^{5x} = 2^6.$$

We can then set the exponents equal and solve:

$$5x = 6$$
$$x = \tfrac{6}{5}$$
$$x = 1.2.$$

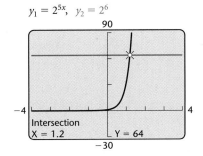

$y_1 = 2^{5x}, \ y_2 = 2^6$

We use the following property.

Base–Exponent Property

For any $a > 0$, $a \neq 1$,

$$a^x = a^y \longleftrightarrow x = y.$$

ONE–TO–ONE FUNCTIONS

REVIEW SECTION 4.1.

This property follows from the fact that for any $a > 0$, $a \neq 1$, $f(x) = a^x$ is a one-to-one function. If $a^x = a^y$, then $f(x) = f(y)$. Then since f is one-to-one, it follows that $x = y$. Conversely, if $x = y$, it follows that $a^x = a^y$, since we are raising a to the same power.

EXAMPLE 1   Solve: $2^{3x-7} = 32$.

Algebraic Solution

Note that $32 = 2^5$. Thus we can write each side as a power of the same number:

$$2^{3x-7} = 2^5.$$

Since the bases are the same number, 2, we can use the base–exponent property and set the exponents equal:

$$3x - 7 = 5$$
$$3x = 12$$
$$x = 4.$$

CHECK: $2^{3x-7} = 32$

$2^{3(4)-7}$? 32

2^{12-7}

2^5

32 | 32 TRUE

The solution is 4.

Graphical Solution

We will use the Intersect method. We graph

$$y_1 = 2^{3x-7} \quad \text{and} \quad y_2 = 32$$

to find the coordinates of the point of intersection.

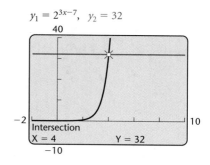

The first coordinate of this point is the solution of the equation $y_1 = y_2$, or $2^{3x-7} = 32$. The solution is 4.

We could also write the equation in the form $2^{3x-7} - 32 = 0$ and use the Zero method.

When it does not seem possible to write each side as a power of the same base, we can take the common or natural logarithm on each side and use the power rule for logarithms.

EXAMPLE 2   Solve: $3^x = 20$.

Algebraic Solution

We have

$$3^x = 20$$
$\log 3^x = \log 20$ Taking the common logarithm on both sides
$x \log 3 = \log 20$ Using the power rule
$x = \dfrac{\log 20}{\log 3}.$ Dividing by log 3

This is an exact answer. We cannot simplify further, but we can approximate using a grapher:

$$x = \frac{\log 20}{\log 3} \approx 2.7268.$$

We can check this by finding $3^{2.7268}$. The solution is about 2.7268.

Graphical Solution

We will use the Intersect method. We graph

$$y_1 = 3^x \quad \text{and} \quad y_2 = 20$$

to find the x-coordinate of the point of intersection.

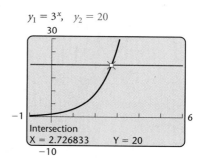

The solution is about 2.7268.

It will make our work easier if we take the natural logarithm when working with equations that have e as a base.

EXAMPLE 3   Solve: $e^{0.08t} = 2500$.

Algebraic Solution

We have

$$e^{0.08t} = 2500$$

$\ln e^{0.08t} = \ln 2500$ **Taking the natural logarithm on both sides**

$0.08t = \ln 2500$ **Finding the logarithm of a base to a power: $\log_a a^x = x$**

$$t = \frac{\ln 2500}{0.08}$$ **Dividing by 0.08**

$$\approx 97.8.$$

The solution is about 97.8.

Graphical Solution

Using the Intersect method, we graph the equations

$$y_1 = e^{0.08x} \quad \text{and} \quad y_2 = 2500$$

and determine the point of intersection.

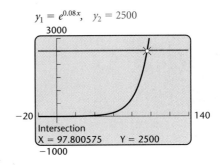

$y_1 = e^{0.08x}, \quad y_2 = 2500$

Intersection
X = 97.800575   Y = 2500

The solution is about 97.8.

EXAMPLE 4   Solve: $e^x + e^{-x} - 6 = 0$.

Algebraic Solution

In this case, we have more than one term with x in the exponent. To get a single expression with x in the exponent, we do the following:

$$e^x + e^{-x} - 6 = 0$$

$$e^x + \frac{1}{e^x} - 6 = 0 \qquad \text{Rewriting } e^{-x} \text{ with a positive exponent}$$

$$e^{2x} + 1 - 6e^x = 0. \qquad \text{Multiplying by } e^x \text{ on both sides}$$

This equation is reducible to quadratic with $u = e^x$:

$$u^2 - 6u + 1 = 0.$$

The coefficients of the reduced quadratic equation are $a = 1$, $b = -6$, and $c = 1$. Using the quadratic formula, we obtain

$$u = e^x = 3 \pm \sqrt{8}.$$

We now take the natural logarithm on both sides:

$$\ln e^x = \ln (3 \pm \sqrt{8})$$
$$x = \ln (3 \pm \sqrt{8}). \qquad \text{Using } \ln e^x = x$$

Approximating each of the solutions, we obtain 1.76 and -1.76.

Graphical Solution

Using the Zero method, we begin by graphing the function

$$y = e^x + e^{-x} - 6.$$

Then we find the zeros of the function.

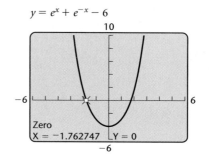

$y = e^x + e^{-x} - 6$

The leftmost zero is about -1.76. Using the ZERO feature one more time, we find that the other zero is about 1.76.

The solutions are about -1.76 and 1.76.

It is possible that when encountering an equation like the one in Example 4, you might not recognize that it could be solved in the algebraic manner shown. This points out the value of the graphical solution.

Solving Logarithmic Equations

Equations containing variables in logarithmic expressions, such as $\log_2 x = 4$ and $\log x + \log (x + 3) = 1$, are called **logarithmic equations.**

> To solve logarithmic equations algebraically, first try to obtain a single logarithmic expression on one side and then write an equivalent exponential equation.

EXAMPLE 5 Solve: $\log_3 x = -2$.

Algebraic Solution

We have

$\log_3 x = -2$

$3^{-2} = x$ Converting to an exponential equation

$\dfrac{1}{3^2} = x$

$\dfrac{1}{9} = x.$

CHECK: $\dfrac{\log_3 x = -2}{}$

$\log_3 \dfrac{1}{9} \ ? \ -2$

$\log_3 3^{-2}$

$-2 \ \bigg| \ -2$ TRUE

The solution is $\frac{1}{9}$.

Graphical Solution

We use the change-of-base formula and graph the equations

$$y_1 = \log_3 x = \frac{\ln x}{\ln 3} \quad \text{and} \quad y_2 = -2.$$

Then we use the Intersect method.

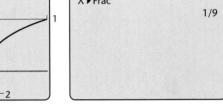

If the solution is a rational number, we usually can find fractional notation for the exact solution by using the FRAC feature from the MATH submenu of the MATH menu.

The solution is $\frac{1}{9}$.

EXAMPLE 6 Solve: $\log x + \log (x + 3) = 1$.

Algebraic Solution

In this case, we have common logarithms. Including the base 10's will help us understand the problem:

$\log_{10} x + \log_{10} (x + 3) = 1$

$\log_{10} [x(x + 3)] = 1$ Using the product rule to obtain a single logarithm

$x(x + 3) = 10^1$ Writing an equivalent exponential equation

$x^2 + 3x = 10$

$x^2 + 3x - 10 = 0$

$(x - 2)(x + 5) = 0$ Factoring

$x - 2 = 0 \quad or \quad x + 5 = 0$

$x = 2 \quad or \quad\quad x = -5.$

CHECK: For 2:

$$\log x + \log (x + 3) = 1$$

$$\log 2 + \log (2 + 3) \; ? \; 1$$
$$\log 2 + \log 5$$
$$\log 10$$
$$1 \quad \mid \quad 1 \quad \text{TRUE}$$

For -5:

$$\log x + \log (x + 3) = 1$$

$$\log (-5) + \log (-5 + 3) \; ? \; 1 \quad \text{FALSE}$$

The number -5 is not a solution because negative numbers do not have real-number logarithms. The solution is 2.

Graphical Solution

We can graph the equations

$$y_1 = \log x + \log (x + 3)$$

and

$$y_2 = 1$$

and use the Intersect method. The first coordinate of the point of intersection is the solution of the equation.

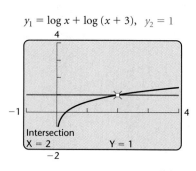

$y_1 = \log x + \log (x + 3), \quad y_2 = 1$

We could also graph the function

$$y = \log x + \log (x + 3) - 1$$

and use the Zero method. The zero of the function is the solution of the equation.

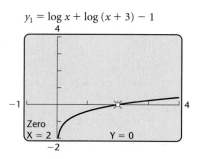

$y_1 = \log x + \log (x + 3) - 1$

With either method, we see that the solution is 2. Note that the graphical solution gives only the one *true* solution.

EXAMPLE 7   Solve: $\log_3 (2x - 1) - \log_3 (x - 4) = 2$.

Algebraic Solution

We have

$\log_3 (2x - 1) - \log_3 (x - 4) = 2$

$\log_3 \dfrac{2x - 1}{x - 4} = 2$ Using the quotient rule

$\dfrac{2x - 1}{x - 4} = 3^2$ Writing an equivalent exponential equation

$\dfrac{2x - 1}{x - 4} = 9$

$2x - 1 = 9(x - 4)$ Multiplying by the LCD, $x - 4$

$2x - 1 = 9x - 36$

$35 = 7x$

$5 = x.$

CHECK: $\dfrac{\log_3 (2x - 1) - \log_3 (x - 4) = 2}{\rule{6cm}{0.4pt}}$

$\log_3 (2 \cdot 5 - 1) - \log_3 (5 - 4)\ ?\ 2$

$\log_3 9 - \log_3 1\ \bigg|$

$2 - 0\ \bigg|$

$2\ \bigg|\ 2$ TRUE

The solution is 5.

Graphical Solution

Here we use the Intersect method to find the solutions of the equation. We use the change-of-base formula and graph the equations

$y_1 = \dfrac{\ln (2x - 1)}{\ln 3} - \dfrac{\ln (x - 4)}{\ln 3}$

and

$y_2 = 2.$

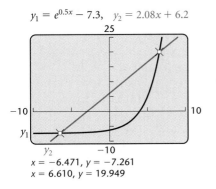

The solution is 5.

Sometimes we encounter equations for which an algebraic solution seems difficult or impossible.

EXAMPLE 8   Solve: $e^{0.5x} - 7.3 = 2.08x + 6.2$.

$y_1 = e^{0.5x} - 7.3,\quad y_2 = 2.08x + 6.2$

$x = -6.471, y = -7.261$
$x = 6.610, y = 19.949$

Graphical Solution

We graph the equations

$y_1 = e^{0.5x} - 7.3$ and $y_2 = 2.08x + 6.2$

and use the Intersect method. (See the window at left.)
    We can also consider the equation

$y = e^{0.5x} - 7.3 - 2.08x - 6.2,$ or $y = e^{0.5x} - 2.08x - 13.5,$

and use the Zero method. The approximate solutions are -6.471 and 6.610.

Exercise Set 4.5

Solve the exponential equation algebraically. Then check using a grapher.

1. $3^x = 81$

2. $2^x = 32$

3. $2^{2x} = 8$

4. $3^{7x} = 27$

5. $2^x = 33$

6. $2^x = 40$

7. $5^{4x-7} = 125$

8. $4^{3x-5} = 16$

9. $27 = 3^{5x} \cdot 9^{x^2}$

10. $3^{x^2+4x} = \frac{1}{27}$

11. $84^x = 70$

12. $28^x = 10^{-3x}$

13. $e^t = 1000$

14. $e^{-t} = 0.04$

15. $e^{-0.03t} = 0.08$

16. $1000e^{0.09t} = 5000$

17. $3^x = 2^{x-1}$

18. $5^{x+2} = 4^{1-x}$

19. $(3.9)^x = 48$

20. $250 - (1.87)^x = 0$

21. $e^x + e^{-x} = 5$

22. $e^x - 6e^{-x} = 1$

23. $\dfrac{e^x + e^{-x}}{e^x - e^{-x}} = 3$

24. $\dfrac{5^x - 5^{-x}}{5^x + 5^{-x}} = 8$

Solve the logarithmic equation algebraically. Then check using a grapher.

25. $\log_5 x = 4$

26. $\log_2 x = -3$

27. $\log x = -4$

28. $\log x = 1$

29. $\ln x = 1$

30. $\ln x = -2$

31. $\log_2 (10 + 3x) = 5$

32. $\log_5 (8 - 7x) = 3$

33. $\log x + \log (x - 9) = 1$

34. $\log_2 (x + 1) + \log_2 (x - 1) = 3$

35. $\log_8 (x + 1) - \log_8 x = 2$

36. $\log x - \log (x + 3) = -1$

37. $\log_4 (x + 3) + \log_4 (x - 3) = 2$

38. $\ln (x + 1) - \ln x = \ln 4$

39. $\log (2x + 1) - \log (x - 2) = 1$

40. $\log_5 (x + 4) + \log_5 (x - 4) = 2$

Use only a grapher. Find approximate solutions of the equation or approximate the point(s) of intersection of the pair of equations.

41. $e^{7.2x} = 14.009$

42. $0.082e^{0.05x} = 0.034$

43. $xe^{3x} - 1 = 3$

44. $5e^{5x} + 10 = 3x + 40$

45. $4 \ln (x + 3.4) = 2.5$

46. $\ln x^2 = -x^2$

47. $\log_8 x + \log_8 (x + 2) = 2$

48. $\log_3 x + 7 = 4 - \log_5 x$

49. $\log_5 (x + 7) - \log_5 (2x - 3) = 1$

50. $y = \ln 3x, \quad y = 3x - 8$

51. $2.3x + 3.8y = 12.4, \quad y = 1.1 \ln (x - 2.05)$

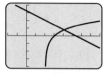

52. $y = 2.3 \ln (x + 10.7), \quad y = 10e^{-0.07x^2}$

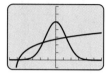

53. $y = 2.3 \ln (x + 10.7), \quad y = 10e^{-0.007x^2}$

Discussion and Writing

54. In Example 3, we took the natural logarithm on both sides of the equation. What would have happened had we taken the common logarithm? Explain which approach seems better to you and why.

55. Explain how Exercises 29 and 30 could be solved using the graph of $f(x) = \ln x$.

Skill Maintenance

In Exercises 56–59:

a) *Find the vertex.*

b) *Find the line of symmetry.*

c) *Determine whether there is a maximum or minimum value and find that value.*

56. $f(x) = -x^2 + 6x - 8$

57. $g(x) = x^2 - 6$

58. $H(x) = 3x^2 - 12x + 16$

59. $G(x) = -2x^2 - 4x - 7$

Synthesis

Solve using any method.

60. $\ln (\ln x) = 2$

61. $\ln (\log x) = 0$

62. $\ln \sqrt[4]{x} = \sqrt{\ln x}$

63. $\sqrt{\ln x} = \ln \sqrt{x}$

64. $\log_3 (\log_4 x) = 0$

65. $(\log_3 x)^2 - \log_3 x^2 = 3$

66. $(\log x)^2 - \log x^2 = 3$

67. $\ln x^2 = (\ln x)^2$

68. $e^{2x} - 9 \cdot e^x + 14 = 0$

69. $5^{2x} - 3 \cdot 5^x + 2 = 0$

70. $x \left(\ln \frac{1}{6} \right) = \ln 6$

71. $\log_3 |x| = 2$

72. $x^{\log x} = \dfrac{x^3}{100}$

73. $\ln x^{\ln x} = 4$

74. $\dfrac{(e^{3x+1})^2}{e^4} = e^{10x}$

75. $\dfrac{\sqrt{(e^{2x} \cdot e^{-5x})^{-4}}}{e^x \div e^{-x}} = e^7$

76. $|\log_a x| = \log_a |x|, \ a > 1$

77. $\ln (x - 2) > 4$

78. $e^x < \dfrac{4}{5}$

79. $|\log_5 x| + 3 \log_5 |x| = 4$

80. $|2^{x^2} - 8| = 3$

81. Given that $a = \log_8 225$ and $b = \log_2 15$, express a as a function of b.

82. Given that $a = (\log_{125} 5)^{\log_5 125}$, find the value of $\log_3 a$.

83. Given that
$$\log_2 [\log_3 (\log_4 x)] = \log_3 [\log_2 (\log_4 y)]$$
$$= \log_4 [\log_3 (\log_2 z)]$$
$$= 0,$$
find $x + y + z$.

84. Given that $f(x) = e^x - e^{-x}$, find $f^{-1}(x)$ if it exists.

Applications and Models: Growth and Decay

4.6

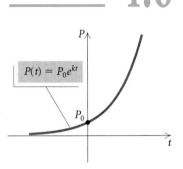

$P(t) = P_0 e^{kt}$

• *Solve applied problems involving exponential growth and decay.*
• *Find models involving exponential and logarithmic functions.*

Exponential and logarithmic functions with base e are rich in applications to many fields such as business, science, psychology, and sociology. In this section, we consider some basic applications and then use curve fitting to do others.

Population Growth

The function

$$P(t) = P_0 e^{kt}, \quad k > 0$$

is a model of many kinds of population growth, whether it be a population of people, bacteria, cellular phones, or money. In this function, P_0 is the population at time 0, P is the population after time t, and k is called the **exponential growth rate.** The graph of such an equation is shown at left.

EXAMPLE 1 *Population Growth of India.* In 1998, the population of India was about 984 million and the exponential growth rate was 1.8% per year (*Source: Statistical Abstract of the United States*).

a) Find the exponential growth function.

b) Graph the exponential growth function.

c) What will the population be in 2005?

d) After how long will the population be double what it was in 1998?

Solution

a) At $t = 0$ (1998), the population was 984 million. We substitute 984 for P_0 and 1.8%, or 0.018, for k to obtain the exponential growth function

$$P(t) = 984e^{0.018t}.$$

b) Using a grapher, we obtain the graph of the exponential growth function, shown at left.

c) In 2005, $t = 7$; that is, 7 yr have passed since 1998. To find the population in 2005, we substitute 7 for t:

$$P(7) = 984e^{0.018(7)} = 984e^{0.126} \approx 1116.$$

We can also use the VALUE feature on a grapher. (See the window at left.) In 2005, the population of India will be about 1116 million, or 1,116,000,000.

d) We are looking for the time T for which $P(T) = 2 \cdot 984$, or 1968. The number T is called the **doubling time.** To find T, we solve the equation

$$1968 = 984e^{0.018T}.$$

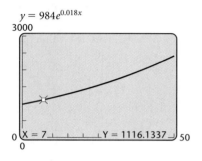

$y = 984e^{0.018x}$

Algebraic Solution

We have

$1968 = 984e^{0.018T}$	**Substituting 1968 for $P(T)$**
$2 = e^{0.018T}$	**Dividing by 984**
$\ln 2 = \ln e^{0.018T}$	
$\ln 2 = 0.018T$	$\ln e^x = x$
$\dfrac{\ln 2}{0.018} = T$	**Dividing by 0.018**
$39 \approx T.$	

The population of India will be double what it was in 1998 about 39 yr after 1998.

Graphical Solution

Using the Intersect method, we graph the equations

$$y_1 = 984e^{0.018x} \quad \text{and} \quad y_2 = 1968$$

and find the first coordinate of their point of intersection.

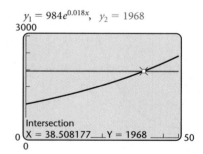

$y_1 = 984e^{0.018x}, \quad y_2 = 1968$

The solution is about 39 yr, so the population of India will be double that of 1998 about 39 yr after 1998.

Interest Compounded Continuously

Here we explore the mathematics behind the concept of **interest compounded continuously.** Suppose that an amount P_0 is invested in a savings account at interest rate k *compounded continuously.* The amount $P(t)$ in the account after t years is given by the exponential function

$$P(t) = P_0 e^{kt}.$$

EXAMPLE 2 *Interest Compounded Continuously.* Suppose that $2000 is invested at interest rate k, compounded continuously, and grows to $2983.65 in 5 yr.

a) What is the interest rate?

b) Find the exponential growth function.

c) What will the balance be after 10 yr?

d) After how long will the $2000 have doubled?

Solution

a) At $t = 0$, $P(0) = P_0 = \$2000$. Thus the exponential growth function is

$$P(t) = 2000 e^{kt}.$$

We know that $P(5) = \$2983.65$. We substitute and solve for k, as shown below.

Algebraic Solution

We have

$$2983.65 = 2000 e^{k(5)} \qquad \text{Substituting 2983.65 for } P(t) \text{ and 5 for } t$$

$$2983.65 = 2000 e^{5k}$$

$$\frac{2983.65}{2000} = e^{5k} \qquad \text{Dividing by 2000}$$

$$\ln \frac{2983.65}{2000} = \ln e^{5k} \qquad \text{Taking the natural logarithm}$$

$$\ln \frac{2983.65}{2000} = 5k \qquad \text{Using } \ln e^x = x$$

$$\frac{\ln \dfrac{2983.65}{2000}}{5} = k \qquad \text{Dividing by 5}$$

$$0.08 \approx k.$$

The interest rate is about 0.08, or 8%.

Graphical Solution

We use the Intersect method. We graph the equations

$$y_1 = 2000 e^{5x} \quad \text{and} \quad y_2 = 2983.65$$

and use the INTERSECT feature to approximate the first coordinate of the point of intersection.

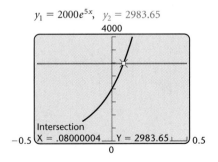

$y_1 = 2000 e^{5x}, \ y_2 = 2983.65$

The solution is about 0.08, or 8%.

b) The exponential growth function is

$$P(t) = 2000e^{0.08t}.$$

c) The balance after 10 yr is

$$P(10) = 2000e^{0.08(10)}$$
$$= 2000e^{0.8}$$
$$\approx \$4451.08.$$

d) We solve using both an algebraic method and a graphical method.

Algebraic Solution

To find the doubling time T, we set $P(T) = 2 \cdot P_0 = \$4000$ and solve for T:

$$4000 = 2000e^{0.08T}$$

$$2 = e^{0.08T} \qquad \text{Dividing by 2000}$$

$$\ln 2 = \ln e^{0.08T} \qquad \text{Taking the natural logarithm}$$

$$\ln 2 = 0.08T \qquad \ln e^x = x$$

$$\frac{\ln 2}{0.08} = T \qquad \text{Dividing by 0.08}$$

$$8.7 \approx T.$$

Thus the original investment of \$2000 will double in about 8.7 yr.

Graphical Solution

The money will have doubled when $P(t) = 2 \cdot P_0 = 4000$, or when $2000e^{0.08t} = 4000$. We use the Zero method. We graph the equation

$$y = 2000e^{0.08x} - 4000$$

and find the zero of the function. The zero of the function is the solution of the equation.

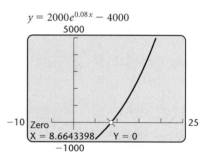
$y = 2000e^{0.08x} - 4000$

The solution is about 8.7, so the original investment of \$2000 will double in about 8.7 yr.

We can find a general expression relating the growth rate k and the doubling time T by solving the following equation:

$$2P_0 = P_0e^{kT} \qquad \text{Substituting } 2P_0 \text{ for } P \text{ and } T \text{ for } t$$

$$2 = e^{kT} \qquad \text{Dividing by } P_0$$

$$\ln 2 = \ln e^{kt} \qquad \text{Taking the natural logarithm}$$

$$\ln 2 = kT \qquad \text{Using } \ln e^x = x$$

$$\frac{\ln 2}{k} = T.$$

Growth Rate and Doubling Time
The **growth rate k** and the **doubling time T** are related by

$$kT = \ln 2, \quad \text{or} \quad k = \frac{\ln 2}{T}, \quad \text{or} \quad T = \frac{\ln 2}{k}.$$

Note that the relationship between k and T does not depend on P_0.

EXAMPLE 3 *World Population Growth.* The population of the world is now doubling every 54.6 yr. What is the exponential growth rate?

Solution We have

$$k = \frac{\ln 2}{T} = \frac{\ln 2}{54.6} \approx 1.3\%.$$

The growth rate of the world population is about 1.3% per year.

Models of Limited Growth

The model $P(t) = P_0 e^{kt}$ has many applications involving unlimited population growth. However, in some populations, there can be factors that prevent a population from exceeding some limiting value—perhaps a limitation on food, living space, or other natural resources. One model of such growth is

$$P(t) = \frac{a}{1 + be^{-kt}},$$

which is called a **logistic function.** This function increases toward a *limiting value a* as $t \to \infty$.

EXAMPLE 4 *Limited Population Growth.* A ship carrying 1000 passengers has the misfortune to be shipwrecked on a small island from which the passengers are never rescued. The natural resources of the island limit the population to 5780. The population gets closer and closer to this limiting value, but never reaches it. The population of the island after time t, in years, is given by the logistic equation

$$P(t) = \frac{5780}{1 + 4.78e^{-0.4t}}.$$

a) Graph the function.

b) Find the population after 0, 1, 2, 5, 10, and 20 yr.

Solution

a) We use a grapher to graph the function. The graph is the S-shaped curve shown at left. Note that this function increases toward a limiting value of 5780.

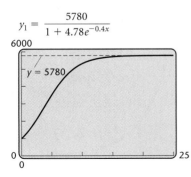

$$y_1 = \frac{5780}{1 + 4.78e^{-0.4x}}$$

b) We can use the TABLE feature on a grapher set in ASK mode to find the function values. (See the window on the left below.) The VALUE feature can also be used. (See the window on the right below.)

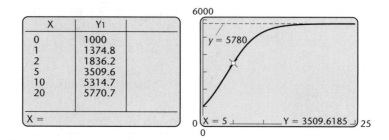

Thus the population will be about 1000 after 0 yr, 1375 after 1 yr, 1836 after 2 yr, 3510 after 5 yr, 5315 after 10 yr, and 5771 after 20 yr.

Another model of limited growth is provided by the function

$$P(t) = L(1 - e^{-kt}), \quad k > 0,$$

which is shown graphed at right. This function also increases toward a limiting value L, as $x \to \infty$.

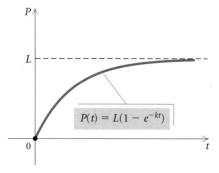

How can scientists determine that an animal bone has lost 30% of its carbon-14? The assumption is that the percentage of carbon-14 in the atmosphere and in living plants and animals is the same. When a plant or an animal dies, the amount of carbon-14 decays exponentially. The scientist burns the animal bone and uses a Geiger counter to determine the percentage of the smoke that is carbon-14. It is the amount that this varies from the percentage in the atmosphere that tells how much carbon-14 has been lost.

The process of carbon-14 dating was developed by the American chemist Willard E. Libby in 1952. It is known that the radioactivity in a living plant is 16 disintegrations per gram per minute. Since the half-life of carbon-14 is 5750 years, an object with an activity of 8 disintegrations per gram per minute is 5750 years old, one with an activity of 4 disintegrations per gram per minute is 11,500 years old, and so on. Carbon-14 dating can be used to measure the age of objects up to 40,000 years old. Beyond such an age, it is too difficult to measure the radioactivity and some other method would have to be used.

Carbon-14 was indeed used to find the age of the Dead Sea Scrolls. It was used recently to refute the authenticity of the Shroud of Turin, presumed to have covered the body of Christ.

Exponential Decay

The function

$$P(t) = P_0 e^{-kt}, \quad k > 0$$

is an effective model of the decline, or decay, of a population. An example is the decay of a radioactive substance. In this case, P_0 is the amount of the substance at time $t = 0$, and $P(t)$ is the amount of the substance left after time t, where k is a positive constant that depends on the situation. The constant k is called the **decay rate.**

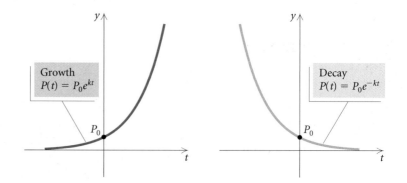

The **half-life** of bismuth is 5 days. This means that half of an amount of bismuth will cease to be radioactive in 5 days. The effect of half-life T is shown in the graph below for nonnegative inputs. The exponential function gets close to 0, but never reaches 0, as t gets very large. Thus, according to an exponential decay model, a radioactive substance never completely decays.

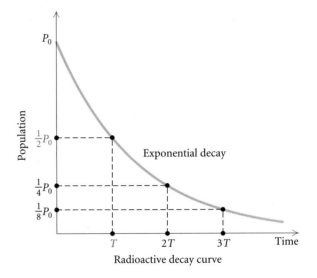

Radioactive decay curve

EXAMPLE 5 *Carbon Dating.* The radioactive element carbon-14 has a half-life of 5750 yr. The percentage of carbon-14 present in the remains of organic matter can be used to determine the age of that organic matter. Archaeologists discovered that the linen wrapping from one of the Dead Sea Scrolls had lost 22.3% of its carbon-14 at the time it was found. How old was the linen wrapping?

Solution   We first find k. When $t = 5750$ (the half-life), $P(t)$ will be half of P_0. We substitute $\frac{1}{2}P_0$ for $P(t)$ and 5750 for t and solve for k. Then

$$\tfrac{1}{2}P_0 = P_0 e^{-k(5750)}$$

or
$$\tfrac{1}{2} = e^{-5750k}.$$

We take the natural logarithm on both sides:

$$\ln \tfrac{1}{2} = \ln e^{-5750k}$$
$$\ln 0.5 = -5750k.$$

Then

$$k = \frac{\ln 0.5}{-5750} \approx 0.00012.$$

We could also solve the equation $\frac{1}{2} = e^{-5750k}$ using a grapher. Now we have the function

$$P(t) = P_0 e^{-0.00012t}.$$

In 1947, a Bedouin youth looking for a stray goat climbed into a cave at Kirbet Qumran on the shores of the Dead Sea near Jericho and came upon earthenware jars containing an incalculable treasure of ancient manuscripts. Shown here are fragments of those so-called Dead Sea Scrolls, a portion of some 600 or so texts found so far and which concern the Jewish books of the Bible. Officials date them before 70 A.D., making them the oldest Biblical manuscripts by 1000 years.

(This equation can be used for any subsequent carbon-dating problem.) If the linen wrapping has lost 22.3% of its carbon-14 from an initial amount P_0, then 77.7%P_0 is the amount present. To find the age t of the wrapping, we solve the following equation for t:

$$77.7\%P_0 = P_0 e^{-0.00012t} \quad \textbf{Substituting 77.7\%}P_0 \textbf{ for } P$$
$$0.777 = e^{-0.00012t}$$
$$\ln 0.777 = \ln e^{-0.00012t}$$
$$\ln 0.777 = -0.00012t \quad \textbf{ln } e^x = x$$
$$\frac{\ln 0.777}{-0.00012} = t$$
$$2103 \approx t.$$

Thus the linen wrapping on the Dead Sea Scrolls was about 2103 yr old when it was found.

Exponential and Logarithmic Curve Fitting

We have added several new functions that can be considered when we fit curves to data. Let's review some of them.

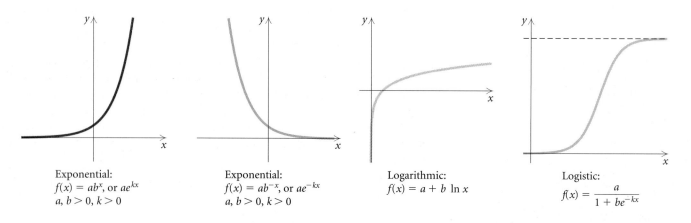

Exponential:
$f(x) = ab^x$, or ae^{kx}
$a, b > 0, k > 0$

Exponential:
$f(x) = ab^{-x}$, or ae^{-kx}
$a, b > 0, k > 0$

Logarithmic:
$f(x) = a + b \ln x$

Logistic:
$f(x) = \dfrac{a}{1 + be^{-kx}}$

Now, when we analyze a set of data for curve fitting, these models can be considered as well as polynomial functions (such as linear, quadratic, cubic, and quartic functions) and rational functions.

EXAMPLE 6   *Credit Card Volume.*   The total credit card volume for Visa, MasterCard, American Express, and Discover has increased dramatically in recent years, as shown in the table on the next page.

a) Use a grapher to fit an exponential function to the data.

b) Graph the function with the scatterplot of the data.

c) Predict the total credit card volume for Visa, MasterCard, American Express, and Discover in 2003.

YEAR, x	CREDIT CARD VOLUME, y (IN BILLIONS)
1988, 0	$261.0
1989, 1	296.3
1990, 2	338.4
1991, 3	361.0
1992, 4	403.1
1993, 5	476.7
1994, 6	584.8
1995, 7	701.2
1996, 8	798.3
1997, 9	885.2

Source: CardWeb Inc.'s CardData

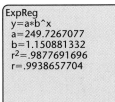

$y = 249.7267077(1.150881332)^x$

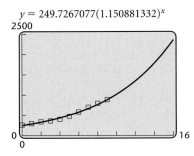

Solution

a) We will fit an equation of the type $y = a \cdot b^x$ to the data, where x is the number of years since 1988. Entering the data into the grapher and carrying out the regression procedure, we find that the equation is

$$y = 249.7267077(1.150881332)^x.$$

The correlation coefficient is very close to 1. This gives us a good indication that the exponential function fits the data well.

b) The graph is shown at left.

c) We evaluate the function found in part (a) for $x = 15$ (2003 − 1988 = 15) and estimate that the total credit card volume in 2003 will be about $2056 billion, or $2,056,000,000,000.

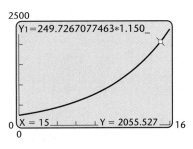

On some graphers, there may be a REGRESSION feature that yields an exponential function, base e. If not, and you wish to find such a function, a conversion can be done using the following.

Converting from Base b to Base e
$$b^x = e^{x(\ln b)}$$

Then, for the equation in Example 6, we have

$$y = 249.7267077(1.150881332)^x$$
$$= 249.7267077e^{x(\ln 1.150881332)}$$
$$= 249.7267077e^{0.1405280246x}.$$

We can prove this conversion formula using properties of logarithms, as follows:

$$e^{x(\ln b)} = e^{\ln b^x} = b^x.$$

Exercise Set 4.6

1. *World Population Growth.* In 1999, the world population was 6.0 billion. The exponential growth rate was 1.3% per year.

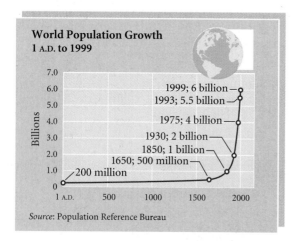

World Population Growth
1 A.D. to 1999

1999; 6 billion
1993; 5.5 billion
1975; 4 billion
1930; 2 billion
1850; 1 billion
1650; 500 million
200 million

Source: Population Reference Bureau

a) Find the exponential growth function.
b) Predict the population of the world in 2005 and 2010.
c) When will the world population be 8 billion?
d) Find the doubling time.

2. *Population Growth of Rabbits.* Under ideal conditions, a population of rabbits has an exponential growth rate of 11.7% per day. Consider an initial population of 100 rabbits.

a) Find the exponential growth function.
b) Graph the function.
c) What will the population be after 7 days?
d) Find the doubling time.

3. *Population Growth.* Complete the following table.

POPULATION	GROWTH RATE, k	DOUBLING TIME, T
a) Mexico	1.9% per year	
b) Japan		346 yr
c) Mozambique	3.3% per year	
d) Norway	0.5% per year	
e) Syria		20.4 yr
f) Philippines		31.5 yr

4. *Female Olympic Athletes.* In 1985, the number of female athletes participating in Summer

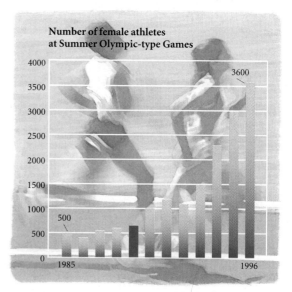

Number of female athletes at Summer Olympic-type Games

3600

500

1985   1996

Olympic-Type Games was 500. In 1996, about 3600 participated in the Summer Olympics in Atlanta. Assuming the exponential model applies:

a) Find the value of k ($P_0 = 500$), and write the function.

b) Estimate the number of female athletes in the Summer Olympics of 2000 and 2004.

5. *Population Growth of Israel.* The population of Israel has a growth rate of 2.6% per year. In 1998, the population was 5,644,000. The land area of Israel is 24,313,062,400 square yards. (*Source: Statistical Abstract of the United States*) Assuming this growth rate continues and is exponential, after how long will there be one person for every square yard of land?

6. *Value of Manhattan Island.* In 1626, Peter Minuit of the Dutch West India Company purchased Manhattan Island from the Indians for $24. Assuming an exponential rate of inflation of 8% per year, how much will Manhattan be worth in 2003?

7. *Interest Compounded Continuously.* Suppose that $10,000 is invested at an interest rate of 5.4% per year, compounded continuously.

a) Find the exponential function that describes the amount in the account after time t, in years.

b) What is the balance after 1 yr? 2 yr? 5 yr? 10 yr?

c) What is the doubling time?

8. *Interest Compounded Continuously.* Complete the following table.

INITIAL INVESTMENT AT $t = 0$, P_0	INTEREST RATE, k	DOUBLING TIME, T	AMOUNT AFTER 5 YR
a) $35,000	6.2%		
b) $5000			$ 7,130.90
c)	8.4%		$11,414.71
d)		11 yr	$17,539.32

9. *Carbon Dating.* A mummy discovered in the pyramid Khufu in Egypt has lost 46% of its carbon-14. Determine its age.

10. *Carbon Dating.* The statue of Zeus at Olympia in Greece is one of the Seven Wonders of the World. It is made of gold and ivory. The ivory was found to

have lost 35% of its carbon-14. Determine the age of the statue.

11. *Radioactive Decay.* Complete the following table.

RADIOACTIVE SUBSTANCE	DECAY RATE, k	HALF-LIFE, T
a) Polonium		3 min
b) Lead		22 yr
c) Iodine-131	9.6% per day	
d) Krypton-85	6.3% per year	
e) Strontium-90		25 yr
f) Uranium-238		4560 yr
g) Plutonium		23,105 yr

12. *Number of Farms.* The number N of farms in the United States has declined continually since 1950. (See Example 5 in Section 3.1.) In 1950, there were 5,647,800 farms, and in 1995 that number had decreased to 2,071,520 (*Source:* U.S. Department of Agriculture). Assuming the number of farms decreased according to the exponential model:

a) Find the value of k, and write an exponential function that describes the number of farms after time t, in years, where t is the number of years since 1950.

b) Estimate the number of farms in 2000, 2005, and 2010.

c) At this decay rate, in what year will only 100,000 farms remain?

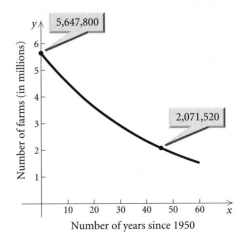

Number of years since 1950

13. *Decline in Beef Consumption.* In 1985, the average annual consumption of beef B was about 80 lb per person. In 1996, it was about 67 lb per person. Assuming consumption is decreasing according to the exponential-decay model:

a) Find the value k, and write an equation that describes beef consumption after time t, in years.

b) Estimate the consumption of beef in 2002.

c) After how many years (theoretically) will the average annual consumption of beef be 20 lb per person?

14. *The Value of Mark McGwire's Baseball Card.* The collecting of baseball cards has become a popular hobby. The card shown here is a photograph of Mark McGwire when he was a member of the Olympic USA Baseball Team in 1984.

In 1987, the value of the card was $8 and in 1997, its value was $20 (*Source: SportsCards,* Joe Clemens, Price Guide Coordinator). Assuming the value of the card has grown exponentially:

a) Find the value k, and determine the exponential growth function, assuming $V_0 = 8$.

b) Using the function found in part (a), estimate the value of the card in 2000.

c) What is the doubling time for the value of the card?

d) After how long will the value of the card be $2000, assuming there is no change in the growth rate?

e) In the Baseball Price Guide in *SportsCards,* April 2000, this card was valued at $200. How does your estimate in part (b) compare with the $200 value? Explain why they differ.

15. *Spread of an Epidemic.* In a town whose population is 2000, a disease creates an epidemic. The number of people N infected t days after the disease has begun is given by the function

$$N(t) = \frac{2000}{1 + 19.9e^{-0.6t}}.$$

a) Graph the function.

b) How many are initially infected with the disease $(t = 0)$?

c) Find the number infected after 2 days, 5 days, 8 days, 12 days, and 16 days.

16. *Acceptance of Seat Belt Laws.* In recent years, many states have passed mandatory seat belt laws. The total number of states N that have passed a seat belt law t years after 1984 is given by the function

$$N(t) = \frac{50}{1 + 22e^{-0.6t}}.$$

(*Source*: National Highway Traffic Safety Administration).

a) Graph the function.

b) How many states had passed the law in 1984? ($t = 0$ corresponds to 1984.)

c) Find the number of states that had passed the law by 1996 and 2002.

d) If the function were to continue to be appropriate, would all 50 states ever pass the law? Explain.

17. *Limited Population Growth in a Lake.* A lake is stocked with 400 fish of a new variety. The size of the lake, the availability of food, and the number of other fish restrict growth in the lake to a *limiting value* of 2500. The population of fish in the lake after time t, in months, is given by the function

$$P(t) = \frac{2500}{1 + 5.25e^{-0.32t}}.$$

a) Graph the function.

b) Find the population after 0, 1, 5, 10, 15, and 20 months.

In Exercises 18–23, determine which, if any, of these functions might be used as a model for the data in the scatterplot.

a) *Quadratic,* $f(x) = ax^2 + bx + c$

b) *Polynomial, not quadratic*

c) *Exponential,* $f(x) = ab^x$, or Be^{kx}, $k > 0$

d) *Exponential,* $f(x) = ab^x$, *or* Be^{-kx}, $k > 0$

e) *Logarithmic,* $f(x) = a + b \ln x$

f) *Logistic,* $f(x) = \dfrac{a}{1 + be^{-kx}}$

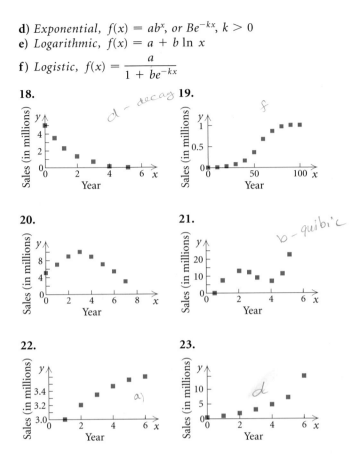

18. *d – decay*

19. *f*

20.

21. *b – quibic*

22. *a)*

23. *d*

24. *Number of Physicians.* The following table contains data regarding the number of physicians in the United States in selected years.

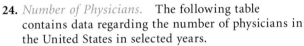

YEAR	TOTAL NUMBER OF PHYSICIANS
1950	219,997
1955	241,711
1960	260,484
1965	292,088
1970	334,028
1975	393,742
1980	467,679
1985	552,716
1990	615,421
1994	684,414
1995	720,325
1996	737,764

Source: American Medical Association

a) Create a scatterplot of these data. Determine whether an exponential function appears to fit the data.

b) Use a grapher to fit the data with an exponential function and determine whether the function is a good fit.

c) Graph the function found in part (b) with a scatterplot of the data.

d) Estimate the number of physicians in 2000 and 2025.

25. *Projected Number of Alzheimer's Patients.* German psychiatrist Alois Alzheimer first described the disease, later called Alzheimer's disease, in 1906. Since life expectancy has significantly increased in the last century, the number of Alzheimer's patients has increased dramatically. The number of patients in the United States reached 4 million in 2000. The following table lists projected data regarding the number of Alzheimer's patients in years beyond 2000.

YEAR	PROJECTED NUMBER OF ALZHEIMER'S PATIENTS IN THE UNITED STATES (IN MILLIONS)
2000	4.0
2010	5.8
2020	6.8
2030	8.7
2040	11.8
2050	14.3

Source: "Alzheimer's, Unlocking the Mystery," by Geoffrey Cowley, *Newsweek*, January 31, 2000

a) Use a grapher to fit the data with an exponential function and determine whether the function is a good fit.

b) Graph the function found in part (a) with a scatterplot of the data.

c) Estimate the number of Alzheimer's patients in 2005, 2025, and 2100.

26. *Forgetting.* In an art class, students were tested at the end of the course on a final exam. Then they were retested with an equivalent test at subsequent time intervals. Their scores after time *t*, in months, are given in the table at the top of the next page.

TIME, t (IN MONTHS)	SCORE, y
1	84.9%
2	84.6%
3	84.4%
4	84.2%
5	84.1%
6	83.9%

a) Use a grapher to fit a logarithmic function $y = a + b \ln x$ to the data.
b) Use the function to predict test scores after 8, 10, 24, and 36 months.
c) After how long will the test scores fall below 82%?

27. *On-line Travel Revenue.* With the explosion of increased Internet use, more and more travelers are booking their travel reservations on-line. The following table lists the total on-line revenue for recent years. Most of the revenue is from airline tickets.

YEAR	ON-LINE TRAVEL REVENUE (IN MILLIONS)
1996	$ 276
1997	827
1998	1900
1999	3200
2000*	4700
2001*	6500
2002*	8900

*Projections
Source: *Travel and Interactive Technology 1999*; Travel Industry Association of America

a) Create a scatterplot of the data. Let $x =$ the number of years since 1996.
b) Use a grapher to fit the data with linear, quadratic, and exponential functions. Determine which function has the best fit.
c) Graph all three functions found in part (b) with the scatterplot in part (a).
d) Use the functions found in part (b) to estimate the on-line travel revenue in 2010. Which function provides the most realistic prediction?

28. *Effect of Advertising.* A company introduces a new software product on a trial run in a city. They advertised the product on television and found the following data relating the percent P of people who bought the product after x ads were run.

NUMBER OF ADS, x	PERCENTAGE WHO BOUGHT, P
0	0.2
10	0.7
20	2.7
30	9.2
40	27
50	57.6
60	83.3
70	94.8
80	98.5
90	99.6

a) Use a grapher to fit a logistic function

$$P(x) = \frac{a}{1 + be^{-kx}}$$

to the data.
b) What percent will buy the product when 55 ads are run? 100 ads?
c) Find an asymptote for the graph. Interpret the asymptote in terms of the advertising situation.

Discussion and Writing

29. Browse through some newspapers or magazines until you find some data and/or a graph that seem as though they can be fit to an exponential function. Make a case for why such a fit is appropriate. Then fit an exponential function to the data and make some predictions.

30. *Atmospheric Pressure.* Atmospheric pressure P at an altitude a is given by

$$P = P_0 e^{-0.00005a},$$

where P_0 is the pressure at sea level ≈ 14.7 lb/in^2 (pounds per square inch). Explain how a barometer, or some device for measuring atmospheric pressure, can be used to find the height of a skyscraper.

Skill Maintenance

Find the slope and the y-intercept of the line.

31. $y = 6$

32. $3x - 10y = 14$

33. $y = 2x - \dfrac{3}{13}$

34. $x = -4$

Synthesis

35. *Present Value.* Following the birth of a child, a parent wants to make an initial investment P_0 that will grow to $50,000 for the child's education at age 18. Interest is compounded continuously at 7%. What should the initial investment be? Such an amount is called the **present value** of $50,000 due 18 yr from now.

36. *Present Value.* Referring to Exercise 35:

a) Solve $P = P_0 e^{kt}$ for P_0.

b) Find the present value of $50,000 due 18 yr from now at interest rate 6.4%.

37. *Supply and Demand.* The supply and demand for the sale of a certain type of VCR are given by

$$S(p) = 480e^{-0.003p} \quad \text{and} \quad D(p) = 150e^{0.004p},$$

where $S(p)$ is the number of VCRs that the company is willing to sell at price p and $D(p)$ is the quantity that the public is willing to buy at price p. Find p, called the **equilibrium price,** such that $D(p) = S(p)$.

38. *Carbon Dating.* Recently, while digging in Chaco Canyon, New Mexico, archeologists found corn pollen that was 4000 yr old (*Source: American Anthropologist*). This was evidence that Native Americans had been cultivating crops in the Southwest centuries earlier than scientists had thought. What percent of the carbon-14 had been lost from the pollen?

39. *Newton's Law of Cooling.* Suppose that a body with temperature T_1 is placed in surroundings with temperature T_0 different from that of T_1. The body will either cool or warm to temperature $T(t)$ after time t, in minutes, where

$$T(t) = T_0 + |T_1 - T_0|e^{-kt}.$$

A cup of coffee with temperature 105°F is placed in a freezer with temperature 0°F. After 5 min, the temperature of the coffee is 70°F. What will its temperature be after 10 min?

40. *When Was the Murder Committed?* The police discover the body of a math professor. Critical to solving the crime is determining when the murder was committed. The coroner arrives at the murder scene at 12:00 P.M. She immediately takes the temperature of the body and finds it to be 94.6°. She then takes the temperature 1 hr later and finds it to be 93.4°. The temperature of the room is 70°. When was the murder committed? (Use Newton's law of cooling in Exercise 39.)

41. *Electricity.* The formula

$$i = \frac{V}{R}[1 - e^{-(R/L)t}]$$

occurs in the theory of electricity. Solve for t.

42. *The Beer–Lambert Law.* A beam of light enters a medium such as water or smog with initial intensity I_0. Its intensity decreases depending on the thickness (or concentration) of the medium. The intensity I at a depth (or concentration) of x units is given by

$$I = I_0 e^{-\mu x}.$$

The constant μ (the Greek letter "mu") is called the **coefficient of absorption,** and it varies with the medium. For sea water, $\mu = 1.4$.

a) What percentage of light intensity I_0 remains at a depth of sea water that is 1 m? 3 m? 5 m? 50 m?

b) Plant life cannot exist below 10 m. What percentage of I_0 remains at 10 m?

43. Given that $y = ae^x$, take the natural logarithm on both sides. Let $Y = \ln y$. Consider Y as a function of x. What kind of function is Y?

44. Given that $y = ax^b$, take the natural logarithm on both sides. Let $Y = \ln y$ and $X = \ln x$. Consider Y as a function of X. What kind of function is Y?

Chapter Summary and Review 4

Important Properties and Formulas

The Composition of Two Functions:	$(f \circ g)(x) = f(g(x))$
One-to-One Function:	$f(a) = f(b) \rightarrow a = b$
Exponential Function:	$f(x) = a^x$
The Number e = 2.7182818284...	
Logarithmic Function:	$f(x) = \log_a x$
A Logarithm is an Exponent:	$\log_a x = y \longleftrightarrow x = a^y$
The Change-of-Base Formula:	$\log_b M = \dfrac{\log_a M}{\log_a b}$
The Product Rule:	$\log_a MN = \log_a M + \log_a N$
The Power Rule:	$\log_a M^p = p \log_a M$
The Quotient Rule:	$\log_a \dfrac{M}{N} = \log_a M - \log_a N$
Other Properties:	$\log_a a = 1, \qquad \log_a 1 = 0,$
	$\log_a a^x = x, \qquad a^{\log_a x} = x$
Base–Exponent Property:	$a^x = a^y \longleftrightarrow x = y$, for $a > 0, a \neq 1$
Exponential Growth Model:	$P(t) = P_0 e^{kt}$
Exponential Decay Model:	$P(t) = P_0 e^{-kt}$
Interest Compounded Continuously:	$P(t) = P_0 e^{kt}$
Limited Growth:	$P(t) = \dfrac{a}{1 + be^{-kt}}$

REVIEW EXERCISES

In Exercises 1 and 2, for the pair of functions:

a) *Find the domain of $f \circ g$ and $g \circ f$.*

b) *Find $(f \circ g)(x)$ and $(g \circ f)(x)$.*

1. $f(x) = \dfrac{4}{x^2}$; $g(x) = 3 - 2x$

2. $f(x) = 3x^2 + 4x$; $g(x) = 2x - 1$

Find $f(x)$ and $g(x)$ such that $h(x) = (f \circ g)x$.

3. $h(x) = \sqrt{5x + 2}$

4. $h(x) = 4(5x - 1)^2 + 9$

5. Find the inverse of the relation

$\{(1.3, -2.7), (8, -3), (-5, 3), (6, -3), (7, -5)\}.$

6. Find an equation of the inverse relation.

a) $y = 3x^2 + 2x - 1$

b) $0.8x^3 - 5.4y^2 = 3x$

In Exercises 7–10, given the function:

a) *Determine whether it is one-to-one, using a grapher if desired.*

b) *If it is one-to-one, find a formula for the inverse.*

7. $f(x) = \sqrt{x - 6}$ **8.** $f(x) = x^3 - 8$

9. $f(x) = 3x^2 + 2x - 1$ **10.** $f(x) = e^x$

11. Find $f(f^{-1}(657))$: $f(x) = \dfrac{4x^5 - 16x^{37}}{119x}$, $x > 1$.

In Exercises 12–17, match the equation with one of figures (a)–(f), which follow. If needed, use a grapher.

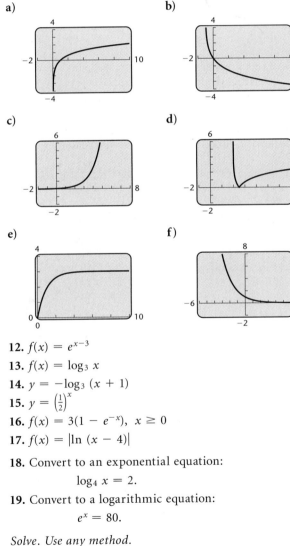

a)   b)

c)   d)

e)   f)

12. $f(x) = e^{x-3}$

13. $f(x) = \log_3 x$

14. $y = -\log_3 (x + 1)$

15. $y = \left(\frac{1}{2}\right)^x$

16. $f(x) = 3(1 - e^{-x})$, $x \geq 0$

17. $f(x) = |\ln (x - 4)|$

18. Convert to an exponential equation:
$$\log_4 x = 2.$$

19. Convert to a logarithmic equation:
$$e^x = 80.$$

Solve. Use any method.

20. $\log_4 x = 2$ **21.** $3^{1-x} = 9^{2x}$

22. $e^x = 80$ **23.** $4^{2x-1} - 3 = 61$

24. $\log_{16} 4 = x$ **25.** $\log_x 125 = 3$

26. $\log_2 x + \log_2 (x - 2) = 3$

27. $\log (x^2 - 1) - \log (x - 1) = 1$

28. $\log x^2 = \log x$ **29.** $e^{-x} = 0.02$

Express as a single logarithm and, if possible, simplify.

30. $3 \log_b x - 4 \log_b y + \frac{1}{2} \log_b z$

31. $\ln (x^3 - 8) - \ln (x^2 + 2x + 4) + \ln (x + 2)$

Express in terms of sums and differences of logarithms.

32. $\ln \sqrt[4]{wr^2}$ **33.** $\log \sqrt[3]{\dfrac{M^2}{N}}$

Given that $\log_a 2 = 0.301$, $\log_a 5 = 0.699$, and $\log_a 6 = 0.778$, find each of the following.

34. $\log_a 3$ **35.** $\log_a 50$

36. $\log_a \frac{1}{5}$ **37.** $\log_a \sqrt[3]{5}$

Simplify.

38. $\ln e^{-5k}$ **39.** $\log_5 5^{-6t}$

40. How long will it take an investment to double itself if it is invested at 8.6%, compounded continuously?

41. The population of a city doubled in 30 yr. What was the exponential growth rate?

42. How old is a skeleton that has lost 27% of its carbon-14?

43. The hydrogen ion concentration of milk is 2.3×10^{-6}. What is the pH?

44. What is the loudness, in decibels, of a sound whose intensity is $1000I_0$?

45. *The Population of Zimbabwe.* The population of Zimbabwe was 11 million in 1998, and the exponential growth rate was 1.2% per year (*Source:* U.S. Bureau of the Census, World Population Profile).

a) Find the exponential growth function.
b) What will the population be in 2004? in 2020?
c) When will the population be 25 million?
d) What is the doubling time?

46. *Toll-free Numbers.* The use of toll-free numbers has grown exponentially. In 1967, there were 7 million such calls, and in 1991, there were 10.2 billion such calls (*Source:* Federal Communication Commission).

a) Find the exponential growth rate k.
b) Find the exponential growth function.
c) Graph the exponential growth function.
d) How many toll-free number calls will be placed in 2002? in 2005?
e) In what year will 1 trillion such calls be placed?

47. *Walking Speed.* The average walking speed w, in feet per second, of a person living in a city of population P, in thousands, is given by the function

$$w(P) = 0.37 \ln P + 0.05.$$

a) The population of Phoenix, Arizona, is 1,200,000. Find the average walking speed.

b) A city's population has an average walking speed of 3.4 ft/sec. Find the population.

48. *Cholesterol Level and the Risk of Heart Attack.* The data in the following table show the relationship of cholesterol level in men to the risk of a heart attack.

CHOLESTEROL LEVEL, x	MEN, PER 10,000, WHO SUFFER A HEART ATTACK, y
100	30
200	65
250	100
275	130
300	180

Source: Nutrition Action Healthletter

a) Use the grapher to fit an exponential function to the data.

b) Graph the function with a scatterplot of the data.

c) Predict the heart attack rate for men with cholesterol levels of 150, 350, and 400.

d) Compare your answers in part (c) with the answer to part (c) in Example 4 of Section 3.1. Which function, power or exponential, is a better fit to the given data?

Discussion and Writing

49. Suppose that you were trying to convince a fellow student that

$$\log_2 (x + 5) \neq \log_2 x + \log_2 5.$$

Give as many explanations as you can.

50. Describe the difference between $f^{-1}(x)$ and $[f(x)]^{-1}$.

Synthesis

Solve.

51. $|\log_4 x| = 3$

52. $\log x = \ln x$

53. $5^{\sqrt{x}} = 625$

54. a) Use a grapher to graph $f(x) = 5e^{-x} \ln x$ in the viewing window $[-1, 10, -5, 5]$.

b) Estimate the relative maximum and the minimum values.

55. Use only a grapher. Determine whether the following functions are inverses of each other:

$$f(x) = \frac{4 + 3x}{x - 2}, \qquad g(x) = \frac{x + 4}{x - 3}.$$

56. Find the domain: $f(x) = \log_3 (\ln x)$.

57. Find the points of intersection of the graphs of the following equations:

$$y = 5x^2 e^{-x}, \qquad y = 2 - e^{-x^2}.$$

Systems of Equations and Matrices 5

APPLICATION

The number of take-out meals purchased per person has surpassed the number of restaurant meals purchased per person in recent years. The functions $r(x) = -0.45x + 68.83$ and $c(x) = 1.70x + 44.15$ can be used to estimate the number of restaurant and take-out meals, respectively, purchased per person x years after 1984. (*Source*: NPD Group CREST Service) We can use these functions to estimate that the number of take-out meals purchased per person was equal to the number of restaurant meals purchased per person about 11.5 yr after 1984.

This problem appears as Exercise 55 in Section 5.1.

It is often desirable or even necessary to use two or more variables to model a situation in a field such as business, science, psychology, engineering, education, or sociology. When this is the case, we write and solve a *system of equations* in order to answer questions about that situation. In this chapter, we study systems of equations and several methods for solving them, including the use of matrices. We also use *systems of inequalities,* along with *linear programming,* to model applications and to find the maximum and minimum values of functions subject to a set of restrictions, or constraints.

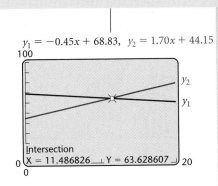

$y_1 = -0.45x + 68.83, \quad y_2 = 1.70x + 44.15$

Intersection
X = 11.486826 Y = 63.628607

359

Systems of Equations in Two Variables

5.1

- Solve a system of two linear equations in two variables by graphing.
- Solve a system of two linear equations in two variables using the substitution and the elimination methods.
- Use systems of two linear equations to solve applied problems.

A **system of equations** is composed of two or more equations considered simultaneously. For example,

$$x + y = 11,$$
$$3x - y = 5$$

is a system of two linear equations in two variables. The solution set of this system consists of all ordered pairs that make *both* equations true. The ordered pair (4, 7) is a solution of the system of equations above. We can verify this by substituting 4 for x and 7 for y in *each* equation.

$$\underline{x + y = 11}$$
$$4 + 7 \; ? \; 11$$
$$11 \; | \; 11 \quad \text{TRUE}$$

$$\underline{3x - y = 5}$$
$$3 \cdot 4 - 7 \; ? \; 5$$
$$12 - 7$$
$$5 \; | \; 5 \quad \text{TRUE}$$

Interactive Discovery

Graph $y_1 = 2x - 5$ and $y_2 = -x + 1$ in the same viewing window. Use the grapher's INTERSECT feature to find the coordinates of the point of intersection of the graphs. Do the same for $y_1 = x + 4$ and $y_2 = -1.25x - 5$ and for $y_1 = 3x - 7$ and $y_2 = -2x + \frac{1}{2}$. What does each of these ordered pairs represent?

Solving Systems of Equations Graphically

Recall that the graph of a linear equation is a line that contains all the ordered pairs in the solution set of the equation. When we graph a system of linear equations, each point at which the equations intersect is a solution of *both* equations and therefore a solution of the system of equations.

Consider the system

$$x + y = 11,$$
$$3x - y = 5.$$

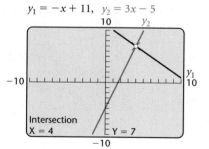

$y_1 = -x + 11, \quad y_2 = 3x - 5$

When we graph these equations on the same set of axes, we see that they intersect at a single point, (4, 7), so (4, 7) is the solution of the system of equations. (Note that when a grapher is used, it might be necessary to write each equation in "$Y = \cdots$" form. If so, we would graph $y_1 = -x + 11$ and $y_2 = 3x - 5$.)

To check this result, we can substitute 4 for x and 7 for y in both equations, as we did above. We can also use the TABLE feature on a grapher to do a check, observing that when $x = 4$, $y_1 = 7$ and $y_2 = 7$.

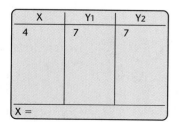

The graphs of most of the systems of equations that we use to model applications intersect at a single point, like the system above. However, it is possible that the graphs will have no points in common or infinitely many points in common. Each of these possibilities is illustrated below.

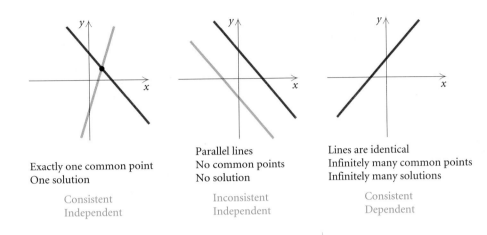

Exactly one common point
One solution

Consistent
Independent

Parallel lines
No common points
No solution

Inconsistent
Independent

Lines are identical
Infinitely many common points
Infinitely many solutions

Consistent
Dependent

If a system of two linear equations in two variables has at least one solution, it is **consistent**. If the system has no solutions, it is **inconsistent**. In addition, if a system of two linear equations in two variables has an infinite number of solutions, the equations are **dependent**. Otherwise, they are **independent**.

The Substitution Method

Solving a system of equations graphically is not always accurate when the solutions are not integers. A pair like $\left(\frac{43}{27}, -\frac{19}{27}\right)$, for instance, will be difficult to determine from a hand-drawn graph. Even when a grapher is used, decimal approximations for the solution of a system of equations are not always successfully converted to fractions.

Algebraic methods for solving systems of equations, when used correctly, always give accurate results. One such technique is the **substitution method.** It is used most often when a variable is alone on one side of an equation or when it is easy to solve for a variable. To apply the substitution method, we begin by using one of the equations to express one variable in terms of the other; then we substitute that expression in the other equation of the system.

EXAMPLE 1   Use the substitution method to solve the system

$$x + y = 11, \qquad (1)$$
$$3x - y = 5. \qquad (2)$$

Solution   First, we solve equation (1) for y. (We could just as well solve for x.)

$$y = 11 - x$$

Then we substitute $11 - x$ for y in equation (2). This gives an equation in one variable, which we know how to solve.

$$3x - (11 - x) = 5 \quad \text{The parentheses are necessary.}$$
$$3x - 11 + x = 5 \quad \text{Removing parentheses}$$
$$4x - 11 = 5$$
$$4x = 16$$
$$x = 4$$

Now we substitute 4 for x in either of the original equations (this is called **back-substitution**) and solve for y. We choose equation (1):

$$4 + y = 11$$
$$y = 7.$$

We have previously checked the pair $(4, 7)$ in both equations both algebraically and graphically. The solution of the system of equations is $(4, 7)$.

The Elimination Method

Another algebraic technique for solving systems of equations is the **elimination method.** With this method, we eliminate a variable by adding two equations. Before we add, it might be necessary to multiply one or both equations by suitable constants in order to find two equations in which coefficients are opposites.

EXAMPLE 2   Solve each of the following systems using the elimination method.

a) $2x + y = 2,$ (1)   **b)** $4x + 3y = 11,$ (1)
     $x - y = 7$ (2) $-5x + 2y = 15$ (2)

Solution

a) Since the y-coefficients are opposites, we can eliminate y by adding the equations:

$$
\begin{array}{ll}
2x + y = 2 & (1) \\
\underline{x - y = 7} & (2) \\
3x = 9 & \text{Adding} \\
x = 3.
\end{array}
$$

We then back-substitute 3 for x in either equation and solve for y. We choose equation (1):

$$2 \cdot 3 + y = 2$$
$$6 + y = 2$$
$$y = -4.$$

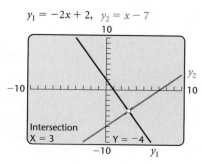

$y_1 = -2x + 2, \quad y_2 = x - 7$

Intersection
X = 3   Y = -4

We can check the pair $(3, -4)$ either algebraically by substituting in both equations or graphically using the INTERSECT feature, as shown at left. The solution is $(3, -4)$.

b) We can obtain x-coefficients that are opposites by multiplying the first equation by 5 and the second equation by 4:

$$\begin{array}{ll} 20x + 15y = 55 & \text{Multiplying equation (1) by 5} \\ \underline{-20x + 8y = 60} & \text{Multiplying equation (2) by 4} \\ \hspace{1.5em} 23y = 115 & \text{Adding} \\ \hspace{2em} y = 5. \end{array}$$

We then back-substitute 5 for y in either equation (1) or (2) and solve for x. We choose equation (1):

$$\begin{array}{ll} 4x + 3 \cdot 5 = 11 & \text{Substituting 5 for } y \text{ in equation (1)} \\ 4x + 15 = 11 & \\ \hspace{1.5em} 4x = -4 & \\ \hspace{2em} x = -1. & \end{array}$$

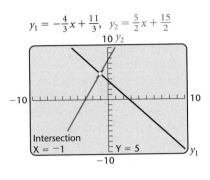

$y_1 = -\frac{4}{3}x + \frac{11}{3}, \quad y_2 = \frac{5}{2}x + \frac{15}{2}$

We can check the pair $(-1, 5)$ either algebraically by substituting in both equations or graphically, as shown at left. The solution is $(-1, 5)$.

In Example 2(b), the two systems

$$\begin{array}{c} 4x + 3y = 11, \\ -5x + 2y = 15 \end{array} \quad \text{and} \quad \begin{array}{c} 20x + 15y = 55, \\ -20x + 8y = 60 \end{array}$$

are **equivalent** because they have exactly the same solutions. When we use the elimination method, we often multiply one or both equations by constants to find equivalent equations that allow us to eliminate a variable by adding.

EXAMPLE 3 Solve each of the following systems using the elimination method.

a) $\begin{array}{ll} x - 3y = 1, & (1) \\ -2x + 6y = 5 & (2) \end{array}$ **b)** $\begin{array}{ll} 2x + 3y = 6, & (1) \\ 4x + 6y = 12 & (2) \end{array}$

Solution

a) We multiply equation (1) by 2 and add:

$$\begin{array}{ll} 2x - 6y = 2 & \text{Multiplying equation (1) by 2} \\ \underline{-2x + 6y = 5} & (2) \\ \hspace{2em} 0 = 7. & \text{Adding} \end{array}$$

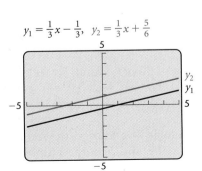

$y_1 = \frac{1}{3}x - \frac{1}{3}, \quad y_2 = \frac{1}{3}x + \frac{5}{6}$

There are no values of x and y for which $0 = 7$ is true, so the system has no solution. The solution set is $\varnothing$. The system of equations is inconsistent.

The graphs of the equations are parallel lines. In fact, we see this when we write the equations in simplified "$y = \cdots$" form to enter them on a grapher. The slopes are the same, $\frac{1}{3}$, but the y-intercepts are different, $\left(0, -\frac{1}{3}\right)$ and $\left(0, \frac{5}{6}\right)$.

b) We multiply equation (1) by -2 and add:

$$\begin{array}{ll} -4x - 6y = -12 & \text{Multiplying equation (1) by } -2 \\ \underline{4x + 6y = 12} & \text{(2)} \\ 0 = 0 \;. & \text{Adding} \end{array}$$

We obtain the equation $0 = 0$, which is true for all values of x and y. This tells us that the equations are dependent, so there are infinitely many solutions. That is, any solution of one equation of the system is also a solution of the other. The graphs of the equations are identical. (In fact, when we write the equations in simplified "$y = \cdots$" form to enter them on a grapher, we get identical equations.)

Using either equation, we can write $y = -\frac{2}{3}x + 2$, so we can write the solutions of the system as ordered pairs of the form $\left(x, -\frac{2}{3}x + 2\right)$. Any real value that we choose for x then gives us a value for y and thus an ordered pair in the solution set. Some of the solutions are $(-3, 4)$, $(0, 2)$, and $(6, -2)$. Similarly, we can write $x = -\frac{3}{2}y + 3$, so the solutions can also be expressed as $\left(-\frac{3}{2}y + 3, y\right)$.

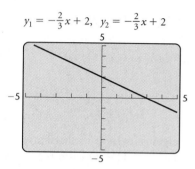

$y_1 = -\frac{2}{3}x + 2, \quad y_2 = -\frac{2}{3}x + 2$

Applications

Frequently the most challenging and time-consuming step in the problem-solving process is translating a situation to mathematical language. However, in many cases, this task is facilitated if we translate to more than one equation in more than one variable.

EXAMPLE 4   *Snack Mixtures.* At Sunda's Snacks, caramel corn worth $2.50 per pound is mixed with honey roasted mixed nuts worth $7.50 per pound in order to get 20 lb of a mixture worth $4.50 per pound. How much of each snack is used?

Solution   We use the five-step problem-solving process.

1. **Familiarize.** Let's begin by making a guess. Suppose 16 lb of caramel corn and 4 lb of nuts were used. Then the total weight of the mixture would be 16 lb + 4 lb, or 20 lb, the desired weight. The total values of these amounts of ingredients are found by multiplying the price per pound by the number of pounds used:

   Caramel corn: $2.50(16) = \$40$

   Nuts: $7.50(4) = \underline{\$30}$

   Total value: $\$70.$

   The desired value of the mixture is $4.50 per pound, so the value of 20 lb would be $4.50(20)$, or $90. Thus we see that our guess, which led to a total of $70, is incorrect. Nevertheless, these calculations will help us to translate.

   We organize the information in a table. We let $x =$ the number of pounds of caramel corn in the mixture and $y =$ the number of pounds of nuts.

	CARAMEL CORN	NUTS	MIXTURE	
PRICE PER POUND	$2.50	$7.50	$4.50	
NUMBER OF POUNDS	x	y	20	$\longrightarrow x + y = 20$
VALUE OF MIXTURE	$2.50x$	$7.50y$	20(4.50), or 90	$\longrightarrow 2.50x + 7.50y = 90$

2. Translate. From the second row of the table, we get one equation:

$$x + y = 20.$$

The last row of the table yields a second equation:

$$2.50x + 7.50y = 90.$$

We can multiply by 10 on both sides of the second equation to clear the decimals. This gives us the following system of equations:

$$x + y = 20, \qquad (1)$$
$$25x + 75y = 900. \qquad (2)$$

3. Carry out. We carry out the solution as follows.

Algebraic Solution

Using the elimination method, we multiply equation (1) by -25 and add it to equation (2):

$$
\begin{array}{r}
-25x - 25y = -500 \\
\underline{25x + 75y = 900} \\
50y = 400 \\
y = 8.
\end{array}
$$

Then we back-substitute to find x:

$$x + 8 = 20 \qquad \textbf{Substituting in equation (1)}$$
$$x = 12.$$

Graphical Solution

We solve each equation for y, getting

$$y = 20 - x,$$
$$y = \frac{900 - 25x}{75}.$$

Next, we graph these equations and find the point of intersection of the graphs.

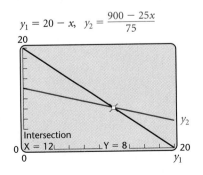

$$y_1 = 20 - x, \quad y_2 = \frac{900 - 25x}{75}$$

4. Check. If 12 lb of caramel corn and 8 lb of nuts are used, the mixture weighs 12 + 8, or 20 lb. The value of the mixture is $2.50(12) + $7.50(8), or $30 + $60, or $90. Since the possible solution yields the desired weight and value of the mixture, our result checks.

5. State. The mixture should consist of 12 lb of caramel corn and 8 lb of honey roasted nuts.

EXAMPLE 5 *Supply and Demand.* Suppose that the price and the supply of the Great Tunes portable CD player are related by the equation

$$y = 90 + 30x,$$

where y is the price, in dollars, at which the seller is willing to supply x thousand units. Also suppose that the price and the demand for the same model of CD player are related by the equation

$$y = 200 - 25x,$$

where y is the price, in dollars, at which the consumer is willing to buy x thousand units.

The **equilibrium point** for this product is the pair (x, y) that is a solution of both equations. The **equilibrium price** is the price at which the amount of the product that the seller is willing to supply is the same as the amount demanded by the consumer. Find the equilibrium point for this product.

Solution

1., 2. Familiarize and **Translate.** We are given a system of equations in the statement of the problem, so no further translation is necessary.

$$y = 90 + 30x, \qquad (1)$$
$$y = 200 - 25x. \qquad (2)$$

We substitute some values for x in each equation to get an idea of the corresponding prices. When $x = 1$,

$$y = 90 + 30 \cdot 1 = 120, \qquad \text{Substituting in equation (1)}$$
$$y = 200 - 25 \cdot 1 = 175. \qquad \text{Substituting in equation (2)}$$

This indicates that the price when 1 thousand units are supplied is lower than the price when 1 thousand units are demanded.
When $x = 4$,

$$y = 90 + 30 \cdot 4 = 210, \qquad \text{Substituting in equation (1)}$$
$$y = 200 - 25 \cdot 4 = 100. \qquad \text{Substituting in equation (2)}$$

In this case, the price related to supply is higher than the price related to demand. It would appear that the x-value we are looking for is between 1 and 4.

3. Carry out. We use the substitution method:

$$200 - 25x = 90 + 30x \qquad \text{Substituting } 200 - 25x \text{ for } y$$
in equation (1)

$$110 = 55x \qquad \text{Adding } 25x \text{ and subtracting } 90 \text{ on both sides}$$

$$2 = x. \qquad \text{Dividing by } 55 \text{ on both sides}$$

We now back-substitute 2 for x in either equation and find y:

$$y = 200 - 25 \cdot 2 \qquad \text{Substituting in equation (2)}$$

$$y = 150.$$

4. Check. We can check algebraically by substituting 2 for x and 150 for y in both equations. Since the equations are given in the statement of the problem, we can also check graphically, as shown at left.

5. State. The equilibrium point is (2, $150). That is, the equilibrium supply is 2 thousand units and the equilibrium price is $150.

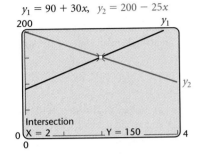

There is another algebraic method for solving systems of two linear equations in two variables known as *Cramer's rule*. It is presented in Appendix B. Systems of equations can be solved on some graphers using the SIMULT feature.

Exercise Set 5.1

In Exercises 1–6, match the system of equations with one of the graphs (a)–(f), which follow.

1. $x + y = -2,$
   $y = x - 8$

2. $x - y = -5,$
   $x = -4y$

3. $x - 2y = -1,$
   $4x - 3y = 6$

4. $2x - y = 1,$
   $x + 2y = -7$

5. $2x - 3y = -1,$
   $-4x + 6y = 2$

6. $4x - 2y = 5,$
   $6x - 3y = -10$

Solve graphically.

7. $x + y = 2,$
   $3x + y = 0$

8. $x + y = 1,$
   $3x + y = 7$

9. $\frac{1}{2}x + y = \frac{3}{2},$
   $2x + 8y = 13$

10. $x + 2y = -\frac{2}{3},$
    $\frac{1}{4}x - \frac{1}{2}y = 1$

11. $y + 1 = 2x,$
    $y - 1 = 2x$

12. $2x - y = 1,$
    $3y = 6x - 3$

Solve using the substitution method. Use a grapher to check your answer.

13. $x + y = 9,$
    $2x - 3y = -2$

14. $3x - y = 5,$
    $x + y = \frac{1}{2}$

15. $x - 2y = 7,$
    $x = y + 4$

16. $x + 4y = 6,$
    $x = -3y + 3$

17. $y = 2x - 6,$
    $5x - 3y = 16$

18. $3x + 5y = 2,$
    $2x - y = -3$

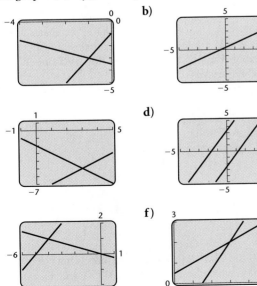

a)

b)

c)

d)

e)

f)

19. $x - 5y = 4,$
$\quad y = 7 - 2x$

20. $5x + 3y = -1,$
$\quad x + y = 1$

21. $2x - 3y = 5,$
$\quad 5x + 4y = 1$

22. $3x + 4y = 6,$
$\quad 2x + 3y = 5$

Solve using the elimination method. Also determine whether each system is consistent or inconsistent and dependent or independent. Use a grapher to check your answer.

23. $x + 2y = 7,$
$\quad x - 2y = -5$

24. $3x + 4y = -2,$
$\quad -3x - 5y = 1$

25. $x - 3y = 2,$
$\quad 6x + 5y = -34$

26. $x + 3y = 0,$
$\quad 20x - 15y = 75$

27. $0.3x - 0.2y = -0.9,$
$\quad 0.2x - 0.3y = -0.6$

28. $0.2x - 0.3y = 0.3,$
$\quad 0.4x + 0.6y = -0.2$

29. $3x - 12y = 6,$
$\quad 2x - 8y = 4$

30. $2x + 6y = 7,$
$\quad 3x + 9y = 10$

31. $\frac{1}{5}x + \frac{1}{2}y = 6,$
$\quad \frac{3}{5}x - \frac{1}{2}y = 2$

32. $\frac{2}{3}x + \frac{3}{5}y = -17,$
$\quad \frac{1}{2}x - \frac{1}{3}y = -1$

33. $2x = 5 - 3y,$
$\quad 4x = 11 - 7y$

34. $7(x - y) = 14,$
$\quad 2x = y + 5$

35. *Sales Promotions.* During a one-month promotional campaign, Video Village gave either a free video rental or a 12-serving box of microwave popcorn to new members. It cost the store $1 for each free rental and $2 for each box of popcorn. In all, 48 new members were signed up and the store's cost for the incentives was $86. How many of each incentive were given away?

36. *Concert Ticket Prices.* One evening 1500 concert tickets were sold for the Fairmont Summer Jazz Festival. Tickets cost $25 for covered pavilion seats and $15 for lawn seats. Total receipts were $28,500. How many of each type of ticket were sold?

37. *Museum Admission Prices.* Admission to the Indianapolis Children's Museum costs twice as much for adults as for children. Admission to the museum for 4 adults and 16 children from the Tiny Tots Day Care Center cost $72. Find the cost of each adult's admission and each child's admission.

38. *Mail-Order Business.* A mail-order lacrosse equipment business shipped 120 packages one day. Customers are charged $3.50 for each standard-delivery package and $7.50 for each express-delivery package. Total shipping charges for the day were $596. How many of each kind of package were shipped?

39. *Supply and Demand.* The supply and demand for a particular model of electronic organizer are related to price by the equations

$$y = 70 + 2x,$$
$$y = 175 - 5x,$$

respectively, where y is the price, in dollars, and x is the number of units, in thousands. Find the equilibrium point for this product.

40. *Supply and Demand.* The supply and demand for a particular model of treadmill are related to price by the equations

$$y = 240 + 40x,$$
$$y = 500 - 25x,$$

respectively, where y is the price, in dollars, and x is the number of units, in thousands. Find the equilibrium point for this product.

The point at which a company's costs equal its revenues is the break-even point. In Exercises 41–44, C represents the production cost, in dollars, of x units of a product and R represents the revenue, in dollars, from the sale of x units. Find the number of units that must be produced and sold in order to break even. That is, find the value of x for which C = R.

41. $C = 14x + 350,$
$\quad R = 16.5x$

42. $C = 8.5x + 75,$
$\quad R = 10x$

43. $C = 15x + 12,000,$
$\quad R = 18x - 6000$

44. $C = 3x + 400,$
$\quad R = 7x - 600$

45. *Nutrition.* A one-cup serving of spaghetti with meatballs contains 260 Cal (calories) and 32 g of carbohydrates. A one-cup serving of chopped iceberg lettuce contains 5 Cal and 1 g of carbohydrates. (*Source: Home and Garden Bulletin No. 72, U.S. Government Printing Office, Washington, D.C. 20402*) How many servings of each would be required to obtain 400 Cal and 50 g of carbohydrates?

46. *Nutrition.* One serving of tomato soup contains 100 Cal and 18 g of carbohydrates. One slice of whole wheat bread contains 70 Cal and 13 g of carbohydrates. (*Source: Home and Garden Bulletin*

No. 72, U.S. Government Printing Office, Washington, D.C. 20402) How many servings of each would be required to obtain 230 Cal and 42 g of carbohydrates?

47. *Motion.* A Leisure Time Cruises riverboat travels 46 km downstream in 2 hr. It travels 51 km upstream in 3 hr. Find the speed of the boat and the speed of the stream.

48. *Motion.* A DC10 travels 3000 km with a tail wind in 3 hr. It travels 3000 km with a head wind in 4 hr. Find the speed of the plane and the speed of the wind.

49. *Investment.* Bernadette inherited $15,000 and invested it in two municipal bonds, which pay 7% and 9% simple interest. The annual interest is $1230. Find the amount invested at each rate.

50. *Tee Shirt Sales.* Mack's Tee Shirt Shack sold 36 shirts one day. All short-sleeved tee shirts cost $12 and all long-sleeved tee shirts cost $18. Total receipts for the day were $522. How many of each kind of shirt were sold?

51. *Coffee Mixtures.* The owner of The Daily Grind coffee shop mixes French roast coffee worth $9.00 per pound with Kenyan coffee worth $7.50 per pound in order to get 10 lb of a mixture worth $8.40 per pound. How much of each type of coffee was used?

52. *Motion.* A Boeing 747 flies the 3000-mi distance from Los Angeles to New York, with a tail wind, in 5 hr. The return trip, against the wind, takes 6 hr. Find the speed of the plane and the speed of the wind.

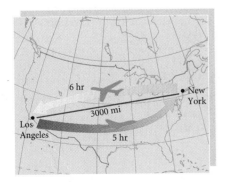

53. *Commissions.* Jackson Manufacturing offers its sales representatives a choice between being paid a commission of 8% of sales or being paid a monthly salary of $1500 plus a commission of 1% of sales. For what monthly sales do the two plans pay the same amount?

54. *Motion.* Two private airplanes travel toward each other from cities that are 780 km apart at speeds of 190 km/h and 200 km/h. They left at the same time. In how many hours will they meet?

55. *Take-out Meals.* The number of take-out meals purchased per person has surpassed the number of restaurant meals purchased per person in recent years, as shown by the data in the following table.

YEAR	NUMBER OF TAKE-OUT MEALS PER PERSON	NUMBER OF RESTAURANT MEALS PER PERSON
1984	43	69
1988	53	68
1992	57	63
1997	66	64

Source: NPD Group's CREST service

a) Find linear regression functions $t(x)$ and $r(x)$ that represent the number of take-out meals purchased per person and the number of restaurant meals purchased per person, respectively, x years after 1984.

b) Use the functions found in part (a) to estimate how many years after 1984 the number of take-out meals purchased was equal to the number of restaurant meals purchased.

56. *U.S. Labor Force.* The percentages of men and women in the civilian labor force in recent years are shown in the table below.

YEAR	PERCENTAGE OF MEN IN THE LABOR FORCE	PERCENTAGE OF WOMEN IN THE LABOR FORCE
1985	76.3	54.5
1990	76.4	57.5
1995	75.0	58.9
1997	75.0	59.8

Source: U.S. Bureau of Labor Statistics

a) Find linear regression functions $m(x)$ and $w(x)$ that represent the percentages of men and women in the civilian labor force, respectively, x years after 1985.

b) Use the functions found in part (a) to predict when the percentages of men and women in the labor force will be equal.

Discussion and Writing

57. Explain in your own words when the elimination method for solving a system of equations is preferable to the substitution method.

58. Cassidy solves the equation $2x + 5 = 3x - 7$ by finding the point of intersection of the graphs of $y_1 = 2x + 5$ and $y_2 = 3x - 7$. She finds the same point when she solves the system of equations

$$y = 2x + 5,$$
$$y = 3x - 7.$$

Explain the difference between the solutions.

Skill Maintenance

Find the zero(s) of the function. Give exact answers.

59. $f(x) = 2x - 5$ **60.** $f(x) = 6 - 3x$

61. $f(x) = x^2 - 4x + 3$ **62.** $f(x) = 3x^2 - x - 5$

Synthesis

63. *Motion.* Nancy jogs and walks to campus each day. She averages 4 km/h walking and 8 km/h jogging. The distance from home to the campus is 6 km and she makes the trip in 1 hr. How far does she jog on each trip?

64. *e-Commerce.* shirts.com advertises a limited-time sale, offering 1 turtleneck for $15 and 2 turtlenecks for $25. A total of 1250 turtlenecks are sold and $16,750 is taken in. How many customers ordered 2 turtlenecks?

65. *Motion.* A train leaves Union Station for Central Station, 216 km away, at 9 A.M. One hour later, a train leaves Central Station for Union Station. They meet at noon. If the second train had started at 9 A.M. and the first train at 10:30 A.M., they would still have met at noon. Find the speed of each train.

66. *Antifreeze Mixtures.* An automobile radiator contains 16 L of antifreeze and water. This mixture is 30% antifreeze. How much of this mixture should be drained and replaced with pure antifreeze so that the final mixture will be 50% antifreeze?

67. Two solutions of the equation $Ax + By = 1$ are $(3, -1)$ and $(-4, -2)$. Find A and B.

68. *Ticket Line.* You are in line at a ticket window. There are 2 more people ahead of you in line than there are behind you. In the entire line, there are three times as many people as there are behind you. How many people are ahead of you?

69. *Gas Mileage.* A 6-cylinder Oldsmobile Aurora gets 19 miles per gallon (mpg) in city driving and 28 mpg in highway driving. The car is driven 405 mi on a full tank of 18 gal of gasoline. How many miles were driven in the city and how many were driven on the highway?

70. *Motion.* Heather is out hiking and is standing on a railroad bridge, as shown in the figure below. A train is approaching from the direction shown by the arrow. If Heather runs at a speed of 10 mph toward the train, she will reach point P on the bridge at the same moment that the train does. If she runs to point Q at the other end of the bridge at a speed of 10 mph, she will reach point Q also at the same moment that the train does. How fast, in miles per hour, is the train traveling?

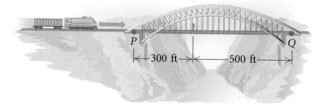

Systems of Equations in Three Variables

5.2

- *Solve systems of linear equations in three variables.*
- *Use systems of three equations to solve applied problems.*
- *Model a situation using a quadratic function.*

A **linear equation in three variables** is an equation equivalent to one of the form $Ax + By + Cz = D$, where A, B, C, and D are real numbers and A, B, and C are not all 0. A **solution of a system of three equations in three variables** is an ordered triple that makes all three equations true. For example, the triple $(2, -1, 0)$ is a solution of the system of equations

$$4x + 2y + 5z = 6,$$
$$2x - y + z = 5,$$
$$3x + 2y - z = 4.$$

We can verify this by substituting 2 for x, -1 for y, and 0 for z in each equation.

Solving Systems of Equations in Three Variables

We will solve systems of equations in three variables using an algebraic method called **Gaussian elimination,** named for the German mathematician Karl Friedrich Gauss (1777–1855). Our goal is to transform the original system to an equivalent one of the form

$$Ax + By + Cz = D,$$
$$Ey + Fz = G,$$
$$Hz = K.$$

Then we solve the third equation for z and back-substitute to find y and then x.

Each of the following operations can be used to transform the original system to an equivalent system in the desired form.

1. Interchange any two equations.
2. Multiply both sides of one of the equations by a nonzero constant.
3. Add a nonzero multiple of one equation to another equation.

EXAMPLE 1 Solve the following system:

$$x - 2y + 3z = 11, \quad (1)$$
$$4x + 2y - 3z = 4, \quad (2)$$
$$3x + 3y - z = 4. \quad (3)$$

Solution We multiply equation (1) by -4 and add it to equation (2). We also multiply equation (1) by -3 and add it to equation (3). These operations produce an equivalent system of equations in which the x-terms have been eliminated from both the second and the third equations.

$$-4x + 8y - 12z = -44 \longleftarrow \quad x - 2y + 3z = 11, \longrightarrow -3x + 6y - 9z = -33$$

$$\underline{4x + 2y - 3z = 4} \quad \text{Eq. (2)} \quad \boxed{\rightarrow} 10y - 15z = -40, \qquad \underline{3x + 3y - z = 4} \quad \text{Eq. (3)}$$

$$10y - 15z = -40 \quad \text{Eq. (4)} \longrightarrow \quad 9y - 10z = -29 \longleftarrow \quad 9y - 10z = -29 \quad \text{Eq. (5)}$$

Now we multiply the last equation by 10 to make the y-coefficient a multiple of the y-coefficient in the equation above it:

$$
\begin{aligned}
x - 2y + 3z &= 11, && (1) \\
10y - 15z &= -40, && (4) \\
90y - 100z &= -290. && (5)
\end{aligned}
$$

Next, we multiply equation (4) by -9 and add it to equation (5).

$$-90y + 135z = 360 \longleftarrow \quad x - 2y + 3z = 11, \qquad (1)$$

$$\underline{90y - 100z = -290} \quad \text{Eq. (5)} \quad \boxed{} -10y - 15z = -40, \qquad (4)$$

$$35z = 70 \longrightarrow 35z = 70. \qquad (6)$$

Now we solve equation (6) for z:

$$
\begin{aligned}
35z &= 70 \\
z &= 2.
\end{aligned}
$$

Then we back-substitute 2 for z in equation (4) and solve for y:

$$
\begin{aligned}
10y - 15 \cdot 2 &= -40 \\
10y - 30 &= -40 \\
10y &= -10 \\
y &= -1.
\end{aligned}
$$

Finally, we back-substitute -1 for y and 2 for z in equation (1) and solve for x:

$$
\begin{aligned}
x - 2(-1) + 3 \cdot 2 &= 11 \\
x + 2 + 6 &= 11 \\
x &= 3.
\end{aligned}
$$

We can check the triple $(3, -1, 2)$ in each of the three original equations. Since it makes all three equations true, the solution is $(3, -1, 2)$. ▬

EXAMPLE 2 Solve the following system:

$$
\begin{aligned}
x + y + z &= 7, && (1) \\
3x - 2y + z &= 3, && (2) \\
x + 6y + 3z &= 25. && (3)
\end{aligned}
$$

Solution We multiply equation (1) by -3 and add it to equation (2). We also multiply equation (1) by -1 and add it to equation (3).

$$
\begin{aligned}
x + y + z &= 7, & (1) \\
-5y - 2z &= -18, & (4) \\
5y + 2z &= 18 & (5)
\end{aligned}
$$

Next, we add equation (4) to equation (5):

$$
\begin{aligned}
x + y + z &= 7, & (1) \\
-5y - 2z &= -18, & (4) \\
0 &= 0.
\end{aligned}
$$

The equation $0 = 0$ tells us that equation (3) of the original system is dependent on the first two equations. Thus the original system is equivalent to

$$
\begin{aligned}
x + y + z &= 7, & (1) \\
3x - 2y + z &= 3. & (2)
\end{aligned}
$$

In this particular case, the original system has infinitely many solutions. (In some cases, the system composed of equations (1) and (2) could be inconsistent.) To find an expression for these solutions, we first solve equation (4) for either y or z. We choose to solve for y:

$$
\begin{aligned}
-5y - 2z &= -18 & (4) \\
-5y &= 2z - 18 \\
y &= -\tfrac{2}{5}z + \tfrac{18}{5}.
\end{aligned}
$$

Then we back-substitute in equation (1) to find an expression for x in terms of z:

$$
\begin{aligned}
x - \tfrac{2}{5}z + \tfrac{18}{5} + z &= 7 \\
x + \tfrac{3}{5}z + \tfrac{18}{5} &= 7 \\
x + \tfrac{3}{5}z &= \tfrac{17}{5} \\
x &= -\tfrac{3}{5}z + \tfrac{17}{5}.
\end{aligned}
$$

The solutions of the system of equations are ordered triples of the form $\left(-\tfrac{3}{5}z + \tfrac{17}{5},\ -\tfrac{2}{5}z + \tfrac{18}{5},\ z\right)$, where z can be any real number. Any real number that we use for z then gives us values for x and y and thus an ordered triple in the solution set. For example, if we choose $z = 0$, we have the solution $\left(\tfrac{17}{5}, \tfrac{18}{5}, 0\right)$. If we choose $z = -1$, we have $(4, 4, -1)$.

If we get a false equation, such as $0 = -5$, at some stage of the elimination process, we conclude that the original system is inconsistent. That is, it has no solutions.

Just as with systems of equations in two variables, systems of three linear equations in three variables can also be solved using Cramer's rule. See Appendix B.

Although systems of three linear equations in three variables do not lend themselves well to graphical solutions, it is of interest to picture some possible solutions. The graph of a linear equation in three variables is a plane. Thus the solution set of such a system is the intersection of three planes. Some possibilities are shown below.

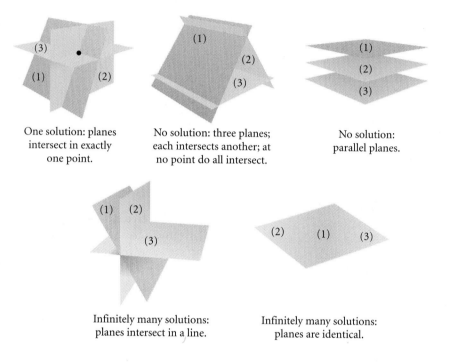

One solution: planes intersect in exactly one point.

No solution: three planes; each intersects another; at no point do all intersect.

No solution: parallel planes.

Infinitely many solutions: planes intersect in a line.

Infinitely many solutions: planes are identical.

If you have access to a grapher that graphs equations in three variables, use it to graph the systems in Examples 1 and 2.

Applications

Systems of equations in three or more variables allow us to solve many problems in fields such as business, the social and natural sciences, and engineering.

EXAMPLE 3   *Investment.*   Moira inherited $15,000 and invested part of it in a money market account, part in municipal bonds, and part in a mutual fund. After 1 yr, she received a total of $730 in simple interest from the three investments. The money market account paid 4% annually, the bonds paid 5% annually, and the mutual fund paid 6% annually. There was $2000 more invested in the mutual fund than in bonds. Find the amount that Moira invested in each category.

Solution

1. **Familiarize.** We let x, y, and z represent the amounts invested in the money market account, the bonds, and the mutual fund, respec-

tively. Then the amounts of income produced annually by each investment are given by 4%x, 5%y, and 6%z, or $0.04x$, $0.05y$, and $0.06z$.

2. **Translate.** The fact that a total of $15,000 is invested gives us one equation:

$$x + y + z = 15,000.$$

Since the total interest is $730, we have a second equation:

$$0.04x + 0.05y + 0.06z = 730.$$

Another statement in the problem gives us a third equation.

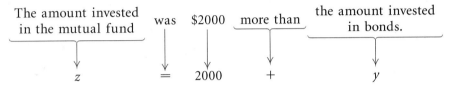

The amount invested in the mutual fund	was	$2000	more than	the amount invested in bonds.
z	$=$	2000	$+$	y

We now have a system of three equations:

$$
\begin{aligned}
x + y + z &= 15,000, \\
0.04x + 0.05y + 0.06z &= 730, \quad \text{or} \\
z &= 2000 + y;
\end{aligned}
\qquad
\begin{aligned}
x + y + z &= 15,000, \\
4x + 5y + 6z &= 73,000, \\
-y + z &= 2000.
\end{aligned}
$$

3. **Carry out.** Solving the system of equations, we get

(7000, 3000, 5000).

4. **Check.** The sum of the numbers is 15,000. The income produced is

$$0.04(7000) + 0.05(3000) + 0.06(5000) = 280 + 150 + 300, \quad \text{or} \quad \$730.$$

Also the amount invested in the mutual fund, $5000, is $2000 more than the amount invested in bonds, $3000. Our solution checks in the original problem.

5. **State.** Moira invested $7000 in a money market account, $3000 in municipal bonds, and $5000 in a mutual fund.

Mathematical Models and Applications

In a situation in which a quadratic function will serve as a mathematical model, we may wish to find an equation, or formula, for the function. For a linear model, we can find an equation if we know two data points. For a quadratic function, we need three data points.

EXAMPLE 4 *The Cost of Operating an Automobile at Various Speeds.* Under certain conditions, it is found that the cost of operating an automobile as a function of speed is approximated by a quadratic function. Use the data shown below to find an

equation of the function. Then use the equation to determine the cost of operating the automobile at 60 mph and at 80 mph.

SPEED (IN MILES PER HOUR)	OPERATING COST PER MILE (IN CENTS)
10	22
20	20
50	20

Solution   Letting x = the speed and $f(x)$ = the cost, we use the three data points (10, 22), (20, 20), and (50, 20) to find a, b, and c in the equation $f(x) = ax^2 + bx + c$. First we substitute:

For (10, 22): $22 = a \cdot 10^2 + b \cdot 10 + c$;

For (20, 20): $20 = a \cdot 20^2 + b \cdot 20 + c$;

For (50, 20): $20 = a \cdot 50^2 + b \cdot 50 + c$.

We now have a system of equations in the variables a, b, and c:

$$100a + 10b + c = 22,$$
$$400a + 20b + c = 20,$$
$$2500a + 50b + c = 20.$$

Solving this system of equations, we obtain $(0.005, -0.35, 25)$. Thus,

$$f(x) = 0.005x^2 - 0.35x + 25.$$

To determine the cost of operating an automobile at 60 mph, we find $f(60)$:

$$f(60) = 0.005(60)^2 - 0.35(60) + 25$$
$$= 22¢.$$

To find the cost of operating an automobile at 80 mph, we find $f(80)$:

$$f(80) = 0.005(80)^2 - 0.35(80) + 25$$
$$= 29¢.$$

$y_1 = 0.005x^2 - 0.35x + 25$

Y₁(60)	
	22
Y₁(80)	
	29

The function in Example 4 can also be found using the QUADRATIC REGRESSION feature on a grapher as we did in Section 2.5. Note that the method of Example 4 works when we have exactly three data points, whereas the QUADRATIC REGRESSION feature on a grapher can be used for three or more points.

Exercise Set 5.2

Solve each of the following systems.

1. $x + y + z = 2,$
$6x - 4y + 5z = 31,$
$5x + 2y + 2z = 13$

2. $x + 6y + 3z = 4,$
$2x + y + 2z = 3,$
$3x - 2y + z = 0$

3. $x - y + 2z = -3,$
$x + 2y + 3z = 4,$
$2x + y + z = -3$

4. $x + y + z = 6,$
$2x - y - z = -3,$
$x - 2y + 3z = 6$

5. $x + 2y - z = 5,$
$2x - 4y + z = 0,$
$3x + 2y + 2z = 3$

6. $2x + 3y - z = 1,$
$x + 2y + 5z = 4,$
$3x - y - 8z = -7$

7. $x + 2y - z = -8,$
$2x - y + z = 4,$
$8x + y + z = 2$

8. $x + 2y - z = 4,$
$4x - 3y + z = 8,$
$5x - y = 12$

9. $2x + y - 3z = 1,$
$x - 4y + z = 6,$
$4x - 7y - z = 13$

10. $x + 3y + 4z = 1,$
$3x + 4y + 5z = 3,$
$x + 8y + 11z = 2$

11. $4a + 9b = 8,$
$8a + 6c = -1,$
$6b + 6c = -1$

12. $3p + 2r = 11,$
$q - 7r = 4,$
$p - 6q = 1$

13. $w + x + y + z = 2,$
$w + 2x + 2y + 4z = 1,$
$-w + x - y - z = -6,$
$-w + 3x + y - z = -2$

14. $w + x - y + z = 0,$
$-w + 2x + 2y + z = 5,$
$-w + 3x + y - z = -4,$
$-2w + x + y - 3z = -7$

15. *e-Commerce.* computerwarehouse.com charges $3 for shipping orders up to 10 lb, $5 for orders from 10 lb up to 15 lb, and $7.50 for orders of 15 lb or more. One day shipping charges for 150 orders totaled $680. The number of orders under 10 lb was

three times the number of orders weighing 15 lb or more. Find the number of packages shipped at each rate.

16. *Mail-Order Business.* Natural Fibers Clothing charges $4 for shipping orders of $30 or less, $6 for orders from $30 to $70, and $7 for orders over $70. One week shipping charges for 600 orders totaled $3340. Eighty more orders for $30 or less were shipped than orders for more than $70. Find the number of orders shipped at each rate.

17. *Nutrition.* A hospital dietician must plan a lunch menu that provides 485 Cal, 41.5 g of carbohydrates, and 35 mg of calcium. A 3-oz serving of broiled ground beef contains 245 Cal, 0 g of carbohydrates, and 9 mg of calcium. One baked potato contains 145 Cal, 34 g of carbohydrates, and 8 mg of calcium. A one-cup serving of strawberries contains 45 Cal, 10 g of carbohydrates, and 21 mg of calcium. (*Source: Home and Garden Bulletin No. 72*, U.S. Government Printing Office, Washington, D.C. 20402) How many servings of each are required to provide the desired nutritional values?

18. *Nutrition.* A diabetic patient wishes to prepare a meal consisting of roasted chicken breast, mashed potatoes, and peas. A 3-oz serving of roasted, skinless chicken breast contains 140 Cal, 27 g of protein, and 64 mg of sodium. A one-cup serving of mashed potatoes contains 160 Cal, 4 g of protein, and 636 mg of sodium, and a one-cup serving of peas contains 125 Cal, 8 g of protein, and 139 mg of sodium. (*Source: Home and Garden Bulletin No. 72*, U.S. Government Printing Office, Washington, D.C. 20402) How many servings of each should be used if the meal is to contain 415 Cal, 50.5 g of protein, and 553 mg of sodium?

19. *Investment.* Jamal earns a year-end bonus of $5000 and puts it in 3 one-year investments that pay $302 in simple interest. Part is invested at 4%, part at 6%, and part at 7%. There is $1500 more invested at 7% than at 4%. Find the amount invested at each rate.

20. *Investment.* Casey receives $336 per year in simple interest from three investments. Part is invested at 8%, part at 9%, and part at 10%. There is $500 more invested at 9% than at 8%. The amount invested at 10% is three times the amount invested at 9%. Find the amount invested at each rate.

21. *Price Increases.* Orange juice, a raisin bagel, and a cup of coffee from Kelly's Koffee Kart cost a total of $3. Kelly posts a notice announcing that, effective next week, the price of orange juice will increase

50% and the price of bagels will increase 20%. After the increase, the same purchase will cost a total of $3.75, and orange juice will cost twice as much as coffee. Find the price of each item before the increase.

22. *Cost of Snack Food.* Martin and Eva pool their loose change to buy snacks on their coffee break. One day, they spent $1.85 on 1 carton of milk, 2 donuts, and 1 cup of coffee. The next day, they spent $2.30 on 3 donuts and 2 cups of coffee. The third day, they bought 1 carton of milk, 1 donut, and 2 cups of coffee and spent $1.75. On the fourth day, they have a total of $1.80 left. Is this enough to buy 2 cartons of milk and 2 donuts?

23. *Passenger Transportation.* The total volume of passenger traffic on domestic airways, by bus (excluding school buses and urban transit buses), and by railroads in the United States in a recent year was 405 billion passenger-miles. (One passenger-mile is the transportation of one passenger the distance of one mile.) The volume of bus traffic was 10 billion passenger-miles more than the volume of railroad traffic. The total volume of railroad traffic was 329 billion passenger-miles less than the volume of traffic on domestic airways. (*Source*: *Transportation in America,* Eno Transportation Foundation, Inc., Landsdowne, VA) What was the volume of each type of passenger traffic?

24. *Cheese Consumption.* The total per capita consumption (the average amount consumed per person) of cheddar, mozzarella, and Swiss cheese in the United States in a recent year was 18.1 lb. The total consumption of mozzarella and Swiss cheese was 0.3 lb less than that of cheddar cheese. The consumption of mozzarella cheese was 6.5 lb more than that of Swiss cheese. (*Source*: U.S. Department of Agriculture, Economic Research Service, *Food Consumption, Prices, and Expenditures*) What was the per capita consumption of each type of cheese?

25. *Golf.* On an 18-hole golf course, there are par-3 holes, par-4 holes, and par-5 holes. A golfer who shoots par on every hole has a score of 72. The sum of the number of par-3 holes and the number of par-5 holes is 8. How many of each type of hole are there on the golf course?

26. *Golf.* On an 18-hole golf course, there are par-3 holes, par-4 holes, and par-5 holes. A golfer who shoots par on every hole has a score of 70. There are twice as many par-4 holes as there are par-5 holes.

How many of each type of hole are there on the golf course?

27. *Number of Marriages.* The following table shows the number of marriages, in thousands, in California, represented as years since 1980.

YEAR, x	NUMBER OF MARRIAGES (IN THOUSANDS)
1980, 0	211
1990, 10	237
1996, 16	203

Source: U.S. National Center for Health Statistics, *Vital Statistics of the United States,* annual, Monthly Vital Statistics Reports

a) Fit a quadratic function $f(x) = ax^2 + bx + c$ to the data.
b) Use the function to predict the number of marriages in California in 2005.

28. *Coffee Consumption.* The following table shows per capita coffee consumption, in gallons, in the United States, represented as years since 1992.

YEAR, x	COFFEE CONSUMPTION (IN GALLONS)
1992, 0	26
1994, 2	21
1996, 4	23

a) Fit a quadratic function $f(x) = ax^2 + bx + c$ to the data.
b) Use the function to predict the per capita consumption of coffee in 2003.

29. *Milk Consumption.* The following table shows per capita milk consumption, in pounds, in the United States, represented as years since 1985.

YEAR, x	MILK CONSUMPTION (IN POUNDS)
1985, 0	594
1990, 5	567
1996, 11	576

Source: U.S. Department of Agriculture, Economic Research Service, *Food Consumption, Prices, and Expenditures*

a) Fit a quadratic function $f(x) = ax^2 + bx + c$ to the data.
b) Use the function to predict the per capita consumption of milk in 2005.

30. *Crude Steel Production.* The following table shows the world production of crude steel, in millions of metric tons, represented as years since 1990.

YEAR, x	CRUDE STEEL PRODUCTION (IN MILLIONS OF METRIC TONS)
1990, 0	771
1995, 5	755
1997, 7	773

Source: U.S. Geological Survey

a) Fit a quadratic function $f(x) = ax^2 + bx + c$ to the data.
b) Use the function to predict the world production of crude steel in 2007.

Discussion and Writing

31. Given two linear equations in three variables, $Ax + By + Cz = D$ and $Ex + Fy + Gz = H$, explain how you would find a third equation such that the system contains dependent equations.

32. Write a problem for a classmate to solve that can be translated to a system of three equations in three variables.

Skill Maintenance

Simplify. Write answers in the form $a + bi$, where a and b are real numbers.

33. $(3 - 4i) - (-2 - i)$ **34.** $(5 + 2i) + (1 - 4i)$

35. $(1 - 2i)(6 + 2i)$ **36.** $\dfrac{3 + i}{4 - 3i}$

Synthesis

In Exercises 37 and 38, let u represent $1/x$, v represent $1/y$, and w represent $1/z$. Solve first for u, v, and w. Then solve the system.

37. $\dfrac{2}{x} - \dfrac{1}{y} - \dfrac{3}{z} = -1,$

$\dfrac{2}{x} - \dfrac{1}{y} + \dfrac{1}{z} = -9,$

$\dfrac{1}{x} + \dfrac{2}{y} - \dfrac{4}{z} = 17$

38. $\dfrac{2}{x} + \dfrac{2}{y} - \dfrac{3}{z} = 3,$

$\dfrac{1}{x} - \dfrac{2}{y} - \dfrac{3}{z} = 9,$

$\dfrac{7}{x} - \dfrac{2}{y} + \dfrac{9}{z} = -39$

39. Find the sum of the angle measures at the tips of the star.

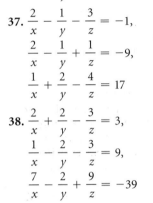

40. *Transcontinental Railroad.* Use the following facts to find the year in which the first U.S. transcontinental railroad was completed. The sum of the digits in the year is 24. The units digit is 1 more than the hundreds digit. Both the tens and the units digits are multiples of three.

In Exercises 41 and 42, three solutions of an equation are given. Use a system of three equations in three variables to find the constants and write the equation.

41. $Ax + By + Cz = 12;$
$\left(1, \frac{3}{4}, 3\right), \left(\frac{4}{3}, 1, 2\right),$ and $(2, 1, 1)$

42. $y = B - Mx - Nz$;
$(1, 1, 2), (3, 2, -6),$ and $\left(\frac{3}{2}, 1, 1\right)$

In Exercises 43 and 44, four solutions of the equation
$y = ax^3 + bx^2 + cx + d$ *are given. Use a system of four equations in four variables to find the constants and write the equation.*

43. $(-2, 59), (-1, 13), (1, -1),$ and $(2, -17)$

44. $(-2, -39), (-1, -12), (1, -6),$ and $(3, 16)$

45. *Theater Attendance.*   A performance at the Bingham Performing Arts Center was attended by 100 people. The audience consisted of adults, students, and children. The ticket prices were $10 for adults, $3 for students, and 50 cents for children. The total amount of money taken in was $100. How many adults, students, and children were in attendance? Does there seem to be some information missing? Do some careful reasoning.

Matrices and Systems of Equations

5.3

• *Solve systems of equations using matrices.*

Matrices and Row-Equivalent Operations

In this section, we consider additional techniques for solving systems of equations. You have probably observed that when we solve a system of equations, we perform computations with the coefficients and the constants and continually rewrite the variables. We can streamline the solution process by omitting the variables until a solution is found. For example, the system

$$2x - 3y = 7,$$
$$x + 4y = -2$$

can be written more simply as

$$\begin{bmatrix} 2 & -3 & | & 7 \\ 1 & 4 & | & -2 \end{bmatrix}.$$

The vertical line replaces the equals signs.

A rectangular array of numbers like the one above is called a **matrix** (pl., **matrices**). The matrix above is called an **augmented matrix** for the given system of equations, because it contains not only the coefficients but also the constant terms. The matrix

$$\begin{bmatrix} 2 & -3 \\ 1 & 4 \end{bmatrix}$$

is called the **coefficient matrix** of the system.

The **rows** of a matrix are horizontal, and the **columns** are vertical. The augmented matrix above has 2 rows and 3 columns, and the coefficient matrix has 2 rows and 2 columns. A matrix with m rows and n columns is said to be of **order** $m \times n$. Thus the order of the augmented matrix above is 2×3, and the order of the coefficient matrix is 2×2. When $m = n$, a matrix is said to be **square**. The coefficient matrix above is a square matrix. The numbers 2 and 4 lie on the **main diagonal** of the coefficient matrix. The numbers in a matrix are called **entries**.

Gaussian Elimination with Matrices

In Section 5.2, we described a series of operations that can be used to transform a system of equations to an equivalent system. Each of these operations corresponds to one that can be used to produce *row-equivalent matrices*.

Row-Equivalent Operations

1. Interchange any two rows.

2. Multiply each entry in a row by the same nonzero constant.

3. Add a nonzero multiple of one row to another row.

We can use these operations on the augmented matrix of a system of equations to solve the system.

EXAMPLE 1   Solve the following system:

$$
\begin{aligned}
2x - \ y + \ 4z &= -3, \\
x - 2y - 10z &= -6, \\
3x \qquad + \ 4z &= 7.
\end{aligned}
$$

Solution   First, we write the augmented matrix, writing 0 for the missing y-term in the last equation:

$$
\left[\begin{array}{rrr|r}
2 & -1 & 4 & -3 \\
1 & -2 & -10 & -6 \\
3 & 0 & 4 & 7
\end{array}\right].
$$

Our goal is to find a row-equivalent matrix of the form

$$
\left[\begin{array}{ccc|c}
1 & a & b & c \\
0 & 1 & d & e \\
0 & 0 & 1 & f
\end{array}\right].
$$

The variables can then be reinserted to form equations from which we can complete the solution. This is done by working from the bottom equation to the top and using back-substitution.

The first step is to multiply and/or interchange rows so that each number in the first column below the first number is a multiple of that number. In this case, we interchange the first and second rows to obtain a 1 in the upper left-hand corner.

$$
\left[\begin{array}{rrr|r}
1 & -2 & -10 & -6 \\
2 & -1 & 4 & -3 \\
3 & 0 & 4 & 7
\end{array}\right]
\qquad
\begin{array}{l}
\text{New row 1} = \text{row 2} \\
\text{New row 2} = \text{row 1}
\end{array}
$$

Next, we multiply the first row by -2 and add it to the second row. We also multiply the first row by -3 and add it to the third row.

$$\begin{bmatrix} 1 & -2 & -10 & | & -6 \\ 0 & 3 & 24 & | & 9 \\ 0 & 6 & 34 & | & 25 \end{bmatrix} \quad \begin{array}{l} \text{New row 2} = -2(\text{row 1}) + \text{row 2} \\ \text{New row 3} = -3(\text{row 1}) + \text{row 3} \end{array}$$

Now we multiply the second row by $\frac{1}{3}$ to get a 1 in the second row, second column.

$$\begin{bmatrix} 1 & -2 & -10 & | & -6 \\ 0 & 1 & 8 & | & 3 \\ 0 & 6 & 34 & | & 25 \end{bmatrix} \quad \text{New row 2} = \tfrac{1}{3}(\text{row 2})$$

Then we multiply the second row by -6 and add it to the third row.

$$\begin{bmatrix} 1 & -2 & -10 & | & -6 \\ 0 & 1 & 8 & | & 3 \\ 0 & 0 & -14 & | & 7 \end{bmatrix} \quad \text{New row 3} = -6(\text{row 2}) + \text{row 3}$$

Finally, we multiply the third row by $-\frac{1}{14}$ to get a 1 in the third row, third column.

$$\begin{bmatrix} 1 & -2 & -10 & | & -6 \\ 0 & 1 & 8 & | & 3 \\ 0 & 0 & 1 & | & -\frac{1}{2} \end{bmatrix} \quad \text{New row 3} = -\tfrac{1}{14}(\text{row 3})$$

Now we can write the system of equations that corresponds to the last matrix above:

$$\begin{aligned} x - 2y - 10z &= -6, & (1) \\ y + 8z &= 3, & (2) \\ z &= -\tfrac{1}{2}. & (3) \end{aligned}$$

We back-substitute $-\frac{1}{2}$ for z in equation (2) and solve for y:

$$\begin{aligned} y + 8\left(-\tfrac{1}{2}\right) &= 3 \\ y - 4 &= 3 \\ y &= 7. \end{aligned}$$

Next, we back-substitute 7 for y and $-\frac{1}{2}$ for z in equation (1) and solve for x:

$$\begin{aligned} x - 2 \cdot 7 - 10\left(-\tfrac{1}{2}\right) &= -6 \\ x - 14 + 5 &= -6 \\ x - 9 &= -6 \\ x &= 3. \end{aligned}$$

The triple $\left(3, 7, -\frac{1}{2}\right)$ checks in the original system of equations, so it is the solution.

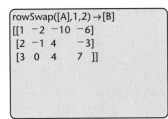

```
rowSwap([A],1,2)→[B]
[[1  −2  −10  −6]
 [2  −1   4   −3]
 [3   0   4    7 ]]
```

STUDY TIP

The *Graphing Calculator Manual* that accompanies this text contains the keystrokes for performing all the row-equivalent operations in Example 1.

Row-equivalent operations can be performed on a grapher. For example, to interchange the first and second rows of the augmented matrix, as we did in the first step above, we enter the matrix as matrix **A** and select "rowSwap" from the MATRIX MATH menu. Some graphers will not automatically store the matrix produced using a row-equivalent operation, so when several operations are to be performed in succession, it is helpful to store the result of each operation as it is produced. In the window at left, we see both the matrix produced by the rowSwap operation and the indication that this matrix is stored as matrix **B**.

The procedure followed in Example 1 is called **Gaussian elimination with matrices.** The last matrix on p. 382 is in **row-echelon form.** To be in this form, a matrix must have the following properties.

Row-Echelon Form

1. If a row does not consist entirely of 0's, then the first nonzero element in the row is a 1 (called a **leading 1**).
2. For any two successive nonzero rows, the leading 1 in the lower row is farther to the right than the leading 1 in the higher row.
3. All the rows consisting entirely of 0's are at the bottom of the matrix.

If a fourth property is also satisfied, a matrix is said to be in **reduced row-echelon form:**

4. Each column that contains a leading 1 has 0's everywhere else.

EXAMPLE 2 Which of the following matrices are in row-echelon form? Which, if any, are in reduced row-echelon form?

a) $\begin{bmatrix} 1 & -3 & 5 & | & -2 \\ 0 & 1 & -4 & | & 3 \\ 0 & 0 & 1 & | & 10 \end{bmatrix}$ b) $\begin{bmatrix} 0 & -1 & | & 2 \\ 0 & 1 & | & 5 \end{bmatrix}$ c) $\begin{bmatrix} 1 & -2 & -6 & 4 & | & 7 \\ 0 & 3 & 5 & -8 & | & -1 \\ 0 & 0 & 1 & 9 & | & 2 \end{bmatrix}$

d) $\begin{bmatrix} 1 & 0 & 0 & | & -2.4 \\ 0 & 1 & 0 & | & 0.8 \\ 0 & 0 & 1 & | & 5.6 \end{bmatrix}$ e) $\begin{bmatrix} 1 & 0 & 0 & 0 & | & \frac{2}{3} \\ 0 & 1 & 0 & 0 & | & -\frac{1}{4} \\ 0 & 0 & 1 & 0 & | & \frac{6}{7} \\ 0 & 0 & 0 & 0 & | & 0 \end{bmatrix}$ f) $\begin{bmatrix} 1 & -4 & 2 & | & 5 \\ 0 & 0 & 0 & | & 0 \\ 0 & 1 & -3 & | & -8 \end{bmatrix}$

Solution The matrices in (a), (d), and (e) satisfy the row-echelon criteria and, thus, are in row-echelon form. In (b) and (c), the first nonzero elements of the first and second rows, respectively, are not 1. In (f), the row consisting entirely of 0's is not at the bottom of the matrix. Thus the matrices in (b), (c), and (f) are not in row-echelon form. In (d) and (e), not only are the row-echelon criteria met but each column that contains a leading 1 also has 0's elsewhere, so these matrices are in reduced row-echelon form.

Gauss–Jordan Elimination

We have seen that with Gaussian elimination we perform row-equivalent operations on a matrix to obtain a row-equivalent matrix in row-echelon form. When we continue to apply these operations until we have a matrix in *reduced* row-echelon form, we are using **Gauss–Jordan elimination.** This method is named for Karl Friedrich Gauss and Wilhelm Jordan (1842–1899).

EXAMPLE 3   Use Gauss–Jordan elimination to solve the system of equations in Example 1.

Solution   Using Gaussian elimination in Example 1, we obtained the matrix

$$\begin{bmatrix} 1 & -2 & -10 & -6 \\ 0 & 1 & 8 & 3 \\ 0 & 0 & 1 & -\frac{1}{2} \end{bmatrix}.$$

We continue to perform row-equivalent operations until we have a matrix in reduced row-echelon form. We multiply the third row by 10 and add it to the first row. We also multiply the third row by -8 and add it to the second row.

$$\begin{bmatrix} 1 & -2 & 0 & -11 \\ 0 & 1 & 0 & 7 \\ 0 & 0 & 1 & -\frac{1}{2} \end{bmatrix}$$ New row 1 = 10(row 3) + row 1
New row 2 = -8(row 3) + row 2

Next, we multiply the second row by 2 and add it to the first row.

$$\begin{bmatrix} 1 & 0 & 0 & 3 \\ 0 & 1 & 0 & 7 \\ 0 & 0 & 1 & -\frac{1}{2} \end{bmatrix}$$ New row 1 = 2(row 2) + row 1

Writing the system of equations that corresponds to this matrix, we have

$$x = 3,$$
$$y = 7,$$
$$z = -\tfrac{1}{2}.$$

We can actually read the solution, $\left(3, 7, -\frac{1}{2}\right)$, directly from the last column of the reduced row-echelon matrix.

EXAMPLE 4   Solve the following system:

$$3x - 4y - z = 6,$$
$$2x - y + z = -1,$$
$$4x - 7y - 3z = 13.$$

Solution We write the augmented matrix and use Gauss–Jordan elimination.

$$\left[\begin{array}{rrr|r} 3 & -4 & -1 & 6 \\ 2 & -1 & 1 & -1 \\ 4 & -7 & -3 & 13 \end{array}\right]$$

We begin by multiplying the second and third rows by 3 so that each number in the first column below the first number, 3, is a multiple of that number.

$$\left[\begin{array}{rrr|r} 3 & -4 & -1 & 6 \\ 6 & -3 & 3 & -3 \\ 12 & -21 & -9 & 39 \end{array}\right] \quad \begin{array}{l} \text{New row 2} = 3(\text{row 2}) \\ \text{New row 3} = 3(\text{row 3}) \end{array}$$

Next, we multiply the first row by -2 and add it to the second row. We also multiply the first row by -4 and add it to the third row.

$$\left[\begin{array}{rrr|r} 3 & -4 & -1 & 6 \\ 0 & 5 & 5 & -15 \\ 0 & -5 & -5 & 15 \end{array}\right] \quad \begin{array}{l} \text{New row 2} = -2(\text{row 1}) + \text{row 2} \\ \text{New row 3} = -4(\text{row 1}) + \text{row 3} \end{array}$$

Now we add the second row to the third row.

$$\left[\begin{array}{rrr|r} 3 & -4 & -1 & 6 \\ 0 & 5 & 5 & -15 \\ 0 & 0 & 0 & 0 \end{array}\right] \quad \text{New row 3} = \text{row 2} + \text{row 3}$$

We can stop at this stage because we have a row consisting entirely of 0's. The last row of the matrix corresponds to the equation $0 = 0$, which is true for all values of x, y, and z. Consequently, the equations are dependent and the system is equivalent to

$$\begin{aligned} 3x - 4y - z &= 6, \\ 5y + 5z &= -15. \end{aligned}$$

This particular system has infinitely many solutions. (A system containing dependent equations could be inconsistent.)

Solving the second equation for y gives us

$$y = -z - 3.$$

Substituting for y in the first equation, we get

$$\begin{aligned} 3x - 4(-z - 3) - z &= 6 \\ x &= -z - 2. \end{aligned}$$

Then the solutions of this system are of the form

$$(-z - 2, -z - 3, z),$$

where z can be any real number.

Similarly, if we obtain a row whose only nonzero entry occurs in the last column, we have an inconsistent system of equations. For example, in the matrix

$$\left[\begin{array}{ccc|c} 1 & 0 & 3 & -2 \\ 0 & 1 & 5 & 4 \\ 0 & 0 & 0 & 6 \end{array}\right],$$

the last row corresponds to the false equation $0 = 6$, so we know the original system of equations has no solution.

Exercise Set 5.3

Determine the order of the matrix.

1. $\begin{bmatrix} 1 & -6 \\ -3 & 2 \\ 0 & 5 \end{bmatrix}$

2. $\begin{bmatrix} 7 \\ -5 \\ -1 \\ 3 \end{bmatrix}$

3. $[2 \ -4 \ 0 \ 9]$

4. $[-8]$

5. $\begin{bmatrix} 1 & -5 & -8 \\ 6 & 4 & -2 \\ -3 & 0 & 7 \end{bmatrix}$

6. $\begin{bmatrix} 13 & 2 & -6 & 4 \\ -1 & 18 & 5 & -12 \end{bmatrix}$

Write the augmented matrix for the system of equations.

7. $2x - y = 7,$
$\quad x + 4y = -5$

8. $3x + 2y = 8,$
$\quad 2x - 3y = 15$

9. $\quad x - 2y + 3z = 12,$
$\quad 2x \qquad - 4z = 8,$
$\qquad 3y + z = 7$

10. $\quad x + y - z = 7,$
$\qquad 3y + 2z = 1,$
$\quad -2x - 5y \qquad = 6$

Write the system of equations that corresponds to the augmented matrix.

11. $\left[\begin{array}{cc|c} 3 & -5 & 1 \\ 1 & 4 & -2 \end{array}\right]$

12. $\left[\begin{array}{cc|c} 1 & 2 & -6 \\ 4 & 1 & -3 \end{array}\right]$

13. $\left[\begin{array}{ccc|c} 2 & 1 & -4 & 12 \\ 3 & 0 & 5 & -1 \\ 1 & -1 & 1 & 2 \end{array}\right]$

14. $\left[\begin{array}{ccc|c} -1 & -2 & 3 & 6 \\ 0 & 4 & 1 & 2 \\ 2 & -1 & 0 & 9 \end{array}\right]$

Solve the system of equations using Gaussian elimination or Gauss–Jordan elimination.

15. $4x + 2y = 11,$
$\quad 3x - y = 2$

16. $2x + y = 1,$
$\quad 3x + 2y = -2$

17. $5x - 2y = -3,$
$\quad 2x + 5y = -24$

18. $2x + y = 1,$
$\quad 3x - 6y = 4$

19. $\quad 3x + 4y = 7,$
$\quad -5x + 2y = 10$

20. $5x - 3y = -2,$
$\quad 4x + 2y = 5$

21. $3x + 2y = 6,$
$\quad 2x - 3y = -9$

22. $\quad x - 4y = 9,$
$\quad 2x + 5y = 5$

23. $\quad x - 3y = 8,$
$\quad -2x + 6y = 3$

24. $4x - 8y = 12,$
$\quad -x + 2y = -3$

25. $-2x + 6y = 4,$
$\quad 3x - 9y = -6$

26. $\quad 6x + 2y = -10,$
$\quad -3x - y = 6$

27. $\quad x + 2y - 3z = 9,$
$\quad 2x - y + 2z = -8,$
$\quad 3x - y - 4z = 3$

28. $\quad x - y + 2z = 0,$
$\quad x - 2y + 3z = -1,$
$\quad 2x - 2y + z = -3$

29. $4x - y - 3z = 1,$
$\quad 8x + y - z = 5,$
$\quad 2x + y + 2z = 5$

30. $3x + 2y + 2z = 3,$
$\quad x + 2y - z = 5,$
$\quad 2x - 4y + z = 0$

31. $\quad x - 2y + 3z = -4,$
$\quad 3x + y - z = 0,$
$\quad 2x + 3y - 5z = 1$

32. $\quad 2x - 3y + 2z = 2,$
$\quad x + 4y - z = 9,$
$\quad -3x + y - 5z = 5$

33. $2x - 4y - 3z = 3,$
$\quad x + 3y + z = -1,$
$\quad 5x + y - 2z = 2$

34. $\quad x + y - 3z = 4,$
$\quad 4x + 5y + z = 1,$
$\quad 2x + 3y + 7z = -7$

35. $\quad p + q + r = 1,$
$\quad p + 2q + 3r = 4,$
$\quad 4p + 5q + 6r = 7$

36. $\quad m + n + t = 9,$
$\quad m - n - t = -15,$
$\quad 3m + n + t = 2$

37. $\begin{aligned} a + b - c &= 7, \\ a - b + c &= 5, \\ 3a + b - c &= -1 \end{aligned}$ **38.** $\begin{aligned} a - b + c &= 3, \\ 2a + b - 3c &= 5, \\ 4a + b - c &= 11 \end{aligned}$

39. $\begin{aligned} -2w + 2x + 2y - 2z &= -10, \\ w + x + y + z &= -5, \\ 3w + x - y + 4z &= -2, \\ w + 3x - 2y + 2z &= -6 \end{aligned}$

40. $\begin{aligned} -w + 2x - 3y + z &= -8, \\ -w + x + y - z &= -4, \\ w + x + y + z &= 22, \\ -w + x - y - z &= -14 \end{aligned}$

Use Gaussian elimination or Gauss–Jordan elimination in Exercises 41–44.

41. *Time of Return.* The Houlihans pay their babysitter $5 per hour before 11 P.M. and $7.50 per hour after 11 P.M. One evening they went out for 5 hr and paid the sitter $30. What time did they come home?

42. *Advertising Expense.* eAuction.com spent a total of $11 million on advertising in fiscal years 1998, 1999, and 2000. The amount spent in 2000 was three times the amount spent in 1998. The amount spent in 1999 was $3 million less than the amount spent in 2000. How much was spent on advertising each year?

43. *Borrowing.* Gonzalez Manufacturing borrowed $30,000 to buy a new piece of equipment. Part of the money was borrowed at 8%, part at 10%, and part at 12%. The annual interest was $3040, and the total amount borrowed at 8% and 10% was twice the amount borrowed at 12%. How much was borrowed at each rate?

44. *Stamp Purchase.* Ricardo spent $15.84 on 33¢ and 22¢ stamps. He bought a total of 60 stamps. How many of each type did he buy?

Discussion and Writing

45. Solve the following system of equations using matrices and Gaussian elimination. Then solve it again using Gauss–Jordan elimination. Do you prefer one method over the other? Why or why not?

$$\begin{aligned} 3x + 4y + 2z &= 0, \\ x - y - z &= 10, \\ 2x + 3y + 3z &= -10 \end{aligned}$$

46. Explain in your own words why the augmented matrix below represents a system of dependent equations.

$$\begin{bmatrix} 1 & -3 & 2 & | & -5 \\ 0 & 1 & -4 & | & 8 \\ 0 & 0 & 0 & | & 0 \end{bmatrix}$$

Skill Maintenance

Solve.

47. $2x^2 + x = 7$

48. $\dfrac{1}{x + 1} - \dfrac{6}{x - 1} = 1$

49. $\sqrt{2x + 1} - 1 = \sqrt{2x - 4}$

50. $x - \sqrt{x} - 6 = 0$

Synthesis

In Exercises 51 and 52, three solutions of the equation $y = ax^2 + bx + c$ are given. Use a system of three equations in three variables and Gaussian elimination or Gauss–Jordan elimination to find the constants and write the equation.

51. $(-3, 12)$, $(-1, -7)$, and $(1, -2)$

52. $(-1, 0)$, $(1, -3)$, and $(3, -22)$

53. Find two different row-echelon forms of

$$\begin{bmatrix} 1 & 5 \\ 3 & 2 \end{bmatrix}.$$

54. Consider the system of equations

$$\begin{aligned} x - y + 3z &= -8, \\ 2x + 3y - z &= 5, \\ 3x + 2y + 2kz &= -3k. \end{aligned}$$

For what value(s) of k, if any, will the system have

a) no solution?

b) exactly one solution?

c) infinitely many solutions?

Solve using matrices.

55. $y = x + z,$
$\quad 3y + 5z = 4,$
$\quad x + 4 = y + 3z$

56. $x + y = 2z,$
$\quad 2x - 5z = 4,$
$\quad x - z = y + 8$

57. $\quad x - 4y + 2z = 7,$
$\quad 3x + y + 3z = -5$

58. $\quad x - y - 3z = 3,$
$\quad -x + 3y + z = -7$

59. $\quad 4x + 5y = 3,$
$\quad -2x + y = 9,$
$\quad 3x - 2y = -15$

60. $2x - 3y = -1,$
$\quad -x + 2y = -2,$
$\quad 3x - 5y = 1$

Matrix Operations

5.4

- *Add, subtract, and multiply matrices when possible.*
- *Write a matrix equation equivalent to a system of equations.*

In Section 5.3, we used matrices to solve systems of equations. Matrices are useful in many other types of applications as well. In this section, we study matrices and some of their properties.

A capital letter is generally used to name a matrix, and lower-case letters with double subscripts generally denote its entries. For example, a_{47}, read "a sub four seven," indicates the entry in the fourth row and the seventh column. A general term is represented by a_{ij}. The notation a_{ij} indicates the entry in row i and column j. In general, we can write a matrix as

$$\mathbf{A} = [a_{ij}] = \begin{bmatrix} a_{11} & a_{12} & a_{13} & \cdots & a_{1n} \\ a_{21} & a_{22} & a_{23} & \cdots & a_{2n} \\ a_{31} & a_{32} & a_{33} & \cdots & a_{3n} \\ \cdot & \cdot & \cdot & & \cdot \\ \cdot & \cdot & \cdot & & \cdot \\ \cdot & \cdot & \cdot & & \cdot \\ a_{m1} & a_{m2} & a_{m3} & \cdots & a_{mn} \end{bmatrix}.$$

The matrix above has m rows and n columns. That is, its order is $m \times n$.

Two matrices are **equal** if they have the same order and corresponding entries are equal.

Matrix Addition and Subtraction

To add or subtract matrices, we add or subtract their corresponding entries. The matrices must have the same order for this to be possible.

Addition and Subtraction of Matrices

Given two $m \times n$ matrices $\mathbf{A} = [a_{ij}]$ and $\mathbf{B} = [b_{ij}]$, their sum is

$$\mathbf{A} + \mathbf{B} = [a_{ij} + b_{ij}]$$

and their difference is

$$\mathbf{A} - \mathbf{B} = [a_{ij} - b_{ij}].$$

Addition of matrices is both commutative and associative.

EXAMPLE 1   Find $\mathbf{A} + \mathbf{B}$ for each of the following.

a) $\mathbf{A} = \begin{bmatrix} -5 & 0 \\ 4 & \frac{1}{2} \end{bmatrix}, \quad \mathbf{B} = \begin{bmatrix} 6 & -3 \\ 2 & 3 \end{bmatrix}$

b) $\mathbf{A} = \begin{bmatrix} 1 & 3 \\ -1 & 5 \\ 6 & 0 \end{bmatrix}, \quad \mathbf{B} = \begin{bmatrix} -1 & -2 \\ 1 & -2 \\ -3 & 1 \end{bmatrix}$

Solution   Each pair of matrices has the same order. We add the corresponding entries.

a) $\mathbf{A} + \mathbf{B} = \begin{bmatrix} -5 & 0 \\ 4 & \frac{1}{2} \end{bmatrix} + \begin{bmatrix} 6 & -3 \\ 2 & 3 \end{bmatrix}$

$\qquad = \begin{bmatrix} -5 + 6 & 0 + (-3) \\ 4 + 2 & \frac{1}{2} + 3 \end{bmatrix}$

$\qquad = \begin{bmatrix} 1 & -3 \\ 6 & 3\frac{1}{2} \end{bmatrix}$

```
[A]+[B]
          [[1 -3 ]
           [6  3.5]]
```

We can also enter $\mathbf{A}$ and $\mathbf{B}$ in a grapher and then find $\mathbf{A} + \mathbf{B}$.

b) $\mathbf{A} + \mathbf{B} = \begin{bmatrix} 1 & 3 \\ -1 & 5 \\ 6 & 0 \end{bmatrix} + \begin{bmatrix} -1 & -2 \\ 1 & -2 \\ -3 & 1 \end{bmatrix}$

$\qquad = \begin{bmatrix} 1 + (-1) & 3 + (-2) \\ -1 + 1 & 5 + (-2) \\ 6 + (-3) & 0 + 1 \end{bmatrix}$

$\qquad = \begin{bmatrix} 0 & 1 \\ 0 & 3 \\ 3 & 1 \end{bmatrix}$

```
[A]+[B]
          [[0  1]
           [0  3]
           [3  1]]
```

This sum can also be found on a grapher after $\mathbf{A}$ and $\mathbf{B}$ have been entered.

EXAMPLE 2   Find $\mathbf{C} - \mathbf{D}$ for each of the following.

a) $\mathbf{C} = \begin{bmatrix} 1 & 2 \\ -2 & 0 \\ -3 & -1 \end{bmatrix}, \quad \mathbf{D} = \begin{bmatrix} 1 & -1 \\ 1 & 3 \\ 2 & 3 \end{bmatrix}$

b) $\mathbf{C} = \begin{bmatrix} 5 & -6 \\ -3 & 4 \end{bmatrix}, \quad \mathbf{D} = \begin{bmatrix} -4 \\ 1 \end{bmatrix}$

Solution

a) We subtract corresponding entries:

$$\mathbf{C} - \mathbf{D} = \begin{bmatrix} 1 & 2 \\ -2 & 0 \\ -3 & -1 \end{bmatrix} - \begin{bmatrix} 1 & -1 \\ 1 & 3 \\ 2 & 3 \end{bmatrix}$$

$$= \begin{bmatrix} 1 - 1 & 2 - (-1) \\ -2 - 1 & 0 - 3 \\ -3 - 2 & -1 - 3 \end{bmatrix} = \begin{bmatrix} 0 & 3 \\ -3 & -3 \\ -5 & -4 \end{bmatrix}.$$

This subtraction can also be done using a grapher.

b) The matrices do not have the same order, so we cannot subtract. ▬

The **opposite**, or **additive inverse,** of a matrix is obtained by replacing each entry with its opposite.

```
[C]−[D]
          [[0    3]
           [−3 −3]
           [−5 −4]]
```

EXAMPLE 3   Find $-\mathbf{A}$ and $\mathbf{A} + (-\mathbf{A})$ for

$$\mathbf{A} = \begin{bmatrix} 1 & 0 & 2 \\ 3 & -1 & 5 \end{bmatrix}.$$

Solution   To find $-\mathbf{A}$, we replace each entry of $\mathbf{A}$ with its opposite.

$$-\mathbf{A} = \begin{bmatrix} -1 & 0 & -2 \\ -3 & 1 & -5 \end{bmatrix},$$

$$\mathbf{A} + (-\mathbf{A}) = \begin{bmatrix} 1 & 0 & 2 \\ 3 & -1 & 5 \end{bmatrix} + \begin{bmatrix} -1 & 0 & -2 \\ -3 & 1 & -5 \end{bmatrix}$$

$$= \begin{bmatrix} 0 & 0 & 0 \\ 0 & 0 & 0 \end{bmatrix} \qquad ▬$$

```
[A]+(−[A])
          [[0 0 0]
           [0 0 0]]
```

A matrix having 0's for all its entries is called a **zero matrix.** When a zero matrix is added to a second matrix of the same order, the second matrix is unchanged. Thus a zero matrix is an **additive identity.** For example,

$$\begin{bmatrix} 0 & 0 & 0 \\ 0 & 0 & 0 \end{bmatrix}$$

is the additive identity for any 2×3 matrix.

Scalar Multiplication

When we find the product of a number and a matrix, we obtain a *scalar product.*

> ### Scalar Product
> The **scalar product** of a number k and a matrix **A** is the matrix denoted $k\mathbf{A}$, obtained by multiplying each entry of **A** by the number k. The number k is called a **scalar**.

EXAMPLE 4   Find $3\mathbf{A}$ and $(-1)\mathbf{A}$ for

$$\mathbf{A} = \begin{bmatrix} -3 & 0 \\ 4 & 5 \end{bmatrix}.$$

Solution   We have

$$3\mathbf{A} = 3\begin{bmatrix} -3 & 0 \\ 4 & 5 \end{bmatrix} = \begin{bmatrix} 3(-3) & 3 \cdot 0 \\ 3 \cdot 4 & 3 \cdot 5 \end{bmatrix} = \begin{bmatrix} -9 & 0 \\ 12 & 15 \end{bmatrix},$$

$$(-1)\mathbf{A} = -1\begin{bmatrix} -3 & 0 \\ 4 & 5 \end{bmatrix} = \begin{bmatrix} -1(-3) & -1 \cdot 0 \\ -1 \cdot 4 & -1 \cdot 5 \end{bmatrix} = \begin{bmatrix} 3 & 0 \\ -4 & -5 \end{bmatrix}.$$

These scalar products can also be found on a grapher after **A** has been entered. ▬

```
3[A]
           [[-9  0]
            [12 15]]
(-1)[A]
           [[3   0 ]
            [-4 -5]]
```

EXAMPLE 5   *Production.* Mitchell Fabricators, Inc., manufactures three styles of bicycle frames in its two plants. The following table shows the number of each style produced at each plant in April.

	MOUNTAIN BIKE	RACING BIKE	TOURING BIKE
NORTH PLANT	150	120	100
SOUTH PLANT	180	90	130

a) Write a 2×3 matrix **A** that represents the information in the table.

b) The manufacturer increased production by 20% in May. Find a matrix **M** that represents the increased production figures.

c) Find the matrix $\mathbf{A} + \mathbf{M}$ and tell what it represents.

Solution

a) Write the entries in the table in a 2×3 matrix **A**.

$$\mathbf{A} = \begin{bmatrix} 150 & 120 & 100 \\ 180 & 90 & 130 \end{bmatrix}$$

b) The production in May will be represented by $\mathbf{A} + 20\%\mathbf{A}$, or

$\mathbf{A} + 0.2\mathbf{A}$, or $1.2\mathbf{A}$. Thus,

$$\mathbf{M} = (1.2)\begin{bmatrix} 150 & 120 & 100 \\ 180 & 90 & 130 \end{bmatrix} = \begin{bmatrix} 180 & 144 & 120 \\ 216 & 108 & 156 \end{bmatrix}.$$

c) $\mathbf{A} + \mathbf{M} = \begin{bmatrix} 150 & 120 & 100 \\ 180 & 90 & 130 \end{bmatrix} + \begin{bmatrix} 180 & 144 & 120 \\ 216 & 108 & 156 \end{bmatrix}$

$\qquad = \begin{bmatrix} 330 & 264 & 220 \\ 396 & 198 & 286 \end{bmatrix}$

The matrix $\mathbf{A} + \mathbf{M}$ represents the total production of each of the three types of frames at each plant in April and May.

The properties of matrix addition and scalar multiplication are similar to the properties of addition and multiplication of real numbers.

Properties of Matrix Addition and Scalar Multiplication

For any $m \times n$ matrices $\mathbf{A}$, $\mathbf{B}$, and $\mathbf{C}$ and any scalars k and l:

$\mathbf{A} + \mathbf{B} = \mathbf{B} + \mathbf{A}$. — *Commutative Property of Addition*

$\mathbf{A} + (\mathbf{B} + \mathbf{C}) = (\mathbf{A} + \mathbf{B}) + \mathbf{C}$. — *Associative Property of Addition*

$(kl)\mathbf{A} = k(l\mathbf{A})$. — *Associative Property of Scalar Multiplication*

$k(\mathbf{A} + \mathbf{B}) = k\mathbf{A} + k\mathbf{B}$. — *Distributive Property*

$(k + l)\mathbf{A} = k\mathbf{A} + l\mathbf{A}$. — *Distributive Property*

There exists a unique matrix $\mathbf{0}$ such that:

$\mathbf{A} + \mathbf{0} = \mathbf{0} + \mathbf{A} = \mathbf{A}$. — *Additive Identity Property*

There exists a unique matrix $-\mathbf{A}$ such that:

$\mathbf{A} + (-\mathbf{A}) = -\mathbf{A} + \mathbf{A} = \mathbf{0}$. — *Additive Inverse Property*

Products of Matrices

Matrix multiplication is defined in such a way that it can be used in solving systems of equations and in many applications.

Matrix Multiplication

For an $m \times n$ matrix $\mathbf{A} = [a_{ij}]$ and an $n \times p$ matrix $\mathbf{B} = [b_{ij}]$, the **product** $\mathbf{AB} = [c_{ij}]$ is an $m \times p$ matrix, where

$$c_{ij} = a_{i1} \cdot b_{1j} + a_{i2} \cdot b_{2j} + a_{i3} \cdot b_{3j} + \cdots + a_{in} \cdot b_{nj}.$$

In other words, the entry c_{ij} in **AB** is obtained by multiplying the entries in row i of **A** by the corresponding entries in column j of **B** and adding the results.

> Note that we can multiply two matrices only when the number of columns in the first matrix is equal to the number of rows in the second matrix.

EXAMPLE 6 For

$$\mathbf{A} = \begin{bmatrix} 3 & 1 & -1 \\ 2 & 0 & 3 \end{bmatrix}, \qquad \mathbf{B} = \begin{bmatrix} 1 & 6 \\ 3 & -5 \\ -2 & 4 \end{bmatrix}, \quad \text{and} \quad \mathbf{C} = \begin{bmatrix} 4 & -6 \\ 1 & 2 \end{bmatrix},$$

find each of the following.

a) **AB**   **b)** **BA**

c) **BC**   **d)** **AC**

Solution

a) **A** is a 2 × 3 matrix and **B** is a 3 × 2 matrix, so **AB** will be a 2 × 2 matrix.

$$\mathbf{AB} = \begin{bmatrix} 3 & 1 & -1 \\ 2 & 0 & 3 \end{bmatrix} \begin{bmatrix} 1 & 6 \\ 3 & -5 \\ -2 & 4 \end{bmatrix}$$

$$= \begin{bmatrix} 3 \cdot 1 + 1 \cdot 3 + (-1)(-2) & 3 \cdot 6 + 1(-5) + (-1)(4) \\ 2 \cdot 1 + 0 \cdot 3 + 3(-2) & 2 \cdot 6 + 0(-5) + 3 \cdot 4 \end{bmatrix}$$

$$= \begin{bmatrix} 8 & 9 \\ -4 & 24 \end{bmatrix}$$

```
[A] [B]
          [[8   9 ]
           [-4 24]]
[B] [A]
          [[15  1 17 ]
           [-1  3 -18]
           [2  -2 14 ]]
```

Matrix multiplication can also be done on a grapher.

b) **B** is a 3 × 2 matrix and **A** is a 2 × 3 matrix, so **BA** will be a 3 × 3 matrix.

$$\mathbf{BA} = \begin{bmatrix} 1 & 6 \\ 3 & -5 \\ -2 & 4 \end{bmatrix} \begin{bmatrix} 3 & 1 & -1 \\ 2 & 0 & 3 \end{bmatrix}$$

$$= \begin{bmatrix} 1 \cdot 3 + 6 \cdot 2 & 1 \cdot 1 + 6 \cdot 0 & 1(-1) + 6 \cdot 3 \\ 3 \cdot 3 + (-5)(2) & 3 \cdot 1 + (-5)(0) & 3(-1) + (-5)(3) \\ -2 \cdot 3 + 4 \cdot 2 & -2 \cdot 1 + 4 \cdot 0 & -2(-1) + 4 \cdot 3 \end{bmatrix}$$

$$= \begin{bmatrix} 15 & 1 & 17 \\ -1 & 3 & -18 \\ 2 & -2 & 14 \end{bmatrix}$$

c) **B** is a 3 × 2 matrix and **C** is a 2 × 2 matrix, so **BC** will be a 3 × 2 matrix.

$$\mathbf{BC} = \begin{bmatrix} 1 & 6 \\ 3 & -5 \\ -2 & 4 \end{bmatrix} \begin{bmatrix} 4 & -6 \\ 1 & 2 \end{bmatrix}$$

$$= \begin{bmatrix} 1 \cdot 4 + 6 \cdot 1 & 1(-6) + 6 \cdot 2 \\ 3 \cdot 4 + (-5)(1) & 3(-6) + (-5)(2) \\ -2 \cdot 4 + 4 \cdot 1 & -2(-6) + 4 \cdot 2 \end{bmatrix}$$

$$= \begin{bmatrix} 10 & 6 \\ 7 & -28 \\ -4 & 20 \end{bmatrix}$$

d) The product **AC** is not defined because the number of columns of **A**, 3, is not equal to the number of rows of **C**, 2.

When **AC** is entered on a grapher, an ERROR message is returned, indicating that the dimensions of the matrices are mismatched. ▬

Note in Example 6 that **AB** ≠ **BA**. Multiplication of matrices is generally not commutative.

EXAMPLE 7   *Dairy Profit.* Dalton's Dairy produces no-fat ice cream and frozen yogurt. The following table shows the number of gallons of each product that are sold at the dairy's three retail outlets one week.

	STORE 1	STORE 2	STORE 3
NO-FAT ICE CREAM	100	80	120
FROZEN YOGURT	160	120	100

On each gallon of no-fat ice cream, the dairy's profit is $4, and on each gallon of frozen yogurt, it is $3. Use matrices to find the total profit on these items at each store for the given week.

Solution   We can write the table showing the distribution of the products as a 2 × 3 matrix:

$$\mathbf{D} = \begin{bmatrix} 100 & 80 & 120 \\ 160 & 120 & 100 \end{bmatrix}.$$

The profit per gallon for each product can also be written as a matrix:

$$\mathbf{P} = [4 \quad 3].$$

Then the total profit at each store is given by the matrix product **PD**:

$$\mathbf{PD} = [4 \quad 3]\begin{bmatrix} 100 & 80 & 120 \\ 160 & 120 & 100 \end{bmatrix}$$

$$= [4 \cdot 100 + 3 \cdot 160 \quad 4 \cdot 80 + 3 \cdot 120 \quad 4 \cdot 120 + 3 \cdot 100]$$

$$= [880 \quad 680 \quad 780].$$

The total profit on no-fat ice cream and frozen yogurt for the given week was $880 at store 1, $680 at store 2, and $780 at store 3.

A matrix that consists of a single row, like **P** in Example 7, is called a **row matrix.** Similarly, a matrix that consists of a single column, like

$$\begin{bmatrix} 8 \\ -3 \\ 5 \end{bmatrix},$$

is called a **column matrix.**

We have already seen that matrix multiplication is generally not commutative. Nevertheless, matrix multiplication does have some properties that are similar to those for multiplication of real numbers.

Properties of Matrix Multiplication

For matrices **A**, **B**, and **C**, assuming that the indicated operation is possible:

$$\mathbf{A}(\mathbf{BC}) = (\mathbf{AB})\mathbf{C}. \qquad \textit{Associative Property of Multiplication}$$

$$\mathbf{A}(\mathbf{B} + \mathbf{C}) = \mathbf{AB} + \mathbf{AC}. \qquad \textit{Distributive Property}$$

$$(\mathbf{B} + \mathbf{C})\mathbf{A} = \mathbf{BA} + \mathbf{CA}. \qquad \textit{Distributive Property}$$

Matrix Equations

We can write a matrix equation equivalent to a system of equations.

EXAMPLE 8 Write a matrix equation equivalent to the following system of equations:

$$\begin{aligned} 4x + 2y - z &= 3, \\ 9x + z &= 5, \\ 4x + 5y - 2z &= 1. \end{aligned}$$

Solution We write the coefficients on the left in a matrix. We then write the product of that matrix and the column matrix containing the variables, and set the result equal to the column matrix containing the con-

STUDY TIP

The AWL Math Tutor Center provides *free* tutoring to students using this text. Assisted by qualified mathematics instructors via telephone, fax, or e-mail, you can receive live tutoring on examples and exercises. This valuable resource provides you with immediate assistance when additional instruction is needed.

stants on the right:

$$\begin{bmatrix} 4 & 2 & -1 \\ 9 & 0 & 1 \\ 4 & 5 & -2 \end{bmatrix} \begin{bmatrix} x \\ y \\ z \end{bmatrix} = \begin{bmatrix} 3 \\ 5 \\ 1 \end{bmatrix}.$$

If we let

$$\mathbf{A} = \begin{bmatrix} 4 & 2 & -1 \\ 9 & 0 & 1 \\ 4 & 5 & -2 \end{bmatrix}, \quad \mathbf{X} = \begin{bmatrix} x \\ y \\ z \end{bmatrix}, \quad \text{and} \quad \mathbf{B} = \begin{bmatrix} 3 \\ 5 \\ 1 \end{bmatrix},$$

we can write this matrix equation as $\mathbf{AX} = \mathbf{B}$.

In the next section, we will solve systems of equations using a matrix equation like the one in Example 8.

Exercise Set 5.4

Find x, y, and z.

1. $[5 \quad x] = [y \quad -3]$

2. $\begin{bmatrix} 6x \\ 25 \end{bmatrix} = \begin{bmatrix} -9 \\ 5y \end{bmatrix}$

3. $\begin{bmatrix} 3 & 2x \\ y & -8 \end{bmatrix} = \begin{bmatrix} 3 & -2 \\ 1 & -8 \end{bmatrix}$

4. $\begin{bmatrix} x-1 & 4 \\ y+3 & -7 \end{bmatrix} = \begin{bmatrix} 0 & 4 \\ -2 & -7 \end{bmatrix}$

For Exercises 5–20, let

$$\mathbf{A} = \begin{bmatrix} 1 & 2 \\ 4 & 3 \end{bmatrix}, \qquad \mathbf{B} = \begin{bmatrix} -3 & 5 \\ 2 & -1 \end{bmatrix},$$

$$\mathbf{C} = \begin{bmatrix} 1 & -1 \\ -1 & 1 \end{bmatrix}, \qquad \mathbf{D} = \begin{bmatrix} 1 & 1 \\ 1 & 1 \end{bmatrix},$$

$$\mathbf{E} = \begin{bmatrix} 1 & 3 \\ 2 & 6 \end{bmatrix}, \qquad \mathbf{F} = \begin{bmatrix} 3 & 3 \\ -1 & -1 \end{bmatrix},$$

$$\mathbf{O} = \begin{bmatrix} 0 & 0 \\ 0 & 0 \end{bmatrix}, \qquad \mathbf{I} = \begin{bmatrix} 1 & 0 \\ 0 & 1 \end{bmatrix}.$$

Find each of the following. Use a grapher to check your answers.

5. $\mathbf{A} + \mathbf{B}$ **6.** $\mathbf{B} + \mathbf{A}$

7. $\mathbf{E} + \mathbf{O}$ **8.** $2\mathbf{A}$

9. $3\mathbf{F}$ **10.** $(-1)\mathbf{D}$

11. $3\mathbf{F} + 2\mathbf{A}$ **12.** $\mathbf{A} - \mathbf{B}$

13. $\mathbf{B} - \mathbf{A}$ **14.** $\mathbf{AB}$

15. $\mathbf{BA}$ **16.** $\mathbf{OF}$

17. $\mathbf{CD}$ **18.** $\mathbf{EF}$

19. $\mathbf{AI}$ **20.** $\mathbf{IA}$

21. *Budget.* For the month of June, Nelia budgets $150 for food, $80 for clothes, and $40 for entertainment.

  a) Write a 1×3 matrix $\mathbf{B}$ that represents these items.

  b) After receiving a raise, Nelia increases the amount budgeted for each item in July by 5%. Find a matrix $\mathbf{R}$ that represents the new amounts.

  c) Find $\mathbf{B} + \mathbf{R}$ and tell what the entries represent.

22. *Produce.* The produce manager at Dugan's Market orders 40 lb of tomatoes, 20 lb of zucchini, and 30 lb of onions from a local farmer one week.

  a) Write a 1×3 matrix $\mathbf{A}$ that represents these items.

  b) The following week the produce manager increases her order by 10%. Find a matrix $\mathbf{B}$ that represents this order.

  c) Find $\mathbf{A} + \mathbf{B}$ and tell what the entries represent.

23. *Nutrition.* A 3-oz serving of roasted, skinless chicken breast contains 140 Cal, 27 g of protein, 3 g of fat, 13 mg of calcium, and 64 mg of sodium. One-half cup of potato salad contains 180 Cal, 4 g of protein, 11 g of fat, 24 mg of calcium, and 662 mg of sodium. One broccoli spear contains 50 Cal, 5 g of protein, 1 g of fat, 82 mg of calcium, and 20 mg of sodium. (*Source*: *Home and Garden Bulletin No. 72,* U.S. Government Printing Office, Washington, D.C. 20402)

a) Write 1×5 matrices **C**, **P**, and **B** that represent the nutritional values of each food.

b) Find **C** + 2**P** + 3**B** and tell what the entries represent.

24. *Nutrition.* One slice of cheese pizza contains 290 Cal, 15 g of protein, 9 g of fat, and 39 g of carbohydrates. One-half cup of gelatin dessert contains 70 Cal, 2 g of protein, 0 g of fat, and 17 g of carbohydrates. One cup of whole milk contains 150 Cal, 8 g of protein, 8 g of fat, and 11 g of carbohydrates. (*Source*: *Home and Garden Bulletin No. 72,* U.S. Government Printing Office, Washington, D.C. 20402)

a) Write 1×4 matrices **P**, **G**, and **M** that represent the nutritional values of each food.

b) Find 3**P** + 2**G** + 2**M** and tell what the entries represent.

Use a grapher to find the product, if possible.

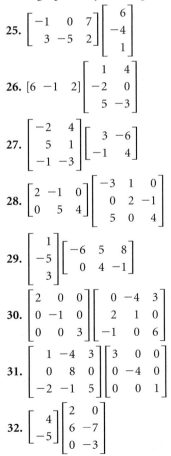

25. $\begin{bmatrix} -1 & 0 & 7 \\ 3 & -5 & 2 \end{bmatrix} \begin{bmatrix} 6 \\ -4 \\ 1 \end{bmatrix}$

26. $\begin{bmatrix} 6 & -1 & 2 \end{bmatrix} \begin{bmatrix} 1 & 4 \\ -2 & 0 \\ 5 & -3 \end{bmatrix}$

27. $\begin{bmatrix} -2 & 4 \\ 5 & 1 \\ -1 & -3 \end{bmatrix} \begin{bmatrix} 3 & -6 \\ -1 & 4 \end{bmatrix}$

28. $\begin{bmatrix} 2 & -1 & 0 \\ 0 & 5 & 4 \end{bmatrix} \begin{bmatrix} -3 & 1 & 0 \\ 0 & 2 & -1 \\ 5 & 0 & 4 \end{bmatrix}$

29. $\begin{bmatrix} 1 \\ -5 \\ 3 \end{bmatrix} \begin{bmatrix} -6 & 5 & 8 \\ 0 & 4 & -1 \end{bmatrix}$

30. $\begin{bmatrix} 2 & 0 & 0 \\ 0 & -1 & 0 \\ 0 & 0 & 3 \end{bmatrix} \begin{bmatrix} 0 & -4 & 3 \\ 2 & 1 & 0 \\ -1 & 0 & 6 \end{bmatrix}$

31. $\begin{bmatrix} 1 & -4 & 3 \\ 0 & 8 & 0 \\ -2 & -1 & 5 \end{bmatrix} \begin{bmatrix} 3 & 0 & 0 \\ 0 & -4 & 0 \\ 0 & 0 & 1 \end{bmatrix}$

32. $\begin{bmatrix} 4 \\ -5 \end{bmatrix} \begin{bmatrix} 2 & 0 \\ 6 & -7 \\ 0 & -3 \end{bmatrix}$

33. *Food Service Management.* The food service manager at a large hospital is concerned about maintaining reasonable food costs. The table below shows the cost per serving, in cents, for items on four menus.

MENU	MEAT	POTATO	VEGETABLE	SALAD	DESSERT
1	45.29	6.63	10.94	7.42	8.01
2	53.78	4.95	9.83	6.16	12.56
3	47.13	8.47	12.66	8.29	9.43
4	51.64	7.12	11.57	9.35	10.72

On a particular day, a dietician orders 65 meals from menu 1, 48 from menu 2, 93 from menu 3, and 57 from menu 4.

a) Write the information in the table as a 4×5 matrix **M**.

b) Write a row matrix **N** that represents the number of each menu ordered.

c) Find the product **NM**.

d) State what the entries of **NM** represent.

34. *Food Service Management.* A college food service manager uses a table like the one below to show the number of units of ingredients, by weight, required for various menu items.

	WHITE CAKE	BREAD	COFFEE CAKE	SUGAR COOKIES
FLOUR	1	2.5	0.75	0.5
MILK	0	0.5	0.25	0
EGGS	0.75	0.25	0.5	0.5
BUTTER	0.5	0	0.5	1

The cost per unit of each ingredient is 15 cents for flour, 28 cents for milk, 54 cents for eggs, and 83 cents for butter.

a) Write the information in the table as a 4×4 matrix **M**.

b) Write a row matrix **C** that represents the cost per unit of each ingredient.

c) Find the product **CM**.

d) State what the entries of **CM** represent.

35. *Production Cost.* Karin supplies two small campus coffee shops with homemade chocolate chip cookies, oatmeal cookies, and peanut butter cookies. The table below shows the number of each type of cookie, in dozens, that Karin sold in one week.

	MUGSY'S COFFEE SHOP	THE COFFEE CLUB
CHOCOLATE CHIP	8	15
OATMEAL	6	10
PEANUT BUTTER	4	3

Karin spends $3 for the ingredients for one dozen chocolate chip cookies, $1.50 for the ingredients for one dozen oatmeal cookies, and $2 for the ingredients for one dozen peanut butter cookies.

a) Write the information in the table as a 3 × 2 matrix **S**.
b) Write a row matrix **C** that represents the cost, per dozen, of the ingredients for each type of cookie.
c) Find the product **CS**.
d) State what the entries of **CS** represent.

36. *Profit.* A manufacturer produces exterior plywood, interior plywood, and fiberboard, which are shipped to two distributors. The table below shows the number of units of each type of product that are shipped to each warehouse.

	DISTRIBUTOR 1	DISTRIBUTOR 2
EXTERIOR PLYWOOD	900	500
INTERIOR PLYWOOD	450	1000
FIBERBOARD	600	700

The profits from each unit of exterior plywood, interior plywood, and fiberboard are $5, $8, and $4, respectively.

a) Write the information in the table as a 3 × 2 matrix **M**.
b) Write a row matrix **P** that represents the profit, per unit, of each type of product.

c) Find the product **PM**.
d) State what the entries of **PM** represent.

37. *Profit.* In Exercise 35, suppose that Karin's profits on one dozen chocolate chip, oatmeal, and peanut butter cookies are $6, $4.50, and $5.20, respectively.

a) Write a row matrix **P** that represents this information.
b) Use the matrices **S** and **P** to find Karin's total profit from each coffee shop.

38. *Production Cost.* In Exercise 36, suppose that the manufacturer's production costs for each unit of exterior plywood, interior plywood, and fiberboard are $20, $25, and $15, respectively.

a) Write a row matrix **C** that represents this information.
b) Use the matrices **M** and **C** to find the total production cost for the products shipped to each distributor.

Write a matrix equation equivalent to the system of equations.

39. $2x - 3y = 7,$
$x + 5y = -6$

40. $-x + y = 3,$
$5x - 4y = 16$

41. $x + y - 2z = 6,$
$3x - y + z = 7,$
$2x + 5y - 3z = 8$

42. $3x - y + z = 1,$
$x + 2y - z = 3,$
$4x + 3y - 2z = 11$

43. $3x - 2y + 4z = 17,$
$2x + y - 5z = 13$

44. $3x + 2y + 5z = 9,$
$4x - 3y + 2z = 10$

45. $-4w + x - y + 2z = 12,$
$w + 2x - y - z = 0,$
$-w + x + 4y - 3z = 1,$
$2w + 3x + 5y - 7z = 9$

46. $12w + 2x + 4y - 5z = 2,$
$-w + 4x - y + 12z = 5,$
$2w - x + 4y = 13,$
$2x + 10y + z = 5$

Discussion and Writing

47. Is it true that if $\mathbf{AB} = \mathbf{0}$, for matrices **A** and **B**, then $\mathbf{A} = \mathbf{0}$ or $\mathbf{B} = \mathbf{0}$? Why or why not?

48. Explain how Karin could use the matrix products found in Exercises 35 and 37 in making business decisions.

Skill Maintenance

In Exercises 49–52:

a) *Find the vertex.*
b) *Find the line of symmetry.*
c) *Determine whether there is a maximum or minimum value and find that value.*

49. $f(x) = x^2 - 3x - 10$ **50.** $f(x) = 2x^2 - 5x - 3$

51. $f(x) = -x^2 - 3x + 5$ **52.** $f(x) = -3x^2 + 4x + 1$

Synthesis

For Exercises 53–56, let

$$\mathbf{A} = \begin{bmatrix} -1 & 0 \\ 2 & 1 \end{bmatrix} \quad and \quad \mathbf{B} = \begin{bmatrix} 1 & -1 \\ 0 & 2 \end{bmatrix}.$$

53. Show that
$$(\mathbf{A} + \mathbf{B})(\mathbf{A} - \mathbf{B}) \neq \mathbf{A}^2 - \mathbf{B}^2,$$
where
$$\mathbf{A}^2 = \mathbf{AA} \quad and \quad \mathbf{B}^2 = \mathbf{BB}.$$

54. Show that
$$(\mathbf{A} + \mathbf{B})(\mathbf{A} + \mathbf{B}) \neq \mathbf{A}^2 + 2\mathbf{AB} + \mathbf{B}^2.$$

55. Show that
$$(\mathbf{A} + \mathbf{B})(\mathbf{A} - \mathbf{B}) = \mathbf{A}^2 + \mathbf{BA} - \mathbf{AB} - \mathbf{B}^2.$$

56. Show that
$$(\mathbf{A} + \mathbf{B})(\mathbf{A} + \mathbf{B}) = \mathbf{A}^2 + \mathbf{BA} + \mathbf{AB} + \mathbf{B}^2.$$

In Exercises 57–61, let

$$\mathbf{A} = \begin{bmatrix} a_{11} & a_{12} & a_{13} & \cdots & a_{1n} \\ a_{21} & a_{22} & a_{23} & \cdots & a_{2n} \\ a_{31} & a_{32} & a_{33} & \cdots & a_{3n} \\ \vdots & \vdots & \vdots & & \vdots \\ a_{m1} & a_{m2} & a_{m3} & \cdots & a_{mn} \end{bmatrix},$$

$$\mathbf{B} = \begin{bmatrix} b_{11} & b_{12} & b_{13} & \cdots & b_{1n} \\ b_{21} & b_{22} & b_{23} & \cdots & b_{2n} \\ b_{31} & b_{32} & b_{33} & \cdots & b_{3n} \\ \vdots & \vdots & \vdots & & \vdots \\ b_{m1} & b_{m2} & b_{m3} & \cdots & b_{mn} \end{bmatrix},$$

$$and\ \mathbf{C} = \begin{bmatrix} c_{11} & c_{12} & c_{13} & \cdots & c_{1n} \\ c_{21} & c_{22} & c_{23} & \cdots & c_{2n} \\ c_{31} & c_{32} & c_{33} & \cdots & c_{3n} \\ \vdots & \vdots & \vdots & & \vdots \\ c_{m1} & c_{m2} & c_{m3} & \cdots & c_{mn} \end{bmatrix},$$

and let k and l be any scalars.

57. Prove that $\mathbf{A} + \mathbf{B} = \mathbf{B} + \mathbf{A}$.

58. Prove that $\mathbf{A} + (\mathbf{B} + \mathbf{C}) = (\mathbf{A} + \mathbf{B}) + \mathbf{C}$.

59. Prove that $(kl)\mathbf{A} = k(l\mathbf{A})$.

60. Prove that $k(\mathbf{A} + \mathbf{B}) = k\mathbf{A} + k\mathbf{B}$.

61. Prove that $(k + l)\mathbf{A} = k\mathbf{A} + l\mathbf{A}$.

Inverses of Matrices

5.5

- *Find the inverse of a square matrix, if it exists.*
- *Use inverses of matrices to solve systems of equations.*

In this section, we continue our study of matrix algebra, finding the **multiplicative inverse,** or simply **inverse**, of a square matrix, if it exists. Then we use such inverses to solve systems of equations.

Interactive Discovery

Enter the following matrices on a grapher.

$$\mathbf{A} = \begin{bmatrix} 2 & -3 \\ -1 & 4 \end{bmatrix}, \quad \mathbf{B} = \begin{bmatrix} 1 & 0 \\ 0 & 1 \end{bmatrix}, \quad \mathbf{C} = \begin{bmatrix} 4 & 0 & -2 \\ 10 & -4 & -6 \\ -1 & 5 & 3 \end{bmatrix}, \quad \mathbf{D} = \begin{bmatrix} 1 & 0 & 0 \\ 0 & 1 & 0 \\ 0 & 0 & 1 \end{bmatrix}$$

Use the grapher to find **AB**, **BA**, **CD**, and **DC**. What is the effect of multiplying by **B**? by **D**?

The Identity Matrix

Recall that, for real numbers, $a \cdot 1 = 1 \cdot a = a$; 1 is the multiplicative identity. A multiplicative identity matrix is very similar to the number 1.

Identity Matrix

For any positive integer n, the $n \times n$ **identity matrix** is an $n \times n$ matrix with 1's on the main diagonal and 0's elsewhere and is denoted by

$$\mathbf{I} = \begin{bmatrix} 1 & 0 & 0 & \cdots & 0 \\ 0 & 1 & 0 & \cdots & 0 \\ 0 & 0 & 1 & \cdots & 0 \\ \vdots & \vdots & \vdots & & \vdots \\ 0 & 0 & 0 & \cdots & 1 \end{bmatrix}.$$

Then $\mathbf{AI} = \mathbf{IA} = \mathbf{A}$, for any $n \times n$ matrix $\mathbf{A}$.

EXAMPLE 1 For

$$\mathbf{A} = \begin{bmatrix} 4 & -7 \\ -3 & 2 \end{bmatrix} \quad \text{and} \quad \mathbf{I} = \begin{bmatrix} 1 & 0 \\ 0 & 1 \end{bmatrix},$$

find each of the following.

a) AI   **b) IA**

Solution

a) $\mathbf{AI} = \begin{bmatrix} 4 & -7 \\ -3 & 2 \end{bmatrix} \begin{bmatrix} 1 & 0 \\ 0 & 1 \end{bmatrix}$

$$= \begin{bmatrix} 4 \cdot 1 - 7 \cdot 0 & 4 \cdot 0 - 7 \cdot 1 \\ -3 \cdot 1 + 2 \cdot 0 & -3 \cdot 0 + 2 \cdot 1 \end{bmatrix} = \begin{bmatrix} 4 & -7 \\ -3 & 2 \end{bmatrix} = \mathbf{A}$$

b) $\mathbf{IA} = \begin{bmatrix} 1 & 0 \\ 0 & 1 \end{bmatrix} \begin{bmatrix} 4 & -7 \\ -3 & 2 \end{bmatrix}$

$$= \begin{bmatrix} 1 \cdot 4 + 0(-3) & 1(-7) + 0 \cdot 2 \\ 0 \cdot 4 + 1(-3) & 0(-7) + 1 \cdot 2 \end{bmatrix} = \begin{bmatrix} 4 & -7 \\ -3 & 2 \end{bmatrix} = \mathbf{A}$$

These products can also be found on a grapher after $\mathbf{A}$ and $\mathbf{I}$ have been entered.

```
[A] [I]
           [[4  -7]
            [-3  2]]
[I] [A]
           [[4  -7]
            [-3  2]]
```

The Inverse of a Matrix

Recall that for every nonzero real number a, there is a multiplicative inverse $1/a$ such that $a(1/a) = (1/a)a = 1$. The multiplicative inverse of a matrix behaves in a similar manner.

> ### Inverse of a Matrix
> For an $n \times n$ matrix $\mathbf{A}$, if there is a matrix $\mathbf{A}^{-1}$ for which $\mathbf{A}^{-1} \cdot \mathbf{A} = \mathbf{I} = \mathbf{A} \cdot \mathbf{A}^{-1}$, then $\mathbf{A}^{-1}$ is the **inverse** of $\mathbf{A}$.

Note that $\mathbf{A}^{-1}$ is read "$\mathbf{A}$ inverse." Also note that not every matrix has an inverse.

EXAMPLE 2 Verify that

$$\mathbf{B} = \begin{bmatrix} 4 & -3 \\ 3 & -2 \end{bmatrix} \quad \text{is the inverse of} \quad \mathbf{A} = \begin{bmatrix} -2 & 3 \\ -3 & 4 \end{bmatrix}.$$

Solution We show that $\mathbf{BA} = \mathbf{I} = \mathbf{AB}$.

$$\mathbf{BA} = \begin{bmatrix} 4 & -3 \\ 3 & -2 \end{bmatrix}\begin{bmatrix} -2 & 3 \\ -3 & 4 \end{bmatrix} = \begin{bmatrix} 1 & 0 \\ 0 & 1 \end{bmatrix}$$

$$\mathbf{AB} = \begin{bmatrix} -2 & 3 \\ -3 & 4 \end{bmatrix}\begin{bmatrix} 4 & -3 \\ 3 & -2 \end{bmatrix} = \begin{bmatrix} 1 & 0 \\ 0 & 1 \end{bmatrix}$$

We can find the inverse of a square matrix, if it exists, by using row-equivalent operations as in the Gauss–Jordan elimination method. For example, consider the matrix

$$\mathbf{A} = \begin{bmatrix} -2 & 3 \\ -3 & 4 \end{bmatrix}.$$

To find its inverse, we first form an **augmented matrix** consisting of $\mathbf{A}$ on the left side and the 2×2 identity matrix on the right side:

$$\begin{bmatrix} -2 & 3 & 1 & 0 \\ -3 & 4 & 0 & 1 \end{bmatrix}.$$

The 2×2 The 2×2
matrix A   identity matrix

Then we attempt to transform the augmented matrix to one of the form

$$\begin{bmatrix} 1 & 0 & a & b \\ 0 & 1 & c & d \end{bmatrix}.$$

The 2×2 The matrix $\mathbf{A}^{-1}$
identity matrix

If we can do this, the matrix on the right, $\begin{bmatrix} a & b \\ c & d \end{bmatrix}$, is $\mathbf{A}^{-1}$.

EXAMPLE 3   Find $\mathbf{A}^{-1}$, where

$$\mathbf{A} = \begin{bmatrix} -2 & 3 \\ -3 & 4 \end{bmatrix}.$$

Solution   First, we write the augmented matrix. Then we transform it to the desired form.

$$\begin{bmatrix} -2 & 3 & | & 1 & 0 \\ -3 & 4 & | & 0 & 1 \end{bmatrix}$$

$$\begin{bmatrix} 1 & -\frac{3}{2} & | & -\frac{1}{2} & 0 \\ -3 & 4 & | & 0 & 1 \end{bmatrix} \qquad \text{New row 1} = -\frac{1}{2}(\text{row 1})$$

$$\begin{bmatrix} 1 & -\frac{3}{2} & | & -\frac{1}{2} & 0 \\ 0 & -\frac{1}{2} & | & -\frac{3}{2} & 1 \end{bmatrix} \qquad \text{New row 2} = 3(\text{row 1}) + \text{row 2}$$

$$\begin{bmatrix} 1 & -\frac{3}{2} & | & -\frac{1}{2} & 0 \\ 0 & 1 & | & 3 & -2 \end{bmatrix} \qquad \text{New row 2} = -2(\text{row 2})$$

$$\begin{bmatrix} 1 & 0 & | & 4 & -3 \\ 0 & 1 & | & 3 & -2 \end{bmatrix} \qquad \text{New row 1} = \frac{3}{2}(\text{row 2}) + \text{row 1}$$

Thus,

$$\mathbf{A}^{-1} = \begin{bmatrix} 4 & -3 \\ 3 & -2 \end{bmatrix},$$

which we verified in Example 2.

The inverse of a matrix can also be found using a grapher.   ▬

```
[A]⁻¹
            [[4  -3]
             [3  -2]]
```

If a matrix has an inverse, we say that it is **invertible**, or **nonsingular**. When we cannot obtain the identity matrix on the left using the Gauss–Jordan method, then no inverse exists. This occurs when we obtain a row consisting entirely of 0's in either of the two matrices in the augmented matrix. In this case, we say that **A** is a **singular matrix.** A grapher will produce an error message similar to "ERR: SINGULAR MATRIX" in this situation.

Solving Systems of Equations

MATRIX EQUATIONS

REVIEW SECTION 5.4.

We can write a system of n linear equations in n variables as a matrix equation $\mathbf{AX} = \mathbf{B}$. If $\mathbf{A}$ has an inverse, then the system of equations has a unique solution that can be found by solving for $\mathbf{X}$, as follows:

$$\mathbf{AX} = \mathbf{B}$$
$$\mathbf{A}^{-1}(\mathbf{AX}) = \mathbf{A}^{-1}\mathbf{B} \qquad \text{Multiplying by } \mathbf{A}^{-1} \text{ on the left on both sides}$$
$$(\mathbf{A}^{-1}\mathbf{A})\mathbf{X} = \mathbf{A}^{-1}\mathbf{B} \qquad \text{Using the associative property of matrix multiplication}$$
$$\mathbf{IX} = \mathbf{A}^{-1}\mathbf{B} \qquad \mathbf{A}^{-1}\mathbf{A} = \mathbf{I}$$
$$\mathbf{X} = \mathbf{A}^{-1}\mathbf{B}. \qquad \mathbf{IX} = \mathbf{X}$$

Matrix Solutions of Systems of Equations

For a system of n linear equations in n variables, $\mathbf{AX} = \mathbf{B}$, if $\mathbf{A}$ is an invertible matrix, then the unique solution of the system is given by

$$\mathbf{X} = \mathbf{A}^{-1}\mathbf{B}.$$

Since matrix multiplication is not commutative in general, care must be taken to multiply *on the left* by $\mathbf{A}^{-1}$.

EXAMPLE 4 Use an inverse matrix to solve the following system of equations:

$$x + 2y - z = -2,$$
$$3x + 5y + 3z = 3,$$
$$2x + 4y + 3z = 1.$$

Solution We write an equivalent matrix equation, $\mathbf{AX} = \mathbf{B}$:

$$\begin{bmatrix} 1 & 2 & -1 \\ 3 & 5 & 3 \\ 2 & 4 & 3 \end{bmatrix} \begin{bmatrix} x \\ y \\ z \end{bmatrix} = \begin{bmatrix} -2 \\ 3 \\ 1 \end{bmatrix}.$$

$$\quad\ \mathbf{A} \qquad\quad \cdot\ \mathbf{X}\ =\ \mathbf{B}$$

```
[A]⁻¹[B]
           [[5 ]
            [−3]
            [1  ]]
```

Then we find $\mathbf{A}^{-1}\mathbf{B}$ using a grapher, as shown at left.

Thus the solution is $x = 5$, $y = -3$, and $z = 1$, or $(5, -3, 1)$.

Note that when a grapher is used, it is not actually necessary to enter the matrix $\mathbf{A}^{-1}$. After the matrices $\mathbf{A}$ and $\mathbf{B}$ are entered on a grapher, the notation $[\mathbf{A}]^{-1}[\mathbf{B}]$ is entered and only the result is displayed.

Exercise Set 5.5

Determine whether $\mathbf{B}$ is the inverse of $\mathbf{A}$.

1. $\mathbf{A} = \begin{bmatrix} 1 & -3 \\ -2 & 7 \end{bmatrix}$, $\mathbf{B} = \begin{bmatrix} 7 & 3 \\ 2 & 1 \end{bmatrix}$

2. $\mathbf{A} = \begin{bmatrix} 3 & 2 \\ 4 & 3 \end{bmatrix}$, $\mathbf{B} = \begin{bmatrix} 3 & -2 \\ -4 & 3 \end{bmatrix}$

3. $\mathbf{A} = \begin{bmatrix} -1 & -1 & 6 \\ 1 & 0 & -2 \\ 1 & 0 & -3 \end{bmatrix}$, $\mathbf{B} = \begin{bmatrix} 2 & 3 & 2 \\ 3 & 3 & 4 \\ 1 & 1 & 1 \end{bmatrix}$

4. $\mathbf{A} = \begin{bmatrix} -2 & 0 & -3 \\ 5 & 1 & 7 \\ -3 & 0 & 4 \end{bmatrix}$, $\mathbf{B} = \begin{bmatrix} 4 & 0 & -3 \\ 1 & 1 & 1 \\ -3 & 0 & 2 \end{bmatrix}$

Use the Gauss–Jordan method to find $\mathbf{A}^{-1}$, if it exists. Check your answers using a grapher by finding $\mathbf{A}^{-1}\mathbf{A}$ and $\mathbf{AA}^{-1}$.

5. $\mathbf{A} = \begin{bmatrix} 3 & 2 \\ 5 & 3 \end{bmatrix}$

6. $\mathbf{A} = \begin{bmatrix} 3 & 5 \\ 1 & 2 \end{bmatrix}$

7. $\mathbf{A} = \begin{bmatrix} 6 & 9 \\ 4 & 6 \end{bmatrix}$

8. $\mathbf{A} = \begin{bmatrix} -4 & -6 \\ 2 & 3 \end{bmatrix}$

9. $A = \begin{bmatrix} 3 & 1 & 0 \\ 1 & 1 & 1 \\ 1 & -1 & 2 \end{bmatrix}$ **10.** $A = \begin{bmatrix} 1 & 0 & 1 \\ 2 & 1 & 0 \\ 1 & -1 & 1 \end{bmatrix}$

11. $A = \begin{bmatrix} 1 & -4 & 8 \\ 1 & -3 & 2 \\ 2 & -7 & 10 \end{bmatrix}$ **12.** $A = \begin{bmatrix} -2 & 5 & 3 \\ 4 & -1 & 3 \\ 7 & -2 & 5 \end{bmatrix}$

Use a grapher to find A^{-1}, *if it exists.*

13. $A = \begin{bmatrix} 4 & -3 \\ 1 & -2 \end{bmatrix}$ **14.** $A = \begin{bmatrix} 0 & -1 \\ 1 & 0 \end{bmatrix}$

15. $A = \begin{bmatrix} 2 & 3 & 2 \\ 3 & 3 & 4 \\ -1 & -1 & -1 \end{bmatrix}$ **16.** $A = \begin{bmatrix} 1 & 2 & 3 \\ 2 & -1 & -2 \\ -1 & 3 & 3 \end{bmatrix}$

17. $A = \begin{bmatrix} 1 & 2 & -1 \\ -2 & 0 & 1 \\ 1 & -1 & 0 \end{bmatrix}$ **18.** $A = \begin{bmatrix} 7 & -1 & -9 \\ 2 & 0 & -4 \\ -4 & 0 & 6 \end{bmatrix}$

19. $A = \begin{bmatrix} 1 & 3 & -1 \\ 0 & 2 & -1 \\ 1 & 1 & 0 \end{bmatrix}$ **20.** $A = \begin{bmatrix} -1 & 0 & -1 \\ -1 & 1 & 0 \\ 0 & 1 & 1 \end{bmatrix}$

21. $A = \begin{bmatrix} 1 & 2 & 3 & 4 \\ 0 & 1 & 3 & -5 \\ 0 & 0 & 1 & -2 \\ 0 & 0 & 0 & -1 \end{bmatrix}$

22. $A = \begin{bmatrix} -2 & -3 & 4 & 1 \\ 0 & 1 & 1 & 0 \\ 0 & 4 & -6 & 1 \\ -2 & -2 & 5 & 1 \end{bmatrix}$

23. $A = \begin{bmatrix} 1 & -14 & 7 & 38 \\ -1 & 2 & 1 & -2 \\ 1 & 2 & -1 & -6 \\ 1 & -2 & 3 & 6 \end{bmatrix}$

24. $A = \begin{bmatrix} 10 & 20 & -30 & 15 \\ 3 & -7 & 14 & -8 \\ -7 & -2 & -1 & 2 \\ 4 & 4 & -3 & 1 \end{bmatrix}$

Solve the system of equations using the inverse of the coefficient matrix of the equivalent matrix equation.

25. $4x + 3y = 2,$
$x - 2y = 6$

26. $2x - 3y = 7,$
$4x + y = -7$

27. $5x + y = 2,$
$3x - 2y = -4$

28. $x - 6y = 5,$
$-x + 4y = -5$

29. $x + z = 1,$
$2x + y = 3,$
$x - y + z = 4$

30. $x + 2y + 3z = -1,$
$2x - 3y + 4z = 2,$
$-3x + 5y - 6z = 4$

31. $2x + 3y + 4z = 2,$
$x - 4y + 3z = 2,$
$5x + y + z = -4$

32. $x + y = 2,$
$3x + 2z = 5,$
$2x + 3y - 3z = 9$

33. $2w - 3x + 4y - 5z = 0,$
$3w - 2x + 7y - 3z = 2,$
$w + x - y + z = 1,$
$-w - 3x - 6y + 4z = 6$

34. $5w - 4x + 3y - 2z = -6,$
$w + 4x - 2y + 3z = -5,$
$2w - 3x + 6y - 9z = 14,$
$3w - 5x + 2y - 4z = -3$

35. *Sales.* Stefan sold a total of 145 Italian sausages and hot dogs from his curbside pushcart and collected $242.05. He sold 45 more hot dogs than sausages. How many of each did he sell?

36. *Price of School Supplies.* Miranda bought 4 lab record books and 3 highlighters for $13.93. Victor bought 3 lab record books and 2 highlighters for $10.25. Find the price of each item.

37. *Cost.* Evergreen Landscaping bought 4 tons of topsoil, 3 tons of mulch, and 6 tons of pea gravel for $2825. The next week the firm bought 5 tons of topsoil, 2 tons of mulch, and 5 tons of pea gravel for $2663. Pea gravel costs $17 less per ton than topsoil. Find the price per ton for each item.

38. *Investment.* Selena receives $537.75 per year in simple interest from three investments totaling $8500. Part is invested at 5.15%, part at 6.05%, and the rest at 7.2%. There is $1500 more invested at 7.2% than at 5.15%. Find the amount invested at each rate.

Discussion and Writing

39. For square matrices A and B, is it true, in general, that $(A + B)^{-1} = A^{-1} + B^{-1}$? Explain.

40. For square matrices A and B, is it true, in general, that $(AB)^{-1} = A^{-1}B^{-1}$? Explain.

Skill Maintenance

Use synthetic division to find the function values.

41. $f(x) = x^3 - 6x^2 + 4x - 8$; find $f(-2)$

42. $f(x) = 2x^4 - x^3 + 5x^2 + 6x - 4$; find $f(3)$

Factor the polynomial $f(x)$.

43. $f(x) = x^3 - 3x^2 - 6x + 8$

44. $f(x) = x^4 + 2x^3 - 16x^2 - 2x + 15$

Synthesis

State the conditions under which $\mathbf{A}^{-1}$ exists. Then find a formula for $\mathbf{A}^{-1}$.

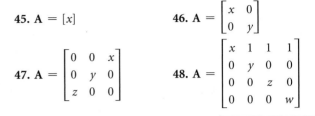

45. $\mathbf{A} = [x]$

46. $\mathbf{A} = \begin{bmatrix} x & 0 \\ 0 & y \end{bmatrix}$

47. $\mathbf{A} = \begin{bmatrix} 0 & 0 & x \\ 0 & y & 0 \\ z & 0 & 0 \end{bmatrix}$

48. $\mathbf{A} = \begin{bmatrix} x & 1 & 1 & 1 \\ 0 & y & 0 & 0 \\ 0 & 0 & z & 0 \\ 0 & 0 & 0 & w \end{bmatrix}$

Systems of Inequalities and Linear Programming

5.6

- Graph linear inequalities.
- Graph systems of linear inequalities.
- Solve linear programming problems.

A graph of an inequality is a drawing that represents its solutions. We have already seen that an inequality in one variable can be graphed on a number line. An inequality in two variables can be graphed on a coordinate plane.

Graphs of Linear Inequalities

A statement like $5x - 4y < 20$ is a linear inequality in two variables.

Linear Inequality in Two Variables

A **linear inequality in two variables** is an inequality that can be written in the form

$$Ax + By < C,$$

where A, B, and C are real numbers and A and B are not both zero. The symbol $<$ may be replaced with $\leq$, $>$, or $\geq$.

A solution of a linear inequality in two variables is an ordered pair (x, y) for which the inequality is true. For example, $(1, 3)$ is a solution of $5x - 4y < 20$ because $5 \cdot 1 - 4 \cdot 3 < 20$, or $-7 < 20$, is true. On the other hand, $(2, -6)$ is not a solution of $5x - 4y < 20$ because $5 \cdot 2 - 4 \cdot (-6) \not< 20$, or $34 \not< 20$.

The **solution set** of an inequality is the set of all the ordered pairs that make it true. The **graph of an inequality** represents its solution set.

EXAMPLE 1 Graph: $y < x + 3$.

Solution We begin by graphing the **related equation** $y = x + 3$. We

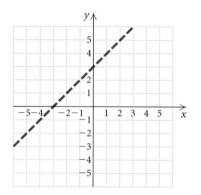

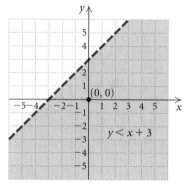

use a dashed line because the inequality symbol is $<$. This indicates that the line itself is not in the solution set of the inequality.

Note that the line divides the coordinate plane into two regions called **half-planes,** one of which satisfies the inequality. Either *all* points in a half-plane are in the solution set of the inequality or *none* is.

To determine which half-plane satisfies the inequality, we try a test point in either region. The point $(0, 0)$ is usually a convenient choice so long as it does not lie on the line.

$$\frac{y < x + 3}{\begin{array}{c} 0 \; ? \; 0 + 3 \\ 0 \; | \; 3 \qquad \text{TRUE} \end{array}}$$

Since $(0, 0)$ satisfies the inequality, so do all points in the half-plane that contains $(0, 0)$. We shade this region to show the solution set of the inequality.

In general, we use the following procedure to graph linear inequalities in two variables.

To graph a linear inequality in two variables:

1. Replace the inequality symbol with an equals sign and graph this related equation. If the inequality symbol is $<$ or $>$, draw the line dashed. If the inequality symbol is $\leq$ or $\geq$, draw the line solid.

2. The graph consists of a half-plane on one side of the line and, if the line is solid, the line as well. To determine which half-plane to shade, test a point not on the line in the original inequality. If that point is a solution, shade the half-plane containing that point. If not, shade the opposite half-plane.

EXAMPLE 2 Graph: $3x + 4y \geq 12$.

Solution

1. First, we graph the related equation $3x + 4y = 12$. We use a solid line because the inequality symbol is $\geq$. This indicates that the line is included in the solution set.

2. To determine which half-plane to shade, we test a point in either region. We choose $(0, 0)$.

$$\frac{3x + 4y \geq 12}{\begin{array}{c} 3 \cdot 0 + 4 \cdot 0 \; ? \; 12 \\ 0 \; | \; 12 \qquad \text{FALSE} \end{array}}$$

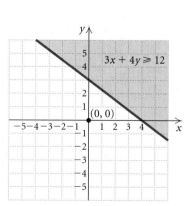

Because $(0, 0)$ is *not* a solution, all the points in the half-plane that does *not* contain $(0, 0)$ are solutions. We shade that region, as shown in the figure at the bottom of p. 406.

This inequality can also be graphed using the "shade above" graph style on a grapher.

$$y = \frac{-3x + 12}{4}$$

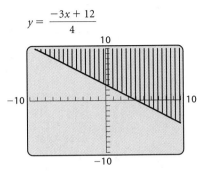

EXAMPLE 3   Graph $x > -3$ on a plane.

Solution

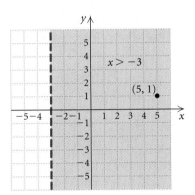

1. First, we graph the related equation $x = -3$. We use a dashed line because the inequality symbol is $>$. This indicates that the line is not included in the solution set.

2. The inequality tells us that all points (x, y) for which $x > -3$ are solutions. These are the points to the right of the line. We can also use a test point to determine the solutions. We choose $(5, 1)$.

$$\frac{x > -3}{5 \ ? \ -3 \quad \text{TRUE}}$$

Because $(5, 1)$ is a solution, we shade the region containing that point—that is, the region to the right of the dashed line.

EXAMPLE 4   Graph $y \leq 4$ on a plane.

Solution

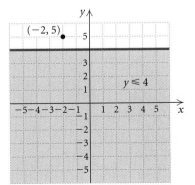

1. First, we graph the related equation $y = 4$. We use a solid line because the inequality symbol is $\leq$.

2. The inequality tells us that all points (x, y) for which $y \leq 4$ are solutions of the inequality. These are the points on or below the line. We can also use a test point to determine the solutions. We choose $(-2, 5)$.

$$\frac{y \leq 4}{5 \ ? \ 4 \quad \text{FALSE}}$$

Because $(-2, 5)$ is not a solution, we shade the half-plane that does not contain that point.

We can also graph this inequality using a grapher.

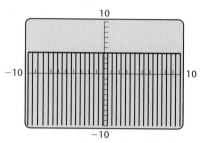

Systems of Linear Inequalities

A system of inequalities in two variables consists of two or more inequalities in two variables considered simultaneously. For example,

$$x + y \leq 4,$$
$$x - y \geq 2$$

is a system of two *linear* inequalities in two variables.

A solution of a system of inequalities is an ordered pair that is a solution of each inequality in the system. To graph a system of linear inequalities, we graph each inequality and determine the region that is common to all the solution sets.

EXAMPLE 5   Graph the solution set of the system

$$x + y \leq 4,$$
$$x - y \geq 2.$$

Solution   We graph $x + y \leq 4$ by first graphing the equation $x + y = 4$ using a solid line. Next, we choose $(0, 0)$ as a test point and find that it is a solution of $x + y \leq 4$, so we shade the half-plane containing $(0, 0)$ using red. Next, we graph $x - y = 2$ using a solid line. We find that $(0, 0)$ is not a solution of $x - y \geq 2$, so we shade the half-plane that does not contain $(0, 0)$ using green. The arrows at the ends of each line help to indicate the half-plane that contains each solution set.

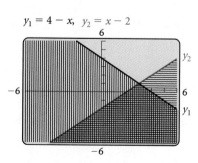

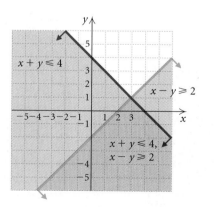

The solution set of the system of equations is the region shaded both red and green, or brown, including parts of the lines $x + y = 4$ and $x - y = 2$.

A system of inequalities may have a graph that consists of a polygon and its interior. As we will see later in this section, it is important in many applications to be able to find the vertices of such a polygon.

EXAMPLE 6 Graph the following system of inequalities and find the coordinates of any vertices formed:

$$3x - y \leq 6, \qquad (1)$$
$$y - 3 \leq 0, \qquad (2)$$
$$x + y \geq 0. \qquad (3)$$

Solution We graph the related equations $3x - y = 6$, $y - 3 = 0$, and $x + y = 0$ using solid lines. The half-plane containing the solution set for each inequality is indicated by the arrows at the ends of each line. We shade the region common to all three solution sets.

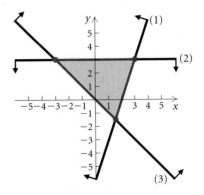

To find the vertices, we solve three systems of equations. The system of equations from inequalities (1) and (2) is

$$3x - y = 6,$$
$$y - 3 = 0.$$

Solving, we obtain the vertex $(3, 3)$.

The system of equations from inequalities (1) and (3) is

$$3x - y = 6,$$
$$x + y = 0.$$

Solving, we obtain the vertex $\left(\frac{3}{2}, -\frac{3}{2}\right)$.

The system of equations from inequalities (2) and (3) is

$$y - 3 = 0,$$
$$x + y = 0.$$

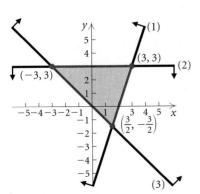

Solving, we obtain the vertex $(-3, 3)$.

We could also graph the system of related equations on a grapher and use the INTERSECT feature to find the vertices. However, some graphers are not capable of shading a region determined by three or more graphs like the one above.

Applications: Linear Programming

In many applications, we want to find a maximum or minimum value. In business, for example, we might want to maximize profit and minimize cost. **Linear programming** can tell us how to do this.

In our study of linear programming, we will consider linear functions of two variables that are to be maximized or minimized subject to several conditions, or **constraints**. These constraints are expressed as inequalities. The solution set of the system of inequalities made up of the constraints contains all the **feasible solutions** of a linear programming problem. The function that we want to maximize or minimize is called the **objective function.**

It can be shown that the maximum and minimum values of the objective function occur at a vertex of the region of feasible solutions. Thus we have the following procedure.

Linear Programming Procedure

To find the maximum or minimum value of a linear objective function subject to a set of constraints:

1. Graph the region of feasible solutions.
2. Determine the coordinates of the vertices of the region.
3. Evaluate the objective function at each vertex. The largest and smallest of those values are the maximum and minimum values of the function, respectively.

EXAMPLE 7 *Maximizing Profit.* Dovetail Carpentry Shop makes bookcases and desks. Each bookcase requires 5 hr of woodworking and 4 hr of finishing. Each desk requires 10 hr of woodworking and 3 hr of finishing. Each month the shop has 600 hr of labor available for woodworking and 240 hr for finishing. The profit on each bookcase is $40 and on each desk is $75. How many of each product should be made each month in order to maximize profit?

Solution We let x = the number of bookcases to be produced and y = the number of desks. Then the profit P is given by the function

$$P = 40x + 75y.$$ To emphasize that P is a function of two variables, we sometimes write $P(x, y) = 40x + 75y$.

We know that x bookcases require $5x$ hr of woodworking and y desks require $10y$ hr of woodworking. Since there is no more than 600 hr of labor available for woodworking, we have one constraint:

$$5x + 10y \leq 600.$$

Similarly, the bookcases and desks require $4x$ hr and $3y$ hr of finishing, respectively. There is no more than 240 hr of labor available for finishing, so we have a second constraint:

$$4x + 3y \leq 240.$$

We also know that $x \geq 0$ and $y \geq 0$ because the carpentry shop cannot make a negative number of either product.

Thus we want to maximize the objective function

$$P = 40x + 75y$$

subject to the constraints

$$5x + 10y \leq 600,$$
$$4x + 3y \leq 240,$$
$$x \geq 0,$$
$$y \geq 0.$$

We graph the system of inequalities and determine the vertices.

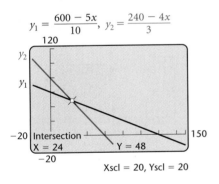

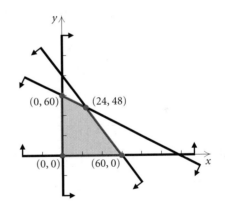

Next, we evaluate the objective function P at each vertex.

VERTICES (x, y)	PROFIT $P = 40x + 75y$	
$(0, 0)$	$P = 40 \cdot 0 + 75 \cdot 0 = 0$	
$(60, 0)$	$P = 40 \cdot 60 + 75 \cdot 0 = 2400$	
$(24, 48)$	$P = 40 \cdot 24 + 75 \cdot 48 = 4560$	⟵ Maximum
$(0, 60)$	$P = 40 \cdot 0 + 75 \cdot 60 = 4500$	

The carpentry shop will make a maximum profit of $4560 when 24 bookcases and 48 desks are produced.

Exercise Set 5.6

In Exercises 1–8, match the inequality with one of the graphs (a)–(h), which follow.

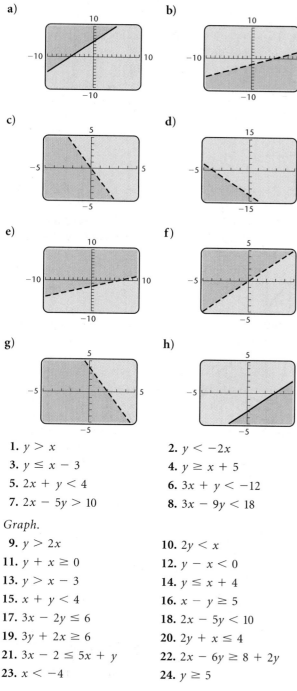

a)

b)

c)

d)

e)

f)

g)

h)

1. $y > x$

2. $y < -2x$

3. $y \leq x - 3$

4. $y \geq x + 5$

5. $2x + y < 4$

6. $3x + y < -12$

7. $2x - 5y > 10$

8. $3x - 9y < 18$

Graph.

9. $y > 2x$

10. $2y < x$

11. $y + x \geq 0$

12. $y - x < 0$

13. $y > x - 3$

14. $y \leq x + 4$

15. $x + y < 4$

16. $x - y \geq 5$

17. $3x - 2y \leq 6$

18. $2x - 5y < 10$

19. $3y + 2x \geq 6$

20. $2y + x \leq 4$

21. $3x - 2 \leq 5x + y$

22. $2x - 6y \geq 8 + 2y$

23. $x < -4$

24. $y \geq 5$

25. $y > -3$

26. $x \leq 5$

27. $-4 < y < -1$
   (*Hint:* Think of this as $-4 < y$ and $y < -1$.)

28. $-3 \leq x \leq 3$
   (*Hint:* Think of this as $-3 \leq x$ and $x \leq 3$.)

29. $y \geq |x|$

30. $y \leq |x + 2|$

In Exercises 31–36, match the system of inequalities with one of the graphs (a)–(f), which follow.

a)

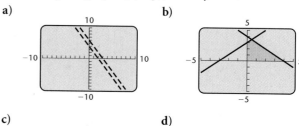

b)

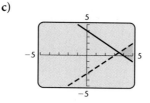

c)

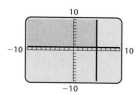

d)

e)

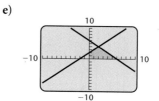

f)

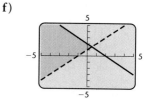

31. $y > x + 1,$
   $y \leq 2 - x$

32. $y < x - 3,$
   $y \geq 4 - x$

33. $2x + y < 4,$
   $4x + 2y > 12$

34. $x \leq 5,$
   $y \geq 1$

35. $x + y \leq 4,$
   $x - y \geq -3,$
   $x \geq 0,$
   $y \geq 0$

36. $x - y \geq -2,$
   $x + y \leq 6,$
   $x \geq 0,$
   $y \geq 0$

Graph the system of inequalities. Then find the coordinates of the vertices.

37. $y \leq x,$
   $y \geq 3 - x$

38. $y \geq x,$
   $y \leq x - 5$

39. $y \geq x,$
   $y \leq x - 4$

40. $y \geq x,$
   $y \leq 2 - x$

41. $y \geq -3,$
   $x \geq 1$

42. $y \leq -2,$
   $x \geq 2$

43. $x \leq 3,$
   $y \geq 2 - 3x$

44. $x \geq -2,$
   $y \leq 3 - 2x$

45. $x + y \leq 1,$
 $\quad x - y \leq 2$

46. $y + 3x \geq 0,$
 $\quad y + 3x \leq 2$

47. $2y - x \leq 2,$
 $\quad y + 3x \geq -1$

48. $\quad y \leq 2x + 1,$
 $\quad\quad y \geq -2x + 1,$
 $\quad\quad x - 2 \leq 0$

49. $x - y \leq 2,$
 $\quad x + 2y \geq 8,$
 $\quad y - 4 \leq 0$

50. $x + 2y \leq 12,$
 $\quad 2x + y \leq 12,$
 $\quad\quad x \geq 0,$
 $\quad\quad y \geq 0$

51. $4y - 3x \geq -12,$
 $\quad 4y + 3x \geq -36,$
 $\quad\quad y \leq 0,$
 $\quad\quad x \leq 0$

52. $8x + 5y \leq 40,$
 $\quad x + 2y \leq 8,$
 $\quad\quad x \geq 0,$
 $\quad\quad y \geq 0$

53. $3x + 4y \geq 12,$
 $\quad 5x + 6y \leq 30,$
 $\quad\quad 1 \leq x \leq 3$

54. $y - x \geq 1,$
 $\quad y - x \leq 3,$
 $\quad 2 \leq x \leq 5$

Find the maximum and the minimum values of the function and the values of x and y for which they occur.

55. $P = 17x - 3y + 60,$ subject to

$$6x + 8y \leq 48,$$
$$0 \leq y \leq 4,$$
$$0 \leq x \leq 7.$$

56. $Q = 28x - 4y + 72,$ subject to

$$5x + 4y \geq 20,$$
$$0 \leq y \leq 4,$$
$$0 \leq x \leq 3.$$

57. $F = 5x + 36y,$ subject to

$$5x + 3y \leq 34,$$
$$3x + 5y \leq 30,$$
$$x \geq 0,$$
$$y \geq 0.$$

58. $G = 16x + 14y,$ subject to

$$3x + 2y \leq 12,$$
$$7x + 5y \leq 29,$$
$$x \geq 0,$$
$$y \geq 0.$$

59. *Maximizing Income.* Golden Harvest Foods makes jumbo biscuits and regular biscuits. The oven can cook at most 200 biscuits per day. Each jumbo biscuit requires 2 oz of flour, each regular biscuit requires 1 oz of flour, and there is 300 oz of flour available. The income from each jumbo biscuit is $0.10 and from each regular biscuit is $0.08. How many of each size biscuit should be made in order to maximize income? What is the maximum income?

60. *Maximizing Mileage.* Omar owns a car and a moped. He can afford 12 gal of gasoline to be split between the car and the moped. Omar's car gets 20 mpg and holds at most 10 gal of gas. His moped gets 100 mpg and holds at most 3 gal of gas. How many gallons of gasoline should each vehicle use if Omar wants to travel as far as possible? What is the maximum number of miles that he can travel?

61. *Maximizing Profit.* Norris Mill can convert logs into lumber and plywood. In a given week, the mill can turn out 400 units of production, of which 100 units of lumber and 150 units of plywood are required by regular customers. The profit is $20 per unit of lumber and $30 per unit of plywood. How many units of each should the mill produce in order to maximize the profit?

62. *Maximizing Profit.* Sunnydale Farm includes 240 acres of cropland. The farm owner wishes to plant this acreage in corn and oats. The profit per acre in corn production is $40 and in oats is $30. A total of 320 hr of labor is available. Each acre of corn requires 2 hr of labor, whereas each acre of oats requires 1 hr of labor. How should the land be divided between corn and oats in order to yield the maximum profit? What is the maximum profit?

63. *Minimizing Cost.* An animal feed to be mixed from soybean meal and oats must contain at least 120 lb of protein, 24 lb of fat, and 10 lb of mineral ash. Each 100-lb sack of soybean meal costs $15 and contains 50 lb of protein, 8 lb of fat, and 5 lb of mineral ash. Each 100-lb sack of oats costs $5 and contains 15 lb of protein, 5 lb of fat, and 1 lb of mineral ash. How many sacks of each should be used to satisfy the minimum requirements at minimum cost?

64. *Minimizing Cost.* Suppose that in the preceding problem the oats were replaced by alfalfa, which costs $8 per 100 lb and contains 20 lb of protein, 6 lb of fat, and 8 lb of mineral ash. How much of each is now required in order to minimize the cost?

65. *Maximizing Income.* Clayton is planning to invest up to $40,000 in corporate and municipal bonds. The least he is allowed to invest in corporate bonds is $6000, and he does not want to invest more than $22,000 in corporate bonds. He also does not want to invest more than $30,000 in municipal bonds. The interest is 8% on corporate bonds and $7\frac{1}{2}\%$ on municipal bonds. This is simple interest for one year. How much should he invest in each type of bond in order to maximize his income? What is the maximum income?

66. *Maximizing Income.* Margaret is planning to invest up to $22,000 in certificates of deposit at City Bank and People's Bank. She wants to invest at least $2000 but no more than $14,000 at City Bank. People's Bank does not insure more than a $15,000 investment, so she will invest no more than that in People's Bank. The interest is 6% at City Bank and $6\frac{1}{2}$% at People's Bank. This is simple interest for one year. How much should she invest in each bank in order to maximize her income? What is the maximum income?

67. *Minimizing Transportation Cost.* An airline with two types of airplanes, P_1 and P_2, has contracted with a tour group to provide transportation for a minimum of 2000 first-class, 1500 tourist-class, and 2400 economy-class passengers. For a certain trip, airplane P_1 costs $12 thousand to operate and can accommodate 40 first-class, 40 tourist-class, and 120 economy-class passengers, whereas airplane P_2 costs $10 thousand to operate and can accommodate 80 first-class, 30 tourist-class, and 40 economy-class passengers. How many of each type of airplane should be used in order to minimize the operating cost?

68. *Minimizing Transportation Cost.* Suppose that in the preceding problem a new airplane P_3 becomes available, having an operating cost for the same trip of $15 thousand and accommodating 40 first-class, 40 tourist-class, and 80 economy-class passengers. If airplane P_1 were replaced by airplane P_3, how many of P_2 and P_3 should be used in order to minimize the operating cost?

69. *Maximizing Profit.* It takes Just Sew 2 hr of cutting and 4 hr of sewing to make a knit suit. It takes 4 hr of cutting and 2 hr of sewing to make a

worsted suit. At most 20 hr per day is available for cutting and at most 16 hr per day is available for sewing. The profit is $34 on a knit suit and $31 on a worsted suit. How many of each kind of suit should be made each day in order to maximize profit? What is the maximum profit?

70. *Maximizing Profit.* Cambridge Metal Works manufactures two sizes of gears. The smaller gear requires 4 hr of machining and 1 hr of polishing and yields a profit of $25. The larger gear requires 1 hr of machining and 1 hr of polishing and yields a profit of $10. The firm has available at most 24 hr per day for machining and 9 hr per day for polishing. How many of each type of gear should be produced each day in order to maximize profit? What is the maximum profit?

71. *Minimizing Nutrition Cost.* Suppose that it takes 12 units of carbohydrates and 6 units of protein to satisfy Jacob's minimum weekly requirements. A particular type of meat contains 2 units of carbohydrates and 2 units of protein per pound. A particular cheese contains 3 units of carbohydrates and 1 unit of protein per pound. The meat costs $3.50 per pound and the cheese costs $4.60 per pound. How many pounds of each are needed in order to minimize the cost and still meet the minimum requirements?

72. *Minimizing Salary Cost.* The Spring Hill school board is analyzing education costs for Hill Top School. It wants to hire teachers and teacher's aides to make up a faculty that satisfies its needs at minimum cost. The average annual salary for a teacher is $35,000 and for a teacher's aide is $18,000. The school building can accommodate a faculty of no more than 50 but needs at least 20 faculty members to function properly. The school must have at least 12 aides, but the number of teachers must be at least twice the number of aides in order to accommodate the expectations of the community. How many teachers and teacher's aides should be hired in order to minimize salary costs?

73. *Maximizing Animal Support in a Forest.* A certain area of forest is populated by two species of animals, which scientists refer to as A and B for simplicity. The forest supplies two kinds of food, referred to as F_1 and F_2. For one year, species A requires 1 unit of F_1 and 0.5 unit of F_2. Species B requires 0.2 unit of F_1 and 1 unit of F_2. The forest can normally supply at most 600 units of F_1 and 525 units of F_2 per year. What is the maximum total number of these animals that the forest can support?

74. *Maximizing Animal Support in a Forest.* Refer to Exercise 73. If there is a wet spring, then supplies of food increase to 1080 units of F_1 and 810 units of F_2. In this case, what is the maximum total number of these animals that the forest can support?

Discussion and Writing

75. Write an applied linear programming problem for a classmate to solve. Devise your problem so that the answer is "The bakery will make a maximum profit when 5 dozen pies and 12 dozen cookies are baked."

76. Describe how the graph of a linear inequality differs from the graph of a linear equation.

Skill Maintenance

Solve.

77. $-5 \leq x + 2 < 4$

78. $|x - 3| \geq 2$

79. $x^2 - 2x \leq 3$

80. $\dfrac{x - 1}{x + 2} > 4$

Synthesis

Graph the system of inequalities.

81. $y \geq x^2 - 2,$
$\quad y \leq 2 - x^2$

82. $y < x + 1,$
$\quad y \geq x^2$

Graph the inequality.

83. $|x + y| \leq 1$

84. $|x| + |y| \leq 1$

85. $|x| > |y|$

86. $|x - y| > 0$

87. *Allocation of Resources.* Significant Sounds manufactures two types of speaker assemblies. The less expensive assembly, which sells for \$350, consists of one midrange speaker and one tweeter. The more expensive speaker assembly, which sells for \$600, consists of one woofer, one midrange speaker, and two tweeters. The manufacturer has in stock 44 woofers, 60 midrange speakers, and 90 tweeters. How many of each type of speaker assembly should be made in order to maximize income? What is the maximum income?

88. *Allocation of Resources.* Sitting Pretty Furniture produces chairs and sofas. The chairs require 20 ft of wood, 1 lb of foam rubber, and 2 yd^2 of fabric. The sofas require 100 ft of wood, 50 lb of foam rubber, and 20 yd^2 of fabric. The manufacturer has in stock 1900 ft of wood, 500 lb of foam rubber, and 240 yd^2 of fabric. The chairs can be sold for \$80 each and the sofas for \$300 each. How many of each should be produced in order to maximize income? What is the maximum income?

Partial Fractions

5.7

• *Decompose rational expressions into partial fractions.*

There are situations in calculus in which it is useful to write a rational expression as a sum of two or more simpler rational expressions. For example, in the equation

$$\frac{4x - 13}{2x^2 + x - 6} = \frac{3}{x + 2} + \frac{-2}{2x - 3},$$

each fraction on the right side is called a **partial fraction**. The expression on the right side is the **partial fraction decomposition** of the rational expression on the left side. In this section, we learn how such decompositions are created.

Partial Fraction Decompositions

The procedure for finding the partial fraction decomposition of a rational expression involves factoring its denominator into linear and quadratic factors.

Procedure for Decomposing a Rational Expression into Partial Fractions

Consider any rational expression $P(x)/Q(x)$ such that $P(x)$ and $Q(x)$ have no common factor other than 1 or -1.

1. If the degree of $P(x)$ is greater than or equal to the degree of $Q(x)$, divide to express $P(x)/Q(x)$ as a quotient + remainder/$Q(x)$ and follow steps (2)–(5) to decompose the resulting rational expression.

2. If the degree of $P(x)$ is less than the degree of $Q(x)$, factor $Q(x)$ into linear factors of the form $(px + q)^n$ and/or quadratic factors of the form $(ax^2 + bx + c)^m$. Any quadratic factor $ax^2 + bx + c$ must be *irreducible*, meaning that it cannot be factored into linear factors with real coefficients.

3. Assign to each linear factor $(px + q)^n$ the sum of n partial fractions:

$$\frac{A_1}{px + q} + \frac{A_2}{(px + q)^2} + \cdots + \frac{A_n}{(px + q)^n}.$$

4. Assign to each quadratic factor $(ax^2 + bx + c)^m$ the sum of m partial fractions:

$$\frac{B_1 x + C_1}{ax^2 + bx + c} + \frac{B_2 x + C_2}{(ax^2 + bx + c)^2} + \cdots + \frac{B_m x + C_m}{(ax^2 + bx + c)^m}.$$

5. Apply algebraic methods, as illustrated in the following examples, to find the constants in the numerators of the partial fractions.

EXAMPLE 1   Decompose into partial fractions:

$$\frac{4x - 13}{2x^2 + x - 6}.$$

Solution   The degree of the numerator is less than the degree of the denominator. We begin by factoring the denominator: $(x + 2)(2x - 3)$. We know that there are constants A and B such that

$$\frac{4x - 13}{(x + 2)(2x - 3)} = \frac{A}{x + 2} + \frac{B}{2x - 3}.$$

To determine A and B, we add the expressions on the right:

$$\frac{4x - 13}{(x + 2)(2x - 3)} = \frac{A(2x - 3) + B(x + 2)}{(x + 2)(2x - 3)}.$$

Next, we equate the numerators:

$$4x - 13 = A(2x - 3) + B(x + 2).$$

Since the last equation containing A and B is true for all x, we can substitute any value of x and still have a true equation. In order to have $2x - 3 = 0$, we choose $x = \frac{3}{2}$. This gives us

$$4\left(\tfrac{3}{2}\right) - 13 = A\left(2 \cdot \tfrac{3}{2} - 3\right) + B\left(\tfrac{3}{2} + 2\right)$$
$$-7 = 0 + \tfrac{7}{2}B.$$

Solving, we obtain $B = -2$.

In order to have $x + 2 = 0$, we choose $x = -2$, which gives us

$$4(-2) - 13 = A[2(-2) - 3] + B(-2 + 2).$$

Solving, we obtain $A = 3$.

The decomposition is as follows:

$$\frac{4x - 13}{2x^2 + x - 6} = \frac{3}{x + 2} + \frac{-2}{2x - 3}, \quad \text{or} \quad \frac{3}{x + 2} - \frac{2}{2x - 3}.$$

To check, we can add to see if we get the expression on the left. We can also use the TABLE feature, comparing values of

$$y_1 = \frac{4x - 13}{2x^2 + x - 6} \quad \text{and} \quad y_2 = \frac{3}{x + 2} - \frac{2}{2x - 3}$$

for the same values of x. Since $y_1 = y_2$ for the given values of x as we scroll through the table, the decomposition appears to be correct.

X	Y₁	Y₂
−1	3.4	3.4
0	2.1667	2.1667
1	3	3
2	−1.25	−1.25
3	−.0667	−.0667
4	.1	.1
5	.14286	.14286

X = −1

EXAMPLE 2 Decompose into partial fractions:

$$\frac{7x^2 - 29x + 24}{(2x - 1)(x - 2)^2}.$$

Solution The degree of the numerator is less than the degree of the denominator. The decomposition has the following form:

$$\frac{7x^2 - 29x + 24}{(2x - 1)(x - 2)^2} = \frac{A}{2x - 1} + \frac{B}{x - 2} + \frac{C}{(x - 2)^2}.$$

As in Example 1, we add and equate the numerators. This gives us

$$7x^2 - 29x + 24 = A(x - 2)^2 + B(2x - 1)(x - 2) + C(2x - 1).$$

Since the equation containing A, B, and C is true for all x, we can substitute any value of x and still have a true equation. In order to have $2x - 1 = 0$, we let $x = \frac{1}{2}$. This gives us

$$7\left(\tfrac{1}{2}\right)^2 - 29 \cdot \tfrac{1}{2} + 24 = A\left(\tfrac{1}{2} - 2\right)^2 + 0.$$

Solving, we obtain $A = 5$.

In order to have $x - 2 = 0$, we let $x = 2$. Substituting gives us

$$7(2)^2 - 29(2) + 24 = 0 + C(2 \cdot 2 - 1).$$

Solving, we obtain $C = -2$.

To find B, we choose any value for x except $\frac{1}{2}$ or 2 and replace A with

5 and C with -2. We let $x = 1$:

$$7 \cdot 1^2 - 29 \cdot 1 + 24 = 5(1 - 2)^2 + B(2 \cdot 1 - 1)(1 - 2)$$
$$+ (-2)(2 \cdot 1 - 1)$$
$$2 = 5 - B - 2$$
$$B = 1.$$

The decomposition is as follows:

$$\frac{7x^2 - 29x + 24}{(2x - 1)(x - 2)^2} = \frac{5}{2x - 1} + \frac{1}{x - 2} - \frac{2}{(x - 2)^2}.$$

We can check the result using a table of values. We let

$$y_1 = \frac{7x^2 - 29x + 24}{(2x - 1)(x - 2)^2} \quad \text{and} \quad y_2 = \frac{5}{2x - 1} + \frac{1}{x - 2} - \frac{2}{(x - 2)^2}.$$

X	Y₁	Y₂
−5	−.6382	−.6382
−4	−.7778	−.7778
−3	−.9943	−.9943
−2	−1.375	−1.375
−1	−2.222	−2.222
0	−6	−6
1	2	2

X = −5

Since $y_1 = y_2$ for given values of x as we scroll through the table, the decomposition appears to be correct.

EXAMPLE 3   Decompose into partial fractions:

$$\frac{6x^3 + 5x^2 - 7}{3x^2 - 2x - 1}.$$

Solution   The degree of the numerator is greater than that of the denominator. Thus we divide and find an equivalent expression:

$$
\begin{array}{r}
2x + 3 \\
3x^2 - 2x - 1 \overline{)6x^3 + 5x^2 - 7} \\
\underline{6x^3 - 4x^2 - 2x} \\
9x^2 + 2x - 7 \\
\underline{9x^2 - 6x - 3} \\
8x - 4
\end{array}
$$

The original expression is thus equivalent to

$$2x + 3 + \frac{8x - 4}{3x^2 - 2x - 1}.$$

We decompose the fraction to get

$$\frac{8x - 4}{(3x + 1)(x - 1)} = \frac{5}{3x + 1} + \frac{1}{x - 1}.$$

The final result is

$$2x + 3 + \frac{5}{3x + 1} + \frac{1}{x - 1}.$$

Systems of equations can be used to decompose rational expressions. Let's reconsider Example 2.

EXAMPLE 4 Decompose into partial fractions:

$$\frac{7x^2 - 29x + 24}{(2x - 1)(x - 2)^2}.$$

Solution The decomposition has the following form:

$$\frac{A}{2x - 1} + \frac{B}{x - 2} + \frac{C}{(x - 2)^2}.$$

We first add:

$$\frac{7x^2 - 29x + 24}{(2x - 1)(x - 2)^2} = \frac{A}{2x - 1} + \frac{B}{x - 2} + \frac{C}{(x - 2)^2}$$

$$= \frac{A(x - 2)^2}{(2x - 1)(x - 2)^2} + \frac{B(2x - 1)(x - 2)}{(2x - 1)(x - 2)^2}$$

$$+ \frac{C(2x - 1)}{(2x - 1)(x - 2)^2}.$$

Then we equate numerators:

$$7x^2 - 29x + 24$$
$$= A(x - 2)^2 + B(2x - 1)(x - 2) + C(2x - 1)$$
$$= A(x^2 - 4x + 4) + B(2x^2 - 5x + 2) + C(2x - 1)$$
$$= Ax^2 - 4Ax + 4A + 2Bx^2 - 5Bx + 2B + 2Cx - C,$$

or

$$7x^2 - 29x + 24$$
$$= (A + 2B)x^2 + (-4A - 5B + 2C)x + (4A + 2B - C).$$

Next, we equate corresponding coefficients:

$7 = A + 2B,$	The coefficients of the x^2-terms must be the same.
$-29 = -4A - 5B + 2C,$	The coefficients of the x-terms must be the same.
$24 = 4A + 2B - C.$	The constant terms must be the same.

We now have a system of three equations. You should confirm that the solution of the system is

$$A = 5, \qquad B = 1, \quad \text{and} \quad C = -2.$$

The decomposition is as follows:

$$\frac{7x^2 - 29x + 24}{(2x - 1)(x - 2)^2} = \frac{5}{2x - 1} + \frac{1}{x - 2} - \frac{2}{(x - 2)^2}.$$

EXAMPLE 5 Decompose into partial fractions:

$$\frac{11x^2 - 8x - 7}{(2x^2 - 1)(x - 3)}.$$

STUDY TIP

The review icons in the text margins provide references to earlier sections where you can find content related to the concept at hand. Reviewing this earlier content will add to your understanding of the current concept.

SYSTEMS OF EQUATIONS IN THREE VARIABLES
REVIEW SECTION 5.2.

Solution The decomposition has the following form:

$$\frac{11x^2 - 8x - 7}{(2x^2 - 1)(x - 3)} = \frac{Ax + B}{2x^2 - 1} + \frac{C}{x - 3}.$$

Adding and equating numerators, we get

$$11x^2 - 8x - 7 = (Ax + B)(x - 3) + C(2x^2 - 1)$$
$$= Ax^2 - 3Ax + Bx - 3B + 2Cx^2 - C,$$

or $\quad 11x^2 - 8x - 7 = (A + 2C)x^2 + (-3A + B)x + (-3B - C).$

We then equate corresponding coefficients:

$$11 = A + 2C, \qquad \text{The coefficients of the } x^2\text{-terms}$$
$$-8 = -3A + B, \qquad \text{The coefficients of the } x\text{-terms}$$
$$-7 = -3B - C. \qquad \text{The constant terms}$$

We solve this system of three equations and obtain

$$A = 3, \qquad B = 1, \quad \text{and} \quad C = 4.$$

The decomposition is as follows:

$$\frac{11x^2 - 8x - 7}{(2x^2 - 1)(x - 3)} = \frac{3x + 1}{2x^2 - 1} + \frac{4}{x - 3}.$$

Exercise Set 5.7

Decompose into partial fractions. Check your answers using a grapher.

1. $\dfrac{x + 7}{(x - 3)(x + 2)}$

2. $\dfrac{2x}{(x + 1)(x - 1)}$

3. $\dfrac{7x - 1}{6x^2 - 5x + 1}$

4. $\dfrac{13x + 46}{12x^2 - 11x - 15}$

5. $\dfrac{3x^2 - 11x - 26}{(x^2 - 4)(x + 1)}$

6. $\dfrac{5x^2 + 9x - 56}{(x - 4)(x - 2)(x + 1)}$

7. $\dfrac{9}{(x + 2)^2(x - 1)}$

8. $\dfrac{x^2 - x - 4}{(x - 2)^3}$

9. $\dfrac{2x^2 + 3x + 1}{(x^2 - 1)(2x - 1)}$

10. $\dfrac{x^2 - 10x + 13}{(x^2 - 5x + 6)(x - 1)}$

11. $\dfrac{x^4 - 3x^3 - 3x^2 + 10}{(x + 1)^2(x - 3)}$

12. $\dfrac{10x^3 - 15x^2 - 35x}{x^2 - x - 6}$

13. $\dfrac{-x^2 + 2x - 13}{(x^2 + 2)(x - 1)}$

14. $\dfrac{26x^2 + 208x}{(x^2 + 1)(x + 5)}$

15. $\dfrac{6 + 26x - x^2}{(2x - 1)(x + 2)^2}$

16. $\dfrac{5x^3 + 6x^2 + 5x}{(x^2 - 1)(x + 1)^3}$

17. $\dfrac{6x^3 + 5x^2 + 6x - 2}{2x^2 + x - 1}$

18. $\dfrac{2x^3 + 3x^2 - 11x - 10}{x^2 + 2x - 3}$

19. $\dfrac{2x^2 - 11x + 5}{(x - 3)(x^2 + 2x - 5)}$

20. $\dfrac{3x^2 - 3x - 8}{(x - 5)(x^2 + x - 4)}$

21. $\dfrac{-4x^2 - 2x + 10}{(3x + 5)(x + 1)^2}$

22. $\dfrac{26x^2 - 36x + 22}{(x - 4)(2x - 1)^2}$

23. $\dfrac{36x + 1}{12x^2 - 7x - 10}$

24. $\dfrac{-17x + 61}{6x^2 + 39x - 21}$

25. $\dfrac{-4x^2 - 9x + 8}{(3x^2 + 1)(x - 2)}$

26. $\dfrac{11x^2 - 39x + 16}{(x^2 + 4)(x - 8)}$

Discussion and Writing

27. Describe the two methods used to find the constants in a partial fraction decomposition.

28. What would you say to a classmate who tells you that the partial fraction decomposition of

$$\frac{3x^2 - 8x + 9}{(x + 3)(x^2 - 5x + 6)}$$

is

$$\frac{2}{x + 3} + \frac{x - 1}{x^2 - 5x + 6}?$$

Explain.

29. Explain the error in the following process.

$$\frac{x^2 + 4}{(x + 2)(x + 1)} = \frac{A}{x + 2} + \frac{B}{x + 1}$$
$$= \frac{A(x + 1) + B(x + 2)}{(x + 2)(x + 1)}$$

Then

$$x^2 + 4 = A(x + 1) + B(x + 2).$$

When $x = -1$:

$$(-1)^2 + 4 = A(-1 + 1) + B(-1 + 2)$$
$$5 = B.$$

When $x = -2$:

$$(-2)^2 + 4 = A(-2 + 1) + B(-2 + 2)$$
$$8 = -A$$
$$-8 = A.$$

Thus,

$$\frac{x^2 + 4}{(x + 2)(x + 1)} = \frac{-8}{x + 2} + \frac{5}{x + 1}.$$

Skill Maintenance

Find the zeros of the polynomial function.

30. $f(x) = x^3 - 3x^2 + x - 3$

31. $f(x) = x^3 + x^2 - 3x - 2$

32. $f(x) = x^4 - x^3 - 5x^2 - x - 6$

33. $f(x) = x^3 + 5x^2 + 5x - 3$

Synthesis

Decompose into partial fractions.

34. $\dfrac{9x^3 - 24x^2 + 48x}{(x - 2)^4(x + 1)}$

[*Hint*: Let the expression equal

$$\frac{A}{x + 1} + \frac{P(x)}{(x - 2)^4}$$

and find $P(x)$.]

35. $\dfrac{x}{x^4 - a^4}$

36. $\dfrac{1}{e^{-x} + 3 + 2e^x}$

37. $\dfrac{1 + \ln x^2}{(\ln x + 2)(\ln x - 3)^2}$

Chapter Summary and Review 5

Important Properties and Formulas

Row-Equivalent Operations

1. Interchange any two rows.
2. Multiply each entry in a row by the same nonzero constant.
3. Add a nonzero multiple of one row to another row.

Row-Echelon Form

1. If a row does not consist entirely of 0's, then the first nonzero element in the row is a 1 (called a leading 1).
2. For any two successive nonzero rows, the leading 1 in the lower row is farther to the right than the leading 1 in the higher row.

(continued)

3. All the rows consisting entirely of 0's are at the bottom of the matrix.

If a fourth property is also satisfied, a matrix is said to be in reduced row-echelon form:

4. Each column that contains a leading 1 has 0's everywhere else.

Properties of Matrix Addition and Scalar Multiplication

For any $m \times n$ matrices **A**, **B**, and **C** and any scalars k and l:

Commutative Property of Addition:
$$\mathbf{A} + \mathbf{B} = \mathbf{B} + \mathbf{A}.$$

Associative Property of Addition:
$$\mathbf{A} + (\mathbf{B} + \mathbf{C}) = (\mathbf{A} + \mathbf{B}) + \mathbf{C}.$$

Associative Property of Scalar Multiplication:
$$(kl)\mathbf{A} = k(l\mathbf{A}).$$

Additive Identity Property:
There exists a unique matrix **0** such that
$$\mathbf{A} + \mathbf{0} = \mathbf{0} + \mathbf{A} = \mathbf{A}.$$

Additive Inverse Property:
There exists a unique matrix $-\mathbf{A}$ such that
$$\mathbf{A} + (-\mathbf{A}) = -\mathbf{A} + \mathbf{A} = \mathbf{0}.$$

Distributive Properties:
$$k(\mathbf{A} + \mathbf{B}) = k\mathbf{A} + k\mathbf{B},$$
$$(k + l)\mathbf{A} = k\mathbf{A} + l\mathbf{A}.$$

Properties of Matrix Multiplication

For matrices **A**, **B**, and **C**, assuming that the indicated operation is possible:

Associative Property of Multiplication:
$$\mathbf{A}(\mathbf{B}\mathbf{C}) = (\mathbf{A}\mathbf{B})\mathbf{C}.$$

Distributive Properties:
$$\mathbf{A}(\mathbf{B} + \mathbf{C}) = \mathbf{A}\mathbf{B} + \mathbf{A}\mathbf{C},$$
$$(\mathbf{B} + \mathbf{C})\mathbf{A} = \mathbf{B}\mathbf{A} + \mathbf{C}\mathbf{A}.$$

Matrix Solutions of Systems of Equations

For a system of n linear equations in n variables, $\mathbf{A}\mathbf{X} = \mathbf{B}$, if **A** is an invertible matrix, then the unique solution of the system is given by $\mathbf{X} = \mathbf{A}^{-1}\mathbf{B}$.

To Graph a Linear Inequality in Two Variables:

1. Replace the inequality symbol with an equals sign and graph this related equation. If the inequality symbol is $<$ or $>$, draw the line dashed. If the inequality symbol is $\leq$ or $\geq$, draw the line solid.

2. The graph consists of a half-plane on one side of the line and, if the line is solid, the line as well. To determine which half-plane to shade, test a point not on the line in the original inequality. If that point is a solution, shade the half-plane containing that point. If not, shade the opposite half-plane.

Linear Programming Procedure

To find the maximum or minimum value of a linear objective function subject to a set of constraints:

1. Graph the region of feasible solutions.

2. Determine the coordinates of the vertices of the region.

3. Evaluate the objective function at each vertex. The largest and smallest of those values are the maximum and minimum values of the function, respectively.

Procedure for Decomposing a Rational Expression into Partial Fractions

Consider any rational expression $P(x)/Q(x)$ such that $P(x)$ and $Q(x)$ have no common factor other than 1 or -1.

1. If the degree of $P(x)$ is greater than or equal to the degree of $Q(x)$, divide to express $P(x)/Q(x)$ as a quotient $+$ remainder$/Q(x)$ and follow steps (2)–(5) to decompose the resulting rational expression.

2. If the degree of $P(x)$ is less than the degree of $Q(x)$, factor $Q(x)$ into linear factors of the form $(px + q)^n$ and/or quadratic factors of the form $(ax^2 + bx + c)^m$. Any quadratic factor $ax^2 + bx + c$ must be irreducible, meaning that it cannot be factored into linear factors with real coefficients.

3. Assign to each linear factor $(px + q)^n$ the sum of n partial fractions:

$$\frac{A_1}{px + q} + \frac{A_2}{(px + q)^2} + \cdots + \frac{A_n}{(px + q)^n}.$$

4. Assign to each quadratic factor $(ax^2 + bx + c)^m$ the sum of m partial

fractions:

$$\frac{B_1x + C_1}{ax^2 + bx + c} + \frac{B_2x + C_2}{(ax^2 + bx + c)^2} + \cdots +$$

$$\frac{B_mx + C_m}{(ax^2 + bx + c)^m}.$$

5. Apply algebraic methods to find the constants in the numerators of the partial fractions.

REVIEW EXERCISES

In Exercises 1–8, match the equations or inequalities with one of the graphs (a)–(h), which follow.

a)

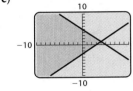

b)

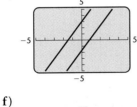

c)

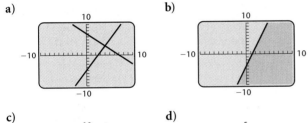

d)

e)

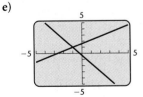

f)

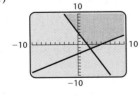

g)

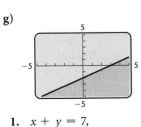

h)

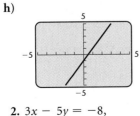

1. $x + y = 7,$
   $2x - y = 5$

2. $3x - 5y = -8,$
   $4x + 3y = -1$

3. $y = 2x - 1,$
   $4x - 2y = 2$

4. $6x - 3y = 5,$
   $y = 2x + 3$

5. $y \leq 3x - 4$

6. $2x - 3y \geq 6$

7. $x - y \leq 3,$
   $x + y \leq 5$

8. $2x + y \geq 4,$
   $3x - 5y \leq 15$

Solve.

9. $5x - 3y = -4,$
   $3x - y = -4$

10. $2x + 3y = 2,$
    $5x - y = -29$

11. $x + 5y = 12,$
    $5x + 25y = 12$

12. $x - y = -2,$
    $-3x + 3y = 6$

13. $2x - 4y + 3z = -3,$
    $-5x + 2y - z = 7,$
    $3x + 2y - 2z = 4$

14. $x + 5y + 3z = 0,$
    $3x - 2y + 4z = 0,$
    $2x + 3y - z = 0$

15. $x - y = 5,$
    $y - z = 6,$
    $z - w = 7,$
    $x + w = 8$

16. Classify each of the systems in Exercises 9–15 as consistent or inconsistent.

17. Classify each of the systems in Exercises 9–15 as dependent or independent.

Solve the system of equations using Gaussian elimination or Gauss–Jordan elimination.

18. $x + 2y = 5,$
    $2x - 5y = -8$

19. $3x + 4y + 2z = 3,$
    $5x - 2y - 13z = 3,$
    $4x + 3y - 3z = 6$

20. $3x + 5y + z = 0,$
    $2x - 4y - 3z = 0,$
    $x + 3y + z = 0$

21.
$$w + x + y + z = -2,$$
$$-3w - 2x + 3y + 2z = 10,$$
$$2w + 3x + 2y - z = -12,$$
$$2w + 4x - y + z = 1$$

22. *Coins.*   The value of 75 coins, consisting of nickels and dimes, is $5.95. How many of each kind are there?

23. *Investment.*   The Mendez family invested $5000, part at 10% and the remainder at 10.5%. The annual income from both investments is $517. What is the amount invested at each rate?

24. *Nutrition.*   A dietician must plan a breakfast menu that provides 460 Cal, 9 g of fat, and 55 mg of calcium. One plain bagel contains 200 Cal, 2 g of fat, and 29 mg of calcium. A one-tablespoon serving of cream cheese contains 100 Cal, 10 g of fat, and 24 mg of calcium. One banana contains 105 Cal, 1 g of fat, and 7 mg of calcium. (*Source: Home and Garden Bulletin No. 72,* U.S. Government Printing Office, Washington, D.C. 20402) How many servings of each are required to provide the desired nutritional values?

25. *Test Scores.*   A student has a total of 225 on three tests. The sum of the scores on the first and second tests exceeds the score on the third test by 61. The first score exceeds the second by 6. Find the three scores.

26. *Ice Milk Consumption.*   The table below shows the per capita ice milk consumption, in pounds, in the United States, represented as years since 1990.

YEAR, x	ICE MILK CONSUMPTION (IN POUNDS)
1990, 0	7.7
1993, 3	6.9
1996, 6	7.6

Source: U.S. Department of Agriculture, Economic Research Service, *Food Consumption, Prices, and Expenditures,* annual

a) Use a system of equations to fit a quadratic function $f(x) = ax^2 + bx + c$ to the data.
b) Use the function to predict the per capita ice milk consumption in 2005.

For Exercises 27–34, let

$$\mathbf{A} = \begin{bmatrix} 1 & -1 & 0 \\ 2 & 3 & -2 \\ -2 & 0 & 1 \end{bmatrix},$$

$$\mathbf{B} = \begin{bmatrix} -1 & 0 & 6 \\ 1 & -2 & 0 \\ 0 & 1 & -3 \end{bmatrix},$$

and

$$\mathbf{C} = \begin{bmatrix} -2 & 0 \\ 1 & 3 \end{bmatrix}.$$

Find each of the following, if possible.

27. A + B   **28. −3A**

29. −A   **30. AB**

31. B + C   **32. A − B**

33. 2A − B   **34. A + 3B**

35. *Food Service Management.*   The table below shows the cost per serving, in cents, for items on four menus that are served at an elder-care facility.

MENU	MEAT	POTATO	VEGETABLE	SALAD	DESSERT
1	46.1	5.9	10.1	8.5	11.4
2	54.6	4.6	9.6	7.6	10.6
3	48.9	5.5	12.7	9.4	9.3
4	51.3	4.8	11.3	6.9	12.7

On a particular day, a dietician orders 32 meals from menu 1, 19 from menu 2, 43 from menu 3, and 38 from menu 4.

a) Write the information in the table as a 4 × 5 matrix **M**.
b) Write a row matrix **N** that represents the number of each menu ordered.
c) Find the product **NM**.
d) State what the entries of **NM** represent.

Find $\mathbf{A}^{-1}$*, if it exists.*

36. A $= \begin{bmatrix} -2 & 0 \\ 1 & 3 \end{bmatrix}$

37. A $= \begin{bmatrix} 0 & 0 & 3 \\ 0 & -2 & 0 \\ 4 & 0 & 0 \end{bmatrix}$

38. A $= \begin{bmatrix} 1 & 0 & 0 & 0 \\ 0 & 4 & -5 & 0 \\ 0 & 2 & 2 & 0 \\ 0 & 0 & 0 & 1 \end{bmatrix}$

39. Write a matrix equation equivalent to this system of equations:

$$\begin{aligned} 3x - 2y + 4z &= 13, \\ x + 5y - 3z &= 7, \\ 2x - 3y + 7z &= -8. \end{aligned}$$

Solve the system of equations using the inverse of the coefficient matrix of the equivalent matrix equation.

40. $2x + 3y = 5,$
$3x + 5y = 11$

41. $5x - y + 2z = 17,$
$3x + 2y - 3z = -16,$
$4x - 3y - z = 5$

42. $w - x - y + z = -1,$
$2w + 3x - 2y - z = 2,$
$-w + 5x + 4y - 2z = 3,$
$3w - 2x + 5y + 3z = 4$

Graph.

43. $y \leq 3x + 6$

44. $4x - 3y \geq 12$

45. Graph this system of inequalities and find the coordinates of any vertices formed.

$$\begin{aligned} 2x + y &\geq 9, \\ 4x + 3y &\geq 23, \\ x + 3y &\geq 8, \\ x &\geq 0, \\ y &\geq 0 \end{aligned}$$

46. Find the maximum and minimum values of $T = 6x + 10y$ subject to

$$\begin{aligned} x + y &\leq 10, \\ 5x + 10y &\geq 50, \\ x &\geq 2, \\ y &\geq 0. \end{aligned}$$

47. *Maximizing a Test Score.* Marita is taking a test that contains questions in group A worth 7 points each and questions in group B worth 12 points each. The total number of questions answered must be at least 8. If Marita knows that group A questions take 8 min each and group B questions take 10 min each and the maximum time for the test is 80 min, how many questions from each group must she answer correctly in order to maximize her score? What is the maximum score?

Decompose into partial fractions.

48. $\dfrac{5}{(x + 2)^2(x + 1)}$

49. $\dfrac{-8x + 23}{2x^2 + 5x - 12}$

Discussion and Writing

50. Write a problem for a classmate to solve that can be translated to a system of equations. Devise the problem so that the solution is "The caterer sold 20 cheese trays and 35 seafood trays."

51. For square matrices **A** and **B**, is it true, in general, that $(\mathbf{AB})^2 = \mathbf{A}^2\mathbf{B}^2$? Explain.

Synthesis

52. One year, Don invested a total of $40,000, part at 12%, part at 13%, and the rest at $14\frac{1}{2}$%. The total amount of interest received on the investments was $5370. The interest received on the $14\frac{1}{2}$% investment was $1050 more than the interest received on the 13% investment. How much was invested at each rate?

Solve.

53. $\dfrac{2}{3x} + \dfrac{4}{5y} = 8,$
$\dfrac{5}{4x} - \dfrac{3}{2y} = -6$

54. $\dfrac{3}{x} - \dfrac{4}{y} + \dfrac{1}{z} = -2,$
$\dfrac{5}{x} + \dfrac{1}{y} - \dfrac{2}{z} = 1,$
$\dfrac{7}{x} + \dfrac{3}{y} + \dfrac{2}{z} = 19$

Graph.

55. $|x| - |y| \leq 1$

56. $|xy| > 1$

Conic Sections 6

APPLICATION

Expenditures from the Social Security Trust Fund are increasing at a greater rate than the fund's income. Expenditures can be modeled with an exponential function and income can be modeled with a linear function. We can use these functions to predict that expenditures will equal income about 25 yr after 1985, or in 2010. (*Source*: Social Security Administration)

This problem appears as Exercise 65 in Section 6.4.

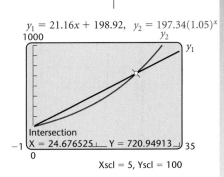

$y_1 = 21.16x + 198.92$, $y_2 = 197.34(1.05)^x$

Intersection
X = 24.676525   Y = 720.94913

Xscl = 5, Yscl = 100

I n this chapter, we study *conic sections*. These curves are formed by the intersection of a cone and a plane. Conic sections and their properties were first studied by the Greeks. We also study systems of equations in which at least one equation is nonlinear. Both conic sections and nonlinear systems of equations have many applications, as we will see.

6.1   The Parabola

6.2   The Circle and the Ellipse

6.3   The Hyperbola

6.4   Nonlinear Systems of Equations

SUMMARY AND REVIEW

427

The Parabola

6.1

- *Given an equation of a parabola, complete the square, if necessary, and then find the vertex, the focus, and the directrix and graph the parabola.*

A **conic section** is formed when a right circular cone with two parts, called *nappes*, is intersected by a plane. One of four types of curves can be formed: a parabola, a circle, an ellipse, or a hyperbola.

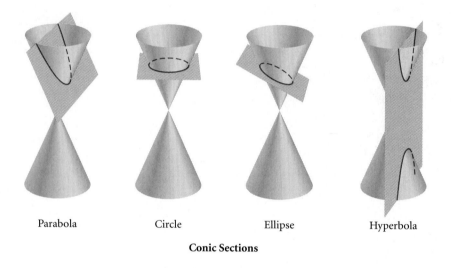

Parabola   Circle   Ellipse   Hyperbola

Conic Sections

Parabolas

Conic sections can be defined algebraically using second-degree equations of the form $Ax^2 + Bxy + Cy^2 + Dx + Ey + F = 0$. In addition, they can be defined geometrically as a set of points that satisfy certain conditions.

In Section 2.4, we saw that the graph of the quadratic function $f(x) = ax^2 + bx + c$, $a \neq 0$, is a parabola. A parabola can be defined geometrically.

Parabola

A **parabola** is the set of all points in a plane equidistant from a fixed line (the **directrix**) and a fixed point not on the line (the **focus**).

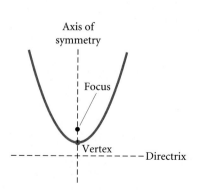

The line that is perpendicular to the directrix and contains the focus is the **axis of symmetry.** The **vertex** is the midpoint of the segment between the focus and the directrix. (See the figure at left.)

Let's derive the standard equation of a parabola with vertex $(0, 0)$ and directrix $y = -p$, where $p > 0$. We place the coordinate axes as shown in the figure at the top of the following page. The y-axis is the axis of symmetry and contains the focus F. The distance from the focus to the vertex is the same as the distance from the vertex to the directrix. Thus the coordinates of F are $(0, p)$.

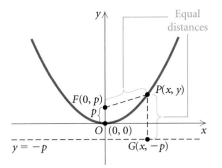

Let $P(x, y)$ be any point of the parabola and consider $\overline{PG}$ perpendicular to the line $y = -p$. The coordinates of G are $(x, -p)$. By the definition of a parabola,

$$PF = PG.$$ The distance from P to the focus is the same as the distance from P to the directrix.

Then using the distance formula, we have

$$\sqrt{(x - 0)^2 + (y - p)^2} = \sqrt{(x - x)^2 + [y - (-p)]^2}$$
$$x^2 + y^2 - 2py + p^2 = y^2 + 2py + p^2 \qquad \text{Squaring both sides and squaring the binomials}$$
$$x^2 = 4py.$$

We have shown that if $P(x, y)$ is on the parabola shown above, then its coordinates satisfy this equation. The converse is also true, but we will not prove it here.

Note that if $p > 0$, as above, the graph opens up. If $p < 0$, the graph opens down.

The equation of a parabola with vertex $(0, 0)$ and directrix $x = -p$ is derived similarly. Such a parabola opens either right ($p > 0$) or left ($p < 0$).

Standard Equation of a Parabola with Vertex at the Origin

The standard equation of a parabola with vertex $(0, 0)$ and directrix $y = -p$ is

$$x^2 = 4py.$$

The focus is $(0, p)$ and the y-axis is the axis of symmetry.

The standard equation of a parabola with vertex $(0, 0)$ and directrix $x = -p$ is

$$y^2 = 4px.$$

The focus is $(p, 0)$ and the x-axis is the axis of symmetry.

EXAMPLE 1 Find the focus and the directrix of the parabola $y = -\frac{1}{12}x^2$. Then graph the parabola.

Solution We write $y = -\frac{1}{12}x^2$ in the form $x^2 = 4py$:

$$-\frac{1}{12}x^2 = y \qquad \text{Given equation}$$
$$x^2 = -12y \qquad \text{Multiplying by } -12 \text{ on both sides}$$
$$x^2 = 4(-3)y. \qquad \text{Standard form}$$

Thus, $p = -3$, so the focus is $(0, p)$, or $(0, -3)$. The directrix is $y = -p = -(-3) = 3$.

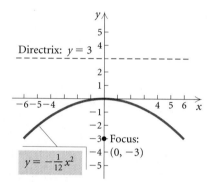

EXAMPLE 2 Find an equation of the parabola with vertex $(0, 0)$ and focus $(5, 0)$. Then graph the parabola.

Solution The focus is on the x-axis so the line of symmetry is the x-axis. Thus the equation is of the type

$$y^2 = 4px.$$

Since the focus is 5 units to the right of the vertex, $p = 5$ and the equation is

$$y^2 = 4(5)x, \quad \text{or}$$
$$y^2 = 20x.$$

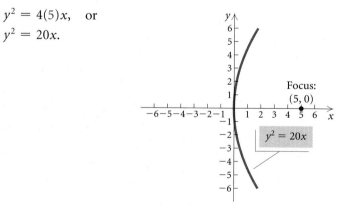

SQUARING THE WINDOW

REVIEW SECTION 1.7.

We can check the graph on a grapher using a squared viewing window. But first it might be necessary to solve for y:

$$y^2 = 20x$$
$$y = \pm\sqrt{20x}. \qquad \text{Using the principle of square roots}$$

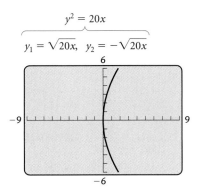

$y^2 = 20x$

$y_1 = \sqrt{20x}, \quad y_2 = -\sqrt{20x}$

We now graph $y_1 = \sqrt{20x}$ and $y_2 = -\sqrt{20x}$. On some graphers, it is possible to graph $y_1 = \sqrt{20x}$ and $y_2 = -y_1$ by using the Y-VARS menu. The hand-drawn graph appears to be correct. ▬

Finding Standard Form by Completing the Square

If a parabola with vertex at the origin is translated horizontally $|h|$ units and vertically $|k|$ units, it has an equation as follows.

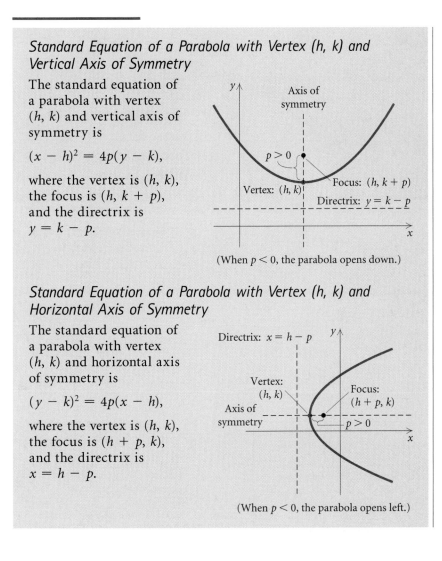

Standard Equation of a Parabola with Vertex (h, k) and Vertical Axis of Symmetry

The standard equation of a parabola with vertex (h, k) and vertical axis of symmetry is

$$(x - h)^2 = 4p(y - k),$$

where the vertex is (h, k), the focus is $(h, k + p)$, and the directrix is $y = k - p$.

(When $p < 0$, the parabola opens down.)

Standard Equation of a Parabola with Vertex (h, k) and Horizontal Axis of Symmetry

The standard equation of a parabola with vertex (h, k) and horizontal axis of symmetry is

$$(y - k)^2 = 4p(x - h),$$

where the vertex is (h, k), the focus is $(h + p, k)$, and the directrix is $x = h - p$.

(When $p < 0$, the parabola opens left.)

COMPLETING THE SQUARE

REVIEW SECTION 2.3.

We can complete the square on equations of the form

$$y = ax^2 + bx + c \quad \text{or} \quad x = ay^2 + by + c$$

in order to write them in standard form.

EXAMPLE 3   For the parabola

$$x^2 + 6x + 4y + 5 = 0,$$

find the vertex, the focus, and the directrix. Then draw the graph.

Solution   We first complete the square:

$$x^2 + 6x + 4y + 5 = 0$$

$$\begin{array}{ll} x^2 + 6x \qquad = -4y - 5 & \text{Subtracting } 4y \text{ and } 5 \text{ on both} \\ & \text{sides} \end{array}$$

$$\begin{array}{ll} x^2 + 6x + 9 = -4y - 5 + 9 & \text{Adding } 9 \text{ on both sides to com-} \\ & \text{plete the square on the left side} \end{array}$$

$$x^2 + 6x + 9 = -4y + 4$$

$$\begin{array}{ll} (x + 3)^2 = -4(y - 1) & \text{Factoring} \end{array}$$

$$\begin{array}{ll} [(x - (-3)]^2 = 4(-1)(y - 1). & \text{Writing standard form:} \\ & (x - h)^2 = 4p(y - k) \end{array}$$

We now have the following:

Vertex (h, k): $(-3, 1)$;

Focus $(h, k + p)$: $(-3, 1 + (-1))$, or $(-3, 0)$;

Directrix $y = k - p$: $y = 1 - (-1)$, or $y = 2$.

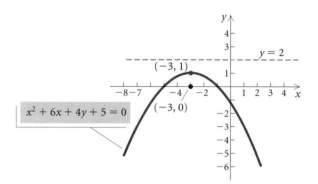

We can check the graph on a grapher using a squared viewing window. It might be necessary to solve for y first:

$$x^2 + 6x + 4y + 5 = 0$$

$$4y = -x^2 - 6x - 5$$

$$y = \tfrac{1}{4}(-x^2 - 6x - 5).$$

The hand-drawn graph appears to be correct.

$y = \tfrac{1}{4}(-x^2 - 6x - 5)$

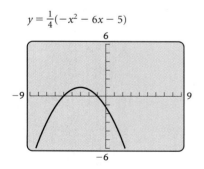

EXAMPLE 4   For the parabola

$$y^2 - 2y - 8x - 31 = 0,$$

find the vertex, the focus, and the directrix. Then draw the graph.

Solution We first complete the square:

$$y^2 - 2y - 8x - 31 = 0$$

$$y^2 - 2y = 8x + 31$$ Adding $8x$ and 31 on both sides

$$y^2 - 2y + 1 = 8x + 31 + 1$$ Adding 1 on both sides to complete the square on the left side

$$y^2 - 2y + 1 = 8x + 32$$

$$(y - 1)^2 = 8(x + 4)$$

$$(y - 1)^2 = 4(2)[x - (-4)].$$ Writing standard form: $(y - k)^2 = 4p(x - h)$

We now have the following:

Vertex (h, k): $(-4, 1)$;

Focus $(h + p, k)$: $(-4 + 2, 1)$, or $(-2, 1)$;

Directrix $x = h - p$: $x = -4 - 2$, or $x = -6$.

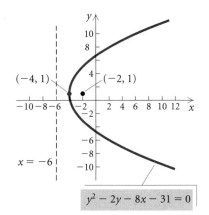

$y^2 - 2y - 8x - 31 = 0$

We can check the graph on a grapher using a squared window. We solve the original equation for y using the quadratic formula:

$$y^2 - 2y - 8x - 31 = 0$$

$$y^2 - 2y + (-8x - 31) = 0$$

$$a = 1, \quad b = -2, \quad c = -8x - 31$$

$$y = \frac{-(-2) \pm \sqrt{(-2)^2 - 4 \cdot 1(-8x - 31)}}{2 \cdot 1}$$

$$y = \frac{2 \pm \sqrt{32x + 128}}{2}.$$

We now graph

$$y_1 = \frac{2 + \sqrt{32x + 128}}{2} \quad \text{and} \quad y_2 = \frac{2 - \sqrt{32x + 128}}{2}.$$

The hand-drawn graph appears to be correct.

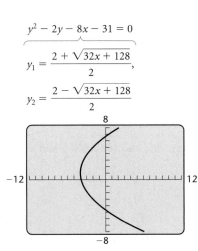

$y^2 - 2y - 8x - 31 = 0$

$$y_1 = \frac{2 + \sqrt{32x + 128}}{2},$$

$$y_2 = \frac{2 - \sqrt{32x + 128}}{2}$$

Applications

Parabolas have many applications. For example, cross sections of car headlights, flashlights, and searchlights are parabolas. The bulb is located at the focus and light from that point is reflected outward parallel to the axis of symmetry. Satellite dishes and field microphones used at sporting events often have parabolic cross sections. Incoming radio waves or sound waves parallel to the axis are reflected into the focus. Cables hung between structures in suspension bridges, such as the Golden Gate Bridge, form parabolas. When a cable supports only its own weight, however, it forms a curve called a *catenary* rather than a parabola.

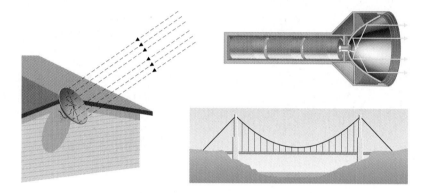

Exercise Set **6.1**

In Exercises 1–6, match the equation with one of the graphs (a)–(f), which follow.

a)

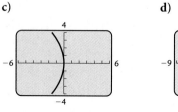

b)

c)

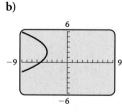

d)

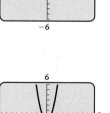

e)

f)

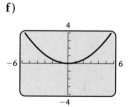

1. $x^2 = 8y$

2. $y^2 = -10x$

3. $(y - 2)^2 = -3(x + 4)$

4. $(x + 1)^2 = 5(y - 2)$

5. $13x^2 - 8y - 9 = 0$

6. $41x + 6y^2 = 12$

Find the vertex, the focus, and the directrix. Then draw the graph.

7. $x^2 = 20y$ **8.** $x^2 = 16y$

9. $y^2 = -6x$ **10.** $y^2 = -2x$

11. $x^2 - 4y = 0$ **12.** $y^2 + 4x = 0$

13. $x = 2y^2$ **14.** $y = \frac{1}{2}x^2$

Find an equation of a parabola satisfying the given conditions.

15. Focus $(4, 0)$, directrix $x = -4$

16. Focus $\left(0, \frac{1}{4}\right)$, directrix $y = -\frac{1}{4}$

17. Focus $(0, -\pi)$, directrix $y = \pi$

18. Focus $(-\sqrt{2}, 0)$, directrix $x = \sqrt{2}$

19. Focus $(3, 2)$, directrix $x = -4$

20. Focus $(-2, 3)$, directrix $y = -3$

Find the vertex, the focus, and the directrix. Then draw the graph.

21. $(x + 2)^2 = -6(y - 1)$

22. $(y - 3)^2 = -20(x + 2)$

23. $x^2 + 2x + 2y + 7 = 0$

24. $y^2 + 6y - x + 16 = 0$

25. $x^2 - y - 2 = 0$

26. $x^2 - 4x - 2y = 0$

27. $y = x^2 + 4x + 3$

28. $y = x^2 + 6x + 10$

29. $y^2 - y - x + 6 = 0$

30. $y^2 + y - x - 4 = 0$

31. *Satellite Dish.* An engineer designs a satellite dish with a parabolic cross section. The dish is 15 ft wide at the opening and the focus is placed 4 ft from the vertex.

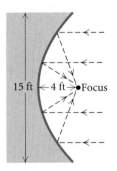

a) Position a coordinate system with the origin at the vertex and the x-axis on the parabola's axis of symmetry and find an equation of the parabola.

b) Find the depth of the satellite dish at the vertex.

32. *Headlight Mirror.* A car headlight mirror has a parabolic cross section with diameter 6 in. and depth 1 in.

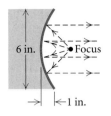

a) Position a coordinate system with the origin at the vertex and the x-axis on the parabola's axis of symmetry and find an equation of the parabola.

b) How far from the vertex should the bulb be positioned if it is to be placed at the focus?

33. *Spotlight.* A spotlight has a parabolic cross section that is 4 ft wide at the opening and 1.5 ft deep at the vertex. How far from the vertex is the focus?

34. *Field Microphone.* A field microphone used at a football game has a parabolic cross section and is 18 in. deep. The focus is 4 in. from the vertex. Find the width of the microphone at the opening.

Discussion and Writing

35. Is a parabola always the graph of a function? Why or why not?

36. Explain how the distance formula is used to find the standard equation of a parabola.

Skill Maintenance

Complete the square and write the result as the square of a binomial.

37. $x^2 + 10x$

38. $y^2 - 9y$

Find the center and the radius of the circle.

39. $(x - 1)^2 + (y + 2)^2 = 9$

40. $(x + 3)^2 + (y - 5)^2 = 36$

Synthesis

41. Find an equation of the parabola with a vertical axis of symmetry and vertex $(-1, 2)$ and containing the point $(-3, 1)$.

42. Find an equation of a parabola with a horizontal axis of symmetry and vertex $(-2, 1)$ and containing the point $(-3, 5)$.

Use a grapher to find the vertex, the focus, and the directrix of each of the following.

43. $4.5x^2 - 7.8x + 9.7y = 0$

44. $134.1y^2 + 43.4x - 316.6y - 122.4 = 0$

45. *Suspension Bridge.* The cables of a suspension bridge are 50 ft above the roadbed at the ends of the bridge and 10 ft above it in the center of the bridge. The roadbed is 200 ft long. Vertical cables are to be spaced every 20 ft along the bridge. Calculate the lengths of these vertical cables.

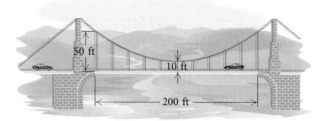

The Circle and the Ellipse

6.2

- *Given an equation of a circle, complete the square, if necessary, and then find the center and the radius and graph the circle.*
- *Given an equation of an ellipse, complete the square, if necessary, and then find the center, the vertices, and the foci and graph the ellipse.*

Circles

We can define a circle geometrically.

Circle

A **circle** is the set of all points in a plane that are at a fixed distance from a fixed point (the **center**) in the plane.

Circles were introduced in Section 1.7. Recall the standard equation of a circle with center (h, k) and radius r.

CIRCLES
REVIEW SECTION 1.7.

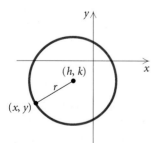

Standard Equation of a Circle

The standard equation of a circle with center (h, k) and radius r is

$$(x - h)^2 + (y - k)^2 = r^2.$$

EXAMPLE 1 For the circle

$$x^2 + y^2 - 16x + 14y + 32 = 0,$$

find the center and the radius. Then graph the circle.

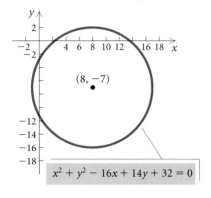

$$x^2 + y^2 - 16x + 14y + 32 = 0$$

Solution   First, we complete the square twice:

$$x^2 + y^2 - 16x + 14y + 32 = 0$$
$$x^2 - 16x \quad + y^2 + 14y \quad = -32$$
$$x^2 - 16x + 64 + y^2 + 14y + 49 = -32 + 64 + 49$$

> **Adding 64 and 49 on both sides to complete the square twice on the left side**

$$(x - 8)^2 + (y + 7)^2 = 81$$
$$(x - 8)^2 + [y - (-7)]^2 = 9^2. \quad \textbf{Writing standard form}$$

The center is $(8, -7)$ and the radius is 9.

To use a grapher to graph the circle, it might be necessary to solve for y first. The original equation can be solved using the quadratic formula, or the standard form of the equation can be solved using the principle of square roots. The second alternative is illustrated here:

$$(x - 8)^2 + (y + 7)^2 = 81$$
$$(y + 7)^2 = 81 - (x - 8)^2$$
$$y + 7 = \pm\sqrt{81 - (x - 8)^2} \quad \textbf{Using the principle of square roots}$$
$$y = -7 \pm \sqrt{81 - (x - 8)^2}.$$

Then we graph

$$y_1 = -7 + \sqrt{81 - (x - 8)^2}$$

and

$$y_2 = -7 - \sqrt{81 - (x - 8)^2}$$

in a squared viewing window.

The hand-drawn graph appears to be correct.   ▬

$$(x - 8)^2 + (y + 7)^2 = 81$$
$$y_1 = -7 + \sqrt{81 - (x - 8)^2},$$
$$y_2 = -7 - \sqrt{81 - (x - 8)^2}$$

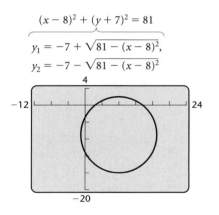

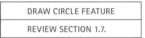

DRAW CIRCLE FEATURE
REVIEW SECTION 1.7.

Some graphers have a DRAW feature that provides a quick way to graph a circle when the center and the radius are known.

Ellipses

We have studied two conic sections, the parabola and the circle. Now we turn our attention to a third, the *ellipse*.

Ellipse

An **ellipse** is the set of all points in a plane, the sum of whose distances from two fixed points (the **foci**) is constant. The **center** of an ellipse is the midpoint of the segment between the foci.

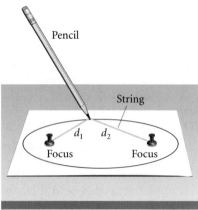

Pencil

String

d_1 d_2

Focus   Focus

We can draw an ellipse by first placing two thumbtacks in a piece of cardboard. These are the foci (singular, *focus*). We then attach a piece of string to the tacks. Its length is the constant sum of the distances $d_1 + d_2$ from the foci to any point on the ellipse. Next, we trace a curve with a pencil held tight against the string. The figure traced is an ellipse.

Let's first consider the ellipse shown below with center at the origin. The points F_1 and F_2 are the foci. The segment $\overline{A'A}$ is the **major axis,** and the points A' and A are the **vertices**. The segment $\overline{B'B}$ is the **minor axis,** and the points B and B' are the **y-intercepts**. Note that the major axis of an ellipse is longer than the minor axis.

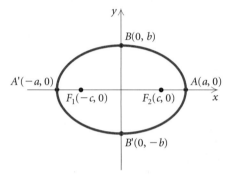

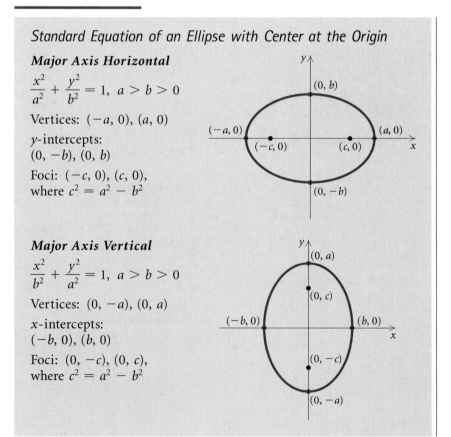

Standard Equation of an Ellipse with Center at the Origin

Major Axis Horizontal

$$\frac{x^2}{a^2} + \frac{y^2}{b^2} = 1, \ a > b > 0$$

Vertices: $(-a, 0)$, $(a, 0)$

y-intercepts:
$(0, -b)$, $(0, b)$

Foci: $(-c, 0)$, $(c, 0)$,
where $c^2 = a^2 - b^2$

Major Axis Vertical

$$\frac{x^2}{b^2} + \frac{y^2}{a^2} = 1, \ a > b > 0$$

Vertices: $(0, -a)$, $(0, a)$

x-intercepts:
$(-b, 0)$, $(b, 0)$

Foci: $(0, -c)$, $(0, c)$,
where $c^2 = a^2 - b^2$

EXAMPLE 2   Find the standard equation of the ellipse with vertices $(-5, 0)$ and $(5, 0)$ and foci $(-3, 0)$ and $(3, 0)$. Then graph the ellipse.

Solution   Since the foci are on the x-axis and the origin is the midpoint of the segment between them, the major axis is horizontal and $(0, 0)$ is the center of the ellipse. Thus the equation is of the form

$$\frac{x^2}{a^2} + \frac{y^2}{b^2} = 1.$$

Since the vertices are $(-5, 0)$ and $(5, 0)$ and the foci are $(-3, 0)$ and $(3, 0)$, we know that $a = 5$ and $c = 3$. These values can be used to find b^2:

$$c^2 = a^2 - b^2$$
$$3^2 = 5^2 - b^2$$
$$9 = 25 - b^2$$
$$b^2 = 16.$$

Thus the equation of the ellipse is

$$\frac{x^2}{25} + \frac{y^2}{16} = 1.$$

To graph the ellipse, we plot the vertices $(-5, 0)$ and $(5, 0)$. Since $b^2 = 16$, we know that $b = 4$ and the y-intercepts are $(0, -4)$ and $(0, 4)$. We plot these points as well and connect the four points we have plotted with a smooth curve.

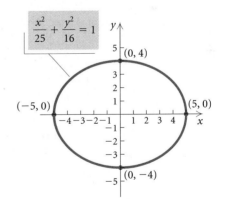

To draw the graph using a grapher, it might be necessary to solve for y first:

$$y = \pm \sqrt{\frac{400 - 16x^2}{25}}.$$

Then we graph

$$y_1 = -\sqrt{\frac{400 - 16x^2}{25}}$$

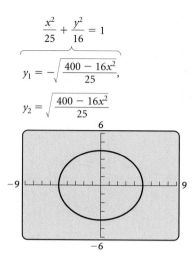

$$\frac{x^2}{25} + \frac{y^2}{16} = 1$$

$$y_1 = -\sqrt{\frac{400 - 16x^2}{25}},$$

$$y_2 = \sqrt{\frac{400 - 16x^2}{25}}$$

and

$$y_2 = \sqrt{\frac{400 - 16x^2}{25}}$$

or

$$y_1 = -\sqrt{\frac{400 - 16x^2}{25}} \quad \text{and} \quad y_2 = -y_1$$

in a squared viewing window.

EXAMPLE 3   For the ellipse

$$9x^2 + 4y^2 = 36,$$

find the vertices and the foci. Then draw the graph.

Solution   We first find standard form:

$$9x^2 + 4y^2 = 36$$

$$\frac{9x^2}{36} + \frac{4y^2}{36} = \frac{36}{36} \qquad \text{Dividing by 36 on both sides to get 1 on the right side}$$

$$\frac{x^2}{4} + \frac{y^2}{9} = 1$$

$$\frac{x^2}{2^2} + \frac{y^2}{3^2} = 1. \qquad \text{Writing standard form}$$

Thus, $a = 3$ and $b = 2$. The major axis is vertical, so the vertices are $(0, -3)$ and $(0, 3)$. Since we know that $c^2 = a^2 - b^2$, we have $c^2 = 3^2 - 2^2 = 5$, so $c = \sqrt{5}$ and the foci are $(0, -\sqrt{5})$ and $(0, \sqrt{5})$.

To graph the ellipse, we plot the vertices. Note also that since $b = 2$, the x-intercepts are $(-2, 0)$ and $(2, 0)$. We plot these points as well and connect the four points we have plotted with a smooth curve.

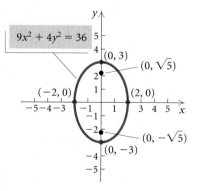

If the center of an ellipse is not at the origin but at some point (h, k), then we can think of an ellipse with center at the origin being translated left or right $|h|$ units and up or down $|k|$ units.

Standard Equation of an Ellipse with Center at (h, k)

Major Axis Horizontal

$$\frac{(x - h)^2}{a^2} + \frac{(y - k)^2}{b^2} = 1, \quad a > b > 0$$

Vertices: $(h - a, k)$, $(h + a, k)$

Length of minor axis: $2b$

Foci: $(h - c, k)$, $(h + c, k)$, where $c^2 = a^2 - b^2$

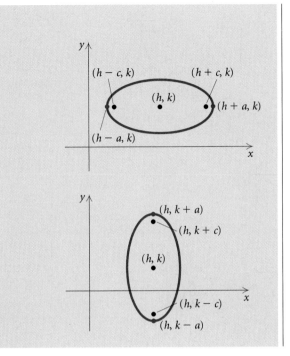

Major Axis Vertical

$$\frac{(x - h)^2}{b^2} + \frac{(y - k)^2}{a^2} = 1, \quad a > b > 0$$

Vertices: $(h, k - a)$, $(h, k + a)$

Length of minor axis: $2b$

Foci: $(h, k - c)$, $(h, k + c)$, where $c^2 = a^2 - b^2$

EXAMPLE 4   For the ellipse

$$4x^2 + y^2 + 24x - 2y + 21 = 0,$$

find the center, the vertices, and the foci. Then draw the graph.

Solution   First, we complete the square to get standard form:

$$4x^2 + y^2 + 24x - 2y + 21 = 0$$

$$4(x^2 + 6x \quad\quad) + (y^2 - 2y \quad\quad) = -21$$

$$4(x^2 + 6x + 9) + (y^2 - 2y + 1) = -21 + 4 \cdot 9 + 1$$

> Completing the square twice by adding $4 \cdot 9$ and 1 on both sides

$$4(x + 3)^2 + (y - 1)^2 = 16$$

$$\frac{1}{16}[4(x + 3)^2 + (y - 1)^2] = \frac{1}{16} \cdot 16$$

$$\frac{(x + 3)^2}{4} + \frac{(y - 1)^2}{16} = 1$$

$$\frac{[x - (-3)]^2}{2^2} + \frac{(y - 1)^2}{4^2} = 1.$$

> Writing standard form:
> $$\frac{(x - h)^2}{b^2} + \frac{(y - k)^2}{a^2} = 1$$

The center is $(-3, 1)$. Note that $a = 4$ and $b = 2$. The major axis is vertical, so the vertices are 4 units above and below the center:

$$(-3, 1 + 4) \text{ and } (-3, 1 - 4), \quad \text{or} \quad (-3, 5) \text{ and } (-3, -3).$$

We know that $c^2 = a^2 - b^2$, so $c^2 = 4^2 - 2^2 = 12$ and $c = \sqrt{12}$, or $2\sqrt{3}$. Then the foci are $2\sqrt{3}$ units above and below the center:

$$(-3, 1 + 2\sqrt{3}) \quad \text{and} \quad (-3, 1 - 2\sqrt{3}).$$

To graph the ellipse, we plot the vertices. Note also that since $b = 2$, two other points on the graph are the endpoints of the minor axis, 2 units right and left of the center:

$$(-3 + 2, 1) \text{ and } (-3 - 2, 1),$$

or

$$(-1, 1) \text{ and } (-5, 1).$$

We plot these points as well and connect the four points with a smooth curve.

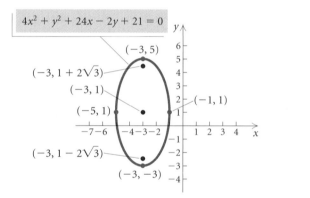

Applications

One of the most exciting recent applications of an ellipse is a medical device called a *lithotripter*. This machine uses underwater shock waves to crush kidney stones. The waves originate at one focus of an ellipse and are reflected to the kidney stone, which is positioned at the other focus. Recovery time following the use of this technique is much shorter than with conventional surgery and the mortality rate is far lower.

A room with an ellipsoidal ceiling is known as a *whispering gallery*. In such a room, a word whispered at one focus can be clearly heard at the other. Whispering galleries are found in the rotunda of the Capitol Building in Washington, D.C., and in the Mormon Tabernacle in Salt Lake City.

Ellipses have many other applications. Planets travel around the sun in elliptical orbits with the sun at one focus, for example, and satellites travel around the earth in elliptical orbits as well.

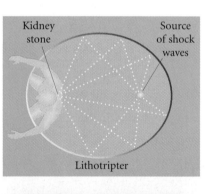

Lithotripter

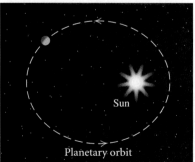

Planetary orbit

Exercise Set 6.2

In Exercises 1–6, match the equation with one of the graphs (a)–(f), which follow.

a)

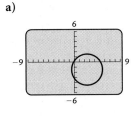

b)

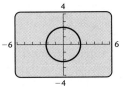

c)

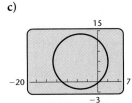

d)

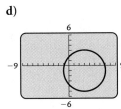

e)

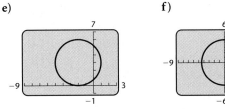

f)

1. $x^2 + y^2 = 5$

2. $y^2 = 20 - x^2$

3. $x^2 + y^2 - 6x + 2y = 6$

4. $x^2 + y^2 + 10x - 12y = 3$

5. $x^2 + y^2 - 5x + 3y = 0$

6. $x^2 + 4x - 2 = 6y - y^2 - 6$

Find the center and the radius of the circle with the given equation. Then draw the graph.

7. $x^2 + y^2 - 14x + 4y = 11$

8. $x^2 + y^2 + 2x - 6y = -6$

9. $x^2 + y^2 + 4x - 6y - 12 = 0$

10. $x^2 + y^2 - 8x - 2y - 19 = 0$

11. $x^2 + y^2 + 6x - 10y = 0$

12. $x^2 + y^2 - 7x - 2y = 0$

13. $x^2 + y^2 - 9x = 7 - 4y$

14. $y^2 - 6y - 1 = 8x - x^2 + 3$

In Exercises 15–18, match the equation with one of the graphs (a)–(d), which follow.

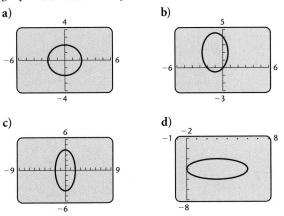

a)

b)

c)

d)

15. $16x^2 + 4y^2 = 64$

16. $4x^2 + 5y^2 = 20$

17. $x^2 + 9y^2 - 6x + 90y = -225$

18. $9x^2 + 4y^2 + 18x - 16y = 11$

Find the vertices and the foci of the ellipse with the given equation. Then draw the graph.

19. $\dfrac{x^2}{4} + \dfrac{y^2}{1} = 1$

20. $\dfrac{x^2}{25} + \dfrac{y^2}{36} = 1$

21. $16x^2 + 9y^2 = 144$

22. $9x^2 + 4y^2 = 36$

23. $2x^2 + 3y^2 = 6$

24. $5x^2 + 7y^2 = 35$

25. $4x^2 + 9y^2 = 1$

26. $25x^2 + 16y^2 = 1$

Find an equation of an ellipse satisfying the given conditions.

27. Vertices: $(-7, 0)$ and $(7, 0)$;
foci: $(-3, 0)$ and $(3, 0)$

28. Vertices: $(0, -6)$ and $(0, 6)$;
foci: $(0, -4)$ and $(0, 4)$

29. Vertices: $(0, -8)$ and $(0, 8)$;
length of minor axis: 10

30. Vertices: $(-5, 0)$ and $(5, 0)$;
length of minor axis: 6

31. Foci: $(-2, 0)$ and $(2, 0)$;
length of major axis: 6

32. Foci: $(0, -3)$ and $(0, 3)$;
length of major axis: 10

Find the center, the vertices, and the foci of the ellipse. Then draw the graph.

33. $\dfrac{(x - 1)^2}{9} + \dfrac{(y - 2)^2}{4} = 1$

34. $\dfrac{(x - 1)^2}{1} + \dfrac{(y - 2)^2}{4} = 1$

35. $\dfrac{(x + 3)^2}{25} + \dfrac{(y - 5)^2}{36} = 1$

36. $\dfrac{(x - 2)^2}{16} + \dfrac{(y + 3)^2}{25} = 1$

37. $3(x + 2)^2 + 4(y - 1)^2 = 192$

38. $4(x - 5)^2 + 3(y - 4)^2 = 48$

39. $4x^2 + 9y^2 - 16x + 18y - 11 = 0$

40. $x^2 + 2y^2 - 10x + 8y + 29 = 0$

41. $4x^2 + y^2 - 8x - 2y + 1 = 0$

42. $9x^2 + 4y^2 + 54x - 8y + 49 = 0$

*The **eccentricity** of an ellipse is defined as $e = c/a$. For an ellipse, $0 < c < a$, so $0 < e < 1$. When e is close to 0, an ellipse appears to be nearly circular. When e is close to 1, an ellipse is very flat.*

43. Observe the shapes of the ellipses in Examples 2 and 4. Which ellipse has the smaller eccentricity? Confirm your answer by computing the eccentricity of each ellipse.

44. Which ellipse below has the smaller eccentricity? (Assume that the coordinate systems have the same scale.)

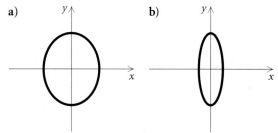

45. Find an equation of an ellipse with vertices $(0, -4)$ and $(0, 4)$ and $e = \frac{1}{4}$.

46. Find an equation of an ellipse with vertices $(-3, 0)$ and $(3, 0)$ and $e = \frac{7}{10}$.

47. *Bridge Supports.* The bridge support shown in the figure below is the top half of an ellipse. Assuming that a coordinate system is superimposed on the drawing in such a way that the center of the ellipse is at point Q, find an equation of the ellipse.

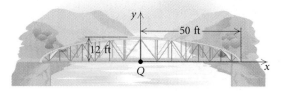

48. *The Ellipse.* In Washington, D.C., there is a large grassy area south of the White House known as the Ellipse. It is actually an ellipse with major axis of length 1048 ft and minor axis of length 898 ft. Assuming that a coordinate system is superimposed on the area in such a way that the center is at the origin and the major and minor axes are on the *x*- and *y*-axes of the coordinate system, respectively, find an equation of the ellipse.

49. *The Earth's Orbit.* The maximum distance of the earth from the sun is 9.3×10^7 miles. The minimum distance is 9.1×10^7 miles. The sun is at one focus of the elliptical orbit. Find the distance from the sun to the other focus.

50. *Carpentry.* A carpenter is cutting a 3-ft by 4-ft elliptical sign from a 3-ft by 4-ft piece of plywood. The ellipse will be drawn using a string attached to the board at the foci of the ellipse.

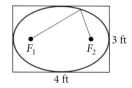

a) How far from the ends of the board should the string be attached?

b) How long should the string be?

Discussion and Writing

51. Explain why function notation is not used in this section.

52. Would you prefer to graph the ellipse

$$\dfrac{(x - 3)^2}{4} + \dfrac{(y + 5)^2}{9} = 1$$

by hand or using a grapher? Explain why you answered as you did.

Skill Maintenance

In Exercises 53–56, given the function:
a) *Determine whether it is one-to-one.*
b) *If it is one-to-one, find a formula for the inverse.*

53. $f(x) = 2x - 3$

54. $f(x) = x^3 + 2$

55. $f(x) = \dfrac{5}{x - 1}$

56. $f(x) = \sqrt{x + 4}$

Synthesis

Find an equation of an ellipse satisfying the given conditions.

57. Vertices: $(3, -4)$, $(3, 6)$;
endpoints of minor axis: $(1, 1)$, $(5, 1)$

58. Vertices: $(-1, -1)$, $(-1, 5)$;
endpoints of minor axis: $(-3, 2)$, $(1, 2)$

59. Vertices: $(-3, 0)$ and $(3, 0)$; passing through $\left(2, \frac{22}{3}\right)$

60. Center: $(-2, 3)$; major axis vertical;
length of major axis: 4;
length of minor axis: 1

Use a grapher to find the center and the vertices of each of the following.

61. $4x^2 + 9y^2 - 16.025x + 18.0927y - 11.346 = 0$

62. $9x^2 + 4y^2 + 54.063x - 8.016y + 49.872 = 0$

63. *Bridge Arch.* A bridge with a semielliptical arch spans a river as shown below. What is the clearance 6 ft from the riverbank?

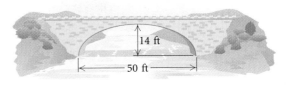

14 ft

50 ft

The Hyperbola

6.3

- *Given an equation of a hyperbola, complete the square, if necessary, and then find the center, the vertices, and the foci and graph the hyperbola.*

The last type of conic section that we will study is the *hyperbola*.

Hyperbola

A **hyperbola** is the set of all points in a plane for which the absolute value of the difference of the distances from two fixed points (the **foci**) is constant. The midpoint of the segment between the foci is the **center** of the hyperbola.

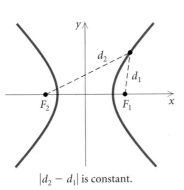

$|d_2 - d_1|$ is constant.

Standard Equations of Hyperbolas

We first consider the equation of a hyperbola with center at the origin. In the figure below, F_1 and F_2 are the foci. The segment $\overline{V_2 V_1}$ is the **transverse axis** and the points V_2 and V_1 are the **vertices**.

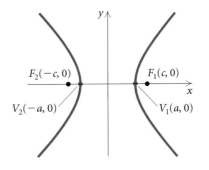

$F_2(-c, 0)$

$F_1(c, 0)$

$V_2(-a, 0)$

$V_1(a, 0)$

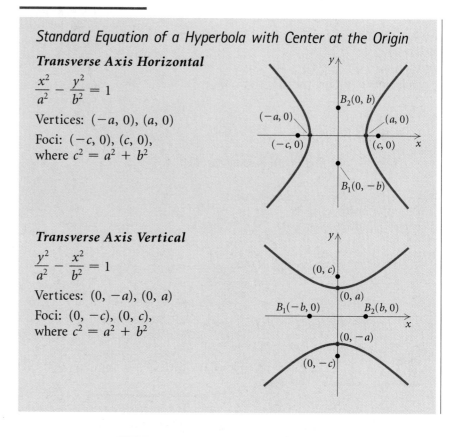

Standard Equation of a Hyperbola with Center at the Origin

Transverse Axis Horizontal

$$\frac{x^2}{a^2} - \frac{y^2}{b^2} = 1$$

Vertices: $(-a, 0)$, $(a, 0)$

Foci: $(-c, 0)$, $(c, 0)$,
where $c^2 = a^2 + b^2$

Transverse Axis Vertical

$$\frac{y^2}{a^2} - \frac{x^2}{b^2} = 1$$

Vertices: $(0, -a)$, $(0, a)$

Foci: $(0, -c)$, $(0, c)$,
where $c^2 = a^2 + b^2$

The segment $\overline{B_1 B_2}$ is the **conjugate axis** of the hyperbola.

To graph a hyperbola with a horizontal transverse axis, it is helpful to begin by graphing the lines $y = -(b/a)x$ and $y = (b/a)x$. These are the **asymptotes** of the hyperbola. For a hyperbola with a vertical transverse axis, the asymptotes are $y = -(a/b)x$ and $y = (a/b)x$. As $|x|$ gets larger and larger, the graph of the hyperbola gets closer and closer to the asymptotes.

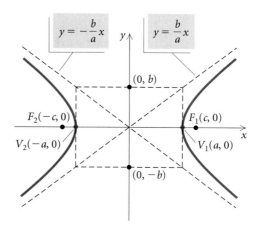

EXAMPLE 1   Find an equation of the hyperbola with vertices $(0, -4)$ and $(0, 4)$ and foci $(0, -6)$ and $(0, 6)$.

Solution   We know that $a = 4$ and $c = 6$. We find b^2.

$$c^2 = a^2 + b^2$$
$$6^2 = 4^2 + b^2$$
$$36 = 16 + b^2$$
$$20 = b^2$$

Since the vertices and the foci are on the y-axis, we know that the transverse axis is vertical. We can now write the equation of the hyperbola:

$$\frac{y^2}{a^2} - \frac{x^2}{b^2} = 1$$
$$\frac{y^2}{16} - \frac{x^2}{20} = 1.$$

EXAMPLE 2   For the hyperbola given by

$$9x^2 - 16y^2 = 144,$$

find the vertices, the foci, and the asymptotes. Then graph the hyperbola.

Solution   First, we find standard form:

$$9x^2 - 16y^2 = 144$$
$$\frac{1}{144}(9x^2 - 16y^2) = \frac{1}{144} \cdot 144 \qquad \text{Multiplying by } \tfrac{1}{144} \text{ to get 1 on the right side}$$
$$\frac{x^2}{16} - \frac{y^2}{9} = 1$$
$$\frac{x^2}{4^2} - \frac{y^2}{3^2} = 1. \qquad \text{Writing standard form}$$

The hyperbola has a horizontal transverse axis, so the vertices are $(-a, 0)$ and $(a, 0)$, or $(-4, 0)$ and $(4, 0)$. From the standard form of the equation, we know that $a^2 = 4^2$, or 16, and $b^2 = 3^2$, or 9. We find the foci:

$$c^2 = a^2 + b^2$$
$$c^2 = 16 + 9$$
$$c^2 = 25$$
$$c = 5.$$

Thus the foci are $(-5, 0)$ and $(5, 0)$.

Next, we find the asymptotes:

$$y = \frac{b}{a}x = \frac{3}{4}x$$

and

$$y = -\frac{b}{a}x = -\frac{3}{4}x.$$

To draw the graph, we sketch the asymptotes first. This is easily done by drawing the rectangle with horizontal sides passing through $(0, 3)$ and $(0, -3)$ and vertical sides through $(4, 0)$ and $(-4, 0)$. Then we draw and extend the diagonals of this rectangle. The two extended diagonals are the asymptotes of the hyperbola. Next, we plot the vertices and draw the branches of the hyperbola outward from the vertices toward the asymptotes.

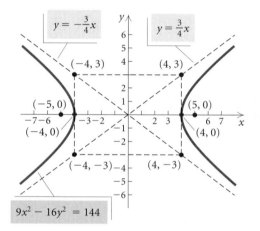

We can use a grapher as a check. It might be necessary to solve for y first and then graph the top and bottom halves of the hyperbola in the same squared viewing window.

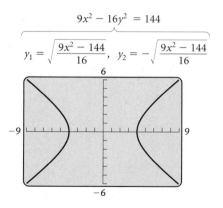

If a hyperbola with center at the origin is translated left or right $|h|$ units and up or down $|k|$ units, the center is at the point (h, k).

Standard Equation of a Hyperbola with Center (h, k)

Transverse Axis Horizontal

$$\frac{(x - h)^2}{a^2} - \frac{(y - k)^2}{b^2} = 1$$

Vertices: $(h - a, k)$, $(h + a, k)$

Asymptotes: $y - k = \dfrac{b}{a}(x - h)$, $y - k = -\dfrac{b}{a}(x - h)$

Foci: $(h - c, k)$, $(h + c, k)$, where $c^2 = a^2 + b^2$

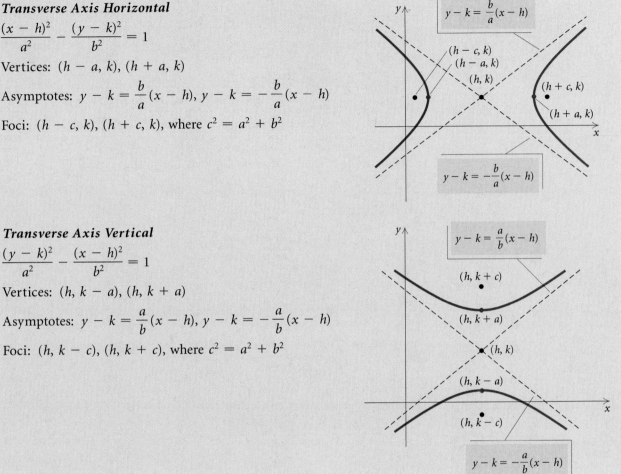

Transverse Axis Vertical

$$\frac{(y - k)^2}{a^2} - \frac{(x - h)^2}{b^2} = 1$$

Vertices: $(h, k - a)$, $(h, k + a)$

Asymptotes: $y - k = \dfrac{a}{b}(x - h)$, $y - k = -\dfrac{a}{b}(x - h)$

Foci: $(h, k - c)$, $(h, k + c)$, where $c^2 = a^2 + b^2$

EXAMPLE 3 For the hyperbola given by

$$4y^2 - x^2 + 24y + 4x + 28 = 0,$$

find the center, the vertices, and the foci. Then draw the graph.

Solution First, we complete the square to get standard form:

$$4y^2 - x^2 + 24y + 4x + 28 = 0$$
$$4(y^2 + 6y \qquad) - (x^2 - 4x \qquad) = -28$$
$$4(y^2 + 6y + 9 - 9) - (x^2 - 4x + 4 - 4) = -28$$
$$4(y^2 + 6y + 9) - (x^2 - 4x + 4) = -28 + 36 - 4.$$

Then

$$4(y + 3)^2 - (x - 2)^2 = 4$$

$$\frac{(y + 3)^2}{1} - \frac{(x - 2)^2}{4} = 1 \qquad \text{Dividing by 4}$$

$$\frac{[y - (-3)]^2}{1^2} - \frac{(x - 2)^2}{2^2} = 1. \qquad \text{Standard form}$$

The center is $(2, -3)$. Note that $a = 1$ and $b = 2$. The transverse axis is vertical, so the vertices are 1 unit below and above the center:

$$(2, -3 - 1) \text{ and } (2, -3 + 1), \quad \text{or} \quad (2, -4) \text{ and } (2, -2).$$

We know that $c^2 = a^2 + b^2$, so $c^2 = 1^2 + 2^2 = 1 + 4 = 5$ and $c = \sqrt{5}$. Thus the foci are $\sqrt{5}$ units below and above the center:

$$(2, -3 - \sqrt{5}) \quad \text{and} \quad (2, -3 + \sqrt{5}).$$

The asymptotes are

$$y - (-3) = \frac{1}{2}(x - 2) \quad \text{and} \quad y - (-3) = -\frac{1}{2}(x - 2),$$

or

$$y + 3 = \frac{1}{2}(x - 2) \quad \text{and} \quad y + 3 = -\frac{1}{2}(x - 2).$$

We sketch the asymptotes, plot the vertices, and draw the graph.

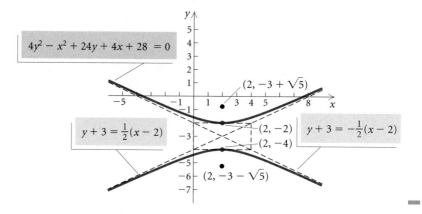

CLASSIFYING EQUATIONS OF CONIC SECTIONS

EQUATION	TYPE OF CONIC SECTION	GRAPH
$x - 4 + 4y = y^2$	Only one variable is squared, so this cannot be a circle, an ellipse, or a hyperbola. Find an equivalent equation: $$x = (y - 2)^2.$$ This is an equation of a parabola.	
$3x^2 + 3y^2 = 75$	Both variables are squared, so this cannot be a parabola. The squared terms are added, so this cannot be a hyperbola. Divide by 3 on both sides to find an equivalent equation: $$x^2 + y^2 = 25.$$ This is an equation of a circle.	
$y^2 = 16 - 4x^2$	Both variables are squared, so this cannot be a parabola. Add $4x^2$ on both sides to find an equivalent equation: $4x^2 + y^2 = 16$. The squared terms are added, so this cannot be a hyperbola. The coefficients of x^2 and y^2 are not the same, so this is not a circle. Divide by 16 on both sides to find an equivalent equation: $$\frac{x^2}{4} + \frac{y^2}{16} = 1.$$ This is an equation of an ellipse.	
$x^2 = 4y^2 + 36$	Both variables are squared, so this cannot be a parabola. Subtract $4y^2$ on both sides to find an equivalent equation: $x^2 - 4y^2 = 36$. The squared terms are not added, so this cannot be a circle or an ellipse. Divide by 36 on both sides to find an equivalent equation: $$\frac{x^2}{36} - \frac{y^2}{9} = 1.$$ This is an equation of a hyperbola.	

Applications

Some comets travel in hyperbolic paths with the sun at one focus. Such comets pass by the sun only one time unlike those with elliptical orbits, which reappear at intervals. A cross section of an amphitheater might be one branch of a hyperbola. A cross section of a nuclear cooling tower might also be a hyperbola.

One other application of hyperbolas is in the long range navigation system LORAN. This system uses transmitting stations in three locations to send out simultaneous signals to a ship or aircraft. The difference in the arrival times of the signals from one pair of transmitters is recorded on the ship or aircraft. This difference is also recorded for signals from another pair of transmitters. For each pair, a computation is performed to determine the difference in the distances from each member of the pair to the ship or aircraft. If each pair of differences is kept constant, two hyperbolas can be drawn. Each has one of the pairs of transmitters as foci and the ship or aircraft lies on the intersection of two of their branches.

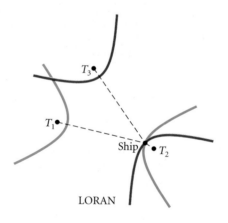

LORAN

Exercise Set 6.3

In Exercises 1–6, match the equation with one of the graphs (a)–(f), which follow.

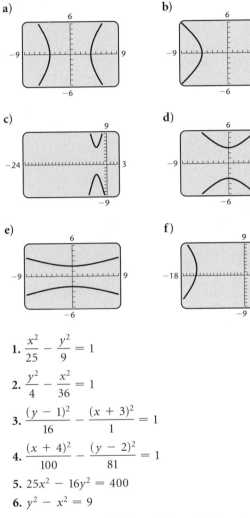

a)

b)

c)

d)

e)

f)

1. $\dfrac{x^2}{25} - \dfrac{y^2}{9} = 1$

2. $\dfrac{y^2}{4} - \dfrac{x^2}{36} = 1$

3. $\dfrac{(y - 1)^2}{16} - \dfrac{(x + 3)^2}{1} = 1$

4. $\dfrac{(x + 4)^2}{100} - \dfrac{(y - 2)^2}{81} = 1$

5. $25x^2 - 16y^2 = 400$

6. $y^2 - x^2 = 9$

Find an equation of a hyperbola satisfying the given conditions.

7. Vertices at $(0, 3)$ and $(0, -3)$; foci at $(0, 5)$ and $(0, -5)$

8. Vertices at $(1, 0)$ and $(-1, 0)$; foci at $(2, 0)$ and $(-2, 0)$

9. Asymptotes $y = \frac{3}{2}x$, $y = -\frac{3}{2}x$; one vertex $(2, 0)$

10. Asymptotes $y = \frac{5}{4}x$, $y = -\frac{5}{4}x$; one vertex $(0, 3)$

Find the center, the vertices, the foci, and the asymptotes. Then draw the graph.

11. $\dfrac{x^2}{4} - \dfrac{y^2}{4} = 1$

12. $\dfrac{x^2}{1} - \dfrac{y^2}{9} = 1$

13. $\dfrac{(x - 2)^2}{9} - \dfrac{(y + 5)^2}{1} = 1$

14. $\dfrac{(x - 5)^2}{16} - \dfrac{(y + 2)^2}{9} = 1$

15. $\dfrac{(y + 3)^2}{4} - \dfrac{(x + 1)^2}{16} = 1$

16. $\dfrac{(y + 4)^2}{25} - \dfrac{(x + 2)^2}{16} = 1$

17. $x^2 - 4y^2 = 4$

18. $4x^2 - y^2 = 16$

19. $9y^2 - x^2 = 81$

20. $y^2 - 4x^2 = 4$

21. $x^2 - y^2 = 2$

22. $x^2 - y^2 = 3$

23. $y^2 - x^2 = \frac{1}{4}$

24. $y^2 - x^2 = \frac{1}{9}$

Find the center, the vertices, the foci, and the asymptotes of the hyperbola. Then draw the graph.

25. $x^2 - y^2 - 2x - 4y - 4 = 0$

26. $4x^2 - y^2 + 8x - 4y - 4 = 0$

27. $36x^2 - y^2 - 24x + 6y - 41 = 0$

28. $9x^2 - 4y^2 + 54x + 8y + 41 = 0$

29. $9y^2 - 4x^2 - 18y + 24x - 63 = 0$

30. $x^2 - 25y^2 + 6x - 50y = 41$

31. $x^2 - y^2 - 2x - 4y = 4$

32. $9y^2 - 4x^2 - 54y - 8x + 41 = 0$

33. $y^2 - x^2 - 6x - 8y - 29 = 0$

34. $x^2 - y^2 = 8x - 2y - 13$

The **eccentricity** of a hyperbola is defined as $e = c/a$. For a hyperbola, $c > a > 0$, so $e > 1$. When e is close to 1, a hyperbola appears to be very narrow. As the eccentricity increases, the hyperbola becomes "wider."

35. Observe the shapes of the hyperbolas in Examples 2 and 3. Which hyperbola has the larger eccentricity? Confirm your answer by computing the eccentricity of each hyperbola.

36. Which hyperbola below has the larger eccentricity? (Assume that the coordinate systems have the same scale.)

a)

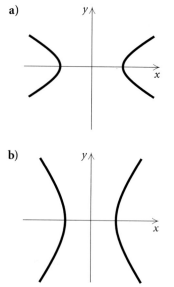

b)

37. Find an equation of a hyperbola with vertices $(3, 7)$ and $(-3, 7)$ and $e = \frac{5}{3}$.

38. Find an equation of a hyperbola with vertices $(-1, 3)$ and $(-1, 7)$ and $e = 4$.

39. *Hyperbolic Mirror.* Certain telescopes contain both a parabolic and a hyperbolic mirror. In the telescope shown in the figure below, the parabola and the hyperbola share focus F_1, which is 14 m above the vertex of the parabola. The hyperbola's second focus F_2 is 2 m above the parabola's vertex. The vertex of the hyperbolic mirror is 1 m below F_1. Position a coordinate system with the origin at the center of the hyperbola and with the foci on the y-axis. Then find the equation of the hyperbola.

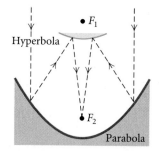

40. *Nuclear Cooling Tower.* A cross section of a

nuclear cooling tower is a hyperbola with equation

$$\frac{x^2}{90^2} - \frac{y^2}{130^2} = 1.$$

The tower is 450 ft tall and the distance from the top of the tower to the center of the hyperbola is half the distance from the base of the tower to the center of the hyperbola. Find the diameter of the top and the base of the tower.

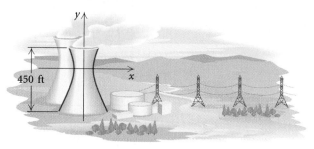

Discussion and Writing

41. How does the graph of a parabola differ from the graph of one branch of a hyperbola?

42. Would you prefer to graph the hyperbola

$$\frac{(y + 5)^2}{16} + \frac{(x - 4)^2}{36} = 1$$

by hand or using a grapher? Explain why you answered as you did.

Skill Maintenance

Solve.

43. $x + y = 5,$
   $x - y = 7$

44. $3x - 2y = 5,$
   $5x + 2y = 3$

45. $2x - 3y = 7,$
   $3x + 5y = 1$

46. $3x + 2y = -1,$
   $2x + 3y = 6$

Synthesis

Find an equation of a hyperbola satisfying the given conditions.

47. Vertices at $(3, -8)$ and $(3, -2)$;
   asymptotes $y = 3x - 14$, $y = -3x + 4$

48. Vertices at $(-9, 4)$ and $(-5, 4)$;
   asymptotes $y = 3x + 25$, $y = -3x - 17$

Use a grapher to find the center, the vertices, and the asymptotes.

49. $5x^2 - 3.5y^2 + 14.6x - 6.7y + 3.4 = 0$

50. $x^2 - y^2 - 2.046x - 4.088y - 4.228 = 0$

51. *Navigation.* Two radio transmitters positioned 300 mi apart send simultaneous signals to a ship that is 200 mi offshore, sailing parallel to the shoreline. The signal from transmitter S reaches the ship 200 microseconds later than the signal from transmitter T. The signals travel at a speed of 186,000 miles per second, or 0.186 mile per microsecond. Find the equation of the hyperbola with foci S and T on which the ship is located.

(*Hint*: For any point on the hyperbola, the absolute value of the difference of its distances from the foci is $2a$.)

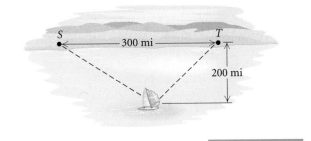

Nonlinear Systems of Equations

6.4

• *Solve a nonlinear system of equations.*
• *Use nonlinear systems of equations to solve applied problems.*

The systems of equations that we have studied so far have been composed of linear equations. Now we consider systems of two equations in two variables in which at least one equation is not linear.

Nonlinear Systems of Equations

The graph of a nonlinear system of equations can have no point of intersection or one or more points of intersection. The coordinates of each point of intersection are a solution of the system of equations. When no point of intersection exists, the system of equations has no real-number solution.

Real-number solutions of nonlinear systems of equations can be found algebraically, using the substitution or elimination method, or graphically. However, since a graph shows *only* real-number solutions, solutions involving imaginary numbers must always be found algebraically.

When we are solving algebraically, the substitution method is preferable for a system consisting of one linear and one nonlinear equation. The elimination method is preferable in most, but not all, cases when both equations are nonlinear.

EXAMPLE 1   Solve the following system of equations:

$$x^2 + y^2 = 25, \qquad (1) \qquad \textbf{(The graph is a circle.)}$$

$$3x - 4y = 0. \qquad (2) \qquad \textbf{(The graph is a line.)}$$

Algebraic Solution

We use the substitution method. First, we solve equation (2) for x:

$$x = \tfrac{4}{3}y. \qquad (3) \qquad \text{We could have solved for } y \text{ instead.}$$

Next, we substitute $\tfrac{4}{3}y$ for x in equation (1) and solve for y:

$$\left(\tfrac{4}{3}y\right)^2 + y^2 = 25$$
$$\tfrac{16}{9}y^2 + y^2 = 25$$
$$\tfrac{25}{9}y^2 = 25$$
$$y^2 = 9 \qquad \text{Multiplying by } \tfrac{9}{25}$$
$$y = \pm 3.$$

Now we substitute these numbers for y in equation (3) and solve for x:

$$x = \tfrac{4}{3}(3) = 4, \qquad \text{The pair } (4, 3) \text{ appears to be a solution.}$$
$$x = \tfrac{4}{3}(-3) = -4. \qquad \text{The pair } (-4, -3) \text{ appears to be a solution.}$$

CHECK: For (4, 3):

$$
\begin{array}{c|c}
x^2 + y^2 = 25 & \\
\hline
4^2 + 3^2 \; ? \; 25 & \\
16 + 9 & \\
25 & 25 \quad \text{TRUE}
\end{array}
\qquad
\begin{array}{c|c}
3x - 4y = 0 & \\
\hline
3(4) - 4(3) \; ? \; 0 & \\
12 - 12 & \\
0 & 0 \quad \text{TRUE}
\end{array}
$$

For (−4, −3):

$$
\begin{array}{c|c}
x^2 + y^2 = 25 & \\
\hline
(-4)^2 + (-3)^2 \; ? \; 25 & \\
16 + 9 & \\
25 & 25 \quad \text{TRUE}
\end{array}
\qquad
\begin{array}{c|c}
3x - 4y = 0 & \\
\hline
3(-4) - 4(-3) \; ? \; 0 & \\
-12 + 12 & \\
0 & 0 \quad \text{TRUE}
\end{array}
$$

The pairs (4, 3) and (−4, −3) check, so they are the solutions.

Graphical Solution

We graph both equations in the same viewing window and note that there are two points of intersection. We can find their coordinates using the INTERSECT feature.

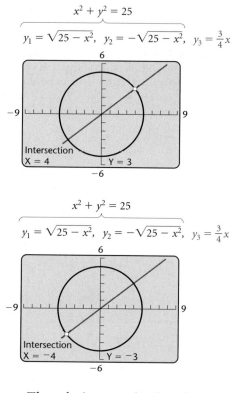

The solutions are (4, 3) and (−4, −3).

The graph can also be used to check the solutions that we found algebraically.

In the algebraic solution in Example 1, suppose that to find x, we had substituted 3 and −3 in equation (1) rather than equation (3). If $y = 3$, $y^2 = 9$, and if $y = -3$, $y^2 = 9$, so both substitutions can be performed at the same time:

$$x^2 + y^2 = 25 \qquad (1)$$
$$x^2 + (\pm 3)^2 = 25$$
$$x^2 + 9 = 25$$
$$x^2 = 16$$
$$x = \pm 4.$$

Thus, if $y = 3$, $x = 4$ or $x = -4$, and if $y = -3$, $x = 4$ or $x = -4$. The possible solutions are $(4, 3)$, $(-4, 3)$, $(4, -3)$, and $(-4, -3)$. A check reveals that $(4, -3)$ and $(-4, 3)$ are not solutions of equation (2). Had we graphed the system of equations before solving algebraically, it would have been clear that there are only two real-number solutions.

EXAMPLE 2   Solve the following system of equations:

$$x + y = 5, \quad (1) \qquad \text{(The graph is a line.)}$$
$$y = 3 - x^2. \quad (2) \qquad \text{(The graph is a parabola.)}$$

Algebraic Solution

We use the substitution method, substituting $3 - x^2$ for y in equation (1):

$$x + 3 - x^2 = 5$$
$$-x^2 + x - 2 = 0 \qquad \text{Subtracting 5 and rearranging}$$
$$x^2 - x + 2 = 0. \qquad \text{Multiplying by } -1$$

Next, we use the quadratic formula:

$$x = \frac{-b \pm \sqrt{b^2 - 4ac}}{2a} = \frac{-(-1) \pm \sqrt{(-1)^2 - 4(1)(2)}}{2(1)}$$
$$= \frac{1 \pm \sqrt{1 - 8}}{2} = \frac{1 \pm \sqrt{-7}}{2} = \frac{1 \pm i\sqrt{7}}{2} = \frac{1}{2} \pm \frac{\sqrt{7}}{2}i.$$

Now, we substitute these values for x in equation (1) and solve for y:

$$\frac{1}{2} + \frac{\sqrt{7}}{2}i + y = 5$$
$$y = 5 - \frac{1}{2} - \frac{\sqrt{7}}{2}i = \frac{9}{2} - \frac{\sqrt{7}}{2}i$$

and $\quad \dfrac{1}{2} - \dfrac{\sqrt{7}}{2}i + y = 5$

$$y = 5 - \frac{1}{2} + \frac{\sqrt{7}}{2}i = \frac{9}{2} + \frac{\sqrt{7}}{2}i.$$

The solutions are

$$\left(\frac{1}{2} + \frac{\sqrt{7}}{2}i, \frac{9}{2} - \frac{\sqrt{7}}{2}i\right) \quad \text{and} \quad \left(\frac{1}{2} - \frac{\sqrt{7}}{2}i, \frac{9}{2} + \frac{\sqrt{7}}{2}i\right).$$

There are no real-number solutions.

Graphical Solution

We graph both equations in the same viewing window.

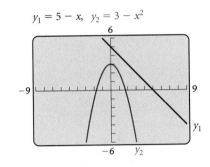

$y_1 = 5 - x, \quad y_2 = 3 - x^2$

Note that there are no points of intersection. This indicates that there are no real-number solutions. Algebra must be used, as at left, to find the imaginary-number solutions.

EXAMPLE 3 Solve the following system of equations:

$$2x^2 + 5y^2 = 39, \quad (1) \qquad \text{(The graph is an ellipse.)}$$
$$3x^2 - y^2 = -1. \quad (2) \qquad \text{(The graph is a hyperbola.)}$$

Algebraic Solution

We use the elimination method. First, we multiply equation (2) by 5 and add to eliminate the y^2-term:

$$
\begin{array}{ll}
2x^2 + 5y^2 = 39 & (1) \\
\underline{15x^2 - 5y^2 = -5} & \text{Multiplying (2) by 5} \\
17x^2 \qquad\;\; = 34 & \text{Adding} \\
\qquad x^2 = 2 & \\
\qquad x = \pm\sqrt{2}. &
\end{array}
$$

If $x = \sqrt{2}$, $x^2 = 2$, and if $x = -\sqrt{2}$, $x^2 = 2$. Thus substituting $\sqrt{2}$ or $-\sqrt{2}$ for x in equation (2) gives us

$$
\begin{aligned}
3(\pm\sqrt{2})^2 - y^2 &= -1 \\
3 \cdot 2 - y^2 &= -1 \\
6 - y^2 &= -1 \\
-y^2 &= -7 \\
y^2 &= 7 \\
y &= \pm\sqrt{7}.
\end{aligned}
$$

Thus, for $x = \sqrt{2}$, we have $y = \sqrt{7}$ or $y = -\sqrt{7}$, and for $x = -\sqrt{2}$, we have $y = \sqrt{7}$ or $y = -\sqrt{7}$. The possible solutions are $(\sqrt{2}, \sqrt{7})$, $(\sqrt{2}, -\sqrt{7})$, $(-\sqrt{2}, \sqrt{7})$, and $(-\sqrt{2}, -\sqrt{7})$. All four pairs check, so they are the solutions.

Graphical Solution

We graph both equations in the same viewing window and note that there are four points of intersection. We can use the INTERSECT feature to find their coordinates.

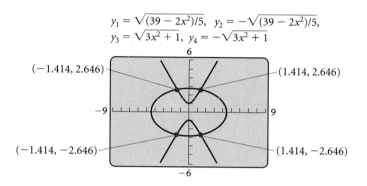

$y_1 = \sqrt{(39 - 2x^2)/5}, \; y_2 = -\sqrt{(39 - 2x^2)/5},$
$y_3 = \sqrt{3x^2 + 1}, \; y_4 = -\sqrt{3x^2 + 1}$

$(-1.414, 2.646)$ $(1.414, 2.646)$

$(-1.414, -2.646)$ $(1.414, -2.646)$

Note that the algebraic method yields exact solutions whereas the graphical method yields decimal approximations of the solutions.

The solutions are approximately $(1.414, 2.646)$, $(1.414, -2.646)$, $(-1.414, 2.646)$, and $(-1.414, -2.646)$.

EXAMPLE 4   Solve the following system of equations:

$$x^2 - 3y^2 = 6, \qquad (1)$$
$$xy = 3. \qquad (2)$$

Algebraic Solution

We use the substitution method. First, we solve equation (2) for y:

$$xy = 3$$
$$y = \frac{3}{x}. \qquad (3)$$

Next, we substitute $3/x$ for y in equation (1) and solve for x:

$$x^2 - 3\left(\frac{3}{x}\right)^2 = 6$$

$$x^2 - 3 \cdot \frac{9}{x^2} = 6$$

$$x^2 - \frac{27}{x^2} = 6$$

$$x^4 - 27 = 6x^2 \qquad \text{Multiplying by } x^2$$

$$x^4 - 6x^2 - 27 = 0$$

$$u^2 - 6u - 27 = 0 \qquad \text{Letting } u = x^2$$

$$(u - 9)(u + 3) = 0 \qquad \text{Factoring}$$

$$u = 9 \quad or \quad u = -3 \qquad \text{Principle of zero products}$$

$$x^2 = 9 \quad or \quad x^2 = -3 \qquad \text{Substituting } x^2 \text{ for } u$$

$$x = \pm 3 \quad or \quad x = \pm i\sqrt{3}.$$

Since $y = 3/x$,

when $x = 3$, $\qquad y = \dfrac{3}{3} = 1;$

when $x = -3$, $\qquad y = \dfrac{3}{-3} = -1;$

when $x = i\sqrt{3}$, $\qquad y = \dfrac{3}{i\sqrt{3}} = \dfrac{3}{i\sqrt{3}} \cdot \dfrac{-i\sqrt{3}}{-i\sqrt{3}} = -i\sqrt{3};$

when $x = -i\sqrt{3}$, $\quad y = \dfrac{3}{-i\sqrt{3}} = \dfrac{3}{-i\sqrt{3}} \cdot \dfrac{i\sqrt{3}}{i\sqrt{3}} = i\sqrt{3}.$

The pairs $(3, 1)$, $(-3, -1)$, $(i\sqrt{3}, -i\sqrt{3})$, and $(-i\sqrt{3}, i\sqrt{3})$ check, so they are the solutions.

Graphical Solution

We graph both equations in the same viewing window and find the coordinates of their points of intersection.

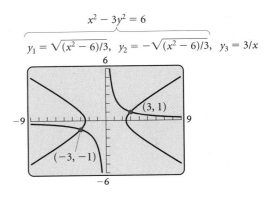

$x^2 - 3y^2 = 6$

$y_1 = \sqrt{(x^2 - 6)/3}, \quad y_2 = -\sqrt{(x^2 - 6)/3}, \quad y_3 = 3/x$

Again, note that the graphical method yields only the real-number solutions of the system of equations. The algebraic method must be used to find *all* the solutions.

Modeling and Problem Solving

EXAMPLE 5 *Dimensions of a Piece of Land.* For a student recreation building at Southport Community College, an architect wants to lay out a rectangular piece of land that has a perimeter of 204 m and an area of 2565 m^2. Find the dimensions of the piece of land.

Solution

1. **Familiarize.** We make a drawing and label it, letting l = the length of the piece of land, in meters, and w = the width, in meters.

2. **Translate.** We now have the following:

   Perimeter: $2w + 2l = 204$,   (1)

   Area: $lw = 2565$.   (2)

3. **Carry out.** We solve the system of equations both algebraically and graphically.

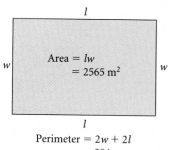

Area = lw
= 2565 m^2

Perimeter = $2w + 2l$
= 204 m

Algebraic Solution

We solve the system of equations

$$2w + 2l = 204,$$
$$lw = 2565.$$

Solving the second equation for l gives us $l = 2565/w$. We then substitute $2565/w$ for l in equation (1) and solve for w:

$$2w + 2\left(\frac{2565}{w}\right) = 204$$

$$2w^2 + 2(2565) = 204w \qquad \text{Multiplying by } w$$

$$2w^2 - 204w + 2(2565) = 0$$

$$w^2 - 102w + 2565 = 0 \qquad \text{Multiplying by } \tfrac{1}{2}$$

$$(w - 57)(w - 45) = 0$$

$$w = 57 \quad or \quad w = 45.$$

Principle of zero products

If $w = 57$, then $l = 2565/w = 2565/57 = 45$. If $w = 45$, then $l = 2565/w = 2565/45 = 57$. Since length is generally considered to be longer than width, we have the solution $l = 57$ and $w = 45$, or (57, 45).

Graphical Solution

We replace l with x and w with y, graph $y_1 = (204 - 2x)/2$ and $y_2 = 2565/x$, and find the point(s) of intersection of the graphs.

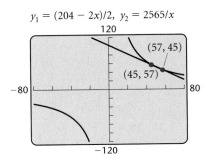

$$y_1 = (204 - 2x)/2, \ y_2 = 2565/x$$

As in the algebraic solution, we have two possible solutions: $(45, 57)$ and $(57, 45)$. Since length, x, is generally considered to be longer than width, y, we have the solution $(57, 45)$.

4. **Check.** If $l = 57$ and $w = 45$, the perimeter is $2 \cdot 57 + 2 \cdot 45$, or 204. The area is $57 \cdot 45$, or 2565. The numbers check.

5. **State.** The length of the piece of land is 57 m and the width is 45 m.

Exercise Set **6.4**

In Exercises 1–6, match the system of equations with one of the graphs (a)–(f), which follow.

a)

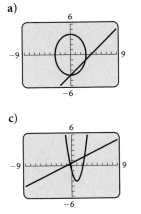

b)

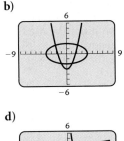

c)

d)

e)

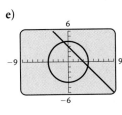

f)

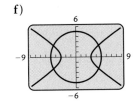

1. $x^2 + y^2 = 16,$,
   $x + y = 3$

2. $16x^2 + 9y^2 = 144,$
   $x - y = 4$

3. $y = x^2 - 4x - 2,$
   $2y - x = 1$

4. $4x^2 - 9y^2 = 36,$
   $x^2 + y^2 = 25$

5. $y = x^2 - 3,$
   $x^2 + 4y^2 = 16$

6. $y^2 - 2y = x + 3,$
   $xy = 4$

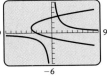

Solve.

7. $x^2 + y^2 = 25,$
$y - x = 1$

8. $x^2 + y^2 = 100,$
$y - x = 2$

9. $4x^2 + 9y^2 = 36,$
$3y + 2x = 6$

10. $9x^2 + 4y^2 = 36,$
$3x + 2y = 6$

11. $x^2 + y^2 = 25,$
$y^2 = x + 5$

12. $y = x^2,$
$x = y^2$

13. $x^2 + y^2 = 9,$
$x^2 - y^2 = 9$

14. $y^2 - 4x^2 = 4,$
$4x^2 + y^2 = 4$

15. $y^2 - x^2 = 9,$
$2x - 3 = y$

16. $x + y = -6,$
$xy = -7$

17. $y^2 = x + 3,$
$2y = x + 4$

18. $y = x^2,$
$3x = y + 2$

19. $x^2 + y^2 = 25,$
$xy = 12$

20. $x^2 - y^2 = 16,$
$x + y^2 = 4$

21. $x^2 + y^2 = 4,$
$16x^2 + 9y^2 = 144$

22. $x^2 + y^2 = 25,$
$25x^2 + 16y^2 = 400$

23. $x^2 + 4y^2 = 25,$
$x + 2y = 7$

24. $y^2 - x^2 = 16,$
$2x - y = 1$

25. $x^2 - xy + 3y^2 = 27,$
$x - y = 2$

26. $2y^2 + xy + x^2 = 7,$
$x - 2y = 5$

27. $x^2 + y^2 = 16,$
$y^2 - 2x^2 = 10$

28. $x^2 + y^2 = 14,$
$x^2 - y^2 = 4$

29. $x^2 + y^2 = 5,$
$xy = 2$

30. $x^2 + y^2 = 20,$
$xy = 8$

31. $3x + y = 7,$
$4x^2 + 5y = 56$

32. $2y^2 + xy = 5,$
$4y + x = 7$

33. $a + b = 7,$
$ab = 4$

34. $p + q = -4,$
$pq = -5$

35. $x^2 + y^2 = 13,$
$xy = 6$

36. $x^2 + 4y^2 = 20,$
$xy = 4$

37. $x^2 + y^2 + 6y + 5 = 0,$
$x^2 + y^2 - 2x - 8 = 0$

38. $2xy + 3y^2 = 7,$
$3xy - 2y^2 = 4$

39. $2a + b = 1,$
$b = 4 - a^2$

40. $4x^2 + 9y^2 = 36,$
$x + 3y = 3$

41. $a^2 + b^2 = 89,$
$a - b = 3$

42. $xy = 4,$
$x + y = 5$

43. $xy - y^2 = 2,$
$2xy - 3y^2 = 0$

44. $4a^2 - 25b^2 = 0,$
$2a^2 - 10b^2 = 3b + 4$

45. $m^2 - 3mn + n^2 + 1 = 0,$
$3m^2 - mn + 3n^2 = 13$

46. $ab - b^2 = -4,$
$ab - 2b^2 = -6$

47. $x^2 + y^2 = 5,$
$x - y = 8$

48. $4x^2 + 9y^2 = 36,$
$y - x = 8$

49. $a^2 + b^2 = 14,$
$ab = 3\sqrt{5}$

50. $x^2 + xy = 5,$
$2x^2 + xy = 2$

51. $x^2 + y^2 = 25,$
$9x^2 + 4y^2 = 36$

52. $x^2 + y^2 = 1,$
$9x^2 - 16y^2 = 144$

53. $5y^2 - x^2 = 1,$
$xy = 2$

54. $x^2 - 7y^2 = 6,$
$xy = 1$

55. *Picture Frame Dimensions.*   Frank's Frame Shop is building a frame for a rectangular oil painting with a perimeter of 28 cm and a diagonal of 10 cm. Find the dimensions of the painting.

56. *Landscaping.*   Green Leaf Landscaping is planting a rectangular wildflower garden with a perimeter of 6 m and a diagonal of $\sqrt{5}$ m. Find the dimensions of the garden.

57. *Pamphlet Design.*   A graphic artist is designing a rectangular advertising pamphlet that is to have an area of 20 in^2 and a perimeter of 18 in. Find the dimensions of the pamphlet.

58. *Sign Dimensions.*   Peden's Advertising is building a rectangular sign with an area of 2 yd^2 and a perimeter of 6 yd. Find the dimensions of the sign.

59. *Banner Design.*   A rectangular banner with an area of $\sqrt{3}$ m^2 is being designed to advertise an exhibit at the Madison Art League. The length of a diagonal is 2 m. Find the dimensions of the banner.

Area = $\sqrt{3}$ m^2

60. *Carpentry.* A carpenter wants to make a rectangular tabletop with an area of $\sqrt{2}$ m^2 and a diagonal of $\sqrt{3}$ m. Find the dimensions of the tabletop.

61. *Fencing.* It will take 210 yd of fencing to enclose a rectangular dog run. The area of the run is 2250 yd^2. What are the dimensions of the run?

62. *Office Dimensions.* The diagonal of the floor of a rectangular office cubicle is 1 ft longer than the length of the cubicle and 3 ft longer than twice the width. Find the dimensions of the cubicle.

63. *Seed Test Plots.* The Burton Seed Company has two square test plots. The sum of their areas is 832 ft^2 and the difference of their areas is 320 ft^2. Find the length of a side of each plot.

64. *Investment.* Jenna made an investment for 1 yr that earned $7.50 simple interest. If the principal had been $25 more and the interest rate 1% less, the interest would have been the same. Find the principal and the rate.

65. *Social Security Trust Fund.* Income and expenditures of the Social Security Trust Fund are shown in the table below.

YEAR	INCOME (IN BILLIONS)	EXPENDITURES (IN BILLIONS)
1985	$203.5	$190.6
1990	315.4	253.1
1991	329.7	274.2
1992	342.6	291.9
1993	355.6	308.8
1994	381.1	323.0
1995	399.7	339.8
1996	424.5	353.6
1997	457.7	369.1
1998	484.3	382.9
1999	503.7	396.3

Source: Social Security Administration

a) Find a linear function $I(x)$ that models the income data. Let x represent the number of years after 1985.

b) Find an exponential function $E(x)$ that models the expenditure data. Let x represent the number of years after 1985.

c) Use the functions found in parts (a) and (b) to determine when income will equal expenditures.

66. *Higher Education.* The percent of associate's, bachelor's, master's, first professional, and doctor's degrees conferred on men and women in the United States in recent years are shown in the table below.

YEAR	PERCENT MALE	PERCENT FEMALE
1960	65.8	34.2
1965	61.5	38.5
1970	59.2	40.8
1975	56.0	44.0
1980	51.1	48.9
1985	49.3	50.7
1990	46.6	53.4
1995	44.9	55.1

Source: U.S. National Center for Education Statistics

a) Find an exponential function $M(x)$ that models the percent of degrees that were earned by men. Let x represent the number of years after 1960.

b) Find a linear function $F(x)$ that models the percent of degrees that were earned by women. Let x represent the number of years after 1960.

c) Use the functions found in parts (a) and (b) to determine when equal percents of men and women earned degrees.

Discussion and Writing

67. What would you say to a classmate who tells you that any nonlinear system of equations can be solved graphically?

68. Write a problem that can be translated to a nonlinear system of equations, and ask a classmate to solve it. Devise the problem so that the solution is "The dimensions of the rectangle are 6 ft by 8 ft."

Skill Maintenance

Solve.

69. $2^{3x} = 64$

70. $5^x = 27$

71. $\log_3 x = 4$

72. $\log(x - 3) + \log x = 1$

Synthesis

73. Find an equation of the circle that passes through the points (2, 4) and (3, 3) and whose center is on the line $3x - y = 3$.

74. Find an equation of the circle that passes through the points (−2, 3) and (−4, 1) and whose center is on the line $5x + 8y = -2$.

75. Find an equation of an ellipse centered at the origin that passes through the points $(1, \sqrt{3}/2)$ and $(\sqrt{3}, 1/2)$.

76. Find an equation of a hyperbola of the type

$$\frac{x^2}{b^2} - \frac{y^2}{a^2} = 1$$

that passes through the points $(-3, -3\sqrt{5}/2)$ and $(-3/2, 0)$.

77. Find an equation of the circle that passes through the points (4, 6), (−6, 2), and (1, 3).

78. Find an equation of the circle that passes through the points (2, 3), (4, 5), and (0, −3).

79. Show that a hyperbola does not intersect its asymptotes. That is, solve the system of equations

$$\frac{x^2}{a^2} - \frac{y^2}{b^2} = 1,$$

$$y = \frac{b}{a}x \quad \left(\text{or } y = -\frac{b}{a}x\right).$$

80. *Numerical Relationship.* Find two numbers whose product is 2 and the sum of whose reciprocals is $\frac{33}{8}$.

81. *Numerical Relationship.* The square of a number exceeds twice the square of another number by $\frac{1}{8}$. The sum of their squares is $\frac{5}{16}$. Find the numbers.

82. *Box Dimensions.* Four squares with sides 5 in. long are cut from the corners of a rectangular metal sheet that has an area of 340 in². The edges are bent up to form an open box with a volume of 350 in³. Find the dimensions of the box.

83. *Numerical Relationship.* The sum of two numbers is 1, and their product is 1. Find the sum of their cubes. There is a method to solve this problem that is easier than solving a nonlinear system of equations. Can you discover it?

84. Solve for x and y:

$$x^2 - y^2 = a^2 - b^2,$$
$$x - y = a - b.$$

Solve.

85. $x^3 + y^3 = 72,$
$x + y = 6$

86. $a + b = \dfrac{5}{6},$

$\dfrac{a}{b} + \dfrac{b}{a} = \dfrac{13}{6}$

87. $p^2 + q^2 = 13,$
$\dfrac{1}{pq} = -\dfrac{1}{6}$

88. $x^2 + y^2 = 4,$
$(x - 1)^2 + y^2 = 4$

89. $5^{x+y} = 100,$
$3^{2x-y} = 1000$

90. $e^x - e^{x+y} = 0,$
$e^y - e^{x-y} = 0$

Solve using a grapher.

91. $y - \ln x = 2,$
$y = x^2$

92. $y = \ln(x + 4),$
$x^2 + y^2 = 6$

93. $e^x - y = 1,$
$3x + y = 4$

94. $y - e^{-x} = 1,$
$y = 2x + 5$

95. $y = e^x,$
$x - y = -2$

96. $y = e^{-x},$
$x + y = 3$

97. $x^2 + y^2 = 19{,}380{,}510.36,$
$27{,}942.25x - 6.125y = 0$

98. $2x + 2y = 1660,$
$xy = 35{,}325$

99. $14.5x^2 - 13.5y^2 - 64.5 = 0,$
$5.5x - 6.3y - 12.3 = 0$

100. $13.5xy + 15.6 = 0,$
$5.6x - 6.7y - 42.3 = 0$

101. $0.319x^2 + 2688.7y^2 = 56{,}548,$
$0.306x^2 - 2688.7y^2 = 43{,}452$

102. $18.465x^2 + 788.723y^2 = 6408,$
$106.535x^2 - 788.723y^2 = 2692$

Chapter Summary and Review 6

Important Properties and Formulas

Standard Equation of a Parabola with Vertex at the Origin

The standard equation of a parabola with vertex $(0, 0)$ and directrix $y = -p$ is

$$x^2 = 4py.$$

The focus is $(0, p)$ and the y-axis is the axis of symmetry.

The standard equation of a parabola with vertex $(0, 0)$ and directrix $x = -p$ is

$$y^2 = 4px.$$

The focus is $(p, 0)$ and the x-axis is the axis of symmetry.

Standard Equation of a Parabola with Vertex (h, k) and Vertical Axis of Symmetry

The standard equation of a parabola with vertex (h, k) and vertical axis of symmetry is

$$(x - h)^2 = 4p(y - k),$$

where the vertex is (h, k), the focus is $(h, k + p)$, and the directrix is $y = k - p$.

Standard Equation of a Parabola with Vertex (h, k) and Horizontal Axis of Symmetry

The standard equation of a parabola with vertex (h, k) and horizontal axis of symmetry is

$$(y - k)^2 = 4p(x - h),$$

where the vertex is (h, k), the focus is $(h + p, k)$, and the directrix is $x = h - p$.

Standard Equation of a Circle

The standard equation of a circle with center (h, k) and radius r is

$$(x - h)^2 + (y - k)^2 = r^2.$$

Standard Equation of an Ellipse with Center at the Origin

Major axis horizontal

$$\frac{x^2}{a^2} + \frac{y^2}{b^2} = 1, \ a > b > 0$$

Vertices: $(-a, 0), (a, 0)$
y-intercepts: $(0, -b), (0, b)$
Foci: $(-c, 0), (c, 0)$, where $c^2 = a^2 - b^2$

Major axis vertical

$$\frac{x^2}{b^2} + \frac{y^2}{a^2} = 1, \ a > b > 0$$

Vertices: $(0, -a), (0, a)$
x-intercepts: $(-b, 0), (b, 0)$
Foci: $(0, -c), (0, c)$, where $c^2 = a^2 - b^2$

Standard Equation of an Ellipse with Center at (h, k)

Major axis horizontal

$$\frac{(x - h)^2}{a^2} + \frac{(y - k)^2}{b^2} = 1, \ a > b > 0$$

Vertices: $(h - a, k), (h + a, k)$
Length of minor axis: $2b$
Foci: $(h - c, k), (h + c, k)$, where $c^2 = a^2 - b^2$

Major axis vertical

$$\frac{(x - h)^2}{b^2} + \frac{(y - k)^2}{a^2} = 1, \ a > b > 0$$

Vertices: $(h, k - a), (h, k + a)$
Length of minor axis: $2b$
Foci: $(h, k - c), (h, k + c)$, where $c^2 = a^2 - b^2$

(continued)

Standard Equation of a Hyperbola with Center at the Origin

Transverse axis horizontal

$$\frac{x^2}{a^2} - \frac{y^2}{b^2} = 1$$

Vertices: $(-a, 0)$, $(a, 0)$

Asymptotes: $y = -\dfrac{b}{a}x$, $y = \dfrac{b}{a}x$

Foci: $(-c, 0)$, $(c, 0)$, where $c^2 = a^2 + b^2$

Transverse axis vertical

$$\frac{y^2}{a^2} - \frac{x^2}{b^2} = 1$$

Vertices: $(0, -a)$, $(0, a)$

Asymptotes: $y = -\dfrac{a}{b}x$, $y = \dfrac{a}{b}x$

Foci: $(0, -c)$, $(0, c)$, where $c^2 = a^2 + b^2$

Standard Equation of a Hyperbola with Center at (h, k)

Transverse axis horizontal

$$\frac{(x - h)^2}{a^2} - \frac{(y - k)^2}{b^2} = 1$$

Vertices: $(h - a, k)$, $(h + a, k)$

Asymptotes: $y - k = \dfrac{b}{a}(x - h)$,

$$y - k = -\frac{b}{a}(x - h)$$

Foci: $(h - c, k)$, $(h + c, k)$, where $c^2 = a^2 + b^2$

Transverse axis vertical

$$\frac{(y - k)^2}{a^2} - \frac{(x - h)^2}{b^2} = 1$$

Vertices: $(h, k - a)$, $(h, k + a)$

Asymptotes: $y - k = \dfrac{a}{b}(x - h)$,

$$y - k = -\frac{a}{b}(x - h)$$

Foci: $(h, k - c)$, $(h, k + c)$, where $c^2 = a^2 + b^2$

REVIEW EXERCISES

In Exercises 1–8, match the equation with one of the graphs (a)–(h), which follow.

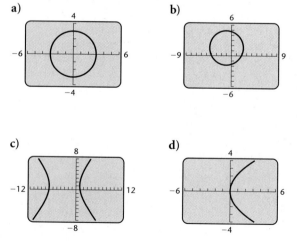

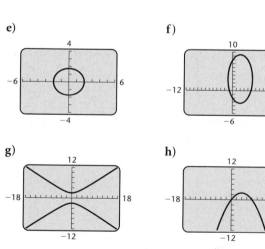

1. $y^2 = 5x$

2. $y^2 = 9 - x^2$

3. $3x^2 + 4y^2 = 12$

4. $9y^2 - 4x^2 = 36$

5. $x^2 + y^2 + 2x - 3y = 8$

6. $4x^2 + y^2 - 16x - 6y = 15$

7. $x^2 - 8x + 6y = 0$

8. $\dfrac{(x+3)^2}{16} - \dfrac{(y-1)^2}{25} = 1$

9. Find an equation of the parabola with directrix $y = \frac{3}{2}$ and focus $\left(0, -\frac{3}{2}\right)$.

10. Find the focus, the vertex, and the directrix of the parabola given by
$$y^2 = -12x.$$

11. Find the vertex, the focus, and the directrix of the parabola given by
$$x^2 + 10x + 2y + 9 = 0.$$

12. Find the center, the vertices, and the foci of the ellipse given by
$$16x^2 + 25y^2 - 64x + 50y - 311 = 0.$$
Then draw the graph.

13. Find an equation of the ellipse having vertices $(0, -4)$ and $(0, 4)$ with minor axis of length 6.

14. Find the center, the vertices, the foci, and the asymptotes of the hyperbola given by
$$x^2 - 2y^2 + 4x + y - \tfrac{1}{8} = 0.$$

15. *Spotlight.* A spotlight has a parabolic cross section that is 2 ft wide at the opening and 1.5 ft deep at the vertex. How far from the vertex is the focus?

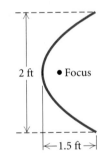

2 ft | • Focus

←1.5 ft→

Solve.

16. $x^2 - 16y = 0,$
$x^2 - y^2 = 64$

17. $4x^2 + 4y^2 = 65,$
$6x^2 - 4y^2 = 25$

18. $x^2 - y^2 = 33,$
$x + y = 11$

19. $x^2 - 2x + 2y^2 = 8,$
$2x + y = 6$

20. $x^2 - y = 3,$
$2x - y = 3$

21. $x^2 + y^2 = 25,$
$x^2 - y^2 = 7$

22. $x^2 - y^2 = 3,$
$y = x^2 - 3$

23. $x^2 + y^2 = 18,$
$2x + y = 3$

24. $x^2 + y^2 = 100,$
$2x^2 - 3y^2 = -120$

25. $x^2 + 2y^2 = 12,$
$xy = 4$

26. *Numerical Relationship.* The sum of two numbers is 11 and the sum of their squares is 65. Find the numbers.

27. *Dimensions of a Rectangle.* A rectangle has a perimeter of 38 m and an area of 84 m². What are the dimensions of the rectangle?

28. *Numerical Relationship.* Find two positive integers whose sum is 12 and the sum of whose reciprocals is $\frac{3}{8}$.

29. *Perimeter.* The perimeter of a square is 12 cm more than the perimeter of another square. The area of the first square exceeds the area of the other by 39 cm². Find the perimeter of each square.

30. *Radius of a Circle.* The sum of the areas of two circles is 130π ft². The difference of the areas is 112π ft². Find the radius of each circle.

Discussion and Writing

31. What would you say to a classmate who tells you that an algebraic solution of a nonlinear system of equations is always preferable to a graphical solution?

32. Is a circle a special type of ellipse? Why or why not?

Synthesis

33. Find an equation of the ellipse containing the point $(-1/2, 3\sqrt{3}/2)$ and with vertices $(0, -3)$ and $(0, 3)$.

34. Find two numbers whose product is 4 and the sum of whose reciprocals is $\frac{65}{56}$.

35. Find an equation of the circle that passes through the points $(10, 7)$, $(-6, 7)$, and $(-8, 1)$.

36. *Navigation.* Two radio transmitters positioned 400 mi apart send simultaneous signals to a ship that is 250 mi offshore, sailing parallel to the shoreline. The signal from transmitter A reaches the ship 300 microseconds before the signal from transmitter B. The signals travel at a speed of 186,000 miles per second, or 0.186 mile per microsecond. Find the equation of the hyperbola with foci A and B on which the ship is located. (*Hint*: For any point on the hyperbola, the absolute value of the difference of its distances from the foci is $2a$.)

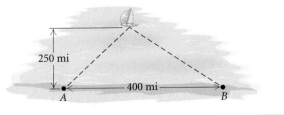

250 mi

←————— 400 mi —————→
A B

23 √

Sequences, Series, and Combinatorics 7

We begin this chapter with a study of a special type of function whose domain is the set of positive integers, a *sequence*. When we add the terms of a sequence, we get a *series*. We also study a method of proof known as *mathematical induction* and we find a method for expanding binomials $(a + b)^n$, where n is a natural number. In addition, we then study *combinatorics*, a method for counting the arrangements and combinations of objects in a set and, finally, we calculate the *probability* that an event will occur.

469

تر تسب, تؤالی

Sequences
and Series

گحریہ ۔رتالم

7.1

• *Find terms of sequences given the nth term.*
• *Look for a pattern in a sequence and try to determine a general term.*
• *Convert between sigma notation and other notation for a series.*
• *Construct the terms of a recursively defined sequence.*

In this section, we discuss sets or lists of numbers, considered in order, and their sums.

Sequences

دونای ترکیب

Suppose that $1000 is invested at 6%, compounded annually. The amounts to which the account will grow after 1 yr, 2 yr, 3 yr, 4 yr, and so on, form the following sequence of numbers:

$$(1) \qquad (2) \qquad (3) \qquad (4)$$
$$\downarrow \qquad \downarrow \qquad \downarrow \qquad \downarrow$$
$$\$1060.00, \quad \$1123.60, \quad \$1191.02, \quad \$1262.48, \ldots.$$

X	Y₁
1	1060
2	1123.6
3	1191
4	1262.5
5	1338.2
6	1418.5
7	1503.6
X = 7	

(We found these numbers using the compound interest formula, $A = P(1 + i)^n$, expressed as $y_1 = 1000(1.06)^x$, in order to create the table on a grapher. Note the rounding error in the table at left.) We can think of this as a function that pairs 1 with $1060.00, 2 with $1123.60, 3 with $1191.02, and so on. A **sequence** is thus a *function,* where the domain is a set of consecutive positive integers beginning with 1.

مساوی ۔ پی درپی   اعداد مست دمتی دصفر

If we continue to compute the amounts of money in the account forever, we obtain an **infinite sequence** with function values

$$\$1060.00, \$1123.60, \$1191.02, \$1262.48, \$1338.23, \$1418.52, \ldots.$$

The dots at the end "..." indicate that the sequence goes on without stopping. If we stop after a certain number of years, we obtain a **finite sequence:**

$$\$1060.00, \$1123.60, \$1191.02, \$1262.48.$$

> ### Sequences
> An **infinite sequence** is a function having for its domain the set of positive integers, $\{1, 2, 3, 4, 5, \ldots\}$.
> A **finite sequence** is a function having for its domain a set of positive integers, $\{1, 2, 3, 4, 5, \ldots, n\}$, for some positive integer n.

Consider the sequence given by the formula

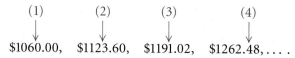

$$a(n) = 2^n, \quad \text{or} \quad a_n = 2^n.$$

Some of the function values, also known as the **terms** of the sequence, follow:

$$a_1 = 2^1 = 2,$$
$$a_2 = 2^2 = 4,$$
$$a_3 = 2^3 = 8,$$
$$a_4 = 2^4 = 16,$$
$$a_5 = 2^5 = 32.$$

The first term of the sequence is denoted as a_1, the fifth term as a_5, and the nth term, or **general term,** as a_n. This sequence can also be denoted as

$$2, 4, 8, \ldots, \quad \text{or as} \quad 2, 4, 8, \ldots, 2^n, \ldots.$$

EXAMPLE 1   Find the first 4 terms and the 23rd term of the sequence whose general term is given by $a_n = (-1)^n n^2$.

Solution   We have $a_n = (-1)^n n^2$, so

$$a_1 = (-1)^1 \cdot 1^2 = -1,$$
$$a_2 = (-1)^2 \cdot 2^2 = 4,$$
$$a_3 = (-1)^3 \cdot 3^2 = -9,$$
$$a_4 = (-1)^4 \cdot 4^2 = 16,$$
$$a_{23} = (-1)^{23} \cdot 23^2 = -529.$$

We can also use a grapher to find the desired terms. We enter $y_1 = (-1)^x x^2$. We then set up a table in ASK mode and enter 1, 2, 3, 4, and 23 as values for x.

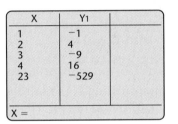

Note in Example 1 that the power $(-1)^n$ causes the signs of the terms to alternate between positive and negative, depending on whether n is even or odd. This kind of sequence is called an **alternating sequence.**

EXAMPLE 2   Use a grapher to find the first 5 terms of the sequence whose general term is given by $a_n = n/(n + 1)$.

Solution   Although we could use a table, we will use the SEQ feature. The grapher will write the terms horizontally as a list. The list can also be written in fractional notation.

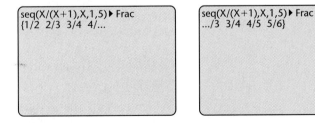

We use $\boxed{\triangleright}$ to view the two items that do not initially appear on the screen. The first 5 terms of the sequence are 1/2, 2/3, 3/4, 4/5, and 5/6.

Finding the General Term

When only the first few terms of a sequence are known, we do not know for sure what the general term is, but we might be able to make a prediction by looking for a pattern.

EXAMPLE 3   For each of the following sequences, predict the general term.

a) $1, \sqrt{2}, \sqrt{3}, 2, \ldots$

b) $-1, 3, -9, 27, -81, \ldots$

c) $2, 4, 8, \ldots$

Solution

a) These are square roots of consecutive integers, so the general term may be $\sqrt{n}$.

b) These are powers of 3 with alternating signs, so the general term may be $(-1)^n 3^{n-1}$.

c) If we see the pattern of powers of 2, we will see 16 as the next term and guess 2^n for the general term. Then the sequence could be written with more terms as

$$2, 4, 8, 16, 32, 64, 128, \ldots .$$

If we see that we can get the second term by adding 2, the third term by adding 4, and the next term by adding 6, and so on, we will see 14 as the next term. A general term for the sequence is then $n^2 - n + 2$, and the sequence can be written with more terms as

$$2, 4, 8, 14, 22, 32, 44, 58, \ldots .$$

━

Example 3(c) illustrates that, in fact, you can never be certain about the general term. The fewer the given terms, the greater the uncertainty.

Graphing Sequences

The domain of a sequence is a set of positive integers, so the graph of a sequence is a set of points that are not connected. Thus we use DOT mode to graph a sequence whose general term is known. In addition to selecting DOT mode, we also select SEQUENCE mode. In this mode, the variable is n and functions are named $u(n)$, $v(n)$, and $w(n)$ rather than y_1, y_2, and y_3.

EXAMPLE 4   Graph the sequence whose general term is given by $a_n = n/(n + 1)$.

Solution   With the grapher set in DOT and SEQUENCE modes, we enter $u(n) = n/(n + 1)$. All the function values will be positive numbers that are less than 1, so we choose the window $[0, 10, 0, 1]$ and we also choose $n\text{Min} = 1$, $n\text{Max} = 10$, PlotStart $= 1$, and PlotStep $= 1$.

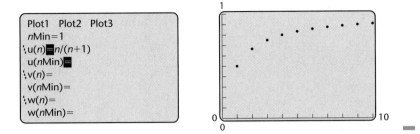

Sums and Series

> ### Series
>
> Given the infinite sequence
>
> $$a_1, a_2, a_3, a_4, \ldots, a_n, \ldots,$$
>
> the sum of the terms
>
> $$a_1 + a_2 + a_3 + \cdots + a_n + \cdots$$
>
> is called an **infinite series.** A **partial sum** is the sum of the first n terms:
>
> $$a_1 + a_2 + a_3 + \cdots + a_n.$$
>
> A partial sum is also called a **finite series,** or ***n*th partial sum,** and is denoted S_n.

EXAMPLE 5 For the sequence $-2, 4, -6, 8, -10, 12, -14, \ldots$, find each of the following.

a) S_1 **b)** S_4 **c)** S_5

Solution

a) $S_1 = -2$

b) $S_4 = -2 + 4 + (-6) + 8 = 4$

c) $S_5 = -2 + 4 + (-6) + 8 + (-10) = -6$

We can also use a grapher to find partial sums of a sequence when a formula for the general term is known.

EXAMPLE 6 Use a grapher to find S_1, S_2, S_3, and S_4 for the sequence whose general term is given by $a_n = n^2 - 3$.

Solution   We can use the CUMSUM feature. The grapher will write the partial sums as a list.

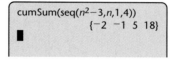

cumSum(seq($n^2-3,n,1,4$))
{−2 −1 5 18}

We have $S_1 = -2$, $S_2 = -1$, $S_3 = 5$, and $S_4 = 18$.

Sigma Notation

The Greek letter Σ (sigma) can be used to simplify notation when the general term of a sequence is a formula. For example, the sum of the first four terms of the sequence 3, 5, 7, 9, ..., $2k + 1$, ... can be named as follows, using what is called **sigma notation**, or **summation notation**:

$$\sum_{k=1}^{4} (2k + 1).$$

This is read "the sum as k goes from 1 to 4 of $(2k + 1)$." The letter k is called the **index of summation.** Sometimes the index of summation starts at a number other than 1, and sometimes letters other than k are used.

EXAMPLE 7   Find and evaluate each of the following sums.

a) $\displaystyle\sum_{k=1}^{5} k^3$

b) $\displaystyle\sum_{k=0}^{4} (-1)^k 5^k$

c) $\displaystyle\sum_{i=8}^{11} \left(2 + \frac{1}{i}\right)$

Solution

a) We replace k with 1, 2, 3, 4, and 5. Then we add the results.

$$\sum_{k=1}^{5} k^3 = 1^3 + 2^3 + 3^3 + 4^3 + 5^3$$

$$= 1 + 8 + 27 + 64 + 125$$

$$= 225$$

We can also find the result using the SUM feature on a grapher.

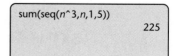

sum(seq($n^{3,}n$,1,5))
225

b) $\displaystyle\sum_{k=0}^{4} (-1)^k 5^k = (-1)^0 5^0 + (-1)^1 5^1 + (-1)^2 5^2 + (-1)^3 5^3 + (-1)^4 5^4$

$$= 1 - 5 + 25 - 125 + 625 = 521$$

c) $\displaystyle\sum_{i=8}^{11} \left(2 + \frac{1}{i}\right) = \left(2 + \frac{1}{8}\right) + \left(2 + \frac{1}{9}\right) + \left(2 + \frac{1}{10}\right) + \left(2 + \frac{1}{11}\right)$

$$= 8\frac{1691}{3960}$$

EXAMPLE 8   Write sigma notation for each sum.

a) $1 + 2 + 4 + 8 + 16 + 32 + 64$

b) $-2 + 4 - 6 + 8 - 10$

c) $x + \dfrac{x^2}{2} + \dfrac{x^3}{3} + \dfrac{x^4}{4} + \cdots$

Solution

a) $1 + 2 + 4 + 8 + 16 + 32 + 64$

This is a sum of powers of 2, beginning with 2^0, or 1, and ending with 2^6, or 64. Sigma notation is $\sum_{k=0}^{6} 2^k$.

b) $-2 + 4 - 6 + 8 - 10$

Disregarding the alternating signs, we see that this is the sum of the first 5 even integers. Note that $2k$ is a formula for the kth positive even integer, and $(-1)^k = -1$ when k is odd and $(-1)^k = 1$ when k is even. Thus the general term is $(-1)^k(2k)$. The sum begins with $k = 1$ and ends with $k = 5$, so sigma notation is $\sum_{k=1}^{5} (-1)^k(2k)$.

c) $x + \dfrac{x^2}{2} + \dfrac{x^3}{3} + \dfrac{x^4}{4} + \cdots$

The general term is x^k/k, beginning with $k = 1$. This is also an infinite series. We use the symbol ∞ for infinity and write the series using sigma notation: $\sum_{k=1}^{\infty} (x^k/k)$.

Recursive Definitions

A sequence may be defined **recursively** or by using a **recursion formula.** Such a definition lists the first term, or the first few terms, and then describes how to determine the remaining terms from the given terms.

EXAMPLE 9   Find the first 5 terms of the sequence defined by

$$a_1 = 5, \qquad a_{k+1} = 2a_k - 3, \quad \text{for } k \geq 1.$$

Solution

$$a_1 = 5,$$

$$a_2 = 2a_1 - 3 = 2 \cdot 5 - 3 = 7,$$

$$a_3 = 2a_2 - 3 = 2 \cdot 7 - 3 = 11,$$

$$a_4 = 2a_3 - 3 = 2 \cdot 11 - 3 = 19,$$

$$a_5 = 2a_4 - 3 = 2 \cdot 19 - 3 = 35.$$

Many graphers have the capability to work with recursively defined sequences. In Example 9, for instance, the function could be entered as $u(n) = 2 * u(n - 1) - 3$ with $u(n\text{Min}) = 5$. We can read the terms of the sequence from a table.

```
Plot1  Plot2  Plot3
nMin=1
\u(n)▬2*u(n-1)-3

u(nMin)▬{5}
\v(n)=
 v(nMin)=
\w(n)=
```

n	u(n)
1	5
2	7
3	11
4	19
5	35
6	67
7	131

n = 1

Exercise Set 7.1

In each of the following, the nth term of a sequence is given. Find the first 4 terms, a_{10}, and a_{15}.

1. $a_n = 4n - 1$

2. $a_n = (n - 1)(n - 2)(n - 3)$

3. $a_n = \dfrac{n}{n - 1}$, $n \geq 2$

4. $a_n = n^2 - 1$, $n \geq 3$

5. $a_n = \dfrac{n^2 - 1}{n^2 + 1}$

6. $a_n = \left(-\dfrac{1}{2}\right)^{n-1}$

7. $a_n = (-1)^n n^2$

8. $a_n = (-1)^{n-1}(3n - 5)$

9. $a_n = 5 + \dfrac{(-2)^{n+1}}{2^n}$

10. $a_n = \dfrac{2n - 1}{n^2 + 2n}$

Find the indicated term of the given sequence.

11. $a_n = 5n - 6$; a_8

12. $a_n = (3n - 4)(2n + 5)$; a_7

13. $a_n = (2n + 3)^2$; a_6

14. $a_n = (-1)^{n-1}(4.6n - 18.3)$; a_{12}

15. $a_n = 5n^2(4n - 100)$; a_{11}

16. $a_n = \left(1 + \dfrac{1}{n}\right)^2$; a_{80}

17. $a_n = \ln e^n$; a_{67}

18. $a_n = 2 - \dfrac{1000}{n}$; a_{100}

Predict the general, or nth term, a_n, of the sequence. Answers may vary.

19. 2, 4, 6, 8, 10, ...

20. 3, 9, 27, 81, 243, ...

21. $-2, 6, -18, 54, \ldots$

22. $-2, 3, 8, 13, 18, \ldots$

23. $\frac{2}{3}, \frac{3}{4}, \frac{4}{5}, \frac{5}{6}, \frac{6}{7}, \ldots$

24. $\sqrt{2}, 2, \sqrt{6}, 2\sqrt{2}, \sqrt{10}, \ldots$

25. $1 \cdot 2, 2 \cdot 3, 3 \cdot 4, 4 \cdot 5, \ldots$

26. $-1, -4, -7, -10, -13, \ldots$

27. $0, \log 10, \log 100, \log 1000, \ldots$

28. $\ln e^2, \ln e^3, \ln e^4, \ln e^5, \ldots$

Find the indicated partial sums for the sequence.

29. $1, 2, 3, 4, 5, 6, 7, \ldots;$ S_3 and S_7

30. $1, -3, 5, -7, 9, -11, \ldots;$ S_2 and S_5

31. $2, 4, 6, 8, \ldots;$ S_4 and S_5

32. $1, \frac{1}{4}, \frac{1}{9}, \frac{1}{16}, \frac{1}{25}, \ldots;$ S_1 and S_5

Find and evaluate the sum.

33. $\displaystyle\sum_{k=1}^{5} \frac{1}{2k}$

34. $\displaystyle\sum_{i=1}^{6} \frac{1}{2i + 1}$

35. $\displaystyle\sum_{i=0}^{6} 2^i$

36. $\displaystyle\sum_{k=4}^{7} \sqrt{2k - 1}$

37. $\displaystyle\sum_{k=7}^{10} \ln k$

38. $\displaystyle\sum_{k=1}^{4} \pi k$

39. $\displaystyle\sum_{k=1}^{8} \frac{k}{k + 1}$

40. $\displaystyle\sum_{i=1}^{5} \frac{i - 1}{i + 3}$

41. $\displaystyle\sum_{i=1}^{5} (-1)^i$

42. $\displaystyle\sum_{k=0}^{5} (-1)^{k+1}$

43. $\displaystyle\sum_{k=1}^{8} (-1)^{k+1} 3k$

44. $\displaystyle\sum_{k=0}^{7} (-1)^k 4^{k+1}$

45. $\displaystyle\sum_{k=0}^{6} \frac{2}{k^2 + 1}$

46. $\displaystyle\sum_{i=1}^{10} i(i + 1)$

47. $\displaystyle\sum_{k=0}^{5} (k^2 - 2k + 3)$

48. $\displaystyle\sum_{k=1}^{10} \frac{1}{k(k + 1)}$

49. $\displaystyle\sum_{i=0}^{10} \frac{2^i}{2^i + 1}$

50. $\displaystyle\sum_{k=0}^{3} (-2)^{2k}$

Write sigma notation.

51. $5 + 10 + 15 + 20 + 25 + \cdots$

52. $7 + 14 + 21 + 28 + 35 + \cdots$

53. $2 - 4 + 8 - 16 + 32 - 64$

54. $3 + 6 + 9 + 12 + 15$

55. $-\frac{1}{2} + \frac{2}{3} - \frac{3}{4} + \frac{4}{5} - \frac{5}{6} + \frac{6}{7}$

56. $\frac{1}{1^2} + \frac{1}{2^2} + \frac{1}{3^2} + \frac{1}{4^2} + \frac{1}{5^2}$

57. $4 - 9 + 16 - 25 + \cdots + (-1)^n n^2$

58. $9 - 16 + 25 + \cdots + (-1)^{n+1} n^2$

59. $\frac{1}{1 \cdot 2} + \frac{1}{2 \cdot 3} + \frac{1}{3 \cdot 4} + \frac{1}{4 \cdot 5} + \cdots$

60. $\frac{1}{1 \cdot 2^2} + \frac{1}{2 \cdot 3^2} + \frac{1}{3 \cdot 4^2} + \frac{1}{4 \cdot 5^2} + \cdots$

Find the first 4 terms of the recursively defined sequence.

61. $a_1 = 4, \quad a_{k+1} = 1 + \frac{1}{a_k}$

62. $a_1 = 256, \quad a_{k+1} = \sqrt{a_k}$

63. $a_1 = 6561, \quad a_{k+1} = (-1)^k \sqrt{a_k}$

64. $a_1 = e^Q, \quad a_{k+1} = \ln a_k$

65. $a_1 = 2, \quad a_2 = 3, \quad a_{k+1} = a_k + a_{k-1}$

66. $a_1 = -10, \quad a_2 = 8, \quad a_{k+1} = a_k - a_{k-1}$

Use a grapher to construct a table of values and a graph for the first 10 terms of the sequence.

67. $a_n = \left(1 + \frac{1}{n}\right)^n$

68. $a_n = \sqrt{n + 1} - \sqrt{n}$

69. $a_1 = 2, \quad a_{k+1} = \sqrt{1 + \sqrt{a_k}}$

70. $a_1 = 2, \quad a_{k+1} = \frac{1}{2}\left(a_k + \frac{2}{a_k}\right)$

71. *Compound Interest.* Suppose that $1000 is invested at 6.2%, compounded annually. The value of the investment after n years is given by the sequence model

$$a_n = \$1000(1.062)^n, \quad n = 1, 2, 3, \ldots.$$

   a) Find the first 10 terms of the sequence.
   b) Find the amount of the investment after 20 yr.

72. *Salvage Value.* The value of an office machine is $5200. Its salvage value each year is 75% of its value the year before. Give a sequence that lists the salvage value of the machine for each year of a 10-yr period.

73. *Bacteria Growth.* A single cell of bacteria divides into two every 15 min. Suppose that the same rate of division is maintained for 4 hr. Give a sequence that lists the number of cells after successive 15-min periods.

74. *Salary Sequence.* Torrey is paid $6.30 per hour for working at Red Freight Limited. Each year he receives a $0.40 hourly raise. Give a sequence that lists Torrey's hourly salary over a 10-yr period.

75. *Driver Fatalities by Age.*   The number of licensed drivers per 100,000 who died in motor vehicle accidents at age n can be estimated by the sequence model

$$a_n = 0.018n^2 - 1.7n + 48.7, \quad n = 15, 16, 17, \ldots, 90.$$

a) Make a table or list of the first 10 terms of the sequence.

b) Construct a graph of the sequence.

c) Find the number of fatalities per 100,000 of those drivers whose age is 16 yr, 23 yr, 50 yr, 75 yr, and 85 yr.

76. *Household Discretionary Income by Age.*   The following table contains factual data relating household discretionary income to age.

AGE, x	HOUSEHOLD DISCRETIONARY INCOME, y
25	$10,223
34	15,607
54	21,173
64	18,257
70	13,747

Source: U.S. Bureau of the Census

a) Use a grapher to fit a quadratic sequence regression function

$$a_n = an^2 + bn + c$$

to the data.

b) Find the household discretionary income of people whose age is 16 yr, 20 yr, 40 yr, 60 yr, and 75 yr.

77. *Fibonacci Sequence: Rabbit Population Growth.*   One of the most famous recursively defined sequences is the *Fibonacci sequence*. In 1202, the Italian mathematician Leonardo Fibonacci (also called Leonardo da Pisa) proposed the following model for rabbit population growth. Start with one pair of rabbits, one female and one male. These rabbits mature and produce a new pair, again one female and one male. The first pair lives until each pair can reproduce, and then each produces a new pair. This pattern continues. The population of mature rabbits can be modeled by the following recursively defined sequence:

$$a_1 = 1, \quad a_2 = 1, \quad a_{k+1} = a_k + a_{k-1}, \quad \text{for } k \geq 2,$$

where a_k is the total number of pairs of rabbits after $k - 2$ reproductions. Find the first 7 terms of the Fibonacci sequence.

Discussion and Writing

78. a) Find the first few terms of the sequence $a_n = n^2 - n + 41$ and describe the pattern you observe.

b) Does the pattern you found in part (a) hold for all choices of n? Why or why not?

79. The Fibonacci sequence has intrigued mathematicians for centuries. In fact, a journal called the *Fibonacci Quarterly* is devoted to publishing the results pertaining to such sequences. Do some research on the connection of the Fibonacci sequence to the idea of the "Golden Section."

Skill Maintenance

Solve.

80. $3x - 2y = 3,$
$2x + 3y = -11$

81. *Research and Development.*   Sun Microsystems, Inc., spent a total of $1.19 billion on original research and development and on the purchase of in-process research and development in 1998. The amount spent on original research and development was $0.837 billion more than the amount spent on purchased research and development. (*Source*: Sun Microsystems, Inc., 1998 Annual Report) Find the amount spent on each category.

Find the center and the radius of the circle with the given equation.

82. $x^2 + y^2 - 6x + 4y = 3$

83. $x^2 + y^2 + 5x - 8y = 2$

Synthesis

Find the first 5 terms of the sequence, and then find S_5.

84. $a_n = \dfrac{1}{2^n} \log 1000^n$

85. $a_n = i^n, \quad i = \sqrt{-1}$

86. $a_n = \ln (1 \cdot 2 \cdot 3 \cdots n)$

For each sequence, find a formula for S_n.

87. $a_n = \ln n$

88. $a_n = \dfrac{1}{n} - \dfrac{1}{n + 1}$

Arithmetic Sequences and Series

7.2

- *For any arithmetic sequence, find the nth term when n is given and n when the nth term is given, and given two terms, find the common difference and construct the sequence.*
- *Find the sum of the first n terms of an arithmetic sequence.*

A sequence in which each term after the first is found by adding the same number to the preceding term is an **arithmetic sequence.**

Arithmetic Sequences

The sequence 2, 5, 8, 11, 14, 17, ... is arithmetic because adding 3 to any term produces the next term. In other words, the difference between any term and the preceding one is 3. Arithmetic sequences are also called *arithmetic progressions.*

> ### Arithmetic Sequence
>
> A sequence is **arithmetic** if there exists a number d, called the **common difference,** such that $a_{n+1} = a_n + d$ for any integer $n \geq 1$.

EXAMPLE 1   For each of the following arithmetic sequences, identify the first term, a_1, and the common difference, d.

a) 4, 9, 14, 19, 24, ...
b) 34, 27, 20, 13, 6, −1, −8, ...
c) 2, $2\frac{1}{2}$, 3, $3\frac{1}{2}$, 4, $4\frac{1}{2}$, ...

Solution   The first term, a_1, is the first term listed. To find the common difference, d, we choose any term beyond the first and subtract the preceding term from it.

SEQUENCE	FIRST TERM, a_1	COMMON DIFFERENCE, d
a) 4, 9, 14, 19, 24, ...	4	5 $(9 - 4 = 5)$
b) 34, 27, 20, 13, 6, −1, −8, ...	34	−7 $(27 - 34 = -7)$
c) 2, $2\frac{1}{2}$, 3, $3\frac{1}{2}$, 4, $4\frac{1}{2}$, ...	2	$\frac{1}{2}$ $\left(2\frac{1}{2} - 2 = \frac{1}{2}\right)$

We obtained the common difference by subtracting a_1 from a_2. Had we subtracted a_2 from a_3 or a_3 from a_4, we would have obtained the same values for d. Thus we can check by adding d to each term in a sequence to see if we progress correctly to the next term.

CHECK:

a) $4 + 5 = 9$, $9 + 5 = 14$, $14 + 5 = 19$, $19 + 5 = 24$
b) $34 + (-7) = 27$, $27 + (-7) = 20$, $20 + (-7) = 13$,
    $13 + (-7) = 6$, $6 + (-7) = -1$, $-1 + (-7) = -8$

STUDY   TIP

The best way to prepare for a final exam is to do so over a period of at least two weeks. First review each chapter, studying the formulas, theorems, properties, and procedures in the sections in the chapter Summary and Review. Then take each of your chapter tests again. If you miss any questions, spend extra time reviewing the corresponding topics. Watch the videotapes that accompany the text or use the InterAct Math Tutorial Software. Also consider participating in a study group or attending a tutoring or review session.

c) $2 + \frac{1}{2} = 2\frac{1}{2}$, $2\frac{1}{2} + \frac{1}{2} = 3$, $3 + \frac{1}{2} = 3\frac{1}{2}$, $3\frac{1}{2} + \frac{1}{2} = 4$, $4 + \frac{1}{2} = 4\frac{1}{2}$

To find a formula for the general, or nth, term of any arithmetic sequence, we denote the common difference by d, write out the first few terms, and look for a pattern:

a_1,

$a_2 = a_1 + d$,

$a_3 = a_2 + d = (a_1 + d) + d = a_1 + 2d$ Substituting for a_2

$a_4 = a_3 + d = (a_1 + 2d) + d = a_1 + 3d$ Substituting for a_3

Note that the coefficient of d in each case is 1 less than the subscript.

Generalizing, we obtain the following formula.

nth Term of an Arithmetic Sequence

The **nth term** of an arithmetic sequence is given by
$a_n = a_1 + (n - 1)d$, for any integer $n \geq 1$.

EXAMPLE 2 Find the 14th term of the arithmetic sequence 4, 7, 10, 13, ...

Solution We first note that $a_1 = 4$, $d = 3$, and $n = 14$. Then using the formula for the nth term, we obtain

$$a_n = a_1 + (n - 1)d$$
$$a_{14} = 4 + (14 - 1) \cdot 3 = 4 + 13 \cdot 3 = 4 + 39 = 43.$$

The 14th term is 43.

EXAMPLE 3 In the sequence of Example 2, which term is 301? That is, find n if $a_n = 301$.

Solution We substitute 301 for a_n, 4 for a_1, and 3 for d in the formula for the nth term and solve for n:

$$a_n = a_1 + (n - 1)d$$
$$301 = 4 + (n - 1) \cdot 3 \qquad \text{Substituting}$$
$$301 = 4 + 3n - 3$$
$$301 = 3n + 1$$
$$300 = 3n \qquad \text{Solving for } n$$
$$100 = n.$$

The term 301 is the 100th term of the sequence.

Given two terms and their places in an arithmetic sequence, we can construct the sequence.

EXAMPLE 4 The 3rd term of an arithmetic sequence is 8, and the 16th term is 47. Find a_1 and d and construct the sequence.

Solution We know that $a_3 = 8$ and $a_{16} = 47$. Thus we would have to add d 13 times to get 47 from 8. That is,

$$8 + 13d = 47. \qquad a_3 \text{ and } a_{16} \text{ are 13 terms apart.}$$

Solving $8 + 13d = 47$, we obtain

$$13d = 39$$
$$d = 3.$$

Since $a_3 = 8$, we subtract d twice to get a_1. Thus,

$$a_1 = 8 - 2 \cdot 3 = 2. \qquad a_1 \text{ and } a_3 \text{ are 2 terms apart.}$$

The sequence is 2, 5, 8, 11, … . Note that we could also subtract d 15 times from a_{16} in order to find a_1. ▬

In general, d should be subtracted $n - 1$ times from a_n in order to find a_1.

Interactive Discovery

Graph the first 10 terms of each arithmetic sequence. What pattern do you observe?

$$a_n = 2 + (n - 1)(4),$$
$$a_n = -5 + (n - 1)(1.2),$$
$$a_n = -4 + (n - 1)(-3),$$
$$a_n = 3 + (n - 1)\left(-\frac{5}{2}\right)$$

The pattern shown above holds in general. *The graph of an arithmetic sequence is a set of points that lie on the graph of a linear function.*

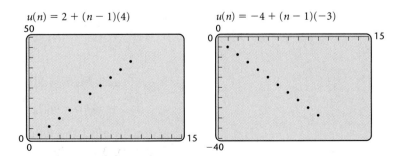

Sum of the First n Terms of an Arithmetic Sequence

Consider the arithmetic sequence

$$3, 5, 7, 9, \ldots .$$

When we add the first 4 terms of the sequence, we get S_4, which is

$$3 + 5 + 7 + 9, \quad \text{or} \quad 24.$$

This sum is called an **arithmetic series.** To find a formula for the sum of the first n terms, S_n, of an arithmetic sequence, we first denote an arithmetic sequence, as follows:

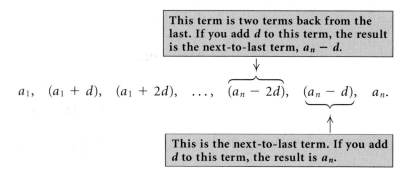

This term is two terms back from the last. If you add d to this term, the result is the next-to-last term, $a_n - d$.

$$a_1, \quad (a_1 + d), \quad (a_1 + 2d), \quad \ldots, \quad (a_n - 2d), \quad (a_n - d), \quad a_n.$$

This is the next-to-last term. If you add d to this term, the result is a_n.

Then S_n is given by

$$S_n = a_1 + (a_1 + d) + (a_1 + 2d) + \cdots + (a_n - 2d)$$
$$+ (a_n - d) + a_n. \tag{1}$$

Reversing the order of the addition gives us

$$S_n = a_n + (a_n - d) + (a_n - 2d) + \cdots + (a_1 + 2d)$$
$$+ (a_1 + d) + a_1. \tag{2}$$

If we add corresponding terms of each side of equations (1) and (2), we get

$$2S_n = [a_1 + a_n] + [(a_1 + d) + (a_n - d)] + [(a_1 + 2d) + (a_n - 2d)]$$
$$+ \cdots + [(a_n - 2d) + (a_1 + 2d)]$$
$$+ [(a_n - d) + (a_1 + d)] + [a_n + a_1]. \qquad \text{There are } n \text{ pairs of square brackets.}$$

This simplifies to

$$2S_n = [a_1 + a_n] + [a_1 + a_n] + [a_1 + a_n] + \cdots + [a_n + a_1]$$
$$+ [a_n + a_1] + [a_n + a_1].$$

Since $a_1 + a_n$ is being added n times, it follows that

$$2S_n = n(a_1 + a_n),$$

from which we get the following formula.

Sum of the First n Terms

The sum of the first n terms of an arithmetic sequence is given by

$$S_n = \frac{n}{2}(a_1 + a_n).$$

EXAMPLE 5   Find the sum of the first 100 natural numbers.

Solution   The sum is

$$1 + 2 + 3 + \cdots + 99 + 100.$$

This is the sum of the first 100 terms of the arithmetic sequence for which

$$a_1 = 1, \qquad a_n = 100, \quad \text{and} \quad n = 100.$$

Thus substituting into the formula

$$S_n = \frac{n}{2}(a_1 + a_n),$$

we get

$$S_{100} = \frac{100}{2}(1 + 100) = 50(101) = 5050.$$

The sum of the first 100 natural numbers is 5050.

EXAMPLE 6   Find the sum of the first 15 terms of the arithmetic sequence 4, 7, 10, 13, ...

Solution   Note that $a_1 = 4$, $d = 3$, and $n = 15$. Before using the formula

$$S_n = \frac{n}{2}(a_1 + a_n),$$

we find the last term, a_{15}:

$$a_{15} = 4 + (15 - 1)3 \qquad \textbf{Substituting into the formula } a_n = a_1 + (n - 1)d$$
$$= 4 + 14 \cdot 3 = 46.$$

Thus,

$$S_{15} = \frac{15}{2}(4 + 46) = \frac{15}{2}(50) = 375.$$

The sum of the first 15 terms is 375.

We can also perform these calculations on a grapher.

```
4+14*3
                    46
(15/2)(4+46)
                   375
```

EXAMPLE 7   Find the sum: $\sum_{k=1}^{130} (4k + 5)$.

Solution   It is helpful to first write out a few terms:

$$9 + 13 + 17 + \cdots.$$

It appears that this is an arithmetic series coming from an arithmetic sequence with $a_1 = 9$, $d = 4$, and $n = 130$. Before using the formula

$$S_n = \frac{n}{2}(a_1 + a_n),$$

we find the last term, a_{130}:

$$a_{130} = 4 \cdot 130 + 5 \qquad \text{The } k\text{th term is } 4k + 5.$$
$$= 520 + 5$$
$$= 525.$$

Thus,

$$S_{130} = \frac{130}{2}(9 + 525) \qquad \text{Substituting into } S_n = \frac{n}{2}(a_1 + a_n)$$
$$= 34{,}710.$$

This sum can also be found on a grapher. It is not necessary to have the grapher set in SEQUENCE mode to do this.   ▬

```
sum(seq(4X+5,X,1,130))
                  34710
```

Applications

The translation of some applications and problem-solving situations may involve arithmetic sequences or series. We consider some examples.

EXAMPLE 8   *Hourly Wages.* Gloria accepts a job, starting with an hourly wage of $14.25, and is promised a raise of 15¢ per hour every 2 months for 5 yr. At the end of 5 yr, what will Gloria's hourly wage be?

Solution   It helps to first write down the hourly wage for several 2-month time periods:

Beginning: $14.25,
After two months: 14.40,
After four months: 14.55,
and so on.

What appears is a sequence of numbers: 14.25, 14.40, 14.55, ... This sequence is arithmetic, because adding $0.15 each time gives us the next term.

We want to find the last term of an arithmetic sequence, so we use the formula $a_n = a_1 + (n - 1)d$. We know that $a_1 = 14.25$ and $d = 0.15$, but what is n? That is, how many terms are in the sequence? Each year there are 6 raises, since Gloria gets a raise every 2 months. There are 5 yr, so the total number of raises will be $5 \cdot 6$, or 30. Thus there will be 31 terms: the original wage and 30 increased rates.

Substituting in the formula $a_n = a_1 + (n - 1)d$ gives us

$$a_{31} = 14.25 + (31 - 1) \cdot 0.15$$
$$= \$18.75.$$

Thus, at the end of 5 yr, Gloria's hourly wage will be $18.75.   —

The calculations in Example 8 could be done in a number of ways. There is often a variety of ways in which a problem can be solved. In this chapter, we will concentrate on the use of sequences and series and their related formulas.

EXAMPLE 9 *Total in a Stack.* A stack of telephone poles has 30 poles in the bottom row. There are 29 poles in the second row, 28 in the next row, and so on. How many poles are in the stack if there are 5 poles in the top row?

Solution   A picture will help in this case. The following figure shows the ends of the poles and the way in which they stack.

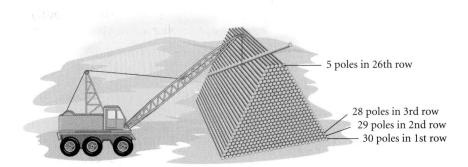

5 poles in 26th row

28 poles in 3rd row
29 poles in 2nd row
30 poles in 1st row

Since the number of poles goes from 30 in a row up to 5 in the top row, there must be 26 rows. We want the sum

$$30 + 29 + 28 + \cdots + 5.$$

Thus we have an arithmetic series. We use the formula

$$S_n = \frac{n}{2}(a_1 + a_n),$$

with $n = 26$, $a_1 = 30$, and $a_{26} = 5$.

Substituting, we get

$$S_{26} = \frac{26}{2}(30 + 5) = 455.$$

There are 455 poles in the stack.   —

Exercise Set 7.2

Find the first term and the common difference.

1. 3, 8, 13, 18, ...

2. $1.08, $1.16, $1.24, $1.32, ...

3. 9, 5, 1, −3, ...

4. −8, −5, −2, 1, 4, ...

5. $\frac{3}{2}, \frac{9}{4}, 3, \frac{15}{4}, ...$

6. $\frac{3}{5}, \frac{1}{10}, -\frac{2}{5}, ...$

7. $316, $313, $310, $307, ...

8. Find the 11th term of the arithmetic sequence 0.07, 0.12, 0.17, ...

9. Find the 12th term of the arithmetic sequence 2, 6, 10, ...

10. Find the 17th term of the arithmetic sequence 7, 4, 1, ...

11. Find the 14th term of the arithmetic sequence $3, \frac{7}{3}, \frac{5}{3}, ...$.

12. Find the 13th term of the arithmetic sequence $1200, $964.32, $728.64, ...

13. Find the 10th term of the arithmetic sequence $2345.78, $2967.54, $3589.30, ...

14. In the sequence of Exercise 9, what term is the number 106?

15. In the sequence of Exercise 8, what term is the number 1.67?

16. In the sequence of Exercise 10, what term is −296?

17. In the sequence of Exercise 11, what term is −27?

18. Find a_{20} when $a_1 = 14$ and $d = -3$.

19. Find a_1 when $d = 4$ and $a_8 = 33$.

20. Find d when $a_1 = 8$ and $a_{11} = 26$.

21. Find n when $a_1 = 25$, $d = -14$, and $a_n = -507$.

22. In an arithmetic sequence, $a_{17} = -40$ and $a_{28} = -73$. Find a_1 and d. Write the first 5 terms of the sequence.

23. In an arithmetic sequence, $a_{17} = \frac{25}{3}$ and $a_{32} = \frac{95}{6}$. Find a_1 and d. Write the first 5 terms of the sequence.

24. Find the sum of the first 14 terms of the series $11 + 7 + 3 + \cdots$.

25. Find the sum of the first 20 terms of the series $5 + 8 + 11 + 14 + \cdots$.

26. Find the sum of the first 300 natural numbers.

27. Find the sum of the first 400 even natural numbers.

28. Find the sum of the odd numbers 1 to 199, inclusive.

29. Find the sum of the multiples of 7 from 7 to 98, inclusive.

30. Find the sum of all multiples of 4 that are between 14 and 523.

31. If an arithmetic series has $a_1 = 2$, $d = 5$, and $n = 20$, what is S_n?

32. If an arithmetic series has $a_1 = 7$, $d = -3$, and $n = 32$, what is S_n?

Find the sum.

33. $\sum_{k=1}^{40} (2k + 3)$

34. $\sum_{k=5}^{20} 8k$

35. $\sum_{k=0}^{19} \frac{k - 3}{4}$

36. $\sum_{k=2}^{50} (2000 - 3k)$

37. $\sum_{k=12}^{57} \frac{7 - 4k}{13}$

38. $\sum_{k=101}^{200} (1.14k - 2.8) - \sum_{k=1}^{5} \left(\frac{k + 4}{10} \right)$

39. *Pole Stacking.* How many poles will be in a stack of telephone poles if there are 50 in the first layer, 49 in the second, and so on, with 6 in the top layer?

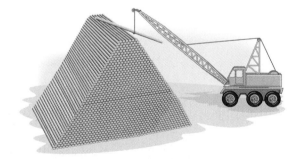

40. *Investment Return.* Max is an investment counselor. He sets up an investment situation for a client that will return $5000 the first year, $6125 the

second year, $7250 the third year, and so on, for 25 yr. How much is received from the investment altogether?

41. *Garden Plantings.* A gardener is making a planting in the shape of a trapezoid. It will have 35 plants in the front row, 31 in the second row, 27 in the third row, and so on. If the pattern is consistent, how many plants will there be in the last row? How many plants are there altogether?

42. *Band Formation.* A formation of a marching band has 14 marchers in the front row, 16 in the second row, 18 in the third row, and so on, for 25 rows. How many marchers are in the last row? How many marchers are there altogether?

43. *Total Savings.* If 10¢ is saved on October 1, 20¢ is saved on October 2, 30¢ on October 3, and so on, how much is saved during the 31 days of October?

44. *Parachutist Free Fall.* It has been found that when a parachutist jumps from an airplane, the distances, in feet, which the parachutist falls in each successive second before pulling the ripcord to release the parachute are as follows:

$$16, 48, 80, 112, 144, \ldots.$$

Is this sequence arithmetic? What is the common difference? What is the total distance fallen after 10 sec?

45. *Theater Seating.* Theaters are often built with more seats per row as the rows move toward the back. Suppose that the first balcony of a theater has 28 seats in the first row, 32 in the second, 36 in the third, and so on, for 20 rows. How many seats are in the first balcony altogether?

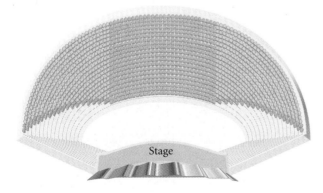

Stage

46. *Small Group Interaction.* In a social science study, Stephan found the following data regarding an

interaction measurement r_n for groups of size n.

n	r_n
3	0.5908
4	0.6080
5	0.6252
6	0.6424
7	0.6596
8	0.6768
9	0.6940
10	0.7112

Source: American Sociological Review, **17** (1952)

Is this sequence arithmetic? What is the common difference?

47. *Raw Material Production.* In an industrial situation, it took 3 units of raw materials to produce 1 unit of a product. The raw material needs thus formed the sequence

$$3, 6, 9, \ldots, 3n, \ldots.$$

Is this sequence arithmetic? What is the common difference?

Discussion and Writing

48. The sum of the first n terms of an arithmetic sequence can be given by

$$S_n = \frac{n}{2}[2a_1 + (n - 1)d].$$

Compare this formula to

$$S_n = \frac{n}{2}(a_1 + a_n).$$

Discuss the reasons for the use of one formula over the other.

49. It is said that as a young child, the mathematician Karl F. Gauss (1777–1855) was able to compute the sum $1 + 2 + 3 + \cdots + 100$ very quickly in his head to the amazement of a teacher. Explain how Gauss might have done this had he possessed some knowledge of arithmetic sequences and series. Then give a formula for the sum of the first n natural numbers.

Skill Maintenance

Solve.

50. $7x - 2y = 4,$
     $x + 3y = 17$

51. $2x + y + 3z = 12,$
     $x - 3y + 2z = -1,$
     $5x + 2y - 4z = -4$

52. Find the vertices and the foci of the ellipse with the equation $9x^2 + 16y^2 = 144$.

53. Find an equation of the ellipse with vertices $(0, -5)$ and $(0, 5)$ and minor axis of length 4.

Synthesis

54. Find three numbers in an arithmetic sequence such that the sum of the first and third is 10 and the product of the first and second is 15.

55. Find a formula for the sum of the first n odd natural numbers:

$$1 + 3 + 5 + \cdots + (2n - 1).$$

56. Find the first 10 terms of the arithmetic sequence for which

$$a_1 = \$8760 \quad \text{and} \quad d = -\$798.23.$$

Then find the sum of the first 10 terms.

57. Find the first term and the common difference for the arithmetic sequence for which

$$a_2 = 40 - 3q \quad \text{and} \quad a_4 = 10p + q.$$

58. The zeros of this polynomial function form an arithmetic sequence. Find them.

$$f(x) = x^4 + 4x^3 - 84x^2 - 176x + 640$$

If p, m, and q form an arithmetic sequence, it can be shown that $m = (p + q)/2$. (See Exercise 65.) The

number m is the **arithmetic mean,** or **average,** of p and q. Given two numbers p and q, if we find k other numbers $m_1, m_2, \ldots, m_k$ such that

$$p, m_1, m_2, \ldots, m_k, q$$

forms an arithmetic sequence, we say that we have "inserted k arithmetic means between p and q."

59. Insert three arithmetic means between 4 and 12.

60. Insert four arithmetic means between 4 and 13.

61. Insert three arithmetic means between -3 and 5.

62. Insert ten arithmetic means between 27 and 300.

63. Insert enough arithmetic means between 1 and 50 so that the sum of the resulting series will be 459.

64. *Straight-Line Depreciation.* A company buys an office machine for $5200 on January 1 of a given year. The machine is expected to last for 8 yr, at the end of which time its **trade-in value,** or **salvage value** will be $1100. If the company's accountant figures the decline in value to be the same each year, then its **book values,** or **salvage values,** after t years, $0 \le t \le 8$, form an arithmetic sequence given by

$$a_t = C - t\left(\frac{C - S}{N}\right),$$

where C is the original cost of the item ($5200), N is the number of years of expected life (8), and S is the salvage value ($1100).

a) Find the formula for a_t for the straight-line depreciation of the office machine.

b) Find the salvage value after 0 yr, 1 yr, 2 yr, 3 yr, 4 yr, 7 yr, and 8 yr.

65. Prove that if p, m, and q form an arithmetic sequence, then

$$m = \frac{p + q}{2}.$$

Geometric Sequences and Series

7.3

- *Identify the common ratio of a geometric sequence, and find a given term and the sum of the first n terms.*
- *Find the sum of an infinite geometric series, if it exists.*

A sequence in which each term after the first is found by multiplying the preceding term by the same number is a **geometric sequence.**

Geometric Sequences

Consider this sequence:

$$2, 6, 18, 54, 162, \ldots.$$

Note that multiplying each term by 3 produces the next term. We call the number 3 the **common ratio** because it can be found by dividing any term by the preceding term. A geometric sequence is also called a *geometric progression*.

Geometric Sequence

A sequence is **geometric** if there is a number r, called the **common ratio,** such that

$$\frac{a_{n+1}}{a_n} = r, \quad \text{or} \quad a_{n+1} = a_n r, \quad \text{for any integer } n \geq 1.$$

EXAMPLE 1   For each of the following geometric sequences, identify the common ratio.

a) 3, 6, 12, 24, 48, ...

b) $1, -\dfrac{1}{2}, \dfrac{1}{4}, -\dfrac{1}{8}, \ldots$

c) $5200, $3900, $2925, $2193.75, ...

d) $1000, $1060, $1123.60, ...

Solution

SEQUENCE	COMMON RATIO
a) 3, 6, 12, 24, 48, ...	2 $\left(\frac{6}{3} = 2, \frac{12}{6} = 2, \text{ and so on} \right)$
b) $1, -\dfrac{1}{2}, \dfrac{1}{4}, -\dfrac{1}{8}, \ldots$	$-\dfrac{1}{2}$ $\left(\dfrac{-\frac{1}{2}}{1} = -\dfrac{1}{2}, \dfrac{\frac{1}{4}}{-\frac{1}{2}} = -\dfrac{1}{2}, \text{ and so on} \right)$
c) $5200, $3900, $2925, $2193.75, ...	0.75 $\left(\dfrac{\$3900}{\$5200} = 0.75, \dfrac{\$2925}{\$3900} = 0.75, \text{ and so on} \right)$
d) $1000, $1060, $1123.60, ...	1.06 $\left(\dfrac{\$1060}{\$1000} = 1.06, \dfrac{\$1123.60}{\$1060} = 1.06, \text{ and so on} \right)$

We now find a formula for the general, or *n*th, term of a geometric sequence. Let a_1 be the first term and r the common ratio. The first few terms are as follows:

$a_1,$

$a_2 = a_1 r,$

$a_3 = a_2 r = (a_1 r)r = a_1 r^2,$ Substituting $a_1 r$ for a_2

$a_4 = a_3 r = (a_1 r^2)r = a_1 r^3.$ Substituting $a_1 r^2$ for a_3

Note that the exponent is 1 less than the subscript.

Generalizing, we obtain the following.

---

nth Term of a Geometric Sequence
The **nth term** of a geometric sequence is given by

$$a_n = a_1 r^{n-1}, \quad \text{for any integer } n \geq 1.$$

EXAMPLE 2 Find the 7th term of the geometric sequence 4, 20, 100, ...

Solution We first note that

$$a_1 = 4 \quad \text{and} \quad n = 7.$$

To find the common ratio, we can divide any term (other than the first) by the preceding term. Since the second term is 20 and the first is 4, we get

$$r = \frac{20}{4}, \quad \text{or} \quad 5.$$

Then using the formula $a_n = a_1 r^{n-1}$, we have

$$a_7 = 4 \cdot 5^{7-1} = 4 \cdot 5^6 = 4 \cdot 15{,}625 = 62{,}500.$$

Thus the 7th term is 62,500.

EXAMPLE 3 Find the 10th term of the geometric sequence 64, −32, 16, −8, ...

Solution We first note that

$$a_1 = 64, \qquad n = 10, \quad \text{and} \quad r = \frac{-32}{64}, \quad \text{or} \quad -\frac{1}{2}.$$

Then using the formula $a_n = a_1 r^{n-1}$, we have

$$a_{10} = 64 \cdot \left(-\frac{1}{2}\right)^{10-1} = 64 \cdot \left(-\frac{1}{2}\right)^9 = 2^6 \cdot \left(-\frac{1}{2^9}\right) = -\frac{1}{8}.$$

Thus the 10th term is $-\frac{1}{8}$.

Interactive Discovery

Graph the first 7 terms of each geometric sequence. What pattern do you observe?

$$a_n = 5 \cdot 3^{n-1}, \qquad a_n = 0.2 \cdot (2.25)^{n-1},$$
$$a_n = \frac{1}{3} \cdot \left(\frac{3}{5}\right)^{n-1}, \qquad a_n = 2(0.95)^{n-1}$$

The pattern shown above holds in general. *The graph of a geometric sequence is a set of points that lie on the graph of an exponential function.*

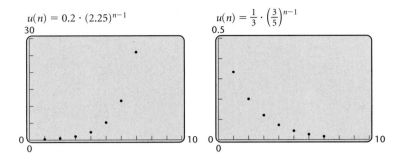

Sum of the First *n* Terms of a Geometric Sequence

Next, we develop a formula for the sum S_n of the first n terms of a geometric sequence:

$$a_1, \; a_1 r, \; a_1 r^2, \; a_1 r^3, \; \ldots, \; a_1 r^{n-1}, \; \ldots .$$

The associated **geometric series** is given by

$$S_n = a_1 + a_1 r + a_1 r^2 + a_1 r^3 + \cdots + a_1 r^{n-1}. \tag{1}$$

We want to find a formula for this sum. If we multiply on both sides of equation (1) by r, we have

$$r S_n = a_1 r + a_1 r^2 + a_1 r^3 + a_1 r^4 + \cdots + a_1 r^n. \tag{2}$$

Subtracting equation (2) from equation (1), we see that the differences of the red terms are 0, leaving

$$S_n - r S_n = a_1 - a_1 r^n,$$

or

$$S_n(1 - r) = a_1(1 - r^n). \qquad \textbf{Factoring}$$

Dividing on both sides by $1 - r$ gives us the following formula.

Sum of the First n Terms

The sum of the first n terms of a geometric sequence is given by

$$S_n = \frac{a_1(1 - r^n)}{1 - r}, \quad \text{for any } r \neq 1.$$

EXAMPLE 4   Find the sum of the first 7 terms of the geometric sequence 3, 15, 75, 375, ...

Solution   We first note that

$$a_1 = 3, \qquad n = 7, \quad \text{and} \quad r = \tfrac{15}{3}, \text{ or } 5.$$

Then using the formula

$$S_n = \frac{a_1(1 - r^n)}{1 - r},$$

we have

$$S_7 = \frac{3(1 - 5^7)}{1 - 5}$$

$$= \frac{3(1 - 78{,}125)}{-4}$$

$$= 58{,}593.$$

Thus the sum of the first 7 terms is 58,593.

EXAMPLE 5　Find the sum: $\displaystyle\sum_{k=1}^{11} (0.3)^k$.

Solution　This is a geometric series with $a_1 = 0.3$, $r = 0.3$, and $n = 11$. Thus,

$$S_{11} = \frac{0.3(1 - 0.3^{11})}{1 - 0.3}$$

$$\approx 0.42857.$$

```
sum(seq(.3^X,X,1,11))
            .4285706694
```

We can also find this sum using a grapher set in either FUNCTION mode or SEQUENCE mode.

Infinite Geometric Series

The sum of the terms of an infinite geometric sequence is an **infinite geometric series.** For some geometric sequences, S_n gets close to a specific number as n gets large. For example, consider the infinite series

$$\frac{1}{2} + \frac{1}{4} + \frac{1}{8} + \frac{1}{16} + \cdots + \frac{1}{2^n} + \cdots.$$

We examine some partial sums. Note that each of the partial sums is less than 1, but S_n gets very close to 1 as n gets large. We say that 1 is the **limit** of S_n and also that 1 is the **sum of the infinite geometric sequence.** The sum of an infinite geometric sequence is denoted S_∞. In this case, $S_\infty = 1$.

n	S_n
1	0.5
5	0.96875
10	0.9990234375
20	0.9999990463
30	0.9999999991

```
sum(seq(1/2^X,X,1,20))
            .9999990463
sum(seq(1/2^X,X,1,30))
            .9999999991
```

Some infinite sequences do not have sums. Consider the infinite geometric series

$$2 + 4 + 8 + 16 + \cdots + 2^n + \cdots.$$

We again examine some partial sums. Note that as n gets large, S_n gets large without bound. This sequence does not have a sum.

n	S_n
1	2
5	62
10	2,046
20	2,097,150
30	2,147,483,646

```
sum(seq(2^X,X,1,20))
                   2097150
sum(seq(2^X,X,1,30))
                2147483646
```

It can be shown (but we will not do so here) that the sum of the terms of an infinite geometric series exists if and only if $|r| < 1$ (that is, the absolute value of the common ratio is less than 1).

To find a formula for the sum of an infinite geometric series, we first consider the sum of the first n terms:

$$S_n = \frac{a_1(1 - r^n)}{1 - r} = \frac{a_1 - a_1 r^n}{1 - r}. \qquad \textbf{Using the distributive law}$$

For $|r| < 1$, values of r^n get close to 0 as n gets large. (See the table at left, where $u(n) = 0.45^n$, for instance.) As r^n gets close to 0, so does $a_1 r^n$. Thus, S_n gets close to $a_1/(1 - r)$.

n	$u(n)$
4	.04101
5	.01845
6	.0083
7	.00374
8	.00168
9	7.6E−4
10	3.4E−4

$n = 10$

Limit or Sum of an Infinite Geometric Series

When $|r| < 1$, the limit or sum of an infinite geometric series is given by

$$S_\infty = \frac{a_1}{1 - r}.$$

EXAMPLE 6 Determine whether each of the following infinite geometric series has a limit. If a limit exists, find it.

a) $1 + 3 + 9 + 27 + \cdots$

b) $-2 + 1 - \frac{1}{2} + \frac{1}{4} - \frac{1}{8} + \cdots$

Solution

a) Here $r = 3$, so $|r| = |3| = 3$. Since $|r| > 1$, the series *does not* have a limit.

b) Here $r = -\frac{1}{2}$, so $|r| = \left|-\frac{1}{2}\right| = \frac{1}{2}$. Since $|r| < 1$, the series *does* have a limit. We find the limit:

$$S_\infty = \frac{a_1}{1-r} = \frac{-2}{1-\left(-\frac{1}{2}\right)} = \frac{-2}{\frac{3}{2}} = -\frac{4}{3}.$$

EXAMPLE 7 Find fractional notation for $0.78787878\ldots$, or $0.\overline{78}$.

Solution We can express this as

$$0.78 + 0.0078 + 0.000078 + \cdots.$$

Then we see that this is an infinite geometric series, where $a_1 = 0.78$ and $r = 0.01$. Since $|r| < 1$, this series has a limit:

$$S_\infty = \frac{a_1}{1-r} = \frac{0.78}{1-0.01} = \frac{0.78}{0.99} = \frac{78}{99}, \quad \text{or} \quad \frac{26}{33}.$$

Thus fractional notation for $0.78787878\ldots$ is $\frac{26}{33}$. You can check this on your calculator.

Applications

The translation of some applications and problem-solving situations may involve geometric sequences or series. Examples 9 and 10 in particular show applications in business and economics.

EXAMPLE 8 *A Daily Doubling Salary.* Suppose someone offered you a job for the month of September (30 days) under the following conditions. You will be paid \$0.01 for the first day, \$0.02 for the second, \$0.04 for the third, and so on, doubling your previous day's salary each day. How much would you earn? (Would you take the job? Make a conjecture before reading further.)

Solution You earn \$0.01 the first day, \$0.01(2) the second day, \$0.01(2)(2) the third day, and so on. The amount earned is the geometric series

$$\$0.01 + \$0.01(2) + \$0.01(2^2) + \$0.01(2^3) + \cdots + \$0.01(2^{29}),$$

where $a_1 = \$0.01$, $r = 2$, and $n = 30$. Using the formula

$$S_n = \frac{a_1(1-r^n)}{1-r},$$

we have

$$S_{30} = \frac{\$0.01(1-2^{30})}{1-2} = \$10,737,418.23.$$

The pay exceeds \$10.7 million for the month. (Most people would probably take the job!)

EXAMPLE 9 *The Amount of an Annuity.* An **annuity** is a sequence of equal payments made at equal time intervals that earn interest. Fixed deposits in a savings account are an example of an annuity. Suppose that to save money to buy a car, Andrea deposits $1000 at the *end* of each of 5 yr in an account that pays 8% interest, compounded annually. The total amount in the account at the end of 5 yr is called the **amount of the annuity.** Find that amount.

Solution   The following time diagram can help visualize the problem. Note that no deposit is made until the end of the first year.

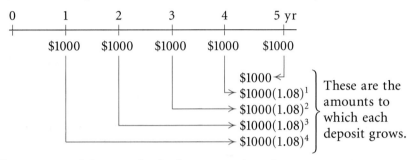

The amount of the annuity is the geometric series

$1000 + \$1000(1.08)^1 + \$1000(1.08)^2 + \$1000(1.08)^3 + \$1000(1.08)^4$,

where $a_1 = \$1000$, $n = 5$, and $r = 1.08$. Using the formula

$$S_n = \frac{a_1(1 - r^n)}{1 - r},$$

we have

$$S_5 = \frac{\$1000(1 - 1.08^5)}{1 - 1.08} \approx \$5866.60.$$

The amount of the annuity is $5866.60.

1000(1−1.08^5)/(1−1.08)
5866.60096

EXAMPLE 10   *The Economic Multiplier.* The finals of the NCAA men's basketball tournament have a significant effect on the economy of the host city. Suppose that 45,000 people visit the city and spend $600 each while there. Then assume that 80% of that money is spent again in the city, and then 80% of that money is spent again, and so on. This is known as the *economic multiplier effect.* Find the total effect on the economy.

Solution   The initial effect is 45,000 · $600, or $27,000,000. The total effect is given by the infinite series

$27,000,000 + \$27,000,000(0.80) + \$27,000,000(0.80^2) + \cdots$.

Using the formula for the sum of an infinite geometric series, we have

$$S_\infty = \frac{a_1}{1 - r} = \frac{27,000,000}{1 - 0.80} = \$135,000,000.$$

Thus the total effect on the economy of the spending from the tournament is $135,000,000.

Exercise Set 7.3

Find the common ratio.

1. 2, 4, 8, 16, ...

2. 18, −6, 2, −$\frac{2}{3}$, ...

3. −1, 1, −1, 1, ...

4. −8, −0.8, −0.08, −0.008, ...

5. $\frac{2}{3}$, −$\frac{4}{3}$, $\frac{8}{3}$, −$\frac{16}{3}$, ...

6. 75, 15, 3, $\frac{3}{5}$, ...

7. 6.275, 0.6275, 0.06275, ...

8. $\frac{1}{x}$, $\frac{1}{x^2}$, $\frac{1}{x^3}$, ...

9. 5, $\dfrac{5a}{2}$, $\dfrac{5a^2}{4}$, $\dfrac{5a^3}{8}$, ...

10. $780, $858, $943.80, $1038.18, ...

Find the indicated term.

11. 2, 4, 8, 16, ...; the 7th term

12. 2, −10, 50, −250, ...; the 9th term

13. 2, 2$\sqrt{3}$, 6, ...; the 9th term

14. 1, −1, 1, −1, ...; the 57th term

15. $\frac{7}{625}$, −$\frac{7}{25}$, ...; the 23rd term

16. $1000, $1060, $1123.60, ...; the 5th term

Find the nth, or general, term.

17. 1, 3, 9, ...

18. 25, 5, 1, ...

19. 1, −1, 1, −1, ...

20. −2, 4, −8, ...

21. $\dfrac{1}{x}$, $\dfrac{1}{x^2}$, $\dfrac{1}{x^3}$, ...

22. 5, $\dfrac{5a}{2}$, $\dfrac{5a^2}{4}$, $\dfrac{5a^3}{8}$, ...

23. Find the sum of the first 7 terms of the geometric series
$$6 + 12 + 24 + \cdots.$$

24. Find the sum of the first 10 terms of the geometric series
$$16 - 8 + 4 - \cdots.$$

25. Find the sum of the first 9 terms of the geometric series
$$\tfrac{1}{18} - \tfrac{1}{6} + \tfrac{1}{2} - \cdots.$$

26. Find the sum of the geometric series
$$-8 + 4 + (-2) + \cdots + \left(-\tfrac{1}{32}\right).$$

Find the sum, if it exists. If you wish, you can also check the results using a grapher.

27. $4 + 2 + 1 + \cdots$

28. $7 + 3 + \frac{9}{7} + \cdots$

29. $25 + 20 + 16 + \cdots$

30. $100 - 10 + 1 - \frac{1}{10} + \cdots$

31. $8 + 40 + 200 + \cdots$

32. $-6 + 3 - \frac{3}{2} + \frac{3}{4} - \cdots$

33. $0.6 + 0.06 + 0.006 + \cdots$

34. $\displaystyle\sum_{k=0}^{10} 3^k$

35. $\displaystyle\sum_{k=1}^{11} 15\left(\frac{2}{3}\right)^k$

36. $\displaystyle\sum_{k=0}^{50} 200(1.08)^k$

37. $\displaystyle\sum_{k=1}^{\infty} \left(\frac{1}{2}\right)^{k-1}$

38. $\displaystyle\sum_{k=1}^{\infty} 2^k$

39. $\displaystyle\sum_{k=1}^{\infty} 12.5^k$

40. $\displaystyle\sum_{k=1}^{\infty} 400(1.0625)^k$

41. $\displaystyle\sum_{k=1}^{\infty} \$500(1.11)^{-k}$

42. $\displaystyle\sum_{k=1}^{\infty} \$1000(1.06)^{-k}$

43. $\displaystyle\sum_{k=1}^{\infty} 16(0.1)^{k-1}$

44. $\displaystyle\sum_{k=1}^{\infty} \frac{8}{3}\left(\frac{1}{2}\right)^{k-1}$

Find fractional notation.

45. $0.131313\ldots$, or $0.\overline{13}$

46. $0.2222\ldots$, or $0.\overline{2}$

47. $8.9999\overline{9}$

48. $6.161\overline{616}$

49. $3.4125\overline{125}$

50. $12.7809\overline{809}$

51. *Bouncing Ping-Pong Ball.* A ping-pong ball is dropped from a height of 16 ft and always rebounds $\frac{1}{4}$ of the distance fallen.

   a) How high does it rebound the 6th time?
   b) Find the total sum of the rebound heights of the ball.

52. *Daily Doubling Salary.* Suppose someone offered you a job for the month of February (28 days) under the following conditions. You will be paid $0.01 the 1st day, $0.02 the 2nd, $0.04 the 3rd, and so on, doubling your previous day's salary each day. How much would you earn altogether?

53. *Bungee Jumping.* A bungee jumper rebounds 60% of the height jumped. A bungee jump is made using a cord that stretches to 200 ft.

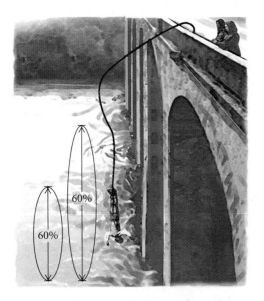

a) After jumping and then rebounding 9 times, how far has a bungee jumper traveled upward (the total rebound distance)?

b) About how far will a jumper have traveled upward (bounced) before coming to rest?

54. *Population Growth.* Gaintown has a present population of 100,000, and the population is increasing by 3% each year.

a) What will the population be in 15 yr?

b) How long will it take for the population to double?

55. *Amount of an Annuity.* To create a college fund, a parent makes a sequence of 18 yearly deposits of $1000 each in a savings account on which interest is compounded annually at 6.2%. Find the amount of the annuity.

56. *Amount of an Annuity.* A sequence of yearly payments of P dollars is invested at the end of each of N years at interest rate i, compounded annually. The total amount in the account, or the amount of the annuity, is V.

a) Show that

$$V = \frac{P[(1 + i)^N - 1]}{i}.$$

b) Suppose that interest is compounded n times per year and deposits are made every compounding period. Show that the formula for V is then given by

$$V = \frac{P\left[\left(1 + \dfrac{i}{n}\right)^{nN} - 1\right]}{i/n}.$$

57. *Loan Repayment.* A family borrows $120,000. The loan is to be repaid in 13 yr at 12% interest, compounded annually. How much will be repaid at the end of 13 yr?

58. *Doubling Paper Folds.* A piece of paper is 0.01 in. thick. It is cut and stacked repeatedly in such a way that its thickness is doubled each time for 20 times. How thick is the result?

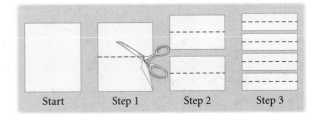

59. *The Economic Multiplier.* The government is making a $13,000,000,000 expenditure for educational improvement. If 85% of this is spent again, and so on, what is the total effect on the economy?

60. *Advertising Effect.* The cereal company Box-o-Vim is about to market a new low-fat great-tasting cereal in a city of 5,000,000 people. They plan an advertising campaign that they think will induce 30% of the people to buy the product. They estimate that if those people like the product, they will induce $30\% \cdot 30\% \cdot 5{,}000{,}000$ more to buy the product, and those will induce $30\% \cdot 30\% \cdot 30\% \cdot 5{,}000{,}000$, and so on. In all, how many people will buy the product as a result of the advertising campaign? What percentage of the population is this?

Discussion and Writing

61. Write a problem for a classmate to solve. Devise the problem so that a geometric series is involved and the solution is "The total amount in the bank is $900(1.08)^{40}$, or about $19,552."

62. The infinite series

$$S_\infty = 2 + \frac{1}{2} + \frac{1}{2 \cdot 3} + \frac{1}{2 \cdot 3 \cdot 4} + \frac{1}{2 \cdot 3 \cdot 4 \cdot 5} + \cdots$$

is not geometric, but it does have a sum. Consider S_1, S_2, S_3, S_4, S_5, and S_6. Construct a table and a graph of the sequence. Expand the sequence of sums, if needed. Make a conjecture about the value of S_∞ and explain your reasoning.

Skill Maintenance

For each pair of functions, find $(f \circ g)(x)$ and $(g \circ f)(x)$.

63. $f(x) = x^2$, $g(x) = 4x + 5$

64. $f(x) = x - 1$, $g(x) = x^2 + x + 3$

Solve.

65. $5^x = 35$ **66.** $\log_2 x = -4$

Synthesis

67. Prove that
$$\sqrt{3} - \sqrt{2}, \quad 4 - \sqrt{6}, \quad \text{and} \quad 6\sqrt{3} - 2\sqrt{2}$$
form a geometric sequence.

68. Consider the sequence
$$4, 20.4, 104.04, 531.6444, \ldots .$$
What is the error in using $a_{277} = 4(5.1)^{276}$ to find the 277th term?

69. Consider the sequence
$$x + 3, x + 7, 4x - 2, \ldots .$$
a) If the sequence is arithmetic, find x and then determine each of the 3 terms and the 4th term.
b) If the sequence is geometric, find x and then determine each of the 3 terms and the 4th term.

70. Find the sum of the first n terms of
$$1 + x + x^2 + \cdots .$$

71. Find the sum of the first n terms of
$$x^2 - x^3 + x^4 - x^5 + \cdots .$$

In Exercises 72 and 73, assume that a_1, a_2, a_3, ... is a geometric sequence.

72. Prove that a_1^2, a_2^2, a_3^2, ..., is a geometric sequence.

73. Prove that $\ln a_1$, $\ln a_2$, $\ln a_3$, ..., is an arithmetic sequence.

74. Prove that 5^{a_1}, 5^{a_2}, 5^{a_3}, ..., is a geometric sequence, if a_1, a_2, a_3, ..., is an arithmetic sequence.

75. The sides of a square are 16 cm long. A second square is inscribed by joining the midpoints of the sides, successively. In the second square, we repeat the process, inscribing a third square. If this process is continued indefinitely, what is the sum of all the areas of all the squares? (*Hint*: Use an infinite geometric series.)

Mathematical Induction

7.4

- *List the statements of an infinite sequence that is defined by a formula.*
- *Do proofs by mathematical induction.*

In this section, we learn to prove a sequence of mathematical statements using a procedure called *mathematical induction.*

Sequences of Statements

Infinite sequences of statements occur often in mathematics. In an infinite sequence of statements, there is a statement for each natural number. For example, consider the sequence of statements represented by the following:

"For each x between 0 and 1, $0 < x^n < 1$."

Let's think of this as $S(n)$, or S_n. Substituting natural numbers for n gives a sequence of statements. We list a few of them.

Statement 1, S_1:   For x between 0 and 1, $0 < x^1 < 1$.

Statement 2, S_2:   For x between 0 and 1, $0 < x^2 < 1$.

Statement 3, S_3:   For x between 0 and 1, $0 < x^3 < 1$.

Statement 4, S_4:   For x between 0 and 1, $0 < x^4 < 1$.

In this context, the symbols S_1, S_2, S_3, and so on, do not represent sums.

EXAMPLE 1   List the first four statements in the sequence obtainable from each of the following.

a) $\log n < n$

b) $1 + 3 + 5 + \cdots + (2n - 1) = n^2$

Solution

a) This time, S_n is "$\log n < n$."

S_1: $\log 1 < 1$

S_2: $\log 2 < 2$

S_3: $\log 3 < 3$

S_4: $\log 4 < 4$

b) This time, S_n is "$1 + 3 + 5 + \cdots + (2n - 1) = n^2$."

S_1: $1 = 1^2$

S_2: $1 + 3 = 2^2$

S_3: $1 + 3 + 5 = 3^2$

S_4: $1 + 3 + 5 + 7 = 4^2$

Proving Infinite Sequences of Statements

We now develop a method of proof, called **mathematical induction,** which we can use to try to prove that all statements in an infinite sequence of statements are true. The statements usually have the form:

"For all natural numbers n, S_n",

where S_n is some mathematical sentence such as those of the preceding examples. Of course, we cannot prove each statement of an infinite sequence individually. Instead, we try to show that "whenever S_k holds, then S_{k+1} must hold." We abbreviate this as $S_k \to S_{k+1}$. (This is also read "If S_k, then S_{k+1}," or "S_k implies S_{k+1}.") Suppose that we could somehow establish that this holds for all natural numbers k. Then we would have the following:

$S_1 \longrightarrow S_2$ meaning "if S_1 is true, then S_2 is true";

$S_2 \longrightarrow S_3$ meaning "if S_2 is true, then S_3 is true";

$S_3 \longrightarrow S_4$ meaning "if S_3 is true, then S_4 is true";

and so on, indefinitely.

Even knowing that $S_k \rightarrow S_{k+1}$, we would still not be certain whether there is *any* k for which S_k is true. All we would know is that "if S_k is true, then S_{k+1} is true." Suppose now that S_k is true for some k, say, $k = 1$. We then must have the following.

S_1 is true.	We have verified, or proved, this.
$S_1 \longrightarrow S_2$	This means that whenever S_1 is true, S_2 is true.
Therefore, S_2 is true.	
$S_2 \longrightarrow S_3$	This means that whenever S_2 is true, S_3 is true.
Therefore, S_3 is true.	
and so on.	

We conclude that S_n is true for all natural numbers n.

This leads us to the principle of mathematical induction, which we use to prove the types of statements considered here.

The Principle of Mathematical Induction

We can prove an infinite sequence of statements S_n by showing the following.

(1) *Basis step.* S_1 is true.

(2) *Induction step.* For all natural numbers k, $S_k \rightarrow S_{k+1}$.

Mathematical induction is analogous to lining up a sequence of dominoes. The induction step tells us that if any one domino is knocked over, then the one next to it will be hit and knocked over. The basis step tells us that the first domino can indeed be knocked over. Note that in order for all dominoes to fall, *both* conditions must be satisfied.

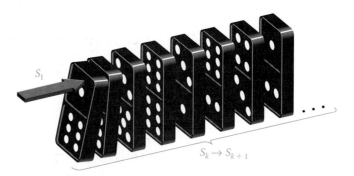

S_1

$S_k \rightarrow S_{k+1}$

When you are learning to do proofs by mathematical induction, it is helpful to first write out S_n, S_1, S_k, and S_{k+1}. This helps to identify what is to be assumed and what is to be deduced.

EXAMPLE 2 Prove: For every natural number n,

$$1 + 3 + 5 + \cdots + (2n - 1) = n^2.$$

Proof We first list S_n, S_1, S_k, and S_{k+1}.

S_n: $1 + 3 + 5 + \cdots + (2n - 1) = n^2$

S_1: $1 = 1^2$

S_k: $1 + 3 + 5 + \cdots + (2k - 1) = k^2$

S_{k+1}: $1 + 3 + 5 + \cdots + (2k - 1) + [2(k + 1) - 1] = (k + 1)^2$

(1) *Basis step.* S_1, as listed, is true.

(2) *Induction step.* We let k be any natural number. We assume S_k to be true and try to show that it implies that S_{k+1} is true. Now S_k is

$$1 + 3 + 5 + \cdots + (2k - 1) = k^2.$$

Starting with the left side of S_{k+1} and substituting k^2 for $1 + 3 + 5 + \cdots + (2k - 1)$, we have

$$\underbrace{1 + 3 + \cdots + (2k - 1)}_{\downarrow} + [2(k + 1) - 1]$$

$$= k^2 + [2(k + 1) - 1] = k^2 + 2k + 1 = (k + 1)^2.$$

We have derived S_{k+1} from S_k. Thus we have shown that for all natural numbers k, $S_k \rightarrow S_{k+1}$. This completes the induction step. It and the basis step tell us that the proof is complete.   —

EXAMPLE 3 Prove: For every natural number n,

$$\frac{1}{2} + \frac{1}{4} + \frac{1}{8} + \cdots + \frac{1}{2^n} = \frac{2^n - 1}{2^n}.$$

Proof We first list S_n, S_1, S_k, and S_{k+1}.

S_n: $\dfrac{1}{2} + \dfrac{1}{4} + \dfrac{1}{8} + \cdots + \dfrac{1}{2^n} = \dfrac{2^n - 1}{2^n}$

S_1: $\dfrac{1}{2} = \dfrac{2^1 - 1}{2^1}$

S_k: $\dfrac{1}{2} + \dfrac{1}{4} + \dfrac{1}{8} + \cdots + \dfrac{1}{2^k} = \dfrac{2^k - 1}{2^k}$

S_{k+1}: $\dfrac{1}{2} + \dfrac{1}{4} + \dfrac{1}{8} + \cdots + \dfrac{1}{2^k} + \dfrac{1}{2^{k+1}} = \dfrac{2^{k+1} - 1}{2^{k+1}}$

(1) *Basis step.* We show S_1 to be true as follows:

$$\frac{2^1 - 1}{2^1} = \frac{2 - 1}{2} = \frac{1}{2}.$$

(2) *Induction step.* We let k be any natural number. We assume S_k to be true and try to show that it implies that S_{k+1} is true. Now S_k is

$$\frac{1}{2} + \frac{1}{4} + \frac{1}{8} + \cdots + \frac{1}{2^k} = \frac{2^k - 1}{2^k}.$$

Starting with the left side of S_{k+1} and substituting

$$\frac{2^k - 1}{2^k} \quad \text{for} \quad \frac{1}{2} + \frac{1}{4} + \cdots + \frac{1}{2^k},$$

we have

$$\underbrace{\frac{1}{2} + \frac{1}{4} + \frac{1}{8} + \cdots + \frac{1}{2^k}}_{} + \frac{1}{2^{k+1}}$$

$$= \frac{2^k - 1}{2^k} + \frac{1}{2^{k+1}} = \frac{2^k - 1}{2^k} \cdot \frac{2}{2} + \frac{1}{2^{k+1}} = \frac{(2^k - 1) \cdot 2 + 1}{2^{k+1}}$$

$$= \frac{2^{k+1} - 2 + 1}{2^{k+1}} = \frac{2^{k+1} - 1}{2^{k+1}}.$$

We have derived S_{k+1} from S_k. Thus we have shown that for all natural numbers k, $S_k \rightarrow S_{k+1}$. This completes the induction step. It and the basis step tell us that the proof is complete.   ▬

EXAMPLE 4   Prove: For every natural number n, $n < 2^n$.

Proof   We first list S_n, S_1, S_k, and S_{k+1}.

S_n: $n < 2^n$

S_1: $1 < 2^1$

S_k: $k < 2^k$

S_{k+1}: $k + 1 < 2^{k+1}$

(1) *Basis step.* S_1, as listed, is true since $2^1 = 2$ and $1 < 2$.

(2) *Induction step.* We let k be any natural number. We assume S_k to be true and try to show that it implies that S_{k+1} is true. Now

$k < 2^k$	This is S_k.
$2k < 2 \cdot 2^k$	Multiplying by 2 on both sides
$2k < 2^{k+1}$	Adding exponents on the right
$k + k < 2^{k+1}$.	Rewriting $2k$ as $k + k$

Since k is any natural number, we know that $1 \leq k$. Thus,

$$k + 1 \leq k + k. \qquad \text{Adding } k \text{ on both sides}$$

Putting the results $k + 1 \leq k + k$ and $k + k < 2^{k+1}$ together gives us

$$k + 1 < 2^{k+1}. \qquad \text{This is } S_{k+1}.$$

We have derived S_{k+1} from S_k. Thus we have shown that for all natural numbers k, $S_k \rightarrow S_{k+1}$. This completes the induction step. It and the basis step tell us that the proof is complete.   ▬

Exercise Set 7.4

List the first five statements in the sequence obtainable from each of the following. Determine whether each of the statements is true or false.

1. $n^2 < n^3$

2. $n^2 - n + 41$ is prime. Find a value for n for which the statement is false.

3. A polygon of n sides has $[n(n - 3)]/2$ diagonals.

4. The sum of the angles of a polygon of n sides is $(n - 2) \cdot 180°$.

Use mathematical induction to prove each of the following.

5. $2 + 4 + 6 + \cdots + 2n = n(n + 1)$

6. $4 + 8 + 12 + \cdots + 4n = 2n(n + 1)$

7. $1 + 5 + 9 + \cdots + (4n - 3) = n(2n - 1)$

8. $3 + 6 + 9 + \cdots + 3n = \dfrac{3n(n + 1)}{2}$

9. $2 + 4 + 8 + \cdots + 2^n = 2(2^n - 1)$

10. $2 \leq 2^n$

11. $n < n + 1$

12. $3^n < 3^{n+1}$

13. $2n \leq 2^n$

14. $\dfrac{1}{1 \cdot 2} + \dfrac{1}{2 \cdot 3} + \cdots + \dfrac{1}{n(n + 1)} = \dfrac{n}{n + 1}$

15. $\dfrac{1}{1 \cdot 2 \cdot 3} + \dfrac{1}{2 \cdot 3 \cdot 4} + \dfrac{1}{3 \cdot 4 \cdot 5} + \cdots$
$+ \dfrac{1}{n(n + 1)(n + 2)} = \dfrac{n(n + 3)}{4(n + 1)(n + 2)}$

16. If x is any real number greater than 1, then for any natural number n, $x \leq x^n$.

The following formulas can be used to find sums of powers of natural numbers. Use mathematical induction to prove each of the following.

17. $1 + 2 + 3 + \cdots + n = \dfrac{n(n + 1)}{2}$

18. $1^2 + 2^2 + 3^2 + \cdots + n^2 = \dfrac{n(n + 1)(2n + 1)}{6}$

19. $1^3 + 2^3 + 3^3 + \cdots + n^3 = \dfrac{n^2(n + 1)^2}{4}$

20. $1^4 + 2^4 + 3^4 + \cdots + n^4$
$= \dfrac{n(n + 1)(2n + 1)(3n^2 + 3n - 1)}{30}$

21. $1^5 + 2^5 + 3^5 + \cdots + n^5$
$= \dfrac{n^2(n + 1)^2(2n^2 + 2n - 1)}{12}$

Use mathematical induction to prove each of the following.

22. $\displaystyle\sum_{i=1}^{n} (3i - 1) = \dfrac{n(3n + 1)}{2}$

23. $\displaystyle\sum_{i=1}^{n} i(i + 1) = \dfrac{n(n + 1)(n + 2)}{3}$

24. $\left(1 + \dfrac{1}{1}\right)\left(1 + \dfrac{1}{2}\right)\left(1 + \dfrac{1}{3}\right) \cdots \left(1 + \dfrac{1}{n}\right) = n + 1$

25. The sum of n terms of an arithmetic sequence:
$a_1 + (a_1 + d) + (a_1 + 2d) + \cdots + [a_1 + (n - 1)d]$
$= \dfrac{n}{2}[2a_1 + (n - 1)d]$

Discussion and Writing

26. Write an explanation of the idea behind mathematical induction for a fellow student.

27. Find two statements not considered in this section that are not true for all natural numbers. Then try to find where a proof by mathematical induction fails.

Skill Maintenance

Solve.

28. $2x - 3y = 1,$
$3x - 4y = 3$

29. $x + y + z = 3,$
$2x - 3y - 2z = 5,$
$3x + 2y + 2z = 8$

30. *e-Commerce.* ebooks.com ran a one-day promotion offering a hardback title for $24.95 and a paperback title for $9.95. A total of 80 books were sold and $1546 was taken in. How many of each type of book were sold?

31. *Investment.* Martin received $376 in simple interest one year from three investments. Part is invested at 6%, part at 8%, and part at 10%. The amount invested at 8% is twice the amount invested at 6%. There is $400 more invested at 10% than at 8%. Find the amount invested at each rate.

Synthesis

Use mathematical induction to prove each of the following.

32. The sum of n terms of a geometric sequence:

$$a_1 + a_1 r + a_1 r^2 + \cdots + a_1 r^{n-1} = \frac{a_1 - a_1 r^n}{1 - r}$$

33. $x + y$ is a factor of $x^{2n} - y^{2n}$

Prove each of the following using mathematical induction. Do the basis step for $n = 2$.

34. For every natural number $n \geq 2$,

$$2n + 1 < 3^n.$$

35. For every natural number $n \geq 2$,

$$\log_a (b_1 b_2 \cdots b_n)$$
$$= \log_a b_1 + \log_a b_2 + \cdots + \log_a b_n.$$

Prove each of the following for any complex numbers $z_1, z_2, \ldots, z_n$, where $i^2 = -1$ and $\bar{z}$ is the conjugate of z (see Section 2.2).

36. $\overline{z^n} = \bar{z}^n$

37. $\overline{z_1 + z_2 + \cdots + z_n} = \overline{z_1} + \overline{z_2} + \cdots + \overline{z_n}$

38. $\overline{z_1 z_2 \cdots z_n} = \overline{z_1} \cdot \overline{z_2} \cdots \overline{z_n}$

39. i^n is either 1, -1, i, or $-i$.

For any integers a and b, b is a factor of a if there exists an integer c such that $a = bc$. Prove each of the following for any natural number n.

40. 2 is a factor of $n^2 + n$.

41. 3 is a factor of $n^3 + 2n$.

42. *The Tower of Hanoi Problem.* There are three pegs on a board. On one peg are n disks, each smaller than the one on which it rests. The problem is to move this pile of disks to another peg. The final order must be the same, but you can move only one disk at a time and can never place a larger disk on a smaller one.

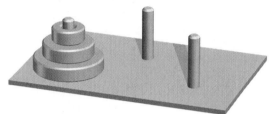

a) What is the *smallest* number of moves needed to move 3 disks? 4 disks? 2 disks? 1 disk?

b) Conjecture a formula for the *smallest* number of moves needed to move n disks. Prove it by mathematical induction.

Combinatorics: Permutations

7.5

- *Evaluate factorial and permutation notation and solve related applied problems.*

In order to study probability, it is first necessary to learn about **combinatorics**, the theory of counting.

Permutations

In this section, we will consider the part of combinatorics called *permutations*.

> The study of permutations involves *order* and *arrangements*.

EXAMPLE 1 How many 3-letter code symbols can be formed with the letters A, B, C *without* repetition (that is, using each letter only once)?

Solution Consider placing the letters in these boxes.

We can select any of the 3 letters for the first letter in the symbol. Once this letter has been selected, the second must be selected from the 2 remaining letters. After this, the third letter is already determined, since only 1 possibility is left. That is, we can place any of the 3 letters in the first box, either of the remaining 2 letters in the second box, and the only remaining letter in the third box. The possibilities can be arrived at using a **tree diagram,** as shown below.

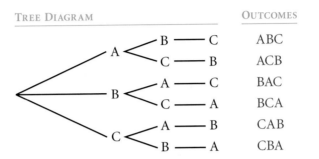

TREE DIAGRAM   OUTCOMES

Each outcome represents one permutation of the letters A, B, C.

We see that there are 6 possibilities. The set of all the possibilities is

{ABC, ACB, BAC, BCA, CAB, CBA}.

Suppose that we perform an experiment such as selecting letters (as in the preceding example), flipping a coin, or drawing a card. The results are called **outcomes**. An **event** is a set of outcomes. The following concerns the counting of actions that occur together, or are combined to form an event.

The Fundamental Counting Principle

Given a combined action, or *event*, in which the first action can be performed in n_1 ways, the second action can be performed in n_2 ways, and so on, the total number of ways in which the combined action can be performed is the product

$$n_1 \cdot n_2 \cdot n_3 \cdot \cdots \cdot n_k.$$

Thus, in Example 1, there are 3 choices for the first letter, 2 for the second letter, and 1 for the third letter, making a total of $3 \cdot 2 \cdot 1$, or 6 possibilities.

EXAMPLE 2   How many 3-letter code symbols can be formed with the letters A, B, C, D, and E *with* repetition (that is, allowing letters to be repeated)?

Solution   Since repetition is allowed, there are 5 choices for the first letter, 5 choices for the second, and 5 for the third. Thus, by the fundamental counting principle, there are $5 \cdot 5 \cdot 5$, or 125 code symbols.

Permutation

A **permutation** of a set of n objects is an ordered arrangement of all n objects.

Consider, for example, a set of 4 objects

{A, B, C, D}.

To find the number of ordered arrangements of the set, we select a first letter: There are 4 choices. Then we select a second letter: There are 3 choices. Then we select a third letter: There are 2 choices. Finally, there is 1 choice for the last selection. Thus, by the fundamental counting principle, there are $4 \cdot 3 \cdot 2 \cdot 1$, or 24, permutations of a set of 4 objects.

We can find a formula for the total number of permutations of all objects in a set of n objects. We have n choices for the first selection, $n - 1$ choices for the second, $n - 2$ for the third, and so on. For the nth selection, there is only 1 choice.

A Formula for the Total Number of Permutations of n Objects

The total number of permutations of n objects, denoted $_nP_n$, is given by

$$_nP_n = n(n - 1)(n - 2) \cdots 3 \cdot 2 \cdot 1.$$

EXAMPLE 3 Find each of the following.

a) $_4P_4$

b) $_7P_7$

Solution

Start with 4.

a) $_4P_4 = 4 \cdot 3 \cdot 2 \cdot 1 = 24$

4 factors

b) $_7P_7 = 7 \cdot 6 \cdot 5 \cdot 4 \cdot 3 \cdot 2 \cdot 1 = 5040$

4 nPr 4
                        24
7 nPr 7
                      5040

We can also find permutations using a grapher, as shown at left.

EXAMPLE 4 In how many different ways can 9 packages be placed in 9 mailboxes, one package in a box?

Solution We have

$$_9P_9 = 9 \cdot 8 \cdot 7 \cdot 6 \cdot 5 \cdot 4 \cdot 3 \cdot 2 \cdot 1 = 362{,}880.$$

Factorial Notation

We will use products such as $7 \cdot 6 \cdot 5 \cdot 4 \cdot 3 \cdot 2 \cdot 1$ so often that it is convenient to adopt a notation for them. For the product

$$7 \cdot 6 \cdot 5 \cdot 4 \cdot 3 \cdot 2 \cdot 1,$$

we write 7!, read "7 factorial."

We now define factorial notation for natural numbers and for 0.

Factorial Notation

For any natural number n,

$$n! = n(n-1)(n-2) \cdots 3 \cdot 2 \cdot 1.$$

For the number 0,

$$0! = 1.$$

We define 0! as 1 so that certain formulas can be stated concisely and with a consistent pattern.

Here are some examples.

$$7! = 7 \cdot 6 \cdot 5 \cdot 4 \cdot 3 \cdot 2 \cdot 1 = 5040$$
$$6! = 6 \cdot 5 \cdot 4 \cdot 3 \cdot 2 \cdot 1 = 720$$
$$5! = 5 \cdot 4 \cdot 3 \cdot 2 \cdot 1 = 120$$
$$4! = 4 \cdot 3 \cdot 2 \cdot 1 = 24$$
$$3! = 3 \cdot 2 \cdot 1 = 6$$
$$2! = 2 \cdot 1 = 2$$
$$1! = 1 = 1$$
$$0! = 1 = 1$$

Factorial notation can also be evaluated using a grapher, as shown at left.

We now see that the following statement is true.

$_nP_n = n!$

We will often need to manipulate factorial notation. For example, note that

$$8! = 8 \cdot 7 \cdot 6 \cdot 5 \cdot 4 \cdot 3 \cdot 2 \cdot 1$$
$$= 8 \cdot (7 \cdot 6 \cdot 5 \cdot 4 \cdot 3 \cdot 2 \cdot 1) = 8 \cdot 7!.$$

Generalizing, we get the following.

For any natural number n, $n! = n(n-1)!$.

Grapher display (at left):

7!	
	5040
6!	
	720
5!	
	120

By using this result repeatedly, we can further manipulate factorial notation.

EXAMPLE 5 Rewrite 7! with a factor of 5!.

Solution We have

$$7! = 7 \cdot 6 \cdot 5!.$$

In general, we have the following.

For any natural numbers k and n, with $k < n$,

$$n! = \underbrace{n(n-1)(n-2) \cdots [n-(k-1)]}_{k \text{ factors}} \cdot \underbrace{(n-k)!}_{n-k \text{ factors}}$$

STUDY TIP

Prepare for class. Review the material that was covered in the previous class and read the portion of the text that will be covered in the next class. When you are prepared, you will be able to follow the lecture more easily and derive the greatest benefit from the time you spend in class.

Permutations of n Objects Taken k at a Time

Consider a set of 5 objects

$$\{A, B, C, D, E\}.$$

How many ordered arrangements can be formed using 3 objects without repetition? Examples of such an arrangement are EBA, CAB, and BCD. There are 5 choices for the first object, 4 choices for the second, and 3 choices for the third. By the fundamental counting principle, there are

$$5 \cdot 4 \cdot 3,$$

or

60 *permutations* of a set of 5 objects taken 3 at a time.

Note that

$$5 \cdot 4 \cdot 3 = \frac{5 \cdot 4 \cdot 3 \cdot 2 \cdot 1}{2 \cdot 1}, \quad \text{or} \quad \frac{5!}{2!}.$$

Permutation of n Objects Taken k at a Time

A **permutation** of a set of n objects taken k at a time is an ordered arrangement of k objects taken from the set.

Consider a set of n objects and the selection of an ordered arrangement of k of them. There would be n choices for the first object. Then there would remain $n - 1$ choices for the second, $n - 2$ choices for the third, and so on. We make k choices in all, so there are k factors in the product. By the fundamental counting principle, the total number of permutations is

$$\underbrace{n(n - 1)(n - 2) \cdots [n - (k - 1)]}_{k \text{ factors}}.$$

We can express this in another way by multiplying by 1, as follows:

$$n(n - 1)(n - 2) \cdots [n - (k - 1)] \cdot \frac{(n - k)!}{(n - k)!}$$

$$= \frac{n(n - 1)(n - 2) \cdots [n - (k - 1)](n - k)!}{(n - k)!}$$

$$= \frac{n!}{(n - k)!}.$$

This gives us the following.

Formulas for the Number of Permutations of n Objects Taken k at a Time

The number of permutations of a set of n objects taken k at a time, denoted $_nP_k$, is given by

$$_nP_k = \underbrace{n(n - 1)(n - 2) \cdots [n - (k - 1)]}_{k \text{ factors}} \qquad (1)$$

$$= \frac{n!}{(n - k)!}. \qquad (2)$$

EXAMPLE 6   Compute $_8P_4$ using both forms of the formula.

SOLUTION   Using form (1), we have

The 8 tells where to start.

$$_8P_4 = \underbrace{8 \cdot 7 \cdot 6 \cdot 5}_{} = 1680.$$

The 4 tells how many factors.

Using form (2), we have

$$_8P_4 = \frac{8!}{(8 - 4)!}$$

$$= \frac{8!}{4!}$$

$$= \frac{8 \cdot 7 \cdot 6 \cdot 5 \cdot 4 \cdot 3 \cdot 2 \cdot 1}{4 \cdot 3 \cdot 2 \cdot 1}$$

$$= 8 \cdot 7 \cdot 6 \cdot 5 = 1680.$$

We can also evaluate $_8P_4$ using a grapher, as shown at left.

```
8 nPr 4
                    1680
```

EXAMPLE 7   *Flags of Nations.* The flags of many nations consist of three vertical stripes similar to the one shown here. For example, the flag of Ireland, shown below, has its first stripe green, second white, and third orange.

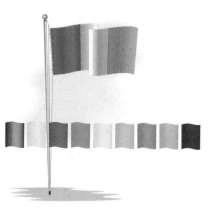

Suppose that the following 9 colors are available:

{black, yellow, red, blue, white, gold, orange, pink, purple}.

How many different flags of 3 colors can be made without repetition of colors? This assumes that the order in which the stripes appear is considered.

Solution   We are determining the number of permutations of 9 objects taken 3 at a time. There is no repetition of colors. Using form (1), we get

$$_9P_3 = 9 \cdot 8 \cdot 7 = 504.$$

EXAMPLE 8   *Batting Orders.* A baseball manager arranges the batting order as follows: The 4 infielders will bat first. Then the 3 outfielders, the catcher, and the pitcher will follow, not necessarily in that order. How many different batting orders are possible?

Solution   The infielders can bat in $_4P_4$ different ways, the rest in $_5P_5$ different ways. Then by the fundamental counting principle, we have

$$_4P_4 \cdot {}_5P_5 = 4! \cdot 5!, \quad \text{or} \quad 2880 \text{ possible batting orders.}$$

If we allow repetition, a situation like the following can occur.

EXAMPLE 9   How many 5-letter code symbols can be formed with the letters A, B, C, and D if we allow a letter to occur more than once?

Solution   We can select each of the 5 letters in 4 ways. That is, we can select the first letter in 4 ways, the second in 4 ways, and so on. Thus there are 4^5, or 1024 arrangements.

The number of distinct arrangements of n objects taken k at a time, allowing repetition, is n^k.

Permutations of Sets with Nondistinguishable Objects

Consider a set of 7 marbles, 4 of which are blue and 3 of which are red. Although the marbles are all different, when they are lined up, one red marble will look just like any other red marble. In this sense, we say that the red marbles are nondistinguishable and, similarly, the blue marbles are nondistinguishable.

We know that there are 7! permutations of this set. Many of them will look alike, however. We develop a formula for finding the number of distinguishable permutations.

Consider a set of n objects in which n_1 are of one kind, n_2 are of a second kind, . . . , n_k are of a kth kind. The total number of permutations of the set is $n!$, but this includes many that are indistinguishable. Let N be the total number of distinguishable permutations. For each of these N permutations, there are $n_1!$ actual permutations, obtained by permuting the objects of the first kind. For each of these $N \cdot n_1!$ permutations, there are $n_2!$ nondistinguishable permutations, obtained by permuting the objects of the second kind, and so on. By the fundamental counting principle, the total number of permutations, including those that are nondistinguishable, is

$$N \cdot n_1! \cdot n_2! \cdot \cdots \cdot n_k!.$$

Then we have $N \cdot n_1! \cdot n_2! \cdot \cdots \cdot n_k! = n!$. Solving for N, we obtain

$$N = \frac{n!}{n_1! \cdot n_2! \cdot \cdots \cdot n_k!}.$$

Now, to finish our problem with the marbles, we have

$$N = \frac{7!}{4!\,3!} = \frac{7 \cdot 6 \cdot 5 \cdot 4 \cdot 3 \cdot 2 \cdot 1}{4 \cdot 3 \cdot 2 \cdot 1 \cdot 3 \cdot 2 \cdot 1}$$

$$= \frac{7 \cdot 5}{1}, \quad \text{or} \quad 35$$

distinguishable permutations of the marbles.

In general:

For a set of n objects in which n_1 are of one kind, n_2 are of another kind, . . . , n_k are of a kth kind, the number of distinguishable permutations is

$$\frac{n!}{n_1! \cdot n_2! \cdot \cdots \cdot n_k!}.$$

EXAMPLE 10   In how many distinguishable ways can the letters of the word CINCINNATI be arranged?

Solution   There are 2 C's, 3 I's, 3 N's, 1 A, and 1 T for a total of 10 letters. Thus,

$$N = \frac{10!}{2! \cdot 3! \cdot 3! \cdot 1! \cdot 1!}, \quad \text{or} \quad 50{,}400.$$

The letters of the word CINCINNATI can be arranged in 50,400 distinguishable ways.

Exercise Set 7.5

Evaluate. Do your work by hand and then check it using a grapher.

1. $_6P_6$

2. $_4P_3$

3. $_{10}P_7$

4. $_{10}P_3$

5. $5!$

6. $7!$

7. $0!$

8. $1!$

9. $\dfrac{9!}{5!}$

10. $\dfrac{9!}{4!}$

11. $(8 - 3)!$

12. $(8 - 5)!$

13. $\dfrac{10!}{7!\,3!}$

14. $\dfrac{7!}{(7 - 2)!}$

15. $_8P_0$

16. $_{13}P_1$

Evaluate. Use a grapher.

17. $_{52}P_4$

18. $_{52}P_5$

Evaluate.

19. $_nP_3$

20. $_nP_2$

21. $_nP_1$

22. $_nP_0$

In each of the following exercises, give your answer using permutation notation, factorial notation, or other operations. Then evaluate.

How many permutations are there of the letters in each of the following words, if all the letters are used without repetition?

23. MARVIN

24. JUDY

25. UNDERMOST

26. COMBINES

27. How many permutations are there of the letters of the word UNDERMOST if the letters are taken 4 at a time?

28. How many permutations are there of the letters of the word COMBINES if the letters are taken 5 at a time?

29. How many 5-digit numbers can be formed using the digits 2, 4, 6, 8, and 9 without repetition? with repetition?

30. In how many ways can 7 athletes be arranged in a straight line?

31. How many distinguishable code symbols can be formed from the letters of the word BUSINESS? BIOLOGY? MATHEMATICS?

32. A professor is going to grade her 24 students on a curve. She will give 3 A's, 5 B's, 9 C's, 4 D's, and 3 F's. In how many ways can she do this?

33. *Phone Numbers.*   How many 7-digit phone numbers can be formed with the digits 0, 1, 2, 3, 4, 5, 6, 7, 8, and 9, assuming that the first number cannot be 0 or 1? Accordingly, how many telephone numbers can there be within a given area code, before the area needs to be split with a new area code?

34. *Program Planning.*   A program is planned to have 5 rock numbers and 4 speeches. In how many ways can this be done if a rock number and a speech are to alternate and a rock number is to come first?

35. Suppose the expression $a^2b^3c^4$ is rewritten without exponents. In how many ways can this be done?

36. *Coin Arrangements.*   A penny, a nickel, a dime, and a quarter are arranged in a straight line.

   a) Considering just the coins, in how many ways can they be lined up?
   b) Considering the coins and heads and tails, in how many ways can they be lined up?

37. How many code symbols can be formed using 5 out of 6 letters of A, B, C, D, E, F if the letters:

   a) are not repeated?

b) can be repeated?

c) are not repeated but must begin with D?

d) are not repeated but must begin with DE?

38. *License Plates.* A state forms its license plates by first listing a number that corresponds to the county in which the owner of the car resides (the names of the counties are alphabetized and the number is its location in that order). Then the plate lists a letter of the alphabet, and this is followed by a number from 1 to 9999. How many such plates are possible if there are 80 counties?

39. *Zip Codes.* A U.S. postal zip code is a five-digit number.

a) How many zip codes are possible if any of the digits 0 to 9 can be used?

b) If each post office has its own zip code, how many possible post offices can there be?

40. *Zip-Plus-4 Codes.* A zip-plus-4 postal code uses a 9-digit number like 75247-5456. How many 9-digit zip-plus-4 postal codes are possible?

41. *Social Security Numbers.* A social security number is a 9-digit number like 243-47-0825.

a) How many different social security numbers can there be?

b) There are about 275 million people in the United States. Can each person have a unique social security number?

Discussion and Writing

42. How "long" is 15!? You own 15 different books and decide to actually make up all the possible arrangements of the books on a shelf. About how long, in years, would it take you if you make one arrangement per second? Write out the reasoning you used for this problem in the form of a paragraph.

43. *Circular Arrangements.* In how many ways can the numbers on a clock face be arranged? See if you can derive a formula for the number of distinct circular arrangements of n objects. Explain your reasoning.

Skill Maintenance

Find the zero(s) of the function.

44. $f(x) = 4x - 9$

45. $f(x) = x^2 + x - 6$

46. $f(x) = 2x^2 - 3x - 1$

47. $f(x) = x^3 - 4x^2 - 7x + 10$

Synthesis

Solve for n.

48. $_nP_5 = 7 \cdot {}_nP_4$

49. $_nP_4 = 8 \cdot {}_{n-1}P_3$

50. $_nP_5 = 9 \cdot {}_{n-1}P_4$

51. $_nP_4 = 8 \cdot {}_nP_3$

52. Show that $n! = n(n - 1)(n - 2)(n - 3)!$.

53. *Single-Elimination Tournaments.* In a single-elimination sports tournament consisting of n teams, a team is eliminated when it loses one game. How many games are required to complete the tournament?

54. *Double-Elimination Tournaments.* In a double-elimination softball tournament consisting of n teams, a team is eliminated when it loses two games. At most, how many games are required to complete the tournament?

Combinatorics: Combinations

7.6

• *Evaluate combination notation and solve related applied problems.*

We now consider counting techniques in which order is not considered.

Combinations

We sometimes make a selection from a set *without regard to order.* Such a selection is called a *combination.* If you play cards, for example, you know that in most situations the *order* in which you hold cards is not important. That is,

The hand

is "equivalent" to these hands.

Each hand contains the same combination of three cards.

EXAMPLE 1   Find all the combinations of 3 letters taken from the set of 5 letters {A, B, C, D, E}.

Solution   The combinations are

$$\{A, B, C\}, \quad \{A, B, D\},$$
$$\{A, B, E\}, \quad \{A, C, D\},$$
$$\{A, C, E\}, \quad \{A, D, E\},$$
$$\{B, C, D\}, \quad \{B, C, E\},$$
$$\{B, D, E\}, \quad \{C, D, E\}.$$

There are 10 combinations of the 5 letters taken 3 at a time.

When we find all the combinations from a set of 5 objects taken 3 at a time, we are finding all the 3-element subsets. When a set is named, the order of the listing is *not* considered. Thus,

$$\{A, C, B\} \quad \text{names the same set as} \quad \{A, B, C\}.$$

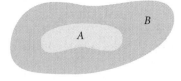

Subset

The set A is a subset of B, denoted $A \subseteq B$, if every element of A is an element of B.

The elements of a subset are not ordered. When thinking of *combinations,* do *not* think about order!

Combination

A **combination** containing k objects is a subset containing k objects.

We want to develop a formula for computing the number of combinations of n objects taken k at a time without actually listing the combinations or subsets.

Combination Notation

The number of combinations of n objects taken k at a time is denoted $_nC_k$.

We call $_nC_k$ **combination notation.** We want to derive a general formula for $_nC_k$ for any $k \leq n$. First, it is true that $_nC_n = 1$, because a set with n objects has only 1 subset with n objects, the set itself. Second, $_nC_1 = n$, because a set with n objects has n subsets with 1 element each. Finally, $_nC_0 = 1$, because a set with n objects has only one subset with 0 elements, namely the empty set $\varnothing$. To consider other possibilities, let's return to Example 1 and compare the number of combinations with the number of permutations.

COMBINATIONS		**PERMUTATIONS**				

$$_5C_3 \text{ of these} \left\{ \begin{array}{l} \{A, B, C\} \longrightarrow ABC \quad BCA \quad CAB \quad CBA \quad BAC \quad ACB \\ \{A, B, D\} \longrightarrow ABD \quad BDA \quad DAB \quad DBA \quad BAD \quad ADB \\ \{A, B, E\} \longrightarrow ABE \quad BEA \quad EAB \quad EBA \quad BAE \quad AEB \\ \{A, C, D\} \longrightarrow ACD \quad CDA \quad DAC \quad DCA \quad CAD \quad ADC \\ \{A, C, E\} \longrightarrow ACE \quad CEA \quad EAC \quad ECA \quad CAE \quad AEC \\ \{A, D, E\} \longrightarrow ADE \quad DEA \quad EAD \quad EDA \quad DAE \quad AED \\ \{B, C, D\} \longrightarrow BCD \quad CDB \quad DBC \quad DCB \quad CBD \quad BDC \\ \{B, C, E\} \longrightarrow BCE \quad CEB \quad EBC \quad ECB \quad CBE \quad BEC \\ \{B, D, E\} \longrightarrow BDE \quad DEB \quad EBD \quad EDB \quad DBE \quad BED \\ \{C, D, E\} \longrightarrow CDE \quad DEC \quad ECD \quad EDC \quad DCE \quad CED \end{array} \right\} 3! \cdot {_5C_3} \text{ of these}$$

Note that each combination of 3 objects yields 6, or 3!, permutations.

$$3! \cdot {_5C_3} = 60 = {_5P_3} = 5 \cdot 4 \cdot 3,$$

so

$$_5C_3 = \frac{_5P_3}{3!} = \frac{5 \cdot 4 \cdot 3}{3 \cdot 2 \cdot 1} = 10.$$

In general, the number of combinations of n objects taken k at a time, ${}_nC_k$, times the number of permutations of these objects, $k!$, must equal the number of permutations of n objects taken k at a time:

$$k! \cdot {}_nC_k = {}_nP_k$$

$$_nC_k = \frac{{}_nP_k}{k!}$$

$$= \frac{1}{k!} \cdot {}_nP_k$$

$$= \frac{1}{k!} \cdot \frac{n!}{(n-k)!} = \frac{n!}{k!\,(n-k)!}.$$

Combinations of n Objects Taken k at a Time

The total number of combinations of n objects taken k at a time, denoted ${}_nC_k$, is given by

$$_nC_k = \frac{n!}{k!\,(n-k)!}, \tag{1}$$

or

$$_nC_k = \frac{{}_nP_k}{k!} = \frac{n(n-1)(n-2)\cdots[n-(k-1)]}{k!}. \tag{2}$$

Another kind of notation for ${}_nC_k$ is **binomial coefficient notation.** The reason for such terminology will be seen later.

Binomial Coefficient Notation

$$\binom{n}{k} = {}_nC_k$$

You should be able to use either notation and either form of the formula.

EXAMPLE 2 Evaluate $\dbinom{7}{5}$, using forms (1) and (2).

Solution

a) By form (1),

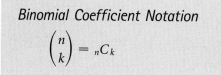

$$\binom{7}{5} = \frac{7!}{5!\,2!} = \frac{7 \cdot 6 \cdot 5 \cdot 4 \cdot 3 \cdot 2 \cdot 1}{5 \cdot 4 \cdot 3 \cdot 2 \cdot 1 \cdot 2 \cdot 1} = \frac{7 \cdot 6}{2 \cdot 1} = 21.$$

b) By form (2),

The 7 tells where to start.

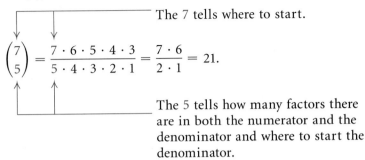

$$\binom{7}{5} = \frac{7 \cdot 6 \cdot 5 \cdot 4 \cdot 3}{5 \cdot 4 \cdot 3 \cdot 2 \cdot 1} = \frac{7 \cdot 6}{2 \cdot 1} = 21.$$

The 5 tells how many factors there are in both the numerator and the denominator and where to start the denominator.

We can find combinations using a grapher as well, as shown below.

7 nCr 5

21

Be sure to keep in mind that $\binom{n}{k}$ does not mean $n \div k$, or n/k.

EXAMPLE 3 Evaluate $\binom{n}{0}$ and $\binom{n}{2}$.

Solution We use form (1) for the first expression and form (2) for the second. Then

$$\binom{n}{0} = \frac{n!}{0!\,(n-0)!} = \frac{n!}{1 \cdot n!} = 1,$$

using form (1), and

$$\binom{n}{2} = \frac{n(n-1)}{2!} = \frac{n(n-1)}{2}, \quad \text{or} \quad \frac{n^2 - n}{2},$$

using form (2).

Note that

$$\binom{7}{2} = \frac{7 \cdot 6}{2 \cdot 1} = 21,$$

so that using the result of Example 2 gives us

$$\binom{7}{5} = \binom{7}{2}.$$

This says that the number of 5-element subsets of a set of 7 objects is the same as the number of 2-element subsets of a set of 7 objects. When 5 elements are chosen from a set, one also chooses *not* to include 2 elements.

To see this, consider such a set:

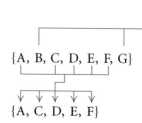

Each time we form a subset with 5 elements, we leave behind a subset with 2 elements, and vice versa.

In general, we have the following.

Subsets of Size k and of Size n − k

$$\binom{n}{k} = \binom{n}{n-k} \quad \text{and} \quad {}_nC_k = {}_nC_{n-k}$$

The number of subsets of size k of a set with n objects is the same as the number of subsets of size $n - k$. The number of combinations of n objects taken k at a time is the same as the number of combinations of n objects taken $n - k$ at a time.

This result provides an alternative way to compute combinations. We now solve problems involving combinations.

EXAMPLE 4   *Michigan Lotto.* The state of Michigan runs a 6-out-of-49-number lotto twice a week that pays at least $2 million. You purchase a card for $1 and pick any 6 numbers from 1 to 49. If your numbers match those that the state draws, you win.

a) How many 6-number combinations are there for drawing?

b) Suppose that it takes 10 min to pick your numbers and buy a ticket. How many tickets can you buy in 4 days?

c) How many people would you have to hire to buy tickets with all the possible combinations and ensure that you win?

Solution

a) No order is implied here. You pick any 6 numbers from 1 to 49. Thus the number of combinations is

$$
\begin{aligned}
{}_{49}C_6 = \binom{49}{6} &= \frac{49!}{6!\,43!} \\
&= \frac{49 \cdot 48 \cdot 47 \cdot 46 \cdot 45 \cdot 44}{6 \cdot 5 \cdot 4 \cdot 3 \cdot 2 \cdot 1} \\
&= 13{,}983{,}816.
\end{aligned}
$$

```
49 nCr 6
                13983816
```

b) In 4 days, there are $4 \cdot 24 \cdot 60$, or 5760 min, so you could buy 5760/10, or 576 tickets in that entire time period.

c) You would need to hire 13,983,816/576, or about 24,278 people to buy tickets with all the possible combinations and ensure a win. (This presumes lottery tickets can be bought 24 hours a day, which is questionable.)

EXAMPLE 5 How many committees can be formed from a group of 5 governors and 7 senators if each committee consists of 3 governors and 4 senators?

Solution The 3 governors can be selected in $_5C_3$ ways and the 4 senators can be selected in $_7C_4$ ways. If we use the fundamental counting principle, it follows that the number of possible committees is

$$_5C_3 \cdot _7C_4 = 10 \cdot 35 = 350.$$

```
5 nCr 3*7 nCr 4
                350
```

CONNECTING THE CONCEPTS

PERMUTATIONS AND COMBINATIONS

PERMUTATIONS

Permutations involve order and arrangements of objects.

Given 5 books, we can arrange 3 of them on a shelf in $_5P_3$, or 60 ways.

Placing the books in different orders produces different arrangements.

COMBINATIONS

Combinations do not involve the order or arrangement of objects.

Given 5 books, we can select 3 of them in $_5C_3$, or 10 ways.

The order in which the books are chosen does not matter.

Exercise Set 7.6

Evaluate. Do your work by hand and then check it using a grapher.

1. $_{13}C_2$

2. $_9C_6$

3. $\binom{13}{11}$

4. $\binom{9}{3}$

5. $\binom{7}{1}$

6. $\binom{8}{8}$

7. $\dfrac{_5P_3}{3!}$

8. $\dfrac{_{10}P_5}{5!}$

9. $\binom{6}{0}$

10. $\binom{6}{1}$

11. $\binom{6}{2}$

12. $\binom{6}{3}$

13. $\binom{7}{0} + \binom{7}{1} + \binom{7}{2} + \binom{7}{3} + \binom{7}{4} + \binom{7}{5} + \binom{7}{6} + \binom{7}{7}$

14. $\binom{6}{0} + \binom{6}{1} + \binom{6}{2} + \binom{6}{3} + \binom{6}{4} + \binom{6}{5} + \binom{6}{6}$

Evaluate. Use a grapher.

15. $_{52}C_4$

16. $_{52}C_5$

17. $\binom{27}{11}$

18. $\binom{37}{8}$

Evaluate.

19. $\binom{n}{1}$

20. $\binom{n}{3}$

21. $\binom{m}{m}$

22. $\binom{t}{4}$

In each of the following exercises, give an expression for the answer using permutation notation, combination notation, factorial notation, or other operations. Then evaluate.

23. *Fraternity Officers.* There are 23 students in a fraternity. How many sets of 4 officers can be selected?

24. *League Games.* How many games can be played in a 9-team sports league if each team plays all other teams once? twice?

25. *Test Options.* On a test, a student is to select 10 out of 13 questions. In how many ways can this be done?

26. *Test Options.* Of the first 10 questions on a test, a student must answer 7. Of the second 5 questions, the student must answer 3. In how many ways can this be done?

27. *Lines and Triangles from Points.* How many lines are determined by 8 points, no 3 of which are collinear? How many triangles are determined by the same points?

28. *Senate Committees.* Suppose the Senate of the United States consists of 58 Republicans and 42 Democrats. How many committees can be formed consisting of 6 Republicans and 4 Democrats?

29. *Poker Hands.* How many 5-card poker hands are possible with a 52-card deck?

30. *Bridge Hands.* How many 13-card bridge hands are possible with a 52-card deck?

31. *Baskin-Robbins Ice Cream.* Baskin-Robbins, a national firm, sells ice cream in 31 flavors.

   a) How many 2-dip cones are possible if order of flavors is to be considered and no flavor is repeated?

   b) How many 2-dip cones are possible if order is to be considered and a flavor can be repeated?

   c) How many 2-dip cones are possible if order is not considered and no flavor is repeated?

Discussion and Writing

32. Explain why a "combination" lock should really be called a "permutation" lock.

33. Give an explanation that you might use with a fellow student to explain that

$$\binom{n}{k} = \binom{n}{n-k}.$$

Skill Maintenance

Solve.

34. $3x - 7 = 5x + 10$

35. $2x^2 - x = 3$

36. $x^2 + 5x + 1 = 0$

37. $x^3 + 3x^2 - 10x = 24$

Synthesis

38. *Full House.* A full house in poker consists of a pair (two of a kind) and three of a kind. How many full houses are there that consist of 3 aces and 2 queens? (See Section 7.8 for a description of a 52-card deck.)

39. *Flush.* A flush in poker consists of a 5-card hand with all cards of the same suit. How many 5-card hands (flushes) are there that consist of all diamonds?

40. There are n points on a circle. How many quadrilaterals can be inscribed with these points as vertices?

41. *League Games.* How many games are played in a league with n teams if each team plays each other team once? twice?

Solve for n.

42. $\dbinom{n + 1}{3} = 2 \cdot \dbinom{n}{2}$

43. $\dbinom{n}{n - 2} = 6$

44. $\dbinom{n}{3} = 2 \cdot \dbinom{n - 1}{2}$

45. $\dbinom{n + 2}{4} = 6 \cdot \dbinom{n}{2}$

46. How many line segments are determined by the n vertices of an n-agon? Of these, how many are

diagonals? Use mathematical induction to prove the result for the diagonals.

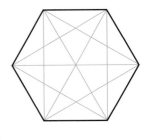

47. Prove that

$$\binom{n}{k - 1} + \binom{n}{k} = \binom{n + 1}{k}$$

for any natural numbers n and k, $k \le n$.

The Binomial Theorem

7.7

- Expand a power of a binomial using Pascal's triangle or factorial notation.
- Find a specific term of a binomial expansion.
- Find the total number of subsets of a set of n objects.

In this section, we consider ways of expanding a binomial $(a + b)^n$.

Binomial Expansions Using Pascal's Triangle

Consider the following expanded powers of $(a + b)^n$, where $a + b$ is any binomial and n is a whole number. Look for patterns.

$$(a + b)^0 = 1$$
$$(a + b)^1 = a + b$$
$$(a + b)^2 = a^2 + 2ab + b^2$$
$$(a + b)^3 = a^3 + 3a^2b + 3ab^2 + b^3$$
$$(a + b)^4 = a^4 + 4a^3b + 6a^2b^2 + 4ab^3 + b^4$$
$$(a + b)^5 = a^5 + 5a^4b + 10a^3b^2 + 10a^2b^3 + 5ab^4 + b^5$$

Each expansion is a polynomial. There are some patterns to be noted.

1. There is one more term than the power of the exponent, n. That is, there are $n + 1$ terms in the expansion of $(a + b)^n$.

2. In each term, the sum of the exponents is n, the power to which the binomial is raised.

3. The exponents of a start with n, the power of the binomial, and decrease to 0. The last term has no factor of a. The first term has no factor of b, so powers of b start with 0 and increase to n.

4. The coefficients start at 1 and increase through certain values about "half"-way and then decrease through these same values back to 1. Let's explore the coefficients further.

Suppose that we want to find an expansion of $(a + b)^6$. The patterns we noted above indicate that there are 7 terms in the expansion:

$$a^6 + c_1a^5b + c_2a^4b^2 + c_3a^3b^3 + c_4a^2b^4 + c_5ab^5 + b^6.$$

How can we determine the value of each coefficient, c_i? We can do so in two different ways. The first method involves writing the coefficients in a triangular array, as follows. This is known as **Pascal's triangle:**

$(a + b)^0$:   1

$(a + b)^1$:   1   1

$(a + b)^2$:   1   2   1

$(a + b)^3$:   1   3   3   1

$(a + b)^4$:   1   4   6   4   1

$(a + b)^5$:   1   5   10   10   5   1

There are many patterns in the triangle. Find as many as you can.

Perhaps you discovered a way to write the next row of numbers, given the numbers in the row above it. There are always 1's on the outside. Each remaining number is the sum of the two numbers above it. Let's try to find an expansion for $(a + b)^6$ by adding another row using the patterns we have discovered:

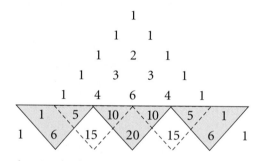

We see that in the last row

the 1st and last numbers are **1**;

the 2nd number is $1 + 5$, or **6**;

the 3rd number is $5 + 10$, or **15**;

the 4th number is $10 + 10$, or **20**;

the 5th number is $10 + 5$, or **15**; and

the 6th number is $5 + 1$, or **6**.

Thus the expansion for $(a + b)^6$ is

$$(a + b)^6 = 1a^6 + 6a^5b + 15a^4b^2 + 20a^3b^3 + 15a^2b^4 + 6ab^5 + 1b^6.$$

To find an expansion for $(a + b)^8$, we complete two more rows of Pascal's triangle:

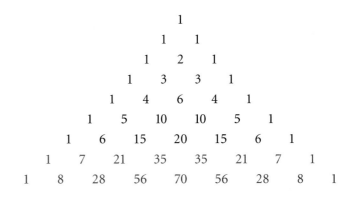

```
                        1
                     1     1
                  1     2     1
               1     3     3     1
            1     4     6     4     1
         1     5    10    10     5     1
      1     6    15    20    15     6     1
   1     7    21    35    35    21     7     1
1     8    28    56    70    56    28     8     1
```

Thus the expansion of $(a + b)^8$ is

$$(a + b)^8 = a^8 + 8a^7b^1 + 28a^6b^2 + 56a^5b^3 + 70a^4b^4 + 56a^3b^5$$
$$+ 28a^2b^6 + 8a^1b^7 + b^8.$$

We can generalize our results as follows.

The Binomial Theorem Using Pascal's Triangle

For any binomial $a + b$ and any natural number n,

$$(a + b)^n = c_0a^nb^0 + c_1a^{n-1}b^1 + c_2a^{n-2}b^2 + \cdots + c_{n-1}a^1b^{n-1}$$
$$+ c_na^0b^n,$$

where the numbers $c_0, c_1, c_2, \ldots, c_{n-1}, c_n$ are from the $(n + 1)$st row of Pascal's triangle.

EXAMPLE 1   Expand: $(u - v)^5$.

Solution   We have $(a + b)^n$, where $a = u$, $b = -v$, and $n = 5$. We use the 6th row of Pascal's triangle:

$$1 \quad 5 \quad 10 \quad 10 \quad 5 \quad 1$$

Then we have

$$(u - v)^5 = [u + (-v)]^5$$
$$= 1(u)^5 + 5(u)^4(-v)^1 + 10(u)^3(-v)^2 + 10(u)^2(-v)^3$$
$$+ 5(u)(-v)^4 + 1(-v)^5$$
$$= u^5 - 5u^4v + 10u^3v^2 - 10u^2v^3 + 5uv^4 - v^5.$$

Note that the signs of the terms alternate between $+$ and $-$. When the power of $-v$ is odd, the sign is $-$.

EXAMPLE 2 Expand: $\left(2t + \dfrac{3}{t}\right)^4$.

Solution We have $(a + b)^n$, where $a = 2t$, $b = 3/t$, and $n = 4$. We use the 5th row of Pascal's triangle:

$$1 \qquad 4 \qquad 6 \qquad 4 \qquad 1$$

Then we have

$$\left(2t + \frac{3}{t}\right)^4 = (2t)^4 + 4(2t)^3\left(\frac{3}{t}\right)^1 + 6(2t)^2\left(\frac{3}{t}\right)^2 + 4(2t)^1\left(\frac{3}{t}\right)^3 + \left(\frac{3}{t}\right)^4$$

$$= 16t^4 + 4(8t^3)\left(\frac{3}{t}\right) + 6(4t^2)\left(\frac{9}{t^2}\right) + 4(2t)\left(\frac{27}{t^3}\right) + \left(\frac{81}{t^4}\right)$$

$$= 16t^4 + 96t^2 + 216 + 216t^{-2} + 81t^{-4}.$$

Binomial Expansion Using Factorial Notation

Suppose that we want to find the expansion of $(a + b)^{11}$. The disadvantage in using Pascal's triangle is that we must compute all the preceding rows of the table to obtain the row needed for the expansion. The following method avoids this. It also enables us to find a specific term—say, the 8th term—without computing all the other terms of the expansion. This method is useful in such courses as finite mathematics, calculus, and statistics, and uses the *binomial coefficient notation* $\dbinom{n}{k}$ developed in Section 7.6.

We can restate the binomial theorem as follows.

The Binomial Theorem Using Factorial Notation
For any binomial $a + b$ and any natural number n,

$$(a + b)^n = \binom{n}{0}a^n b^0 + \binom{n}{1}a^{n-1}b^1 + \binom{n}{2}a^{n-2}b^2 + \cdots$$

$$+ \binom{n}{n-1}a^1 b^{n-1} + \binom{n}{n}a^0 b^n$$

$$= \sum_{k=0}^{n}\binom{n}{k}a^{n-k}b^k.$$

The binomial theorem can be proved by mathematical induction, but we will not do so here. This form shows why $\dbinom{n}{k}$ is called a *binomial coefficient.*

EXAMPLE 3 Expand: $(x^2 - 2y)^5$.

Solution We have $(a + b)^n$, where $a = x^2$, $b = -2y$, and $n = 5$. Then using the binomial theorem, we have

$$(x^2 - 2y)^5 = \binom{5}{0}(x^2)^5 + \binom{5}{1}(x^2)^4(-2y) + \binom{5}{2}(x^2)^3(-2y)^2$$

$$+ \binom{5}{3}(x^2)^2(-2y)^3 + \binom{5}{4}x^2(-2y)^4 + \binom{5}{5}(-2y)^5$$

$$= \frac{5!}{0!\,5!}x^{10} + \frac{5!}{1!\,4!}x^8(-2y) + \frac{5!}{2!\,3!}x^6(4y^2) + \frac{5!}{3!\,2!}x^4(-8y^3)$$

$$+ \frac{5!}{4!\,1!}x^2(16y^4) + \frac{5!}{5!\,0!}(-32y^5)$$

$$= x^{10} - 10x^8y + 40x^6y^2 - 80x^4y^3 + 80x^2y^4 - 32y^5.$$ ▬

EXAMPLE 4 Expand: $\left(\dfrac{2}{x} + 3\sqrt{x}\right)^4$.

Solution We have $(a + b)^n$, where $a = 2/x$, $b = 3\sqrt{x}$, and $n = 4$. Then using the binomial theorem, we have

$$\left(\frac{2}{x} + 3\sqrt{x}\right)^4 = \binom{4}{0}\left(\frac{2}{x}\right)^4 + \binom{4}{1}\left(\frac{2}{x}\right)^3(3\sqrt{x}) + \binom{4}{2}\left(\frac{2}{x}\right)^2(3\sqrt{x})^2$$

$$+ \binom{4}{3}\left(\frac{2}{x}\right)(3\sqrt{x})^3 + \binom{4}{4}(3\sqrt{x})^4$$

$$= \frac{4!}{0!\,4!}\left(\frac{2}{x}\right)^4 + \frac{4!}{1!\,3!}\left(\frac{2}{x}\right)^3(3\sqrt{x}) + \frac{4!}{2!\,2!}\left(\frac{2}{x}\right)^2(3\sqrt{x})^2$$

$$+ \frac{4!}{3!\,1!}\left(\frac{2}{x}\right)(3\sqrt{x})^3 + \frac{4!}{4!\,0!}(3\sqrt{x})^4$$

$$= \frac{16}{x^4} + \frac{96}{x^{5/2}} + \frac{216}{x} + 216\sqrt{x} + 81x^2.$$ ▬

Finding a Specific Term

Suppose that we want to determine only a particular term of an expansion. The method we have developed will allow us to find such a term without computing all the rows of Pascal's triangle or all the preceding coefficients.

Note that in the binomial theorem, $\binom{n}{0}a^n b^0$ gives us the 1st term, $\binom{n}{1}a^{n-1}b^1$ gives us the 2nd term, $\binom{n}{2}a^{n-2}b^2$ gives us the 3rd term, and so on. This can be generalized as follows.

Finding the $(k + 1)$st Term

The $(k + 1)$st term of $(a + b)^n$ is $\binom{n}{k}a^{n-k}b^k$.

EXAMPLE 5   Find the 5th term in the expansion of $(2x - 5y)^6$.

Solution   First, we note that $5 = 4 + 1$. Thus, $k = 4$, $a = 2x$, $b = -5y$, and $n = 6$. Then the 5th term of the expansion is

$$\binom{6}{4}(2x)^{6-4}(-5y)^4, \quad \text{or} \quad \frac{6!}{4!\,2!}(2x)^2(-5y)^4, \quad \text{or} \quad 37{,}500x^2y^4.$$

EXAMPLE 6   Find the 8th term in the expansion of $(3x - 2)^{10}$.

Solution   First, we note that $8 = 7 + 1$. Thus, $k = 7$, $a = 3x$, $b = -2$, and $n = 10$. Then the 8th term of the expansion is

$$\binom{10}{7}(3x)^{10-7}(-2)^7, \quad \text{or} \quad \frac{10!}{7!\,3!}(3x)^3(-2)^7, \quad \text{or} \quad -414{,}720x^3.$$

Total Number of Subsets

Suppose that a set has n objects. The number of subsets containing k elements is $\binom{n}{k}$ by a result of Section 7.6. The total number of subsets of a set is the number of subsets with 0 elements, plus the number of subsets with 1 element, plus the number of subsets with 2 elements, and so on. The total number of subsets of a set with n elements is

$$\binom{n}{0} + \binom{n}{1} + \binom{n}{2} + \cdots + \binom{n}{n}.$$

Now consider the expansion of $(1 + 1)^n$:

$$(1 + 1)^n = \binom{n}{0} \cdot 1^n + \binom{n}{1} \cdot 1^{n-1} \cdot 1^1 + \binom{n}{2} \cdot 1^{n-2} \cdot 1^2$$

$$+ \cdots + \binom{n}{n} \cdot 1^n$$

$$= \binom{n}{0} + \binom{n}{1} + \binom{n}{2} + \cdots + \binom{n}{n}.$$

Thus the total number of subsets is $(1 + 1)^n$, or 2^n. We have proved the following.

Total Number of Subsets

The total number of subsets of a set with n elements is 2^n.

EXAMPLE 7   The set {A, B, C, D, E} has how many subsets?

Solution   The set has 5 elements, so the number of subsets is 2^5, or 32.

EXAMPLE 8   Wendy's, a national restaurant firm, offers the following condiments for its hamburgers:

> {*catsup, mustard, mayonnaise, tomato,*
>
> *lettuce, onions, pickle, relish, cheese*}.

How many different kinds of hamburgers can Wendy's serve, excluding size of hamburger or number of patties?

Solution   The condiments on each hamburger are the elements of a subset of the set of all possible condiments, the empty set being a plain hamburger. The total number of possible hamburgers is

$$\binom{9}{0} + \binom{9}{1} + \binom{9}{2} + \cdots + \binom{9}{9} = 2^9 = 512.$$

Thus Wendy's serves hamburgers in 512 different ways.

Exercise Set 7.7

Expand.

1. $(x + 5)^4$
2. $(x - 1)^4$
3. $(x - 3)^5$
4. $(x + 2)^9$
5. $(x - y)^5$
6. $(x + y)^8$
7. $(5x + 4y)^6$
8. $(2x - 3y)^5$
9. $\left(2t + \dfrac{1}{t}\right)^7$
10. $\left(3y - \dfrac{1}{y}\right)^4$
11. $(x^2 - 1)^5$
12. $(1 + 2q^3)^8$
13. $(\sqrt{5} + t)^6$
14. $(x - \sqrt{2})^6$
15. $\left(a - \dfrac{2}{a}\right)^9$
16. $(1 + 3)^n$
17. $(\sqrt{2} + 1)^6 - (\sqrt{2} - 1)^6$
18. $(1 - \sqrt{2})^4 + (1 + \sqrt{2})^4$

19. $(x^{-2} + x^2)^4$
20. $\left(\dfrac{1}{\sqrt{x}} - \sqrt{x}\right)^6$

Find the indicated term of the binomial expansion.

21. 3rd; $(a + b)^7$
22. 6th; $(x + y)^8$
23. 6th; $(x - y)^{10}$
24. 5th; $(p - 2q)^9$
25. 12th; $(a - 2)^{14}$
26. 11th; $(x - 3)^{12}$
27. 5th; $(2x^3 - \sqrt{y})^8$
28. 4th; $\left(\dfrac{1}{b^2} + \dfrac{b}{3}\right)^7$
29. Middle; $(2u - 3v^2)^{10}$
30. Middle two; $(\sqrt{x} + \sqrt{3})^5$

Determine the number of subsets of each of the following.

31. A set of 7 elements
32. A set of 6 members
33. The set of letters of the Greek alphabet, which contains 24 letters

34. The set of letters of the English alphabet, which contains 26 letters

35. What is the degree of $(x^5 + 3)^4$?

36. What is the degree of $(2 - 5x^3)^7$?

Expand each of the following, where $i^2 = -1$.

37. $(3 + i)^5$

38. $(1 + i)^6$

39. $(\sqrt{2} - i)^4$

40. $\left(\dfrac{\sqrt{3}}{2} - \dfrac{1}{2}i\right)^{11}$

41. Find a formula for $(a - b)^n$. Use sigma notation.

42. Expand and simplify:
$$\frac{(x + h)^{13} - x^{13}}{h}.$$

43. Expand and simplify:
$$\frac{(x + h)^n - x^n}{h}.$$
Use sigma notation.

Discussion and Writing

44. Discuss the pros and cons of each method of finding a binomial expansion. Give examples of when you might use one method rather than the other.

45. Blaise Pascal (1623–1662) was a French scientist and philosopher who founded the modern theory of probability. Do some research on Pascal and see if you can find out how he discovered his famous "triangle of numbers."

Skill Maintenance

Given that $f(x) = x^2 + 1$ and $g(x) = 2x - 3$, find each of the following.

46. $(f + g)(x)$

47. $(fg)(x)$

48. $(f \circ g)(x)$

49. $(g \circ f)(x)$

Synthesis

Solve for x.

50. $\displaystyle\sum_{k=0}^{8} \binom{8}{k} x^{8-k} 3^k = 0$

51. $\displaystyle\sum_{k=0}^{4} \binom{4}{k} 5^{4-k} x^k = 64$

52. $\displaystyle\sum_{k=0}^{5} \binom{5}{k} (-1)^k x^{5-k} 3^k = 32$

53. $\displaystyle\sum_{k=0}^{4} \binom{4}{k} (-1)^k x^{4-k} 6^k = 81$

54. Find the term of
$$\left(\frac{3x^2}{2} - \frac{1}{3x}\right)^{12}$$
that does not contain x.

55. Find the middle term of $(x^2 - 6y^{3/2})^6$.

56. Find the ratio of the 4th term of
$$\left(p^2 - \frac{1}{2} p \sqrt[3]{q}\right)^5$$
to the 3rd term.

57. Find the term of
$$\left(\sqrt[3]{x} - \frac{1}{\sqrt{x}}\right)^7$$
containing $1/x^{1/6}$.

58. *Money Combinations.* A money clip contains one each of the following bills: \$1, \$2, \$5, \$10, \$20, \$50, and \$100. How many different sums of money can be formed using the bills?

Find the sum.

59. $_{100}C_0 + {}_{100}C_1 + \cdots + {}_{100}C_{100}$

60. $_nC_0 + {}_nC_1 + \cdots + {}_nC_n$

Simplify.

61. $\displaystyle\sum_{k=0}^{23} \binom{23}{k} (\log_a x)^{23-k} (\log_a t)^k$

62. $\displaystyle\sum_{k=0}^{15} \binom{15}{k} i^{30-2k}$

63. Use mathematical induction and the property
$$\binom{n}{r - 1} + \binom{n}{r} = \binom{n + 1}{r}$$
to prove the binomial theorem.

Probability

7.8

- *Compute the probability of a simple event.*

When a coin is tossed, we can reason that the chance, or likelihood, that it will fall heads is 1 out of 2, or the **probability** that it will fall heads is $\frac{1}{2}$. Of course, this does not mean that if a coin is tossed 10 times it will necessarily fall heads 5 times. If the coin is a "fair coin" and it is tossed a great many times, however, it will fall heads very nearly half of the time. Here we give an introduction to two kinds of probability, **experimental** and **theoretical**.

Experimental and Theoretical Probability

If we toss a coin a great number of times—say, 1000—and count the number of times it falls heads, we can determine the probability that it will fall heads. If it falls heads 503 times, we would calculate the probability of its falling heads to be

$$\frac{503}{1000}, \quad \text{or} \quad 0.503.$$

This is an **experimental** determination of probability. Such a determination of probability is discovered by the observation and study of data and is quite common and very useful. Here, for example, are some probabilities that have been determined *experimentally*:

1. The probability that a woman will get breast cancer in her lifetime is $\frac{1}{11}$.
2. If you kiss someone who has a cold, the probability of your catching a cold is 0.07.
3. A person who has just been released from prison has an 80% probability of returning.

If we consider a coin and reason that it is just as likely to fall heads as tails, we would calculate the probability that it will fall heads to be $\frac{1}{2}$. This is a **theoretical** determination of probability. Here are some other probabilities that have been determined *theoretically*, using mathematics:

1. If there are 30 people in a room, the probability that two of them have the same birthday (excluding year) is 0.706.
2. While on a trip, you meet someone, and after a period of conversation, discover that you have a common acquaintance. The typical reaction, "It's a small world!", is actually not appropriate, because the probability of such an occurrence is quite high—just over 22%.

In summary, experimental probabilities are determined by making observations and gathering data. Theoretical probabilities are determined by reasoning mathematically. Examples of experimental and theoretical probability like those above, especially those we do not expect, lead us to see the value of a study of probability. You might ask, "What is the

STUDY TIP

It is always best to study for a final exam over a period of at least two weeks. If you have only one or two days of study time, however, begin by reviewing each chapter, studying the formulas, theorems, properties, and procedures in the sections and in the chapter Summary and Review. Then do the review exercises in each chapter Summary and Review. Also attend a review session if one is available.

true probability?" In fact, there is none. Experimentally, we can determine probabilities within certain limits. These may or may not agree with the probabilities that we obtain theoretically. There are situations in which it is much easier to determine one of these types of probabilities than the other. For example, it would be quite difficult to arrive at the probability of catching a cold using theoretical probability.

Computing Experimental Probabilities

We first consider experimental determination of probability. The basic principle we use in computing such probabilities is as follows.

Principle P (Experimental)

An experiment is performed in which n observations are made. If a situation, or event, E occurs m times out of n observations, then we say that the *experimental probability* of the event, $P(E)$, is given by

$$P(E) = \frac{m}{n}.$$

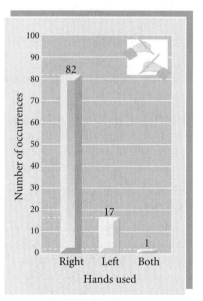

EXAMPLE 1 *Sociological Survey.* The authors of this text conducted an experiment to determine the number of people who are left-handed, right-handed, or both. The results are shown in the graph at left.

a) Determine the probability that a person is right-handed.

b) Determine the probability that a person is left-handed.

c) Determine the probability that a person is ambidextrous (uses both hands with equal ability).

d) For most tournaments held by the Professional Bowlers Association, there are 120 bowlers. On the basis of the data in this experiment, how many of the bowlers would you expect to be left-handed?

Solution

a) The number of people who are right-handed is 82, the number who are left-handed is 17, and the number who are ambidextrous is 1. The total number of observations is $82 + 17 + 1$, or 100. Thus the probability that a person is right-handed is

$$P = \frac{82}{100}, \quad \text{or} \quad 0.82, \quad \text{or} \quad 82\%.$$

b) The probability that a person is left-handed is P, where

$$P = \frac{17}{100}, \quad \text{or} \quad 0.17, \quad \text{or} \quad 17\%.$$

c) The probability that a person is ambidextrous is P, where

$$P = \frac{1}{100}, \quad \text{or} \quad 0.01, \quad \text{or} \quad 1\%.$$

d) There are 120 bowlers, and from part (b) we can expect 17% to be left-handed. Since

$$17\% \text{ of } 120 = 0.17 \cdot 120 = 20.4,$$

we can expect that about 20 of the bowlers will be left-handed. ▬

EXAMPLE 2 *Quality Control.* It is very important for a manufacturer to maintain the quality of its products. In fact, companies hire quality control inspectors to ensure this process. The goal is to produce as few defective products as possible. But since a company is producing thousands of products every day, it cannot afford to check every product to see if it is defective. To find out what percentage of its products are defective, the company checks a smaller sample.

The U.S. Department of Agriculture requires that 80% of the seeds that a company produces must sprout. To find out about the quality of the seeds it produces, a company takes 500 seeds from those it has produced and plants them. It finds that 417 of the seeds sprout.

a) What is the probability that a seed will sprout?

b) Did the seeds pass government standards?

Solution

a) We know that 500 seeds were planted and 417 sprouted. The probability of a seed sprouting is P, where

$$P = \frac{417}{500} = 0.834, \quad \text{or} \quad 83.4\%.$$

b) Since the percentage of seeds exceeded the 80% requirement, the company determines that it is producing quality seeds. ▬

EXAMPLE 3 *Television Ratings.* Television networks are always concerned about the percentage of homes that have TVs and are watching their programs. A sample of the homes are contacted by attaching an electronic device to the TVs of about 1400 homes across the country. Viewing information is then fed into a computer. The following are the results of a recent survey.

NETWORK	ABC	CBS	NBC	Fox	Other, or not watching
NUMBER OF HOMES WATCHING	118	134	143	99	906

What is the probability that a home was tuned to NBC during the time period? to Fox?

Solution   The probability that a home was tuned to NBC is P, where

$$P = \frac{143}{1400} \approx 0.102 \approx 10.2\%.$$

The probability that a home was tuned to Fox is P, where

$$P = \frac{99}{1400} \approx 0.071 \approx 7.1\%.$$

The percentages found in Example 3 are called *ratings*.

Theoretical Probability

Suppose that we perform an experiment such as flipping a coin, throwing a dart, drawing a card from a deck, or checking an item off an assembly line for quality. The results of such an experiment are called **outcomes**. The set of all possible outcomes is called the **sample space.** An **event** is a set of outcomes, that is, a subset of the sample space.

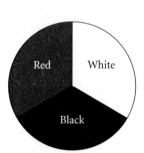

EXAMPLE 4   *Dart Throwing.* Consider this dartboard. Assume that the experiment is "throwing a dart" and that the dart hits the board. Find each of the following.

a) The outcomes

b) The sample space

Solution

a) The outcomes are *hitting black* (B), *hitting red* (R), and *hitting white* (W).

b) The sample space is {*hitting black, hitting red, hitting white*}, which can be simply stated as {B, R, W}.

EXAMPLE 5   *Die Rolling.* A die (pl., dice) is a cube, with six faces, each containing a number of dots from 1 to 6 on each side.

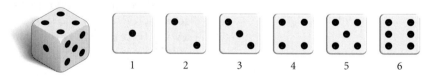

A die is rolled. Find each of the following.

a) The outcomes

b) The sample space

Solution

a) The outcomes are 1, 2, 3, 4, 5, 6.

b) The sample space is {1, 2, 3, 4, 5, 6}. ▬

We denote the probability that an event E occurs as $P(E)$. For example, "a coin falling heads" may be denoted H. Then $P(H)$ represents the probability of the coin falling heads. When all the outcomes of an experiment have the same probability of occurring, we say that they are *equally likely*. To see the distinction between events that are equally likely and those that are not, consider the dartboards shown below.

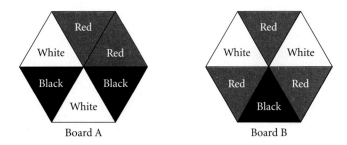

Board A   Board B

For board A, the events *hitting black*, *hitting red*, and *hitting white* are equally likely, because the black, red, and white areas are the same. However, for board B the areas are not the same so these events are not equally likely.

Principle P (Theoretical)

If an event E can occur m ways out of n possible equally likely outcomes of a sample space S, then the **theoretical probability** of the event, $P(E)$, is given by

$$P(E) = \frac{m}{n}.$$

EXAMPLE 6   What is the probability of rolling a 3 on a die?

Solution   On a fair die, there are 6 equally likely outcomes and there is 1 way to roll a 3. By Principle P, $P(3) = \frac{1}{6}$. ▬

EXAMPLE 7   What is the probability of rolling an even number on a die?

Solution   The event is rolling an *even* number. It can occur 3 ways (getting 2, 4, or 6). The number of equally likely outcomes is 6. By Principle P, $P(\text{even}) = \frac{3}{6}$, or $\frac{1}{2}$. ▬

We now use a number of examples related to a standard bridge deck of 52 cards. Such a deck is made up as shown in the following figure.

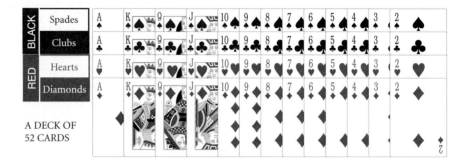

A DECK OF
52 CARDS

EXAMPLE 8 What is the probability of drawing an ace from a well-shuffled deck of cards?

Solution   There are 52 outcomes (the number of cards in the deck), they are equally likely (from a well-shuffled deck), and there are 4 ways to obtain an ace, so by Principle *P*, we have

$$P(\text{drawing an ace}) = \frac{4}{52}, \quad \text{or} \quad \frac{1}{13}.$$

EXAMPLE 9 Suppose that we select, without looking, one marble from a bag containing 3 red marbles and 4 green marbles. What is the probability of selecting a red marble?

Solution   There are 7 equally likely ways of selecting any marble, and since the number of ways of getting a red marble is 3, we have

$$P(\text{selecting a red marble}) = \frac{3}{7}.$$

The following are some results that follow from Principle *P*.

Probability Properties

a) If an event *E* cannot occur, then $P(E) = 0$.

b) If an event *E* is certain to occur, then $P(E) = 1$.

c) The probability that an event *E* will occur is a number from 0 to 1: $0 \leq P(E) \leq 1$.

For example, in coin tossing, the event that a coin will land on its edge has probability 0. The event that a coin falls either heads or tails has probability 1.

In the following examples, we use the combinatorics that we studied in Sections 7.5 and 7.6 to calculate theoretical probabilities.

EXAMPLE 10 Suppose that 2 cards are drawn from a well-shuffled deck of 52 cards. What is the probability that both of them are spades?

Solution The number of ways n of drawing 2 cards from a well-shuffled deck of 52 is $_{52}C_2$. Since 13 of the 52 cards are spades, the number of ways m of drawing 2 spades is $_{13}C_2$. Thus,

$$P(\text{getting 2 spades}) = \frac{m}{n} = \frac{_{13}C_2}{_{52}C_2} = \frac{78}{1326} = \frac{1}{17}.$$

EXAMPLE 11 Suppose that 3 people are selected at random from a group that consists of 6 men and 4 women. What is the probability that 1 man and 2 women are selected?

Solution The number of ways of selecting 3 people from a group of 10 is $_{10}C_3$. One man can be selected in $_6C_1$ ways, and 2 women can be selected in $_4C_2$ ways. By the fundamental counting principle, the number of ways of selecting 1 man and 2 women is $_6C_1 \cdot {_4}C_2$. Thus the probability that 1 man and 2 women are selected is

$$P = \frac{_6C_1 \cdot {_4}C_2}{_{10}C_3} = \frac{3}{10}.$$

EXAMPLE 12 *Rolling Two Dice.* What is the probability of getting a total of 8 on a roll of a pair of dice?

Solution On each die, there are 6 possible outcomes. The outcomes are paired so there are $6 \cdot 6$, or 36, possible ways in which the two can fall. (Assuming that the dice are different—say, one red and one blue—can help in visualizing this.)

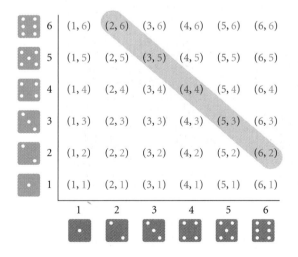

The pairs that total 8 are as shown in the figure above. There are 5 possible ways of getting a total of 8, so the probability is $\frac{5}{36}$.

Calculator display (left margin, Example 10):

```
13 nCr 2/52 nCr 2►Frac
                  1/17
```

Calculator display (left margin, Example 11):

```
6 nCr 1*4 nCr 2/10 nCr
3►Frac
                  3/10
```

Exercise Set 7.8

1. *Select a Number.* In a survey conducted by the authors, 100 people were polled and asked to select a number from 1 to 5. The results are shown in the following table.

NUMBER OF CHOICES	1	2	3	4	5
NUMBER WHO CHOSE THAT NUMBER	18	24	23	23	12

a) What is the probability that the number chosen is 1? 2? 3? 4? 5?

b) What general conclusion might a psychologist make from the experiment?

2. *Mason Dots®.* Made by the Tootsie Industries of Chicago, Illinois, Mason Dots® is a gumdrop candy. A box was opened by the authors and was found to contain the following number of gumdrops:

Strawberry	7
Lemon	8
Orange	9
Cherry	4
Lime	5
Grape	6

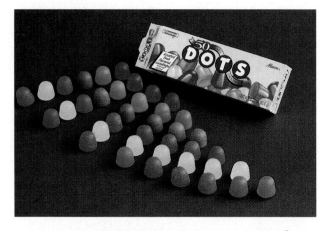

If we take one gumdrop out of the box, what is the probability of getting a lemon? lime? orange? grape? strawberry? licorice?

3. *Junk Mail.* Have you ever wondered why you receive so much junk mail? In experimental studies, the U.S. Postal Service has found that the probability that a piece of advertising is opened and read is 78%. A business sends out 15,000 pieces of advertising. How many of these can the company expect to be opened and read?

4. *Linguistics.* An experiment was conducted by the authors to determine the relative occurrence of various letters of the English alphabet. The front page of a newspaper was considered. In all, there were 9136 letters. The number of occurrences of each letter of the alphabet is listed in the following table.

LETTER	NUMBER OF OCCURRENCES	PROBABILITY
A	853	853/9136 ≈ 9.3%
B	136	
C	273	
D	286	
E	1229	
F	173	
G	190	
H	399	
I	539	
J	21	
K	57	
L	417	
M	231	
N	597	
O	705	
P	238	
Q	4	
R	609	
S	745	
T	789	
U	240	
V	113	
W	127	
X	20	
Y	124	
Z	21	21/9136 ≈ 0.2%

a) Complete the table of probabilities with the percentage, to the nearest tenth of a percent, of the occurrence of each letter.

b) What is the probability of a vowel occurring?

c) What is the probability of a consonant occurring?

5. *Wheel of Fortune®.* The results of the experiment in Exercise 4 can be quite useful to a person playing the popular television game show *Wheel of Fortune.* Players guess letters in order to spell out a phrase, a person, or a thing.

a) What 5 consonants have the greatest probability of occurring?

b) What vowel has the greatest probability of occurring?

c) The winner of the main part of the show plays for a grand prize and at one time was allowed to guess 5 consonants and a vowel in order to discover the secret wording. The 5 consonants R, S, T, L, N, and the vowel E seemed to be chosen most often. Do the results in parts (a) and (b) support such a choice?

6. *Card Drawing.* Suppose we draw a card from a well-shuffled deck of 52 cards.

a) How many equally likely outcomes are there?

What is the probability of drawing each of the following?

b) A queen   **c)** A heart

d) A 7   **e)** A red card

f) A 9 or a king   **g)** A black ace

7. *Marbles.* Suppose we select, without looking, one marble from a bag containing 4 red marbles and 10 green marbles. What is the probability of selecting each of the followng?

a) A red marble

b) A green marble

c) A purple marble

d) A red or a green marble

8. *Production Unit.* The sales force of a business consists of 10 men and 10 women. A production unit of 4 people is set up at random. What is the probability that 2 men and 2 women are chosen?

9. *Coin Drawing.* A sack contains 7 dimes, 5 nickels, and 10 quarters. Eight coins are drawn at random. What is the probability of getting 4 dimes, 3 nickels, and 1 quarter?

10. *Michigan Lotto.* Twice a week, the state of Michigan runs a 6-out-of-49-number lotto that pays at least $2 million. You purchase a card for $1 and pick any 6 numbers from 1 to 49. If your numbers match those that the state draws, you win.

a) How many 6-number combinations are there for drawing?

b) You buy 1 lottery ticket. What is your probability of winning?

Five-Card Poker Hands. Suppose that 5 cards are drawn from a deck of 52 cards. What is the probability of drawing each of the following?

11. 3 sevens and 2 kings

12. 5 aces

13. 5 spades

14. 4 aces and 1 five

15. *Random-Number Generator.* Many graphers have a **random-number generator.** This feature produces a random number in the interval [0, 1]. (Consult your user's manual.) We can use such a feature to simulate coin flipping. A number r such that $0 \le r \le 0.5$ would indicate heads, H. A number r such that $0.5 < r \le 1.0$ would indicate tails, T. Use a random-number generator 100 times.

a) What is the experimental probability of getting heads?

b) What is the experimental probability of getting tails?

16. *Tossing Three Coins.* Three coins are flipped. An outcome might be HTH.

a) Find the sample space.

What is the probability of getting each of the following?

b) Exactly one head

c) At most two tails

d) At least one head

e) Exactly two tails

Roulette. An American roulette wheel contains 38 slots numbered 00, 0, 1, 2, 3, ..., 35, 36. Eighteen of the slots numbered 1–36 are colored red and 18 are colored black. The 00 and 0 slots are uncolored. The wheel is spun, and a ball is rolled around the rim until it falls into a slot. What is the probability that the ball falls in each of the following?

17. A black slot

18. A red slot

19. A red or a black slot

20. The 00 slot

21. The 0 slot

22. Either the 00 or the 0 slot (in this case, the house always wins)

23. An odd-numbered slot

24. The number 24

25. *Dartboard.*   The figure below shows a dartboard. A dart is thrown and hits the board. Find the probabilities

$$P(\text{red}), P(\text{green}), P(\text{blue}), P(\text{yellow}).$$

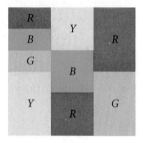

Discussion and Writing

26. *Random Best-Selling Novels.*   Sir Arthur Stanley Eddington, an astronomer, once wrote in a satirical essay that if a monkey were left alone long enough with a typewriter and typed randomly, any great novel could be replicated.

What is the probability that the following passage could have been written by a monkey? Ignore capital letters and punctuation and consider only letters and spaces.

"It was the best of times, it was the worst of times, . . ." (Charles Dickens, 1859). Explain your answer.

27. Find at least one use of probability in today's newspaper. Make a report.

Skill Maintenance

Solve.

28. $3x^2 - 4x = 3$

29. $2x^3 + 5x^2 - 4x - 3 = 0$

30. $x - y = 1,$
$2x - 3y = 6$

31. $2x + y - 3z = 5,$
$3x + 3y - 5z = 4,$
$x - 2y + 2z = 11$

Synthesis

Five-Card Poker Hands.   Suppose that 5 cards are drawn from a deck of 52 cards. For the following exercises, give both a reasoned expression and an answer.

32. *Royal Flush.*   A *royal flush* consists of a 5-card hand with A-K-Q-J-10 of the same suit.

a) How many royal flushes are there?
b) What is the probability of getting a royal flush?

33. *Straight Flush.*   A *straight flush* consists of 5 cards in sequence in the same suit, but excludes royal flushes. An ace can be used low, before a two, or high, following a king.

a) How many straight flushes are there?
b) What is the probability of getting a straight flush?

34. *Four of a Kind.*   A *four-of-a-kind* is a 5-card hand in which 4 of the cards are of the same denomination, such as J-J-J-J-6, 7-7-7-7-A, or 2-2-2-2-5.

a) How many four-of-a-kind hands are there?
b) What is the probability of getting four of a kind?

35. *Full House.*   A *full house* consists of a pair and 3 of a kind, such as Q-Q-Q-4-4.

a) How many full houses are there?
b) What is the probability of getting a full house?

36. *Three of a Kind.*   A *three-of-a-kind* is a 5-card hand in which exactly 3 of the cards are of the same denomination and the other 2 are *not*, such as Q-Q-Q-10-7.

a) How many three-of-a-kind hands are there?
b) What is the probability of getting three of a kind?

37. *Flush.*   An ordinary *flush* is a 5-card hand in which all the cards are of the same suit, but not all in sequence (not a straight flush or royal flush).

a) How many flushes are there?
b) What is the probability of getting a flush?

38. *Two Pairs.*   A hand with *two pairs* is a hand like Q-Q-3-3-A.

a) How many are there?
b) What is the probability of getting two pairs?

39. *Straight.*   An ordinary *straight* is any 5 cards in sequence, but not of the same suit—for example, 4 of spades, 5 of hearts, 6 of diamonds, 7 of hearts, and 8 of clubs.

a) How many straights are there?
b) What is the probability of getting a straight?

Chapter Summary and Review 7

Important Properties and Formulas

Arithmetic Sequences and Series

General term: $a_n = a_{n-1} + d$

$a_n = a_1 + (n - 1)d$

Common difference: d

Sum of the first n terms: $S_n = \dfrac{n}{2}(a_1 + a_n)$

Geometric Sequences and Series

General term: $a_{n+1} = a_n r$

$a_n = a_1 r^{n-1}$

Common ratio: r

Sum of the first n terms: $S_n = \dfrac{a_1(1 - r^n)}{1 - r}$

Sum of an infinite geometric series:

$$S_\infty = \dfrac{a_1}{1 - r}, \quad |r| < 1$$

The Principle of Mathematical Induction

(1) *Basis step*: Prove S_1 is true.

(2) *Induction step*: Prove for all numbers k, $S_k \rightarrow S_{k+1}$.

The Fundamental Counting Principle

The total number of ways in which k actions can be performed together is $n_1 \cdot n_2 \cdot n_3 \cdots n_k$.

Permutations of n Objects Taken n at a Time

$_nP_n = n! = n(n - 1)(n - 2) \cdots 3 \cdot 2 \cdot 1$

Permutations of n Objects Taken k at a Time

$$_nP_k = \underbrace{n(n - 1)(n - 2) \cdots [n - (k - 1)]}_{k \text{ factors}}$$

$$= \dfrac{n!}{(n - k)!}$$

Permutations of Sets With Some Nondistinguishable Objects

$$\dfrac{n!}{n_1! \cdot n_2! \cdots \cdots n_k!}$$

Combinations of n Objects Taken k at a Time

$$_nC_k = \binom{n}{k} = \dfrac{_nP_k}{k!} = \dfrac{n!}{k!\,(n - k)!}$$

$$= \dfrac{n(n - 1)(n - 2) \cdots [n - (k - 1)]}{k!}$$

The Binomial Theorem

$$(a + b)^n = \sum_{k=0}^{n} \binom{n}{k} a^{n-k} b^k$$

The (k + 1)st Term of Binomial Expansion

The $(k + 1)$st term of $(a + b)^n$ is $\binom{n}{k} a^{n-k} b^k$.

The total number of subsets of a set with n elements is 2^n.

Probability Principle P

$$P(E) = \dfrac{m}{n}$$

REVIEW EXERCISES

1. Find the first 4 terms, a_{11}, and a_{23}:

$$a_n = (-1)^n \left(\frac{n^2}{n^4 + 1} \right).$$

2. Predict the general, or nth, term. Answers may vary.

$$2, -5, 10, -17, 26, \ldots$$

3. Find and evaluate:

$$\sum_{k=1}^{4} \frac{(-1)^{k+1} 3^k}{3^k - 1}.$$

4. Use a grapher to construct a table of values and a graph for the first 10 terms of this sequence.

$$a_1 = 0.3, \quad a_{k+1} = 5a_k + 1$$

5. Write sigma notation:

$$0 + 3 + 8 + 15 + 24 + 35 + 48.$$

6. Find the 10th term of the arithmetic sequence

$$\tfrac{3}{4}, \tfrac{13}{12}, \tfrac{17}{12}, \ldots .$$

7. Find the 6th term of the arithmetic sequence

$$a - b, \ a, \ a + b, \ \ldots .$$

8. Find the sum of the first 18 terms of the arithmetic sequence

$$4, 7, 10, \ldots .$$

9. Find the sum of the first 200 natural numbers.

10. The 1st term in an arithmetic sequence is 5, and the 17th term is 53. Find the 3rd term.

11. The common difference in an arithmetic sequence is 3. The 10th term is 23. Find the first term.

12. For a geometric sequence, $a_1 = -2$, $r = 2$, and $a_n = -64$. Find n and S_n.

13. For a geometric sequence, $r = \frac{1}{2}$, $n = 5$, and $S_n = \frac{31}{2}$. Find a_1 and a_n.

Find the sum of each infinite geometric series, if it exists. If you wish, you can also check the results using a grapher.

14. $25 + 27.5 + 30.25 + 33.275 + \cdots$

15. $0.27 + 0.0027 + 0.000027 + \cdots$

16. $\frac{1}{2} - \frac{1}{6} + \frac{1}{18} - \cdots$

17. Find fractional notation for $2.\overline{43}$.

18. Insert four arithmetic means between 5 and 9.

19. *Bouncing Golfball.* A golfball is dropped from a height of 30 ft to the pavement. It always rebounds three fourths of the distance that it drops. How far (up and down) will the ball have traveled when it hits the pavement for the 6th time?

20. *The Amount of an Annuity.* To create a college fund, a parent makes a sequence of 18 yearly deposits of $2000 each in a savings account on which interest is compounded annually at 5.8%. Find the amount of the annuity.

21. *Total Gift.* You receive 10¢ on the first day of the year, 12¢ on the 2nd day, 14¢ on the 3rd day, and so on.

a) How much will you receive on the 365th day?
b) What is the sum of all these 365 gifts?

22. *The Economic Multiplier.* The government is making a $24,000,000,000 expenditure for travel to Mars. If 73% of this amount is spent again, and so on, what is the total effect on the economy?

Use mathematical induction to prove each of the following.

23. For every natural number n,

$$1 + 4 + 7 + \cdots + (3n - 2) = \frac{n(3n - 1)}{2}.$$

24. For every natural number n,

$$1 + 3 + 3^2 + \cdots + 3^{n-1} = \frac{3^n - 1}{2}.$$

25. For every natural number $n \geq 2$,

$$\left(1 - \frac{1}{2}\right)\left(1 - \frac{1}{3}\right) \cdots \left(1 - \frac{1}{n}\right) = \frac{1}{n}.$$

26. *Book Arrangements.* In how many ways can 6 books be arranged on a shelf?

27. *Flag Displays.* If 9 different signal flags are available, how many different displays are possible using 4 flags in a row?

28. *Prize Choices.* The winner of a contest can choose any 8 of 15 prizes. How many different sets of prizes can be chosen?

29. *Fraternity–Sorority Names.* The Greek alphabet contains 24 letters. How many fraternity or sorority names can be formed using 3 different letters?

30. *Letter Arrangements.* In how many distinguishable ways can the letters of the word TENNESSEE be arranged?

31. *Floor Plans.* A manufacturer of houses has 1 floor plan but achieves variety by having 3 different roofs, 4 different ways of attaching the garage, and 3 different types of entrance. Find the number of different houses that can be produced.

32. *Code Symbols.* How many code symbols can be formed using 5 out of 6 of the letters of G, H, I, J, K, L if the letters:

a) cannot be repeated?
b) can be repeated?
c) cannot be repeated but must begin with K?
d) cannot be repeated but must end with IGH?

33. Determine the number of subsets of a set containing 8 members.

Expand.

34. $(m + n)^7$

35. $(x - \sqrt{2})^5$

36. $(x^2 - 3y)^4$

37. $\left(a + \dfrac{1}{a}\right)^8$

38. $(1 + 5i)^6$, where $i^2 = -1$

39. Find the 4th term of $(a + x)^{12}$.

40. Find the 12th term of $(2a - b)^{18}$. Do not multiply out the factorials.

41. *Election Poll.* Before an election, a poll was conducted to see which candidate was favored. Three people were running for a particular office. During the polling, 86 favored candidate A, 97 favored B, and 23 favored C. Assuming that the poll is a valid indicator of the election, what is the probability that the election will be won by A? B? C?

42. *Rolling Dice.* What is the probability of getting a 10 on a roll of a pair of dice? on a roll of 1 die?

43. *Drawing a Card.* From a deck of 52 cards, 1 card is drawn at random. What is the probability that it is a club?

44. *Drawing Three Cards.* From a deck of 52 cards, 3 are drawn at random without replacement. What is the probability that 2 are aces and 1 is a king?

45. *Video Sales.* The table at the top of the next column shows the revenue from video sales in the United States in several recent years.

YEAR	VIDEO SALES REVENUE (IN BILLIONS)
1985	$0.86
1990	3.18
1997	8.24
2000	9.76

Source: Paul Kagan Associates, Inc.

a) Find a linear sequence function $a_n = an + b$ that models the data. Let n represent the number of years after 1985.
b) Use the sequence found in part (a) to predict the revenue in 2005.

Discussion and Writing

46. Write an exercise for a classmate to solve. Design it so that the solution is $_9C_4$.

47. *Chain Business Deals.* Chain letters have been outlawed by the U.S. government. Nevertheless, "chain" business deals still exist and they can be fraudulent. Suppose that a salesperson is charged with the task of hiring 4 new salespersons. Each of them gives half of their profits to the person who hires them. Each of these people hires 4 new salespersons. Each of these gives half of his or her profits to the person who hired them. Half of these profits then go back to the original hiring person. Explain the lure of this business to someone who has managed several sequences of hirings. Explain the fallacy of such a business as well. Keep in mind that there are only 275 million people in the United States.

Synthesis

48. Explain why the following cannot be proved by mathematical induction: For every natural number n,

a) $3 + 5 + \cdots + (2n + 1) = (n + 1)^2$.
b) $1 + 3 + \cdots + (2n - 1) = n^2 + 3$.

49. Suppose that $a_1, a_2, \ldots, a_n$ and $b_1, b_2, \ldots, b_n$ are geometric sequences. Prove that $c_1, c_2, \ldots, c_n$ is a geometric sequence, where $c_n = a_n b_n$.

50. Suppose that $a_1, a_2, \ldots, a_n$ is an arithmetic sequence. Is $b_1, b_2, \ldots, b_n$ an arithmetic sequence if:

a) $b_n = |a_n|$? **b)** $b_n = a_n + 8$?

c) $b_n = 7a_n$? **d)** $b_n = \dfrac{1}{a_n}$?

e) $b_n = \log a_n$? **f)** $b_n = a_n^3$?

51. The zeros of this polynomial function form an arithmetic sequence. Find them.
$$f(x) = x^4 - 4x^3 - 4x^2 + 16x$$

52. Write the first 3 terms of the infinite geometric series with $r = -\frac{1}{3}$ and $S_\infty = \frac{3}{8}$.

53. Simplify:
$$\sum_{k=0}^{10} (-1)^k \binom{10}{k} (\log x)^{10-k} (\log y)^k.$$

Solve for n.

54. $\dbinom{n}{6} = 3 \cdot \dbinom{n-1}{5}$ **55.** $\dbinom{n}{n-1} = 36$

56. Solve for a:
$$\sum_{k=0}^{5} \binom{5}{k} 9^{5-k} a^k = 0.$$

Appendixes

Descartes' Rule of Signs

A

- *Use Descartes' rule of signs to find information about the number of real zeros of a polynomial function with real coefficients.*

The development of a rule that helps determine the number of positive real zeros and the number of negative real zeros of a polynomial function is credited to Descartes. To use the rule, we must have the polynomial arranged in descending or ascending order, with no zero terms written in, the leading coefficient positive, and the *constant term* not 0. The **constant term** is the term that does not contain a variable. Then we determine the number of *variations of sign*, that is, the number of times, in reading through the polynomial, that successive coefficients are of different sign.

EXAMPLE 1 Determine the number of variations of sign in the polynomial function $P(x) = 2x^5 - 3x^2 + x + 4$.

Solution We have

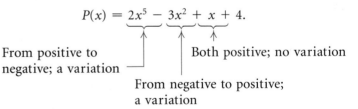

$$P(x) = 2x^5 - 3x^2 + x + 4.$$

From positive to negative; a variation

Both positive; no variation

From negative to positive; a variation

The number of variations of sign is 2.

Note the following:

$$P(-x) = 2(-x)^5 - 3(-x)^2 + (-x) + 4$$
$$= -2x^5 - 3x^2 - x + 4.$$

We see that the number of variations of sign in $P(-x)$ is 1.

We now state Descartes' rule, without proof.

Descartes' Rule of Signs

Let $P(x)$ be a polynomial function with real coefficients and a nonzero constant term. The number of positive real zeros of $P(x)$ is either:

1. The same as the number of variations of sign in $P(x)$, or

2. Less than the number of variations of sign in $P(x)$ by a positive even integer.

The number of negative real zeros of $P(x)$ is either:

3. The same as the number of variations of sign in $P(-x)$, or

4. Less than the number of variations of sign in $P(-x)$ by a positive even integer.

A zero of multiplicity m must be counted m times.

In each of Examples 2–4, what does Descartes' rule of signs tell you about the number of positive real zeros and the number of negative real zeros?

EXAMPLE 2 $P(x) = 2x^5 - 5x^2 - 3x + 6$

Solution   The number of variations of sign in $P(x)$ is 2. Therefore, the number of positive real zeros is either 2 or less than 2 by 2, 4, 6, and so on. Thus the number of positive real zeros is either 2 or 0, since a negative number of zeros has no meaning.

$$P(-x) = -2x^5 - 5x^2 + 3x + 6$$

The number of variations of sign in $P(-x)$ is 1. Thus there is exactly 1 negative real zero. Since nonreal, complex conjugates occur in pairs, we also know the possible ways in which zeros might occur, as summarized in the table shown at left.

TOTAL NUMBER OF ZEROS	5	5
POSITIVE REAL	2	0
NEGATIVE REAL	1	1
NONREAL	2	4

EXAMPLE 3 $P(x) = 5x^4 - 3x^3 + 7x^2 - 12x + 4$

Solution   There are 4 variations of sign. Thus the number of positive real zeros is either

$$4 \quad \text{or} \quad 4 - 2 \quad \text{or} \quad 4 - 4.$$

That is, the number of positive real zeros is 4, 2, or 0.

$$P(-x) = 5x^4 + 3x^3 + 7x^2 + 12x + 4$$

There are 0 changes in sign, so there are no negative real zeros.

EXAMPLE 4 $P(x) = 6x^6 - 2x^2 - 5x$

Solution   As stated, the polynomial does not satisfy the conditions of Descartes' rule of signs because the constant term is 0. But because x is a factor of every term, we know that the polynomial has 0 as a zero. We can then factor as follows:

$$P(x) = x(6x^5 - 2x - 5)$$

Now we analyze $Q(x) = 6x^5 - 2x - 5$ and $Q(-x) = -6x^5 + 2x - 5$. The number of variations of sign in $Q(x)$ is 1. Therefore, there is exactly 1 positive real zero. The number of variations of sign in $Q(-x)$ is 2. Thus the number of negative real zeros is 2 or 0. The same results apply to $P(x)$.

Exercise Set A

What does Descartes' rule of signs tell you about the number of positive real zeros and the number of negative real zeros?

1. $f(x) = 3x^5 - 2x^2 + x - 1$

2. $g(x) = 5x^6 - 3x^3 + x^2 - x$

3. $h(x) = 6x^7 + 2x^2 + 5x + 4$

4. $P(x) = -3x^5 - 7x^3 - 4x - 5$

5. $F(x) = 3p^{18} + 2p^4 - 5p^2 + p + 3$

6. $H(x) = 5t^{12} - 7t^4 + 3t^2 + t + 1$

7. $C(x) = 7x^6 + 3x^4 - x - 10$

8. $g(x) = -z^{10} + 8z^7 + z^3 + 6z - 1$

9. $h(x) = -4t^5 - t^3 + 2t^2 + 1$

10. $P(x) = x^6 + 2x^4 - 9x^3 - 4$

11. $f(x) = y^4 + 13y^3 - y + 5$

12. $Q(x) = x^4 - 2x^2 + 12x - 8$

13. $r(x) = x^4 - 6x^2 + 20x - 24$

14. $f(x) = x^4 - 2x^2 - 8$

15. $R(x) = 3x^4 - 5x^2 - 4$

16. $f(x) = x^4 - 9x^2 - 6x + 4$

17. $f(x) = x^4 - 21x^2 + 4x + 6$

18. $h(x) = x^4 + 3x^2 + 2$

19. $g(x) = x^4 + 5x^2 + 6$

20. $A(x) = x^5 - 14x^3 + 12x^2 + 13x - 12$

21. $P(x) = 3x^9 - 4x^7 - x^5 + 5x^2$

22. $f(x) = 2x^5 + 7x^4 + 10x^2 - x$

Determinants and Cramer's Rule B

- *Evaluate determinants of square matrices.*
- *Use Cramer's rule to solve systems of equations.*

Determinants of Square Matrices

With every square matrix, we associate a number called its *determinant*.

Determinant of a 2 × 2 Matrix

The **determinant** of the matrix $\begin{bmatrix} a & c \\ b & d \end{bmatrix}$ is denoted $\begin{vmatrix} a & c \\ b & d \end{vmatrix}$ and is defined as

$$\begin{vmatrix} a & c \\ b & d \end{vmatrix} = ad - bc.$$

EXAMPLE 1   Evaluate: $\begin{vmatrix} \sqrt{2} & -3 \\ -4 & -\sqrt{2} \end{vmatrix}$.

Solution

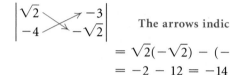   The arrows indicate the products involved.

$$= \sqrt{2}(-\sqrt{2}) - (-4)(-3)$$
$$= -2 - 12 = -14$$

We now consider a way to evaluate determinants of square matrices of order 3×3 or higher.

> ### Minor
>
> For a square matrix $\mathbf{A} = [a_{ij}]$, the **minor** M_{ij} of an element a_{ij} is the determinant of the matrix formed by deleting the ith row and the jth column of $\mathbf{A}$.

EXAMPLE 2   For the matrix

$$[a_{ij}] = \begin{bmatrix} -8 & 0 & 6 \\ 4 & -6 & 7 \\ -1 & -3 & 5 \end{bmatrix},$$

find each of the following.

a) M_{11}

b) M_{23}

Solution

a) We delete the first row and the first column and find the determinant of the 2×2 matrix formed by the remaining elements.

$$\begin{bmatrix} -8 & 0 & 6 \\ 4 & -6 & 7 \\ -1 & -3 & 5 \end{bmatrix} \qquad M_{11} = \begin{vmatrix} -6 & 7 \\ -3 & 5 \end{vmatrix}$$
$$= (-6) \cdot 5 - (-3) \cdot 7$$
$$= -30 - (-21)$$
$$= -30 + 21 = -9$$

b) We delete the second row and the third column and find the determinant of the 2×2 matrix formed by the remaining elements.

$$\begin{bmatrix} -8 & 0 & 6 \\ 4 & -6 & 7 \\ -1 & -3 & 5 \end{bmatrix} \qquad M_{23} = \begin{vmatrix} -8 & 0 \\ -1 & -3 \end{vmatrix}$$
$$= -8(-3) - (-1)0 = 24$$

Cofactor

For a square matrix $\mathbf{A} = [a_{ij}]$, the **cofactor** A_{ij} of an element a_{ij} is given by

$$A_{ij} = (-1)^{i+j}M_{ij},$$

where M_{ij} is the minor of a_{ij}.

EXAMPLE 3 For the matrix given in Example 2, find each of the following.

a) A_{11} **b)** A_{23}

Solution

a) In Example 2, we found that $M_{11} = -9$. Then

$$A_{11} = (-1)^{1+1}(-9) = (1)(-9) = -9.$$

b) In Example 2, we found that $M_{23} = 24$. Then

$$A_{23} = (-1)^{2+3}(24) = (-1)(24) = -24.$$

Evaluating Determinants Using Cofactors

Consider the matrix $\mathbf{A}$ given by

$$\mathbf{A} = \begin{bmatrix} a_{11} & a_{12} & a_{13} \\ a_{21} & a_{22} & a_{23} \\ a_{31} & a_{32} & a_{33} \end{bmatrix}.$$

The determinant of the matrix, denoted $|\mathbf{A}|$, can be found by multiplying each element of the first column by its cofactor and adding:

$$|\mathbf{A}| = a_{11} A_{11} + a_{21} A_{21} + a_{31} A_{31}.$$

Because

$$A_{11} = (-1)^{1+1}M_{11} = M_{11},$$
$$A_{21} = (-1)^{2+1}M_{21} = -M_{21},$$

and $\quad A_{31} = (-1)^{3+1}M_{31} = M_{31},$

we can write

$$|\mathbf{A}| = a_{11} \cdot \begin{vmatrix} a_{22} & a_{23} \\ a_{32} & a_{33} \end{vmatrix} - a_{21} \cdot \begin{vmatrix} a_{12} & a_{13} \\ a_{32} & a_{33} \end{vmatrix} + a_{31} \cdot \begin{vmatrix} a_{12} & a_{13} \\ a_{22} & a_{23} \end{vmatrix}.$$

It can be shown that we can determine $|\mathbf{A}|$ by picking *any* row or column, multiplying each element in that row or column by its cofactor, and adding. This is called *expanding* across a row or down a column. We just expanded down the first column. We now define the determinant of a square matrix of any order.

> **Determinant of Any Square Matrix**
>
> For any square matrix **A** of order $n \times n$ $(n > 1)$, we define the **determinant** of **A**, denoted $|\mathbf{A}|$, as follows. Choose any row or column. Multiply each element in that row or column by its cofactor and add the results. The determinant of a 1×1 matrix is simply the element of the matrix. The value of a determinant will be the same no matter how it is evaluated.

EXAMPLE 4 Evaluate $|\mathbf{A}|$ by expanding across the third row.

$$\mathbf{A} = \begin{bmatrix} -8 & 0 & 6 \\ 4 & -6 & 7 \\ -1 & -3 & 5 \end{bmatrix}$$

Solution We have

$$|\mathbf{A}| = (-1)A_{31} + (-3)A_{32} + 5A_{33}$$

$$= (-1)(-1)^{3+1} \cdot \begin{vmatrix} 0 & 6 \\ -6 & 7 \end{vmatrix} + (-3)(-1)^{3+2} \cdot \begin{vmatrix} -8 & 6 \\ 4 & 7 \end{vmatrix}$$

$$+ 5(-1)^{3+3} \cdot \begin{vmatrix} -8 & 0 \\ 4 & -6 \end{vmatrix}$$

$$= (-1) \cdot 1 \cdot [0 \cdot 7 - (-6)6] + (-3)(-1)[-8 \cdot 7 - 4 \cdot 6]$$

$$+ 5 \cdot 1 \cdot [-8(-6) - 4 \cdot 0]$$

$$= -[36] + 3[-80] + 5[48] = -36 - 240 + 240 = -36.$$

The value of this determinant is -36 no matter which row or column we expand upon.

Determinants can be evaluated on most graphers using the MATRIX package. After entering a matrix on the grapher, we select the determinant operation from the MATRIX MATH menu and enter the name of the matrix. The grapher will return the value of the determinant of the matrix. For example, for

$$\mathbf{A} = \begin{bmatrix} 1 & 6 & -1 \\ -3 & -5 & 3 \\ 0 & 4 & 2 \end{bmatrix}, \qquad \text{we have}$$

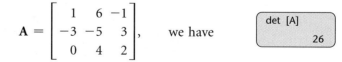

det [A]

    26

Cramer's Rule

Determinants can be used to solve systems of linear equations. Consider a system of two linear equations:

$$a_1 x + b_1 y = c_1,$$
$$a_2 x + b_2 y = c_2.$$

Using the methods of Chapter 5, we obtain

$$x = \frac{c_1 b_2 - c_2 b_1}{a_1 b_2 - a_2 b_1} \quad \text{and} \quad y = \frac{a_1 c_2 - a_2 c_1}{a_1 b_2 - a_2 b_1}.$$

The numerators and denominators of these expressions can be written as determinants:

$$x = \frac{\begin{vmatrix} c_1 & b_1 \\ c_2 & b_2 \end{vmatrix}}{\begin{vmatrix} a_1 & b_1 \\ a_2 & b_2 \end{vmatrix}} \quad \text{and} \quad y = \frac{\begin{vmatrix} a_1 & c_1 \\ a_2 & c_2 \end{vmatrix}}{\begin{vmatrix} a_1 & b_1 \\ a_2 & b_2 \end{vmatrix}}.$$

If we let

$$D = \begin{vmatrix} a_1 & b_1 \\ a_2 & b_2 \end{vmatrix}, \qquad D_x = \begin{vmatrix} c_1 & b_1 \\ c_2 & b_2 \end{vmatrix}, \quad \text{and} \quad D_y = \begin{vmatrix} a_1 & c_1 \\ a_2 & c_2 \end{vmatrix},$$

we have

$$x = \frac{D_x}{D} \quad \text{and} \quad y = \frac{D_y}{D}.$$

This procedure for solving systems of equations is known as *Cramer's rule*.

Cramer's Rule for 2 × 2 Systems

The solution of the system of equations

$$a_1 x + b_1 y = c_1,$$
$$a_2 x + b_2 y = c_2$$

is given by

$$x = \frac{D_x}{D}, \qquad y = \frac{D_y}{D},$$

where

$$D = \begin{vmatrix} a_1 & b_1 \\ a_2 & b_2 \end{vmatrix}, \qquad D_x = \begin{vmatrix} c_1 & b_1 \\ c_2 & b_2 \end{vmatrix},$$

$$D_y = \begin{vmatrix} a_1 & c_1 \\ a_2 & c_2 \end{vmatrix}, \quad \text{and} \quad D \neq 0.$$

Note that the denominator D contains the coefficients of x and y, in the same position as in the original equations. For x, the numerator is obtained by replacing the x-coefficients in D (the a's) with the c's. For y, the numerator is obtained by replacing the y-coefficients in D (the b's) with the c's.

EXAMPLE 5 Solve using Cramer's rule:

$$2x + 5y = 7,$$
$$5x - 2y = -3.$$

Solution We have

$$x = \frac{\begin{vmatrix} 7 & 5 \\ -3 & -2 \end{vmatrix}}{\begin{vmatrix} 2 & 5 \\ 5 & -2 \end{vmatrix}} = \frac{7(-2) - (-3)5}{2(-2) - 5 \cdot 5} = \frac{1}{-29} = -\frac{1}{29},$$

$$y = \frac{\begin{vmatrix} 2 & 7 \\ 5 & -3 \end{vmatrix}}{\begin{vmatrix} 2 & 5 \\ 5 & -2 \end{vmatrix}} = \frac{2(-3) - 5 \cdot 7}{-29} = \frac{-41}{-29} = \frac{41}{29}.$$

The solution is $\left(-\frac{1}{29}, \frac{41}{29}\right)$.

Cramer's rule works only when a system of equations has a unique solution. This occurs when $D \neq 0$. If $D = 0$, $D_x = 0$, and $D_y = 0$, then the equations are dependent. If $D = 0$ and D_x and/or D_y is not 0, then the system is inconsistent.

Cramer's rule can be extended to a system of n linear equations in n variables. We consider a 3×3 system.

Cramer's Rule for 3×3 *Systems*

The solution of the system of equations

$$a_1x + b_1y + c_1z = d_1,$$
$$a_2x + b_2y + c_2z = d_2,$$
$$a_3x + b_3y + c_3z = d_3$$

is given by

$$x = \frac{D_x}{D}, \qquad y = \frac{D_y}{D}, \qquad z = \frac{D_z}{D},$$

where

$$D = \begin{vmatrix} a_1 & b_1 & c_1 \\ a_2 & b_2 & c_2 \\ a_3 & b_3 & c_3 \end{vmatrix}, \qquad D_x = \begin{vmatrix} d_1 & b_1 & c_1 \\ d_2 & b_2 & c_2 \\ d_3 & b_3 & c_3 \end{vmatrix},$$

$$D_y = \begin{vmatrix} a_1 & d_1 & c_1 \\ a_2 & d_2 & c_2 \\ a_3 & d_3 & c_3 \end{vmatrix}, \qquad D_z = \begin{vmatrix} a_1 & b_1 & d_1 \\ a_2 & b_2 & d_2 \\ a_3 & b_3 & d_3 \end{vmatrix}, \quad \text{and } D \neq 0.$$

Note that the determinant D_x is obtained from D by replacing the x-coefficients with d_1, d_2, and d_3. A similar thing happens with D_y and D_z. When $D = 0$, Cramer's rule cannot be used. If $D = 0$ and D_x, D_y, and D_z are 0, the equations are dependent. If $D = 0$ and one of D_x, D_y, or D_z is not 0, then the system is inconsistent.

EXAMPLE 6   Solve using Cramer's rule:

$$x - 3y + 7z = 13,$$
$$x + y + z = 1,$$
$$x - 2y + 3z = 4.$$

Solution   We have

$$D = \begin{vmatrix} 1 & -3 & 7 \\ 1 & 1 & 1 \\ 1 & -2 & 3 \end{vmatrix} = -10, \qquad D_x = \begin{vmatrix} 13 & -3 & 7 \\ 1 & 1 & 1 \\ 4 & -2 & 3 \end{vmatrix} = 20,$$

$$D_y = \begin{vmatrix} 1 & 13 & 7 \\ 1 & 1 & 1 \\ 1 & 4 & 3 \end{vmatrix} = -6, \qquad D_z = \begin{vmatrix} 1 & -3 & 13 \\ 1 & 1 & 1 \\ 1 & -2 & 4 \end{vmatrix} = -24.$$

Then

$$x = \frac{D_x}{D} = \frac{20}{-10} = -2,$$

$$y = \frac{D_y}{D} = \frac{-6}{-10} = \frac{3}{5},$$

$$z = \frac{D_z}{D} = \frac{-24}{-10} = \frac{12}{5}.$$

The solution is $\left(-2, \frac{3}{5}, \frac{12}{5}\right)$. In practice, it is not necessary to evaluate D_z. When we have found values for x and y, we can substitute them into one of the equations to find z.

Exercise Set **B**

Evaluate the determinant algebraically.

1. $\begin{vmatrix} -2 & -\sqrt{5} \\ -\sqrt{5} & 3 \end{vmatrix}$

2. $\begin{vmatrix} \sqrt{5} & -3 \\ 4 & 2 \end{vmatrix}$

3. $\begin{vmatrix} x & 4 \\ x & x^2 \end{vmatrix}$

4. $\begin{vmatrix} y^2 & -2 \\ y & 3 \end{vmatrix}$

5. $\begin{vmatrix} 3 & 1 & 2 \\ -2 & 3 & 1 \\ 3 & 4 & -6 \end{vmatrix}$

6. $\begin{vmatrix} 3 & -2 & 1 \\ 2 & 4 & 3 \\ -1 & 5 & 1 \end{vmatrix}$

7. $\begin{vmatrix} x & 0 & -1 \\ 2 & x & x^2 \\ -3 & x & 1 \end{vmatrix}$ **8.** $\begin{vmatrix} x & 1 & -1 \\ x^2 & x & x \\ 0 & x & 1 \end{vmatrix}$

9. Use a grapher to check your answer to Exercise 1.

10. Use a grapher to check your answer to Exercise 2.

11. Use a grapher to check your answer to Exercise 5.

12. Use a grapher to check your answer to Exercise 6.

Use the following matrix for Exercises 13–20:

$$A = \begin{bmatrix} 7 & -4 & -6 \\ 2 & 0 & -3 \\ 1 & 2 & -5 \end{bmatrix}.$$

13. Find M_{11}, M_{32}, and M_{22}.

14. Find M_{13}, M_{31}, and M_{23}.

15. Find A_{11}, A_{32}, and A_{22}.

16. Find A_{13}, A_{31}, and A_{23}.

17. Evaluate $|A|$ by expanding across the second row.

18. Evaluate $|A|$ by expanding down the second column.

19. Evaluate $|A|$ by expanding down the third column.

20. Evaluate $|A|$ on a grapher.

Use the following matrix for Exercises 21–27:

$$A = \begin{bmatrix} 1 & 0 & 0 & -2 \\ 4 & 1 & 0 & 0 \\ 5 & 6 & 7 & 8 \\ -2 & -3 & -1 & 0 \end{bmatrix}.$$

21. Find M_{41} and M_{33}.

22. Find M_{12} and M_{44}.

23. Find A_{24} and A_{43}.

24. Find A_{22} and A_{34}.

25. Evaluate $|A|$ by expanding across the first row.

26. Evaluate $|A|$ by expanding down the third column.

27. Evaluate $|A|$ on a grapher.

28. Evaluate on a grapher:

$$\begin{vmatrix} 5 & -4 & 2 & -2 \\ 3 & -3 & -4 & 7 \\ -2 & 3 & 2 & 4 \\ -8 & 9 & 5 & -5 \end{vmatrix}.$$

Solve using Cramer's rule.

29. $-2x + 4y = 3,$
$\quad\ 3x - 7y = 1$

30. $5x - 4y = -3,$
$\quad\ 7x + 2y = 6$

31. $2x - y = 5,$
$\quad\ x - 2y = 1$

32. $3x + 4y = -2,$
$\quad\ 5x - 7y = 1$

33. $2x + 9y = -2,$
$\quad\ 4x - 3y = 3$

34. $2x + 3y = -1,$
$\quad\ 3x + 6y = -0.5$

35. $2x + 5y = 7,$
$\quad\ 3x - 2y = 1$

36. $3x + 2y = 7,$
$\quad\ 2x + 3y = -2$

37. $3x + 2y - z = 4,$
$\quad\ 3x - 2y + z = 5,$
$\quad\ 4x - 5y - z = -1$

38. $\quad 3x - y + 2z = 1,$
$\qquad\ x - y + 2z = 3,$
$\quad -2x + 3y + z = 1$

39. $3x + 5y - z = -2,$
$\quad\ x - 4y + 2z = 13,$
$\quad 2x + 4y + 3z = 1$

40. $3x + 2y + 2z = 1,$
$\quad 5x - y - 6z = 3,$
$\quad 2x + 3y + 3z = 4$

41. $\ x - 3y - 7z = 6,$
$\ 2x + 3y + z = 9,$
$\ 4x + y \qquad\ = 7$

42. $\ x - 2y - 3z = 4,$
$\ 3x - \qquad 2z = 8,$
$\ 2x + y + 4z = 13$

43. $\qquad 6y + 6z = -1,$
$\ 8x \qquad + 6z = -1,$
$\ 4x + 9y \qquad\ = 8$

44. $3x + 5y \qquad\ = 2,$
$\ 2x \qquad\ - 3z = 7,$
$\qquad 4y + 2z = -1$

Discussion and Writing

45. Explain why the system of equations

$$a_1 x + b_1 y = c_1,$$
$$a_2 x + b_2 y = c_2$$

is either dependent or inconsistent when

$$\begin{vmatrix} a_1 & b_1 \\ a_2 & b_2 \end{vmatrix} = 0.$$

46. If the lines $a_1 x + b_1 y = c_1$ and $a_2 x + b_2 y = c_2$ are parallel, what can you say about the values of

$$\begin{vmatrix} a_1 & b_1 \\ a_2 & b_2 \end{vmatrix}, \quad \begin{vmatrix} c_1 & b_1 \\ c_2 & b_2 \end{vmatrix}, \quad \text{and} \quad \begin{vmatrix} a_1 & c_1 \\ a_2 & c_2 \end{vmatrix}?$$

Synthesis

Solve.

47. $\begin{vmatrix} x & 5 \\ -4 & x \end{vmatrix} = 24$

48. $\begin{vmatrix} y & 2 \\ 3 & y \end{vmatrix} = y$

49. $\begin{vmatrix} x & -3 \\ -1 & x \end{vmatrix} \geq 0$

50. $\begin{vmatrix} y & -5 \\ -2 & y \end{vmatrix} < 0$

51. $\begin{vmatrix} x+3 & 4 \\ x-3 & 5 \end{vmatrix} = -7$

52. $\begin{vmatrix} m+2 & -3 \\ m+5 & -4 \end{vmatrix} = 3m - 5$

53. $\begin{vmatrix} 2 & x & 1 \\ 1 & 2 & -1 \\ 3 & 4 & -2 \end{vmatrix} = -6$

54. $\begin{vmatrix} x & 2 & x \\ 3 & -1 & 1 \\ 1 & -2 & 2 \end{vmatrix} = -10$

Rewrite the expression using a determinant. Answers may vary.

55. $2L + 2W$

56. $\pi r + \pi h$

57. $a^2 + b^2$

58. $\frac{1}{2} h(a + b)$

59. $2\pi r^2 + 2\pi rh$

60. $x^2 y^2 - Q^2$

Parametric Equations
C

- *Graph parametric equations and determine an equivalent rectangular equation.*
- *Determine parametric equations for a rectangular equation.*

Graphing Parametric Equations

Much of our graphing of curves in this text has been with rectangular equations involving only two variables, x and y. We graphed sets of ordered pairs, (x, y), where y is a function of x. In this section, we introduce a third variable, t, such that x and y are each a function of t.

Consider a point P with coordinates (x, y) in the rectangular coordinate plane. As P moves in the plane, its location at time t is given by two functions

$$x = f(t) \quad \text{and} \quad y = g(t).$$

For example, let the equations be

$$x = \tfrac{1}{2}t \quad \text{and} \quad y = t^2 - 3, \quad t \geq 0.$$

We restrict t to nonnegative values since t represents time. When $t = 2$, we have

$$x = \tfrac{1}{2} \cdot 2 = 1 \quad \text{and} \quad y = 2^2 - 3 = 1.$$

Thus after 2 sec, the coordinates of P are $(1, 1)$. The table below lists other ordered pairs. We plot them and draw the curve for $t \geq 0$.

t	x	y	(x, y)
0	0	-3	$(0, -3)$
1	$\frac{1}{2}$	-2	$\left(\frac{1}{2}, -2\right)$
2	1	1	$(1, 1)$
3	$\frac{3}{2}$	6	$\left(\frac{3}{2}, 6\right)$

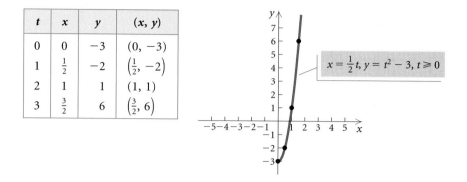

$x = \frac{1}{2}t, y = t^2 - 3, t \geq 0$

This curve appears to be part of a parabola. Let's verify this by finding the equivalent rectangular equation. Solving $x = \frac{1}{2}t$ for t, we get $t = 2x$. Substituting $2x$ for t in $y = t^2 - 3$, we have

$$y = (2x)^2 - 3 = 4x^2 - 3.$$

We know from Section 2.4 that this is a quadratic equation and that the graph is a parabola. Thus the curve is part of the parabola $y = 4x^2 - 3$.

The equations $x = \frac{1}{2}t$ and $y = t^2 - 3$ are called the **parametric equations** for the curve. The variable t is called the **parameter**. The equation $y = 4x^2 - 3$, with the restriction $x \geq 0$, is the equivalent rectangular equation for the curve. Since $t \geq 0$ and $x = \frac{1}{2}t$, the restriction $x \geq 0$ must be included. In this example, t represents time, but t, in general, can represent any real number in a specified interval. One advantage of parametric equations is that the restriction on the parameter t allows us to investigate a portion of the curve.

Parametric Equations

If f and g are continuous functions of t on an interval I, then the set of ordered pairs (x, y) such that $x = f(t)$ and $y = g(t)$ is a **plane curve**. The equations $x = f(t)$ and $y = f(t)$ are **parametric equations** for the curve. The variable t is the **parameter**.

Plane curves described with parametric equations can also be graphed on graphers. Consult your grapher's manual for specific instructions.

EXAMPLE 1   Using a grapher, graph each of the following plane curves given their respective parametric equations and the restriction for the parameter. Then find the equivalent rectangular equation.

a) $x = t^2$, $y = t - 1$; $-1 \leq t \leq 4$
b) $x = \sqrt{t}$, $y = 2t + 3$; $0 \leq t \leq 3$

Solution

a) When using a grapher set in PARAMETRIC mode, we must set minimum and maximum values for x, y, and t.

WINDOW
  Tmin = -1
  Tmax = 4
  Tstep = .1
  Xmin = -2
  Xmax = 18
  Xscl = 1
  Ymin = -4
  Ymax = 4
  Yscl = 1

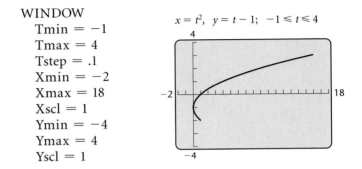

Using the TRACE feature, we can see the T, X, and Y values along the curve. To find an equivalent rectangular equation, we can solve either equation for t. Let's select the equation $y = t - 1$ and solve for t:

$$y = t - 1$$
$$y + 1 = t.$$

We then substitute $y + 1$ for t in $x = t^2$:

$$x = t^2$$
$$x = (y + 1)^2. \qquad \textbf{Parabola}$$

This is an equation of a parabola that opens to the right. Given that $-1 \le t \le 4$, we have the corresponding restrictions on x and y: $0 \le x \le 16$ and $-2 \le y \le 3$. Thus the equivalent rectangular equation is

$$x = (y + 1)^2; \quad 0 \le x \le 16, \quad -2 \le y \le 3.$$

b) Using a grapher set in PARAMETRIC mode, we have

WINDOW
  Tmin = 0
  Tmax = 3
  Tstep = .1
  Xmin = -3
  Xmax = 3
  Xscl = 1
  Ymin = -2
  Ymax = 10
  Yscl = 1

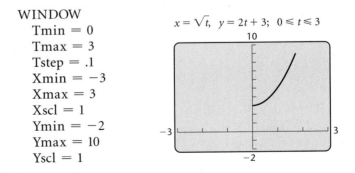

To find an equivalent rectangular equation, we first solve $x = \sqrt{t}$ for t:

$$x = \sqrt{t}$$
$$x^2 = t.$$

Then we substitute x^2 for t in $y = 2t + 3$:

$$y = 2t + 3$$
$$y = 2x^2 + 3. \qquad \textbf{Parabola}$$

When $0 \le t \le 3$, $0 \le x \le \sqrt{3}$ and $3 \le y \le 9$. The equivalent rectangular equation is

$$y = 2x^2 + 3; \quad 0 \le x \le \sqrt{3}, \quad 3 \le y \le 9.$$

Determining Parametric Equations for a Given Rectangular Equation

Many sets of parametric equations can represent the same plane curve. In fact, there are infinitely many such equations.

EXAMPLE 2 Find three sets of parametric equations for the parabola

$$y = 4 - (x + 3)^2.$$

Solution

If $x = t$, then $y = 4 - (t + 3)^2$.

If $x = t - 3$, then $y = 4 - (t - 3 + 3)^2$, or $4 - t^2$.

If $x = \dfrac{t}{3}$, then $y = 4 - \left(\dfrac{t}{3} + 3\right)^2$, or $-\dfrac{t^2}{9} - 2t - 5$.

Exercise Set C

Using a grapher, graph the plane curve given by the set of parametric equations and the restriction for the parameter. Then find the equivalent rectangular equation.

1. $x = \frac{1}{2}t$, $y = 6t - 7$; $-1 \le t \le 6$

2. $x = t$, $y = 5 - t$; $-2 \le t \le 3$

3. $x = t^3$, $y = t - 4$; $-1 \le t \le 10$

4. $x = \sqrt{t}$, $y = 2t + 3$; $0 \le t \le 8$

5. $x = t^2$, $y = \sqrt{t}$; $0 \le t \le 4$

6. $x = t^3 + 1$, $y = t$; $-3 \le t \le 3$

7. $x = t + 3$, $y = \dfrac{1}{t + 3}$; $-2 \le t \le 2$

8. $x = 2t^3 + 1$, $y = 2t^3 - 1$; $-4 \le t \le 4$

9. $x = 2t - 1$, $y = t^2$; $-3 \le t \le 3$

10. $x = \frac{1}{3}t$, $y = t$; $-5 \le t \le 5$

11. $x = e^{-t}$, $y = e^t$; $-\infty < t < \infty$

12. $x = 2 \ln t$, $y = t^2$; $0 < t < \infty$

Find two sets of parametric equations for the rectangular equation.

13. $y = 4x - 3$

14. $y = x^2 - 1$

15. $y = (x - 2)^2 - 6x$

16. $y = x^3 + 3$

Synthesis

17. Determine a set of parametric equations for the line that passes through $(-2, 4)$ and $(-3, 8)$.

18. Determine a set of parametric equations for the circle with center $(-1, 2)$ and radius 5.

Answers

Introduction to Graphs and the Graphing Calculator

1.

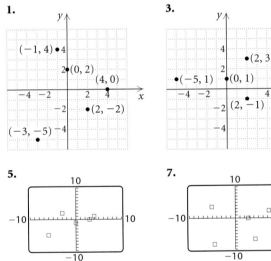

3.

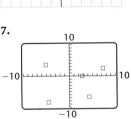

5.

7.

9. Yes; no   **11.** Yes; no   **13.** No; yes   **15.** No; yes

17.

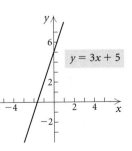

19.

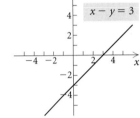

21.

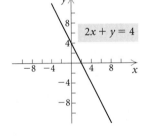

23.

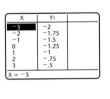

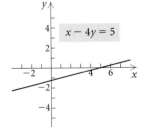

A-1

25.

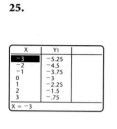 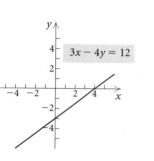

27.

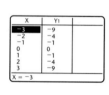 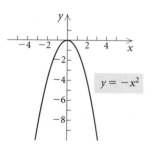

29.

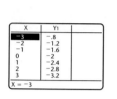 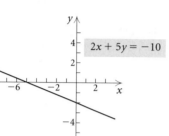

31.

 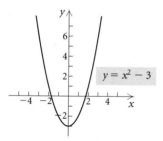

33.

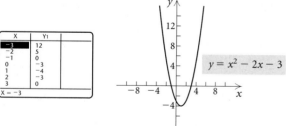

35. (b)   **37.** (a)

39.

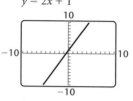

41.

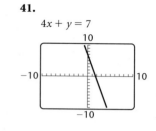

43.

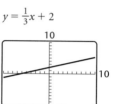

45.

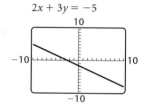

47.

49.

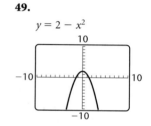

51.

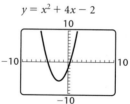

53. $[-25, 25, -25, 25]$ **55.** Standard window
57. $[-1, 1, -0.3, 0.3]$ **59.** $(1, 5)$ **61.** $\left(\frac{5}{3}, \frac{7}{3}\right)$
63. $\left(\frac{11}{6}, \frac{1}{3}\right)$ **65.** $(-2, 4), (2, 4)$
67. $(-0.8284271, 3.1715729), (4.8284271, 8.8284271)$

Chapter R

Exercise Set R.1

1. $\sqrt[3]{8}, 0, 9, \sqrt{25}$ **3.** $\sqrt{7}, 5.242242224\ldots, -\sqrt{14}, \sqrt[5]{5},$
$\sqrt[3]{4}$ **5.** $-12, 5.\overline{3}, -\frac{7}{3}, \sqrt[3]{8}, 0, -1.96, 9, 4\frac{2}{3}, \sqrt{25}, \frac{5}{7}$
7. Answers may vary; $\frac{3}{4}$ **9.** Answers may vary; -5
11. $[-3, 3]$ **13.** $[-14, -11)$ **15.** $(-\infty, -4]$

17. $(-\infty, 3.8)$ **19.** $(-\infty, 7) \cup (7, \infty)$ **21.** $(0, 5)$
23. $[-9, -4)$ **25.** $[x, x + h]$ **27.** (p, ∞) **29.** True
31. False   **33.** True   **35.** False   **37.** False   **39.** True
41. True   **43.** True   **45.** False   **47.** Commutative
property of multiplication   **49.** Multiplicative identity
property   **51.** Associative property of multiplication
53. Commutative property of multiplication
55. Commutative property of addition   **57.** Multiplicative
inverse property   **59.** 7.1   **61.** $\frac{5}{4}$ **63.** 11   **65.** 6
67. 5.4   **69.** Discussion and Writing
71. Answers may vary; $0.124124412444\ldots$
73. Answers may vary; -0.00999
75.

Exercise Set R.2

1. 1   **3.** 5^2, or 25   **5.** 1   **7.** 7^{-1}, or $\frac{1}{7}$ **9.** $6x^5$
11. $15a^{-1}b^5$, or $\dfrac{15b^5}{a}$ **13.** $72x^5$ **15.** b^3 **17.** x^3y^{-3},
or $\dfrac{x^3}{y^3}$ **19.** $8xy^{-5}$, or $\dfrac{8x}{y^5}$ **21.** $8a^3b^6$ **23.** $16x^{12}$
25. $\dfrac{c^2d^4}{25}$ **27.** $\dfrac{8^5a^{20}c^{10}}{b^{25}}$ **29.** 4.05×10^5
31. 3.9×10^{-7} **33.** 1.6×10^{-5} **35.** 0.000083
37. $20{,}700{,}000$ **39.** $11{,}358{,}000{,}000$ **41.** 1.395×10^3
43. 8×10^{-14} **45.** $\$1.19 \times 10^7$ **47.** 1.332×10^{14}
49. 2   **51.** 2048   **53.** 5   **55.** $\$2883.67$ **57.** $\$8763.54$
59. Discussion and Writing   **61.** x^{8t} **63.** t^{8x}
65. $9x^{2a}y^{2b}$ **67.** $\$660.29$ **69.** $\$942.54$ **71.** $x \leq 1$

Exercise Set R.3

1. $-5y^4, 3y^3, 7y^2, -y, -4; 4$ **3.** $3a^4b, -7a^3b^3, 5ab, -2; 6$
5. $3x^2y - 5xy^2 + 7xy + 2$ **7.** $3x + 2y - 2z - 3$
9. $-2x^2 + 6x - 2$ **11.** $x^4 - 3x^3 - 4x^2 + 9x - 3$
13. $2a^4 - 2a^3b - a^2b + 4ab^2 - 3b^3$ **15.** $x^2 + 2x - 15$
17. $2a^2 + 13a + 15$ **19.** Left to the student
21. $4x^2 + 8xy + 3y^2$ **23.** $y^2 + 10y + 25$
25. $25x^2 - 30x + 9$ **27.** Left to the student
29. $4x^2 + 12xy + 9y^2$ **31.** $4x^4 - 12x^2y + 9y^2$
33. $a^2 - 9$ **35.** Left to the student   **37.** $9x^2 - 4y^2$
39. $4x^2 + 12xy + 9y^2 - 16$ **41.** $x^4 - 1$
43. Discussion and Writing   **45.** $a^{2n} - b^{2n}$
47. $a^{2n} + 2a^nb^n + b^{2n}$ **49.** $x^6 - 1$
51. $x^{a^2 - b^2}$ **53.** $a^2 + b^2 + c^2 + 2ab + 2ac + 2bc$

Exercise Set R.4

1. $2(x - 5)$ **3.** $3x^2(x^2 - 3)$ **5.** $4(a^2 - 3a + 4)$
7. $(b - 2)(a + c)$ **9.** $(x + 3)(x^2 + 6)$

11. $(y - 3)(y + 2)(y - 2)$ **13.** Left to the student
15. $(p + 2)(p + 4)$ **17.** $(2n - 7)(n + 8)$
19. $(y^2 - 7)(y^2 + 3)$ **21.** Left to the student
23. $(3x + 5)(3x - 5)$ **25.** $4x(y^2 + z)(y^2 - z)$
27. $(y - 3)^2$ **29.** $(1 - 4x)^2$ **31.** Left to the student
33. $(x + 2)(x^2 - 2x + 4)$ **35.** $(m - 1)(m^2 + m + 1)$
37. Left to the student   **39.** $3ab(6a - 5b)$
41. $(x - 4)(x^2 + 5)$ **43.** $8(x + 2)(x - 2)$ **45.** Prime
47. Left to the student   **49.** $(x + 4)(x + 5)$
51. $(y - 5)(y - 1)$ **53.** $(2a + 1)(a + 4)$
55. $(3x - 1)(2x + 3)$ **57.** $(y - 9)^2$ **59.** $(xy - 7)^2$
61. $4a(x + 7)(x - 2)$ **63.** $3(z - 2)(z^2 + 2z + 4)$
65. $2ab(2a^2 + 3b^2)(4a^4 - 6a^2b^2 + 9b^4)$
67. Left to the student   **69.** Discussion and Writing
71. $(y^2 + 12)(y^2 - 7)$ **73.** $\left(y + \frac{4}{7}\right)\left(y - \frac{2}{7}\right)$
75. $h(3x^2 + 3xh + h^2)$ **77.** $(y + 4)(y - 7)$
79. $(x^n + 8)(x^n - 3)$ **81.** $(x + a)(x + b)$
83. $(5y^m + x^n - 1)(5y^m - x^n + 1)$ **85.** $y(y - 1)^2(y - 2)$

Exercise Set R.5

1. $\{x \mid x \text{ is a real number}\}$ **3.** $\{x \mid x \text{ is a real number } and$
$x \neq 0 \text{ and } x \neq 1\}$ **5.** $\{x \mid x \text{ is a real number } and \, x \neq -2$
$and \, x \neq 2 \text{ and } x \neq -5\}$
7.

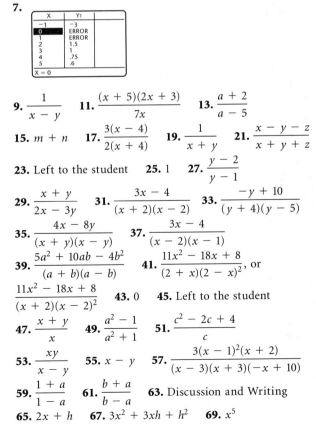

9. $\dfrac{1}{x - y}$ **11.** $\dfrac{(x + 5)(2x + 3)}{7x}$ **13.** $\dfrac{a + 2}{a - 5}$
15. $m + n$ **17.** $\dfrac{3(x - 4)}{2(x + 4)}$ **19.** $\dfrac{1}{x + y}$ **21.** $\dfrac{x - y - z}{x + y + z}$
23. Left to the student   **25.** 1   **27.** $\dfrac{y - 2}{y - 1}$
29. $\dfrac{x + y}{2x - 3y}$ **31.** $\dfrac{3x - 4}{(x + 2)(x - 2)}$ **33.** $\dfrac{-y + 10}{(y + 4)(y - 5)}$
35. $\dfrac{4x - 8y}{(x + y)(x - y)}$ **37.** $\dfrac{3x - 4}{(x - 2)(x - 1)}$
39. $\dfrac{5a^2 + 10ab - 4b^2}{(a + b)(a - b)}$ **41.** $\dfrac{11x^2 - 18x + 8}{(2 + x)(2 - x)^2}$, or
$\dfrac{11x^2 - 18x + 8}{(x + 2)(x - 2)^2}$ **43.** 0   **45.** Left to the student
47. $\dfrac{x + y}{x}$ **49.** $\dfrac{a^2 - 1}{a^2 + 1}$ **51.** $\dfrac{c^2 - 2c + 4}{c}$
53. $\dfrac{xy}{x - y}$ **55.** $x - y$ **57.** $\dfrac{3(x - 1)^2(x + 2)}{(x - 3)(x + 3)(-x + 10)}$
59. $\dfrac{1 + a}{1 - a}$ **61.** $\dfrac{b + a}{b - a}$ **63.** Discussion and Writing
65. $2x + h$ **67.** $3x^2 + 3xh + h^2$ **69.** x^5

71. $\dfrac{(n + 1)(n + 2)(n + 3)}{2 \cdot 3}$

73. $\dfrac{x^3 + 2x^2 + 11x + 20}{2(x + 1)(2 + x)}$

Exercise Set R.6

1. 11 **3.** $4|y|$ **5.** $|b + 1|$ **7.** $-3x$ **9.** $|x - 2|$
11. 2 **13.** $6\sqrt{5}$ **15.** $3\sqrt[3]{2}$ **17.** $8\sqrt{2}|c|d^2$ **19.** Left to
the student **21.** $2x^2y\sqrt{6}$ **23.** $3x\sqrt[3]{4y}$
25. $2(x + 4)\sqrt[3]{(x + 4)^2}$ **27.** $\dfrac{m^2n^4}{2}$ **29.** 2 **31.** $\dfrac{1}{2x}$
33. $\dfrac{4a\sqrt[3]{a}}{3b}$ **35.** $\dfrac{x\sqrt{7x}}{6y^3}$ **37.** $51\sqrt{2}$
39. $-2x\sqrt{2} - 12\sqrt{5x}$ **41.** 1 **43.** $4 + 2\sqrt{3}$
45. Left to the student **47.** About 13,709.5 ft
49. (a) $h = \dfrac{a}{2}\sqrt{3}$; (b) $A = \dfrac{a^2}{4}\sqrt{3}$ **51.** 8
53. $\dfrac{\sqrt{6}}{3}$ **55.** $\dfrac{\sqrt[3]{10}}{2}$ **57.** $\dfrac{2\sqrt[3]{6}}{3}$ **59.** $\dfrac{9 - 3\sqrt{5}}{2}$
61. $\dfrac{6\sqrt{m} + 6\sqrt{n}}{m - n}$ **63.** $\dfrac{6}{5\sqrt{3}}$ **65.** $\dfrac{7}{\sqrt[3]{98}}$ **67.** $\dfrac{11}{\sqrt{33}}$
69. $\dfrac{76}{27 + 3\sqrt{5} - 9\sqrt{3} - \sqrt{15}}$ **71.** $\dfrac{a - b}{3a\sqrt{a} - 3a\sqrt{b}}$
73. $\sqrt[4]{x^3}$ **75.** 8 **77.** $\dfrac{1}{5}$ **79.** $\dfrac{a\sqrt[4]{a}}{\sqrt[4]{b^3}}$, or $a\sqrt[4]{\dfrac{a}{b^3}}$
81. $13\sqrt[4]{13}$ **83.** $20^{2/3}$ **85.** $11^{1/6}$ **87.** $5^{5/6}$ **89.** 4
91. $8a^2$ **93.** $\dfrac{x^3}{3b^{-2}}$, or $\dfrac{x^3b^2}{3}$ **95.** $x\sqrt[3]{y}$ **97.** Left to the
student **99.** $\sqrt[6]{288}$ **101.** $\sqrt[12]{x^{11}y^7}$ **103.** $a\sqrt[6]{a^5}$
105. $(a + x)\sqrt[12]{(a + x)^{11}}$ **107.** Discussion and Writing
109. $\dfrac{(2 + x^2)\sqrt{1 + x^2}}{1 + x^2}$ **111.** $a^{a/2}$
113. (a) 0, 1; (b) $x > 1$; (c) $0 < x < 1$

Exercise Set R.7

1. 4 **3.** 8 **5.** $\dfrac{4}{5}$ **7.** $-\dfrac{3}{2}$ **9.** $\dfrac{2}{3}, \dfrac{3}{2}$ **11.** $-1, \dfrac{2}{3}$
13. $-\sqrt{3}, \sqrt{3}$ **15.** 0, 3 **17.** $-\dfrac{1}{3}, 0, 2$ **19.** $-1, -\dfrac{1}{7}, 1$
21. $b = \dfrac{2A}{h}$ **23.** $w = \dfrac{P - 2l}{2}$ **25.** $h = \dfrac{2A}{b_1 + b_2}$
27. $\pi = \dfrac{3V}{4r^3}$ **29.** $C = \dfrac{5}{9}(F - 32)$ **31.** Discussion and
Writing **33.** $-\dfrac{2}{3}$ **35.** 1 **37.** $-\sqrt{7}, -\dfrac{3}{2}, 0, \dfrac{1}{3}, \sqrt{7}$

Review Exercises, Chapter R

1. [R.1] 12, -3, -1, -19, 31, 0 **2.** [R.1] 12, 31
3. [R.1] -43.89, 12, -3, $-\dfrac{1}{5}$, -1, $-\dfrac{4}{3}$, $7\dfrac{2}{3}$, -19, 31, 0
4. [R.1] All of them **5.** [R.1] $\sqrt{7}$, $\sqrt[3]{10}$
6. [R.1] 12, 31, 0 **7.** [R.1] $[-3, 5)$ **8.** [R.1] 3.5
9. [R.1] 16 **10.** [R.1] 10 **11.** [R.2] 117 **12.** [R.2] -10
13. [R.2] 3,261,000 **14.** [R.2] 0.00041
15. [R.2] 1.432×10^{-2} **16.** [R.2] 4.321×10^4
17. [R.2] 7.8125×10^{-22} **18.** [R.2] 5.46×10^{-32}
19. [R.2] $-14a^{-2}b^7$, or $\dfrac{-14b^7}{a^2}$
20. [R.2] $6x^9y^{-6}z^6$, or $\dfrac{6x^9z^6}{y^6}$ **21.** [R.6] 3 **22.** [R.6] -2
23. [R.5] $\dfrac{b}{a}$ **24.** [R.5] $\dfrac{x + y}{xy}$ **25.** [R.6] -4
26. [R.6] $25x^4 - 10\sqrt{2}x^2 + 2$ **27.** [R.6] $13\sqrt{5}$
28. [R.3] $x^3 + t^3$ **29.** [R.3] $10a^2 - 7ab - 12b^2$
30. [R.3] $8xy^4 - 9xy^2 + 4x^2 + 2y - 7$
31. [R.4] $(x + 2)(x^2 - 3)$
32. [R.4] $3a(2a + 3b^2)(2a - 3b^2)$ **33.** [R.4] $(x + 12)^2$
34. [R.4] $x(9x - 1)(x + 4)$
35. [R.4] $(2x - 1)(4x^2 + 2x + 1)$
36. [R.4] $(3x^2 + 5y^2)(9x^4 - 15x^2y^2 + 25y^4)$
37. [R.4] $6(x + 2)(x^2 - 2x + 4)$
38. [R.4] $(x - 1)(2x + 3)(2x - 3)$
39. [R.4] $(3x - 5)^2$ **40.** [R.4] $3(6x^2 - x + 2)$
41. [R.4] $(ab - 3)(ab + 2)$ **42.** [R.5] 3
43. [R.5] $\dfrac{x - 5}{(x + 5)(x + 3)}$ **44.** [R.6] $y^3\sqrt[6]{y}$
45. [R.6] $\sqrt[3]{(a + b)^2}$ **46.** [R.6] $b\sqrt[5]{b^2}$ **47.** [R.6] $\dfrac{m^4n^2}{3}$
48. [R.6] $\dfrac{x - 2\sqrt{xy} + y}{x - y}$ **49.** [R.6] About 18.8 ft
50. [R.7] 4 **51.** [R.7] $-5, 1$ **52.** [R.7] $-2, \dfrac{4}{3}$
53. [R.7] $-\sqrt{3}, \sqrt{3}$ **54.** [R.7] $-\sqrt{10}, \sqrt{10}$
55. [R.7] $h = \dfrac{V}{lw}$ **56.** [R.7] $s = \dfrac{M - n}{0.3}$
57. Discussion and Writing [R.5], [R.7] Two expressions
are equivalent if they have the same value for all numbers
in both domains. Two equations are equivalent if they
have the same solution set.
58. Discussion and Writing [R.7] If an equation contains
no fractions, using the addition principle before using
the multiplication principle eliminates the need to add
or subtract fractions.
59. [R.3] $x^{2n} + 6x^n - 40$ **60.** [R.3] $t^{2a} + 2 + t^{-2a}$
61. [R.3] $y^{2b} - z^{2c}$
62. [R.3] $a^{3n} - 3a^{2n}b^n + 3a^nb^{2n} - b^{3n}$
63. [R.4] $(y^n + 8)^2$ **64.** [R.4] $(x^t - 7)(x^t + 4)$
65. [R.4] $m^{3n}(m^n - 1)(m^{2n} + m^n + 1)$ **66.** [R.7] 256

Chapter 1

Exercise Set 1.1

1. Yes   **3.** Yes   **5.** No   **7.** Yes   **9.** Yes   **11.** Yes
13. No   **15.** Function; domain: $\{2, 3, 4\}$;
range: $\{10, 15, 20\}$ **17.** Not a function;
domain: $\{-7, -2, 0\}$; range: $\{3, 1, 4, 7\}$
19. Function; domain: $\{-2, 0, 2, 4, -3\}$; range: $\{1\}$
21. $f(-1) = 2; f(0) = 0; f(1) = -2$
23. (a) 1; **(b)** 6; **(c)** 22; **(d)** $3x^2 + 2x + 1$; **(e)** $3t^2 - 4t + 2$
25. (a) 8; **(b)** -8; **(c)** $-x^3$; **(d)** $27y^3$; **(e)** $8 + 12h + 6h^2 + h^3$
27. (a) $\dfrac{1}{8}$; **(b)** 0; **(c)** does not exist; **(d)** $\dfrac{81}{53}$, or

approximately 1.5283; **(e)** $\dfrac{x + h - 4}{x + h + 3}$

29. 0; does not exist; does not exist as a real number; $\dfrac{1}{\sqrt{3}}$,

or $\dfrac{\sqrt{3}}{3}$ **31.** All real numbers, or $(-\infty, \infty)$
33. $\{x \mid x \neq 0\}$, or $(-\infty, 0) \cup (0, \infty)$ **35.** $\{x \mid x \neq 2\}$, or
$(-\infty, 2) \cup (2, \infty)$ **37.** $\{x \mid x \neq -1 \ and \ x \neq 5\}$, or
$(-\infty, -1) \cup (-1, 5) \cup (5, \infty)$ **39.** Domain: all real
numbers; range: $[0, \infty)$ **41.** Domain: $[-3, 3]$; range: $[0, 3]$
43. Domain: all real numbers; range: all real numbers
45. Domain: $(-\infty, 7]$; range: $[0, \infty)$
47. Domain: all real numbers; range: $(-\infty, 3]$ **49.** No
51. Yes   **53.** Yes   **55.** No   **57.** Function; domain: $[0, 5]$;
range: $[0, 3]$ **59.** Function; domain: $[-2\pi, 2\pi]$;
range: $[-1, 1]$ **61.** 645 m; 0 m   **63.** 0.4 acre; 20.4 acres;
50.6 acres; 416.9 acres; 1033.6 acres   **65.** Discussion and
Writing   **67.** $f(x) = x, g(x) = x + 1$ **69.** 35

Exercise Set 1.2

1. (a) Yes; **(b)** yes; **(c)** yes   **3. (a)** Yes; **(b)** no; **(c)** no
5. $-\dfrac{3}{5}$ **7.** 0   **9.** $\dfrac{1}{5}$ **11.** $-\dfrac{5}{3}$ **13.** Not defined
15. $2a + h$ **17.** 6.7% grade; $y = 0.067x$
19. $100 per year   **21.** $\dfrac{1}{6}$ km per minute   **23.** $\dfrac{3}{5}$; $(0, -7)$
25. $-\dfrac{1}{2}$; $(0, 5)$ **27.** $-\dfrac{3}{2}$; $(0, 5)$ **29.** $\dfrac{4}{3}$; $(0, -7)$
31. $y = \dfrac{2}{9}x + 4$ **33.** $y = -4x - 7$ **35.** $y = -4.2x + \dfrac{3}{4}$
37. $y = \dfrac{2}{9}x + \dfrac{19}{3}$ **39.** $y = 3x - 5$ **41.** $y = -\dfrac{3}{5}x - \dfrac{17}{5}$
43. $y = -3x + 2$ **45.** $y = -\dfrac{1}{2}x + \dfrac{7}{2}$ **47.** Parallel
49. Perpendicular   **51.** $y = \dfrac{2}{7}x + \dfrac{29}{7}$; $y = -\dfrac{7}{2}x + \dfrac{31}{2}$
53. $y = -0.3x - 2.1$; $y = \dfrac{10}{3}x + \dfrac{70}{3}$
55. $y = -\dfrac{3}{4}x + \dfrac{1}{4}$; $y = \dfrac{4}{3}x - 6$ **57.** $x = 3$; $y = -3$

59. (a) $W(h) = 3.5h - 110$; **(b)**

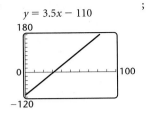

$y = 3.5x - 110$

(c) 107 lb; **(d)** $\{h \mid h > 31.43\}$, or $(31.43, \infty)$

61. (a)

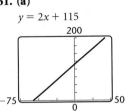

$y = 2x + 115$

(b) 115, 75, 135, 179;
(c) Below $-57.5°$, stopping
distance is negative; above
$32°$, ice doesn't form.

63. (a) $\dfrac{11}{10}$. For each mile per hour faster that the car
travels, it takes $\dfrac{11}{10}$ ft longer to stop;
(b)

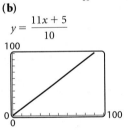

$y = \dfrac{11x + 5}{10}$

(c) 6, 11.5, 22.5, 55.5, 72; **(d)** $\{r \mid r > 0\}$, or $(0, \infty)$. If r is
allowed to be 0, the function says that a stopped car
travels $\dfrac{1}{2}$ ft before stopping.

65.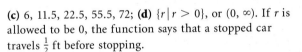

Height of corn (in feet) / Number of weeks after planting

67. $C(t) = 60 + 40t$;
$C(6) = \$300$

$y = 60 + 40x$

69. $C(x) = 800 + 3x$;
$C(75) = \$1025$

$y = 800 + 3x$

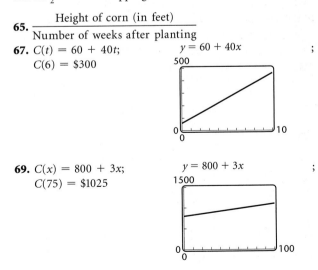

71. Discussion and Writing   **73.** [1.1] 40   **74.** [1.1] 10
75. [1.1] $a^2 + 3a$ **76.** [1.1] $a^2 + 2ah + h^2 - 3a - 3h$
77. -7.75 **79.** $F(C) = \frac{9}{5}C + 32$; $C(F) = \frac{5}{9}(F - 32)$
81. False   **83.** False   **85.** $f(x) = x + b$

Exercise Set 1.3

1. Yes   **3.** No   **5.** No   **7.** Yes
9. (a) $y = 1304.844444x + 21950.01111$; 54,571;
(b) $r = 0.9890$, a good fit
11. (a) $y = 645.7x + 9799.2$; \$14,319.10; \$15,610.50;
\$18,193.30; **(b)** $r = 0.9994$, a good fit; **(c)** The president's
assertion is about \$1000, or 6%, higher than the estimate
from the model.
13. (a) $M = 0.2H + 156$; **(b)** 164, 169, 171, 173; **(c)** $r = 1$;
the regression line fits the data perfectly and should be a
good predictor.
15. Discussion and Writing   **17.** [1.2] -1
18. [1.2] Not defined   **19.** [1.2] $y = \frac{3}{4}x - 5$
20. [1.2] $y = -10$
21. (a) $H = -0.4020073552x + 813.785167$; 33, 32, 32, 29,
29; -0.3668; not a good predictor; **(b)** no;
(c) $R = -3.353892124x + 6646.484217$; 130, 127, 123, 100,
96; -0.7817; would be an above-average predictor, but not
high enough to earn confidence in its predictability; **(d)** yes

Exercise Set 1.4

1. (a) $(-5, 1)$; **(b)** $(3, 5)$; **(c)** $(1, 3)$
3. (a) $(-3, -1)$, $(3, 5)$; **(b)** $(1, 3)$; **(c)** $(-5, -3)$
5. Increasing: $(0, \infty)$; decreasing: $(-\infty, 0)$; relative
minimum: 0 at $x = 0$
7. Increasing: $(-\infty, 0)$; decreasing: $(0, \infty)$; relative
maximum: 5 at $x = 0$
9. Increasing: $(3, \infty)$; decreasing: $(-\infty, 3)$; relative
minimum: 1 at $x = 3$
11. Increasing: $(1, 3)$; decreasing: $(-\infty, 1)$, $(3, \infty)$; relative
maximum: -4 at $x = 3$; relative minimum: -8 at $x = 1$
13. Increasing: $(-1.552, 0)$, $(1.552, \infty)$; decreasing:
$(-\infty, -1.552)$, $(0, 1.552)$; relative maximum: 4.07 at $x = 0$;
relative minima: -2.314 at $x = -1.552$, -2.314 at $x = 1.552$
15. (a)

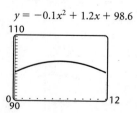

$y = -0.1x^2 + 1.2x + 98.6$

(b) 102.2; **(c)** 6 days after the illness began, the
temperature is 102.2°F.

17. Increasing: $(-1, 1)$; decreasing: $(-\infty, -1)$, $(1, \infty)$
19. Increasing: $(-1.414, 1.414)$; decreasing: $(-2, -1.414)$,
$(1.414, 2)$ **21.** $d(t) = \sqrt{(120t)^2 + (400)^2}$
23. $A(x) = 24x - x^2$ **25.** $A(w) = 10w - \dfrac{w^2}{2}$
27. $d(s) = \dfrac{14}{s}$ **29. (a)** $A(x) = x(30 - x)$, or $30x - x^2$;
(b) $\{x \mid 0 < x < 30\}$;
(c)   ;

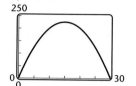

$y = 30x - x^2$

(d) 15 ft by 15 ft
31. (a) $V(x) = x(12 - 2x)(12 - 2x)$, or $4x(6 - x)^2$;
(b) $\{x \mid 0 < x < 6\}$;
(c)   ;

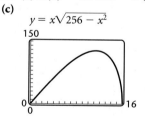

$y = 4x(6 - x)^2$

(d) 8 cm by 8 cm by 2 cm
33. (a) $A(x) = x\sqrt{256 - x^2}$; **(b)** $\{x \mid 0 < x < 16\}$;
(c)

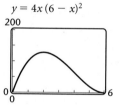

$y = x\sqrt{256 - x^2}$

(d) 11.314 ft by 11.314 ft
35.   **37.**

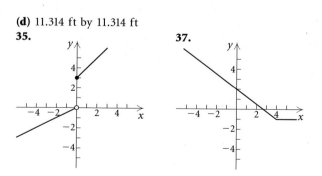

39.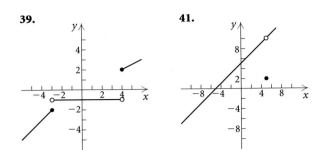

41.

43.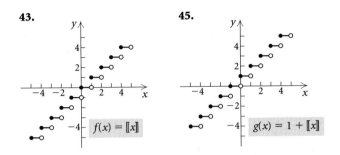

$f(x) = [\![x]\!]$

45.

$g(x) = 1 + [\![x]\!]$

47.
$$y = \begin{cases} \sqrt[3]{x}, & \text{for } x \le -1, \\ x^2 - 3x, & \text{for } -1 < x < 4, \\ \sqrt{x - 4}, & \text{for } x \ge 4 \end{cases}$$

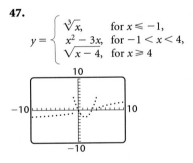

49. 33   **51.** −1   **53.** 5   **55.** 2   **57.** 0   **59.** 0

61. (a) Domain of f, all real numbers; domain of g, $[-4, \infty)$; domain of $f + g$, $f - g$, and fg, $[-4, \infty)$; domain of ff, all real numbers; domain of f/g, $(-4, \infty)$; domain of g/f, $[-4, 3) \cup (3, \infty)$; **(b)** $(f + g)(x) = x - 3 + \sqrt{x + 4}$; $(f - g)(x) = x - 3 - \sqrt{x + 4}$; $(fg)(x) = (x - 3)\sqrt{x + 4}$; $(ff)(x) = x^2 - 6x + 9$; $(f/g)(x) = (x - 3)/\sqrt{x + 4}$; $(g/f)(x) = \sqrt{x + 4}/(x - 3)$
63. (a) Domain of f, g, $f + g$, $f - g$, fg, and ff, all real numbers; domain of f/g, all real numbers except -3 and $\frac{1}{2}$; domain of g/f, all real numbers except 0;
(b) $(f + g)(x) = x^3 + 2x^2 + 5x - 3$;
$(f - g)(x) = x^3 - 2x^2 - 5x + 3$;
$(fg)(x) = 2x^5 + 5x^4 - 3x^3$; $(ff)(x) = x^6$;
$(f/g)(x) = x^3/(2x^2 + 5x - 3)$;
$(g/f)(x) = (2x^2 + 5x - 3)/x^3$
65. $(f + g)(x) = 2x^2 - 2$; $(f - g)(x) = -6$;
$(fg)(x) = x^4 - 2x^2 - 8$; $(f/g)(x) = (x^2 - 4)/(x^2 + 2)$

67. $(f + g)(x) = \sqrt{x - 7} + x^2 - 25$;
$(f - g)(x) = \sqrt{x - 7} - x^2 + 25$;
$(fg)(x) = \sqrt{x - 7}(x^2 - 25)$;
$(f/g)(x) = \sqrt{x - 7}/(x^2 - 25)$
69. Domain of F: $[0, 9]$; domain of G: $[3, 10]$; domain of $F + G$: $[3, 9]$ **71.** $[3, 6) \cup (6, 8) \cup (8, 9)$
73.

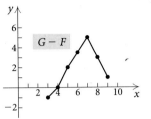

$G - F$

75. (a) $P(x) = -0.4x^2 + 57x - 13$; **(b)** $R(100) = 2000$; $C(100) = 313$; $P(100) = 1687$; **(c)** left to the student
77. $6x + 3h - 2$ **79.** $3x^2 + 3xh + h^2$
81. $\dfrac{7}{(x + h + 3)(x + 3)}$ **83.** Discussion and Writing
85. [1.1] All real numbers, or $(-\infty, \infty)$
86. [1.1] $\{x \mid x \ne 3\}$ **87.** [1.1] $\{x \mid x \ne -5\}$
88. [1.1] All real numbers, or $(-\infty, \infty)$
89. Increasing: $(-5, -2)$, $(4, \infty)$; decreasing: $(-\infty, -5)$, $(-2, 4)$; relative maximum: 560 at $x = -2$; relative minima: 425 at $x = -5$, -304 at $x = 4$
91. (a)

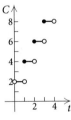

(b) $C(t) = 2([\![t]\!] + 1)$, $t > 0$
93. $\{x \mid -5 \le x < -4 \text{ or } 5 \le x < 6\}$
95. (a) $h(r) = \dfrac{30 - 5r}{3}$; **(b)** $V(r) = \pi r^2 \left(\dfrac{30 - 5r}{3}\right)$;
(c) $V(h) = \pi h \left(\dfrac{30 - 3h}{5}\right)^2$

Exercise Set 1.5

1. x-axis, no; y-axis, yes; origin, no
3. x-axis, yes; y-axis, no; origin, no
5. x-axis, no; y-axis, no; origin, yes
7. x-axis, no; y-axis, yes; origin, no
9. x-axis, no; y-axis, no; origin, no
11. x-axis, no; y-axis, yes; origin, no
13. x-axis, no; y-axis, no; origin, yes
15. x-axis, no; y-axis, no; origin, yes

17. *x*-axis, yes; *y*-axis, yes; origin, yes
19. *x*-axis, no; *y*-axis, yes; origin, no
21. *x*-axis, yes; *y*-axis, yes; origin, yes
23. *x*-axis, no; *y*-axis, no; origin, no
25. *x*-axis, no; *y*-axis, no; origin, yes
27. Even   **29.** Odd   **31.** Neither   **33.** Odd   **35.** Even
37. Odd   **39.** Even   **41.** Even
43. Start with the graph of
$g(x) = x^2$. Shift it up 1 unit.

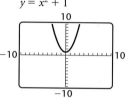
$y = x^2 + 1$

45. Start with the graph of
$f(x) = x^3$. Shift it left 5 units.

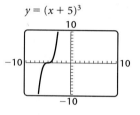
$y = (x + 5)^3$

47. Start with the graph of
$g(x) = x^2$. Reflect it across
the *x*-axis.

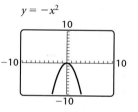
$y = -x^2$

49. Start with the graph of $g(x) = \sqrt{x}$. Stretch it vertically by multiplying each *y*-coordinate by 3, then shift it down 5 units.

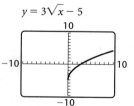
$y = 3\sqrt{x} - 5$

51. Start with the graph of $f(x) = |x|$. Stretch it horizontally by multiplying each *x*-coordinate by 3, then shift it down 4 units.

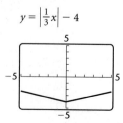
$y = \left|\frac{1}{3}x\right| - 4$

53. Start with the graph of $g(x) = x^2$. Shift it left 5 units, then down 4 units.

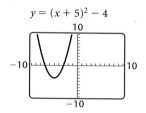
$y = (x + 5)^2 - 4$

55. Start with the graph of $g(x) = x^2$. Shift it right 5 units, shrink it vertically by multiplying each *y*-coordinate by $\frac{1}{4}$, and reflect it across the *x*-axis.

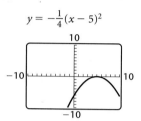
$y = -\frac{1}{4}(x - 5)^2$

57. Start with the graph of $g(x) = 1/x$. Shift it left 3 units, then up 2 units. Use DOT mode.

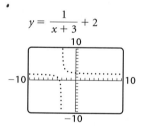
$y = \dfrac{1}{x + 3} + 2$

59. $f(x) = -(x - 8)^2$ **61.** $f(x) = |x + 7| + 2$
63. $f(x) = \dfrac{1}{2x} - 3$ **65.** $f(x) = -(x - 3)^2 + 4$
67. $f(x) = \sqrt{-(x + 2)} - 1$ **69.** $f(x) = 0.83(x + 4)^3$
71.

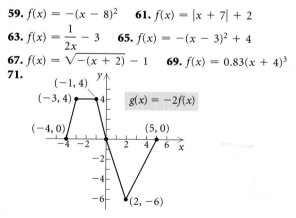

$\left(-\dfrac{m}{a}\right) = -cx$

$\dfrac{1}{a} = c$

73.

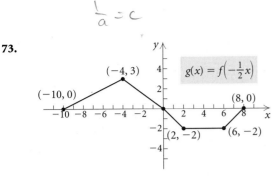

75.

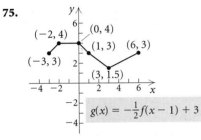

29. $y = xz$ **31.** $y = \dfrac{3}{10}xz^2$ **33.** $y = \dfrac{xz}{5wp}$ **35.** 2.5 m

37. 36 mph **39.** About 106

41. Discussion and Writing

43. [1.4]

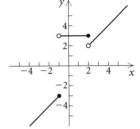

44. [1.5] x-axis, no; y-axis, yes; origin, no
45. [1.5] x-axis, yes; y-axis, no; origin, no
46. [1.5] x-axis, no; y-axis, no; origin, yes

47. \$7.20 **49.** $\dfrac{\pi}{4}$

77. (f) **79.** (f) **81.** (d) **83.** (c)
85. $f(-x) = 2(-x)^4 - 35(-x)^3 + 3(-x) - 5 =$
$2x^4 + 35x^3 - 3x - 5 = g(x)$ **87.** $g(x) = x^3 - 3x^2 + 2$
89. $k(x) = (x + 1)^3 - 3(x + 1)^2$ **91.** Discussion and
Writing **93.** Discussion and Writing
94. [1.1] **(a)** 22; **(b)** -22; **(c)** $4a^3 - 5a$; **(d)** $-4a^3 + 5a$
95. [1.1] **(a)** 38; **(b)** 38; **(c)** $5a^2 - 7$; **(d)** $5a^2 - 7$
96. [1.2] Slope is $\frac{2}{9}$; y-intercept is $\left(0, \frac{1}{9}\right)$.
97. [1.2] $y = -\frac{1}{8}x + \frac{7}{8}$ **99.** Odd
101. x-axis, yes; y-axis, yes; origin, yes
103. x-axis, yes; y-axis, no; origin, no
105. Start with the graph of $g(x) = [\![x]\!]$. Shift it right
$\frac{1}{2}$ unit. Domain: all real numbers; range: all integers.
107. (3, 8); (3, 6); $\left(\frac{3}{2}, 4\right)$ **109.** 5 **111.** True
113. $E(-x) = \dfrac{f(-x) + f(-(-x))}{2} = \dfrac{f(-x) + f(x)}{2} = E(x)$

115. (a) $E(x) + O(x) = \dfrac{f(x) + f(-x)}{2} + \dfrac{f(x) - f(-x)}{2} =$
$\dfrac{2f(x)}{2} = f(x)$; **(b)** $f(x) = \dfrac{-22x^2 + \sqrt{x} + \sqrt{-x} - 20}{2} +$
$\dfrac{8x^3 + \sqrt{x} - \sqrt{-x}}{2}$

Exercise Set 1.6

1. 4.5; $y = 4.5x$ **3.** 36; $y = \dfrac{36}{x}$ **5.** 4; $y = 4x$

7. 4; $y = \dfrac{4}{x}$ **9.** $\dfrac{3}{8}$; $y = \dfrac{3}{8}x$ **11.** 0.54; $y = \dfrac{0.54}{x}$

13. 3.5 hr **15.** 90 g **17.** About 686 kg **19.** 40 lb

21. $66\frac{2}{3}$ cm **23.** 1.92 ft **25.** $y = \dfrac{0.0015}{x^2}$ **27.** $y = 15x^2$

Exercise Set 1.7

1. $\sqrt{10}$, 3.162 **3.** $\sqrt{45}$, 6.708 **5.** $\sqrt{14 + 6\sqrt{2}}$, 4.742
7. 6.5 **9.** Yes
11. The distances between the points are $\sqrt{10}$, 10, and
$\sqrt{90}$. Since $(\sqrt{10})^2 + (\sqrt{90})^2 = 10^2$, the three points are
vertices of a right triangle.
13. $\left(\dfrac{9}{2}, \dfrac{15}{2}\right)$ **15.** $\left(\dfrac{15}{2}, 2\right)$ **17.** $\left(\dfrac{\sqrt{3} - \sqrt{6}}{2}, -\dfrac{\sqrt{5}}{2}\right)$

19.

$\left(-\dfrac{1}{2}, \dfrac{3}{2}\right), \left(\dfrac{7}{2}, \dfrac{1}{2}\right), \left(\dfrac{5}{2}, \dfrac{9}{2}\right), \left(-\dfrac{3}{2}, \dfrac{11}{2}\right)$; no

21. $\left(\dfrac{\sqrt{7} + \sqrt{2}}{2}, -\dfrac{1}{2}\right)$

23. Square the window; for example, $[-12, 9, -4, 10]$.
25. $(x - 2)^2 + (y - 3)^2 = \dfrac{25}{9}$
27. $(x + 1)^2 + (y - 4)^2 = 25$
29. $(x - 2)^2 + (y - 1)^2 = 169$
31. $(x + 2)^2 + (y - 3)^2 = 4$

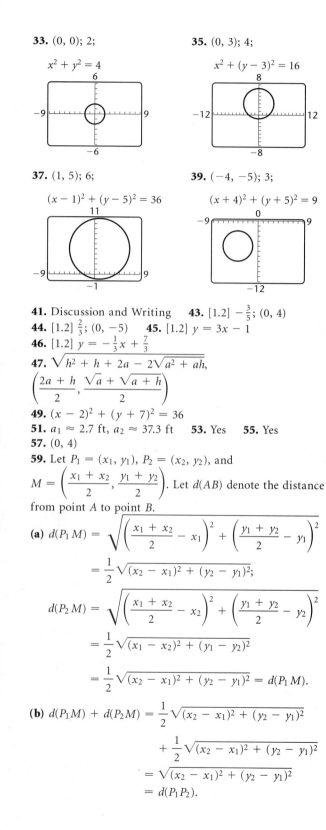

33. $(0, 0)$; 2;

$x^2 + y^2 = 4$

35. $(0, 3)$; 4;

$x^2 + (y - 3)^2 = 16$

37. $(1, 5)$; 6;

$(x - 1)^2 + (y - 5)^2 = 36$

39. $(-4, -5)$; 3;

$(x + 4)^2 + (y + 5)^2 = 9$

41. Discussion and Writing   **43.** [1.2] $-\frac{3}{5}$; $(0, 4)$
44. [1.2] $\frac{2}{3}$; $(0, -5)$ **45.** [1.2] $y = 3x - 1$
46. [1.2] $y = -\frac{1}{3}x + \frac{7}{3}$
47. $\sqrt{h^2 + h + 2a - 2\sqrt{a^2 + ah}}$,

$$\left(\frac{2a + h}{2}, \frac{\sqrt{a} + \sqrt{a + h}}{2} \right)$$

49. $(x - 2)^2 + (y + 7)^2 = 36$
51. $a_1 \approx 2.7$ ft, $a_2 \approx 37.3$ ft   **53.** Yes   **55.** Yes
57. $(0, 4)$
59. Let $P_1 = (x_1, y_1)$, $P_2 = (x_2, y_2)$, and
$M = \left(\dfrac{x_1 + x_2}{2}, \dfrac{y_1 + y_2}{2} \right)$. Let $d(AB)$ denote the distance
from point A to point B.

(a) $d(P_1 M) = \sqrt{\left(\dfrac{x_1 + x_2}{2} - x_1 \right)^2 + \left(\dfrac{y_1 + y_2}{2} - y_1 \right)^2}$

$= \dfrac{1}{2}\sqrt{(x_2 - x_1)^2 + (y_2 - y_1)^2}$;

$d(P_2 M) = \sqrt{\left(\dfrac{x_1 + x_2}{2} - x_2 \right)^2 + \left(\dfrac{y_1 + y_2}{2} - y_2 \right)^2}$

$= \dfrac{1}{2}\sqrt{(x_1 - x_2)^2 + (y_1 - y_2)^2}$

$= \dfrac{1}{2}\sqrt{(x_2 - x_1)^2 + (y_2 - y_1)^2} = d(P_1 M)$.

(b) $d(P_1 M) + d(P_2 M) = \dfrac{1}{2}\sqrt{(x_2 - x_1)^2 + (y_2 - y_1)^2}$

$+ \dfrac{1}{2}\sqrt{(x_2 - x_1)^2 + (y_2 - y_1)^2}$

$= \sqrt{(x_2 - x_1)^2 + (y_2 - y_1)^2}$

$= d(P_1 P_2)$.

Review Exercises, Chapter 1

1. [1.1] Not a function; domain: {3, 5, 7}; range: {1, 3, 5, 7}
2. [1.1] Function; domain: {−2, 0, 1, 2, 7}; range: {−7, −4, −2, 2, 7}   **3.** [1.1] No   **4.** [1.1] Yes   **5.** [1.1] No
6. [1.1] Yes   **7.** [1.1] All real numbers
8. [1.1] $\{x \mid x \neq 0\}$ **9.** [1.1] $\{x \mid x \neq 5 \text{ and } x \neq 1\}$
10. [1.1] $\{x \mid x \neq -4 \text{ and } x \neq 4\}$
11. [1.1] Domain: $[-5, 3]$; range: $[0, 4]$
12. [1.1] Domain: all real numbers; range: $[-5, \infty)$
13. [1.1] Domain: all real numbers; range: all real numbers
14. [1.1] Domain: all real numbers; range: $[-19, \infty)$
15. [1.1] **(a)** -3; **(b)** 9; **(c)** $a^2 - 3a - 1$; **(d)** $2x + h - 1$
16. [1.2] **(a)** Yes; **(b)** no; **(c)** no, strictly speaking, but data might be modeled by a linear regression function.
17. [1.2] **(a)** Yes; **(b)** yes; **(c)** yes   **18.** [1.2] -2; $(0, -7)$
19. [1.2] $y + 4 = -\frac{2}{3}(x - 0)$
20. [1.2] $y + 1 = 3(x + 2)$
21. [1.2] $y - 1 = \frac{1}{3}(x - 4)$, or $y + 1 = \frac{1}{3}(x + 2)$
22. [1.2] Parallel   **23.** [1.2] Neither
24. [1.2] Perpendicular   **25.** [1.2] $y = -\frac{2}{3}x - \frac{1}{3}$
26. [1.2] $y = \frac{3}{2}x - \frac{5}{2}$
27. [1.2] $C(t) = 25 + 20t$; $145   **28.** [1.2] **(a)** 70°C, 220°C, 10,020°C;

(b)

$y = 25 + 20x$

$y = 10x + 20$

(c) $[0, 5600]$
29. [1.3] **(a)** $y = 0.9362739726x + 9.656712329$, 56.6 million; **(b)** $r = 0.9946$, a good fit
30. [1.4]   **31.** [1.4]

32. [1.4]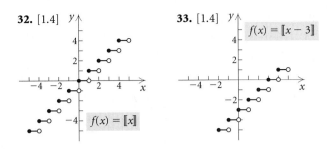

33. [1.4]

60. [1.5]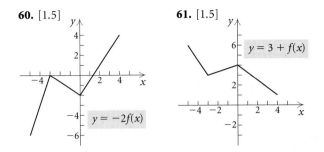

61. [1.5]

34. [1.4] **(a)** Increasing: $(-\infty, -0.860)$, $(6.339, \infty)$; decreasing: $(-0.860, 6.339)$; **(b)** relative maximum: $(-0.860, 710.094)$; relative minimum: $(6.339, 504.836)$
35. [1.4] **(a)** Increasing: $(-\infty, -4)$; decreasing: $(-4, \infty)$; **(b)** relative maximum: $(-4, 15)$
36. [1.4] **(a)** Increasing: $(-3.589, 2.089)$; decreasing: $(-5, -3.589)$, $(2.089, 3)$; **(b)** relative maximum: $(2.089, 4.247)$; relative minimum: $(-3.589, -8.755)$
37. [1.4] $A(x) = 2x\sqrt{4 - x^2}$

38. [1.4] **(a)** $A(x) = x^2 + \dfrac{432}{x}$; **(b)** $(0, \infty)$;

(c)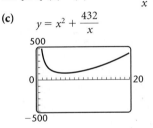
$$y = x^2 + \frac{432}{x}$$

(d) $x = 6$ in.; height $= 3$ in.

39. [1.5] x-axis, yes; y-axis, yes; origin, yes
40. [1.5] x-axis, yes; y-axis, yes; origin, yes
41. [1.5] x-axis, no; y-axis, no; origin, no
42. [1.5] x-axis, no; y-axis, yes; origin, no
43. [1.5] x-axis, no; y-axis, no; origin, yes
44. [1.5] x-axis, no; y-axis, yes; origin, no
45. [1.5] Even   **46.** [1.5] Even   **47.** [1.5] Odd
48. [1.5] Even   **49.** [1.5] Even   **50.** [1.5] Neither
51. [1.5] Odd   **52.** [1.5] Even   **53.** [1.5] Even
54. [1.5] Odd   **55.** [1.5] $f(x) = (x + 3)^2$
56. [1.5] $f(x) = -\sqrt{x - 3} + 4$
57. [1.5] $f(x) = 2|x - 3|$
58. [1.5]

59. [1.5]

62. [1.4] **(a)** Domain of f, $\{x \,|\, x \neq 0\}$; domain of g, all real numbers; domain of $f + g$, $f - g$, and fg, $\{x \,|\, x \neq 0\}$; domain of f/g, $\left\{ x \,\middle|\, x \neq 0 \ and \ x \neq \dfrac{3}{2} \right\}$
(b) $(f + g)(x) = \dfrac{4}{x^2} + 3 - 2x$; $(f - g)(x) = \dfrac{4}{x^2} - 3 + 2x$; $fg(x) = \dfrac{12}{x^2} - \dfrac{8}{x}$; $(f/g)(x) = \dfrac{4}{x^2(3 - 2x)}$
63. [1.4] **(a)** Domain of f, g, $f + g$, $f - g$, and fg, all real numbers; domain of f/g, $\left\{ x \,\middle|\, x \neq \dfrac{1}{2} \right\}$;
(b) $(f + g)(x) = 3x^2 + 6x - 1$; $(f - g)(x) = 3x^2 + 2x + 1$; $fg(x) = 6x^3 + 5x^2 - 4x$; $(f/g)(x) = \dfrac{3x^2 + 4x}{2x - 1}$
64. [1.4] $P(x) = -0.5x^2 + 105x - 6$
65. [1.6] $y = 4x$ **66.** [1.6] $y = \dfrac{2500}{x}$ **67.** [1.6] 20 min
68. [1.6] 500 watts   **69.** [1.6] About 78
70. [1.7] $\sqrt{34} \approx 5.831$ **71.** [1.7] $\left(\frac{1}{2}, \frac{11}{2}\right)$
72. [1.7] $(x + 2)^2 + (y - 6)^2 = 13$ **73.** [1.7] $(-1, 3)$; 4
74. [1.7] $(x - 2)^2 + (y - 4)^2 = 26$
75. Discussion and Writing [1.1], [1.5] **(a)** $4x^3 - 2x + 9$; **(b)** $4x^3 + 24x^2 + 46x + 35$; **(c)** $4x^3 - 2x + 42$. **(a)** Adds 2 to each function value; **(b)** adds 2 to each input before finding a function value; **(c)** adds the output for 2 to the output for x
76. Discussion and Writing [1.5] In the graph of $y = f(cx)$, the constant c stretches or shrinks the graph of $y = f(x)$ horizontally. The constant c in $y = cf(x)$ stretches or shrinks the graph of $y = f(x)$ vertically. For $y = f(cx)$, the x-coordinates of $y = f(x)$ are divided by c; for $y = cf(x)$, the y-coordinates of $y = f(x)$ are multiplied by c.
77. Discussion and Writing [1.5] **(a)** To draw the graph of y_2 from the graph of y_1, reflect across the x-axis the portions of the graph for which the y-coordinates are negative. **(b)** To draw the graph of y_2 from the graph of y_1, draw the portion of the graph of y_1 to the right of the y-axis; then draw its reflection across the y-axis.

78. [1.1] $\{x \mid x < 0\}$
79. [1.1] $\{x \mid x \neq -3 \text{ and } x \neq 0 \text{ and } x \neq 3\}$
80. [1.4], [1.5] Let $f(x)$ and $g(x)$ be odd functions. Then by definition, $f(-x) = -f(x)$, or $f(x) = -f(-x)$, and $g(-x) = -g(x)$, or $g(x) = -g(-x)$. Thus, $(f + g)(x) = f(x) + g(x) = -f(-x) + [-g(-x)] = -[f(-x) + g(-x)] = -(f + g)(-x)$ and $f + g$ is odd.
81. [1.5] Reflect the graph of $y = f(x)$ across the x-axis and then across the y-axis.

Chapter 2

Exercise Set 2.1

1. (a) $(4, 0)$; (b) 4   **3.** -5 **5.** 18   **7.** 16   **9.** -12
11. 6   **13.** 20   **15.** 6   **17.** 15   **19.** $-\frac{3}{4}$ **21.** -4
23. $-\frac{2}{3}$ **25.** 1.6   **27.** 6   **29.** 0.5   **31.** $-\frac{25}{8}$, or -3.125
33. 5   **35.** 47.1 million   **37.** 125   **39.** 1.93 million
41. Length: 93 m; width: 68 m   **43.** $26°, 130°, 24°$
45. About 280,526   **47.** Freight: 66 mph; passenger: 80 mph   **49.** 67.5 lb   **51.** About 5.5 million   **53.** 23
55. 660 yr   **57.** Discussion and Writing
59. [1.2] $y = -\frac{3}{4}x + \frac{13}{4}$ **60.** [1.3] 76.0 million
61. [1.4] $x^2 + 2x - 6$ **62.** [1.4] 12   **63.** Yes   **65.** No

Exercise Set 2.2

1. $2 + 11i$ **3.** $5 - 12i$ **5.** $5 + 9i$ **7.** $5 + 4i$
9. $5 + 7i$ **11.** $13 - i$ **13.** $-11 + 16i$ **15.** $35 + 14i$
17. $-14 + 23i$ **19.** 41   **21.** $12 + 16i$ **23.** $-45 - 28i$
25. $\dfrac{-4\sqrt{3} + 10}{41} + \dfrac{5\sqrt{3} + 8}{41}i$ **27.** $-\dfrac{14}{13} + \dfrac{5}{13}i$
29. $\frac{15}{146} + \frac{33}{146}i$ **31.** $-\frac{1}{2} + \frac{1}{2}i$ **33.** $-\frac{1}{2} - \frac{13}{2}i$ **35.** $-i$
37. $-i$ **39.** 1   **41.** i **43.** 625   **45.** Discussion and Writing   **47.** [1.2] $y = -2x + 1$ **48.** [1.4] $x^2 - 3x - 1$
49. [1.4] $\frac{8}{11}$ **50.** [1.4] $2x + h - 3$ **51.** True   **53.** True
55. $a^2 + b^2$

Exercise Set 2.3

1. (a) $(-4, 0), (2, 0)$; (b) $-4, 2$ **3.** $-7, 1$ **5.** $4 \pm \sqrt{7}$
7. $-4 \pm 3i$ **9.** $-2, \frac{1}{3}$ **11.** $-3, 5$ **13.** $-1, \frac{2}{5}$
15. $\dfrac{5 \pm \sqrt{7}}{3}$ **17.** $-\dfrac{1}{2} \pm \dfrac{\sqrt{7}}{2}i$ **19.** $\dfrac{4 \pm \sqrt{31}}{5}$
21. $\dfrac{5}{6} \pm \dfrac{\sqrt{23}}{6}i$ **23.** No   **25.** Yes   **27.** No   **29.** 2, 6
31. 0.143, 6   **33.** $-0.151, 1.651$ **35.** $-0.637, 3.137$
37. $-5, -1$ **39.** $\dfrac{3 \pm \sqrt{21}}{2}$ **41.** $\dfrac{5 \pm \sqrt{21}}{2}$
43. $-3.449, 1.449$ **45.** $-1.535, 0.869$ **47.** $-0.347, 1.181$

49. $\pm 1, \pm \sqrt{2}$ **51.** 16   **53.** $-8, 64$ **55.** $\frac{5}{2}, 3$
57. $-\frac{3}{2}, -1, \frac{1}{2}, 1$ **59.** About 11.5 sec   **61.** Length: 4 ft; width: 3 ft   **63.** 4 and 9; -9 and -4 **65.** 2 cm
67. Linear   **69.** Quadratic   **71.** Linear   **73.** Discussion and Writing   **75.** [1.5] x-axis: yes; y-axis: yes; origin: yes
76. [1.5] x-axis: no; y-axis: yes; origin: no   **77.** [1.5] Odd
78. [1.5] Neither   **79.** $\dfrac{-1 \pm \sqrt{1 + 4\sqrt{2}}}{2}$ **81.** $3 \pm \sqrt{5}$

83. 19   **85.** $-2 \pm \sqrt{2}, \dfrac{1}{2} \pm \dfrac{\sqrt{7}}{2}i$

87. $t = \dfrac{-v_0 \pm \sqrt{v_0^2 - 2ax_0}}{a}$

Exercise Set 2.4

1. (a) $(-0.5, -2.25)$; (b) $x = -0.5$; (c) minimum: -2.25
3. (a) $(4, -4)$; (b) $x = 4$; (c) minimum: -4
5. (a) $\left(\frac{7}{2}, -\frac{1}{4}\right)$; (b) $x = \frac{7}{2}$; (c) minimum: $-\frac{1}{4}$
7. (a) $\left(-\frac{3}{2}, \frac{7}{2}\right)$; (b) $x = -\frac{3}{2}$; (c) minimum: $\frac{7}{2}$
9. (a) $\left(\frac{1}{2}, \frac{3}{2}\right)$; (b) $x = \frac{1}{2}$; (c) maximum: $\frac{3}{2}$
11. (f)   **13.** (b)   **15.** (h)   **17.** (c)
19. (a) $(3, -4)$; (b) minimum: -4; (c) $[-4, \infty)$;
(d) increasing: $(3, \infty)$; decreasing: $(-\infty, 3)$
21. (a) $(-1, -18)$; (b) minimum: -18; (c) $[-18, \infty)$;
(d) increasing: $(-1, \infty)$; decreasing: $(-\infty, -1)$
23. (a) $\left(5, \frac{9}{2}\right)$; (b) maximum: $\frac{9}{2}$; (c) $\left(-\infty, \frac{9}{2}\right]$;
(d) increasing: $(-\infty, 5)$; decreasing: $(5, \infty)$
25. (a) $(-1, 2)$; (b) minimum: 2; (c) $[2, \infty)$;
(d) increasing: $(-1, \infty)$; decreasing: $(-\infty, -1)$
27. 4.5 in.   **29.** Base: 10 cm; height: 10 cm   **31.** 350.6 ft
33. 350   **35.** $797; 40   **37.** 4800 yd^2
39. Discussion and Writing   **41.** Discussion and Writing
42. [1.4] 3   **43.** [1.4] $4x + 2h - 1$
44. [1.5]   **45.** [1.5]

47. (a) 2; (b) $\frac{11}{2}$ **49.** (a) 2; (b) $1 - i$ **51.** -236.25

53.
$$y = (|x| - 5)^2 - 3$$

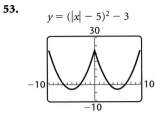

Exercise Set 2.5

1. (c)   **3.** (d)   **5.** (a)

7. (a)

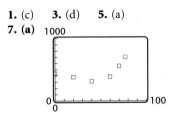

(b) quadratic;
(c) $f(x) = 0.1729980538x^2 - 10.70013387x + 465.7917864;$
(d)

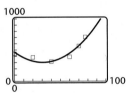

(e) 788; 844

9. (a)

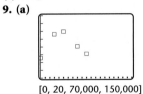

[0, 20, 70,000, 150,000]

(b) quadratic;
(c) $f(x) = -1226.47072x^2 + 12,279.05135x + 97,260.77349;$
(d)

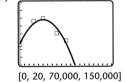

[0, 20, 70,000, 150,000]

(e) \$49,615 million; \$5491 million

11. (a)

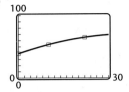

(b) Answers may vary; quadratic;
(c) $f(x) = -0.0223484848x^2 + 1.673484848x + 39;$
(d)

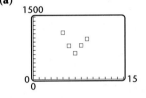

(e) 68.3%; 70.2%
13. (a)

(b) quadratic;
(c) $f(x) = 93.28571429x^2 - 1336x + 5460.828571;$
(d)

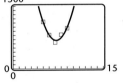

(e) 688; 1429
15. (a) $f(x) = -0.8630952381x^2 + 12.125x + 36.76785714;$
(b) 77 yr; 77 yr
17. Discussion and Writing   **19.** [1.1] −1
20. [1.1] Does not exist   **21.** [1.1] 0
22. [1.1] $\{x \mid x \neq 1\}$, or $(-\infty, 1) \cup (1, \infty)$
23. (a)

350

0   30
0

(b) quadratic;
(c) $f(x) = 2.031904548x^2 - 59.04179598x + 527.2818092$;
(d)

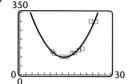

(e) \$20.9 million

Exercise Set 2.6

1. $\frac{20}{9}$ **3.** 286   **5.** 6   **7.** No solution   **9.** No solution
11. No solution   **13.** $\{x \mid x$ is a real number *and* $x \neq 0$
and $x \neq 6\}$ **15.** 2   **17.** $\frac{5}{3}$ **19.** $\pm\sqrt{2}$ **21.** No solution
23. -6 **25.** 7   **27.** No solution   **29.** 1   **31.** 3, 7
33. -8 **35.** 81   **37.** $-7, 7$ **39.** No solution
41. $-3, 5$ **43.** $-\frac{1}{3}, \frac{1}{3}$ **45.** 0   **47.** $-1, -\frac{1}{3}$
49. $-24, 44$ **51.** $-2, 4$ **53.** $T_1 = \dfrac{P_1 V_1 T_2}{P_2 V_2}$
55. $R_2 = \dfrac{RR_1}{R_1 - R}$ **57.** $p = \dfrac{Fm}{m - F}$ **59.** Discussion and
Writing   **60.** [2.1] 3   **61.** [2.1] 7.5   **62.** [2.1] 10.4%
63. [2.1] 481,000   **65.** $3 \pm 2\sqrt{2}$ **67.** -1

Exercise Set 2.7

1. $\{x \mid x > 3$, or $(3, \infty)$ **3.** $\left\{x \mid x \geq -\frac{5}{12}\right\}$, or $\left[-\frac{5}{12}, \infty\right)$
5. $\left\{y \mid y \geq \frac{22}{13}\right\}$, or $\left[\frac{22}{13}, \infty\right)$ **7.** $\left\{x \mid x \leq \frac{15}{34}\right\}$, or $\left(-\infty, \frac{15}{34}\right]$
9. $\{x \mid x < 1\}$, or $(-\infty, 1)$ **11.** Left to the student
13. $[-3, 3)$ **15.** $[8, 10]$ **17.** $[-7, -1]$ **19.** $\left(-\frac{3}{2}, 2\right)$
21. $(1, 5]$ **23.** $\left(-\frac{11}{3}, \frac{13}{3}\right]$ **25.** $(-\infty, -2] \cup (1, \infty)$
27. $\left(-\infty, -\frac{7}{2}\right] \cup \left[\frac{1}{2}, \infty\right)$ **29.** $(-\infty, 9.6) \cup (10.4, \infty)$
31. $\left(-\infty, -\frac{57}{4}\right] \cup \left[-\frac{55}{4}, \infty\right)$ **33.** Left to the student
35. $(-7, 7)$;

37. $(-\infty, -4.5] \cup [4.5, \infty)$;

39. $(-17, 1)$ **41.** $(-\infty, -17] \cup [1, \infty)$ **43.** $\left(-\frac{1}{4}, \frac{3}{4}\right)$
45. $\left(-\frac{1}{3}, \frac{1}{3}\right)$ **47.** $[-6, 3]$ **49.** $(-\infty, 4.9) \cup (5.1, \infty)$
51. $\left[-\frac{1}{2}, \frac{7}{2}\right]$ **53.** $\left[-\frac{7}{3}, 1\right]$ **55.** $(-\infty, -8) \cup (7, \infty)$
57. $\varnothing$ **59.** Left to the student
61. Discussion and Writing   **63.** Discussion and Writing
64. [2.2] $2 + i$ **65.** [2.2] $10 - 7i$ **66.** [2.2] $13 - 13i$

67. [2.2] $\frac{7}{25} + \frac{26}{25}i$ **69.** $\left(-\frac{1}{4}, \frac{5}{9}\right]$ **71.** $\left(-\infty, \frac{1}{2}\right)$
73. No solution

Review Exercises, Chapter 2

1. [2.1] 3   **2.** [2.1] 4   **3.** [2.1] 0.2, or $\frac{1}{5}$ **4.** [2.1] 4
5. [2.3] 1   **6.** [2.3] $-5, 3$ **7.** [2.3] $\dfrac{1 \pm \sqrt{41}}{4}$
8. [2.3] $\dfrac{-1 \pm \sqrt{10}}{3}$ **9.** [2.1] -1 **10.** [2.3] $-\frac{5}{2}, \frac{1}{3}$
11. [2.6] $\frac{27}{7}$ **12.** [2.6] $-\frac{1}{2}, \frac{9}{4}$ **13.** [2.6] 0, 3
14. [2.6] 5   **15.** [2.6] 1, 7   **16.** [2.6] $-8, 1$
17. [2.7] $\left[-\frac{4}{3}, \frac{4}{3}\right]$ **18.** [2.7] $\left(\frac{2}{5}, 2\right]$ **19.** [2.7] $(-\infty, \infty)$
20. [2.7] $\left(-\infty, -\frac{5}{3}\right] \cup [1, \infty)$ **21.** [2.7] $\left(-\frac{2}{3}, 1\right)$
22. [2.7] $(-\infty, -6] \cup [-2, \infty)$ **23.** [2.6] $h = \dfrac{v^2}{2g}$
24. [2.6] $t = \dfrac{ab}{a + b}$ **25.** [2.2] $-2\sqrt{10}i$
26. [2.2] $-4\sqrt{15}$ **27.** [2.2] $-\frac{7}{8}$ **28.** [2.2] $-18 - 26i$
29. [2.2] $\frac{11}{10} + \frac{3}{10}i$ **30.** [2.2] $1 - 4i$ **31.** [2.2] $2 - i$
32. [2.2] $-i$ **33.** [2.2] 3^{28}
34. [2.3] $x^2 - 3x + \frac{9}{4} = 18 + \frac{9}{4}$; $\left(x - \frac{3}{2}\right)^2 = \frac{81}{4}$; $x = \frac{3}{2} \pm \frac{9}{2}$;
$-3, 6$ **35.** [2.3] $x^2 - 4x = 2$; $x^2 - 4x + 4 = 2 + 4$;
$(x - 2)^2 = 6$; $x = 2 \pm \sqrt{6}$; $2 - \sqrt{6}, 2 + \sqrt{6}$
36. [2.3] $-2, \frac{4}{3}$ **37.** [2.3] $1 - 3i, 1 + 3i$
38. [2.3] $-3, 6$ **39.** [2.3] 1
40. [2.3] $\pm\sqrt{\dfrac{3 \pm \sqrt{5}}{2}}$ **41.** [2.3] $-\sqrt{3}, 0, \sqrt{3}$
42. [2.3] $-2, -\frac{2}{3}, 3$ **43.** [2.3] $-5, -2, 2$
44. [2.4] **(a)** $\left(\frac{3}{8}, -\frac{7}{16}\right)$; **(b)** $x = \frac{3}{8}$; **(c)** maximum: $-\frac{7}{16}$
45. [2.4] **(a)** $(1, -2)$; **(b)** $x = 1$; **(c)** minimum: -2
46. [2.3] **(d)**   **47.** [2.4] **(c)**   **48.** [2.4] **(b)**   **49.** [2.4] **(a)**
50. [2.3] 30 ft, 40 ft   **51.** [2.6] 6 mph
52. [2.3] 80 km/h
53. [2.3] $35 - 5\sqrt{33}$ ft, or about 6.3 ft
54. [2.3] 6 ft by 6 ft
55. [2.3] $\dfrac{15 - \sqrt{115}}{2}$ cm, or about 2.1 cm
56. [2.3] 2.1 cm   **57.** [2.5] **(a)** $f(x) = 0.0030322581x^2 -$
$0.5764516129x + 57.53225806$; **(b)** 227; 312
58. Discussion and Writing [2.3] $ax^2 + bx + c =$
$\left(x - \dfrac{-b + \sqrt{b^2 - 4ac}}{2a}\right)\left(x - \dfrac{-b - \sqrt{b^2 - 4ac}}{2a}\right)$
59. Discussion and Writing [2.4] You can conclude that
$|a_1| = |a_2|$ since these constants determine how wide the
parabolas are. Nothing can be concluded about the h's and
the k's.
60. [2.6] $-7, 9$ **61.** [2.6] $4 \pm \sqrt[4]{243}$, or 0.052, 7.948
62. [2.3] -1 **63.** [2.3] $-\frac{1}{4}, 2$ **64.** [2.4] ± 6
65. [2.3] 9%

Chapter 3

Exercise Set 3.1

1. Cubic; $\frac{1}{2}x^3$; $\frac{1}{2}$; 3   **3.** Linear; 0.9x; 0.9; 1
5. Quartic; $305x^4$; 305; 4   **7.** (D)   **9.** (F)
11. (B)   **13.** −1.532, −0.347, 1.879
15. −1.414, 0, 1.414   **17.** −1, 0, 1
19. −10.153, −1.871, −0.821, −0.303, 0.098, 0.535, 1.219,
3.297   **21.** −1.386; relative maximum: 1.506 at $x =$
−0.632, relative minimum: 0.494 at $x = 0.632$; $(-\infty, \infty)$
23. −0.720, 4.089; relative maxima: 6.59375 at $x = 0.5$,
10.5 at $x = 3$, relative minimum: 6.5 at $x = 1$; $(-\infty, 10.5]$
25. −1.249, 1.249; relative minimum: −3.8 at $x = 0$, no
relative maxima; $[-3.8, \infty)$
27. 3.822; no relative maxima or relative minima; $(-\infty, \infty)$
29. −1.697, 0, 1.856; relative maximum: 11.012 at $x =$
1.258, relative minimum: −8.183 at $x = -1.116$; $(-\infty, \infty)$
31. −1.932, −0.518, 0, 0.518, 1.932; relative maxima: 0.195
at $x = -0.294$, 4.414 at $x = 1.521$, relative minima:
−4.414 at $x = -1.521$, −0.195 at $x = 0.294$; $(-\infty, \infty)$
33. $f(-5) = -18$ and $f(-4) = 7$. By the intermediate
value theorem, since $f(-5)$ and $f(-4)$ have opposite signs,
then $f(x)$ has a zero between −5 and −4.   **35.** 0.866 sec
37. $3240   **39. (a)** 18.75%; **(b)** 10%   **41.** (b)
43. (c)   **45.** (a)   **47.** (b)
49. (a)

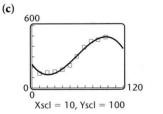

(b) cubic: $y = -0.0016285158x^3 + 0.2897410836x^2 -$
10.07476028x + 227.1101167;
(c)   ;

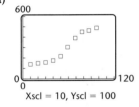

(d) linear; 510, 557; answers may vary
51. (a)

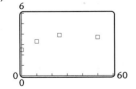

(b) cubic: $y = 0.000008x^3 - 0.00188x^2 + 0.098x + 2.5$;
(c)

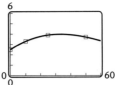

(d) cubic; 2.65; answers may vary
53. Discussion and Writing   **55.** Discussion and Writing
56. [1.7] $6\sqrt{2}$ **57.** [1.7] 5
58. [1.7] Center: $(-4, 3)$; radius: $2\sqrt{2}$
59. [1.7] Center: $(3, -5)$; radius: 7

Exercise Set 3.2

1. Yes; no; no   **3.** (a)
5. $P(x) = (x + 2)(x^2 - 2x + 4) - 16$
7. $P(x) = (x^2 + 4)(x^2 + 5) + 0$
9. $Q(x) = 2x^3 + x^2 - 3x + 10, R(x) = -42$
11. $Q(x) = x^2 - 4x + 8, R(x) = -24$
13. $Q(x) = x^3 + x^2 + x + 1, R(x) = 0$
15. $Q(x) = 2x^3 + x^2 + \frac{7}{2}x + \frac{7}{4}, R(x) = -\frac{1}{8}$
17. $Q(x) = x^3 + x^2y + xy^2 + y^3, R(x) = 0$
19. 0; −60; 0   **21.** 5,935,988; −772
23. 0; 0; 65; $1 - 12\sqrt{2}$ **25.** Yes; no   **27.** No; no
29. $f(x) = (x - 1)(x + 2)(x + 3)$; 1, −2, −3
31. $f(x) = (x - 2)(x - 5)(x + 1)$; 2, 5, −1
33. $f(x) = (x - 2)(x - 3)(x + 4)$; 2, 3, −4
35. $f(x) = (x - 1)(x - 2)(x - 3)(x + 5)$; 1, 2, 3, −5
37. Discussion and Writing   **39.** [2.1] −5
40. [2.3] $\dfrac{5 \pm i\sqrt{71}}{4}$ **41.** [2.3] $-1, \dfrac{3}{7}$
42. [2.3] $b = 15$ in., $h = 15$ in.   **43. (a)** $x + 4, x + 3,$
$x - 2, x - 5$; **(b)** $P(x) = (x + 4)(x + 3)(x - 2)(x - 5)$;
(c) yes; two examples are $f(x) = c \cdot P(x)$ for any nonzero
constant c and $g(x) = (x - a)P(x)$; **(d)** no
45. $\frac{14}{3}$ **47.** 0, 0.9636   **49.** $-1 \pm \sqrt{7}$
51. Answers can vary. One possibility is $P(x) = x^{15} - x^{14}$.
53. $x - 3 + i, R\ 6 - 3i$

Exercise Set 3.3

1. −3, multiplicity 2; 1, multiplicity 1
3. 0, multiplicity 3; 1, multiplicity 2; −4, multiplicity 1
5. $\pm\sqrt{3}$, ±1, each has multiplicity 1
7. −3, −1, 1, each has multiplicity 1
9. $f(x) = x^3 - 6x^2 - x + 30$
11. $f(x) = x^3 + 3x^2 + 4x + 12$

13. $f(x) = x^3 - \sqrt{3}x^2 - 2x + 2\sqrt{3}$
15. $f(x) = x^5 + 2x^4 - 2x^2 - x$
17. $-3 - 4i, 4 + \sqrt{5}$
19. $f(x) = x^3 - 4x^2 + 6x - 4$
21. $f(x) = x^3 - 5x^2 + 16x - 80$
23. $f(x) = x^4 + 4x^2 - 45$
25. $i, 2, 3$ **27.** $1 + 2i, 1 - 2i$ **29.** ± 1
31. $\pm 1, \pm 2, \pm \frac{1}{3}, \pm \frac{1}{5}, \pm \frac{2}{3}, \pm \frac{2}{5}, \pm \frac{1}{15}, \pm \frac{2}{15}$
33. (a) Rational: -3; other: $\pm\sqrt{2}$;
(b) $f(x) = (x + 3)(x + \sqrt{2})(x - \sqrt{2})$
35. (a) Rational: $-2, 1$; other: none;
(b) $f(x) = (x + 2)(x - 1)^2$
37. (a) Rational: -1; other: $3 \pm 2\sqrt{2}i$;
(b) $f(x) = (x + 1)(x - 3 - 2\sqrt{2}i)(x - 3 + 2\sqrt{2}i)$
39. (a) Rational: $-\frac{1}{5}, 1$; other: $\pm 2i$;
(b) $f(x) = (5x + 1)(x + 1)(x + 2i)(x - 2i)$
41. (a) Rational: $-2, -1$; other: $3 \pm \sqrt{13}$;
(b) $f(x) = (x + 2)(x + 1)(x - 3 - \sqrt{13})(x - 3 + \sqrt{13})$
43. (a) Rational: 2; other: $1 \pm \sqrt{3}$;
(b) $f(x) = (x - 2)(x - 1 - \sqrt{3})(x - 1 + \sqrt{3})$
45. (a) Rational: -2; other: $1 \pm \sqrt{3}i$;
(b) $f(x) = (x + 2)(x - 1 - \sqrt{3}i)(x - 1 + \sqrt{3}i)$
47. (a) Rational: $\dfrac{1}{2}$; other: $\dfrac{1 \pm \sqrt{5}}{2}$;
(b) $f(x) = \dfrac{1}{3}\left(x - \dfrac{1}{2}\right)\left(x - \dfrac{1 + \sqrt{5}}{2}\right)\left(x - \dfrac{1 - \sqrt{5}}{2}\right)$
49. No rational zeros   **51.** No rational zeros
53. No rational zeros   **55.** $-2, 1, 2$ **57.** Discussion and
Writing   **59.** [2.4] **(a)** $(4, -6)$; **(b)** $x = 4$; **(c)** minimum:
-6 at $x = 4$ **60.** [2.4] **(a)** $(-1, 2)$; **(b)** $x = -1$;
(c) maximum: 2 at $x = -1$ **61.** [2.1] 10
62. [2.3] $-3, 11$ **63. (a)** $-1, \frac{1}{2}, 3$; **(b)** $0, \frac{3}{2}, 4$;
(c) $-3, -\frac{3}{2}, 1$; **(d)** $-\frac{1}{2}, \frac{1}{4}, \frac{3}{2}$
65. By the rational zeros theorem, only ± 1 and ± 5 can be
rational solutions of $x^2 - 5 = 0$. Since none of them is a
solution, the equation has no rational solutions. But $\sqrt{5}$ is
a solution, so $\sqrt{5}$ must be irrational.
67. $-8, -\frac{3}{2}, 4, 7, 15$

Exercise Set 3.4

1. (d); $x = 2, x = -2, y = 0$
3. (e); $x = 2, x = -2, y = 0$
5. (c); $x = 2, x = -2, y = 8x$

7. No x-intercepts,
y-intercepts: $\left(0, \frac{1}{3}\right)$;

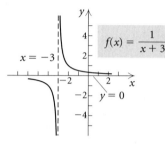

9. No x-intercepts,
y-intercept: $\left(0, \frac{2}{5}\right)$;

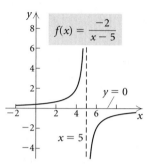

11. x-intercept: $\left(-\frac{1}{2}, 0\right)$,
no y-intercept;

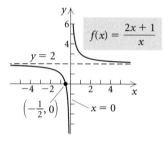

13. No x-intercepts,
y-intercept: $\left(0, \frac{1}{4}\right)$;

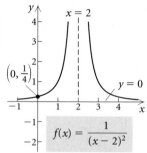

15. No x-intercepts,
no y-intercept;

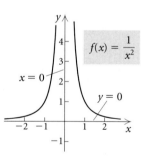

17. No x-intercepts,
y-intercept: $\left(0, \frac{1}{3}\right)$;

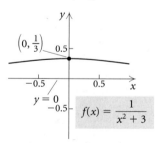

19. x-intercept: $(-2, 0)$,
y-intercept: $(0, 2)$;

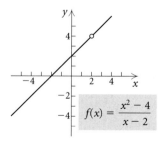

21. x-intercept: $(1, 0)$,
y-intercept: $\left(0, -\frac{1}{2}\right)$;

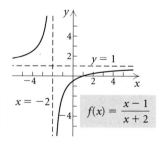

23. x-intercept: $(-3, 0)$, y-intercept: $(0, -1)$;

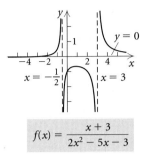

$$f(x) = \frac{x + 3}{2x^2 - 5x - 3}$$

25. x-intercepts: $(-3, 0)$ and $(3, 0)$, y-intercept: $(0, -9)$;

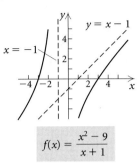

$$f(x) = \frac{x^2 - 9}{x + 1}$$

27. x-intercepts: $(-2, 0)$ and $(1, 0)$, y-intercept: $(0, -2)$;

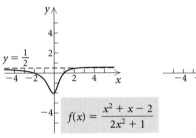

$$f(x) = \frac{x^2 + x - 2}{2x^2 + 1}$$

29. x-intercept: $\left(-\frac{2}{3}, 0\right)$, y-intercept: $(0, 2)$;

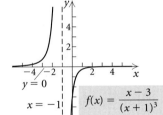

$$g(x) = \frac{3x^2 - x - 2}{x - 1}$$

31. x-intercept: $(1, 0)$, y-intercept: $\left(0, \frac{1}{3}\right)$;

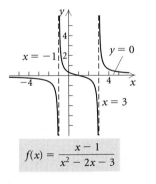

$$f(x) = \frac{x - 1}{x^2 - 2x - 3}$$

33. x-intercepts: $(3, 0)$, y-intercept: $(0, -3)$;

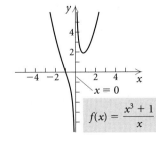

$$f(x) = \frac{x - 3}{(x + 1)^3}$$

35. x-intercept: $(-1, 0)$, no y-intercept;

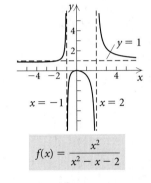

$$f(x) = \frac{x^3 + 1}{x}$$

37. x-intercepts: $(-5, 0)$, $(0, 0)$, and $(3, 0)$, y-intercept: $(0, 0)$;

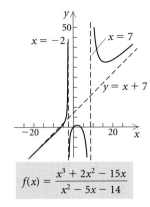

$$f(x) = \frac{x^3 + 2x^2 - 15x}{x^2 - 5x - 14}$$

39. x-intercept: $(0, 0)$, y-intercept: $(0, 0)$;

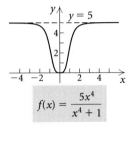

$$f(x) = \frac{5x^4}{x^4 + 1}$$

41. x-intercept: $(0, 0)$, y-intercept: $(0, 0)$;

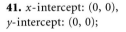

$$f(x) = \frac{x^2}{x^2 - x - 2}$$

43. $f(x) = \dfrac{1}{x^2 - x - 20}$ **45.** $f(x) = \dfrac{3x^2 + 12x + 12}{2x^2 - 2x - 40}$

47. (a)

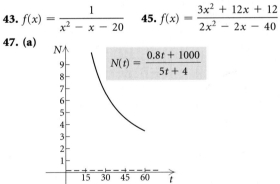

$$N(t) = \frac{0.8t + 1000}{5t + 4}$$

$N(t) \to 0.16$ as $t \to \infty$; **(b)** The medication never completely disappears from the body; a trace amount remains.

49. (a)

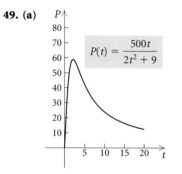

$$P(t) = \frac{500t}{2t^2 + 9}$$

(b) $P(0) = 0$; $P(1) = 45{,}455$; $P(3) = 55{,}556$; $P(8) = 29{,}197$
(c) $P(t) \to 0$ as $t \to \infty$; **(d)** In time, no one lives in Lordsburg. **(e)** 58,926
51. Discussion and Writing   **53.** Discussion and Writing
54. [1.6] $y = 44x$ **55.** [1.6] $y = \dfrac{4}{x}$
56. [1.6] $y = \dfrac{0.0000005xz^2}{w}$ **57.** [1.6] 4 hr
59. $y = x^3 + 4$
61.

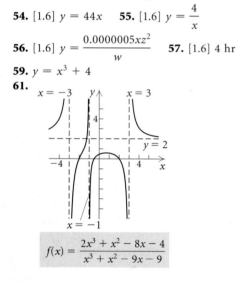

$$f(x) = \frac{2x^3 + x^2 - 8x - 4}{x^3 + x^2 - 9x - 9}$$

63. $(-\infty, -3) \cup (7, \infty)$

Exercise Set 3.5

1. $(-\infty, -5) \cup (-3, 2)$ **3.** $(-2, 0] \cup (2, \infty)$ **5.** $(-4, 1)$
7. $(-\infty, -2] \cup [4, \infty)$ **9.** $(-\infty, -2) \cup (1, \infty)$
11. $(-\infty, -5) \cup (5, \infty)$ **13.** $(-\infty, -2] \cup [2, \infty)$
15. $(-\infty, 3) \cup (3, \infty)$ **17.** $\varnothing$ **19.** $\left(-\infty, -\frac{5}{4}\right] \cup [0, 3]$
21. $[-3, -1] \cup [1, \infty)$ **23.** $(-\infty, -2) \cup (1, 3)$
25. $[-\sqrt{2}, -1] \cup [\sqrt{2}, \infty)$ **27.** $(-\infty, -1] \cup \left[\frac{3}{2}, 2\right]$
29. $(-\infty, 5]$ **31.** $(-\infty, -1.680) \cup (2.154, 5.526)$
33. $(-4, \infty)$ **35.** $\left(-\frac{5}{2}, \infty\right)$ **37.** $\left(-3, -\frac{1}{5}\right] \cup (1, \infty)$
39. $(-\infty, -3) \cup \left[\dfrac{5 - \sqrt{105}}{10}, -\dfrac{1}{3}\right) \cup \left[\dfrac{5 + \sqrt{105}}{10}, \infty\right)$
41. $\left(2, \frac{7}{2}\right]$ **43.** $(1 - \sqrt{2}, 0) \cup (1 + \sqrt{2}, \infty)$
45. $(-\infty, -3) \cup (1, 3) \cup \left[\frac{11}{3}, \infty\right)$ **47.** $(-\infty, \infty)$

49. $\left(-3, \dfrac{1 - \sqrt{61}}{6}\right) \cup \left(-\dfrac{1}{2}, 0\right) \cup \left(\dfrac{1 + \sqrt{61}}{6}, \infty\right)$
51. $(-1, 0) \cup \left(\frac{2}{7}, \frac{7}{2}\right)$
53. $[-6 - \sqrt{33}, -5) \cup [-6 + \sqrt{33}, 1) \cup (5, \infty)$
55. $(0.408, 2.449)$ **57. (a)** $(10, 200)$;
(b) $(0, 10) \cup (200, \infty)$ **59.** $\{n \mid 9 \le n \le 23\}$
61. Discussion and Writing
63. [1.7] $(x + 2)^2 + (y - 4)^2 = 9$
64. [1.7] $x^2 + (y + 3)^2 = \frac{49}{4}$ **65.** [2.4] **(a)** $\left(\frac{3}{4}, -\frac{55}{8}\right)$;
(b) maximum: $-\frac{55}{8}$ when $x = \frac{3}{4}$; **(c)** $\left(-\infty, -\frac{55}{8}\right]$
66. [2.4] **(a)** $(5, -23)$; **(b)** minimum: -23 when $x = 5$;
(c) $[-23, \infty)$ **67.** $\{3\}$ **69.** $(-\infty, \infty)$ **71.** $[-\sqrt{5}, \sqrt{5}]$
73. $\left[-\frac{3}{2}, \frac{3}{2}\right]$ **75.** $\left(-\infty, -\frac{1}{4}\right) \cup \left(\frac{1}{2}, \infty\right)$
77. $(-4, -2) \cup (-1, 1)$
79. $x^2 + x - 12 < 0$; answers may vary

Review Exercises, Chapter 3

1. [3.1] **(a)** $-2.637, 1.137$; **(b)** relative maximum: 7.125 at $x = -0.75$; **(c)** none; **(d)** domain: all real numbers; range: $(-\infty, 7.125]$
2. [3.1] **(a)** $-3, -1.414, 1.414$; **(b)** relative maximum: 2.303 at $x = -2.291$; **(c)** relative minimum: -6.303 at $x = 0.291$; **(d)** domain: all real numbers; range: all real numbers
3. [3.1] **(a)** $0, 1, 2$; **(b)** relative maximum: 0.202 at $x = 0.610$; **(c)** relative minima: 0 at $x = 0$, -0.620 at $x = 1.640$; **(d)** domain: all real numbers; range: $[-0.620, \infty)$
4. [3.1] **(a)** 4%; **(b)** 5%
5. [3.1] **(a)** Linear: $f(x) = 0.5408695652x - 30.30434783$; quadratic: $f(x) = 0.0030322581x^2 - 0.5764516129x + 57.53225806$; cubic: $f(x) = 0.0000247619x^3 - 0.0112857143x^2 + 2.002380952x - 82.14285714$; **(b)** the cubic function; **(c)** 298, 498
6. [3.2] $4x^2 - \frac{14}{5}x - \frac{17}{25}$, R 284/25
7. [3.2] $2x^3 + x + 1$, R $2x + 6$ **8.** [3.2] $\{-5, 1, 2\}$
9. [3.2] $\{-2 - \sqrt{11}, -5, -2 + \sqrt{11}, 4\}$
10. [3.2] $x^2 + 7x + 22$, R 120
11. [3.2] $x^3 + x^2 + x + 1$, R 0   **12.** [3.2] 120
13. [3.2] 0
14. [3.2] -3 is a root; $P(x) = (x + 5)(x + 3)(x - 1)$; other roots: $-5, 1$
15. [3.2] -2 is a root; $P(x) = (x + 2)(x + 1) \cdot [x - (-3 + \sqrt{10})][x - (-3 - \sqrt{10})]$ other roots: $-1, -3 \pm \sqrt{10}$
16. [3.3] $x^3 - x^2 - 10x - 8 = 0$; answers may vary
17. [3.3] $x^4 + 3x^3 - 7x^2 - 9x + 12 = 0$; answers may vary
18. [3.3] $x^6 - 13x^5 + 62x^4 - 132x^3 - 45x^2 + 745x - 858 = 0$; answers may vary
19. [3.3] $\{-1, 2, 3\}$ **20.** [3.3] $\{-5, -3, 1\}$
21. [3.3] $\{-3, 2 - \sqrt{11}, 2 + \sqrt{11}, 3\}$

22. [3.3] **(a)** $\{-10, 1\}$; **(b)**

$y = x^3 + 8x^2 - 19x + 10$

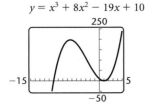

(c) $P(x) = (x - 1)^2(x + 10)$
23. [3.3] **(a)** $\{-4, 0, 3, 4\}$;
(b)

$y = x^6 + x^5 - 28x^4 - 16x^3 + 192x^2$

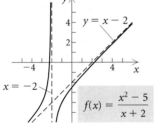

(c) $P(x) = x^2(x + 4)^2(x - 3)(x - 4)$
24. [3.4]

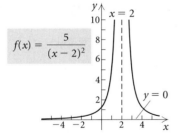

$y = x - 2$

$x = -2$

$f(x) = \dfrac{x^2 - 5}{x + 2}$

25. [3.4]

$f(x) = \dfrac{5}{(x - 2)^2}$

$x = 2$

$y = 0$

26. [3.4]

$y = 1$

$x = -4$ $x = 5$

$f(x) = \dfrac{x^2 + x - 6}{x^2 - x - 20}$

27. [3.4]

$y = 0$

$x = -3$ $x = 5$

$f(x) = \dfrac{x - 2}{x^2 - 2x - 15}$

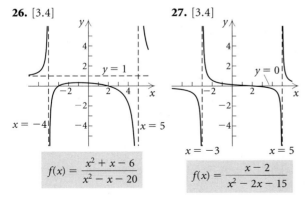

28. [3.4] $f(x) = \dfrac{1}{x^2 - x - 6}$

29. [3.4] $f(x) = \dfrac{4x^2 + 12x}{x^2 - x - 6}$

30. [3.4] **(a)** $N(t)$

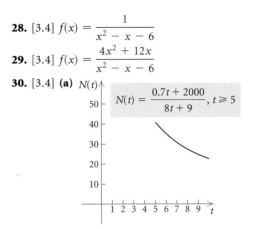

$N(t) = \dfrac{0.7t + 2000}{8t + 9}, t \geq 5$

$N(t) \to 0.0875$ as $t \to \infty$; **(b)** The medication never completely disappears from the body; a trace amount remains.   **31.** [3.5] $(-3, 3)$
32. [3.5] $\left(-\infty, -\frac{1}{2}\right) \cup (2, \infty)$
33. [3.5] $[-4, 1] \cup [2, \infty)$
34. [3.5] $\left(-\infty, -\frac{14}{3}\right) \cup (-3, \infty)$
35. [3.5] $(-5, -3.386] \cup (-2, 2) \cup (4, \infty)$
36. [3.1], [3.5] **(a)** 324 ft, 2.5 sec; **(b)** $t = 7$; **(c)** $(2, 3)$
37. [3.5] $\left[\dfrac{5 - \sqrt{15}}{2}, \dfrac{5 + \sqrt{15}}{2}\right]$
38. Discussion and Writing [3.1], [3.4] A polynomial function is a function that can be defined by a polynomial expression. A rational function is a function that can be defined as a quotient of two polynomials.
39. Discussion and Writing [3.4] Vertical asymptotes occur at any x-values that make the denominator zero. The graph of a rational function does not cross any vertical asymptotes. Horizontal asymptotes occur when the degree of the numerator is less than or equal to the degree of the denominator. Oblique asymptotes occur when the degree of the numerator is 1 greater than the degree of the denominator. Graphs of rational functions may cross horizontal or oblique asymptotes.
40. [3.1] 9%   **41.** [3.5] $(-\infty, -1 - \sqrt{6}] \cup [-1 + \sqrt{6}, \infty)$
42. [3.5] $\left(-\infty, -\frac{1}{2}\right) \cup \left(\frac{1}{2}, \infty\right)$
43. [3.3] $\{1 + i, 1 - i, i, -i\}$ **44.** [3.5] $(-\infty, 2)$
45. [3.3] $(x - 1)\left(x + \dfrac{1}{2} - \dfrac{\sqrt{3}}{2}i\right)\left(x + \dfrac{1}{2} + \dfrac{\sqrt{3}}{2}i\right)$
46. [3.2] 7   **47.** [3.2] -4 **48.** [3.5] $(-\infty, -5] \cup [2, \infty)$
49. [3.5] $(-\infty, 1.1] \cup [2, \infty)$ **50.** [3.5] $\left(-1, \frac{3}{7}\right)$

Chapter 4

Exercise Set 4.1

1. $(f \circ g)(x) = (g \circ f)(x) = x$
3. $(f \circ g)(x) = (g \circ f)(x) = x$
5. $(f \circ g)(x) = 20;\ (g \circ f)(x) = 0.05$
7. $(f \circ g)(x) = |x|;\ (g \circ f)(x) = x$
9. $(f \circ g)(x) = (g \circ f)(x) = x$
11. $(f \circ g)(x) = x^3 - 2x^2 - 4x + 6;$
$(g \circ f)(x) = x^3 - 5x^2 + 3x + 8$
13. $f(x) = x^5;\ g(x) = 4 + 3x$

15. $f(x) = \dfrac{1}{x};\ g(x) = (x - 2)^4$

17. $f(x) = \dfrac{x - 1}{x + 1};\ g(x) = x^3$

19. $f(x) = x^6;\ g(x) = \dfrac{2 + x^3}{2 - x^3}$

21. $f(x) = \sqrt{x};\ g(x) = \dfrac{x - 5}{x + 2}$

23. $f(x) = x^3 - 5x^2 + 3x - 1;\ g(x) = x + 2$
25. $f(x) = 2(x - 20)$
27. $\{(8, 7), (8, -2), (-4, 3), (-8, 8)\}$
29. $\{(-1, -1), (4, -3)\}$ **31.** $x = 4y - 5$
33. $y^3 x = -5$
35.

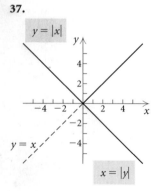

37.

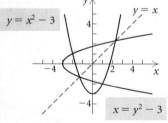

39. Yes   **41.** No   **43.** No   **45.** Yes   **47.** Yes
49. No   **51.** No   **53.** Yes

55. $y_1 = 0.8x + 1.7,$
$y_2 = \dfrac{x - 1.7}{0.8}$

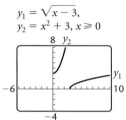

Domain and range of both f and f^{-1}: all real numbers

57. $y_1 = \dfrac{1}{2}x - 4,$
$y_2 = 2x + 8$

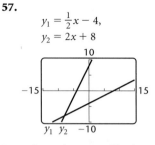

Domain and range of both f and f^{-1}: all real numbers

59. $y_1 = \sqrt{x - 3},$
$y_2 = x^2 + 3, x \geqslant 0$

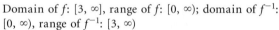

Domain of f: $[3, \infty]$, range of f: $[0, \infty)$; domain of f^{-1}: $[0, \infty)$, range of f^{-1}: $[3, \infty)$

61. $y_1 = x^2 - 4, x \geqslant 0;\ y_2 = \sqrt{4 + x}$

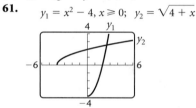

Domain of f: $[0, \infty)$, range of f: $[-4, \infty)$; domain of f^{-1}: $[-4, \infty)$, range of f^{-1}: $[0, \infty)$

63. $y_1 = (3x - 9)^3,\ y_2 = \dfrac{\sqrt[3]{x} + 9}{3}$

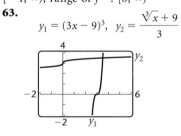

Domain and range of both f and f^{-1}: all real numbers
65. (a) One-to-one; **(b)** $f^{-1}(x) = x - 4$

67. (a) One-to-one; **(b)** $f^{-1}(x) = \dfrac{x + 1}{2}$

69. (a) One-to-one; **(b)** $f^{-1}(x) = \dfrac{4}{x} - 7$

71. (a) One-to-one; **(b)** $f^{-1}(x) = \dfrac{3x + 4}{x - 1}$

73. (a) One-to-one; **(b)** $f^{-1}(x) = \sqrt[3]{x + 1}$
75. (a) Not one-to-one; **(b)** does not have an inverse that is a function

77. (a) One-to-one; **(b)** $f^{-1}(x) = \sqrt{\dfrac{x+2}{5}}, \; x \geq -2$

79. (a) One-to-one; **(b)** $f^{-1}(x) = x^2 - 1, \; x \geq 0$

81. $\frac{1}{3}x$ **83.** $-x$ **85.** $x^3 + 5$

87. $(f^{-1} \circ f)(x) = f^{-1}(f(x))$
$$= f^{-1}\!\left(\tfrac{7}{8}x\right)$$
$$= \tfrac{8}{7}\!\left(\tfrac{7}{8}x\right) = x;$$
$(f \circ f^{-1})(x) = f(f^{-1}(x))$
$$= f\!\left(\tfrac{8}{7}x\right)$$

89. $(f^{-1} \circ f)(x) = f^{-1}\!\left(\dfrac{1-x}{x}\right)$

$$= \dfrac{1}{\dfrac{1-x}{x} + 1} = \dfrac{1}{\dfrac{1}{x}} = x;$$

$(f \circ f^{-1})(x) = f\!\left(\dfrac{1}{x+1}\right)$

$$= \dfrac{1 - \dfrac{1}{x+1}}{\dfrac{1}{x+1}} = \dfrac{\dfrac{x+1-1}{x+1}}{\dfrac{1}{x+1}} = x$$

91. $5; \; a$ **93. (a)** 36, 40, 44, 52, 60; **(b)** $g^{-1}(x) = \dfrac{x}{2} - 12;$

(c) 6, 8, 10, 14, 18

95. (a) 0.5, 11.5, 22.5, 55.5, 72;

(b), (d)

$$y_1 = \dfrac{11x + 5}{10}, \quad y_2 = \dfrac{10x - 5}{11}$$

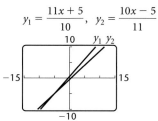

(c) $D^{-1}(r) = \dfrac{10r - 5}{11};$ the speed, in miles per hour, that the car is traveling when the reaction distance is r feet

97. Discussion and Writing

99. [1.2]

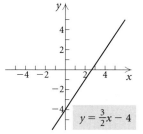

$$y = \tfrac{3}{2}x - 4$$

100. [1.7]

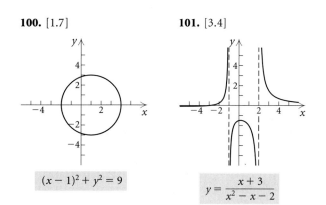

$$(x-1)^2 + y^2 = 9$$

101. [3.4]

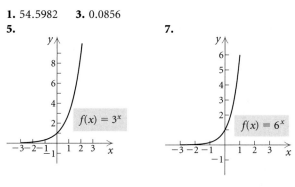

$$y = \dfrac{x+3}{x^2 - x - 2}$$

102. [2.4]

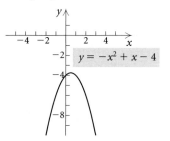

$$y = -x^2 + x - 4$$

103. Yes   **105.** No   **107.** $f(x) = x^2 - 3$, for inputs $x \geq 0$; $f^{-1}(x) = \sqrt{x + 3}$, for inputs $x \geq -3$

109. Answers may vary. $f(x) = 3/x, \; f(x) = 1 - x, \; f(x) = x$

Exercise Set 4.2

1. 54.5982   **3.** 0.0856

5.

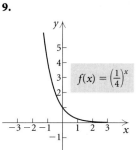

$$f(x) = 3^x$$

7.

$$f(x) = 6^x$$

9.

$$f(x) = \left(\tfrac{1}{4}\right)^x$$

11.

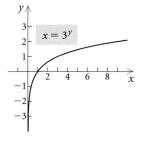

$$x = 3^y$$

13.

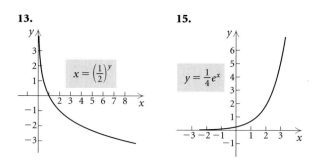

$$x = \left(\tfrac{1}{2}\right)^y$$

15.

$$y = \tfrac{1}{4}e^x$$

17.

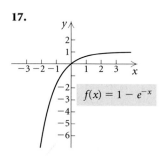

$$f(x) = 1 - e^{-x}$$

19. Shift the graph of $y = 2^x$ left 1 unit.

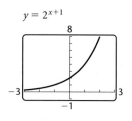

$$y = 2^{x+1}$$

21. Shift the graph of $y = 2^x$ down 3 units.

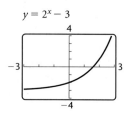

$$y = 2^x - 3$$

23. Reflect the graph of $y = 3^x$ across the y-axis, then across the x-axis, and then shift it up 4 units.

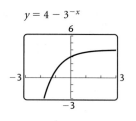

$$y = 4 - 3^{-x}$$

25. Shift the graph of $y = \left(\tfrac{3}{2}\right)^x$ right 1 unit.

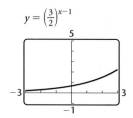

$$y = \left(\tfrac{3}{2}\right)^{x-1}$$

27. Shift the graph of $y = 2^x$ left 3 units, and then down 5 units.

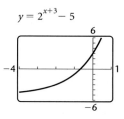

$$y = 2^{x+3} - 5$$

29. Shrink the graph of $y = e^x$ horizontally.

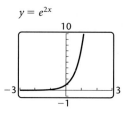

$$y = e^{2x}$$

31. Shift the graph of $y = e^x$ left 1 unit and reflect it across the y-axis.

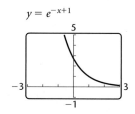

$$y = e^{-x+1}$$

33. Reflect the graph of $y = e^x$ across the y-axis, then across the x-axis, then shift it up 1 unit, and then stretch it vertically.

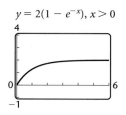

$$y = 2(1 - e^{-x}),\ x > 0$$

35. (a) 799,053; **(b)** 1,363,576;
(c)

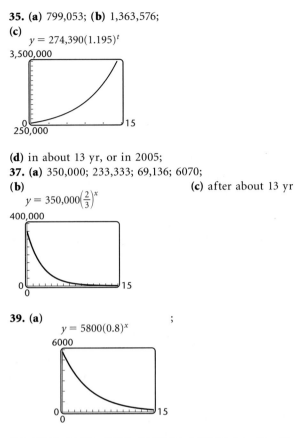

$y = 274,390(1.195)^t$
3,500,000
0
250,000
15

(d) in about 13 yr, or in 2005;
37. (a) 350,000; 233,333; 69,136; 6070;
(b)   **(c)** after about 13 yr

$y = 350,000\left(\frac{2}{3}\right)^x$
400,000
0
0
15

39. (a)   ;

$y = 5800(0.8)^x$
6000
0
0
15

(b) $5800, $4640, $3712, $1900.54, $622.77; **(c)** after 11 yr
41. (a)   ;

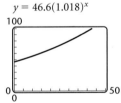

$y = 46.6(1.018)^x$
100
0
0
50

(b) About 65.4 billion ft³, 78.2 billion ft³; **(c)** after about
39 yr
43. (a)   ;

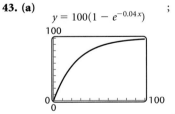

$y = 100(1 - e^{-0.04x})$
100
0
0
100

(b) about 63%; **(c)** after 58 days
45. (c)   **47. (a)**   **49. (l)**   **51. (g)**   **53. (i)**   **55. (k)**
57. (m)   **59.** (1.481, 4.090)   **61.** $(-0.402, -1.662)$,
(1.051, 2.722)   **63.** 4.448   **65.** $(0, \infty)$
67. 2.294, 3.228   **69.** Discussion and Writing
71. Discussion and Writing   **72.** [3.1] -2
73. [3.1] 0.63188461, 1.4252375
74. [3.1] -0.808143, 2.4748096   **75.** [3.1] $-8, 0, 2$
77. $y = 1$
79. (a)

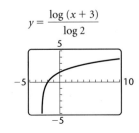

$y = e^{-x^2}$
3
-5 5
-1

(b) none; **(c)** relative maximum: 1 at $x = 0$

Exercise Set 4.3

1.   **3.**

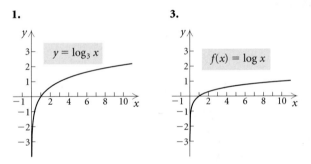

$y = \log_3 x$ $f(x) = \log x$

5. 4   **7.** 3   **9.** -3 **11.** -2 **13.** 0   **15.** 1
17. $\log_{10} 1000 = 3$ **19.** $\log_8 2 = \frac{1}{3}$ **21.** $\log_e t = 3$
23. $\log_e 7.3891 = 2$
25. $\log_p 3 = k$ **27.** $5^1 = 5$ **29.** $10^{-2} = 0.01$
31. $e^{3.4012} = 30$ **33.** $a^{-x} = M$ **35.** $a^x = T^3$
37. 0.4771   **39.** 2.7259   **41.** -0.2441 **43.** Does not
exist   **45.** 0.6931   **47.** 6.6962   **49.** Does not exist
51. 3.3219   **53.** -0.2614 **55.** 0.7384
57. Shift the graph of $y = \log_2 x$ left 3 units. Domain:
$(-3, \infty)$; vertical asymptote: $x = -3$;

$y = \dfrac{\log (x + 3)}{\log 2}$
5
-5 10
-5

59. Shift the graph of $y = \log_3 x$ down 1 unit. Domain: $(0, \infty)$; vertical asymptote: $x = 0$;

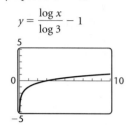

$$y = \frac{\log x}{\log 3} - 1$$

61. Stretch the graph of $y = \ln x$ vertically. Domain: $(0, \infty)$; vertical asymptote: $x = 0$;

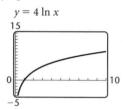

$$y = 4 \ln x$$

63. Reflect the graph of $y = \ln x$ across the x-axis and shift it up 2 units. Domain: $(0, \infty)$; vertical asymptote: $x = 0$;

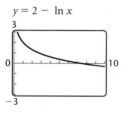

$$y = 2 - \ln x$$

65.   **67.**

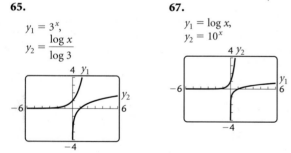

$y_1 = 3^x$,
$y_2 = \dfrac{\log x}{\log 3}$

$y_1 = \log x$,
$y_2 = 10^x$

69. (a) 2.3 ft/sec; **(b)** 3.0 ft/sec; **(c)** 2.2 ft/sec; **(d)** 1.9 ft/sec; **(e)** 1.7 ft/sec

71. (a) 78%; **(b)** 67.5%, 57%;
(c)

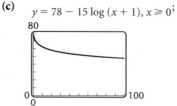

$y = 78 - 15 \log (x + 1), x \geq 0$;   **(d)** after 73 months

73. (a) 10^{-7}; **(b)** 4.0×10^{-6}; **(c)** 6.3×10^{-4}; **(d)** 1.6×10^{-5}
75. (a) 34 decibels; **(b)** 64 decibels; **(c)** 60 decibels;
(d) 90 decibels   **77.** Discussion and Writing
78. [3.2] -4 **79.** [3.2] -280
80. [3.3] $f(x) = x^3 - x^2 + 16x - 16$
81. [3.3] $f(x) = x^3 - 7x$ **83.** 3   **85.** $(0, \infty)$
87. $(-\infty, 0) \cup (0, \infty)$ **89.** $\left(-\frac{5}{2}, -2\right)$ **91.** (d)   **93.** (b)
95. (a)

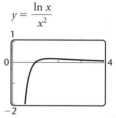

$y = x \ln x$

(b) 1; **(c)** relative minimum: -0.368 at $x = 0.368$
97. (a)

$$y = \frac{\ln x}{x^2}$$

(b) 1; **(c)** relative maximum: 0.184 at $x = 1.649$

Exercise Set 4.4

1. $\log_3 81 + \log_3 27$ **3.** $\log_5 5 + \log_5 125$
5. $\log_t 8 + \log_t Y$ **7.** $3 \log_b t$ **9.** $8 \log y$
11. $-6 \log_c K$ **13.** $\log_t M - \log_t 8$
15. $\log_a x - \log_a y$
17. $\log_a 6 + \log_a x + 5 \log_a y + 4 \log_a z$
19. $2 \log_b p + 5 \log_b q - 4 \log_b m - 9$
21. $3 \log_a x - \frac{5}{2} \log_a p - 4 \log_a q$
23. $2 \log_a m + 3 \log_a n - \frac{3}{4} - \frac{5}{4} \log_a b$
25. $\log_a 150$ **27.** $\log 100 = 2$
29. $\log_a x^{-5/2} y^4$, or $\log_a \dfrac{y^4}{x^{5/2}}$ **31.** $\ln x$ **33.** $\ln (x - 2)$
35. $\ln \dfrac{x}{(x^2 - 25)^3}$ **37.** $\ln \dfrac{2^{11/5} x^9}{y^8}$ **39.** -0.5108
41. -1.6094 **43.** $\frac{1}{2}$ **45.** 2.6094   **47.** 4.3174

49. 3   **51.** $|x - 4|$ **53.** $4x$ **55.** w **57.** $8t$
59. Discussion and Writing   **61.** [2.2] $31 - 22i$
62. [2.2] $\frac{1}{2} - \frac{1}{2}i$ **63.** [2.3] $\left(-\frac{1}{2}, 0\right)$, $(7, 0)$; $-\frac{1}{2}$, 7
64. [3.3] $(1, 0)$; 1   **65.** 4   **67.** $\log_a (x^3 - y^3)$
69. $\frac{1}{2} \log_a (x - y) - \frac{1}{2} \log_a (x + y)$ **71.** 7   **73.** True
75. True   **77.** True   **79.** -2 **81.** 3
83. $\log_a \left(\dfrac{x + \sqrt{x^2 - 5}}{5} \cdot \dfrac{x - \sqrt{x^2 - 5}}{x - \sqrt{x^2 - 5}} \right)$

$$= \log_a \frac{5}{5(x - \sqrt{x^2 - 5})}$$
$$= -\log_a (x - \sqrt{x^2 - 5})$$

Exercise Set 4.5

1. 4   **3.** $\frac{3}{2}$ **5.** 5.044   **7.** $\frac{5}{2}$ **9.** $-3, \frac{1}{2}$ **11.** 0.959
13. 6.908   **15.** 84.191   **17.** -1.710 **19.** 2.844
21. $-1.567, 1.567$ **23.** 0.347   **25.** 625   **27.** 0.0001
29. e **31.** $\frac{22}{3}$ **33.** 10   **35.** $\frac{1}{63}$ **37.** 5   **39.** $\frac{21}{8}$
41. 0.367   **43.** 0.621   **45.** -1.532 **47.** 7.062
49. 2.444   **51.** $(4.093, 0.786)$ **53.** $(7.586, 6.684)$
55. Discussion and Writing
56. [2.4] (a) $(3, 1)$; (b) $x = 3$; (c) maximum: 1 when $x = 3$
57. [2.4] (a) $(0, -6)$; (b) $x = 0$; (c) minimum: -6 when
$x = 0$ **58.** [2.4] (a) $(2, 4)$; (b) $x = 2$; (c) minimum: 4
when $x = 2$ **59.** [2.4] (a) $(-1, -5)$; (b) $x = -1$;
(c) maximum: -5 when $x = -1$ **61.** 10
63. 1, e^4 or 1, 54.598   **65.** $\frac{1}{3}$, 27   **67.** 1, e^2 or 1, 7.389
69. 0, 0.431   **71.** $-9, 9$ **73.** e^{-2}, e^2 or 0.135, 7.389
75. $\frac{7}{4}$ **77.** $(56.598, \infty)$ **79.** 5   **81.** $a = \frac{2}{3}b$ **83.** 88

Exercise Set 4.6

1. (a) $P(t) = 6.0e^{0.013t}$; (b) 6.5 billion, 6.9 billion;
(c) in 22.1 yr; (d) 53.3 yr
3. (a) 36.5 yr; (b) 0.2% per year; (c) 21.0 yr; (d) 138.6 yr;
(e) 3.4% per year; (f) 2.2% per year
5. About 322 yr   **7.** (a) $P(t) = 10,000e^{0.054t}$; (b) \$10,555;
\$11,140; \$13,100; \$17,160; (c) 12.8 yr   **9.** About 5135 yr
11. (a) 23.1% per minute; (b) 3.15% per year; (c) 7.2 days;
(d) 11 yr; (e) 2.8% per year; (f) 0.015% per year;
(g) 0.003% per year   **13.** (a) $k \approx 0.016$; $P(t) = 80e^{-0.016t}$;
(b) about 61 lb per person; (c) 86.6 yr
15. (a)

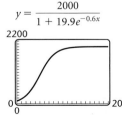

$$y = \frac{2000}{1 + 19.9e^{-0.6x}}$$

(b) 96; (c) 286, 1005, 1719, 1971, 1997

17. (a)

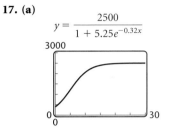

$$y = \frac{2500}{1 + 5.25e^{-0.32x}}$$

(b) 400, 520, 1214, 2059, 2396, 2478
19. (f)   **21.** (b)   **23.** (c)
25. (a) $y = 4.195491964(1.025306189)^x$, or
$y = 4.195491964e^{0.024991289x}$; since $r \approx 0.9954$, the function
is a good fit.
(b)

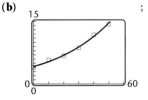

;

(c) 4.8 million, 7.8 million, 51.1 million
27. (a)

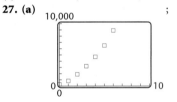

;

(b) linear: $y = 1429.214286x - 530.0714286$, $r^2 = 0.9641$;
quadratic: $y = 158.0952381x^2 + 480.6428571x + 260.4047619$, $r^2 = 0.9995$; $y = 445.8787388(1.736315606)^x$,
$r^2 = 0.9361$; according to r^2, the quadratic has the best fit;
(c)

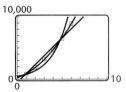

(d) linear: \$19,479 million, or \$19.479 billion; quadratic:
\$37,976 million, or \$37.976 billion; exponential:
\$1,009,295 million, or \$1009.295 billion; the revenue
from the quadratic model seems most realistic. Answers
may vary.
29. Discussion and Writing   **31.** [1.2] 0; $(0, 6)$
32. [1.2] $\frac{3}{10}$; $\left(0, -\frac{7}{5}\right)$ **33.** [1.2] 2; $\left(0, -\frac{3}{13}\right)$
34. [1.2] Slope not defined; no y-intercept
35. \$14,182.70   **37.** \$166.16   **39.** 46.7°F
41. $t = -\dfrac{L}{R} \left[\ln \left(1 - \dfrac{iR}{V} \right) \right]$ **43.** Linear

Review Exercises, Chapter 4

1. [4.1] **(a)** Domain of $f \circ g$, $\left\{x \mid x \neq \frac{3}{2}\right\}$; domain of $g \circ f$, $\left\{x \mid x \neq 0\right\}$; **(b)** $(f \circ g)(x) = \dfrac{4}{(3 - 2x)^2}$; $(g \circ f)(x) = 3 - \dfrac{8}{x^2}$
2. [4.1] **(a)** Domain of $f \circ g$ and $g \circ f$, all real numbers; **(b)** $(f \circ g)(x) = 12x^2 - 4x - 1$; $(g \circ f)(x) = 6x^2 + 8x - 1$
3. [4.1] $f(x) = \sqrt{x}$, $g(x) = 5x + 2$. Answers may vary.
4. [4.1] $f(x) = 4x^2 + 9$, $g(x) = 5x - 1$. Answers may vary.
5. [4.1] $\{(-2.7, 1.3), (-3, 8), (3, -5), (-3, 6), (-5, 7)\}$
6. [4.1] **(a)** $x = 3y^2 + 2y - 1$; **(b)** $0.8y^3 - 5.4x^2 = 3y$
7. [4.1] **(a)** Yes; **(b)** $f^{-1}(x) = x^2 + 6$, $x \geq 0$
8. [4.1] **(a)** Yes; **(b)** $f^{-1}(x) = \sqrt[3]{x + 8}$ **9.** [4.1] **(a)** No
10. [4.1] **(a)** Yes; **(b)** $f^{-1}(x) = \ln x$ **11.** [4.1] 657
12. [4.2] (c)   **13.** [4.3] (a)   **14.** [4.3] (b)   **15.** [4.2] (f)
16. [4.2] (e)   **17.** [4.3] (d)   **18.** [4.3] $4^2 = x$
19. [4.3] $\log_e 80 = x$ **20.** [4.5] 16   **21.** [4.5] $\frac{1}{5}$
22. [4.5] 4.382   **23.** [4.5] 2   **24.** [4.5] $\frac{1}{2}$
25. [4.5] 5   **26.** [4.5] 4   **27.** [4.5] 9   **28.** [4.5] 1
29. [4.5] 3.912   **30.** [4.4] $\log_b \dfrac{x^3\sqrt{z}}{y^4}$
31. [4.4] $\ln(x^2 - 4)$ **32.** [4.4] $\frac{1}{4}\ln w + \frac{1}{2}\ln r$
33. [4.4] $\frac{2}{3}\log M - \frac{1}{3}\log N$ **34.** [4.4] 0.477
35. [4.4] 1.699   **36.** [4.4] -0.699 **37.** [4.4] 0.233
38. [4.4] $-5k$ **39.** [4.4] $-6t$ **40.** [4.6] 8.1 yr
41. [4.6] 2.3%   **42.** [4.6] About 2623 yr   **43.** [4.3] 5.6
44. [4.3] 30 decibels
45. [4.6] **(a)** $P(t) = 11e^{0.012t}$; **(b)** 11.8 million, 14.3 million; **(c)** in 68.4 yr; **(d)** 57.8 yr
46. [4.6] **(a)** $k = 0.304$; **(b)** $P(t) = 7e^{0.304t}$; **(c)**

$$y = 7e^{0.304x}$$

(d) 292.4 billion, 727.9 billion; **(e)** 2006
47. [4.3] **(a)** 2.7 ft/sec; **(b)** 8,553,143
48. [4.6] **(a)** $y = 11.96557466(1.00877703)^x$; **(b)**

$$y = 11.96557466(1.00877703)^x$$

(c) 44, 255, 394; **(d)** Comparing the coefficient of correlation values (power: 0.9739; exponential: 0.9955), we see that the exponential function is the better fit.

49. Discussion and Writing [4.4] By the product rule, $\log_2 x + \log_2 5 = \log_2 5x$, not $\log_2 (x + 5)$. Also, substituting various numbers for x shows that both sides of the inequality are indeed unequal. You could also graph each side and show that the graphs do not coincide.
50. Discussion and Writing [4.1] The inverse of a function $f(x)$ is written $f^{-1}(x)$, whereas $[f(x)]^{-1}$ means $\dfrac{1}{f(x)}$.
51. [4.5] $\frac{1}{64}$, 64   **52.** [4.5] 1   **53.** [4.5] 16
54. [4.2], [4.3] **(a)**

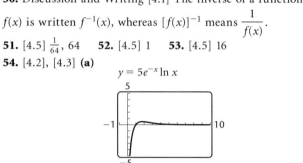
$$y = 5e^{-x}\ln x$$

(b) relative maximum: 0.486 at $x = 1.763$
55. [4.1] No   **56.** [4.3] $(1, \infty)$
57. [4.2] $(-0.393, 1.143)$, $(0.827, 1.495)$, $(3.310, 2.000)$

Chapter 5

Exercise Set 5.1

1. (c)   **3.** (f)   **5.** (b)   **7.** $(-1, 3)$ **9.** $(-0.5, 1.75)$
11. No solution   **13.** $(5, 4)$ **15.** $(1, -3)$ **17.** $(2, -2)$
19. $\left(\frac{39}{11}, -\frac{1}{11}\right)$ **21.** $(1, -1)$
23. $(1, 3)$; consistent, independent
25. $(-4, -2)$; consistent, independent
27. $(-3, 0)$; consistent, independent
29. $(4y + 2, y)$ or $\left(x, \frac{1}{4}x - \frac{1}{2}\right)$; consistent, dependent
31. $(10, 8)$; consistent, independent
33. $(1, 1)$; consistent, independent
35. 10 free rentals, 38 boxes of popcorn
37. Adult: $6, child: $3   **39.** (15, $100)   **41.** 140
43. 6000
45. $1\frac{1}{2}$ servings of spaghetti, 2 servings of lettuce
47. Boat: 20 km/h, stream: 3 km/h
49. $6000 at 7%, $9000 at 9%
51. 6 lb of French roast, 4 lb of Kenyan
53. $21,428.57
55. **(a)** $t(x) = 1.69541779x + 44.15363881$, $r(x) = -0.4528301887x + 68.83018868$; **(b)** 11.5 yr
57. Discussion and Writing   **59.** [2.1] $\frac{5}{2}$, or 2.5
60. [2.1] 2   **61.** [2.3] 1, 3   **62.** [2.3] $\dfrac{1 \pm \sqrt{61}}{6}$
63. 4 km
65. First train: 36 km/h, second train: 54 km/h
67. $A = \frac{1}{10}$, $B = -\frac{7}{10}$ **69.** City: 209; highway: 196

Exercise Set 5.2

1. $(3, -2, 1)$ **3.** $(-3, 2, 1)$ **5.** $\left(2, \frac{1}{2}, -2\right)$

7. No solution   **9.** $\left(\dfrac{11y + 19}{5}, y, \dfrac{9y + 11}{5}\right)$

11. $\left(\frac{1}{2}, \frac{2}{3}, -\frac{5}{6}\right)$ **13.** $(1, -2, 4, -1)$

15. Under 10 lb: 60; 10 lb up to 15 lb: 70; 15 lb or more: 20

17. $1\frac{1}{4}$ servings of beef, 1 baked potato, $\frac{3}{4}$ serving strawberries   **19.** 4%: \$1300; 6%: \$900; 7%: \$2800

21. Orange juice: \$1; bagel: \$1.25; coffee: \$0.75

23. Airways: 351 billion; bus: 32 billion; railroad: 22 billion   **25.** Par-3: 4; par-4: 10; par-5: 4

27. **(a)** $f(x) = -\frac{31}{60}x^2 + \frac{233}{30}x + 211$; **(b)** about 82 thousand

29. **(a)** $f(x) = \frac{69}{110}x^2 - \frac{939}{110}x + 594$; **(b)** 674 lb

31. Discussion and Writing   **33.** [2.2] $5 - 3i$

34. [2.2] $6 - 2i$ **35.** [2.2] $10 - 10i$ **36.** [2.2] $\frac{9}{25} + \frac{13}{25}i$

37. $\left(-1, \frac{1}{5}, -\frac{1}{2}\right)$ **39.** 180°   **41.** $3x + 4y + 2z = 12$

43. $y = -4x^3 + 5x^2 - 3x + 1$

45. Adults: 5; students: 1; children: 94

Exercise Set 5.3

1. 3×2 **3.** 1×4 **5.** 3×3 **7.** $\left[\begin{array}{cc|c} 2 & -1 & 7 \\ 1 & 4 & -5 \end{array}\right]$

9. $\left[\begin{array}{ccc|c} 1 & -2 & 3 & 12 \\ 2 & 0 & -4 & 8 \\ 0 & 3 & 1 & 7 \end{array}\right]$ **11.** $3x - 5y = 1,$
$x + 4y = -2$

13. $2x + y - 4z = 12,$
$3x \qquad + 5z = -1,$
$x - y + z = 2$

15. $\left(\frac{3}{2}, \frac{5}{2}\right)$ **17.** $\left(-\frac{63}{29}, -\frac{114}{29}\right)$ **19.** $\left(-1, \frac{5}{2}\right)$ **21.** $(0, 3)$

23. No solution   **25.** $(3y - 2, y)$ **27.** $(-1, 2, -2)$

29. $\left(\frac{3}{2}, -4, 3\right)$ **31.** $(-1, 6, 3)$

33. $\left(\frac{1}{2}z + \frac{1}{2}, -\frac{1}{2}z - \frac{1}{2}, z\right)$ **35.** $(r - 2, -2r + 3, r)$

37. No solution   **39.** $(1, -3, -2, -1)$ **41.** 1:00 A.M.

43. \$8000 at 8%; \$12,000 at 10%; \$10,000 at 12%

45. Discussion and Writing   **47.** [2.3] $\dfrac{-1 \pm \sqrt{57}}{4}$

48. [2.6] $-3, -2$ **49.** [2.6] 4   **50.** [2.3] 9

51. $y = 3x^2 + \frac{5}{2}x - \frac{15}{2}$ **53.** $\left[\begin{array}{cc} 1 & 5 \\ 0 & 1 \end{array}\right], \left[\begin{array}{cc} 1 & 0 \\ 0 & 1 \end{array}\right]$

55. $\left(-\frac{4}{3}, -\frac{1}{3}, 1\right)$ **57.** $\left(-\frac{14}{13}z - 1, \frac{3}{13}z - 2, z\right)$

59. $(-3, 3)$

Exercise Set 5.4

1. $x = -3, y = 5$ **3.** $x = -1, y = 1$ **5.** $\left[\begin{array}{cc} -2 & 7 \\ 6 & 2 \end{array}\right]$

7. $\left[\begin{array}{cc} 1 & 3 \\ 2 & 6 \end{array}\right]$ **9.** $\left[\begin{array}{cc} 9 & 9 \\ -3 & -3 \end{array}\right]$ **11.** $\left[\begin{array}{cc} 11 & 13 \\ 5 & 3 \end{array}\right]$

13. $\left[\begin{array}{cc} -4 & 3 \\ -2 & -4 \end{array}\right]$ **15.** $\left[\begin{array}{cc} 17 & 9 \\ -2 & 1 \end{array}\right]$ **17.** $\left[\begin{array}{cc} 0 & 0 \\ 0 & 0 \end{array}\right]$

19. $\left[\begin{array}{cc} 1 & 2 \\ 4 & 3 \end{array}\right]$ **21.** **(a)** [150 80 40]; **(b)** [157.5 84 42];

(c) [307.5 164 82], the total budget for each area in June and July   **23.** **(a)** $\mathbf{C} = [140\ \ 27\ \ 3\ \ 13\ \ 64]$,
$\mathbf{P} = [180\ \ 4\ \ 11\ \ 24\ \ 662]$, $\mathbf{B} = [50\ \ 5\ \ 1\ \ 82\ \ 20]$;
(b) [650 50 28 307 1448], the total nutritional values of a meal of 1 serving of chicken, 1 cup of potato salad, and 3 broccoli spears

25. $\left[\begin{array}{c} 1 \\ 40 \end{array}\right]$ **27.** $\left[\begin{array}{cc} -10 & 28 \\ 14 & -26 \\ 0 & -6 \end{array}\right]$ **29.** Not defined

31. $\left[\begin{array}{ccc} 3 & 16 & 3 \\ 0 & -32 & 0 \\ -6 & 4 & 5 \end{array}\right]$

33. **(a)** $\left[\begin{array}{ccccc} 45.29 & 6.63 & 10.94 & 7.42 & 8.01 \\ 53.78 & 4.95 & 9.83 & 6.16 & 12.56 \\ 47.13 & 8.47 & 12.66 & 8.29 & 9.43 \\ 51.64 & 7.12 & 11.57 & 9.35 & 10.72 \end{array}\right]$;

(b) [65 48 93 57];
(c) [12,851.86 1862.1 3019.81 2081.9 2611.56];
(d) the total cost, in cents, for each item for the day's meals

35. **(a)** $\left[\begin{array}{cc} 8 & 15 \\ 6 & 10 \\ 4 & 3 \end{array}\right]$; **(b)** [3 1.50 2]; **(c)** [41 66];

(d) the total cost, in dollars, of ingredients for each coffee shop

37. **(a)** [6 4.50 5.20]; **(b)** $\mathbf{PS} = [95.80\ \ 150.60]$

39. $\left[\begin{array}{cc} 2 & -3 \\ 1 & 5 \end{array}\right]\left[\begin{array}{c} x \\ y \end{array}\right] = \left[\begin{array}{c} 7 \\ -6 \end{array}\right]$

41. $\left[\begin{array}{ccc} 1 & 1 & -2 \\ 3 & -1 & 1 \\ 2 & 5 & -3 \end{array}\right]\left[\begin{array}{c} x \\ y \\ z \end{array}\right] = \left[\begin{array}{c} 6 \\ 7 \\ 8 \end{array}\right]$

43. $\left[\begin{array}{ccc} 3 & -2 & 4 \\ 2 & 1 & -5 \end{array}\right]\left[\begin{array}{c} x \\ y \\ z \end{array}\right] = \left[\begin{array}{c} 17 \\ 13 \end{array}\right]$

45. $\begin{bmatrix} -4 & 1 & -1 & 2 \\ 1 & 2 & -1 & -1 \\ -1 & 1 & 4 & -3 \\ 2 & 3 & 5 & -7 \end{bmatrix} \begin{bmatrix} w \\ x \\ y \\ z \end{bmatrix} = \begin{bmatrix} 12 \\ 0 \\ 1 \\ 9 \end{bmatrix}$

47. Discussion and Writing

49. [2.4] **(a)** $\left(\frac{3}{2}, -\frac{49}{4}\right)$; **(b)** $x = \frac{3}{2}$; **(c)** minimum: $-\frac{49}{4}$

50. [2.4] **(a)** $\left(\frac{5}{4}, -\frac{49}{8}\right)$; **(b)** $x = \frac{5}{4}$; **(c)** minimum: $-\frac{49}{8}$

51. [2.4] **(a)** $\left(-\frac{3}{2}, \frac{29}{4}\right)$; **(b)** $x = -\frac{3}{2}$; **(c)** maximum: $\frac{29}{4}$

52. [2.4] **(a)** $\left(\frac{2}{3}, \frac{7}{3}\right)$; **(b)** $x = \frac{2}{3}$; **(c)** maximum: $\frac{7}{3}$

53. $(\mathbf{A} + \mathbf{B})(\mathbf{A} - \mathbf{B}) = \begin{bmatrix} -2 & 1 \\ 2 & -1 \end{bmatrix}$; $\mathbf{A}^2 - \mathbf{B}^2 = \begin{bmatrix} 0 & 3 \\ 0 & -3 \end{bmatrix}$

55. $(\mathbf{A} + \mathbf{B})(\mathbf{A} - \mathbf{B}) = \begin{bmatrix} -2 & 1 \\ 2 & -1 \end{bmatrix}$
$= \mathbf{A}^2 + \mathbf{B}\mathbf{A} - \mathbf{A}\mathbf{B} - \mathbf{B}^2$

57. $\mathbf{A} + \mathbf{B} =$

$\begin{bmatrix} a_{11} + b_{11} & a_{12} + b_{12} & a_{13} + b_{13} & \cdots & a_{1n} + b_{1n} \\ a_{21} + b_{21} & a_{22} + b_{22} & a_{23} + b_{23} & \cdots & a_{2n} + b_{2n} \\ a_{31} + b_{31} & a_{32} + b_{32} & a_{33} + b_{33} & \cdots & a_{3n} + b_{3n} \\ \vdots & \vdots & \vdots & & \vdots \\ a_{m1} + b_{m1} & a_{m2} + b_{m2} & a_{m3} + b_{m3} & \cdots & a_{mn} + b_{mn} \end{bmatrix}$
$=$
$\begin{bmatrix} b_{11} + a_{11} & b_{12} + a_{12} & b_{13} + a_{13} & \cdots & b_{1n} + a_{1n} \\ b_{21} + a_{21} & b_{22} + a_{22} & b_{23} + a_{23} & \cdots & b_{2n} + a_{2n} \\ b_{31} + a_{31} & b_{32} + a_{32} & b_{33} + a_{33} & \cdots & b_{3n} + a_{3n} \\ \vdots & \vdots & \vdots & & \vdots \\ b_{m1} + a_{m1} & b_{m2} + a_{m2} & b_{m3} + a_{m3} & \cdots & b_{mn} + a_{mn} \end{bmatrix}$
$= \mathbf{B} + \mathbf{A}$

59. $(kl)\mathbf{A} = \begin{bmatrix} (kl)a_{11} & (kl)a_{12} & (kl)a_{13} & \cdots & (kl)a_{1n} \\ (kl)a_{21} & (kl)a_{22} & (kl)a_{23} & \cdots & (kl)a_{2n} \\ (kl)a_{31} & (kl)a_{32} & (kl)a_{33} & \cdots & (kl)a_{3n} \\ \vdots & \vdots & \vdots & & \vdots \\ (kl)a_{m1} & (kl)a_{m2} & (kl)a_{m3} & \cdots & (kl)a_{mn} \end{bmatrix}$

$= \begin{bmatrix} k(la_{11}) & k(la_{12}) & k(la_{13}) & \cdots & k(la_{1n}) \\ k(la_{21}) & k(la_{22}) & k(la_{23}) & \cdots & k(la_{2n}) \\ k(la_{31}) & k(la_{32}) & k(la_{33}) & \cdots & k(la_{3n}) \\ \vdots & \vdots & \vdots & & \vdots \\ k(la_{m1}) & k(la_{m2}) & k(la_{m3}) & \cdots & k(la_{mn}) \end{bmatrix}$

$= k \begin{bmatrix} la_{11} & la_{12} & la_{13} & \cdots & la_{1n} \\ la_{21} & la_{22} & la_{23} & \cdots & la_{2n} \\ la_{31} & la_{32} & la_{33} & \cdots & la_{3n} \\ \vdots & \vdots & \vdots & & \vdots \\ la_{m1} & la_{m2} & la_{m3} & \cdots & la_{mn} \end{bmatrix}$

$= k(l\mathbf{A})$

61. $(k + l)\mathbf{A} =$
$\begin{bmatrix} (k+l)a_{11} & (k+l)a_{12} & (k+l)a_{13} & \cdots & (k+l)a_{1n} \\ (k+l)a_{21} & (k+l)a_{22} & (k+l)a_{23} & \cdots & (k+l)a_{2n} \\ (k+l)a_{31} & (k+l)a_{32} & (k+l)a_{33} & \cdots & (k+l)a_{3n} \\ \vdots & \vdots & \vdots & & \vdots \\ (k+l)a_{m1} & (k+l)a_{m2} & (k+l)a_{m3} & \cdots & (k+l)a_{mn} \end{bmatrix}$
$=$
$\begin{bmatrix} ka_{11} + la_{11} & ka_{12} + la_{12} & ka_{13} + la_{13} & \cdots & ka_{1n} + la_{1n} \\ ka_{21} + la_{21} & ka_{22} + la_{22} & ka_{23} + la_{23} & \cdots & ka_{2n} + la_{2n} \\ ka_{31} + la_{31} & ka_{32} + la_{32} & ka_{33} + la_{33} & \cdots & ka_{3n} + la_{3n} \\ \vdots & \vdots & \vdots & & \vdots \\ ka_{m1} + la_{m1} & ka_{m2} + la_{m2} & ka_{m3} + la_{m3} & \cdots & ka_{mn} + la_{mn} \end{bmatrix}$
$= k\mathbf{A} + l\mathbf{A}$

Exercise Set 5.5

1. Yes **3.** No **5.** $\begin{bmatrix} -3 & 2 \\ 5 & -3 \end{bmatrix}$ **7.** Does not exist

9. $\begin{bmatrix} \frac{3}{8} & -\frac{1}{4} & \frac{1}{8} \\ -\frac{1}{8} & \frac{3}{4} & -\frac{3}{8} \\ -\frac{1}{4} & \frac{1}{2} & \frac{1}{4} \end{bmatrix}$ **11.** Does not exist **13.** $\begin{bmatrix} 0.4 & -0.6 \\ 0.2 & -0.8 \end{bmatrix}$

15. $\begin{bmatrix} -1 & -1 & -6 \\ 1 & 0 & 2 \\ 0 & 1 & 3 \end{bmatrix}$ **17.** $\begin{bmatrix} 1 & 1 & 2 \\ 1 & 1 & 1 \\ 2 & 3 & 4 \end{bmatrix}$ **19.** Does not exist

21. $\begin{bmatrix} 1 & -2 & 3 & 8 \\ 0 & 1 & -3 & 1 \\ 0 & 0 & 1 & -2 \\ 0 & 0 & 0 & -1 \end{bmatrix}$ **23.** $\begin{bmatrix} 0.25 & 0.25 & 1.25 & -0.25 \\ 0.5 & 1.25 & 1.75 & -1 \\ -0.25 & -0.25 & -0.75 & 0.75 \\ 0.25 & 0.5 & 0.75 & -0.5 \end{bmatrix}$

25. $(2, -2)$ **27.** $(0, 2)$ **29.** $(3, -3, -2)$
31. $(-1, 0, 1)$ **33.** $(1, -1, 0, 1)$
35. 50 sausages, 95 hot dogs
37. Topsoil: \$239; mulch: \$179; pea gravel: \$222
39. Discussion and Writing
41. [3.2] -48 **42.** [3.2] 194
43. [3.2] $(x + 2)(x - 1)(x - 4)$
44. [3.2] $(x + 5)(x + 1)(x - 1)(x - 3)$

45. $\mathbf{A}^{-1}$ exists if and only if $x \neq 0$. $\mathbf{A}^{-1} = \begin{bmatrix} \frac{1}{x} \end{bmatrix}$

47. $\mathbf{A}^{-1}$ exists if and only if $xyz \neq 0$. $\mathbf{A}^{-1} = \begin{bmatrix} 0 & 0 & \frac{1}{z} \\ 0 & \frac{1}{y} & 0 \\ \frac{1}{x} & 0 & 0 \end{bmatrix}$

Exercise Set 5.6

1. (f) **3.** (h) **5.** (g)
7. (b)
9.

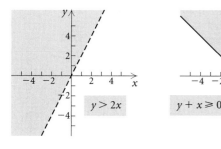

$y > 2x$

11.

$y + x \geq 0$

13.

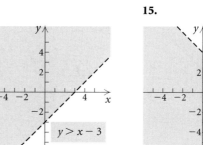

$y > x - 3$

15.

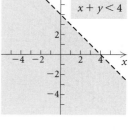

$x + y < 4$

17.

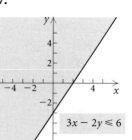

$3x - 2y \leq 6$

19.

$3y + 2x \geq 6$

21.

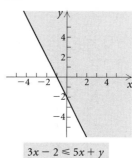

$3x - 2 \leq 5x + y$

23.

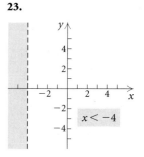

$x < -4$

25.

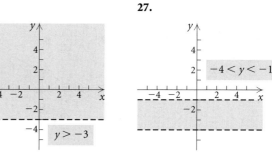

$y > -3$

27.

$-4 < y < -1$

29.

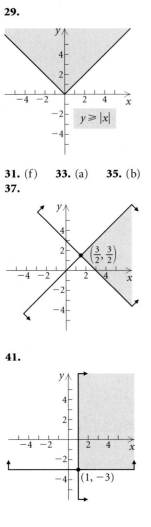

$y \geq |x|$

31. (f)   **33.** (a)   **35.** (b)

37.

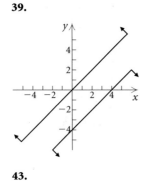

$\left(\frac{3}{2}, \frac{3}{2}\right)$

39.

41.

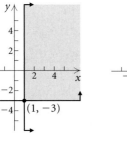

$(1, -3)$

43.

$(3, -7)$

45.

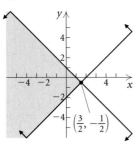

$\left(\frac{3}{2}, -\frac{1}{2}\right)$

47.

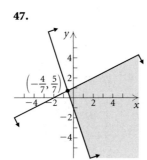

$\left(-\frac{4}{7}, \frac{5}{7}\right)$

49.

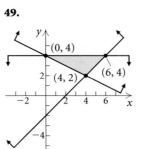

$(0, 4)$ $(6, 4)$ $(4, 2)$

51.

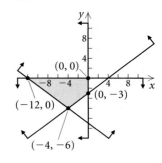

$(0, 0)$ $(-12, 0)$ $(0, -3)$ $(-4, -6)$

53.

$\left(1, \frac{25}{6}\right)$ $\left(1, \frac{9}{4}\right)$ $\left(3, \frac{5}{2}\right)$ $\left(3, \frac{3}{4}\right)$

55. Maximum: 179 when $x = 7$ and $y = 0$; minimum: 48 when $x = 0$ and $y = 4$

57. Maximum: 216 when $x = 0$ and $y = 6$; minimum: 0 when $x = 0$ and $y = 0$

59. Maximum income of \$18 is achieved when 100 of each type of biscuit is made.

61. Maximum profit of \$11,000 is achieved by producing 100 units of lumber and 300 units of plywood.

63. Minimum cost of \$36$\frac{12}{13}$ is achieved by using $1\frac{11}{13}$ sacks of soybean meal and $1\frac{11}{13}$ sacks of oats.

65. Maximum income of \$3110 is achieved when \$22,000 is invested in corporate bonds and \$18,000 is invested in municipal bonds.

67. Minimum cost of \$460 thousand is achieved using 30 P_1's and 10 P_2's.

69. Maximum profit per day of \$192 is achieved when 2 knit suits and 4 worsted suits are made.

71. Minimum weekly cost of \$19.05 is achieved when 1.5 lb of meat and 3 lb of cheese are used.

73. Maximum total number is 800, or 550 of A and 250 of B.   **75.** Discussion and Writing

77. [2.7] $\{x \mid -7 \leq x < 2\}$, or $[-7, 2)$

78. [2.7] $\{x \mid x \leq 1 \text{ or } x \geq 5\}$, or $(-\infty, 1] \cup [5, \infty)$

79. [3.5] $\{x \mid -1 \leq x \leq 3\}$, or $[-1, 3]$

80. [3.5] $\{x \mid -3 < x < -2\}$, or $(-3, -2)$

81.

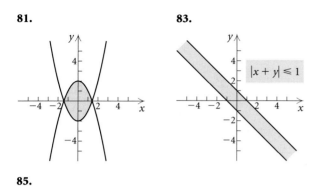

83.

$|x + y| \leqslant 1$

85.

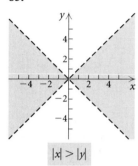

$|x| > |y|$

87. Maximum income of \$28,500 is achieved by making 30 less expensive assemblies and 30 more expensive assemblies.

Exercise Set 5.7

1. $\dfrac{2}{x - 3} - \dfrac{1}{x + 2}$ **3.** $-\dfrac{4}{3x - 1} + \dfrac{5}{2x - 1}$

5. $-\dfrac{3}{x - 2} + \dfrac{2}{x + 2} + \dfrac{4}{x + 1}$

7. $-\dfrac{3}{(x + 2)^2} - \dfrac{1}{x + 2} + \dfrac{1}{x - 1}$ **9.** $\dfrac{3}{x - 1} - \dfrac{4}{2x - 1}$

11. $x - 2 - \dfrac{\frac{11}{4}}{(x + 1)^2} + \dfrac{\frac{17}{16}}{x + 1} - \dfrac{\frac{17}{16}}{x - 3}$

13. $\dfrac{3x + 5}{x^2 + 2} - \dfrac{4}{x - 1}$ **15.** $-\dfrac{2}{x + 2} + \dfrac{10}{(x + 2)^2} + \dfrac{3}{2x - 1}$

17. $3x + 1 + \dfrac{2}{2x - 1} + \dfrac{3}{x + 1}$

19. $-\dfrac{1}{x - 3} + \dfrac{3x}{x^2 + 2x - 5}$

21. $\dfrac{5}{3x + 5} - \dfrac{3}{x + 1} + \dfrac{4}{(x + 1)^2}$ **23.** $\dfrac{8}{4x - 5} + \dfrac{3}{3x + 2}$

25. $\dfrac{2x - 5}{3x^2 + 1} - \dfrac{2}{x - 2}$ **27.** Discussion and Writing

29. Discussion and Writing   **30.** [3.3] $3, \pm i$

31. [3.3] $-2, \dfrac{1 \pm \sqrt{5}}{2}$ **32.** [3.3] $-2, 3, \pm i$

33. [3.3] $-3, -1 \pm \sqrt{2}$

35. $-\dfrac{\frac{1}{2a^2}x}{x^2 + a^2} + \dfrac{\frac{1}{4a^2}}{x - a} + \dfrac{\frac{1}{4a^2}}{x + a}$

37. $-\dfrac{3}{25(\ln x + 2)} + \dfrac{3}{25(\ln x - 3)} + \dfrac{7}{5(\ln x - 3)^2}$

Review Exercises, Chapter 5

1. [5.1] (a)   **2.** [5.1] (e)   **3.** [5.1] (h)   **4.** [5.1] (d)
5. [5.6] (b)   **6.** [5.6] (g)   **7.** [5.6] (c)   **8.** [5.6] (f)
9. [5.1] $(-2, -2)$ **10.** [5.1] $(-5, 4)$
11. [5.1] No solution   **12.** [5.1] $(y - 2, y)$, or $(x, x + 2)$
13. [5.2] No solution   **14.** [5.2] $(0, 0, 0)$
15. [5.2] $(-5, 13, 8, 2)$
16. [5.1], [5.2] Consistent: 9, 10, 12, 14, 15; the others are inconsistent.
17. [5.1], [5.2] Dependent: 12; the others are independent.
18. [5.3] $(1, 2)$ **19.** [5.3] $(-3, 4, -2)$
20. [5.3] $\left(\dfrac{z}{2}, -\dfrac{z}{2}, z\right)$ **21.** [5.3] $(-4, 1, -2, 3)$
22. [5.1] 31 nickels, 44 dimes
23. [5.1] \$1600 at 10%, \$3400 at 10.5%
24. [5.2] 1 bagel, $\frac{1}{2}$ serving cream cheese, 2 bananas
25. [5.2] 74.5, 68.5, 82
26. [5.2] **(a)** $f(x) = \frac{1}{12}x^2 - \frac{31}{60}x + \frac{77}{10}$; **(b)** 18.7 lb

27. [5.4] $\begin{bmatrix} 0 & -1 & 6 \\ 3 & 1 & -2 \\ -2 & 1 & -2 \end{bmatrix}$ **28.** [5.4] $\begin{bmatrix} -3 & 3 & 0 \\ -6 & -9 & 6 \\ 6 & 0 & -3 \end{bmatrix}$

29. [5.4] $\begin{bmatrix} -1 & 1 & 0 \\ -2 & -3 & 2 \\ 2 & 0 & -1 \end{bmatrix}$ **30.** [5.4] $\begin{bmatrix} -2 & 2 & 6 \\ 1 & -8 & 18 \\ 2 & 1 & -15 \end{bmatrix}$

31. [5.4] Not possible   **32.** [5.4] $\begin{bmatrix} 2 & -1 & -6 \\ 1 & 5 & -2 \\ -2 & -1 & 4 \end{bmatrix}$

33. [5.4] $\begin{bmatrix} 3 & -2 & -6 \\ 3 & 8 & -4 \\ -4 & -1 & 5 \end{bmatrix}$ **34.** [5.4] $\begin{bmatrix} -2 & -1 & 18 \\ 5 & -3 & -2 \\ -2 & 3 & -8 \end{bmatrix}$

35. [5.4] **(a)** $\begin{bmatrix} 46.1 & 5.9 & 10.1 & 8.5 & 11.4 \\ 54.6 & 4.6 & 9.6 & 7.6 & 10.6 \\ 48.9 & 5.5 & 12.7 & 9.4 & 9.3 \\ 51.3 & 4.8 & 11.3 & 6.9 & 12.7 \end{bmatrix}$
(b) $[32 \quad 19 \quad 43 \quad 38]$;
(c) $[6564.7 \quad 695.1 \quad 1481.1 \quad 1082.8 \quad 1448.7]$;
(d) the total cost, in cents, for each item for the day's meal

36. [5.5] $\begin{bmatrix} -\frac{1}{2} & 0 \\ \frac{1}{6} & \frac{1}{3} \end{bmatrix}$ **37.** [5.5] $\begin{bmatrix} 0 & 0 & \frac{1}{4} \\ 0 & -\frac{1}{2} & 0 \\ \frac{1}{3} & 0 & 0 \end{bmatrix}$

38. [5.5] $\begin{bmatrix} 1 & 0 & 0 & 0 \\ 0 & \frac{1}{9} & \frac{5}{18} & 0 \\ 0 & -\frac{1}{9} & \frac{2}{9} & 0 \\ 0 & 0 & 0 & 1 \end{bmatrix}$

39. [5.4] $\begin{bmatrix} 3 & -2 & 4 \\ 1 & 5 & -3 \\ 2 & -3 & 7 \end{bmatrix} \begin{bmatrix} x \\ y \\ z \end{bmatrix} = \begin{bmatrix} 13 \\ 7 \\ -8 \end{bmatrix}$ **40.** [5.5] $(-8, 7)$

41. [5.5] $(1, -2, 5)$ **42.** [5.5] $(2, -1, 1, -3)$
43. [5.6]

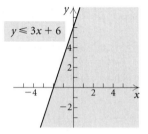
$y \le 3x + 6$

44. [5.6]

45. [5.6]

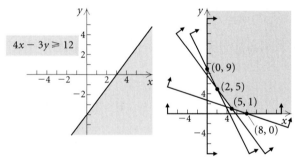
$4x - 3y \ge 12$

$(0, 9)$
$(2, 5)$
$(5, 1)$
$(8, 0)$

46. [5.6] Minimum $= 52$ at $(2, 4)$; maximum $= 92$ at $(2, 8)$
47. [5.6] Maximum score of 96 when 0 group A questions and 8 group B questions are answered

48. [5.7] $\dfrac{5}{x + 1} - \dfrac{5}{x + 2} - \dfrac{5}{(x + 2)^2}$

49. [5.7] $\dfrac{2}{2x - 3} - \dfrac{5}{x + 4}$

50. Discussion and Writing [5.1] During a holiday season, a caterer sold a total of 55 food trays. She sold 15 more seafood trays than cheese trays. How many of each were sold?
51. Discussion and Writing [5.4] In general, $(\mathbf{AB})^2 \ne \mathbf{A}^2\mathbf{B}^2$. $(\mathbf{AB})^2 = \mathbf{ABAB}$ and $\mathbf{A}^2\mathbf{B}^2 = \mathbf{AABB}$. Since matrix multiplication is not commutative, $\mathbf{BA} \ne \mathbf{AB}$, so $(\mathbf{AB})^2 \ne \mathbf{A}^2\mathbf{B}^2$.

52. [5.2] 12%: \$10,000; 13%: \$12,000; $14\frac{1}{2}$%: \$18,000
53. [5.1] $\left(\frac{5}{18}, \frac{1}{7}\right)$ **54.** [5.2] $\left(1, \frac{1}{2}, \frac{1}{3}\right)$
55. [5.6] **56.** [5.6]

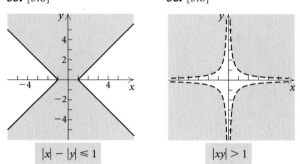
$|x| - |y| \le 1$ $|xy| > 1$

Chapter 6

Exercise Set 6.1

1. (f) **3.** (b) **5.** (d)
7. V: $(0, 0)$; **9.** V: $(0, 0)$; F: $\left(-\frac{3}{2}, 0\right)$;
 F: $(0, 5)$; D: $y = -5$ D: $x = \frac{3}{2}$

$x^2 = 20y$

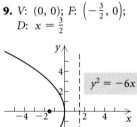
$y^2 = -6x$

11. V: $(0, 0)$; **13.** V: $(0, 0)$; F: $\left(\frac{1}{8}, 0\right)$;
 F: $(0, 1)$; D: $y = -1$ D: $x = -\frac{1}{8}$

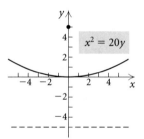
$x^2 - 4y = 0$

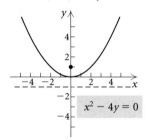
$x = 2y^2$

15. $y^2 = 16x$ **17.** $x^2 = -4\pi y$

19. $(y - 2)^2 = 14\left(x + \frac{1}{2}\right)$

21. V: $(-2, 1)$; F: $\left(-2, -\frac{1}{2}\right)$; D: $y = \frac{5}{2}$

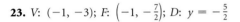

$(x + 2)^2 = -6(y - 1)$

23. V: $(-1, -3)$; F: $\left(-1, -\frac{7}{2}\right)$; D: $y = -\frac{5}{2}$

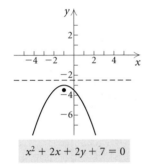

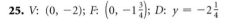

$x^2 + 2x + 2y + 7 = 0$

25. V: $(0, -2)$; F: $\left(0, -1\frac{3}{4}\right)$; D: $y = -2\frac{1}{4}$

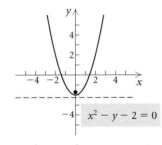

$x^2 - y - 2 = 0$

27. V: $(-2, -1)$; F: $\left(-2, -\frac{3}{4}\right)$; D: $y = -1\frac{1}{4}$

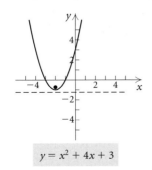

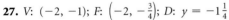

$y = x^2 + 4x + 3$

29. V: $\left(5\frac{3}{4}, \frac{1}{2}\right)$; F: $\left(6, \frac{1}{2}\right)$; D: $x = 5\frac{1}{2}$

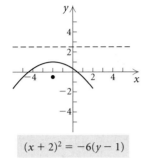

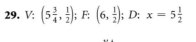

$y^2 - y - x + 6 = 0$

31. (a) $y^2 = 16x$; **(b)** $3\frac{33}{64}$ ft   **33.** $\frac{2}{3}$ ft, or 8 in.

35. Discussion and Writing   **37.** [2.3] $(x + 5)^2$

38. [2.3] $\left(y - \frac{9}{2}\right)^2$ **39.** [1.7] $(1, -2)$; 3

40. [1.7] $(-3, 5)$; 6   **41.** $(x + 1)^2 = -4(y - 2)$

43. V: $(0.867, 0.348)$; F: $(0.867, -0.191)$; D: $y = 0.887$

45. 10 ft, 11.6 ft, 16.4 ft, 24.4 ft, 35.6 ft, 50 ft

Exercise Set 6.2

1. (b)   **3.** (d)   **5.** (a)

7. $(7, -2)$; 8   **9.** $(-2, 3)$; 5

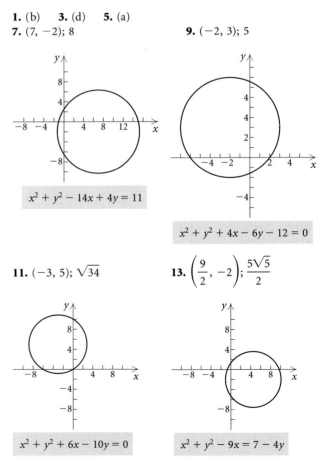

$x^2 + y^2 - 14x + 4y = 11$

$x^2 + y^2 + 4x - 6y - 12 = 0$

11. $(-3, 5)$; $\sqrt{34}$ **13.** $\left(\frac{9}{2}, -2\right)$; $\frac{5\sqrt{5}}{2}$

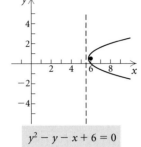

$x^2 + y^2 + 6x - 10y = 0$

$x^2 + y^2 - 9x = 7 - 4y$

15. (c)   **17.** (d)

19. *V*: (2, 0), (−2, 0);
   F: ($\sqrt{3}$, 0), (−$\sqrt{3}$, 0)

21. *V*: (0, 4), (0, −4);
   F: (0, $\sqrt{7}$), (0, −$\sqrt{7}$)

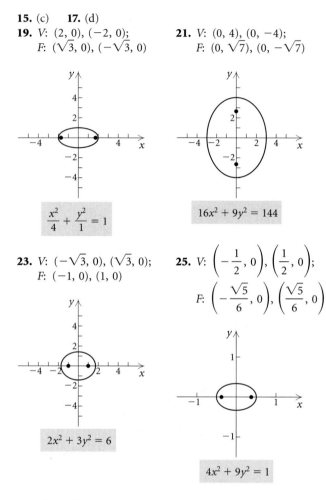

$$\frac{x^2}{4} + \frac{y^2}{1} = 1$$

$$16x^2 + 9y^2 = 144$$

23. *V*: (−$\sqrt{3}$, 0), ($\sqrt{3}$, 0);
   F: (−1, 0), (1, 0)

25. *V*: $\left(-\frac{1}{2}, 0\right)$, $\left(\frac{1}{2}, 0\right)$;

   F: $\left(-\frac{\sqrt{5}}{6}, 0\right)$, $\left(\frac{\sqrt{5}}{6}, 0\right)$

$$2x^2 + 3y^2 = 6$$

$$4x^2 + 9y^2 = 1$$

27. $\dfrac{x^2}{49} + \dfrac{y^2}{40} = 1$ **29.** $\dfrac{x^2}{25} + \dfrac{y^2}{64} = 1$

31. $\dfrac{x^2}{9} + \dfrac{y^2}{5} = 1$

33. *C*: (1, 2); *V*: (4, 2), (−2, 2); *F*: (1 + $\sqrt{5}$, 2),
(1 − $\sqrt{5}$, 2)

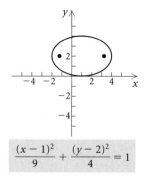

$$\frac{(x-1)^2}{9} + \frac{(y-2)^2}{4} = 1$$

35. *C*: (−3, 5); *V*: (−3, 11), (−3, −1); *F*: (−3, 5 + $\sqrt{11}$),
(−3, 5 − $\sqrt{11}$)

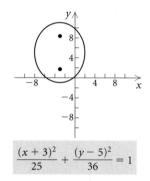

$$\frac{(x+3)^2}{25} + \frac{(y-5)^2}{36} = 1$$

37. *C*: (−2, 1); *V*: (−10, 1), (6, 1); *F*: (−6, 1), (2, 1)

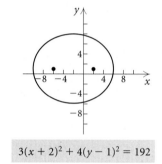

$$3(x+2)^2 + 4(y-1)^2 = 192$$

39. *C*: (2, −1); *V*: (−1, −1), (5, −1); *F*: (2 + $\sqrt{5}$, −1),
(2 − $\sqrt{5}$, −1)

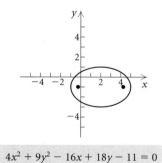

$$4x^2 + 9y^2 - 16x + 18y - 11 = 0$$

41. C: $(1, 1)$; V: $(1, 3)$, $(1, -1)$; F: $(1, 1 + \sqrt{3})$, $(1, 1 - \sqrt{3})$

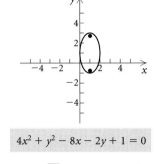

$$4x^2 + y^2 - 8x - 2y + 1 = 0$$

43. Example 2; $\dfrac{3}{5} < \dfrac{\sqrt{12}}{4}$ **45.** $\dfrac{x^2}{15} + \dfrac{y^2}{16} = 1$

47. $\dfrac{x^2}{2500} + \dfrac{y^2}{144} = 1$ **49.** 2×10^6 mi

51. Discussion and Writing

53. [4.1] **(a)** Yes; **(b)** $f^{-1}(x) = \dfrac{x + 3}{2}$

54. [4.1] **(a)** Yes; **(b)** $f^{-1}(x) = \sqrt[3]{x - 2}$

55. [4.1] **(a)** Yes; **(b)** $f^{-1}(x) = \dfrac{5}{x} + 1$, or $\dfrac{5 + x}{x}$

56. [4.1] **(a)** Yes; **(b)** $f^{-1}(x) = x^2 - 4$, $x \geq 0$

57. $\dfrac{(x - 3)^2}{4} + \dfrac{(y - 1)^2}{25} = 1$ **59.** $\dfrac{x^2}{9} + \dfrac{y^2}{484/5} = 1$

61. C: $(2.003, -1.005)$; V: $(-1.017, -1.005)$, $(5.023, -1.005)$ **63.** About 9.1 ft

Exercise Set 6.3

1. (b)   **3.** (c)   **5.** (a)   **7.** $\dfrac{y^2}{9} - \dfrac{x^2}{16} = 1$

9. $\dfrac{x^2}{4} - \dfrac{y^2}{9} = 1$

11. C: $(0, 0)$; V: $(2, 0)$, $(-2, 0)$; F: $(2\sqrt{2}, 0)$, $(-2\sqrt{2}, 0)$; A: $y = x$, $y = -x$

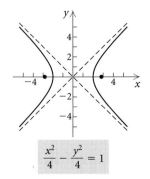

$$\dfrac{x^2}{4} - \dfrac{y^2}{4} = 1$$

13. C: $(2, -5)$; V: $(-1, -5)$, $(5, -5)$; F: $(2 - \sqrt{10}, -5)$, $(2 + \sqrt{10}, -5)$; A: $y = -\dfrac{x}{3} - \dfrac{13}{3}$, $y = \dfrac{x}{3} - \dfrac{17}{3}$

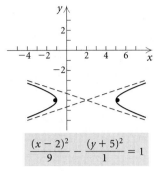

$$\dfrac{(x - 2)^2}{9} - \dfrac{(y + 5)^2}{1} = 1$$

15. C: $(-1, -3)$; V: $(-1, -1)$, $(-1, -5)$; F: $(-1, -3 + 2\sqrt{5})$, $(-1, -3 - 2\sqrt{5})$; A: $y = \frac{1}{2}x - \frac{5}{2}$, $y = -\frac{1}{2}x - \frac{7}{2}$

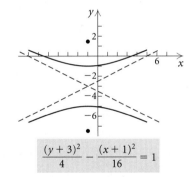

$$\dfrac{(y + 3)^2}{4} - \dfrac{(x + 1)^2}{16} = 1$$

17. C: $(0, 0)$; V: $(-2, 0)$, $(2, 0)$; F: $(-\sqrt{5}, 0)$, $(\sqrt{5}, 0)$; A: $y = -\frac{1}{2}x$, $y = \frac{1}{2}x$

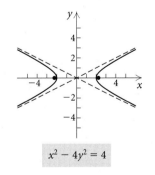

$$x^2 - 4y^2 = 4$$

19. C: $(0, 0)$; V: $(0, -3)$, $(0, 3)$; F: $(0, -3\sqrt{10})$, $(0, 3\sqrt{10})$; A: $y = \frac{1}{3}x$, $y = -\frac{1}{3}x$

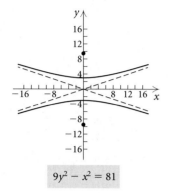

$$9y^2 - x^2 = 81$$

21. C: $(0, 0)$; V: $(-\sqrt{2}, 0)$, $(\sqrt{2}, 0)$; F: $(-2, 0)$, $(2, 0)$; A: $y = x$, $y = -x$

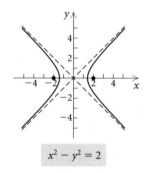

$$x^2 - y^2 = 2$$

23. C: $(0, 0)$; V: $\left(0, -\frac{1}{2}\right)$, $\left(0, \frac{1}{2}\right)$; F: $\left(0, -\frac{\sqrt{2}}{2}\right)$, $\left(0, \frac{\sqrt{2}}{2}\right)$; A: $y = x$, $y = -x$

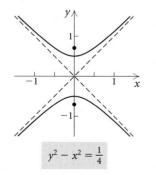

$$y^2 - x^2 = \frac{1}{4}$$

25. C: $(1, -2)$; V: $(0, -2)$, $(2, -2)$; F: $(1 - \sqrt{2}, -2)$, $(1 + \sqrt{2}, -2)$; A: $y = -x - 1$, $y = x - 3$

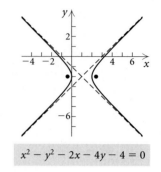

$$x^2 - y^2 - 2x - 4y - 4 = 0$$

27. C: $\left(\frac{1}{3}, 3\right)$; V: $\left(-\frac{2}{3}, 3\right)$, $\left(\frac{4}{3}, 3\right)$; F: $\left(\frac{1}{3} - \sqrt{37}, 3\right)$, $\left(\frac{1}{3} + \sqrt{37}, 3\right)$; A: $y = 6x + 1$, $y = -6x + 5$

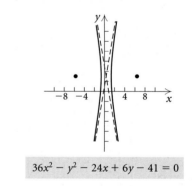

$$36x^2 - y^2 - 24x + 6y - 41 = 0$$

29. C: $(3, 1)$; V: $(3, 3)$, $(3, -1)$; F: $(3, 1 + \sqrt{13})$, $(3, 1 - \sqrt{13})$; A: $y = \frac{2}{3}x - 1$, $y = -\frac{2}{3}x + 3$

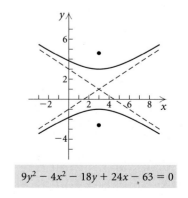

$$9y^2 - 4x^2 - 18y + 24x - 63 = 0$$

31. C: $(1, -2)$; V: $(2, -2)$, $(0, -2)$; F: $(1 + \sqrt{2}, -2)$, $(1 - \sqrt{2}, -2)$; A: $y = x - 3$, $y = -x - 1$

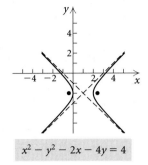

$$x^2 - y^2 - 2x - 4y = 4$$

33. C: $(-3, 4)$; V: $(-3, 10)$, $(-3, -2)$; F: $(-3, 4 + 6\sqrt{2})$, $(-3, 4 - 6\sqrt{2})$; A: $y = x + 7$, $y = -x + 1$

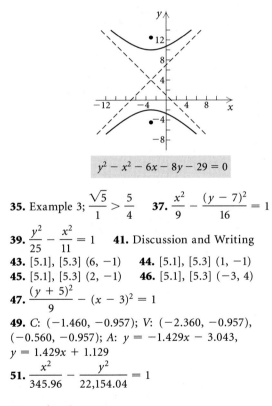

$$y^2 - x^2 - 6x - 8y - 29 = 0$$

35. Example 3; $\dfrac{\sqrt{5}}{1} > \dfrac{5}{4}$ **37.** $\dfrac{x^2}{9} - \dfrac{(y - 7)^2}{16} = 1$

39. $\dfrac{y^2}{25} - \dfrac{x^2}{11} = 1$ **41.** Discussion and Writing

43. [5.1], [5.3] $(6, -1)$ **44.** [5.1], [5.3] $(1, -1)$
45. [5.1], [5.3] $(2, -1)$ **46.** [5.1], [5.3] $(-3, 4)$

47. $\dfrac{(y + 5)^2}{9} - (x - 3)^2 = 1$

49. C: $(-1.460, -0.957)$; V: $(-2.360, -0.957)$, $(-0.560, -0.957)$; A: $y = -1.429x - 3.043$, $y = 1.429x + 1.129$

51. $\dfrac{x^2}{345.96} - \dfrac{y^2}{22{,}154.04} = 1$

Exercise Set 6.4

1. (e)   **3.** (c)   **5.** (b)   **7.** $(-4, -3)$, $(3, 4)$
9. $(0, 2)$, $(3, 0)$ **11.** $(-5, 0)$, $(4, 3)$, $(4, -3)$
13. $(3, 0)$, $(-3, 0)$ **15.** $(0, -3)$, $(4, 5)$
17. $(-2, 1)$ **19.** $(3, 4)$, $(-3, -4)$, $(4, 3)$, $(-4, -3)$
21. $\left(\dfrac{6\sqrt{21}}{7}, \dfrac{4i\sqrt{35}}{7} \right)$, $\left(\dfrac{6\sqrt{21}}{7}, -\dfrac{4i\sqrt{35}}{7} \right)$,

$\left(-\dfrac{6\sqrt{21}}{7}, \dfrac{4i\sqrt{35}}{7} \right)$, $\left(-\dfrac{6\sqrt{21}}{7}, -\dfrac{4i\sqrt{35}}{7} \right)$
23. $(3, 2)$, $\left(4, \dfrac{3}{2} \right)$
25. $\left(\dfrac{5 + \sqrt{70}}{3}, \dfrac{-1 + \sqrt{70}}{3} \right)$, $\left(\dfrac{5 - \sqrt{70}}{3}, \dfrac{-1 - \sqrt{70}}{3} \right)$
27. $(\sqrt{2}, \sqrt{14})$, $(-\sqrt{2}, \sqrt{14})$, $(\sqrt{2}, -\sqrt{14})$, $(-\sqrt{2}, -\sqrt{14})$ **29.** $(1, 2)$, $(-1, -2)$, $(2, 1)$, $(-2, -1)$
31. $\left(\dfrac{15 + \sqrt{561}}{8}, \dfrac{11 - 3\sqrt{561}}{8} \right)$,

$\left(\dfrac{15 - \sqrt{561}}{8}, \dfrac{11 + 3\sqrt{561}}{8} \right)$
33. $\left(\dfrac{7 - \sqrt{33}}{2}, \dfrac{7 + \sqrt{33}}{2} \right)$, $\left(\dfrac{7 + \sqrt{33}}{2}, \dfrac{7 - \sqrt{33}}{2} \right)$
35. $(3, 2)$, $(-3, -2)$, $(2, 3)$, $(-2, -3)$
37. $\left(\dfrac{5 - 9\sqrt{15}}{20}, \dfrac{-45 + 3\sqrt{15}}{20} \right)$,

$\left(\dfrac{5 + 9\sqrt{15}}{20}, \dfrac{-45 - 3\sqrt{15}}{20} \right)$ **39.** $(3, -5)$, $(-1, 3)$

41. $(8, 5)$, $(-5, -8)$ **43.** $(3, 2)$, $(-3, -2)$
45. $(2, 1)$, $(-2, -1)$, $(1, 2)$, $(-1, -2)$
47. $\left(4 + \dfrac{3i\sqrt{6}}{2}, -4 + \dfrac{3i\sqrt{6}}{2} \right)$, $\left(4 - \dfrac{3i\sqrt{6}}{2}, -4 - \dfrac{3i\sqrt{6}}{2} \right)$
49. $(3, \sqrt{5})$, $(-3, -\sqrt{5})$, $(\sqrt{5}, 3)$, $(-\sqrt{5}, -3)$
51. $\left(\dfrac{8\sqrt{5}}{5} i, \dfrac{3\sqrt{105}}{5} \right)$, $\left(\dfrac{8\sqrt{5}}{5} i, -\dfrac{3\sqrt{105}}{5} \right)$,

$\left(-\dfrac{8\sqrt{5}}{5} i, \dfrac{3\sqrt{105}}{5} \right)$, $\left(-\dfrac{8\sqrt{5}}{5} i, -\dfrac{3\sqrt{105}}{5} \right)$
53. $(2, 1)$, $(-2, -1)$, $\left(-i\sqrt{5}, \dfrac{2i\sqrt{5}}{5} \right)$, $\left(i\sqrt{5}, -\dfrac{2i\sqrt{5}}{5} \right)$

55. 6 cm by 8 cm   **57.** 4 in. by 5 in.
59. 1 m by $\sqrt{3}$ m   **61.** 30 yd by 75 yd   **63.** 16 ft, 24 ft
65. **(a)** $I(x) = 21.15508287x + 198.9151934$;
(b) $E(x) = 197.3380928 \cdot 1.053908223^x$; **(c)** about 25 yr
after 1985   **67.** Discussion and Writing   **69.** [4.5] 2
70. [4.5] 2.048   **71.** [4.5] 81   **72.** [4.5] 5

73. $(x - 2)^2 + (y - 3)^2 = 1$ **75.** $\dfrac{x^2}{4} + y^2 = 1$

77. $\left(x + \dfrac{5}{13} \right)^2 + \left(y - \dfrac{32}{13} \right)^2 = \dfrac{5365}{169}$

79. There is no number x such that $\dfrac{x^2}{a^2} - \dfrac{\left(\dfrac{b}{a} x \right)^2}{b^2} = 1$,

because the left side simplifies to $\dfrac{x^2}{a^2} - \dfrac{x^2}{a^2}$, which is 0.
81. $\left(\dfrac{1}{2}, \dfrac{1}{4} \right)$, $\left(\dfrac{1}{2}, -\dfrac{1}{4} \right)$, $\left(-\dfrac{1}{2}, \dfrac{1}{4} \right)$, $\left(-\dfrac{1}{2}, -\dfrac{1}{4} \right)$

83. Factor: $x^3 + y^3 = (x + y)(x^2 - xy + y^2)$. We know that $x + y = 1$, so $(x + y)^2 = x^2 + 2xy + y^2 = 1$, or $x^2 + y^2 = 1 - 2xy$. We also know that $xy = 1$, so $x^2 + y^2 = 1 - 2 \cdot 1 = -1$. Then $x^3 + y^3 = 1 \cdot (-1 - 1) = -2$. **85.** $(2, 4), (4, 2)$
87. $(3, -2), (-3, 2), (2, -3), (-2, 3)$

89. $\left(\dfrac{2 \log 3 + 3 \log 5}{3(\log 3 \cdot \log 5)}, \dfrac{4 \log 3 - 3 \log 5}{3(\log 3 \cdot \log 5)} \right)$

91. $(1.564, 2.448), (0.138, 0.019)$ **93.** $(0.871, 1.388)$
95. $(1.146, 3.146), (-1.841, 0.159)$ **97.** $(0.965, 4402.33),$
$(-0.965, -4402.33)$ **99.** $(2.112, -0.109),$
$(-13.041, -13.337)$ **101.** $(400, 1.431), (-400, 1.431),$
$(400, -1.431), (-400, -1.431)$

Review Exercises, Chapter 6

1. [6.1] (d) **2.** [6.2] (a) **3.** [6.2] (e) **4.** [6.3] (g)
5. [6.2] (b) **6.** [6.2] (f) **7.** [6.1] (h) **8.** [6.3] (c)
9. [6.1] $x^2 = -6y$ **10.** [6.1] F: $(-3, 0)$; V: $(0, 0)$;
D: $x = 3$ **11.** [6.1] V: $(-5, 8)$; F: $\left(-5, \frac{15}{2}\right)$; D: $y = \frac{17}{2}$
12. [6.2] C: $(2, -1)$; V: $(-3, -1), (7, -1)$; F: $(-1, -1),$
$(5, -1)$;

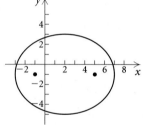

$$16x^2 + 25y^2 - 64x + 50y - 311 = 0$$

13. [6.2] $\dfrac{x^2}{9} + \dfrac{y^2}{16} = 1$

14. [6.3] C: $\left(-2, \dfrac{1}{4}\right)$; V: $\left(0, \dfrac{1}{4}\right) \left(-4, \dfrac{1}{4}\right)$;

F: $\left(-2 + \sqrt{6}, \dfrac{1}{4}\right), \left(-2 - \sqrt{6}, \dfrac{1}{4}\right)$;

A: $y - \dfrac{1}{4} = \dfrac{\sqrt{2}}{2}(x + 2), \; y - \dfrac{1}{4} = -\dfrac{\sqrt{2}}{2}(x + 2)$

15. [6.1] 0.167 ft **16.** [6.4] $(-8\sqrt{2}, 8), (8\sqrt{2}, 8)$

17. [6.4] $\left(3, \dfrac{\sqrt{29}}{2}\right), \left(-3, \dfrac{\sqrt{29}}{2}\right), \left(3, -\dfrac{\sqrt{29}}{2}\right),$

$\left(-3, -\dfrac{\sqrt{29}}{2}\right)$ **18.** [6.4] $(7, 4)$

19. [6.4] $(2, 2), \left(\dfrac{32}{9}, -\dfrac{10}{9}\right)$ **20.** [6.4] $(0, -3), (2, 1)$
21. [6.4] $(4, 3), (4, -3), (-4, 3), (-4, -3)$
22. [6.4] $(-\sqrt{3}, 0), (\sqrt{3}, 0), (-2, 1), (2, 1)$
23. [6.4] $\left(-\dfrac{3}{5}, \dfrac{21}{5}\right), (3, -3)$

24. [6.4] $(6, 8), (6, -8), (-6, 8), (-6, -8)$
25. [6.4] $(2, 2), (-2, -2), (2\sqrt{2}, \sqrt{2}), (-2\sqrt{2}, -\sqrt{2})$
26. [6.4] 7, 4 **27.** [6.4] 7 m by 12 m **28.** [6.4] 4, 8
29. [6.4] 32 cm, 20 cm **30.** [6.4] 11 ft, 3 ft
31. Discussion and Writing [6.4] An algebraic solution of a nonlinear system of equations may be preferable to a graphical solution if there are complex-number solutions or if any of the equations is difficult to enter on a grapher. However, there are nonlinear systems that are difficult or impossible to solve using algebraic methods. Approximations of the real-number solutions of many of those systems may be found using graphical methods.
32. Discussion and Writing [6.2] The equation of a circle can be written as

$$\dfrac{(x - h)^2}{a^2} + \dfrac{(y - k)^2}{b^2} = 1,$$

where $a = b = r$, the radius of the circle. In an ellipse, $a > b$, so a circle is not a special type of ellipse.

33. [6.2] $x^2 + \dfrac{y^2}{9} = 1$ **34.** [6.4] $\dfrac{8}{7}, \dfrac{7}{2}$

35. [6.2], [6.4] $(x - 2)^2 + (y - 1)^2 = 100$

36. [6.3] $\dfrac{x^2}{778.41} - \dfrac{y^2}{39{,}221.59} = 1$

Chapter 7

Exercise Set 7.1

1. 3, 7, 11, 15; 39; 59 **3.** $2, \frac{3}{2}, \frac{4}{3}, \frac{5}{4}; \frac{10}{9}; \frac{15}{14}$
5. $0, \frac{3}{5}, \frac{4}{5}, \frac{15}{17}; \frac{99}{101}; \frac{112}{113}$ **7.** $-1, 4, -9, 16; 100; -225$
9. 7, 3, 7, 3; 3; 7 **11.** 34 **13.** 225 **15.** $-33{,}880$
17. 67 **19.** $2n$ **21.** $(-1)^n \cdot 2 \cdot 3^{n-1}$ **23.** $\dfrac{n + 1}{n + 2}$
25. $n(n + 1)$ **27.** $\log 10^{n-1}$, or $n - 1$ **29.** 6; 28
31. 20; 30 **33.** $\frac{1}{2} + \frac{1}{4} + \frac{1}{6} + \frac{1}{8} + \frac{1}{10} = \frac{137}{120}$
35. $1 + 2 + 4 + 8 + 16 + 32 + 64 = 127$
37. $\ln 7 + \ln 8 + \ln 9 + \ln 10 = \ln (7 \cdot 8 \cdot 9 \cdot 10) = \ln 5040 \approx 8.5252$
39. $\frac{1}{2} + \frac{2}{3} + \frac{3}{4} + \frac{4}{5} + \frac{5}{6} + \frac{6}{7} + \frac{7}{8} + \frac{8}{9} = \frac{15{,}551}{2520}$
41. $-1 + 1 - 1 + 1 - 1 = -1$
43. $3 - 6 + 9 - 12 + 15 - 18 + 21 - 24 = -12$
45. $2 + 1 + \frac{2}{5} + \frac{1}{5} + \frac{2}{17} + \frac{1}{13} + \frac{2}{37} = \frac{157{,}351}{40{,}885}$
47. $3 + 2 + 3 + 6 + 11 + 18 = 43$
49. $\frac{1}{2} + \frac{2}{3} + \frac{4}{5} + \frac{8}{9} + \frac{16}{17} + \frac{32}{33} + \frac{64}{65} + \frac{128}{129} + \frac{256}{257} + \frac{512}{513} + \frac{1024}{1025} \approx 9.736$
51. $\displaystyle\sum_{k=1}^{\infty} 5k$ **53.** $\displaystyle\sum_{k=1}^{6} (-1)^{k+1} 2^k$ **55.** $\displaystyle\sum_{k=1}^{6} (-1)^k \dfrac{k}{k + 1}$
57. $\displaystyle\sum_{k=2}^{n} (-1)^k k^2$ **59.** $\displaystyle\sum_{k=1}^{\infty} \dfrac{1}{k(k + 1)}$ **61.** $4, 1\frac{1}{4}, 1\frac{4}{5}, 1\frac{5}{9}$
63. $6561, -81, 9i, -3\sqrt{i}$ **65.** 2, 3, 5, 8

67.

n	U_n
1	2
2	2.25
3	2.3704
4	2.4414
5	2.4883
6	2.5216
7	2.5465
8	2.5658
9	2.5812
10	2.5937

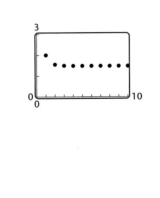

69.

n	U_n
1	2
2	1.5538
3	1.4988
4	1.4914
5	1.4904
6	1.4902
7	1.4902
8	1.4902
9	1.4902
10	1.4902

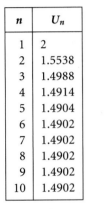

71. (a) 1062, 1127.84, 1197.77, 1272.03, 1350.90, 1434.65, 1523.60, 1618.07, 1718.39, 1824.93; **(b)** $3330.35
73. 1, 2, 4, 8, 16, 32, 64, 128, 256, 512, 1024, 2048, 4096, 8192, 16,384, 32,768, 65,536
75. (a)

n	U_n
15	27.25
16	26.108
17	25.002
18	23.932
19	22.898
20	21.9
21	20.938
22	20.012
23	19.122
24	18.268

(b)

(c) 26, 19, 9, 22, 34

77. 1, 1, 2, 3, 5, 8, 13   **79.** Discussion and Writing
80. [5.1], [5.3], $(-1, -3)$ **81.** [5.1], [5.3] Original:
$1.0135 billion; purchased: $0.1765 billion
82. [6.1] $(3, -2)$; 4   **83.** [6.1] $\left(-\dfrac{5}{2}, 4\right)$; $\dfrac{\sqrt{97}}{2}$
85. $i, -1, -i, 1, i$; i **87.** $\ln(1 \cdot 2 \cdot 3 \cdot \cdots \cdot n)$

Exercise Set 7.2

1. $a_1 = 3$, $d = 5$ **3.** $a_1 = 9$, $d = -4$
5. $a_1 = \dfrac{3}{2}$, $d = \dfrac{3}{4}$ **7.** $a_1 = \$316$, $d = -\$3$ **9.** $a_{12} = 46$
11. $a_{14} = -\dfrac{17}{3}$ **13.** $a_{10} = \$7941.62$ **15.** 33rd
17. 46th   **19.** $a_1 = 5$ **21.** $n = 39$
23. $a_1 = \dfrac{1}{3}$; $d = \dfrac{1}{2}$; $\dfrac{1}{3}, \dfrac{5}{6}, \dfrac{4}{3}, \dfrac{11}{6}, \dfrac{7}{3}$ **25.** 670   **27.** 160,400
29. 735   **31.** 990   **33.** 1760   **35.** $\dfrac{65}{2}$ **37.** $-\dfrac{6026}{13}$
39. 1260   **41.** 3; 171   **43.** 4960¢, or $49.60   **45.** 1320
47. Yes; 3   **49.** Discussion and Writing
50. [5.1], [5.3] $(2, 5)$ **51.** [5.2], [5.3] $(2, -1, 3)$
52. [6.1] $(-4, 0)$, $(4, 0)$; $(-\sqrt{7}, 0)$, $(\sqrt{7}, 0)$
53. [6.1] $\dfrac{x^2}{4} + \dfrac{y^2}{25} = 1$ **55.** n^2 **57.** $a_1 = 60 - 5p - 5q$;
$d = 5p + 2q - 20$ **59.** 6, 8, 10   **61.** $-1, 1, 3$
63. Insert 16 arithmetic a means between 1 and 50 with $d = \dfrac{49}{17}$.
65.
$$m = p + d$$
$$\underline{m = q - d}$$
$$2m = p + q \qquad \textbf{Adding}$$
$$m = \dfrac{p + q}{2}$$

Exercise Set 7.3

1. 2   **3.** -1 **5.** -2 **7.** 0.1   **9.** $\dfrac{a}{2}$ **11.** 128
13. 162   **15.** $7(5)^{40}$ **17.** 3^{n-1} **19.** $(-1)^{n-1}$
21. $\dfrac{1}{x^n}$ **23.** 762   **25.** $\dfrac{4921}{18}$ **27.** 8   **29.** 125
31. Does not exist   **33.** $\dfrac{2}{3}$ **35.** $29\dfrac{38,569}{59,049}$ **37.** 2
39. Does not exist   **41.** $\$4545.\overline{45}$ **43.** $\dfrac{160}{9}$ **45.** $\dfrac{13}{99}$
47. 9   **49.** $\dfrac{34,091}{9990}$ **51. (a)** $\dfrac{1}{256}$ ft; **(b)** $5\dfrac{1}{3}$ ft
53. (a) About 297 ft; **(b)** 300 ft   **55.** $31,497.57
57. $523,619.17   **59.** $86,666,666,667   **61.** Discussion
and Writing   **63.** [4.1] $(f \circ g)(x) = 16x^2 + 40x + 25$;
$(g \circ f)(x) = 4x^2 + 5$ **64.** [4.1] $(f \circ g)(x) = x^2 + x + 2$;
$(g \circ f)(x) = x^2 - x + 3$ **65.** [4.5] 2.209   **66.** [4.5] $\dfrac{1}{16}$
67. $(4 - \sqrt{6})/(\sqrt{3} - \sqrt{2}) = 2\sqrt{3} + \sqrt{2}$,
$(6\sqrt{3} - 2\sqrt{2})/(4 - \sqrt{6}) = 2\sqrt{3} + \sqrt{2}$; there exists a
common ratio, $2\sqrt{3} + \sqrt{2}$; thus the sequence is
geometric.

69. (a) $\frac{13}{3}; \frac{22}{3}, \frac{34}{3}, \frac{46}{3}, \frac{58}{3}$; **(b)** $-\frac{11}{3}$; $-\frac{2}{3}, \frac{10}{3}, -\frac{50}{3}, \frac{250}{3}$ or 5; 8, 12, 18, 27 **71.** $S_n = \dfrac{x^2(1 - (-x)^n)}{x + 1}$

73. $\dfrac{a_{n+1}}{a_n} = r$, so $\ln \dfrac{a_{n+1}}{a_n} = \ln r$. But $\ln \dfrac{a_{n+1}}{a_n} = \ln a_{n+1} -$ $\ln a_n = \ln r$. Thus, $\ln a_1, \ln a_2, \ldots$, is an arithmetic sequence with common difference $\ln r$. **75.** 512 cm^2

Exercise Set 7.4

1. $1^2 < 1^3$, false; $2^2 < 2^3$, true; $3^2 < 3^3$, true; $4^2 < 4^3$, true; $5^2 < 5^3$, true

3. A polygon of 3 sides has $\dfrac{3(3 - 3)}{2}$ diagonals. True; A

5polygon of 4 sides has $\dfrac{4(4 - 3)}{2}$ diagonals. True; A polygon

of 5 sides has $\dfrac{5(5 - 3)}{2}$ diagonals. True; A polygon of

6 sides has $\dfrac{6(6 - 3)}{2}$ diagonals. True; A polygon of 7 sides

has $\dfrac{7(7 - 3)}{2}$ diagonals. True.

5.

S_n: $2 + 4 + 6 + \cdots + 2n = n(n + 1)$
S_1: $2 = 1(1 + 1)$
S_k: $2 + 4 + 6 + \cdots + 2k = k(k + 1)$
S_{k+1}: $2 + 4 + 6 + \cdots + 2k + 2(k + 1)$
            $= (k + 1)(k + 2)$

1. *Basis step:* S_1 true by substitution.
2. *Induction step:* Assume S_k. Deduce S_{k+1}.
   Starting with the left side of S_{k+1}, we have

   $2 + 4 + 6 + \cdots + 2k + 2(k + 1)$
   $= k(k + 1) + 2(k + 1)$ **By S_k**
   $= (k + 1)(k + 2)$.   **Distributive law**

7. S_n: $1 + 5 + 9 + \cdots + (4n - 3) = n(2n - 1)$
    S_1: $1 = 1(2 \cdot 1 - 1)$
    S_k: $1 + 5 + 9 + \cdots + (4k - 3) = k(2k - 1)$
    S_{k+1}: $1 + 5 + 9 + \cdots + (4k - 3) + [4(k + 1) - 3]$
               $= (k + 1)[2(k + 1) - 1]$
               $= (k + 1)(2k + 1)$

1. *Basis step:* S_1 true by substitution.
2. *Induction step:* Assume S_k. Deduce S_{k+1}.
   Starting with the left side of S_{k+1}, we have

   $\underbrace{1 + 5 + 9 + \cdots + (4k - 3)} + [4(k + 1) - 3]$
   $= k(2k - 1) + [4(k + 1) - 3]$ **By S_k**
   $= 2k^2 - k + 4k + 4 - 3$
   $= (k + 1)(2k + 1)$.

9. S_n: $2 + 4 + 8 + \cdots + 2^n = 2(2^n - 1)$
    S_1: $2 = 2(2 - 1)$
    S_k: $2 + 4 + 8 + \cdots + 2^k = 2(2^k - 1)$
    S_{k+1}: $2 + 4 + 8 + \cdots + 2^k + 2^{k+1} = 2(2^{k+1} - 1)$

1. *Basis step:* S_1 is true by substitution.
2. *Induction step:* Assume S_k. Deduce S_{k+1}.
   Starting with the left side of S_{k+1}, we have

   $\underbrace{2 + 4 + 8 + \cdots + 2^k} + 2^{k+1}$
   $= \underbrace{2(2^k - 1)} + 2^{k+1}$ **By S_k**
   $= 2^{k+1} - 2 + 2^{k+1}$
   $= 2 \cdot 2^{k+1} - 2$
   $= 2(2^{k+1} - 1)$.

11. 1. *Basis step:* Since $1 < 1 + 1$, S_1 is true.
    2. *Induction step:* Assume S_k. Deduce S_{k+1}. Now

   $k < k + 1$ **By S_k**
   $k + 1 < k + 1 + 1$;   **Adding 1**
   $k + 1 < k + 2$.

13. 1. *Basis step:* Since $2 = 2$, S_1 is true.
    2. *Induction step:* Let k be any natural number. Assume S_k. Deduce S_{k+1}.

   $2k \le 2^k$ **By S_k**
   $2 \cdot 2k \le 2 \cdot 2^k$ **Multiplying by 2**
   $4k \le 2^{k+1}$

   Since $1 \le k$, $k + 1 \le k + k$, or $k + 1 \le 2k$.
   Then $2(k + 1) \le 4k$.
   Thus, $2(k + 1) \le 4k \le 2^{k+1}$, so $2(k + 1) \le 2^{k+1}$.

15.

S_n: $\dfrac{1}{1 \cdot 2 \cdot 3} + \dfrac{1}{2 \cdot 3 \cdot 4} + \dfrac{1}{3 \cdot 4 \cdot 5} + \cdots$

$+ \dfrac{1}{n(n + 1)(n + 2)}$

$= \dfrac{n(n + 3)}{4(n + 1)(n + 2)}$

S_1: $\dfrac{1}{1 \cdot 2 \cdot 3} = \dfrac{1(1 + 3)}{4 \cdot 2 \cdot 3}$

S_k: $\dfrac{1}{1 \cdot 2 \cdot 3} + \dfrac{1}{2 \cdot 3 \cdot 4} + \cdots + \dfrac{1}{k(k + 1)(k + 2)}$

$= \dfrac{k(k + 3)}{4(k + 1)(k + 2)}$

S_{k+1}: $\dfrac{1}{1 \cdot 2 \cdot 3} + \dfrac{1}{2 \cdot 3 \cdot 4} + \cdots + \dfrac{1}{k(k + 1)(k + 2)}$

$+ \dfrac{1}{(k + 1)(k + 2)(k + 3)}$

$= \dfrac{(k + 1)(k + 1 + 3)}{4(k + 1 + 1)(k + 1 + 2)} = \dfrac{(k + 1)(k + 4)}{4(k + 2)(k + 3)}$

1. *Basis step:* Since $\dfrac{1}{1 \cdot 2 \cdot 3} = \dfrac{1}{6}$ and $\dfrac{1(1 + 3)}{4 \cdot 2 \cdot 3} =$

$\dfrac{1 \cdot 4}{4 \cdot 2 \cdot 3} = \dfrac{1}{6}$, S_1 is true.

2. *Induction step*: Assume S_k. Deduce S_{k+1}.

Add $\dfrac{1}{(k+1)(k+2)(k+3)}$ on both sides of S_k and simplify the right side.
Only the right side is shown here.

$$\frac{k(k+3)}{4(k+1)(k+2)} + \frac{1}{(k+1)(k+2)(k+3)}$$

$$= \frac{k(k+3)(k+3)+4}{4(k+1)(k+2)(k+3)}$$

$$= \frac{k^3 + 6k^2 + 9k + 4}{4(k+1)(k+2)(k+3)}$$

$$= \frac{(k+1)^2(k+4)}{4(k+1)(k+2)(k+3)}$$

$$= \frac{(k+1)(k+4)}{4(k+2)(k+3)}$$

17.

S_n: $1 + 2 + 3 + \cdots + n = \dfrac{n(n+1)}{2}$

S_1: $1 = \dfrac{1(1+1)}{2}$

S_k: $1 + 2 + 3 + \cdots + k = \dfrac{k(k+1)}{2}$

S_{k+1}: $1 + 2 + 3 + \cdots + k + (k+1) = \dfrac{(k+1)(k+2)}{2}$

1. *Basis step*: S_1 true by substitution.
2. *Induction step*: Assume S_k. Deduce S_{k+1}.
   Starting with the left side of S_{k+1}, we have

$$\underbrace{1 + 2 + 3 + \cdots + k} + (k+1)$$

$$= \frac{k(k+1)}{2} + (k+1) \qquad \textbf{By } S_k$$

$$= \frac{k(k+1) + 2(k+1)}{2} \qquad \textbf{Adding}$$

$$= \frac{(k+1)(k+2)}{2}. \qquad \textbf{Distributive law}$$

19. 1. *Basis step*. S_1: $1^3 = \dfrac{1^2(1+1)^2}{4} = 1$. True.

2. *Induction step*: Assume S_k. Deduce S_{k+1}.

S_k: $1^3 + 2^3 + \cdots + k^3 = \dfrac{k^2(k+1)^2}{4}$

$$1^3 + 2^3 + \cdots + (k+1)^3 = \frac{k^2(k+1)^2}{4} + (k+1)^3 \quad \textbf{By } S_k$$

$$= \frac{(k+1)^2}{4}[k^2 + 4(k+1)]$$

$$= \frac{(k+1)^2(k+2)^2}{4}$$

21.

1. *Basis step*. S_1: $1^5 = \dfrac{1^2(1+1)^2(2 \cdot 1^2 + 2 \cdot 1 - 1)}{12}$.
True.

2. *Induction step*. Assume S_k:

$1^5 + 2^5 + \cdots + k^5 = \dfrac{k^2(k+1)^2(2k^2 + 2k - 1)}{12}$.

Then $1^5 + 2^5 + \cdots + k^5 + (k+1)^5$

$$= \frac{k^2(k+1)^2(2k^2 + 2k - 1)}{12} + (k+1)^5$$

$$= \frac{k^2(k+1)^2(2k^2 + 2k - 1) + 12(k+1)^5}{12}$$

$$= \frac{(k+1)^2(2k^4 + 14k^3 + 35k^2 + 36k + 12)}{12}$$

$$= \frac{(k+1)^2(k+2)^2(2k^2 + 6k + 3)}{12}$$

$$= \frac{(k+1)^2(k+1+1)^2(2(k+1)^2 + 2(k+1) - 1)}{12}.$$

23.

1. *Basis step*. S_1: $1(1+1) = \dfrac{1(1+1)(1+2)}{3}$. True.

2. *Induction step*. Assume

S_k: $1(1+1) + 2(2+1) + \cdots + k(k+1)$

$\quad = \dfrac{k(k+1)(k+2)}{3}$.

Then $1(1+1) + 2(2+1) + \cdots + k(k+1)$
$+ (k+1)(k+1+1)$

$$= \frac{k(k+1)(k+2)}{3} + (k+1)(k+2)$$

$$= \frac{(k+1)(k+2)(k+3)}{3}$$

$$= \frac{(k+1)(k+1+1)(k+1+2)}{3}.$$

25. 1. *Basis step*: Since $\frac{1}{2}[2a_1 + (1-1)d] =$
$\frac{1}{2} \cdot 2a_1 = a_1$, S_1 is true.

2. *Induction step*: Assume S_k. Deduce S_{k+1}. Starting with the left side of S_{k+1}, we have

$$\underbrace{a_1 + (a_1 + d) + \cdots + [a_1 + (k-1)d]} + [a_1 + kd]$$

$$= \frac{k}{2}[2a_1 + (k-1)d] \qquad + [a_1 + kd] \qquad \textbf{By } S_k$$

$$= \frac{k[2a_1 + (k-1)d]}{2} + \frac{2[a_1 + kd]}{2}$$

$$= \frac{2ka_1 + k(k-1)d + 2a_1 + 2kd}{2}$$

$$= \frac{2a_1(k+1) + k(k-1)d + 2kd}{2}$$

$$= \frac{2a_1(k+1) + (k-1+2)kd}{2}$$

$$= \frac{2a_1(k+1) + (k+1)kd}{2} = \frac{k+1}{2}[2a_1 + kd].$$

27. Discussion and Writing   **28.** [5.1], [5.3] (5, 3)
29. [5.2], [5.3] (2, −3, 4)   **30.** [5.1], [5.3] 50 hardback, 30 paperback   **31.** [5.2], [5.3] $800 at 6%, $1600 at 8%, $2000 at 10%

33.
1. *Basis step.* S_1: $x + y$ is a factor of $x^2 - y^2$. True.
    S_2: $x + y$ is a factor of $x^4 - y^4$. True.
2. *Induction step.* Assume S_{k-1}: $x + y$ is a factor of
    $x^{2(k-1)} - y^{2(k-1)}$.
    Then $x^{2(k-1)} - y^{2(k-1)} = (x + y)Q(x)$
    for some polynomial Q.
    Assume S_k: $x + y$ is a factor of
    $x^{2k} - y^{2k}$.
    Then $x^{2k} - y^{2k} = (x + y)P(x)$ for some
    polynomial P.

$x^{2(k+1)} - y^{2(k+1)}$
$\quad = (x^{2k} - y^{2k})(x^2 + y^2) - (x^{2(k-1)} - y^{2(k-1)})(x^2 y^2)$
$\quad = (x + y)P(x)(x^2 + y^2) - (x + y)Q(x)(x^2 y^2)$
$\quad = (x + y)[P(x)(x^2 + y^2) - Q(x)(x^2 y^2)]$

so $x + y$ is a factor of $x^{2(k+1)} - y^{2(k+1)}$.

35.
S_2: $\log_a (b_1 b_2) = \log_a b_1 + \log_a b_2$
S_k: $\log_a (b_1 b_2 \cdots b_k) = \log_a b_1 + \log_a b_2 + \cdots + \log_a b_k$
S_{k+1}: $\log_a (b_1 b_2 \cdots b_{k+1}) = \log_a b_1 + \log_a b_2 + \cdots + \log_a b_{k+1}$

1. *Basis step*: S_2 is true by the properties of logarithms.
2. *Induction step*: Let k be a natural number $k \geq 2$.
    Assume S_k. Deduce S_{k+1}.

$\log_a (b_1 b_2 \cdots b_{k+1})$ **Left side of S_{k+1}**
$\quad = \log_a (b_1 b_2 \cdots b_k) + \log_a b_{k+1}$ **By S_2**
$\quad = \log_a b_1 + \log_a b_2 + \cdots + \log_a b_k + \log_a b_{k+1}$

37. S_2: $\overline{z_1 + z_2} = \bar{z}_1 + \bar{z}_2$:
$\overline{(a + bi) + (c + di)} = \overline{(a + c) + (b + d)i}$
$\qquad\qquad = (a + c) - (b + d)i$
$\overline{(a + bi)} + \overline{(c + di)} = a - bi + c - di$
$\qquad\qquad = (a + c) - (b + d)i.$
S_k: $\overline{z_1 + z_2 + \cdots + z_k} = \bar{z}_1 + \bar{z}_2 + \cdots + \bar{z}_k.$
$\overline{(z_1 + z_2 + \cdots + z_k) + z_{k+1}}$
$\qquad = \overline{(z_1 + z_2 + \cdots + z_k)} + \overline{z_{k+1}}$ **By S_2**
$\qquad = \bar{z}_1 + \bar{z}_2 + \cdots + \bar{z}_k + \bar{z}_{k+1}$ **By S_k**

39. S_1: i is either i or -1 or $-i$ or 1.
    S_k: i^k is either i or -1 or $-i$ or 1.
    $i^{k+1} = i^k \cdot i$ is then $i \cdot i = -1$ or $-1 \cdot i = -i$ or $-i \cdot i = 1$ or $1 \cdot i = i$.

41. S_1:   3 is a factor of $1^3 + 2 \cdot 1$.
    S_k: 3 is a factor of $k^3 + 2k$, i.e., $k^3 + 2k = 3 \cdot m$.
    S_{k+1}: 3 is a factor of $(k + 1)^3 + 2(k + 1)$.

Consider
$(k + 1)^3 + 2(k + 1) = k^3 + 3k^2 + 5k + 3$
$\qquad\qquad = (k^3 + 2k) + 3k^2 + 3k + 3$
$\qquad\qquad = 3m + 3(k^2 + k + 1).$
A multiple of 3

Exercise Set 7.5

1. 720   **3.** 604,800   **5.** 120   **7.** 1   **9.** 3024   **11.** 120
13. 120   **15.** 1   **17.** 6,497,400   **19.** $n(n - 1)(n - 2)$
21. n **23.** 6! = 720   **25.** 9! = 362,880   **27.** $_9P_4 = 3024$
29. $_5P_5 = 120$; $5^5 = 3125$ **31.** $\dfrac{8!}{3!} = 6720$; $\dfrac{7!}{2!} = 2520$;
$\dfrac{11!}{2!\,2!\,2!} = 4,989,600$ **33.** $8 \cdot 10^6 = 8,000,000$; 8 million
35. $\dfrac{9!}{2!\,3!\,4!} = 1260$ **37. (a)** $_6P_5 = 720$; **(b)** $6^5 = 7776$;
(c) $1 \cdot {}_5P_4 = 120$; **(d)** $1 \cdot 1 \cdot {}_4P_3 = 24$
39. (a) 10^5, or 100,000; **(b)** 100,000
41. (a) $10^9 = 1,000,000,000$; **(b)** yes
43. Discussion and Writing   **44.** [2.1] $\dfrac{9}{4}$, or 2.25
45. [2.3] $-3, 2$ **46.** [2.3] $\dfrac{3 \pm \sqrt{17}}{4}$ **47.** [3.3] $-2, 1, 5$
49. 8   **51.** 11   **53.** $n - 1$

Exercise Set 7.6

1. 78   **3.** 78   **5.** 7   **7.** 10   **9.** 1   **11.** 15   **13.** 128
15. 270,725   **17.** 13,037,895   **19.** n **21.** 1
23. $_{23}C_4 = 8855$ **25.** $_{13}C_{10} = 286$
27. $\dbinom{8}{2} = 28$; $\dbinom{8}{3} = 56$ **29.** $\dbinom{52}{5} = 2,598,960$
31. (a) $_{31}P_2 = 930$; **(b)** $31^2 = 961$; **(c)** $_{31}C_2 = 465$
33. Discussion and Writing
34. [2.1] $-\dfrac{17}{2}$, or -8.5 **35.** [2.3] $-1, \dfrac{3}{2}$, or $-1, 1.5$
36. [2.3] $\dfrac{-5 \pm \sqrt{21}}{2}$ **37.** [3.3] $-4, -2, 3$
39. $\dbinom{13}{5} = 1287$ **41.** $\dbinom{n}{2}$; $2\dbinom{n}{2}$ **43.** 4   **45.** 7
47. $\dbinom{n}{k - 1} + \dbinom{n}{k}$

$\quad = \dfrac{n!}{(k - 1)!(n - k + 1)!} \cdot \dfrac{k}{k}$
$\qquad + \dfrac{n!}{k!(n - k)!} \cdot \dfrac{(n - k + 1)}{(n - k + 1)}$
$\quad = \dfrac{n!(k + (n - k + 1))}{k!(n - k + 1)!}$
$\quad = \dfrac{(n + 1)!}{k!(n - k + 1)!} = \dbinom{n + 1}{k}$

Exercise Set 7.7

1. $x^4 + 20x^3 + 150x^2 + 500x + 625$

3. $x^5 - 15x^4 + 90x^3 - 270x^2 + 405x - 243$

5. $x^5 - 5x^4y + 10x^3y^2 - 10x^2y^3 + 5xy^4 - y^5$

7. $15{,}625x^6 + 75{,}000x^5y + 150{,}000x^4y^2 +$
$160{,}000x^3y^3 + 96{,}000x^2y^4 + 30{,}720xy^5 + 4096y^6$

9. $128t^7 + 448t^5 + 672t^3 + 560t + 280t^{-1} + 84t^{-3}$
$+ 14t^{-5} + t^{-7}$

11. $x^{10} - 5x^8 + 10x^6 - 10x^4 + 5x^2 - 1$

13. $125 + 150\sqrt{5}t + 375t^2 + 100\sqrt{5}t^3 + 75t^4 +$
$6\sqrt{5}t^5 + t^6$

15. $a^9 - 18a^7 + 144a^5 - 672a^3 + 2016a - 4032a^{-1} +$
$5376a^{-3} - 4608a^{-5} + 2304a^{-7} - 512a^{-9}$

17. $140\sqrt{2}$ **19.** $x^{-8} + 4x^{-4} + 6 + 4x^4 + x^8$

21. $21a^5b^2$ **23.** $-252x^5y^5$ **25.** $-745{,}472a^3$

27. $1120x^{12}y^2$ **29.** $-1{,}959{,}552u^5v^{10}$ **31.** 128

33. 2^{24}, or $16{,}777{,}216$ **35.** 20 **37.** $-12 + 316i$

39. $-7 - 4\sqrt{2}\,i$ **41.** $\displaystyle\sum_{k=0}^{n} \binom{n}{k}(-1)^k a^{n-k}b^k$

43. $\displaystyle\sum_{k=1}^{n} \binom{n}{k}x^{n-k}h^{k-1}$ **45.** Discussion and Writing

46. [1.4] $x^2 + 2x - 2$ **47.** [1.4] $2x^3 - 3x^2 + 2x - 3$

48. [4.1] $4x^2 - 12x + 10$ **49.** [4.1] $2x^2 - 1$

51. $-5 \pm 2\sqrt{2}$, $-5 \pm 2\sqrt{2}\,i$ **53.** $3, 9, 6 \pm 3i$

55. $-4320x^6y^{9/2}$ **57.** $-\dfrac{35}{x^{1/6}}$ **59.** 2^{100} **61.** $[\log_a(xt)]^{23}$

63. 1. *Basis step*: Since $a + b = (a + b)^1$, S_1 is true.
2. *Induction step*: Let S_k be the statement of the
binomial theorem with n replaced by k. Multiply
both sides of S_k by $(a + b)$ to obtain

$$(a + b)^{k+1}$$
$$= \left[a^k + \cdots + \binom{k}{r-1}a^{k-(r-1)}b^{r-1} \right.$$
$$\left. + \binom{k}{r}a^{k-r}b^r + \cdots + b^k \right](a + b)$$
$$= a^{k+1} + \cdots + \left[\binom{k}{r-1} + \binom{k}{r} \right]a^{(k+1)-r}b^r$$
$$+ \cdots + b^{k+1}$$
$$= a^{k+1} + \cdots + \binom{k+1}{r}a^{(k+1)-r}b^r + \cdots + b^{k+1}.$$

This proves S_{k+1}, assuming S_k. Hence S_n is true for
$n = 1, 2, 3, \ldots$.

Exercise Set 7.8

1. (a) 0.18, 0.24, 0.23, 0.23, 0.12; **(b)** Opinions may vary,
but it seems that people tend not to pick the first or last
numbers.   **3.** 11,700   **5. (a)** T, S, R, N, L; **(b)** E; **(c)** yes

7. (a) $\frac{2}{7}$; **(b)** $\frac{5}{7}$; **(c)** 0; **(d)** 1   **9.** $\frac{350}{31{,}977}$ **11.** $\frac{1}{108{,}290}$

13. $\frac{33}{66{,}640}$ **15.** Answers will vary.   **17.** $\frac{9}{19}$ **19.** $\frac{18}{19}$

21. $\frac{1}{38}$ **23.** $\frac{9}{19}$ **25.** $P(\text{red}) = \frac{1}{3}$, $P(\text{green}) = \frac{2}{9}$,
$P(\text{blue}) = \frac{1}{6}$, $P(\text{yellow}) = \frac{5}{18}$ **27.** Discussion and Writing

28. [2.3] $\dfrac{2 \pm \sqrt{13}}{3}$ **29.** [3.3] $-3, -\frac{1}{2}, 1$, or $-3, -0.5, 1$

30. [5.1], [5.3] $(-3, -4)$ **31.** [5.2], [5.3] $(5, -2, 1)$

33. (a) 36; **(b)** 1.39×10^{-5}

35. (a) $(13 \cdot {}_4C_3) \cdot (12 \cdot {}_4C_2) = 3744$; **(b)** 0.00144

37. (a) $4 \cdot \binom{13}{5} - 4 - 36 = 5108$; **(b)** 0.00197

39. (a) $\binom{10}{1}\binom{4}{1}\binom{4}{1}\binom{4}{1}\binom{4}{1}\binom{4}{1} - 4 - 36 = 10{,}200$;

(b) 0.00392

Review Exercises, Chapter 7

1. [7.1] $-\frac{1}{2}, \frac{4}{17}, -\frac{9}{82}, \frac{16}{257}; -\frac{121}{14{,}642}; -\frac{529}{279{,}842}$

2. [7.1] $(-1)^{n+1}(n^2 + 1)$ **3.** [7.1] $\frac{3}{2} - \frac{9}{8} + \frac{27}{26} - \frac{81}{80} = \frac{417}{1040}$

4. [7.1]

n	U_n
1	0.3
2	2.5
3	13.5
4	68.5
5	343.5
6	1718.5
7	8593.5
8	42968.5
9	214843.5
10	1074218.5

5. [7.1] $\displaystyle\sum_{k=1}^{7} k^2 - 1$ **6.** [7.2] $3\frac{3}{4}$ **7.** [7.2] $a + 4b$

8. [7.2] 531   **9.** [7.2] 20,100   **10.** [7.2] 11   **11.** [7.2] -4

12. [7.3] $n = 6$, $S_n = -126$ **13.** [7.3] $a_1 = 8$, $a_5 = \frac{1}{2}$

14. [7.3] Does not exist   **15.** [7.3] $\frac{3}{11}$ **16.** [7.3] $\frac{3}{8}$

17. [7.3] $\frac{241}{99}$ **18.** [7.2] $5\frac{4}{5}, 6\frac{3}{5}, 7\frac{2}{5}, 8\frac{1}{5}$ **19.** [7.3] 167.3 ft

20. [7.3] \$60,653.05   **21.** [7.2] **(a)** \$7.38; **(b)** \$1365.10

22. [7.3] \$88,888,888,889

23. [7.4] S_n: $1 + 4 + 7 + \cdots + (3n - 2) = \dfrac{n(3n - 1)}{2}$

S_1: $1 = \dfrac{1(3 - 1)}{2}$

S_k: $1 + 4 + 7 + \cdots + (3k - 2) = \dfrac{k(3k - 1)}{2}$

S_{k+1}: $1 + 4 + 7 + \cdots + [3(k + 1) - 2]$
$= 1 + 4 + 7 + \cdots + (3k - 2) + (3k + 1)$
$= \dfrac{(k + 1)(3k + 2)}{2}$

1. *Basis step*: $1 = \dfrac{2}{2} = \dfrac{1(3-1)}{2}$ is true.

2. *Induction step*: Assume S_k. Add $(3k+1)$ to both sides.

$$1 + 4 + 7 + \cdots + (3k-2) + (3k+1)$$
$$= \frac{k(3k-1)}{2} + (3k+1)$$
$$= \frac{k(3k-1)}{2} + \frac{2(3k+1)}{2}$$
$$= \frac{3k^2 - k + 6k + 2}{2}$$
$$= \frac{3k^2 + 5k + 2}{2}$$
$$= \frac{(k+1)(3k+2)}{2}$$

24. [7.4] S_1: $\quad 1 = \dfrac{3^1 - 1}{2}$; S_2: $\quad 1 + 3 = \dfrac{3^2 - 1}{2}$

S_k: $\quad 1 + 3 + 3^2 + \cdots + 3^{k-1} = \dfrac{3^k - 1}{2}$

2. *Induction step*: Assume S_k. Add 3^k on both sides.

$$1 + 3 + \cdots + 3^{k-1} + 3^k$$
$$= \frac{3^k - 1}{2} + 3^k = \frac{3^k - 1}{2} + 3^k \cdot \frac{2}{2}$$
$$= \frac{3 \cdot 3^k - 1}{2} = \frac{3^{k+1} - 1}{2}$$

25. [7.4]

S_n: $\quad \left(1 - \dfrac{1}{2}\right)\left(1 - \dfrac{1}{3}\right) \cdots \left(1 - \dfrac{1}{n}\right) = \dfrac{1}{n}$

S_2: $\quad \left(1 - \dfrac{1}{2}\right) = \dfrac{1}{2}$

S_k: $\quad \left(1 - \dfrac{1}{2}\right)\left(1 - \dfrac{1}{3}\right) \cdots \left(1 - \dfrac{1}{k}\right) = \dfrac{1}{k}$

S_{k+1}: $\quad \left(1 - \dfrac{1}{2}\right)\left(1 - \dfrac{1}{3}\right) \cdots \left(1 - \dfrac{1}{k}\right)\left(1 - \dfrac{1}{k+1}\right)$
$$= \frac{1}{k+1}$$

1. *Basis step*: S_2 is true by substitution.
2. *Induction step*: Assume S_k. Deduce S_{k+1}.
   Starting with the left side of S_{k+1}, we have

$$\underbrace{\left(1 - \frac{1}{2}\right)\left(1 - \frac{1}{3}\right) \cdots \left(1 - \frac{1}{k}\right)}\left(1 - \frac{1}{k+1}\right)$$
$$= \frac{1}{k} \cdot \left(1 - \frac{1}{k+1}\right). \qquad \textbf{By } S_k$$

Then
$$= \frac{1}{k} \cdot \left(\frac{k+1-1}{k+1}\right)$$
$$= \frac{1}{k} \cdot \frac{k}{k+1}$$
$$= \frac{1}{k+1}. \qquad \textbf{Simplifying}$$

26. [7.5] $6! = 720$ **27.** [7.5] $9 \cdot 8 \cdot 7 \cdot 6 = 3024$

28. [7.6] $\dbinom{15}{8} = 6435$ **29.** [7.5] $24 \cdot 23 \cdot 22 = 12{,}144$

30. [7.5] $\dfrac{9!}{1!\,4!\,2!\,2!} = 3780$ **31.** [7.5] 36

32. [7.5] **(a)** $_6P_5 = 720$; **(b)** $6^5 = 7776$; **(c)** $_5P_4 = 120$;
(d) $_3P_2 = 6$ **33.** [7.7] 256

34. [7.7] $m^7 + 7m^6n + 21m^5n^2 + 35m^4n^3 + 35m^3n^4 + 21m^2n^5 + 7mn^6 + n^7$

35. [7.7] $x^5 - 5\sqrt{2}\,x^4 + 20x^3 - 20\sqrt{2}\,x^2 + 20x - 4\sqrt{2}$

36. [7.7] $x^8 - 12x^6y + 54x^4y^2 - 108x^2y^3 + 81y^4$

37. [7.7] $a^8 + 8a^6 + 28a^4 + 56a^2 + 70 + 56a^{-2} + 28a^{-4} + 8a^{-6} + a^{-8}$ **38.** [7.7] $-6624 + 16{,}280i$

39. [7.7] $220a^9x^3$ **40.** [7.7] $-\dbinom{18}{11}128a^7b^{11}$

41. [7.8] $\frac{86}{206} \approx 0.42$, $\frac{97}{206} \approx 0.47$, $\frac{23}{206} \approx 0.11$ **42.** [7.8] $\frac{1}{12}$, 0

43. [7.8] $\frac{1}{4}$ **44.** [7.8] $\frac{6}{5525}$

45. [7.1] **(a)** $a_n = 0.6149275362n + 0.5905797101$;
(b) $\$12.89$ billion

46. Discussion and Writing [7.6] A list of 9 candidates for an office is to be narrowed down to 4 candidates. In how many ways can this be done?

47. Discussion and Writing [7.3] Someone who has managed several sequences of hiring has a considerable income from the sales of the people in the lower levels. However, with a finite population, it will not be long before the salespersons in the lowest level have no one to hire and no one to sell to.

48. [7.4] S_1 fails for both (a) and (b).

49. [7.3] $\dfrac{a_{k+1}}{a_k} = r_1$, $\dfrac{b_{k+1}}{b_k} = r_2$, so $\dfrac{a_{k+1}b_{k+1}}{a_kb_k} = r_1r_2$, a constant

50. [7.2] **(a)** No (unless a_n is all positive or all negative); **(b)** yes; **(c)** yes; **(d)** no (unless a_n is constant); **(e)** no (unless a_n is constant); **(f)** no (unless a_n is constant)

51. [7.2] $-2, 0, 2, 4$ **52.** [7.3] $\frac{1}{2}, -\frac{1}{6}, \frac{1}{18}$

53. [7.6] $\left(\log \dfrac{x}{y}\right)^{10}$ **54.** [7.6] 18 **55.** [7.6] 36

56. [7.7] -9

Appendixes

Exercise Set A

1. 3 or 1; 0   **3.** 0; 3 or 1   **5.** 2 or 0; 2 or 0   **7.** 1; 1
9. 1; 0   **11.** 2 or 0; 2 or 0   **13.** 3 or 1; 1   **15.** 1; 1
17. 2 or 0; 2 or 0   **19.** 0; 0   **21.** 2 or 0; 1

Exercise Set B

1. -11 **3.** $x^3 - 4x$ **5.** -109 **7.** $-x^4 + x^2 - 5x$
9. The answer checks.   **11.** The answer checks.
13. $M_{11} = 6$, $M_{32} = -9$, $M_{22} = -29$
15. $A_{11} = 6$, $A_{32} = 9$, $A_{22} = -29$ **17.** -10
19. -10 **21.** $M_{41} = -14$, $M_{33} = 20$
23. $A_{24} = 15$, $A_{43} = 30$ **25.** 110   **27.** 110
29. $\left(-\frac{25}{2}, -\frac{11}{2}\right)$ **31.** $(3, 1)$ **33.** $\left(\frac{1}{2}, -\frac{1}{3}\right)$ **35.** $(1, 1)$
37. $\left(\frac{3}{2}, \frac{13}{14}, \frac{33}{14}\right)$ **39.** $(3, -2, 1)$ **41.** $(1, 3, -2)$
43. $\left(\frac{1}{2}, \frac{2}{3}, -\frac{5}{6}\right)$ **45.** Discussion and Writing   **47.** ± 2
49. $(-\infty, -\sqrt{3}] \cup [\sqrt{3}, \infty)$ **51.** -34 **53.** 4

55. Answers may vary.
$$\begin{vmatrix} L & -W \\ 2 & 2 \end{vmatrix}$$
57. Answers may vary.
$$\begin{vmatrix} a & b \\ -b & a \end{vmatrix}$$
59. Answers may vary.
$$\begin{vmatrix} 2\pi r & 2\pi r \\ -h & r \end{vmatrix}$$

Exercise Set C

1. $y = 12x - 7$, $-\frac{1}{2} \le x \le 3$
3. $y = \sqrt[3]{x} - 4$, $-1 \le x \le 1000$
5. $x = \sqrt[4]{x}$, $0 \le x \le 16$ **7.** $y = \dfrac{1}{x}$, $1 \le x \le 5$
9. $y = \frac{1}{4}(x + 1)^2$, $-7 \le x \le 5$ **11.** $y = \dfrac{1}{x}$, $x > 0$
13. Answers may vary. $x = t$, $y = 4t - 3$; $x = \dfrac{t}{4} + 3$, $y = t + 9$
15. Answers may vary. $x = t$, $y = (t - 2)^2 - 6t$; $x = t + 2$, $y = t^2 - 6t - 12$
17. Answers may vary. $x = \frac{1}{4}t$, $y = -t - 4$

Index

Index of Applications

Geometry

Plane Geometry

Rectangle
Area: $A = lw$
Perimeter: $P = 2l + 2w$

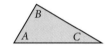

Square
Area: $A = s^2$
Perimeter: $P = 4s$

Triangle
Area: $A = \frac{1}{2}bh$

Sum of Angle Measures
$A + B + C = 180°$

Right Triangle
Pythagorean theorem
(equation):
$a^2 + b^2 = c^2$

Parallelogram
Area: $A = bh$

Trapezoid
Area: $A = \frac{1}{2}h(a + b)$

Circle
Area: $A = \pi r^2$
Circumference:
$C = \pi d = 2\pi r$

Solid Geometry

Rectangular Solid
Volume: $V = lwh$

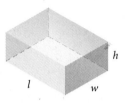

Cube
Volume: $V = s^3$

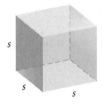

Right Circular Cylinder
Volume: $V = \pi r^2 h$
Lateral surface area:
$L = 2\pi rh$
Total surface area:
$S = 2\pi rh + 2\pi r^2$

Right Circular Cone
Volume: $V = \frac{1}{3}\pi r^2 h$
Lateral surface area:
$L = \pi rs$
Total surface area:
$S = \pi r^2 + \pi rs$
Slant height:
$s = \sqrt{r^2 + h^2}$

Sphere
Volume: $V = \frac{4}{3}\pi r^3$
Surface area: $S = 4\pi r^2$

Algebra

Properties of Real Numbers

Commutative: $a + b = b + a; \quad ab = ba$

Associative: $a + (b + c) = (a + b) + c;$
$a(bc) = (ab)c$

Additive Identity: $a + 0 = 0 + a = a$

Additive Inverse: $-a + a = a + (-a) = 0$

Multiplicative Identity: $a \cdot 1 = 1 \cdot a = a$

Multiplicative Inverse: $a \cdot \dfrac{1}{a} = 1, \ a \neq 0$

Distributive: $a(b + c) = ab + ac$

Exponents and Radicals

$a^m \cdot a^n = a^{m+n}$ $\qquad \dfrac{a^m}{a^n} = a^{m-n}$

$(a^m)^n = a^{mn}$ $\qquad (ab)^m = a^m b^m$

$\left(\dfrac{a}{b}\right)^m = \dfrac{a^m}{b^m}$ $\qquad a^{-n} = \dfrac{1}{a^n}$

If n is even, $\sqrt[n]{a^n} = |a|$.

If n is odd, $\sqrt[n]{a^n} = a$.

$\sqrt[n]{a} \cdot \sqrt[n]{b} = \sqrt[n]{ab}, \ a,b \geq 0$

$\sqrt[n]{\dfrac{a}{b}} = \dfrac{\sqrt[n]{a}}{\sqrt[n]{b}}$

$\sqrt[n]{a^m} = (\sqrt[n]{a})^m = a^{m/n}$

Special-Product Formulas

$(a + b)(a - b) = a^2 - b^2$

$(a + b)^2 = a^2 + 2ab + b^2$

$(a - b)^2 = a^2 - 2ab + b^2$

$(a + b)^3 = a^3 + 3a^2b + 3ab^2 + b^3$

$(a - b)^3 = a^3 - 3a^2b + 3ab^2 - b^3$

$(a + b)^n = \displaystyle\sum_{k=0}^{n} \binom{n}{k} a^{n-k} b^k, \quad$ where

$\binom{n}{k} = \dfrac{n!}{k! \, (n - k)!}$

$\qquad = \dfrac{n(n - 1)(n - 2) \cdots [n - (k - 1)]}{k!}$

Factoring Formulas

$a^2 - b^2 = (a + b)(a - b)$

$a^2 + 2ab + b^2 = (a + b)^2$

$a^2 - 2ab + b^2 = (a - b)^2$

$a^3 + b^3 = (a + b)(a^2 - ab + b^2)$

$a^3 - b^3 = (a - b)(a^2 + ab + b^2)$

Interval Notation

$(a, b) = \{x | a < x < b\}$

$[a, b] = \{x | a \leq x \leq b\}$

$(a, b] = \{x | a < x \leq b\}$

$[a, b) = \{x | a \leq x < b\}$

$(-\infty, a) = \{x | x < a\}$

$(a, \infty) = \{x | x > a\}$

$(-\infty, a] = \{x | x \leq a\}$

$[a, \infty) = \{x | x \geq a\}$

Absolute Value

$|a| \geq 0$

For $a > 0$,

$\quad |X| = a \rightarrow X = -a \quad \text{or} \quad X = a,$

$\quad |X| < a \rightarrow -a < X < a,$

$\quad |X| > a \rightarrow X < -a \quad \text{or} \quad X > a.$

Equation-Solving Principles

$a = b \rightarrow a + c = b + c$

$a = b \rightarrow ac = bc$

$a = b \rightarrow a^n = b^n$

$ab = 0 \leftrightarrow a = 0 \quad \text{or} \quad b = 0$

$x^2 = k \rightarrow x = \sqrt{k} \quad \text{or} \quad x = -\sqrt{k}$

Inequality-Solving Principles

$a < b \rightarrow a + c < b + c$

$a < b \text{ and } c > 0 \rightarrow ac < bc$

$a < b \text{ and } c < 0 \rightarrow ac > bc$

(Algebra continued)

Algebra (continued)

The Distance Formula

The distance from (x_1, y_1) to (x_2, y_2) is given by
$$d = \sqrt{(x_2 - x_1)^2 + (y_2 - y_1)^2}.$$

Midpoint Formula

The midpoint of the line segment from (x_1, y_1) to (x_2, y_2) is given by
$$\left(\frac{x_1 + x_2}{2}, \frac{y_1 + y_2}{2} \right).$$

Formulas Involving Lines

The slope of the line containing points (x_1, y_1) to (x_2, y_2) is given by
$$m = \frac{y_2 - y_1}{x_2 - x_1}.$$

Slope–intercept equation: $\quad y = f(x) = mx + b$

Horizontal line: $\quad y = b \quad$ or $\quad f(x) = b$

Vertical line: $\quad x = a$

Point–slope equation: $\quad y - y_1 = m(x - x_1)$

The Quadratic Formula

The solutions of $ax^2 + bx + c = 0$, $a \neq 0$, are given by
$$x = \frac{-b \pm \sqrt{b^2 - 4ac}}{2a}.$$

Compound Interest Formulas

Compounded n times per year: $\quad A = P\left(1 + \dfrac{i}{n}\right)^{nt}$

Compounded continuously: $\quad P(t) = P_0 e^{kt}$

Properties of Exponential and Logarithmic Functions

$\log_a x = y \leftrightarrow x = a^y \qquad\qquad a^x = a^y \leftrightarrow x = y$

$\log_a MN = \log_a M + \log_a N \qquad \log_a M^p = p \log_a M$

$\log_a \dfrac{M}{N} = \log_a M - \log_a N$

$\log_b M = \dfrac{\log_a M}{\log_a b}$

$\log_a a = 1 \qquad\qquad\qquad\qquad \log_a 1 = 0$

$\log_a a^x = x \qquad\qquad\qquad\qquad a^{\log_a x} = x$

Conic Sections

Circle: $\qquad (x - h)^2 + (y - k)^2 = r^2$

Ellipse: $\qquad \dfrac{(x - h)^2}{a^2} + \dfrac{(y - k)^2}{b^2} = 1,$

$\qquad\qquad \dfrac{(x - h)^2}{b^2} + \dfrac{(y - k)^2}{a^2} = 1$

Parabola: $\qquad (x - h)^2 = 4p(y - k),$

$\qquad\qquad (y - k)^2 = 4p(x - h)$

Hyperbola: $\qquad \dfrac{(x - h)^2}{a^2} - \dfrac{(y - k)^2}{b^2} = 1,$

$\qquad\qquad \dfrac{(y - k)^2}{a^2} - \dfrac{(x - h)^2}{b^2} = 1$

Arithmetic Sequences and Series

$a_1, a_1 + d, a_1 + 2d, a_1 + 3d, \ldots$

$a_{n+1} = a_n + d \qquad\qquad a_n = a_1 + (n - 1)d$

$S_n = \dfrac{n}{2}(a_1 + a_n)$

Geometric Sequences and Series

$a_1, a_1 r, a_1 r^2, a_1 r^3, \ldots$

$a_{n+1} = a_n r \qquad\qquad a_n = a_1 r^{n-1}$

$S_n = \dfrac{a_1(1 - r^n)}{1 - r} \qquad S_\infty = \dfrac{a_1}{1 - r}, |r| < 1$